The theory of generalised functions

The theory of generalised functions

D.S.JONES
Ivory Professor of Mathematics, University of Dundee

CAMBRIDGE UNIVERSITY PRESS
Cambridge
London New York New Rochelle
Melbourne Sydney

CAMBRIDGE UNIVERSITY PRESS
Cambridge, New York, Melbourne, Madrid, Cape Town, Singapore, São Paulo, Delhi

Cambridge University Press
The Edinburgh Building, Cambridge CB2 8RU, UK

Published in the United States of America by Cambridge University Press, New York

www.cambridge.org
Information on this title: www.cambridge.org/9780521237239

This publication is a substantially revised and completely reset edition of
Generalised Functions, published by McGraw-Hill 1966
First published by Cambridge University Press 1982
This digitally printed version 2008

A catalogue record for this publication is available from the British Library

ISBN 978-0-521-23723-9 hardback
ISBN 978-0-521-10004-5 paperback

To Katie, Kim and Corrie
for the pleasure they have given
Ivy and myself

Contents

Preface

For some years I have been offering lectures on generalised functions to undergraduate and postgraduate students. The undergraduate course was based originally on M.J. Lighthill's stimulating book *An Introduction to Fourier Analysis and Generalised Functions* which contains a simplified version of a theory evolved by G. Temple to make generalised functions more readily accessible and intelligible to students. It is an approach to the theory of generalised functions which permits early introduction in student courses while retaining the power and practical utility of the methods. At the same time it can be developed so as to include the more advanced aspects appropriate to postgraduate instruction. This book has grown from the courses which I have given expounding the ramifications of the Lighthill–Temple theory to various groups of students. It is arranged so that sections can be chosen relevant to any level of course.

Much of the material was originally contained in my book *Generalised Functions*, published by McGraw-Hill in 1966, but this book differs from the earlier version in several major respects. The treatment and definitions of the special generalised functions which are powers of the single variable x have been completely changed as well as those of the powers of the radial distance in higher dimensions. A different definition of δ-functions, whose support is on a surface, has also been introduced. The properties of the hyperbolic and ultrahyperbolic distances have been tackled in another way, with consequences for the general quadratic form. Numerous subsequent formulae are thereby altered. Further, a section has been added on the Fourier transform of weak functions and ultradistributions.

The purpose of the first chapter is to summarise some of the basic theorems of analysis which are required in subsequent chapters. It is anticipated that most readers will have met this material in one form or another before reading this book. For this reason explanation and argument have been cut to a minimum and, consequently, this chapter is not a suitable first reading for those who have not met several of the analytical ideas before. Since the chapter is self-contained some readers will, I hope, find it a useful introduction to the notions and terminology employed in other books where the approach to the subject has a more topological character. Many readers will find it profitable, on a first reading, to start at Chapter 2 and read onwards, referring back to Chapter 1 only for notation and statements of theorems.

In Chapter 2 the properties of good functions are given. Generalised functions of a single variable are introduced in Chapter 3 via sequences of good functions. After an examination of the derivative, Fourier transform and limit, the general structure of a generalised function is determined.

Chapter 4 is concerned with some special generalised functions, their Fourier transforms and the evaluation of certain integrals which are too singular to be embraced by classical analysis. The final section contains a brief discussion of generalised functions on a half-line.

Chapter 5 is devoted to series of generalised functions and shows, in particular, that any generalised function can be represented as a series of Hermite functions. There is also a detailed investigation of expansions in Fourier series, many theorems being much simpler than in classical analysis.

The problem of multiplication and division is dealt with in Chapter 6; the properties of the convolution product are also derived.

Generalised functions of several variables are introduced in Chapter 7. Most of the results are obvious generalisations of those for a single variable but new features are the direct product and the Fourier transform with respect to one of several variables. The last sections deal with spherically symmetric generalised functions and integration with respect to a parameter.

Chapter 8 treats the difficult problem of changing variables in a generalised function. This leads naturally to δ-functions on a hypersurface and the meaning to be attached to powers of the hyperbolic distance and its generalisations.

The asymptotic evaluation of Fourier integrals and the method of stationary phase in several variables comprise Chapter 9.

Applications of generalised functions are considered in Chapter 10. Particular reference is made to integral equations, ordinary and partial differential equations, as well as correlation theory.

Chapter 11 brings in the notion of a weak function, which is not so restricted at infinity as the generalised function. The significance of weak functions in solving integral equations, ordinary differential equations and in the justification of the operational method is shown. The Fourier transform of a weak function and ultradistributions are discussed, as well as the relation between weak functions and distributions.

The Laplace transform of a weak function is defined in Chapter 12 and a number of applications is given.

Exercises are given at various stages throughout the chapters. Most of them are to enable the reader to become thoroughly familiar with the theory, though some are extensions of theorems in the text. There are also some exercises which are worded so that they could be used as topics for minor theses. It is hoped that this variety will provide instructors with plenty of flexibility.

The author takes this opportunity of expressing his thanks to Mrs D. Ross for turning his manuscript into legible typescript despite a certain obscurity about the way it was organised.

The author's gratitude to his wife Ivy, who manages to display nonchalance and good cheer whatever burden is imposed on her, is immeasurable.

University of Dundee D.S. Jones
October 1980

1
Convergence

This chapter is concerned primarily with deriving certain theorems on convergence which will be required in subsequent chapters. It is expected that most readers will be familiar with the notions involved so that much of the material is given in a condensed manner. However, an attempt has been made to make the chapter self-contained. Some readers may find the chapter a helpful introduction to the ideas and terminology employed in other books on generalised functions. The reader who does not have a good background in analysis is strongly advised to go straight to Chapter 2 and to just refer to Chapter 1 for the theorems that are needed.

1.1. Preliminary definitions

A *set* is a collection of elements. A set containing no elements is called a *null* or *empty set*. There is no restriction on what an element is: it may be a number or a point or a vector and so on. Usually we shall call the elements points and take all sets to be sets of points in a fixed non-empty set Ω, which will be called a *space*. The empty set will be denoted by $\varnothing$ and the capitals $A, B, \ldots$ will denote sets. If ω is a point of A, we write $\omega \in A$; if ω is not a point of A, we write $\omega \notin A$. Another useful notation is $\{\omega | P\}$ for the set of points satisfying condition P ; for example, the set of points common to both sets A and B can be written $\{\omega | \omega \in A \text{ and } \omega \in B\}$.

A set of sets is called a *class*. The class of all sets in Ω is called the *space of sets* in Ω. A class of sets in Ω is a set in this space of sets so that all set theories apply to classes considered as sets in the corresponding space of sets. Classes will be denoted by the script capitals $\mathscr{A}, \mathscr{B}, \ldots$

If all the points of A are points of B we write $A \subset B$ or, equivalently, $B \supset A$. Obviously, $A \subset A$ and $\varnothing \subset A \subset \Omega$. If $A \subset B$ and $B \subset C$ then $A \subset C$. If $A \subset B$ and $B \subset A$ we write $A = B$.

The *intersection* $A \cap B$ is the set of all points common to A and B, i.e. if $\omega \in A$ and $\omega \in B$ then $\omega \in A \cap B$ and conversely. The *union* $A \cup B$ is the set of all points which belong to at least one of the sets A or B, i.e. if $\omega \in A$ or $\omega \in B$ then $\omega \in A \cup B$ and conversely. If $A \cap B = \varnothing$ the sets A and B are said to be *disjoint* and their union may then be called a *sum* and written as $A + B$, i.e. if $A \cap B = \varnothing$ then $A \cup B = A + B$.

The *difference* $A - B$ is the set of all points of A which are not in B, i.e. if $\omega \in A$ and $\omega \notin B$ then $\omega \in A - B$ and conversely. The difference $\Omega - A$ is called the *complement* of A and denoted by A^c; it is the set of all points which do not belong to A.

The following *commutative, associative* and *distributive* laws are valid, i.e.

$$A \cup B = B \cup A, \qquad A \cap B = B \cap A;$$

$$(A \cup B) \cup C = A \cup (B \cup C),$$
$$(A \cap B) \cap C = A \cap (B \cap C);$$

$$(A \cup B) \cup C = (A \cap C) \cup (B \cap C),$$
$$(A \cup B) \cap (A \cup C) = A \cup (B \cap C).$$

Relations between sets and their complements are:

$$\Omega^c = \varnothing, \qquad \varnothing^c = \Omega, \qquad A \cap A^c = \varnothing, \qquad A + A^c = \Omega;$$
$$A - B = A \cap B^c, \qquad (A \cup B)^c = A^c \cap B^c,$$
$$(A \cap B)^c = A^c \cup B^c;$$

if $A \subset B$ then $A^c \supset B^c$.

The operations of union and intersection can be extended to arbitrary classes. Let I be a set, not necessarily in Ω, and corresponding to each $i \in I$ choose a set $A_i \subset \Omega$. The class of sets so chosen will be denoted by $\{A_i | i \in I\}$. For obvious reasons I is called an *index set.* The *intersection* of $\{A_i | i \in I\}$ is the set of all points which belong to every A_i and is denoted by $\bigcap_{i \in I} A_i$, i.e.

$$\bigcap_{i \in I} A_i = \{\omega | \omega \in A_i \text{ for every } i \in I\}.$$

The *union* $\bigcup_{i \in I} A_i$ is the set of all points which belong to at least one A_i, i.e.

$$\bigcup_{i \in I} A_i = \{\omega | \omega \in A_i \text{ for some } i \in I\}.$$

If $A_i \cap A_j = \varnothing$ for all $i, j \in I$, $i \neq j$, the class $\{A_i | i \in I\}$ is said to be a *disjoint class* and the union of its sets may be called a sum and denoted by $\sum A_i$.

If $\omega \notin A_i$ then $\omega \in A_i^c$ and conversely. Consequently

$$\left(\bigcup_{i \in I} A_i\right)^c = \bigcap_{i \in I} A_i^c, \quad \left(\bigcap_{i \in I} A_i\right)^c = \bigcup_{i \in I} A_i^c. \tag{1}$$

By convention

$$\bigcup_{i \in \varnothing} A_i = \varnothing, \quad \bigcap_{i \in \varnothing} A_i = \Omega. \tag{2}$$

It will be observed that the following *principle of duality* holds: *any relation between sets involving unions and intersections becomes a valid relation by replacing* $\cup, \cap, \varnothing, \Omega$ *by* $\cap, \cup, \Omega, \varnothing$ *respectively.*

Finally we introduce the notion of equivalence class. Suppose we have a rule R which places the sets A and B in one-to-one correspondence, which we denote by ARB. The relation is *reflexive*, ARA; *symmetric*, ARB implies BRA; *transitive*, ARB and BRC imply ARC. A reflexive, symmetric and transitive relation is called an *equivalence relation*. The class $\{B | BRA\}$ is called the *equivalence class* corresponding to A. In essence an equivalence class is determined by any one of its members.

A class or set is said to be *finite* if its elements can be put in one-to-one correspondence with the first n positive integers, for some n. It is said to be *denumerable* if it can be put in one-to-one correspondence with all the positive integers. It is said to be *countable* if it is either finite or denumerable.

1.2. Sequences

For each value of $n (= 1, 2, \ldots)$ take a corresponding set A_n. The ordered denumerable class $A_1, A_2, \ldots$ is called a *sequence* and is denoted by $\{A_n\}$. It is not necessary that $A_m \neq A_n$. The *limit superior* $\overline{\lim}_n A_n$ is defined by

$$\overline{\lim_n} A_n = \bigcap_{k=1}^{\infty} \bigcup_{n=k}^{\infty} A_n;$$

it consists of the set of all those points which belong to infinitely many A_n. The *limit inferior* $\underline{\lim}_n A_n$ is defined by

$$\underline{\lim}_n A_n = \bigcup_{k=1}^{\infty} \bigcap_{n=k}^{\infty} A_n;$$

it consists of the set of all those points which belong to all but a finite number of A_n. Every point which belongs to all but a finite number of A_n belongs to infinitely many A_n so that

$$\underline{\lim}\, A_n \subset \overline{\lim}\, A_n.$$

If $\underline{\lim}\, A_n \supset \overline{\lim}\, A_n$ then $\underline{\lim}\, A_n = \overline{\lim}\, A_n$ and if this common set be denoted by A the sequence $\{A_n\}$ is said to *converge* to A.

A sequence is said to be *non-decreasing* if $A_n \subset A_{n+1}$ for each n; *non-increasing* if $A_{n+1} \subset A_n$ for each n. A *monotone* sequence is one which is either non-decreasing or non-increasing. *Every monotone sequence converges and, if it is non-decreasing,* $\lim A_n = \bigcup_{k=1}^{\infty} A_k$, *whereas if it is non-increasing,* $\lim A_n = \bigcap_{k=1}^{\infty} A_k$. This follows at once from the definitions.

The idea of sequence occurs in other ways; thus the *sequence* $\{\omega_n\}$ is the ordered denumerable set of points $\omega_1, \omega_2, \ldots$ A *subsequence* is obtained by selecting a sequence $\{n_i\}$ of positive integers with $n_i > n_j$ when $i > j$ and selecting the terms ω_{n_i} of the original sequence; the result is a sequence $\{\omega_{n_i}\}$ whose ith term is the n_ith term of the original sequence.

Many sequences involve real numbers, whose properties we now briefly review. A set X of real numbers is *bounded above* by the real number b if $x \leq b$ for every $x \in X$; b is called an *upper bound* for X. If b is an upper bound for X, if c is any other upper bound and if $b \leq c$ whatever c is then b is the smallest possible upper bound; in that case b is known as the *least upper bound* or *supremum* of X and written sup X. Sometimes the notation l.u.b. X is used. By reversing the inequalities in these definitions we define *bounded below, lower bound, greatest lower bound* or *infimum* of X (written inf X).

A fundamental postulate is: *every non-empty set of real numbers which is bounded above possesses a real supremum.* If the non-empty set X of real numbers is bounded below, the set $\{-x \mid x \in X\}$ is bounded above and hence possesses a real supremum. Therefore X has a real infimum, i.e. a non-empty set bounded above and below possesses both a real supremum and a real infimum.

The supremum of a sequence $\{x_n\}$ is denoted by $\sup_n x_n$. The *limit superior* is defined by

$$\overline{\lim_n}\, x_n = \inf_k \sup_{n \geq k} x_n$$

and the *limit inferior* by

$$\underline{\lim_n}\, x_n = \sup_k \inf_{n \geq k} x_n .$$

If a sequence is bounded above and below, it possesses both a real supremum and a real infimum and so has both a limit superior and a limit inferior.

The ordinary number system consists of *finite numbers*; the *extended real number system* is obtained by adding the *infinite numbers* ∞ and $-\infty$. These symbols have the properties:

$$x + (\pm\infty) = (\pm\infty) + x = \pm\infty, \qquad \frac{x}{\pm\infty} = 0 \quad \text{if} \quad -\infty < x < \infty;$$

$$x(\pm\infty) = (\pm\infty)x = \begin{cases} \pm\infty & \text{if } 0 < x \leq \infty \\ \mp\infty & \text{if } -\infty \leq x < 0. \end{cases}$$

The expression $\infty - \infty$ is meaningless so that if one of the sum of two numbers be $\pm\infty$ the other must not be $\mp\infty$ for the sum to exist.

Any set of extended real numbers has both a supremum (which may be infinite) and an infimum. Consequently *every sequence of extended real numbers has a limit superior and a limit inferior.* Moreover, if inclusion, union and intersection of numbers be identified with $x \leq y$, $\sup_{i \in I} x_i$, $\inf_{i \in I} x_i$ respectively these operations have the properties of the corresponding set operations; thus *monotone sequences of extended real numbers* (i.e. $x_{n+1} \geq x_n$ or $x_{n+1} \leq x_n$ for all n) *always converge (possibly to infinity).*

The set of all finite numbers $-\infty < x < \infty$ is the *real line* R_1 or $(-\infty, \infty)$; the set $-\infty \leq x \leq \infty$ is the *extended real line* $\bar{R}_1$ or $[-\infty, \infty]$.

1.3. Functions

If a rule is provided which associates with each $\omega \in \Omega$ a point $\omega' \in \Omega'$ we say that a *function f on Ω* or a *function from Ω to Ω'* is defined. The space Ω is called the *domain* of f. The point ω' which corresponds to ω is called the *value of f at ω* and is denoted by $f(\omega)$. The subset of Ω' comprising the values of f is called the *range* of f. We shall suppose that each point of the range corresponds to only one point of the domain so that the correspondence is one-to-one and that a function is always *single-valued*. Multiple-valued functions which occur frequently in analysis can be subsumed under the preceding definition by giving a rule specifying the branch to be employed.

Note that a sequence can be regarded as a function whose domain is the set of positive integers. However we shall continue to use the notation f_n for the value at the nth integer rather than $f(n)$.

Sometimes it is convenient to use the notation $f(A)$ for the set of values of f for all $\omega \in A$; $f(A)$ is called the *image* of A. Similarly $f(\mathscr{A})$ is the class of images $f(A)$ for $A \in \mathscr{A}$. The *inverse image* of $A' \subset \Omega'$ is the set of all points such that $f(\omega) \in A'$. The *inverse function* f^{-1} of f is defined by assigning to every A' its inverse image, i.e.

$$f^{-1}(A') = \{\omega \mid f(\omega) \in A'\}.$$

If A' consists of the single point ω' we write $f^{-1}(\omega')$ for $\{\omega \mid f(\omega) = \omega'\}$. The inverse is defined from the class of all subsets of Ω' to the class of all subsets of Ω. If A' does not contain a point of the range of f then $f^{-1}(A') = \varnothing$. Since f is single-valued the inverse images of disjoint sets of Ω' are themselves disjoint. Hence

$$f^{-1}(A' - B') = f^{-1}(A') - f^{-1}(B'),$$

$$f^{-1}\left(\bigcup_{i \in I} A'_i\right) = \bigcup_{i \in I} f^{-1}(A'_i),$$

$$f^{-1}\left(\bigcap_{i \in I} A'_i\right) = \bigcap_{i \in I} f^{-1}(A'_i)$$

so that *inverse functions preserve inclusion and all set and class operations.*

Another notation for functions arises in connection with product spaces. If A_1 and A_2 are two arbitrary sets the *product set* $A_1 \times A_2$ is defined as the set of all ordered pairs (ω_1, ω_2) where $\omega_1 \in A_1, \omega_2 \in A_2$. If $A_1, B_1, \ldots$ are sets in Ω_1 and $A_2, B_2, \ldots$ sets in Ω_2 then $A_1 \times A_2$, $B_1 \times B_2, \ldots$ are sets in the *product space* $\Omega_1 \times \Omega_2$. If we are given a rule which associates $\omega' \in \Omega'$ with $\omega_1 \in \Omega_1$ and $\omega_2 \in \Omega_2$ then we can regard it as a function from $\Omega_1 \times \Omega_2$ to Ω' and include it in the above notation. Sometimes, however, it is convenient to indicate the connection with Ω_1 and Ω_2 separately and write $f(\omega_1, \omega_2)$ for its value at (ω_1, ω_2). Another interpretation is that, for each fixed $\omega_2 \in \Omega_2$, f determines a function from Ω_1 to Ω', i.e. f may be thought of as a function from Ω_2 to the set of all functions from Ω_1 to Ω'; in this case its value at ω_2 is denoted by $f(\ ,\omega_2)$. Similarly for fixed ω_1, the function from Ω_2 to Ω' is written $f(\omega_1,\ )$.

1.4. Topological space

A space is provided with a *topology* when the class of *open sets* is defined. What is to be meant by an open set is at our disposal subject only to the restriction that arbitrary unions and finite intersections of open sets must also be open sets. Because of the convention (2) the class of open sets contains $\varnothing$ and Ω. The complement of an open set is called a *closed set*. By (1) arbitrary intersections and finite unions of closed sets are closed sets and the class contains Ω and $\varnothing$.

Two different topologies on the real line are obtained by regarding a single point as an open set or as a closed set respectively.

A space in which a topology has been supplied is called a *topological space* and will be denoted by $\mathscr{X}$, points of it by x and the class of open sets by $\mathscr{O}$. If A is a set in $\mathscr{X}$ it can be regarded as a space in its own right and supplied with its own topology or it can be provided with the topology of the intersections of A with the open sets of $\mathscr{X}$; in the latter case the *induced topology* is said to be chosen.

Any set which contains a non-empty open set is a *neighbourhood* of any point x of this open set. A neighbourhood of x will be denoted by $N(x)$. The *punctured* or *deleted neighbourhood* $\underline{N}(x)$ consists of a neighbourhood with x removed. Sometimes neighbourhoods and open sets are identified and a topology provided by specifying neighbourhoods in a space.

The *interior* A° of A is the union of all open sets in A; if A is open then $A = A^\circ$. The point x belongs to A°, i.e. is interior to A if A is a neighbourhood $N(x)$ of x. The *adherence* $\bar{A}$ of A is the intersection of all closed sets containing A; if A is closed then $A = \bar{A}$. The point x belongs to $\bar{A}$, i.e. is *adherent* to A, if no $N(x)$ is disjoint from A. Clearly

$$(A^\circ)^c = \overline{A^c}, \qquad (A^c)^\circ = (\bar{A})^c.$$

A function f from a topological domain $\mathscr{X}$ to a topological range space $\mathscr{X}'$ is said to be *continuous* at x if the inverse images of neighbourhoods of $f(x)$ are neighbourhoods of x. If f is continuous at every $x \in \mathscr{X}$ then f is said to be *continuous* on $\mathscr{X}$. Since inverse functions preserve all set operations f is continuous if, and only if, the inverse images of open sets are open. By taking complements we could replace the word 'open' by 'closed'. The spaces $\mathscr{X}$ and $\mathscr{X}'$ are *topologically equivalent* if, and only if, there exists a one-to-one correspondence f on $\mathscr{X}$ to $\mathscr{X}'$ such that f and f^{-1} are continuous.

Before introducing the notion of limit we consider the ordering of sets. Let I be a set of points denoted variously by i, j, k. A *partial ordering* $\prec$ ($i \prec j$ means i precedes j, $j \succ i$ means j follows i) is a relation which is reflexive, $i \prec i$; transitive, $i \prec j$ and $j \prec k$ imply $i \prec k$; and such that $i \prec j$ and $j \prec i$ imply $i = j$. A typical example is the inclusion relation of sets. I is said to be a *direction* if it is partially ordered and if every pair i, j is followed by some k. An example is the neighbourhoods of a point x. I is said to be *linearly ordered* if every pair i, j is ordered so that either $i \prec j$ or $j \prec i$; I is then also a direction. The set of integers is linearly ordered by the relation $\leq$.

The indexed set $\{x_i | i \in I\}$ or $\{x_i\}$ for short is called a *directed set* if I is a direction; a sequence is a particular example. If for every $N(x)$ there is a j such that $x_i \in N(x)$ for all $i \succ j$, we say that the directed set $\{x_i\}$ is *convergent* and write $x_i \to x$, calling x the *limit* of the directed set. Although we say x is the limit there is nothing to prevent a directed set having more than one limit. Topological spaces in which directed sets have no more than one limit are called *Hausdorff* or *separated* spaces.

Definition 1.1. *A topological space is called a Hausdorff or separated space if every directed set has at most one limit.*

A separated space has the following important property.

Theorem 1.1. *The two definitions*

(i) *every pair of distinct points has disjoint neighbourhoods,*

(ii) *the intersection of all closed sets containing a point is the point, are equivalent and, moreover, are equivalent to Definition* 1.1.

Proof. Suppose $x_i \to x$ and $x_i \to y$ where $x \neq y$. Then there are j and k such that $x_i \in N(x)$ for all $i \succ j$ and $x_i \in N(y)$ for all $i \succ k$. Hence

$x_i \in N(x) \cap N(y)$ for all those i which follow both j and k. Such i exist because I is a direction. Hence no pair of $N(x)$ and $N(y)$ is disjoint so that (i) is not true if a directed set has more than one limit. On the other hand, if no pair $N(x)$, $N(y)$ is disjoint there are points $z \in N(x) \cap N(y)$ and we can make the pairs form a direction by saying that $N(x) \supset N_1(x)$ and $N(y) \supset N_1(y)$ implies $(N(x), N(y)) \prec (N_1(x), N_1(y))$. The points z then form a directed set converging to both x and y. Thus if distinct points do not have disjoint neighbourhoods a directed set can have more than one limit and the proof that (i) and Definition 1.1 are equivalent is complete.

Turning now to (ii) we note that, if (i) holds, for every $y \neq x$ there exists a $N(x)$ such that $y \notin \overline{N(x)}$ and therefore (ii) holds. Conversely, if (ii) holds there exists, for every $y \neq x$, a $N(x)$ such that $y \notin \overline{N(x)}$. Then $y \in (\overline{N(x)})^c$ and this is an open set which is a neighbourhood of y disjoint from $N(x)$, i.e. (ii) implies (i) and the proof of the theorem is complete. □

Closely related to the concept of limit is that of point of accumulation. A point x is a *point of accumulation* or a *limit point* of the directed set $\{x_i\}$ if for every $N(x)$ and i there exists some $j \succ i$ such that $x_j \in N(x)$. The connection between limit and point of accumulation can be expressed in terms of the sets $A_i = \{x_j | j \succ i\}$. For $x_i \to x$ if, and only if, for every $N(x)$ there exists an $A_i \subset N(x)$, whereas x is a point of accumulation if, and only if, no pair A_i, $N(x)$ is disjoint. Obviously the set of all points of accumulation of $\{x_i\}$ consists of the intersection of all $\bar{A}_i$, and if $x_i \to x$ this set comprises the single point x.

1.5. Compactness

If a directed set possesses at least one point of accumulation it is said to have the *Bolzano–Weierstrass property*. Spaces in which all directed sets have the Bolzano–Weierstrass property are called *compact*.

Definition 1.2. *A separated space is compact if every directed set has at least one point of accumulation.*

A set is compact if it is compact in its induced topology. A class

$\mathscr{C}$ of open sets is called an *open covering* of a set A if, for every $x \in A$, there is some $C \subset \mathscr{C}$ such that $x \in C$.

Theorem 1.2. *A separated space is compact if, and only if, every open covering of the space contains a finite subclass which is also a covering of the space.*

For this reason a compact space is often said to have the *Heine–Borel property.*

Proof. If every open covering contains a finite covering we see, by taking complements, that every class of closed sets whose intersection is empty contains a finite subclass whose intersection is empty. Hence every class of closed sets, all of whose finite subclasses have non-empty intersections, has a non-empty intersection.

If $\{x_i\}$ is a directed set let $A_i = \{x_j | j \succ i\}$. The $\bar{A}_i$ form a class of closed sets whose finite subclasses have non-empty intersections and hence the intersection of all $\bar{A}_i$ is non-empty, i.e. $\{x_i\}$ has a point of accumulation.

Conversely, consider a class of closed sets all of whose finite subclasses have non-empty intersections. From this class together with the non-empty intersections we can form a direction by the relation of inclusion and then, by selecting a point from every set, we obtain a directed set $\{x_i\}$. If the space is compact $\{x_i\}$ has a point of accumulation which must belong to every set of the class. Hence the intersection of the class is non-empty and reversing the argument of the first paragraph shows that every open covering contains a finite covering. The proof of the theorem is complete. □

Theorem 1.3. *Every compact set is closed; in a compact space every closed set is compact.*

Proof. Let A be compact and let $x \in A$, $y \in A^c$. Let $N(x)$ be a neighbourhood of x and $N(x, y)$ a neighbourhood of y disjoint from $N(x)$. As x goes through every point of A the $N(x)$ form an open covering of A which by Theorem 1.2 contains a finite covering $\bigcup_{k=1}^{n} N(x_k)$. Then $y \in \bigcap_{k=1}^{n} N(x_k, y)$ which is disjoint from the finite covering so that $y \notin \bar{A}$. Since y was chosen arbitrarily in A^c it follows that $\bar{A}$ and A^c are disjoint, i.e. A is closed.

The second half of the theorem can be deduced from the first paragraph of Theorem 1.2. □

Theorem 1.4. *In a compact space a directed set $\{x_i\}$ is convergent if, and only if, it has a unique point of accumulation.*

Proof. Suppose x is the unique point of accumulation of $\{x_i\}$ but that x is not the limit of the set. Then, if $A_i = \{x_j | j \succ i\}$, there exists a $N(x)$ such that no A_i is disjoint from $(N(x))^c$. Thus for every i we can choose a $k \succ i$ such that $x_k \in A_i \cap (N(x))^c$. Since the space is compact the directed set $\{x_k\}$ has a point of accumulation $x' \in (N(x))^c$. Hence $\{x_i\}$ has x' as a point of accumulation and, since $x' \neq x$, we have a contradiction of x being the unique point of accumulation. Hence $\{x_i\}$ converges. The 'only if' assertion has already been proved at the end of § 1.4. □

Theorem 1.5. *The range of a continuous function on a compact domain is compact.*

Proof. The inverse image of every open covering of the range is an open covering of the domain because the function is continuous. The open covering of the domain contains a finite covering which is the inverse image of a finite covering of the range. The theorem now follows from Theorem 1.2. □

A separated space is said to be *locally compact* if every point has a compact neighbourhood; then every neighbourhood contains a compact one. Clearly, a compact space is locally compact but the converse is not necessarily true. However, it is not difficult to make a locally compact space compact.

Suppose $\mathscr{X}$ is locally compact but not compact. Consider a new space $\mathscr{X}_\infty$ which consists of all the points of $\mathscr{X}$ and one point, chosen arbitrarily, not in $\mathscr{X}$ which will be denoted by ∞. Provide $\mathscr{X}_\infty$ with a topology in which any open set in $\mathscr{X}$ is an open set in $\mathscr{X}_\infty$ and in which an open set containing ∞ is ∞ together with any open set in $\mathscr{X}$ whose complement is compact. The induced topology of $\mathscr{X}$ is then its original topology. $\mathscr{X}_\infty$ is separated (Theorem 1.1) because two distinct points in $\mathscr{X}$ have disjoint neighbourhoods in $\mathscr{X}_\infty$ and ∞ always has a disjoint neighbourhood from any $x \in \mathscr{X}$, by choosing, for example, a compact $N(x) \subset \mathscr{X}$, so that $\infty \in (N(x))^c$. Further, every open covering of $\mathscr{X}_\infty$ contains one open set consisting of ∞ and O where O^c is compact and so, by Theorem 1.2, contains a finite covering of O^c; the one open set together with the finite covering of O^c forms a finite covering of $\mathscr{X}_\infty$. Hence

Theorem 1.6. *The space $\mathscr{X}_\infty$ is compact.*

1.6. Metric spaces

A set A is *dense in* B if $A \subset B \subset \bar{A}$. For example, the set of rational numbers is dense in the real number system.

It is sometimes convenient to use the terminology that there is a *countable base at* x when there is a countable class $\{N_m(x) | m = 1, 2, \ldots\}$ of neighbourhoods of x such that every neighbourhood of x contains an $N_m(x)$. A space has a *countable base* $\{N_m | m = 1, 2, \ldots\}$ if, for every x, a subclass of N_ms is a base at x. In order to have a countable base a space must be separable.

A *metric space* is a space in which there exists a *distance function d*, i.e. a finite real-valued function from $\mathscr{X} \times \mathscr{X}$ to R_1 which has the property $d(x, x) = 0$ and the *triangle property*

$$d(x, y) \leq d(z, x) + d(z, y)$$

for all x, y, z in $\mathscr{X}$. A typical example is the real number system with $d(x, y) = |x - y|$. Putting $y = x$ in the triangle property we have $d(z, x) \geq 0$; putting $z = y$ in the triangle property gives $d(x, y) \leq d(y, x)$ which, being true for all x and y, implies $d(x, y) = d(y, x)$. The relation $d(x, y) = 0$ is therefore reflexive, transitive and symmetric, and partitions $\mathscr{X}$ into equivalence classes. By working with this space of equivalence classes we can say $d(x, y) = 0$ implies $x = y$. In other words we identify all y such that $d(x, y) = 0$.

The set of points y such that $d(x, y) < a$ is called a *sphere* with centre x and radius a. The *diameter* of a set A in a metric space is defined by

$$\operatorname{diam} A = \sup\{d(x, y) | x \in A, y \in A\}.$$

There may or may not be points x and y in A such that $d(x, y) = \operatorname{diam} A$. The *distance between a point and a set* $d(x, A)$ and *the distance between two sets* $d(A, B)$ are defined by

$$d(x, A) = \inf\{d(x, y) \mid y \in A\},$$
$$d(A, B) = \inf\{d(x, y) \mid x \in A, y \in B\}.$$

A *metric space* is provided with a topology by saying that a set A is *open* if, for every $x \in A$, there exists a sphere with centre x which is contained in A. Every sphere is then open, and generally neighbourhoods can be thought of as spheres so that the notation $N(x, a)$ for a sphere of centre x and radius a can be used. Since $d(x, y) > 0$ when $x \neq y$ spheres with centres x and y are disjoint if their radii are less than $\frac{1}{2}d(x, y)$ and it follows from Theorem 1.1 that the space is separated with the topology provided. The statement $x_i \to x$ means that $d(x_i, x) \to 0$. Consequently, *sequences cannot converge to more than one point.*

Furthermore if the sequence $\{x_n\}$ in A converges to x then $x \in \bar{A}$. The converse is also true, for we may choose $x_n \in A \cap N(x, 1/n)$. Thus *a set is closed if, and only if, it contains the limits of all convergent sequences of its points.*

If f is a function from a metric space $\mathscr{X}$ with metric d to a metric space $\mathscr{X}'$ with metric d' and if $f(y) \to f(x)$ as $y \to x$ then clearly $f(y_n) \to f(x)$ for every sequence $\{y_n\}$ converging to x. On the other hand, if $f(y)$ does not converge to $f(x)$ as $y \to x$ there exist $x_n \in N(x, 1/n)$ such that $x_n \to x$ but $f(x_n)$ does not converge to $f(x)$. Hence $f(y) \to f(x)$ *as* $y \to x$ *if, and only if,* $f(x_n) \to f(x)$ *for every sequence* $\{x_n\}$ *converging to* x.

A sequence $\{x_n\}$ is called a *Cauchy sequence* if, for each $\varepsilon > 0$, there exists a positive integer n_0 such that $d(x_m, x_n) < \varepsilon$ for $m > n_0, n > n_0$; more briefly, $d(x_m, x_n) \to 0$ as $m, n \to \infty$. If $x_n \to x$ then

$$d(x_m, x_n) \leq d(x, x_m) + d(x, x_n) \to 0$$

and so *any convergent sequence is a Cauchy sequence.* The converse is not necessarily true but if every Cauchy sequence converges to some limit which is a point of the space, the space is said to be *complete.*

Theorem 1.7. *If every Cauchy sequence contains a convergent subsequence the space is complete.*

Proof. If the Cauchy sequence $\{x_n\}$ contains a subsequence $\{x_m\}$ such that $x_m \to x$ then $d(x_n, x) \leq d(x_m, x_n) + d(x_m, x) \to 0$ as $m, n \to \infty$. Therefore $x_n \to x$. □

Theorem 1.8. *In a metric space the two statements*

(i) *the space is compact,*

(ii) *every sequence contains a convergent subsequence,*

are equivalent. Both imply that the space is complete.

Proof. If (i) is true, (ii) follows at once and then Theorem 1.7 shows that the space is complete. It remains to prove that (ii) implies (i).

When (ii) is true the space can be covered by a finite number of spheres of radius $1/n$, n being a positive integer. Otherwise we could select a sequence of points, each of which was distant at least $1/n$ from all the others and would therefore contain no convergent subsequence; if there were only a finite number of such points then spheres centred at these points would cover the space. The union of these finite coverings for $n = 1, 2, \ldots$ is a countable covering $\mathcal{N}$ of the space such that, if C is any open set, there is an $N_m \in \mathcal{N}$ and $N_m \subset C$.

Given an open covering of the space every one of its sets contains a N_m so that, for every N_m, we select one set C_m of the covering containing it. The countable class of C_ms is an open covering of the space. If no finite union of the C_ms covers the space then, for every n, there exists a point $x_n \notin \bigcup_{m=1}^{n} C_m$ and the sequence $\{x_n\}$ must contain a convergent subsequence. Since the space is complete this sequence has a point of accumulation x which must belong to some $C_{m'}$. But this implies that, for some $n > m'$, $x_n \in C_{m'} \subset \bigcup_{m=1}^{n} C_m$ and we have a contradiction. Thus a finite union of C_ms covers the space and (i) follows from Theorem 1.2. □

One method of *completing* a metric space which is not complete is to form all Cauchy sequences $\{x_n\}$ and construct a metric $d_1(s, s')$, where s is a Cauchy sequence, by $d_1(s, s') = \lim d(x_n, x'_n)$. The relation $d_1(s, s') = 0$ provides a space of equivalence classes $s = s'$ which is a complete metric space. The one-to-one correspondence between $\mathscr{X}$ and the set of classes of equivalence of sequences $\{x\}$ maintains distances.

1.7. Function spaces

A considerable body of analysis is concerned with spaces whose elements are functions defined over some fixed set. Take, for example, the space of all real-valued continuous functions on $0 \leq t \leq 1$; this is provided with a metric $d(x, y) = \sup_{0 \leq t \leq 1} |x(t) - y(t)|$. An algebraic structure can be defined by taking $z = x + y$ and $w = ax$ to mean that, for each t in the domain of the functions, $z(t) = x(t) + y(t)$ and $w(t) = ax(t)$. Here, and in the following, a, b, c are scalars, i.e. arbitrary complex numbers in general, although in some spaces they are restricted to real numbers.

$\mathscr{X}$ is a *linear space* if there are defined the operations of 'addition' and 'multiplication by scalars', together with an equivalence relation '$=$', from $\mathscr{X}$ to $\mathscr{X}$ such that

$$x + y = y + x; \qquad x + (y + z) = (x + y) + z;$$

if

$$x + y = x + z$$

then

$$y = z;$$

$$1.x = x; \qquad a(x + y) = ax + ay;$$

$$(a + b)x = ax + bx; \qquad a(bx) = (ab)x.$$

'Subtraction' is defined by taking $x - y = x + (-1).y$. Choose some $y \in \mathscr{X}$ and put $\theta = 0.y$. Then $y + \theta = (1 + 0).y = y$ and for any $x \in \mathscr{X}$

$$x + y = x + (y + \theta) = (x + \theta) + y$$

which implies $x = x + \theta$. Thus θ is a zero point and it is unique because $x + \theta_1 = x = x + \theta_2$ implies $\theta_1 = \theta_2$. A set in a linear space generates a *linear variety* or *linear subspace* or *linear closure* by adding to its points $x, y, \ldots$ all points of the form $ax + by + \ldots$

A *linear metric space* is a linear space with a metric d in which $d(x, y) = 0$ implies $x = y$ (in the same sense as defined in § 1.6) and in which

$$d(x, y) = d(x - y, \theta);$$

$$\text{if } x_n \to \theta \text{ then } ax_n \to \theta;$$

$$\text{if } a_n \to 0 \text{ then } a_n x \to \theta.$$

A *normed linear space* is a linear space on which is defined a norm

$\|x\|$ such that $\|x\| \geq 0$ and

$$\|x+y\| \leq \|x\| + \|y\|, \qquad \|ax\| = |a| . \|x\|,$$
$$\text{if } \|x\| = 0 \text{ then } x = \theta \text{ and conversely.}$$

It becomes a linear metric space on choosing $d(x, y) = \|x - y\|$. On the other hand a linear space with a metric such that $d(x, y) = d(x - y, \theta)$ and $d(ax, \theta) = |a| d(x, \theta)$ is a linear metric space and is a normed linear space on taking $\|x\| = d(x, \theta)$.

A normed linear space which is complete in the metric of the norm is called a *Banach space*, *real* or *complex* according as multiplication by real or complex scalars is defined. For example, the space of real-valued continuous functions on [0, 1] is a Banach space with

$$\|x\| = \sup_{0 \leq t \leq 1} |x(t)|.$$

If $\|x+y\|^2 + \|x-y\|^2 = 2\|x\|^2 + 2\|y\|^2$ the Banach space becomes a *Hilbert space* with a scalar product (x, y) defined by

$$4(x, y) = \|x+y\|^2 - \|x-y\|^2$$

or

$$4(x, y) = \|x+y\|^2 - \|x-y\|^2 + \mathrm{i}\|x+\mathrm{i}y\|^2 - \mathrm{i}\|x-\mathrm{i}y\|^2$$

according as the space is real or complex.

In general, a *scalar product* is a function from $\mathscr{X} \times \mathscr{X}$ to the scalars of the linear space such that $(x, y) = \overline{(y, x)}$ (the complex conjugate), $(x, x) > 0$ if $x \neq \theta$ and $(ax + by, z) = a(x, z) + b(y, z)$. Usually a Hilbert space is determined first by a scalar product and then a norm is defined by $\|x\| = (x, x)^{1/2}$. It is immediately obvious that this definition has the last two properties of a norm. To show that it has the first we use the *Schwarz inequality*

$$|(x, y)| \leq \|x\| . \|y\|. \tag{3}$$

If $(x, y) = 0$ this inequality is trivial and if $(x, y) \neq 0$ we note that $(x - ay, x - ay) \geq 0$ or $\|x\|^2 + |a|^2\|y\|^2 - a(y, x) - \bar{a}\overline{(y, x)} \geq 0$ and (3) follows by choosing $a = \|x\|^2/(y, x)$. Hence

$$\begin{aligned}\|x+y\|^2 = (x+y, x+y) &= \|x\|^2 + \|y\|^2 + (x, y) + (y, x)\\ &\leq \|x\|^2 + \|y\|^2 + 2\|x\| . \|y\|\end{aligned}$$

from (3) and the first norm property is verified.

One of the most important ideas in normed linear spaces is that of a *functional*. A functional F is a function from a normed linear space to the space of scalars. It is called *linear* if $F(ax+by) = aF(x) + bF(y)$. It is said to be *continuous at* x if $F(x_n) \to F(x)$ as $x_n \to x$, i.e. $\|x_n - x\| \to 0$ and to be a *continuous functional* if this is true for all x. It is said to be *bounded* if $|F(x)| \le M\|x\|$ where $M < \infty$ is independent of x. For a bounded linear functional a norm can be defined by

$$\|F\| = \sup_{x \neq \theta} |F(x)|/\|x\|.$$

A typical example of a linear bounded continuous functional is the scalar product (x, y) for fixed y.

If F is linear then $F(\theta) = 0$; thus a linear F is continuous if it is continuous at θ. Generally

Theorem 1.9. *A linear functional over a normed linear space is continuous if it is bounded, and conversely.*

Proof. If F is bounded

$$|F(x_n) - F(x)| = |F(x_n - x)| \le M\|x_n - x\| \to 0$$

as $x_n \to x$ and F is continuous.

If F is not bounded, for each n, there exists an x_n such that $|F(x_n)| > n\|x_n\|$ and then, if $y_n = x_n/n\|x_n\|$, $F(y_n) > 1$ while $\|y_n\| = 1/n \to 0$ so that F is not continuous and the converse is proved. □

Theorem 1.10. *The space of all bounded linear functionals over a normed linear space is a Banach space with norm* $\|F\|$.

Proof. Obviously the space is linear; it is also normed because

$$\|F_1 + F_2\| = \sup_{x \neq \theta} |F_1(x) + F_2(x)|/\|x\| \le \|F_1\| + \|F_2\|,$$

$\|aF\| = |a|\,\|F\|$ and $\|F\| = 0$ implies $|F| = 0$, and conversely. It remains to show that the space is complete.

If $\|F_m - F_n\| \to 0$ as $m, n \to \infty$ there exists, for every $\varepsilon > 0$, n_0 such that $\|F_m - F_n\| < \varepsilon$ for $m, n > n_0$. Hence $|F_m(x) - F_n(x)| < \varepsilon\|x\|$ for every x. Consequently, since the space of scalars is complete, there is an $F(x)$ such that $F_n(x) \to F(x)$ and F is linear. Letting $m \to \infty$, $|F(x) - F_n(x)| < \varepsilon\|x\|$ for $n > n_0$ and every x, i.e. $\|F\| < \|F_n\| + \varepsilon$ and $\|F - F_n\| < \varepsilon$. Hence F is bounded and $F_n \to F$ in the metric of the space; the proof is complete. □

If $\mathscr{B}$ is a Banach space, the space of all bounded linear functionals over $\mathscr{B}$ is a Banach space, called the *dual* of $\mathscr{B}$ and denoted by $\mathscr{B}^*$.

Theorem 1.11. (Hahn–Banach) *If F is a bounded linear functional over a linear variety A of a normed linear space, there is a bounded linear functional F_1 over the whole space which agrees with F over A and has the same norm.*

F_1 is called an *extension* of F; F is the *restriction* of F_1 to A.

Proof. Let $x_1, x_2 \in A$ and $x_0 \notin A$. We can take $\|F\| = 1$ without loss of generality. Assume firstly that the scalars are real. Then

$$F(x_1) - F(x_2) = F(x_1 - x_2) \leq \|x_1 - x_2\| \leq \|x_1 - x_0\| + \|x_2 - x_0\|$$

so that $F(x_1) - \|x_1 - x_0\| \leq F(x_2) + \|x_2 - x_0\|$.

Thus as x goes through the points of A the quantities $F(x) - \|x - x_0\|$ and $F(x) + \|x - x_0\|$ form, for fixed x_0, two classes of real numbers of which the first is situated to the left of the second. Therefore there is at least one number x' such that

$$F(x) - \|x - x_0\| \leq x' \leq F(x) + \|x - x_0\|$$

for all $x \in A$. Since A is a linear variety $-x \in A$ and so

$$|F(x) + x'| \leq \|x + x_0\|. \tag{4}$$

Now define $F_1(x + ax_0)$ for $x \in A$ and arbitrary real a by

$$F_1(x + ax_0) = F(x) + ax'.$$

It is clearly a linear functional, agreeing with F for $a = 0$, and bounded with $\|F_1\| = \|F\|$ because of (4).

The set of $x + ax_0$ is a linear variety $A' \supset A$. If A' is not the whole space a further extension can be obtained in the same way by choosing a point outside A', and clearly the process can be continued. The class $A, A', \ldots$ is linearly ordered by taking $A \prec A'$ when $A \subset A'$ and so possesses a supremum in the family of all possible extensions of F to linear functionals without change of norm. Since this is true of every linearly ordered class in the family, it follows from Zorn's theorem that the family has a supremum which is a member of the family. It must have the whole space as domain since otherwise it could be extended further as above. The proof is complete for real scalars.

For complex scalars let $G(x) = \text{Re}\, F(x)$. Then G is a real-valued bounded linear functional with $\|G\| \leq 1$. It may first be extended to all points $x + ax_0$, a real, as above, and then, since $\text{i}x_0 \notin A$, to all points $x + (a + \text{i}b)x_0$, b real, by understanding them to be $(x + ax_0) + b(\text{i}x_0)$. The required extension of F is now provided by $F_1(x) = G_1(x) - \text{i}G_1(\text{i}x)$, G_1 being the extension of G just described. Clearly F_1 is a linear functional which agrees with F for $x \in A$ and is bounded because, if $r \geq 0$, α be real and $F_1(x) = r\text{e}^{\text{i}\alpha}$,

$$|F_1(x)| = \text{e}^{-\text{i}\alpha} F_1(x) = F_1(x\text{e}^{-\text{i}\alpha}) = G_1(x\text{e}^{\text{i}\alpha}) \leq \|x\|.$$

The argument now proceeds as in the preceding paragraph and the theorem is proved. □

Theorem 1.12. *If A is a linear variety of a normed linear space and x_0 is a point such that $d(x_0, A) > 0$, there exist linear functionals F, F_0 over the space such that $F(x_0) = \|x_0\|$, $\|F\| = 1$ and $F_0(x) = 0$ on A, $F_0(x_0) = d(x_0, A)$.*

Proof. Put $F(ax_0) = a\|x_0\|$ and $F_0(x + ax_0) = ad(x_0, A)$, $x \in A$, and extend by Theorem 1.11. □

Let x be a fixed point of a Banach space $\mathscr{B}$. Then $F(x)$, as F varies in $\mathscr{B}^*$, will be a bounded linear functional over $\mathscr{B}^*$ and therefore an element of $\mathscr{B}^{**} = (\mathscr{B}^*)^*$. Since $|F(x)| \leq \|F\|.\|x\|$ its norm does not exceed $\|x\|$ and since, by Theorem 1.12, there is an F such that $F(x) = \|x\|$, $\|F\| = 1$ the norm in $\mathscr{B}^{**}$ equals $\|x\|$. Therefore, to each $x \in \mathscr{B}$ corresponds a $\Phi(x) \in \mathscr{B}^{**}$ such that $\Phi(ax + by) = a\Phi(x) + b\Phi(y)$, $\|\Phi(x)\| = \|x\|$. In particular $\|\Phi(x) - \Phi(y)\| = \|x - y\|$ so that $\Phi(x)$ and $\Phi(y)$ differ unless $x = y$. Thus the metric is preserved and we can identify $\Phi(x)$ with x thereby mapping $\mathscr{B}$ into $\mathscr{B}^{**}$, i.e. $\mathscr{B} \subset \mathscr{B}^{**}$. If it turns out that $\mathscr{B} = \mathscr{B}^{**}$, $\mathscr{B}$ is said to be *reflexive*.

The *anti-dual* is the space generated by the complex conjugates $\overline{F(x)}$ of linear functionals. The anti-dual and dual of a Hilbert space coincide.

1.8. Numbers

A *numerical function* is one which assigns as its value at each $\omega \in \Omega$ a real number, i.e. it is a function from Ω to $\bar{R}_1$. It is a *finite* or *bounded function* if infinite values do not occur. Unless we state otherwise, R_1 will be understood to have the metric $d(x, y) = |x - y|$. The *p-dimensional real space* R_p is the set of all ordered p-tuples $(x_1, x_2, \ldots, x_p)$ or, equivalently, the product space $R_1 \times R_1 \times \ldots \times R_1$ containing p factors. If each R_1 is replaced by $\bar{R}_1$ the *extended p-dimensional real space* $\bar{R}_p$ is obtained. The metric in R_p will always be taken as $d(x, y) = \{\sum_{i=1}^{p} (x_i - y_i)^2\}^{1/2}$ unless we state otherwise.

In R_1 the set of all points x such that $a < x < b$ will be denoted by (a, b); the set $a \leq x \leq b$ by $[a, b]$; the set $a \leq x < b$ by $[a, b)$; the set $a < x \leq b$ by $(a, b]$. The same notation may be employed in R_p with the understanding that $a < b$ means $a_i < b_i\ (i = 1, \ldots, p)$ and similarly for $a \leq b$. The sets (a, b) and $[a, b]$ are open and closed respectively; $[a, b)$ and $(a, b]$ are neither.

Theorem 1.13. *R_1 is a complete separated locally compact metric space.*

Proof. That a metric space is separated has been proved in § 1.6.

Let A be any bounded set containing an infinite number of points and let B be the set of points each of which is greater than an infinite number of points of A. By the fundamental postulate (§ 1.2) $\inf B = x_0$ (say) exists and any interval $(x_0 - \varepsilon, x_0 + \varepsilon)$ for $\varepsilon > 0$ contains an infinite number of points of A. Therefore x_0 is a point of accumulation of A. If A is a subset of a closed set C, $x_0 \in C$ and so C is compact.

If $\{x_n\}$ is a Cauchy sequence, $|x_m - x_n| < \varepsilon$ for $m > n_0, n > n_0$. Keeping ε and m fixed we see that x_n is bounded; therefore $\{x_n\}$ has both a finite limit superior and finite limit inferior. If these differed by $a > 0$ we could find arbitrarily large integers j and k such that $|x_j - x_k| > \frac{1}{2}a$, contrary to $\{x_n\}$ being a Cauchy sequence. Hence $\overline{\lim}_n x_n = \underline{\lim}_n x_n$ and the common value is the limit of the sequence. The proof is complete. □

It follows from Theorem 1.6 that $\bar{R}_1$ is compact. Theorem 1.3 then shows that every bounded closed set in R_1 is compact.

In R_p, x has components $(x_1, x_2, \ldots, x_p)$; in a sequence $\{x_n\}$, x_n has components $(x_{n1}, x_{n2}, \ldots, x_{np})$.

Theorem 1.14. *R_p is a complete, separated, locally compact metric space.*

Proof. The proof is so similar to that of Theorem 1.13, apart from the complication of components, that we just prove completeness and leave the rest to the reader.

If $\{x_n\}$ is a Cauchy sequence in R_p, $d(x_m, x_n) < \varepsilon$ for $m, n > n_0$. Therefore $|x_{mi} - x_{ni}| \leq d(x_m, x_n) < \varepsilon$ for each i, and $\{x_{mi}\}$ is a Cauchy sequence in R_1. By Theorem 1.13 this sequence has a limit x_{0i} and hence $\{x_n\}$ converges to x_0 in R_p, i.e. R_p is complete. □

From now on we shall be primarily concerned with metric spaces and shall obtain a number of results that will be needed later.

1.9. Limits

Let f be a function from the metric space $\mathscr{X}$ to the metric space Ξ with metric δ. Let $\xi \in \Xi$ and let x_0 be a point of accumulation of $\mathscr{X}$. Then we have the following definition.

Definition 1.3. *If for any given* $\varepsilon > 0$, *there exists* $\eta > 0$ *such that if* $d(x, x_0) < \eta$, *where* $x \neq x_0$, *then* $\delta(f(x), \xi) < \varepsilon$, *we define* $\lim_{x \to x_0} f(x) = \xi$.

First note that if the set of positive integers together with ∞ is provided with the metric $d(m, n) = |1/m - 1/n|$, $d(m, \infty) = 1/m$ a metric space is obtained with ∞ as a point of accumulation. Then a sequence $\{x_n\}$ which converges to x_0 comes under the above definition and we may write $\lim_{n \to \infty} x_n = x_0$.

As in §1.6 a definition equivalent to Definition 1.3 is that $\lim_{n \to \infty} f(x_n) = \xi$ for every sequence $\{x_n\}$ such that $\lim_{n \to \infty} x_n = x_0$.

If $x_0 \in \mathscr{X}$ then $\lim_{x \to x_0} f(x) = f(x_0)$ implies continuity of f at x_0, and conversely; if $\lim_{x \to x_0} f(x) \neq f(x_0)$, f is said to be *discontinuous* at x_0.

Theorem 1.15. *If* Ξ *is a complete metric space, a necessary and sufficient condition that there exists* $\xi \in \Xi$ *such that* $\lim_{x \to x_0} f(x) = \xi$ *is that* $\{f(x_n)\}$ *is a Cauchy sequence for every sequence* $\{x_n\}$ *such that* $\lim_{n \to \infty} x_n = x_0$.

Proof. Necessity follows from the preceding paragraph.

With regard to sufficiency any Cauchy sequence in a complete metric space determines a limit. If the limits of $\{f(x_n)\}$ and $\{f(y_n)\}$

were different, the sequence obtained by taking terms alternately from the two sequences would not be a Cauchy sequence. □

Let f be a function from the product space $\mathscr{X} \times \mathscr{Y}$ to Ξ. It may be that, for a given $y \in \mathscr{Y}$, $\lim_{x \to x_0} f(x, y)$ exists; as y varies over $\mathscr{Y}$ we obtain a function from $\mathscr{Y}$. This leads to

Definition 1.4. *If* $\lim_{x \to x_0} f(x, y) = f_0(y)$, *the function* f *is said to converge pointwise on* $\mathscr{Y}$ *to* f_0 *as* $x \to x_0$.

If f converges pointwise to f_0, there exists $\eta > 0$ such that, for any given $y \in \mathscr{Y}$ and $\varepsilon > 0$, $\delta(f(x, y), f_0(y)) < \varepsilon$ when $d(x, x_0) < \eta$. In general η will depend on y and ε. However, it may happen that η is independent of y; in that case we have

Definition 1.5. *If for any given* $\varepsilon > 0$ *and every* $y \in \mathscr{Y}$ *there exists* $\eta > 0$ *such that* $\delta(f(x, y), f_0(y)) < \varepsilon$ *when* $d(x, x_0) < \eta$, $x \neq x_0$, *we say that* f *converges uniformly to* f_0.

For example, let $f(x, y) = y/(x + y)$ on $0 < x \leq 1$ and $0 < y < 1$ and let Ξ be R_1. Then $x = 0$ is a point of accumulation and $|1 - y/(x + y)| = x/(x + y) < \varepsilon$ if $x < \varepsilon y/(1 - \varepsilon)$ when $\varepsilon < 1$. Hence, taking $\eta = \varepsilon y/(1 - \varepsilon)$ we have $\lim_{x \to 0} f(x, y) = 1$, i.e. f converges pointwise on $0 < y < 1$. However, this η depends on y and no η independent of y is satisfactory for the whole interval so that the convergence is not uniform. If the interval for y were $0 < a \leq y < 1$ we could choose $\eta = \varepsilon a/(1 - \varepsilon)$ and then the convergence would be uniform.

By choosing $\mathscr{X}$ to be the set of positive integers and ∞, with the metric described above, we obtain results for sequences. Thus $\lim_{n \to \infty} f_n(y) = f_0(y)$ if there is an n_0 (which may depend on y and ε) such that $\delta(f_n(y), f_0(y)) < \varepsilon$ for $n > n_0$. The convergence is uniform if n_0 is independent of y.

Theorem 1.16. *In order that* $\lim_{x \to x_0} f(x, y) = f_0(y)$ *uniformly on* $\mathscr{Y}$ *it is necessary and sufficient that for every sequence* $\{x_n\}$ *such that* $\lim_{n \to \infty} x_n = x_0$, $\lim_{n \to \infty} \delta(f(x_n, y_n), f_0(y_n)) = 0$ *for every sequence* $\{y_n\}$.

Proof. If the convergence is uniform there is an η independent of y such that $\delta(f(x, y), f_0(y)) < \varepsilon$ for $d(x, x_0) < \eta$. Since $\lim_{n \to \infty} x_n = x_0$ there is an n_0 such that $d(x_n, x_0) < \eta$ for $n > n_0$. Consequently, x and

y can be replaced by x_n and y_n respectively in the inequality on δ and necessity is proved.

Uniform convergence will fail if for every $\eta > 0$ there is some $y \in \mathscr{Y}$ and an $\varepsilon > 0$ such that $\delta(f(x, y), f_0(y)) \geq \varepsilon$ for some x with $d(x, x_0) < \eta$. The set X of x for which this is true has x_0 as a point of accumulation. Now the set $X_n = X \cap \underline{N}(x_0, 1/n) - \underline{N}(x_0, 1/(n+1))$ is non-empty for an infinite number of $n \geq k$ otherwise $X \cap N(x_0, 1/k) = \varnothing$ contrary to x_0 being a point of accumulation. Choose a subsequence $\{X_{n_i}\}$ of non-empty sets and define $\{x_i\}$ by taking $x_i \in X_{n_i}$. Then, for $n_i > 1/\varepsilon$, $x_i \in \underline{N}(x_0, \varepsilon)$ and $\lim_{i \to \infty} x_i = x_0$. Thus there is a sequence such that $\delta(f(x_n, y_n), f_0(y_n)) \geq \varepsilon$ contrary to assumption. Sufficiency of the condition is therefore proved. □

When $\mathscr{Y}$ is a metric space the possibility of $\lim_{y \to y_0} f_0(y)$ can be considered. This process gives naturally the *iterated limit*

$$\lim_{y \to y_0} \lim_{x \to x_0} f(x, y).$$

Similarly the iterated limit

$$\lim_{x \to x_0} \lim_{y \to y_0} f(x, y)$$

can be considered. Another type of limit occurs from regarding $\mathscr{X} \times \mathscr{Y}$ as a metric space. This can be done by defining the distance function between (x_1, y_1) and (x_2, y_2) by

$$d[(x_1, y_1), (x_2, y_2)] = \max[d_1(x_1, x_2), d_2(y_1, y_2)],$$

d_1 and d_2 being those of $\mathscr{X}$ and $\mathscr{Y}$ respectively. Now the limit as $(x, y) \to (x_0, y_0)$ is available from Definition 1.3. This limit, known as the *double limit*, is written

$$\lim_{\substack{x \to x_0 \\ y \to y_0}} f(x, y).$$

For example, let f be defined on $0 < x \leq 1, 0 < y \leq 1$ by $f(x, y) = xy/(x^2 + y^2)$. Then $\lim_{x \to 0} f(x, y) = 0$ so that $\lim_{y \to 0} \lim_{x \to 0} f(x, y) = 0$. Similarly $\lim_{x \to 0} \lim_{y \to 0} f(x, y) = 0$. However, if we take $y_n = ax_n$ where $\lim_{n \to \infty} x_n = 0, f(x_n, y_n) = a/(1 + a^2)$ and $d[(x_n, y_n), (0, 0)] \to 0$. Since a is arbitrary $\lim_{\substack{x \to 0 \\ y \to 0}} f(x, y)$ does not exist.

This example shows that iterated limits can exist when the double limit does not exist. Similarly the double limit can exist when iterated limits do not. A theorem concerning the existence of limits is

Theorem 1.17. *Let f be defined from the product metric space $\mathscr{X} \times \mathscr{Y}$ to the complete metric space Ξ. Suppose $\lim_{x \to x_0} f(x, y) = f(x_0, y)$ for each $y \neq y_0$ and suppose $\lim_{y \to y_0} f(x, y) = f(x, y_0)$ uniformly on $\mathscr{X}$ with $x \neq x_0$. Then the double limit and iterated limits all exist at (x_0, y_0) and are all equal.*

Proof. For any given $\varepsilon > 0$ there is, since $\lim_{y \to y_0} f(x, y) = f(x, y_0)$ uniformly, an η_1 such that, if $d_2(y_0, y_1) < \eta_1$ and $d_2(y_0, y_2) < \eta_1$, $\delta(f(x, y_1), f(x, y_2)) < \varepsilon$ for every $x \neq x_0$. Let y_3 be fixed in $\underline{N}(y_0, \eta_1)$. Since $\lim_{x \to x_0} f(x, y) = f(x_0, y)$ for $y \neq y_0$, there is an η_2 such that $\delta(f(x_1, y_3), f(x_2, y_3)) < \varepsilon$ for $d_1(x_0, x_1) < \eta_2$, $d_1(x_0, x_2) < \eta_2$. Let $\eta = \min(\eta_1, \eta_2)$ and let $(x, y) \in \mathscr{X} \times \mathscr{Y}$ be such that $0 < d[(x, y), (x_0, y_0)] < \eta$; let (x', y') be another such point. Then $d_1(x, x_0) < \eta \leq \eta_2$ and $d_2(y, y_0) < \eta \leq \eta_1$ so that

$$\begin{aligned}\delta(f(x, y), f(x', y')) &\leq \delta(f(x, y_3), f(x, y)) + \delta(f(x, y_3), f(x', y')) \\ &\leq \delta(f(x, y_3), f(x, y)) + \delta(f(x', y_3), f(x, y_3)) \\ &\quad + \delta(f(x', y_3), f(x', y')) \\ &< 3\varepsilon\end{aligned}$$

from the preceding provided that $x \neq x_0, x' \neq x_0$. If $x = x_0$ then $y \neq y_0$ and

$$\begin{aligned}\delta(f(x, y), f(x', y')) \leq \delta(f(x_3, y), f(x, y)) + \delta(f(x_3, y'), f(x_3, y)) \\ + \delta(f(x_3, y'), f(x', y'))\end{aligned}$$

where x_3 is fixed in $\underline{N}(x_0, \eta_2)$. The first two terms are each less than ε. The third term is less than ε from the preceding result on δ unless $x' = x_0$ but then $y' \neq y_0$ and the earlier inequality may be employed. In all cases $\delta(f(x, y), f(x', y')) < 3\varepsilon$ and so it follows from Theorem 1.15 that $\lim_{\substack{x \to x_0 \\ y \to y_0}} f(x, y)$ exists.

Denote the double limit by ξ. Let $y \in \underline{N}(y_0, \eta)$; then $0 < d[(x_0, y), (x_0, y_0)] < \eta$ and so $\delta(f(x_0, y), \xi) < \varepsilon$. Therefore $\lim_{y \to y_0} f(x_0, y) = \xi$, i.e. $\lim_{y \to y_0} \lim_{x \to x_0} f(x, y) = \xi$. The other iterated limit may be dealt with similarly and the proof is complete. □

Corollary 1.17. *If $\{f_n\}$ is a sequence of real-valued continuous functions on $\mathscr{X}$ such that* $\lim_{n\to\infty} f_n(x) = f_0(x)$ *uniformly on $\mathscr{X}$, then f_0 is continuous.*

Proof. Since $f_n(x)$ is continuous, $\lim_{x\to x_0} f_n(x) = f_n(x_0)$ for every finite n. Therefore, by Theorem 1.17, $\lim_{x\to x_0} \lim_{n\to\infty} f_n(x) = \lim_{n\to\infty} \lim_{x\to x_0} f_n(x)$, i.e. $\lim_{x\to x_0} f_0(x) = \lim_{n\to\infty} f_n(x_0) = f_0(x_0)$. □

A function is said to be *uniformly continuous on* $\mathscr{X}$ if for every $\varepsilon > 0$ there is an η, which depends only on ε, such that $\delta(f(x), f(y)) < \varepsilon$ whenever $d(x, y) < \eta$.

Theorem 1.18. *A continuous function from a compact metric space $\mathscr{X}$ to a metric space Ξ is uniformly continuous.*

Proof. Since f is continuous, for every $x \in \mathscr{X}$ there is an $\eta(x)$ such that $\delta(f(x), f(y)) < \varepsilon$ for $d(x, y) < 2\eta(x)$. Since the domain is compact it can be covered by a finite number of spheres; let their centres and radii be $x_k, \eta(x_k)$ respectively, $k = 1, \ldots, n$. Let η be the smallest radius. Suppose x is in the kth sphere. Then, if $d(x, y) < \eta$, $d(x_k, y) < 2\eta(x_k)$ and so

$$\delta(f(x), f(y)) \leq \delta(f(x_k), f(x)) + \delta(f(x_k), f(y)) < 2\varepsilon$$

and the theorem is proved. □

Theorem 1.19. *A real-valued function, continuous on an interval in R_1, attains every value between its supremum and infimum.*

For, if the range consisted of two (or more) disjoint closed sets so would the domain.

1.10. Series

Let $\{a_n\}$ be a sequence of real numbers and let $S_n = a_1 + a_2 + \ldots + a_n$. If the sequence $\{S_n\}$ converges to a limit S in R_1 the series $a_1 + a_2 + \ldots$ *is said to converge to sum* S and we write

$$\sum_{n=1}^{\infty} a_n = S = \lim_{n\to\infty} S_n.$$

If $\{S_n\}$ does not converge the series is said to be *divergent*. A consequence of Theorems 1.13 and 1.15 is

Theorem 1.20. *A necessary and sufficient condition for an infinite series to converge is that for every* $\varepsilon > 0$ *there is an* n_0 *such that* $|S_m - S_n| < \varepsilon$ *for* $m > n_0$ *and* $n > n_0$.

It follows at once, by taking $m = n + 1$, that a necessary condition for convergence is $\lim_{n \to \infty} a_n = 0$. This is not sufficient since the series with $a_n = 1/n$ is divergent.

The series $a_1 + a_2 + \ldots$ is said to be *absolutely convergent* when the series $|a_1| + |a_2| + \ldots$ is convergent.

Theorem 1.21. (Comparison test) *If* $|a_n| \leq b_n$ *for all n, and* $b_1 + b_2 + \ldots$ *is convergent then* $a_1 + a_2 + \ldots$ *is absolutely convergent.*

Proof. If $\sigma_n = \sum_{m=1}^{n} b_m$ there is an n_0 such that $|\sigma_{n+p} - \sigma_n| < \varepsilon$ for $n > n_0, p \geq 0$. Hence $|S_{n+p} - S_n| \leq |a_{n+1}| + |a_{n+2}| + \ldots + |a_{n+p}| \leq |\sigma_{n+p} - \sigma_n| < \varepsilon$ which demonstrates the theorem and also shows, by Theorem 1.20, that $a_1 + a_2 + \ldots$ is convergent. By taking $b_n = a_n$ we see that an *absolutely convergent series is convergent.* □

If $|a_{n+1}| \leq a|a_n|$ for $n \geq n_0$, then $|a_{n+1}| \leq a^{n+1-n_0}|a_{n_0}|$ and the series converges absolutely if $a < 1$. On the other hand, if $|a_{n+1}| \geq a|a_n|$ and $a > 1$, a_n cannot tend to zero as $n \to \infty$ and the series cannot converge. Hence we have

Corollary 1.21. (Ratio test) *The series is absolutely convergent or divergent according as* $\lim_{n \to \infty} |a_{n+1}/a_n| \lessgtr 1$.

Some other test is necessary if the limit is unity.

Theorem 1.22. *The sum of an absolutely convergent series is not affected by altering the order of its terms.*

Proof. Let S'_n be the sum of the first n terms in a rearrangement. Given $\varepsilon > 0$, there is an n_0 such that $|S_{n_0+p} - S_{n_0}| < \varepsilon$ for all $p \geq 0$. Suppose that the first n_0 terms of S are included in the first m_0 terms of S'. Then, for $m > m_0$, $|S'_m - S| \leq |S_{n_0} - S| + |\text{terms with suffixes} > n_0| < 2\varepsilon$ which shows that $\lim_{m \to \infty} S'_m = S$ and proves the theorem. □

Series of complex numbers, where a typical term is $a_n + ic_n$, are dealt with by saying that the series converges when the $\{a_n\}$ and $\{c_n\}$

series converge separately. It is evident that Theorems 1.20–1.22 remain valid with a_n replaced by $a_n + ic_n$.

Let $\{f_n(x)\}$ be a sequence of complex-valued functions and let $S_n(x) = \sum_{m=1}^{n} f_m(x)$. Then the series is said to *converge pointwise (uniformly) on $\mathscr{X}$ to $S(x)$ if the sequence $\{S_n(x)\}$ converges pointwise (uniformly) to $S(x)$.* The series *converges absolutely* if $\sum |f_m(x)|$ converges pointwise. It is apparent that Theorems 1.20–1.22 hold for pointwise convergence. Theorem 1.20 is also true for uniform convergence provided that n_0 can be chosen independent of x.

Theorem 1.23. (Weierstrass's M-test.) *If for every $x \in \mathscr{X}$, $|f_n(x)| \leq M_n$ where $\sum_{n=1}^{\infty} M_n$ is a convergent series of positive real numbers, the series $\sum f_n(x)$ is absolutely and uniformly convergent on $\mathscr{X}$.*

Proof. Absolute convergence follows from Theorem 1.21. Also there is an n_0, depending only on ε, such that $M_{n+1} + M_{n+2} + \dots + M_{n+p} < \varepsilon$ for $n > n_0$. Hence $|S_{n+p}(x) - S_n(x)| < \varepsilon$ and uniform convergence follows. □

In general, absolute convergence does not imply uniform convergence nor is the converse true. The series $\sum_{n=1}^{\infty} x^2/(1+x^2)^{n+1}$ converges absolutely for $x \geq 0$ but not uniformly. In fact

$$S_n(x) = 1 + x^2 - (1 + x^2)^{1-n}$$

and $\lim_{n\to\infty} S_n(x) = 1 + x^2$ if $x \neq 0$, but $\lim_{n\to\infty} S_n(0) = 0$. On the other hand the series $\sum_{n=1}^{\infty} (-1)^n/(n + x^2)$ converges uniformly but not absolutely.

An immediate consequence of Corollary 1.17 is

Theorem 1.24. *The sum of a uniformly convergent series of continuous functions is continuous.*

One application of the preceding theory is to construct an extension of a real-valued continuous function f on the closed set A to $\mathscr{X}$. Let $M_1 = \sup_{x\in A} |f(x)|$ and let A_1, B_1 be the sets for which $f(x) \leq -\frac{1}{3}M_1$ and $f(x) \geq \frac{1}{3}M_1$ respectively. Define g_1 by

$$g_1(x) = \tfrac{2}{3}M_1 \frac{d(x, A_1)}{d(x, A_1) + d(x, B_1)} - \tfrac{1}{3}M_1$$

so that $g_1(x) = -\frac{1}{3}M_1$ for $x \in A_1$, $g_1(x) = \frac{1}{3}M_1$ for $x \in B_1$. Also g_1 is continuous on $\mathscr{X}$ and $|g_1(x)| \leq \frac{1}{3}M_1$ for every $x \in \mathscr{X}$. Define f_2 on A by $f_2 = f - g_1$ and construct g_2 from f_2 in the same way as g_1 was constructed from f. Continuing the process we obtain a sequence $\{g_n\}$ of real-valued continuous functions such that $M_n = \sup_{x \in A}|f_n(x)| = \sup_{x \in A}|f_{n-1}(x) - g_{n-1}(x)| \leq \frac{2}{3}M_{n-1}$. Therefore, by Theorem 1.23, $\sum_{n=1}^{\infty} g_n(x)$ is uniformly convergent and, by Theorem 1.24, is continuous; since $\sup|f(x) - \sum_{n=1}^{m-1} g_n(x)| = M_m \to 0$ on A, the series equals f on A and an extension has been found.

1.11. Derivatives

In this section x will denote a point of R_1 and f a numerical function unless we state otherwise. The *derivative* $\mathrm{d}f/\mathrm{d}x$ of f at x is defined by

$$\frac{\mathrm{d}f(x)}{\mathrm{d}x} = \lim_{h \to 0} \frac{f(x+h) - f(x)}{h}.$$

The notation $f'(x)$, sometimes employed, is reserved for later use. The *upper and lower right derived numbers* are the limit superior and limit inferior respectively of $(1/h)\{f(x+h) - f(x)\}$ as $h \to 0$ with $h > 0$. The left derived numbers are defined similarly with $h < 0$. The derived numbers always have a meaning (being finite or infinite) even when a derivative does not exist; in fact, a finite derivative exists if, and only if, all four derived numbers have the same finite value. If the two right derived numbers are finite and equal the function possesses a *right-hand derivative*; there is a similar definition for the *left-hand derivative.*

Usually it is not difficult to decide whether a derivative exists for most of the functions which occur in practice, but there is a theorem which covers a wide class of functions. To prove it we first introduce the notion of a *set of measure zero*, which is a set of points which can be enclosed in a countable number of intervals whose total length is arbitrarily small. Clearly a subset of such a set is a set of measure zero; so is the union of a countable number of sets of measure zero (enclose the nth with a set of intervals whose total length is less than $\varepsilon/2^n$).

Relations which hold at all points except those of a set of measure zero are said to be true *almost everywhere.*

A *monotone* function is either *non-increasing* $f(x) \le f(y)$, $x > y$ or *non-decreasing* $f(x) \ge f(y)$, $x > y$. If $f(x)$ is finite, non-decreasing and continuous from the left, i.e. $\lim_{x \to x_0 - 0} f(x) = f(x_0)$ the function $f(b) - f(a)$, $b \ge a$, defines a (Lebesgue–Stieltjes) *measure* $\mu[a, b)$, for intervals $[a, b]$ with the properties $0 \le \mu[a, b) < \infty$, $\mu[a, b) \to 0$ as $a \to b$.

Theorem 1.25. *A monotone function possesses a finite derivative almost everywhere.*

Proof. Let the derived numbers be $n_{rs}, n_{ri}, n_{ls}, n_{li}$, the suffixes r, l, s, i referring to right, left, superior and inferior respectively. We have to show that these four numbers are equal and finite. Suppose that f is non-decreasing in $a \le x \le b$; the proof when f is non-increasing is similar.

Assume firstly that f is continuous. The set of points where $n_{rs} = \infty$ is contained in the set A for which $n_{rs} > c_1$, where c_1 is an arbitrary real number. Now $n_{rs} > c_1$ at x implies that there is a $y > x$ such that $f(y) - f(x) > c_1(y - x)$, i.e. $g(y) > g(x)$ where $g(x) = f(x) - c_1 x$. The function g is continuous on $[a, b]$.

Let B be the set of points x interior to the interval such that there exists $y > x$ with $g(y) > g(x)$. B is open since the inequality, if valid for one x, is valid for nearby x because of the continuity of g. Let (a_k, b_k) be a typical interval of B and let $a_k < x < b_k$. Let x_1, between x and b_k, be the point nearest to the last for which $g(x_1) \ge g(x)$. Then, if $x_1 \ne b_k$, there is a y_1 beyond b_k since $b_k \notin B$ such that $g(x_1) < g(y_1) \le g(b_k) < g(x_1)$. In view of this contradiction $x_1 = b_k$ and $g(x) \le g(b_k)$; allowing $x \to a_k$ we have $g(a_k) \le g(b_k)$. B comprises a countable union of disjoint intervals of this type (countable, because there is only a finite number of disjoint intervals of length greater than $(b - a)/n$).

Now $A \subset B$ and therefore $f(b_k) - f(a_k) \ge c_1(b_k - a_k)$; hence

$$c_1 \sum_k (b_k - a_k) \le \sum_k \{f(b_k) - f(a_k)\} \le f(b) - f(a)$$

which shows that the total length of the intervals (a_k, b_k) can be made arbitrarily small by choosing c_1 large enough. Hence the set of points where $n_{rs} = \infty$ is of measure zero, i.e. $n_{rs} < \infty$ almost everywhere.

Let $c_2 < c_1$ and let B_1 be the set of intervals of the last but one paragraph for the function $g(x)=f(-x)+c_2x$; then B_1 contains those x for which $n_{\text{li}}<c_2$. Let B_2 be the set (a_{kj}, b_{kj}) corresponding to $f(x)-c_1x$, but considered only in the interior of the intervals (a_k, b_k) of B_1; then $f(b_k)-f(a_k)\le c_2(b_k-a_k)$ and $f(b_{kj})-f(a_{kj})\ge c_1(b_{kj}-a_{kj})$. Hence $c_1 l_2 \le c_2 l_1$ where l_1, l_2 are the total lengths of B_1 and B_2. In the interior of B_2 we form the set B_3 corresponding to $f(-x)+c_2x$ and, proceeding in this way, obtain a sequence such that $c_1 l_{2n}\le c_2 l_{2n-1}$; therefore $l_{2n}\le (c_2/c_1)^n l_1 \to 0$.

This shows that the points where $n_{\text{rs}}>c_1$ and $n_{\text{li}}<c_2$, obviously in B_n, form a set of measure zero. Each point x such that $n_{\text{rs}}>n_{\text{li}}$ belongs to such a set; in fact c_1 and c_2 can be taken as rational numbers, since two rational numbers can always be interposed between two different real numbers. The union of such sets is countable since the rationals are countable and so is of measure zero. Hence $n_{\text{rs}}\le n_{\text{li}}$ almost everywhere.

Applying this to $-f(-x)$ we have $n_{\text{ls}}\le n_{\text{ri}}$ almost everywhere. Therefore $n_{\text{rs}}\le n_{\text{li}}\le n_{\text{ri}}\le n_{\text{ls}}$, i.e. the four derived numbers are equal almost everywhere. Since n_{rs} is finite almost everywhere, the derivative exists almost everywhere. This completes the proof when f is continuous.

If f is not continuous its points of discontinuity are countable and therefore form a set of measure zero. For the number of points where $|f(x+0)-f(x-0)|>1/n$ must be finite otherwise there would be a convergent subsequence such that one at least of the limits of $f(x+0), f(x-0)$ did not exist, and this is not possible because f is monotone. Now the main place where continuity was used was in the determination of B; in this case take $g_1(x)=\max\{g(x-0), g(x), g(x+0)\}$ and let B be the set of x where there is $y>x$ such that $g(y)>g_1(x)$. It is a countable union of disjoint intervals with $g(a_k+0)\le g_1(b_k)$. The ensuing modifications are slight and will be omitted; there is no change at points of continuity and the points of discontinuity form a set of measure zero. □

This theorem covers many of the cases encountered in applied mathematics since the domain can usually be split into intervals where the function is monotone. One important class is that of the functions of *bounded variation*, i.e. functions which are the difference between two non-decreasing monotone functions. Obviously, a function of bounded variation possesses a derivative almost everywhere. In particular, a measure has a derivative almost everywhere.

If f is defined on R_2 we can consider its derivative for fixed y. In this way we obtain the partial derivative $\partial f/\partial x$ defined by

$$\frac{\partial}{\partial x} f(x, y) = \lim_{h \to 0} \frac{1}{h} \{f(x+h, y) - f(x, y)\}.$$

Similarly for $\partial f/\partial y$. From these we may form the iterated cross derivatives $(\partial/\partial x)(\partial f/\partial y)$ and $(\partial/\partial y)(\partial f/\partial x)$. We may also define a double cross derivative as

$$\lim_{\substack{h \to 0 \\ k \to 0}} \frac{1}{hk} \{f(x+h, y+k) - f(x, y+k) - f(x+h, y) + f(x, y)\}.$$

It follows from Theorem 1.17 that the iterated cross derivatives exist and are equal whenever the double cross derivative exists and the first partial derivatives exist in some neighbourhood of (x, y). The double cross derivative is the natural generalisation of the preceding theory, by working with areas instead of lengths: *a set of measure zero* is a set of points which can be enclosed in a countable number of rectangles whose total area is arbitrarily small. A *measure* $\mu[a, b)$ for $[a, b]$ (see § 1.8 for notation) has the properties $0 \leq \mu[a, b) < \infty$, $\mu[a, b) \to 0$ as $a \to b$, i.e. $a_i \to b_i$ $(i = 1, 2)$ and $\mu[a, b)$ is of the form

$$f(b_1, b_2) - f(a_1, b_2) - f(b_1, a_2) + f(a_1, a_2).$$

A measure possesses a double cross derivative almost everywhere.

The generalisations to functions defined in R_p are obvious.

1.12. Integration

There are several ways in which an integral can be defined; we use the one most convenient for subsequent purposes. As in the preceding section x will denote a point of R_1 and f a numerical function unless we state otherwise. Divide the interval (a, b) into sub-intervals at the points $x_1, x_2, \ldots, x_{n-1}$ with $a \leq x_1 \leq x_2 \leq \ldots \leq x_{n-1} \leq b$ and take $x_0 = a$, $x_n = b$. Form the sums

$$S_n = \sum_{m=1}^{n} (x_m - x_{m-1}) \sup_{x_{m-1} \leq x \leq x_m} f(x),$$

$$s_n = \sum_{m=1}^{n} (x_m - x_{m-1}) \inf_{x_{m-1} \leq x \leq x_m} f(x).$$

If, as $n \to \infty$ and the lengths of the sub-intervals tend uniformly to zero, S_n and s_n have the same limit S then S is called the *Riemann integral* of $f(x)$ and we write $S = \int_a^b f(x)\mathrm{d}x$; f is said to be *integrable*. If $c < a$, $\int_a^c f(x)\mathrm{d}x$ means $-\int_c^a f(x)\mathrm{d}x$.

If f is integrable it is necessary that there is a division of (a, b) into sub-intervals whose lengths do not exceed δ_n such that

$$0 \leq \sum (x_m - x_{m-1}) \left\{ \sup_{x_{m-1} \leq x \leq x_m} f(x) - \inf_{x_{m-1} \leq x \leq x_m} f(x) \right\} < 2^{-n}.$$

Let $F_n(x)$ be the supremum of such sums for all divisions of (a, x) whose sub-intervals have lengths not exceeding δ_n. Clearly $2^{-n} > F_n(y) \geq F_n(x)$ for $y > x$ and therefore $F_n(x)$ has a derivative almost everywhere, by Theorem 1.25. Also $\sum_n F_n(x)$ is convergent (Theorem 1.21) and $\sum_{m=1}^n \mathrm{d}F_m/\mathrm{d}x$, which is finite almost everywhere, satisfies $\sum_{m=1}^n \mathrm{d}F_m/\mathrm{d}x \leq \sum_{m=1}^{n+1} \mathrm{d}F_m/\mathrm{d}x \leq \mathrm{d}\sum_{m=1}^{\infty} F_m/\mathrm{d}x$ and so is convergent almost everywhere (§ 1.2). It is therefore necessary that $\lim_{n \to \infty} \mathrm{d}F_n/\mathrm{d}x = 0$ almost everywhere. Hence, if $h \leq \delta_n$,

$$\sup_{x \leq y \leq x+h} f(y) - \inf_{x \leq y \leq x+h} f(y) \leq \frac{1}{h}\{F_n(x+h) - F_n(x)\} \to 0$$

almost everywhere. Hence, a *necessary condition for f to be integrable is that it be continuous almost everywhere.* It can be shown that this condition is also sufficient when f is bounded but the proof will not be given.

Another way of approaching integration is to define the integral of a *step function* – a function which vanishes except in a finite number of intervals of finite length where it has a constant value which may differ from interval to interval – in an interval (a, b), which may be finite or infinite, as $\sum_i c_i l_i$, l_i being the length of the interval in which the constant value is c_i. From the class $\mathscr{C}_0$ of step functions we form a new class $\mathscr{C}_1$ of functions which are limits almost everywhere of sequences $\{f_n\}$ such that $f_n(x) \leq f_{n+1}(x)$ and $\sum_i c_{ni} l_{ni} \leq B < \infty$.

To show that such sequences have limits we can suppose that $f_n \geq 0$; otherwise we consider the sequence $\{f_n - f_1\}$. The set of points A for which the sequence diverges is contained in the set for which $f_m(x) > B/\varepsilon$ for $m > m_0$ and this set is the union of the sets A_n for which $f_n(x) > B/\varepsilon$. A_n consists of a finite number of intervals whose total length L_n, multiplied by B/ε, cannot exceed the integral. Therefore $L_n B/\varepsilon \leq B$. Since $A_n \subset A_{n+1}$ the total length of A does not exceed ε, i.e. A is of measure zero.

Since the integrals of f_n are bounded and increasing, they converge to a finite limit (§1.2) and we call this limit the integral of $f(=\lim_{n\to\infty} f_n)$, i.e. $\int_a^b f(x)\mathrm{d}x = \lim_{n\to\infty} \int_a^b f_n(x)\mathrm{d}x$.

A new class $\mathscr{C}_2$ is formed by saying that if $f_1 \in \mathscr{C}_1$ and $f_2 \in \mathscr{C}_1$, the integral of $f_1 - f_2$ is defined by

$$\int_a^b \{f_1(x) - f_2(x)\}\mathrm{d}x = \int_a^b f_1(x)\mathrm{d}x - \int_a^b f_2(x)\mathrm{d}x.$$

The integral so defined is the *Lebesgue integral*. It has the property that if $f \in \mathscr{C}_2$ then $|f| \in \mathscr{C}_2$. If (a, b) is a *finite* interval and f has a Riemann integral then it is the limit of s_n, i.e. it is the limit of an integral formed from a step function with values $\inf f$ on the sub-intervals. Since f is continuous almost everywhere $f \in \mathscr{C}_1$ and so the Riemann integral coincides with the Lebesgue integral. Conversely, if $f \in \mathscr{C}_1$ and $-f \in \mathscr{C}_1$, f equals almost everywhere a function which has a Riemann integral which agrees with the Lebesgue integral. The class $\mathscr{C}_2$ provides examples of functions possessing Lebesgue integrals but not Riemann integrals.

To justify the definition of the Lebesgue integral one should show that two different sequences of step functions converging almost everywhere to the same function in $\mathscr{C}_1$ give the same limiting integral. This is true but will not be proved here. Nor will the fact that $\mathscr{C}_2$ is complete, i.e. an increasing sequence from $\mathscr{C}_2$ with bounded integrals always converges to an element of $\mathscr{C}_2$.

The *Stieltjes integral* $\int_a^b f\mathrm{d}\mu$ where μ is a measure and f continuous is defined as the limit of the sum $\sum_m \mu[x_m, x_{m-1}) f(\xi_m)$ $(x_{m-1} \leq \xi_m \leq x_m)$. It can be generalised to other f by various means such as limiting ξ_m only to points of continuity of f. One of the most convenient approaches is to regard the integral as a linear functional which assigns a value to each continuous f (such functions form a Banach space with norm $\|f\| = \max_{a \leq x \leq b} |f(x)|$) and then extend the functional beyond the domain of continuous f. Conversely, the

Riesz representation theorem states that any linear functional over continuous f can be written as a Stieltjes integral. The Stieltjes integral can always be expressed as an ordinary integral by means of the formula

$$\int_a^b f\,\mathrm{d}\mu = \int_{T(a)}^{T(b)} f\frac{\mathrm{d}\mu}{\mathrm{d}T}\mathrm{d}T$$

where $T(x)$ is the total variation of $\mu[x, a)$.

The preceding paragraphs show that the Lebesgue integral can be obtained by a limiting operation from the Riemann integral. It will therefore be sufficient for many purposes to consider all integrals as of the Riemann type. All remarks apply to complex-valued f by dealing with the real and imaginary parts separately.

If $\lim_{x\to\infty}\int_a^x f(t)\,\mathrm{d}t$ exists we say that the infinite integral $\int_a^\infty f(t)\,\mathrm{d}t$ exists and write $\int_a^\infty f(t)\,\mathrm{d}t = \lim_{x\to\infty}\int_a^x f(t)\,\mathrm{d}t$.

(*Note*: In the definition of the Lebesgue integral the interval (a, b) can be infinite so that $\int_a^\infty$ is defined directly; it is not necessarily the same as $\lim_{b\to\infty}\int_a^b$ because $|f|$ must also be integrable. An integrand for which the two differ is $(\sin x)/x$.)

Similarly we define

$$\int_{-\infty}^{\infty} f(t)\,\mathrm{d}t = \lim_{x\to\infty}\int_{-x}^{x} f(t)\,\mathrm{d}t.$$

Observe that if $\lim_{x\to\infty}\int_a^x f(t)\,\mathrm{d}t$ exists and if $\lim_{y\to-\infty}\int_y^a f(t)\,\mathrm{d}t$ exists then the integral on the right-hand side certainly exists, but this limit may exist when the two former do not and so it covers a wider class of functions.

For functions which are unbounded in the interval of integration the definitions are similar. If f is singular at $c > a$ we define

$$\int_a^c f(t)\,\mathrm{d}t = \lim_{x\to c-0}\int_a^x f(t)\,\mathrm{d}t, \qquad \int_c^b f(t)\,\mathrm{d}t = \lim_{x\to c+0}\int_x^b f(t)\,\mathrm{d}t,$$

$$\int_a^b f(t)\,\mathrm{d}t = \lim_{\delta\to+0}\left\{\int_a^{c-\delta} f(t)\,\mathrm{d}t + \int_{c+\delta}^b f(t)\,\mathrm{d}t\right\}$$

whenever the right-hand sides exist. Sometimes the last definition is said to give a *Cauchy principal value*; it may exist when neither $\int_a^c$ nor $\int_c^b$ exist.

It is evident that

$$\left|\int_a^b f(t)\mathrm{d}t\right| \le \int_a^b |f(t)|\,\mathrm{d}t$$

and that, if $\int_a^b |f(t)|\,\mathrm{d}t = 0$, f, being continuous almost everywhere, must vanish almost everywhere.

From Theorem 1.15 it is clear that $\int_a^\infty f(t)\mathrm{d}t$ exists if, and only if, there is x_0 such that, for given $\varepsilon > 0$, $\left|\int_{x_1}^{x_2} f(t)\mathrm{d}t\right| < \varepsilon$ for any $x_2 \ge x_1 \ge x_0$.

If $\int_a^\infty |f(t)|\,\mathrm{d}t < \infty$, $\int_a^\infty f(t)\mathrm{d}t$ is said to be *absolutely convergent*. In a similar manner to Theorem 1.21, can be shown

Theorem 1.26. *If* $|f(t)| \le f_0(t)$ *and* $\int_a^\infty f_0(t)\mathrm{d}t$ *exists,* $\int_a^\infty f(t)\mathrm{d}t$ *is absolutely convergent.*

If $\int_a^b |f(t)|^p\,\mathrm{d}t$ exists we write $f \in L_p(a, b)$; often when no confusion can arise this will be simplified to $f \in L_p$. Since $\int_a^b f_1(t)\overline{f_2(t)}\mathrm{d}t$ has the properties of a scalar product (f_1, f_2) Schwarz's inequality (3) applies and

Theorem 1.27. *If* $f_1 \in L_2$ *and* $f_2 \in L_2$ *then* $f_1 f_2 \in L_1$ *and*

$$\left|\int f_1(t)\overline{f_2(t)}\mathrm{d}t\right|^2 \le \int |f_1(t)|^2\,\mathrm{d}t \int |f_2(t)|^2\,\mathrm{d}t.$$

More generally

Theorem 1.28. (Hölder inequality) *If* $f_1 \in L_p$ *and* $f_2 \in L_q$ *where* $p > 1$ *and* $1/p + 1/q = 1$ *then* $f_1 f_2 \in L_1$ *and*

$$\left|\int f_1(t) f_2(t)\mathrm{d}t\right| \le \left\{\int |f_1(t)|^p\,\mathrm{d}t\right\}^{1/p} \left\{\int |f_2(t)|^q\,\mathrm{d}t\right\}^{1/q}.$$

Proof. If $0 < \mu < 1$ and $t \ge 1$, $t^{\mu-1} \le 1$ so that

$$\int_1^{c/d} \mu t^{\mu-1}\,\mathrm{d}t \le \int_1^{c/d} \mu\,\mathrm{d}t$$

if $c \ge d$, i.e. $c^\mu d^{1-\mu} \le \mu c + (1-\mu)d$. The replacement of μ by $1-\mu$ shows that this inequality is also true for $c < d$. Choose $\mu = 1/p$, $1-\mu = 1/q$, $c = |f_1(t)|^p / \int |f_1(u)|^p\,\mathrm{d}u$ and $d = |f_2(t)|^q / \int |f_2(u)|^q\,\mathrm{d}u$ (we take $\int |f_1(u)|^p\,\mathrm{d}u \ne 0$ and $\int |f_2(u)|^q\,\mathrm{d}u \ne 0$ otherwise there is nothing to prove). Then

$$\frac{|f_1(t)f_2(t)|}{\{\int|f_1(u)|^p\mathrm{d}u\}^{1/p}\{\int|f_2(u)|^q\mathrm{d}u\}^{1/q}} \le \frac{|f_1(t)|^p}{p\int|f_1(u)|^p\mathrm{d}u} + \frac{|f_2(t)|^q}{q\int|f_2(u)|^q\mathrm{d}u}.$$

Integration with respect to t gives

$$\frac{\int|f_1(t)f_2(t)|\mathrm{d}t}{\{\int|f_1(u)|^p\mathrm{d}u\}^{1/p}\{\int|f_2(u)|^q\mathrm{d}u\}^{1/q}} \le \frac{1}{p}+\frac{1}{q} \le 1$$

and the theorem is proved. □

Theorem 1.29. (Minkowski inequality) *If* $f_1 \in L_p$ *and* $f_2 \in L_p (p \ge 1)$ *then*

$$\left\{\int|f_1(t)+f_2(t)|^p\mathrm{d}t\right\}^{1/p} \le \left\{\int|f_1(t)|^p\mathrm{d}t\right\}^{1/p} + \left\{\int|f_2(t)|^p\mathrm{d}t\right\}^{1/p}.$$

Proof. If $p=1$ or $\int|f_1(t)+f_2(t)|^p\mathrm{d}t = 0$ the result is obvious. For $p>1$

$$\begin{aligned}\int|f_1(t)+f_2(t)|^p\mathrm{d}t &= \int|f_1(t)+f_2(t)||f_1(t)+f_2(t)|^{p-1}\mathrm{d}t \\ &\le \int|f_1(t)||f_1(t)+f_2(t)|^{p-1}\mathrm{d}t \\ &\quad + \int|f_2(t)||f_1(t)+f_2(t)|^{p-1}\mathrm{d}t \\ &\le \left\{\int|f_2(t)|^p\mathrm{d}t\right\}^{1/p}\left\{\int|f_1(t)+f_2(t)|^p\mathrm{d}t\right\}^{1/q} \\ &\quad + \left\{\int|f_2(t)|^p\mathrm{d}t\right\}^{1/p}\left\{\int|f_1(t)+f_2(t)|^p\mathrm{d}t\right\}^{1/q}\end{aligned}$$

by Theorem 1.28. Division by $\{\int|f_1(t)+f_2(t)|^p\mathrm{d}t\}^{1/q}$ now proves the theorem. □

The derivative and integral are inverses of one another in a certain sense. For, if f is integrable in (a, b) and continuous at $x\,(a<x<b)$, we can choose h_0 such that $|f(t)-f(x)|<\varepsilon$ for $|t-x|\le h_0$ and then

$$\left|\frac{1}{h}\left\{\int_a^{x+h} f(t)\mathrm{d}t - \int_a^x f(t)\mathrm{d}t\right\} - f(x)\right| = \left|\frac{1}{h}\int_x^{x+h}\{f(t)-f(x)\}\mathrm{d}t\right| \le \varepsilon$$

for $|h| \le h_0$. Hence $(\mathrm{d}/\mathrm{d}x)\int_a^x f(t)\,\mathrm{d}t = f(x)$. Further, if f is continuous on $[a, b]$ and $F(x) = \int_a^x f(t)\,\mathrm{d}t$, $\mathrm{d}F(x)/\mathrm{d}x = f(x)$ which may be expressed as follows. If $\mathrm{d}f(x)/\mathrm{d}x$ is continuous on $[a, b]$ and $a \le a_1 \le b_1 \le b$,

$$\int_{a_1}^{b_1} \frac{\mathrm{d}f(t)}{\mathrm{d}t}\,\mathrm{d}t = f(b_1) - f(a_1).$$

Theorem 1.30. (First mean value theorem) *If f is continuous there is a ξ in $[a, b]$ such that $\int_a^b f(t)\,\mathrm{d}t = (b-a)f(\xi)$.*

Proof. If U and L are the upper and lower bounds of f in $[a, b]$ then $U - f(t) \ge 0$ and $f(t) - L \ge 0$ so that

$$U(b-a) \ge \int_a^b f(t)\,\mathrm{d}t \ge L(b-a).$$

By Theorem 1.19 there is a ξ in $[a, b]$ such that $f(\xi)$ has any given value between U and L. The theorem follows at once. □

Corollary 1.30*a*. *If* $\mathrm{d}f(x)/\mathrm{d}x$ *is continuous in $[a, b]$ there is a ξ in $[a, b]$ such that*

$$f(b) - f(a) = (b-a)\,\mathrm{d}f(\xi)/\mathrm{d}\xi.$$

This follows at once from (4).

Corollary 1.30*b*. *If f is continuous and $f_1 \ge 0$, there is a ξ in $[a, b]$ such that*

$$\int_a^b f(t) f_1(t)\,\mathrm{d}t = f(\xi)\int_a^b f_1(t)\,\mathrm{d}t.$$

Repeat the argument of Theorem 1.30 retaining the factor f_1 in the integrand.

Corollary 1.30c. *If f and its first n derivatives are continuous in $[a, a+h]$ there is a ξ in $[0, 1]$ such that*

$$f(a+h) = \sum_{m=0}^{n-1} \frac{h^m}{m!}\frac{\mathrm{d}^m f(a)}{\mathrm{d}a^m} + \frac{h^n}{n!}\left[\frac{\mathrm{d}^n f(x)}{\mathrm{d}x^n}\right]_{x=a+\xi h}.$$

For

$$f(a+h)-\sum_{m=0}^{n-1}\frac{h^m}{m!}\frac{\mathrm{d}^m f(a)}{\mathrm{d}a^m}=\int_0^1\left\{\frac{\mathrm{d}}{\mathrm{d}t}\sum_{m=0}^{n-1}\frac{(1-t)^m}{m!}\frac{\mathrm{d}^m}{\mathrm{d}t^m}f(a+th)\right\}\mathrm{d}t$$

$$=\int_0^1\frac{(1-t)^{n-1}}{(n-1)!}\frac{\mathrm{d}^n}{\mathrm{d}t^n}f(a+th)\mathrm{d}t$$

and the result follows from Corollary 1.30*b* with $f_1=(1-t)^{n-1}$. □

If $\int_a^x f(t,y)\mathrm{d}t$ converges uniformly on $\mathscr{Y}$ as $x\to\infty$ the integral $\int_a^\infty f(t,y)\mathrm{d}t$ is said to be *uniformly convergent* on $\mathscr{Y}$. There is a similar definition, and analogous theorems to those that follow, for f singular in the interval of integration.

The analogue of Theorem 1.23 (the proof being parallel) is

Theorem 1.31. *If* $|f(t,y)|\le f_1(t)$ *for* $t\ge a$ *and* $y\in\mathscr{Y}$, *and if* $\int_a^\infty f_1(t)\mathrm{d}t<\infty$ *then* $\int_a^\infty f(t,y)\,\mathrm{d}t$ *is uniformly convergent on* $\mathscr{Y}$.

Theorem 1.32. *If* $f(t,y)$ *is continuous at* y_0 *uniformly with respect to* t *and if* $\int_a^\infty f(t,y)\mathrm{d}t$ *is uniformly convergent, then* $\int_a^\infty f(t,y)\mathrm{d}t$ *is continuous at* y_0, *i.e.* $\lim_{y\to y_0}\int_a^\infty f(t,y)\mathrm{d}t=\int_a^\infty f(t,y_0)\mathrm{d}t$.

Proof. Given $\varepsilon>0$, there is an η such that $|f(t,y)-f(t,y_0)|<\varepsilon$ for $|y-y_0|<\eta$ and $a\le t\le b$. Hence

$$\left|\int_a^b\{f(t,y)-f(t,y_0)\}\mathrm{d}t\right|<\varepsilon(b-a)$$

which shows that the integral over any finite interval is continuous. Since $\int_a^\infty f(t,y)\mathrm{d}t=\lim_{n\to\infty}\int_a^n f(t,y)\mathrm{d}t$ the theorem follows from Corollary 1.17. □

Since continuity implies uniform continuity (Theorem 1.18) the first condition in Theorem 1.32 can be replaced by $f(t,y)$ *continuous on* $[a,\infty]\times\mathscr{Y}$.

Theorem 1.33. *If* $\partial f(t,y)/\partial y$ *is continuous on* $[a,\infty]\times\mathscr{Y}$ *and* $\int_a^\infty\partial f(t,y)/\partial y\mathrm{d}t$ *is uniformly convergent on* $\mathscr{Y}$ *then*

$$\frac{\mathrm{d}}{\mathrm{d}y}\int_a^\infty f(t,y)\mathrm{d}t=\int_a^\infty\frac{\partial}{\partial y}f(t,y)\mathrm{d}t.$$

Proof.

$$\frac{\mathrm{d}}{\mathrm{d}y}\int_a^{\infty} f(t,y)\,\mathrm{d}t = \lim_{h\to 0}\int_a^{\infty}\frac{f(t,y+h)-f(t,y)}{h}\,\mathrm{d}t$$

$$= \lim_{h\to 0}\int_a^{\infty}\left[\frac{\partial}{\partial\xi}f(t,\xi)\right]_{\xi=y+\theta h}\mathrm{d}t$$

where $0\le\theta\le 1$, by Corollary 1.30*a*. The result now follows from Theorem 1.32. □

As in Theorem 1.32 the first condition on $\partial f/\partial y$ could be '*If* $\partial f/\partial y$ *is continuous on* $\mathscr{Y}$ *uniformly with respect to t*'.

Theorem 1.34. *If* $\int_a^b\{\int_\alpha^\beta f(t,y)\,\mathrm{d}y\}\,\mathrm{d}t = \int_\alpha^\beta\{\int_a^b f(t,y)\,\mathrm{d}t\}\,\mathrm{d}y$ *for all* $b>a$ *and if* $\int_a^\infty f(t,y)\,\mathrm{d}t$ *is uniformly convergent on* $[\alpha,\beta]$ *then*

$$\int_a^{\infty}\left\{\int_\alpha^\beta f(t,y)\,\mathrm{d}y\right\}\mathrm{d}t = \int_\alpha^\beta\left\{\int_a^\infty f(t,y)\,\mathrm{d}t\right\}\mathrm{d}y.$$

Proof. Let $S_n(y)=\int_a^n f(t,y)\,\mathrm{d}t$; then

$$\int_a^{n}\left\{\int_\alpha^\beta f(t,y)\,\mathrm{d}y\right\}\mathrm{d}t = \int_\alpha^\beta S_n(y)\,\mathrm{d}y$$

$$\to\int_a^\beta\left\{\int_a^\infty f(t,y)\,\mathrm{d}t\right\}\mathrm{d}y$$

since there is n_0 independent of y such that $|S_n(y)-\int_a^\infty f(t,y)\,\mathrm{d}t|<\varepsilon$ for $n>n_0$. □

Theorem 1.35. *If* $\int_a^b\{\int_\alpha^\infty f(t,y)\,\mathrm{d}y\}\,\mathrm{d}t = \int_\alpha^\infty\{\int_a^b f(t,y)\,\mathrm{d}t\}\,\mathrm{d}y$ *for all* $b>a$ *and if* $\int_a^\infty\{\int_\alpha^\beta f(t,y)\,\mathrm{d}y\}\,\mathrm{d}t = \int_\alpha^\beta\{\int_a^\infty f(t,y)\,\mathrm{d}t\}\,\mathrm{d}y$ *for all* $\beta>\alpha$ *then*

$$\int_a^{\infty}\left\{\int_\alpha^\infty f(t,y)\,\mathrm{d}y\right\}\mathrm{d}t = \int_\alpha^\infty\left\{\int_a^\infty f(t,y)\,\mathrm{d}t\right\}\mathrm{d}y \tag{5}$$

provided that either

$$\int_a^{\infty}\left\{\int_\alpha^\infty |f(t,y)|\,\mathrm{d}y\right\}\mathrm{d}t<\infty \quad \textit{or} \quad \int_\alpha^\infty\left\{\int_a^\infty |f(t,y)|\,\mathrm{d}t\right\}\mathrm{d}y<\infty.$$

Proof. Suppose $f(t,y)\ge 0$ and that the left-hand side of (5) exists.

Then

$$\int_{\alpha}^{\beta}\left\{\int_{a}^{\infty} f(t, y)\mathrm{d}t\right\}\mathrm{d}y = \int_{a}^{\infty}\left\{\int_{\alpha}^{\beta} f(t, y)\mathrm{d}y\right\}\mathrm{d}t$$

$$\leq \int_{a}^{\infty}\left\{\int_{\alpha}^{\infty} f(t, y)\mathrm{d}y\right\}\mathrm{d}t.$$

Letting $\beta \to \infty$ we see that the right-hand side of (5) exists. The inequality reversed may be obtained in a similar way and so (5) holds for $f(t, y) \geq 0$. If f has varying sign write it as $\frac{1}{2}(|f|+f) - \frac{1}{2}(|f|-f)$. Each of $|f| \pm f$ is non-negative and each is bounded by $2|f|$ so that the two may be dealt with separately as above. If f is complex deal with the real and imaginary parts separately. □

Theorem 1.36. *If $\sum_{n=1}^{\infty} f_n$ is a uniformly convergent series of continuous functions on $[a, b]$ then*

$$\sum_{n=1}^{\infty} \int_{a}^{b} f_n(t)\mathrm{d}t = \int_{a}^{b} \sum_{n=1}^{\infty} f_n(t)\mathrm{d}t.$$

If this is true for all finite b then $\sum_{n=1}^{\infty} \int_a^{\infty} f_n(t)\mathrm{d}t = \int_a^{\infty} \sum_{n=1}^{\infty} f_n(t)\mathrm{d}t$ provided that either

$$\sum_{n=1}^{\infty} \int_{a}^{\infty} |f_n(t)|\mathrm{d}t < \infty \quad \textit{or} \quad \int_{a}^{\infty} \sum_{n=1}^{\infty} |f_n(t)|\mathrm{d}t < \infty.$$

Proof. For given ε, there is an n_0 such that $|S_n(x) - \sum_{m=1}^{\infty} f_m(x)| < \varepsilon$ for $n > n_0$ and all x in $[a, b]$. Hence

$$\left|\int_{a}^{b}\left\{S_n(t) - \sum_{m=1}^{\infty} f_m(t)\right\}\mathrm{d}t\right| < \varepsilon(b-a)$$

and the first result follows since the integral of the infinite series exists because of Theorem 1.24.

For the second part suppose firstly that $f_n(t) \geq 0$ for all n. Then

$$\int_{a}^{b} \sum_{n=1}^{\infty} f_n(t)\mathrm{d}t = \sum_{n=1}^{\infty} \int_{a}^{b} f_n(t)\mathrm{d}t \leq \sum_{n=1}^{\infty} \int_{a}^{\infty} f_n(t)\mathrm{d}t.$$

Hence

$$\int_a^\infty \sum_{n=1}^\infty f_n(t)\mathrm{d}t \le \sum_{n=1}^\infty \int_a^\infty f_n(t)\mathrm{d}t.$$

On the other hand,

$$\sum_{n=1}^N \int_a^\infty f_n(t)\mathrm{d}t = \int_a^\infty \sum_{n=1}^N f_n(t)\mathrm{d}t \le \int_a^\infty \sum_{n=1}^\infty f_n(t)\mathrm{d}t$$

and letting $N \to \infty$ we obtain the inequality reversed. Hence equality must hold and either of the given conditions makes both sides finite.

For f_n not necessarily positive proceed as in Theorem 1.35. □

Theorem 1.37. *If* $\sum_{n=1} \mathrm{d}f_n(x)/\mathrm{d}x$ *is uniformly convergent and* $\sum_{n=1} f_n$ *is convergent then* $(\mathrm{d}/\mathrm{d}x)\sum_{n=1} f_n(x) = \sum_{n=1} \mathrm{d}f_n(x)/\mathrm{d}x$.

Proof. Along the lines of Theorem 1.33. □

Theorem 1.38. (Second mean value theorem) *If f is absolutely integrable over (a, b) and ϕ is positive, bounded and non-increasing then*

$$\int_a^b f(t)\phi(t)\mathrm{d}t = \phi(a+0)\int_a^\xi f(t)\mathrm{d}t$$

for some ξ in $[a, b]$.

Proof. Let $0 < \varepsilon < \phi(a+0) - \phi(b-0)$. Then, because ϕ is bounded, there is an x_1 such that

$$\phi(a+0) - \phi(x)\begin{cases} < \varepsilon & (a < x < x_1) \\ \ge \varepsilon & (x > x_1). \end{cases}$$

Similarly, there are points $x_2, x_3, \dots$ such that

$$\phi(x_{m-1}+0) - \phi(x)\begin{cases} < \varepsilon & (x_{m-1} < x < x_m) \\ \ge \varepsilon & (x > x_m) \end{cases} \tag{6}$$

unless $\phi(x_{r-1}+0) - \phi(b-0) \le \varepsilon$ when we take $x_r = b$. There is only a finite number of x_m since $\phi(b-0) - \phi(a+0)$ is finite and $\phi(x_{m-1}+0) - \phi(x_r+0) \ge \varepsilon$.

Define $\phi_1(x)$ in $x_m \leq x < x_{m+1}$ by $\phi_1(x) = \phi(x_m + 0)$. Then

$$\int_a^b \phi_1(t) f(t) \mathrm{d}t = \sum_{m=0}^{r-1} \phi(x_m + 0) \int_{x_m}^{x_{m+1}} f(t) \mathrm{d}t$$
$$= \sum_{m=0}^{r-2} \{\phi(x_m + 0) - \phi(x_{m+1} + 0)\} \int_a^{x_{m+1}} f(t) \mathrm{d}t$$
$$+ \phi(x_{r-1} + 0) \int_a^{x_r} f(t) \mathrm{d}t.$$

Hence if U and L are the upper and lower bounds of $\int_a^x f(t) \mathrm{d}t$,

$$L\phi(a + 0) \leq \int_a^b \phi_1(t) f(t) \mathrm{d}t \leq U\phi(a + 0). \tag{7}$$

Also

$$\left| \int_a^b \phi_1(t) f(t) \mathrm{d}t - \int_a^b \phi(t) f(t) \mathrm{d}t \right| = \left| \sum_{m=0}^{r-1} \int_{x_m}^{x_{m+1}} \{\phi(x_m + 0) - \phi(t)\} f(t) \mathrm{d}t \right|$$
$$< \varepsilon \int_a^b |f(t)| \mathrm{d}t$$

by (6). Therefore, making $\varepsilon \to 0$, we can replace ϕ_1 by ϕ in (7). But $\int_a^x f(t) \mathrm{d}t$ is continuous and therefore takes every value between U and and L. In particular there is a ξ such that it has the value $\{1/\phi(a + 0)\} \times \int_a^b \phi(t) f(t) \mathrm{d}t$ and the theorem is proved. □

If ϕ is positive and non-decreasing the corresponding result is

$$\int_a^b f(t) \phi(t) \mathrm{d}t = \phi(b - 0) \int_\xi^b f(t) \mathrm{d}t.$$

The Riemann *double integral* of a function of two variables can be defined in a manner analogous to the single variable case, with areas replacing linear intervals. Thus the common limit (if it exists) as $n, q \to \infty$ of

$$\sum_{m=1}^{n} \sum_{p=1}^{q} (x_m - x_{m-1})(y_p - y_{p-1}) \sup_{\substack{x_{m-1} \leq x \leq x_m \\ y_{p-1} \leq y \leq y_p}} f(x, y)$$

and the same with inf for sup is called the double integral of f and denoted by $\int_{\mathscr{X} \times \mathscr{Y}} f(x, y) \mathrm{d}x \mathrm{d}y$.

If $\mathscr{Y}$ be $[\alpha, \beta]$ and f is continuous we note that

$$\sum_{p=1}^{q} (y_p - y_{p-1}) \sup_{\substack{x_{m-1} \leq x \leq x_m \\ y_{p-1} \leq y \leq y_p}} f(x, y) \geq \sup_{x_{m-1} \leq x \leq x_m} \int_{\alpha}^{\beta} f(x, y) \mathrm{d}y$$

$$\geq \inf_{x_{m-1} \leq x \leq x_m} \int_{\alpha}^{\beta} f(x, y) \mathrm{d}y$$

$$\geq \sum_{p=1}^{q} (y_p - y_{p-1}) \inf_{\substack{x_{m-1} \leq x \leq x_m \\ y_{p-1} \leq y \leq y_p}} f(x, y)$$

from which, after the operation $\sum_m (x_m - x_{m-1})$, we deduce that the double integral equals the *iterated* or *repeated integral*, i.e.

$$\int_{\mathscr{X} \times \mathscr{Y}} f(x, y) \mathrm{d}x \mathrm{d}y = \int_a^b \left\{ \int_{\alpha}^{\beta} f(x, y) \mathrm{d}y \right\} \mathrm{d}x.$$

Similarly

$$\int_{\mathscr{X} \times \mathscr{Y}} f(x, y) \mathrm{d}x \mathrm{d}y = \int_{\alpha}^{\beta} \left\{ \int_a^b f(x, y) \mathrm{d}x \right\} \mathrm{d}y.$$

For an infinite domain of integration proceed along lines analogous to the preceding for single integrals.

The definition of an integral in n-dimensional space is an obvious generalisation of that for the double integral. Such multiple integrals have properties analogous to those for the single integral. For example, the theorem on absolute convergence (1.26) and those on the Schwarz inequality (1.27), the Hölder inequality (1.28) and the Minkowski inequality (1.29) remain unaltered provided that t is interpreted as a point of n-dimensional space and $\mathrm{d}t$ as an element of volume in that space. With a similar interpretation the first mean value theorem (1.30) is also valid. The theorems on uniform convergence and continuity (1.31–1.37) are also true with the understanding that $\mathscr{Y}$ need not have the same dimension as the space of t and that the derivative (where it occurs) may be a partial derivative.

2

Good functions

In this chapter all functions will be complex-valued and defined on R_1. In fact we shall be concerned with functions having special properties and shall describe them in the graphic terminology introduced by M.J. Lighthill in his book *Fourier Analysis and Generalised Functions.*

2.1. Good functions

Definition 2.1. *A function γ is said to be good if it is differentiable infinitely often everywhere on R_1 and if*

$$\lim_{|x|\to\infty}\left|x^r \frac{\mathrm{d}^k}{\mathrm{d}x^k}\gamma(x)\right| = 0$$

for every integer $r \geq 0$ and every integer $k \geq 0$.

Clearly e^{-x^2}, $(1+\mathrm{i})\mathrm{e}^{-x^2}$, $x^p\mathrm{e}^{-x^2}$ (p a positive integer) are all good functions. But e^x is not a good function because it does not behave properly at infinity. For the same reason the function that is equal to 1 for every point of R_1 is not good. On the other hand the function that is equal to $0 (x \leq 0)$, $\mathrm{e}^{-x^2} (x > 0)$ is not good because it does not have an infinite number of derivatives at the origin.

It is evident that if γ_1 and γ_2 are good then so are $\gamma_1 + \gamma_2$ and $\gamma_1\gamma_2$. Also if γ is good then so is $\mathrm{d}\gamma/\mathrm{d}x$. Furthermore, if $\gamma(x)$ is good so is $\gamma(ax+b)$, a and b being real constants. However, it is not necessarily true that the integral of a good function is good, for example $\int_{-\infty}^{x} \mathrm{e}^{-t^2}\,\mathrm{d}t \not\to 0$ as $x \to \infty$.

Exercises

1. Which of the following are good functions:
   $x^2\mathrm{e}^{-x^2+\pi\mathrm{i}/2}$; $\sin x$; $\operatorname{sech} x$; x^6; $1/(x^3+1)$; $\tanh x - 1$;
   $\mathrm{e}^{-x^2}\sin 3x$; $\mathrm{e}^{-|x|}$; 0; $0 (x \leq 0)$ and $\mathrm{e}^{1/x} (x > 0)$;
   $0 (|x| > 1)$ and $\mathrm{e}^{-1/(1-x^2)} (|x| \leq 1)$?
2. If γ is good show that $x^p\gamma(x)$ (p a positive integer) is good.
3. If $\gamma(x)$ is good is $\gamma(x^3)$ good?

Definition 2.2. *A function ψ is said to be fairly good if it is differentiable infinitely often everywhere on R_1 and if there is some fixed N such that*

$$\lim_{|x|\to\infty}\left|x^{-N}\frac{\mathrm{d}^k}{\mathrm{d}x^k}\psi(x)\right|=0$$

for every integer $k \geq 0$.

Examples of a fairly good function are a polynomial and e^{ix}, but e^x is not fairly good.

It is evident that if ψ_1 and ψ_2 are fairly good then so are $\psi_1 + \psi_2$ and $\psi_1\psi_2$. If ψ is fairly good so is $\mathrm{d}\psi/\mathrm{d}x$. If $\psi(x)$ is fairly good so is $\psi(ax+b)$, a and b being real constants.

Any good function is fairly good and

Theorem 2.1. *If γ is good and ψ fairly good then $\psi\gamma$ is good.*

Proof. For

$$\lim_{|x|\to\infty}\left|x^r\frac{\mathrm{d}^s\psi(x)}{\mathrm{d}x^s}\frac{\mathrm{d}^{k-s}\gamma(x)}{\mathrm{d}x^{k-s}}\right| = \lim_{|x|\to\infty}\left|x^{-N}\frac{\mathrm{d}^s\psi(x)}{\mathrm{d}x^s}.x^{r+N}\frac{\mathrm{d}^{k-s}\gamma(x)}{\mathrm{d}x^{k-s}}\right|$$
$$=0$$

by Definitions 2.1 and 2.2. Since this is a typical term in the kth derivative of $\psi\gamma$ the proof is complete. □

Exercises

4. Which of the functions in Exercise 1 are fairly good?
5. Which of the following are fairly good?
   $\sin x$, $\cos 3x$, $\tan 2x$, $(1+x^2)^n$,
   $e^{-x^2}\sin ax$, $\cos x/\cosh x$. Are any of them good?
6. The function $\psi(x+iy)$ of the complex variable $x+iy$ is regular in $y_1-\delta \leq y \leq y_1+\delta\ (\delta>0)$ and $\lim_{|x|\to\infty}|x^{-N}\psi(x+iy)|=0$ in this region. Is ψ a fairly good function of x?

A somewhat wider class than that of Definition 2.2 is the class of *moderately good* functions χ, which possess any number of derivatives and for every integer $k \geq 0$ there is a finite N_k such that $\lim_{|x|\to\infty}|x^{-N_k}\mathrm{d}^k\chi(x)/\mathrm{d}x^k| = 0$. A typical example is e^{ix^2}. Obviously a fairly good function is moderately good and the two classes have similar properties. Indeed Theorem 2.1 remains true if ψ is replaced by χ because it is only necessary to change N to N_s in the proof.

Definition 2.3. *A function ϕ is said to be fine if it is differentiable infinitely often everywhere on R_1 and if it and all its derivatives vanish identically outside some finite interval.*

A typical example is the function which equals $e^{-x^2/(1-x^2)}$ for $|x|<1$ and vanishes for $|x|\geq 1$. The sum and product of two fine functions are fine. A fine function is good. The product of a good and fine function is fine; the same is true if good is replaced by moderately good. The derivative of a fine function is fine.

Exercises

7. Which of the functions in Exercises 1 and 5 are moderately good and which are fine?
8. Is the function which equals $\exp\{1/(x-d)-1/(x-c)\}$ in $[c,d]$ and vanishes elsewhere fine?
9. The function σ is defined by $\sigma(x)=e^{-x^2/(1-x^2)}(|x|<1)$, $\sigma(x)=0(|x|\geq 1)$. The sequence $\{n_i\}$ is an increasing sequence of positive integers. Show that $\sum_{i=1}^{\infty}\sigma(x-n_i)/n_i^i$ is good. (Hint: Theorem 1.24.)
10. The function τ is defined by $\tau(x)=0\ (|x|\geq 1)$,

$$\tau(x)=\int_{|x|}^{1}e^{-1/t(1-t)}dt\bigg/\int_0^1 e^{-1/t(1-t)}dt \quad (|x|<1).$$

Show that $\tau(x)+\tau(x-1)=1$ when $0\leq x\leq 1$ and deduce that $\sum_{m=-n}^{n}\tau(x-m)$ is a fine function which equals 1 for $|x|\leq n$.
11. If γ is good prove that

$$\max_{x\in R_1}(1+x^2)^{k/2}\left|\frac{d^r\gamma(x)}{dx^r}\right|\leq(\tfrac{1}{2}\pi)^n\max_{x\in R_1}(1+x^2)^{n+k/2}\left|\frac{d^{r+n}\gamma(x)}{dx^{r+n}}\right|.$$

(Hint: use $\gamma(x)=\int_{-\infty}^{x}\{d\gamma(t)/dt\}dt$ for $x\leq 0$ and $\gamma(x)=-\int_x^{\infty}\{d\gamma(t)/dt\}dt$ for $x\geq 0$.)

2.2. Fourier transforms

An important notion in what follows will be that of the Fourier transform. In its definition we shall use α to denote a point of R_1. Then

Definition 2.4. *If $F(\alpha)=\int_{-\infty}^{\infty}f(x)e^{-i\alpha x}dx$ exists for every α in R_1, F is called the Fourier transform of f.*

One particular Fourier transform of frequent occurrence is that of $f(x)=e^{-x^2/2}$. This is calculated by considering $\int e^{-z^2/2}dz$ taken round the rectangular contour in the complex z-plane, joining the points $(-X, 0), (X, 0), (X, \alpha), (-X, \alpha)$. Since $e^{-z^2/2}$ has no singularities within this region the integral is zero. Hence

$$\left(\int_{-X}^{X} + \int_{X}^{X+i\alpha} + \int_{X+i\alpha}^{-X+i\alpha} + \int_{-X+i\alpha}^{-X}\right) e^{-z^2/2}\,dz = 0.$$

Now

$$\left|\int_{X}^{X+i\alpha} e^{-z^2/2}\,dz\right| = \left|\int_{0}^{\alpha} ie^{-(X+iu)^2/2}\,du\right|$$
$$\leq e^{-X^2/2}\int_{0}^{|\alpha|} e^{u^2/2}\,du$$

which tends to zero as $X \to \infty$. The same is true of $\int_{-X+i\alpha}^{-X}$ and hence, letting $X \to \infty$,

$$\int_{-\infty}^{\infty} e^{-(x+i\alpha)^2/2}\,dx = \int_{-\infty}^{\infty} e^{-x^2/2}\,dx = 2\int_{0}^{\infty} e^{-x^2/2}\,dx.$$

Since $x = (2y)^{1/2}$ gives

$$\int_{0}^{\infty} e^{-x^2/2}\,dx = 2^{-1/2}\int_{0}^{\infty} y^{-1/2}e^{-y}\,dy = (\tfrac{1}{2}\pi)^{1/2},$$

$$\int_{-\infty}^{\infty} e^{-(x^2/2)-i\alpha x}\,dx = (2\pi)^{1/2}e^{-\alpha^2/2}.$$

For reference purposes it is convenient to put this as

Theorem 2.2. *The Fourier transform of* $e^{-x^2/2}$ *is* $(2\pi)^{1/2}e^{-\alpha^2/2}$.

This particular example illustrates an important property of Fourier transforms. Since $F(\alpha) = (2\pi)^{1/2}e^{-\alpha^2/2}$,

$$\int_{-\infty}^{\infty} F(\alpha)e^{i\alpha x}\,d\alpha = (2\pi)^{1/2}\int_{-\infty}^{\infty} e^{-(\alpha^2/2)+i\alpha x}\,d\alpha = 2\pi e^{-x^2/2}$$

by Theorem 2.2. Hence, for this particular example,

$$\frac{1}{2\pi}\int_{-\infty}^{\infty} F(\alpha)e^{i\alpha x}\,d\alpha = f(x).$$

It is of interest to know whether such reciprocity exists for all

Fourier transforms. We shall not be concerned here with the most general class of f for which it is true but only with one which will be amply sufficient for our subsequent need.

Theorem 2.3. *If f and its derivative* $\mathrm{d}f/\mathrm{d}x$ *are both continuous and* $O(x^{-2})$ *as* $|x| \to \infty$, *then the Fourier transform F of f exists and*

$$\frac{1}{2\pi}\int_{-\infty}^{\infty} F(\alpha)\mathrm{e}^{\mathrm{i}\alpha x}\,\mathrm{d}\alpha = f(x).$$

The notation $O(x^{-2})$ means that there is a constant K such that $|f(x)| \le K/x^2$.

Proof. Since

$$\left|\int_{-\infty}^{\infty} f(x)\mathrm{e}^{-\mathrm{i}\alpha x}\,\mathrm{d}x\right| \le \int_{-\infty}^{\infty} |f(x)|\,\mathrm{d}x < \infty$$

because f is continuous and $O(x^{-2})$ as $|x| \to \infty$ the Fourier transform exists, being absolutely convergent by Theorem 1.26. It is also uniformly convergent for $\alpha \in \mathrm{R}_1$ by Theorem 1.31.

Also

$$\begin{aligned}\frac{1}{2\pi}\int_{-\infty}^{\infty} F(\alpha)\mathrm{e}^{\mathrm{i}\alpha x}\,\mathrm{d}\alpha &= \lim_{A\to\infty}\frac{1}{2\pi}\int_{-A}^{A} F(\alpha)\mathrm{e}^{\mathrm{i}\alpha x}\,\mathrm{d}\alpha\\ &= \lim_{A\to\infty}\frac{1}{2\pi}\int_{-A}^{A} \mathrm{e}^{\mathrm{i}\alpha x}\left\{\int_{-\infty}^{\infty} f(u)\mathrm{e}^{-\mathrm{i}\alpha u}\,\mathrm{d}u\right\}\mathrm{d}\alpha\\ &= \lim_{A\to\infty}\frac{1}{2\pi}\int_{-\infty}^{\infty} f(u)\left\{\int_{-A}^{A} \mathrm{e}^{\mathrm{i}\alpha(x-u)}\,\mathrm{d}\alpha\right\}\mathrm{d}u\end{aligned}$$

by Theorem 1.34 because of the uniform convergence of $\int_{-\infty}^{\infty} f(u)\mathrm{e}^{-\mathrm{i}\alpha u}\,\mathrm{d}u$. Hence

$$\begin{aligned}\frac{1}{2\pi}\int_{-\infty}^{\infty} F(\alpha)\mathrm{e}^{\mathrm{i}\alpha x}\,\mathrm{d}\alpha &= \lim_{A\to\infty}\int_{-\infty}^{\infty} f(u)\frac{\sin A(x-u)}{\pi(x-u)}\,\mathrm{d}u\\ &= \lim_{A\to\infty}\int_{-\infty}^{\infty} f\left(x+\frac{y}{A}\right)\frac{\sin y}{\pi y}\,\mathrm{d}y.\end{aligned}$$

Now $|f(x+y/A)-f(x)| = |\int_x^{x+y/A}(\mathrm{d}f/\mathrm{d}x)\mathrm{d}x| \le Ky/A$ since $\mathrm{d}f/\mathrm{d}x$

is bounded. Therefore

$$\left|\int_{-A^{3/4}}^{A^{3/4}}\left\{f\left(x+\frac{y}{A}\right)-f(x)\right\}\frac{\sin y}{\pi y}\mathrm{d}y\right|\le\int_{-A^{3/4}}^{A^{3/4}}\frac{K\,\mathrm{d}y}{A}\le K/A^{1/4}.$$

Also

$$\left|\int_{A^{3/4}}^{\infty}f\left(x+\frac{y}{A}\right)\frac{\sin y}{y}\mathrm{d}y\right|$$

$$=\left|\left[-\frac{\cos y}{y}f\left(x+\frac{y}{A}\right)\right]_{A^{3/4}}^{\infty}+\int_{A^{3/4}}^{\infty}\frac{\cos y}{y}\left\{\frac{\mathrm{d}f}{\mathrm{d}y}-\frac{f}{y}\right\}\mathrm{d}y\right|$$

$$\le\frac{K}{A^{1/4}}+\int_{A^{3/4}}^{\infty}\left(\frac{KA}{y^3}+\frac{KA^2}{y^4}\right)\mathrm{d}y\to 0$$

because of the behaviour of f and $\mathrm{d}f/\mathrm{d}y$ at infinity, for example,

$$|f(x+y)|\le K\{1+(x+y)^2\}^{-1}\le 4K(1+x^2)/y^2.$$

Hence

$$\frac{1}{2\pi}\int_{-\infty}^{\infty}F(\alpha)\mathrm{e}^{\mathrm{i}\alpha x}\mathrm{d}\alpha=\lim_{A\to\infty}\int_{-A^{3/4}}^{A^{3/4}}f(x)\frac{\sin y}{\pi y}\mathrm{d}y$$

$$=f(x)$$

since $\int_{-\infty}^{\infty}(\sin y/y)\mathrm{d}y=\pi$; the proof is complete. □

The Fourier transform exists under less stringent conditions than those of Theorem 2.3, for example, if $f\in L_1(-\infty,\infty)$ but then the inversion formula in the second half of the theorem is no longer necessarily true. However, it is clear that there are many simple functions such as x and $\sin x$ which do not have a Fourier transform. The way of overcoming this difficulty will be described in the next chapter.

Theorem 2.4. *If $\gamma(x)$ is a good function so is its Fourier transform $\Gamma(\alpha)$.*

Proof. For $p\ge 0$

$$\left|\int_{-\infty}^{\infty}x^p\gamma(x)\mathrm{e}^{-\mathrm{i}\alpha x}\mathrm{d}x\right|\le\int_{-\infty}^{\infty}|x|^p|\gamma(x)|\mathrm{d}x<\infty \qquad (1)$$

since the integrand is bounded on any finite interval and tends to zero faster than any inverse power of $|x|$ as $|x| \to \infty$. By Theorem 1.31 the integral on the left is uniformly convergent for $\alpha \in R_1$. Taking $p = 0$ we see that $\Gamma(\alpha)$ exists.

Taking $p = 1$ and applying Theorem 1.33 we see that

$$d\Gamma(\alpha)/d\alpha = \int_{-\infty}^{\infty} (-ix)\gamma(x)e^{-i\alpha x}dx;$$

continual repetition of the process leads to

$$d^q\Gamma(\alpha)/d\alpha^q = \int_{-\infty}^{\infty} (-ix)^q\gamma(x)e^{-i\alpha x}dx$$

so that all derivatives of Γ exist.

Since $(-ix)^q$ is fairly good $(-ix)^q\gamma(x)$ is a good function $\gamma_1(x)$ by Theorem 2.1. Then integration by parts gives

$$\frac{d^q\Gamma(\alpha)}{d\alpha^q} = \left[\frac{\gamma_1(x)e^{-i\alpha x}}{-i\alpha}\right]_{-\infty}^{\infty} + \frac{1}{i\alpha}\int_{-\infty}^{\infty} e^{-i\alpha x}\frac{d\gamma_1(x)}{dx}dx$$

$$= \frac{1}{i\alpha}\int_{-\infty}^{\infty} e^{-i\alpha x}\frac{d\gamma_1(x)}{dx}dx$$

because of the behaviour of a good function at infinity. Since the derivative of a good function is good the process can be repeated and we are thus led to

$$\frac{d^q\Gamma(\alpha)}{d\alpha^q} = \frac{1}{(i\alpha)^m}\int_{-\infty}^{\infty} e^{-i\alpha x}\frac{d^m\gamma_1(x)}{dx^m}dx.$$

The integral is bounded ((1) with $p = 0$, $\gamma = d^m\gamma_1/dx^m$) and so, for any given r we can find m large enough to make

$$\lim_{|\alpha|\to\infty}\left|\alpha^r\frac{d^q\Gamma(\alpha)}{d\alpha^q}\right| = 0$$

and the proof is complete. □

A fine function is good. Therefore a fine function has a good Fourier transform but this in general is not fine. This distinguishing feature of the Fourier transform, that it does not produce a new class from good functions whereas it does for fine functions, will be important subsequently.

Theorem 2.5. *If $\gamma(x)$ is a good function*

(i) $\gamma(x) = (1/2\pi)\int_{-\infty}^{\infty} \Gamma(\alpha)e^{i\alpha x}\,d\alpha$,
(ii) *the Fourier transform of* $d\gamma/dx$ is $i\alpha\Gamma(\alpha)$.

Proof. Part (i) is an immediate consequence of Theorem 2.3 because γ satisfies the conditions imposed there on f.

With regard to (ii) integration by parts gives

$$\int_{-\infty}^{\infty} \frac{d\gamma(x)}{dx} e^{-i\alpha x}\,dx = [\gamma(x)e^{-i\alpha x}]_{-\infty}^{\infty} + \int_{-\infty}^{\infty} i\alpha\gamma(x)e^{-i\alpha x}\,dx$$

$$= i\alpha \int_{-\infty}^{\infty} \gamma(x)e^{-i\alpha x}\,dx$$

because $\gamma \to 0$ as $|x| \to \infty$ and the theorem is proved. □

Theorem 2.6. *If γ is good and $f \in L_1(-\infty, \infty)$, then*

$$\int_{-\infty}^{\infty} \Gamma(\alpha)F(\alpha)e^{i\alpha y}\,d\alpha = 2\pi \int_{-\infty}^{\infty} f(x)\gamma(y-x)\,dx$$

for $y \in R_1$.

Proof. Since Γ is good (Theorem 2.4) and $f \in L_1$

$$\int_{-\infty}^{\infty}\left\{\int_{-\infty}^{\infty} |\Gamma(\alpha)f(x)e^{i(y-x)}|\,dx\right\}d\alpha = \left\{\int_{-\infty}^{\infty} |\Gamma(\alpha)|\,d\alpha\right\}\left\{\int_{-\infty}^{\infty} |f(x)|\,dx\right\}$$

$$< \infty.$$

Hence, by Theorems 1.34 and 1.35

$$\int_{-\infty}^{\infty} \Gamma(\alpha)e^{i\alpha y}\left\{\int_{-\infty}^{\infty} f(x)e^{-i\alpha x}\,dx\right\}d\alpha$$

$$= \int_{-\infty}^{\infty} f(x)\left\{\int_{-\infty}^{\infty} \Gamma(\alpha)e^{i\alpha(y-x)}\,d\alpha\right\}dx$$

and the theorem follows from Theorem 2.5 (i). □

Exercises

12. Find the Fourier transform of sech $\frac{1}{2}x$ and verify the Fourier inversion formula for it.

13. Find the Fourier transform of each of (i) $(x^2+1)^{-1}$, (ii)$(x^4+1)^{-1}$ and then evaluate $(1/2\pi)\int_{-\infty}^{\infty} F(\alpha)e^{i\alpha x}d\alpha$.

14. If γ is good, with Fourier transform Γ, prove that the Fourier transform of $\gamma(ax+b)$ is $(1/|a|)e^{ib\alpha/a}\Gamma(\alpha/a)$.

15. Determine the Fourier transforms of (i) $xe^{-x^2/2}$, (ii) $(1-x^2)e^{-x^2/2}$, (iii) $e^{-(x-1)^2}$, (iv) $xe^{-(x-1)^2}$.

16. Prove that

$$\int_{-\infty}^{\infty} \frac{e^{-(x-y)^2/2}}{x^2+1}dx = (2\pi)^{1/2}\int_0^{\infty} e^{-(\alpha^2/2)-\alpha}\cos\alpha y\, d\alpha.$$

17. The function f is infinitely differentiable everywhere on R_1 and there is a sequence $\{C_k\}$ of finite constants and a good function γ_1 such that $|d^kf/dx^k| < C_k\gamma_1$. Prove that f is good. Is the converse true?

3

Generalised functions

3.1. Generalised functions

The notion of a good function permits a considerable extension of the idea of a function in the following way. Let γ_n and γ be good functions. Then $\gamma_n\gamma$ is good and $\int_{-\infty}^{\infty}\gamma_n(x)\gamma(x)\mathrm{d}x$ exists.

Definition 3.1. *A sequence $\{\gamma_n\}$ of good functions is said to be regular if, for every given good γ, $\lim_{n\to\infty}\int_{-\infty}^{\infty}\gamma_n(x)\gamma(x)\mathrm{d}x$ exists and is finite. Two regular sequences which give the same limit are said to be equivalent.*

Thus, if $\{\gamma_{1,n}\}$ and $\{\gamma_{2,n}\}$ are regular, they are equivalent if, and only if, $\lim_{n\to\infty}\int_{-\infty}^{\infty}\gamma_{1,n}(x)\gamma(x)\mathrm{d}x = \lim_{n\to\infty}\int_{-\infty}^{\infty}\gamma_{2,n}(x)\gamma(x)\mathrm{d}x$ for every good γ.

The sequence $\{\mathrm{e}^{-x^2/n}\}$ is regular. To show this use the inequality

$$|\mathrm{e}^{-x^2/n} - 1| = \left|\int_0^x \frac{2x\mathrm{e}^{-x^2/n}}{n}\mathrm{d}x\right| < \int_0^x \frac{2x}{n}\mathrm{d}x < \frac{x^2}{n}.$$

Then

$$\int_{-\infty}^{\infty} \mathrm{e}^{-x^2/n}\gamma(x)\mathrm{d}x = \int_{-\infty}^{\infty} (\mathrm{e}^{-x^2/n} - 1)\gamma(x)\mathrm{d}x + \int_{-\infty}^{\infty} \gamma(x)\mathrm{d}x$$

and

$$\left|\int_{-\infty}^{\infty} (\mathrm{e}^{-x^2/n} - 1)\gamma(x)\mathrm{d}x\right| < \int_{-\infty}^{\infty} \frac{x^2|\gamma(x)|}{n}\mathrm{d}x \to 0$$

as $n \to \infty$ since $\int_{-\infty}^{\infty} x^2|\gamma(x)|\mathrm{d}x < \infty$ because γ is good. Hence

$$\lim_{n\to\infty}\int_{-\infty}^{\infty} \mathrm{e}^{-x^2/n}\gamma(x)\mathrm{d}x = \int_{-\infty}^{\infty} \gamma(x)\mathrm{d}x \tag{1}$$

so that the limit exists and is finite for every good γ; the sequence is regular.

By replacing n by $2n$ we see that

$$\lim_{n\to\infty}\int_{-\infty}^{\infty} e^{-x^2/2n}\gamma(x)dx = \int_{-\infty}^{\infty}\gamma(x)dx.$$

Thus $\{e^{-x^2/2n}\}$ is a regular sequence, equivalent to $\{e^{-x^2/n}\}$.

One would expect $\{e^{-x^2/n}\}$ to be regular since $e^{-x^2/n}\to 1$ uniformly in x as $n\to\infty$. Equally the sequence $\{ne^{-x^2}\}$ is not regular. In fact a sequence is not regular if one can find a single good γ for which the limit does not exist. A sequence whose regularity or otherwise is not immediately obvious is $\{n^2 e^{-nx^2}\}$; since $\lim_{n\to\infty} n^2 e^{-nx^2} = 0$ uniformly on any interval excluding $x = 0$ it appears that any departure from regularity will be caused by behaviour near the origin. The reader may find it interesting to show that the sequence is not regular.

The equivalence relation (§1.1) provided by equivalent regular sequences splits the regular sequences into equivalence classes.

Definition 3.2. *An equivalence class of regular sequences is a generalised function.*

A convenient notation is to write, if g is the generalised function associated with the equivalence class of which $\{\gamma_n\}$ is a typical member,

$$\lim_{n\to\infty}\int_{-\infty}^{\infty}\gamma_n(x)\gamma(x)dx = \rightarrow\!\!\!\!\!\!\int_{-\infty}^{\infty} g(x)\gamma(x)dx.$$

The arrow emphasises that a limiting process is involved and that the quantity on the right-hand side is not an ordinary integral. Later on, when certain properties have been established, it will be found reasonable to drop the arrow.

Let $\{\gamma_{1n}\}$ and $\{\gamma_{2n}\}$ (these abbreviations for $\gamma_{1,n}$ and $\gamma_{2,n}$ will be used when there is little likelihood of confusion with a subsequence of $\{\gamma_n\}$) be regular sequences defining the generalised functions g_1 and g_2 respectively. Then

$$\lim_{n\to\infty}\int_{-\infty}^{\infty}\gamma_{1n}(x)\gamma(x)dx = \rightarrow\!\!\!\!\!\!\int_{-\infty}^{\infty} g_1(x)\gamma(x)dx,$$

$$\lim_{n\to\infty}\int_{-\infty}^{\infty}\gamma_{2n}(x)\gamma(x)dx = \rightarrow\!\!\!\!\!\!\int_{-\infty}^{\infty} g_2(x)\gamma(x)dx.$$

If the sequences $\{\gamma_{1n}\}$ and $\{\gamma_{2n}\}$ are equivalent the two left-hand sides, and therefore the two right-hand sides, are equal, and conversely. Hence

$$\int_{-\infty}^{\infty} g_1(x)\gamma(x)\mathrm{d}x = \int_{-\infty}^{\infty} g_2(x)\gamma(x)\mathrm{d}x$$

if, and only if,

$$g_1(x) = g_2(x).$$

Example 1. According to (1) the regular sequence $\{\mathrm{e}^{-x^2/n}\}$ defines a generalised function $I(x)$ such that

$$\int_{-\infty}^{\infty} I(x)\gamma(x)\mathrm{d}x = \int_{-\infty}^{\infty} \gamma(x)\mathrm{d}x.$$

This generalised function is clearly related to the function which is equal to 1 for all x and we shall, in fact, denote it by 1.

Example 2. The sequence $\{(\mathrm{e}^{-x^2/n})/n\}$, obtained from the preceding by dividing the nth term by n, is regular by an analysis similar to that leading to (1) and gives

$$\lim_{n\to\infty} \int_{-\infty}^{\infty} \frac{\mathrm{e}^{-x^2/n}\gamma(x)}{n}\mathrm{d}x = 0.$$

It therefore defines a generalised function which may be denoted by 0, being connected with the function equal to 0 for all x.

Definition 3.3. *The sequence $\{(n/\pi)^{1/2}\mathrm{e}^{-nx^2}\}$ is regular and defines a generalised function, denoted by $\delta(x)$, such that*

$$\int_{-\infty}^{\infty} \delta(x)\gamma(x)\mathrm{d}x = \gamma(0).$$

The statements made in this definition of $\delta(x)$ must be substantiated. Firstly

$$\int_{-\infty}^{\infty} \mathrm{e}^{-nx^2}\gamma(x)\mathrm{d}x = \gamma(0)\int_{-\infty}^{\infty} \mathrm{e}^{-nx^2}\mathrm{d}x + \int_{-\infty}^{\infty} \{\gamma(x) - \gamma(0)\}\mathrm{e}^{-nx^2}\mathrm{d}x$$

and

$$\int_{-\infty}^{\infty} e^{-nx^2}dx = \int_{-\infty}^{\infty} \frac{e^{-x^2/2}dx}{(2n)^{1/2}} = \left(\frac{\pi}{n}\right)^{1/2}$$

from Theorem 2.2. Also

$$|\gamma(x)-\gamma(0)| = \left|\int_0^x \frac{d\gamma(x)}{dx}dx\right| \le M|x|$$

where M is a finite constant since $d\gamma/dx$, being good, is bounded. Hence

$$\left|\int_{-\infty}^{\infty} \{\gamma(x)-\gamma(0)\}e^{-nx^2}dx\right| \le 2M\int_0^{\infty} xe^{-nx^2}dx$$
$$\le (M/n)[-e^{-nx^2}]_0^{\infty} = M/n.$$

Therefore

$$\int_{-\infty}^{\infty} \left(\frac{n}{\pi}\right)^{1/2} e^{-nx^2}\gamma(x)dx = \gamma(0) + O\left(\frac{M}{n^{1/2}}\right)$$

and

$$\lim_{n\to\infty}\int_{-\infty}^{\infty} \left(\frac{n}{\pi}\right)^{1/2} e^{-nx^2}\gamma(x)dx = \gamma(0)$$

which shows that the sequence $\{(n/\pi)^{1/2}e^{-nx^2}\}$ of good functions is regular and that the generalised function has the property stated.

Exercise

1. Which of the following sequences are regular?
(i) $\{e^{-x^6/n}\}$, (ii) $\{e^{-|x|/n}\}$, (iii) $\{(e^{-x^4/n})/n\}$, (iv) $\{ne^{-nx^2}\}$, (v) $\{n^{1/4}e^{-nx^4}\}$.
What generalised functions do the regular sequences define?

We have had some examples of regular sequences and the generalised functions they define but it is not self-evident that the procedure distinguishes between quantities which we would customarily regard as different. Suppose that

$$\lim_{n\to\infty}\int_{-\infty}^{\infty} \gamma_{1n}(x)\gamma(x)dx = \int_{-\infty}^{\infty} f_1(x)\gamma(x)dx \tag{2}$$

where f_1 is an integrable function. One way in which this could happen is that $\gamma_{1n}=\gamma_{11}$ for all n; another important way will be seen later. Let there be a regular sequence $\{\gamma_{2n}\}$ such that

$$\lim_{n\to\infty}\int_{-\infty}^{\infty}\gamma_{2n}(x)\gamma(x)\,dx=\int_{-\infty}^{\infty}f_2(x)\gamma(x)\,dx \tag{3}$$

where f_2 is integrable. Then if the sequences $\{\gamma_{1n}\}$ and $\{\gamma_{2n}\}$ were equivalent it would be expected that f_1 and f_2 were equal in some sense. On the other hand if f_1 and f_2 were not equal it would be desirable that $\{\gamma_{1n}\}$ and $\{\gamma_{2n}\}$ were not equivalent. By subtracting (3) from (2) we can see that the question is: if $\int_{-\infty}^{\infty}f(x)\gamma(x)\,dx=0$ for all good γ does f vanish? Since (§ 1.12) an integrable function is continuous almost everywhere it will be sufficient for our purposes to consider an interval of continuity (a, b) of f and then the answer to the question is contained in

Theorem 3.1. *If $\int_{-\infty}^{\infty}f(x)\gamma(x)\,dx=0$ for all good functions γ which vanish outside the interval of continuity (a, b) of f then $f\equiv 0$ in (a, b).*

Proof. Select (c, d) arbitrarily in (a, b) and then consider the good function

$$\gamma_1(x)=\begin{cases}\exp\left\{-\dfrac{1}{m}\left(\dfrac{1}{x-c}+\dfrac{1}{d-x}\right)\right\} & (a\le c\le x\le d\le b,\ m>0)\\ 0 & (x\notin(c,d)).\end{cases}$$

Given $\varepsilon>0$ we can choose m so large that $|1-\gamma_1(x)|\le\varepsilon$ for $c+\varepsilon\le x\le d-\varepsilon$. Since γ_1 satisfies the conditions of the theorem

$$\int_c^d f(x)\gamma_1(x)\,dx=\int_{-\infty}^{\infty}f(x)\gamma_1(x)\,dx=0.$$

Hence

$$\begin{aligned}\int_c^d f(x)\,dx&=\int_c^d f(x)\{1-\gamma_1(x)\}\,dx\\&=\left(\int_c^{c+\varepsilon}+\int_{c+\varepsilon}^{d-\varepsilon}+\int_{d-\varepsilon}^{d}\right)\{f(x)[1-\gamma_1(x)]\,dx\}.\end{aligned}$$

Since f is continuous it is bounded, by M (say). Then the three integrals do not exceed, in modulus, $2M\varepsilon$, $M\varepsilon$, $2M\varepsilon$ respectively and so

$$\left|\int_c^d f(x)\mathrm{d}x\right| \le 5M\varepsilon.$$

Since ε is arbitrarily small this implies $\int_c^d f(x)\mathrm{d}x = 0$ which implies, because (c, d) is arbitrary, that $f \equiv 0$ in (a, b) and the theorem is proved. □

The definition of generalised functions given by L. Schwartz in *Théorie des distributions* uses a different approach. The space of good functions is denoted by $\mathscr{S}$. $\mathscr{S}$ is provided with a topology by saying that $\{\gamma_n\}$ converges to zero if, for any finite number of a_{lm}, $\sum_{l,m} a_{lm} x^l \mathrm{d}^m\gamma_n/\mathrm{d}x^m$ tends to zero uniformly on R_1. $\mathscr{S}$ is locally convex, complete and has a countable base. The dual $\mathscr{S}'$ of $\mathscr{S}$ is the space of continuous linear functionals $F(\gamma)$ defined for $\gamma\in\mathscr{S}$ and continuous in the sense that the complex numbers $F(\gamma_n)$ tend to zero when $\{\gamma_n\}$ converges to zero in $\mathscr{S}$. $\mathscr{S}'$ is the space of generalised functions. It can be provided with a topology by saying that $\{F_n\}$ converges to zero if the numbers $F_n(\gamma)$ tend to zero for any $\gamma\in\mathscr{S}$ and uniformly for any bounded set of γ. Then $\mathscr{S}$ and $\mathscr{S}'$ are reflexive, i.e. duals of one another. (See also §11.16.)

3.2. Conventional functions as generalised functions

It will now be shown that a large class of conventional functions can be regarded as generalised functions. The notation $f\in L_p$ has been introduced already; it means $\int_{-\infty}^{\infty} |f(x)|^p \mathrm{d}x < \infty$. We now introduce the notation of

Definition 3.4. *If, for some* $N \ge 0$, $\int_{-\infty}^{\infty} |f(x)|^p/(1+x^2)^N \mathrm{d}x < \infty$ *then* $f\in K_p$ *and conversely.*

Note that if

$$\int_{-\infty}^{\infty} \frac{|f(x)|^p}{(1+x^2)^N}\mathrm{d}x < \infty, \quad \int_{-\infty}^{\infty} \frac{|f(x)|^p}{(1+x^2)^{N+m}}\mathrm{d}x \le \int_{-\infty}^{\infty} \frac{|f(x)|^p}{(1+x^2)^N}\mathrm{d}x < \infty$$

for $m \ge 0$.

Obviously if $f\in L_p$ then $f\in K_p$ but the converse is not true. For example, sin x is in K_1 (take $N = 1$) but is not in L_1. On the other hand we have

Theorem 3.2. *If $f \in K_p (p \geq 1)$ then $f \in K_1$.*

Proof. For $p = 1$ there is nothing to prove, and for $p > 1$ the Hölder inequality (Theorem 1.28) gives

$$\int_{-\infty}^{\infty} \frac{|f(x)|\,\mathrm{d}x}{(1+x^2)^{(N/p)+1/q}} \leq \left\{\int_{-\infty}^{\infty} \frac{|f(x)|^p}{(1+x^2)^N}\mathrm{d}x\right\}^{1/p} \left\{\int_{-\infty}^{\infty} \frac{\mathrm{d}x}{1+x^2}\right\}^{1/q}$$

with $1/q = 1 - 1/p$. Both integrals on the right are finite so that the theorem is proved. □

Let $\rho(x)$ denote the particular good function defined by

$$\rho(x) = \begin{cases} \dfrac{\mathrm{e}^{-1/(1-x^2)}}{\int_{-1}^{1} \mathrm{e}^{-1/(1-t^2)}\mathrm{d}t} & (|x| < 1) \\ 0 & (|x| \geq 1). \end{cases}$$

Note that $\int_{-1}^{1} \rho(x)\,\mathrm{d}x = 1$.

Theorem 3.3. *If $f \in K_1$ the sequence $\{\gamma_n\}$, where*

$$\gamma_n(x) = \int_{-\infty}^{\infty} f(u)\rho\{n(u-x)\}\,n\mathrm{e}^{-u^2/n^2}\,\mathrm{d}u,$$

is regular and defines a generalised function g such that

$$\int_{-\infty}^{\infty} g(x)\gamma(x)\,\mathrm{d}x = \int_{-\infty}^{\infty} f(x)\gamma(x)\,\mathrm{d}x.$$

Proof. First it must be shown that γ_n is good. Since ρ is good there is an M, independent of x, such that $|(1+u^2)^N \rho\{n(u-x)\}| < M$. Therefore, because $f \in K_1$, the integral for γ_n is uniformly convergent by Theorem 1.31. Since the derivative of a good function is good the same is true of the integral containing $\partial\rho/\partial x$. Hence, by Theorem 1.33 (see remark at end),

$$\frac{\mathrm{d}^q\gamma_n(x)}{\mathrm{d}x^q} = -\int_{-\infty}^{\infty} f(u)(-n)^{q+1}\rho^{(q)}\{n(u-x)\}\,\mathrm{e}^{-u^2/n^2}\,\mathrm{d}u$$

where $\rho^{(q)}$ is the qth derivative of ρ with respect to its argument. Continuity of $\mathrm{d}^q\gamma_n/\mathrm{d}x^q$ follows from Theorem 1.32. Now $\rho(x)$ and all its derivatives vanish outside $(-1, 1)$. Therefore

$$\frac{\mathrm{d}^q\gamma_n(x)}{\mathrm{d}x^q} = -\int_{x-1/n}^{x+1/n} f(u)(-n)^{q+1}\rho^{(q)}\{n(u-x)\}\,\mathrm{e}^{-u^2/n^2}\,\mathrm{d}u.$$

In the interval of integration $x - 1/n \le u \le x + 1/n$. This implies $x - 1 \le u \le x + 1$ which means $|x| - 1 \le |u| \le |x| + 1$. Hence

$$\mathrm{e}^{-u^2/n^2} \le \mathrm{e}^{-(|x|-1)^2/n^2} \quad \text{and} \quad (1+u^2)^N \le \{1 + (|x|+1)^2\}^N.$$

Since $\rho^{(q)}$ is good it is bounded, by M_1 (say), and so

$$\left|\frac{\mathrm{d}^q\gamma_n(x)}{\mathrm{d}x^q}\right| \le n^{q+1} M_1 \{1 + (|x|+1)^2\}^N \mathrm{e}^{-(|x|-1)^2/n^2} \int_{-\infty}^{\infty} \frac{|f(u)|}{(1+u^2)^N}\,\mathrm{d}u.$$

The integral is finite, because $f \in K_1$, and independent of x so that $\lim_{|x|\to\infty} |x^r \mathrm{d}^q\gamma_n(x)/\mathrm{d}x^q| = 0$. Because r and q are arbitrary, it has been shown that γ_n is good.

It must now be demonstrated that the sequence $\{\gamma_n\}$ is regular. We have

$$\begin{aligned}\int_{-\infty}^{\infty} \gamma_n(x)\gamma(x)\,\mathrm{d}x &= \int_{-\infty}^{\infty} \gamma(x)\left[\int_{-\infty}^{\infty} f(u)\rho\{n(u-x)\}n\mathrm{e}^{-u^2/n^2}\,\mathrm{d}u\right]\mathrm{d}x \\ &= \int_{-\infty}^{\infty} f(u)\mathrm{e}^{-u^2/n^2}\left[\int_{-\infty}^{\infty} \gamma(x)\rho\{n(u-x)\}n\,\mathrm{d}x\right]\mathrm{d}u\end{aligned}$$

by Theorem 1.35. On putting $x = u - y/n$, we have

$$\begin{aligned}\int_{-\infty}^{\infty} \gamma(x)\rho\{n(u-x)\}n\,\mathrm{d}x &= \int_{-\infty}^{\infty} \gamma(u - y/n)\rho(y)\,\mathrm{d}y \\ &= \int_{-1}^{1} \gamma(u-y/n)\rho(y)\,\mathrm{d}y.\end{aligned}$$

Hence

$$\begin{aligned}\int_{-\infty}^{\infty} \gamma_n(x)\gamma(x)\,\mathrm{d}x = \int_{-1}^{1} \rho(y)\,\mathrm{d}y&\left[\int_{-\infty}^{\infty} f(u)\{\gamma(u-y/n) - \gamma(u)\}\mathrm{e}^{-u^2/n^2}\,\mathrm{d}u\right. \\ &\left. + \int_{-\infty}^{\infty} f(u)\gamma(u)\mathrm{e}^{-u^2/n^2}\,\mathrm{d}u\right]\end{aligned}$$

since $\int_{-1}^{1} \rho(y)\,dy = 1$. Now

$$\left|\int_{-\infty}^{\infty} f(u)\gamma(u)e^{-u^2/n^2}\,du - \int_{-\infty}^{\infty} f(u)\gamma(u)\,du\right|$$

$$\leq \int_{-\infty}^{\infty} |f(u)|\,|\gamma(u)|\,|e^{-u^2/n^2} - 1|\,du$$

$$\leq \frac{1}{n^2}\int_{-\infty}^{\infty} |f(u)|\,|\gamma(u)|\,u^2\,du$$

$$\leq \frac{1}{n^2}\int_{-\infty}^{\infty} \frac{f(u)}{(1+u^2)^N}(1+u^2)^{N+1}|\gamma(u)|\,du$$

$$\leq \frac{B}{n^2}\int_{-\infty}^{\infty} \frac{|f(u)|}{(1+u^2)^N}\,du$$

because the good function $(1+u^2)^{N+1}\gamma(u)$ is bounded. Since $f \in K_1$ we see that

$$\lim_{n\to\infty}\int_{-\infty}^{\infty} f(u)\gamma(u)e^{-u^2/n^2}\,du = \int_{-\infty}^{\infty} f(u)\gamma(u)\,du.$$

By Corollary 1.30*a*, $\gamma(u - y/n) - \gamma(u) = (y/n)\,d\gamma(\xi)/d\xi$ for some ξ in $(u - y/n, u)$. For finite u, $d\gamma/d\xi$ is bounded and, for large $|u|$, $d\gamma/d\xi$ tends to zero faster than any inverse power of u, i.e.

$$\max_{u - y/n \leq \xi \leq u} d\gamma(\xi)/d\xi \leq B_1(1+u^2)^{-N}.$$

Hence

$$\left|\int_{-1}^{1} \rho(y)\,dy \int_{-\infty}^{\infty} f(u)\{\gamma(u - y/n) - \gamma(u)\}e^{-u^2/n^2}\,du\right|$$

$$\leq \frac{B_1}{n}\int_{-\infty}^{\infty} \frac{|f(u)|}{(1+u^2)^N}\,du \to 0$$

as $n \to \infty$. Consequently

$$\lim_{n\to\infty}\int_{-\infty}^{\infty} \gamma_n(x)\gamma(x)\,dx = \int_{-\infty}^{\infty} f(u)\gamma(u)\,du$$

and the theorem is proved. □

Theorem 3.3 provides a very wide range of generalised functions. It includes all integrable functions and measures; it also includes functions which, although not integrable over $(-\infty, \infty)$, are such that $|f(x)|/(1+x^2)^N$ is integrable. Integrable functions which are equal almost everywhere give rise to the same generalised function. On the other hand functions which differ on a set of positive measure give different generalised functions on account of Theorem 3.1.

Exercises

2. If $f_1 \in K_p$ and $f_2 \in K_p$, show that $f_1 + f_2 \in K_p$.
3. If $f_1 \in K_p$ $(p > 1)$ and $f_2 \in K_q$, $q = p/(p-1)$, show that $f_1 f_2 \in K_1$.
4. Which of the following functions belongs to K_2 and which to K_1? $1+x^2, 1/|x|^{1/2}, \sin x, \cos x^2, e^x, e^{ix^2}/|x|^{1/4}, |x^2-1|^{1/2} \sin(1/x)$.
5. If $f \in K_p$ show that $f = P_n f_1$ where P_n is a polynomial and $f_1 \in L_p$.
6. If $f \in K_p$ prove that $f \in K_{p'}$ for $1 \leq p' \leq p$.
7. Which of the functions in Exercise 4 gives rise to a generalised function and which of the Bessel functions $J_0(x)$, $J_\nu(x)$ $(\nu > -1)$, $H_\nu^{(2)}(x)$?

So far a name has been given to the generalised functions 0, 1, $\delta(x)$. It is now convenient to introduce the following nomenclature.

Definition 3.5. *If f is a function and g a generalised function such that*

$$⨍_{-\infty}^{\infty} g(x)\gamma(x)\,dx = \int_a^b f(x)\gamma(x)\,dx$$

for every good function γ which vanishes outside (a, b), we say that

$$g(x) = f(x) \qquad (a < x < b)$$

and conversely. If $⨍_{-\infty}^{\infty} g(x)\gamma(x)\,dx = \int_{-\infty}^{\infty} f(x)\gamma(x)\,dx$ for every good γ we say

$$g(x) = f(x)$$

and conversely.

Since integrable functions which are equal almost everywhere give rise to the same generalised functions this definition implies that such functions are to be regarded as the same. For this reason values of a generalised function can be specified only in an interval and not at a point.

Example 3. $\delta(x) = 0$ in $x > 0$ and in $x < 0$. For, if γ vanishes on $(0, \infty)$, $\gamma(0) = 0$ by continuity and $⨍_{-\infty}^{\infty} \delta(x)\gamma(x)\,dx = 0$ by Definition 3.3. This shows that $\delta(x) = 0$ in $x < 0$. The proof for $x > 0$ is similar. Since $\delta(x)$ is non-zero only on a set of measure zero it follows that $\delta(x)$ is not a conventional integrable function otherwise $\int_{-\infty}^{\infty} \delta(x)\gamma(x)\mathrm{d}x$ would vanish for all γ.

Two functions which are of frequent occurrence and which belong to K_1 are the *Heaviside unit function* $\mathrm{H}(x)$ defined by

$$\mathrm{H}(x) = \begin{cases} 1 & (x > 0) \\ 0 & (x \leq 0) \end{cases}$$

and sgn x defined by

$$\operatorname{sgn} x = \begin{cases} 1 & (x > 0) \\ 0 & (x = 0) \\ -1 & (x < 0). \end{cases}$$

Of course, when they are regarded as generalised functions the specification of the value at the origin has no significance.

3.3. The derivative

Let $\{\gamma_{1n}\}, \{\gamma_{2n}\}$ be regular sequences defining the generalised functions g_1 and g_2 respectively. Then

Definition 3.6. *The sequence $\{\gamma_{1n} + \gamma_{2n}\}$ is regular and defines a generalised function which will be denoted by $g_1 + g_2$.*

The statement that $\{\gamma_{1n} + \gamma_{2n}\}$ is regular has to be checked. Now $\gamma_{1n} + \gamma_{2n}$ is good and

$$\begin{aligned} &\lim_{n\to\infty} \int_{-\infty}^{\infty} \{\gamma_{1n}(x) + \gamma_{2n}(x)\}\gamma(x)\mathrm{d}x \\ &\quad = \lim_{n\to\infty} \int_{-\infty}^{\infty} \gamma_{1n}(x)\gamma(x)\mathrm{d}x + \lim_{n\to\infty} \int_{-\infty}^{\infty} \gamma_{2n}(x)\gamma(x)\mathrm{d}x \\ &\quad = \int_{-\infty}^{\infty} g_1(x)\gamma(x)\mathrm{d}x + \int_{-\infty}^{\infty} g_2(x)\gamma(x)\mathrm{d}x. \end{aligned} \tag{4}$$

This shows that the sequence is regular. Furthermore, since the right-hand side is unaltered if $\{\gamma_{1n}\}$ (or $\{\gamma_{2n}\}$) is replaced by an equivalent sequence, all sequences $\{\gamma_{1n} + \gamma_{2n}\}$ obtained from sequences equivalent to $\{\gamma_{1n}\}$ and $\{\gamma_{2n}\}$ are equivalent. Thus the definition is reasonable. Note too that it is consistent with Definition 3.5 and Theorem 3.3 (if $f_1 \in K_1$ and $f_2 \in K_1$ then $f_1 + f_2 \in K_1$ and the sequence in Theorem 3.3 would give the generalised function $f_1 + f_2$). Equation (4) may be restated as

$$\int_{-\infty}^{\infty} \{g_1(x)+g_2(x)\}\gamma(x)\mathrm{d}x = \int_{-\infty}^{\infty} g_1(x)\gamma(x)\mathrm{d}x + \int_{-\infty}^{\infty} g_2(x)\gamma(x)\mathrm{d}x. \tag{5}$$

Let $\{\gamma_n(x)\}$ be a regular sequence defining the generalised function $g(x)$. Then we have

Definition 3.7. *The sequence* $\{\gamma_n(ax+b)\}$ *(a and b real) is a regular sequence and defines a generalised function which will be denoted by* $g(ax+b)$.

To check regularity, note that $\gamma_n(ax+b)$ is good and

$$\lim_{n\to\infty}\int_{-\infty}^{\infty} \gamma_n(ax+b)\gamma(x)\mathrm{d}x = \lim_{n\to\infty}\int_{-\infty}^{\infty} \gamma_n(x)\gamma\left(\frac{x-b}{a}\right)\frac{\mathrm{d}x}{|a|}$$
$$= \int_{-\infty}^{\infty} g(x)\gamma\left(\frac{x-b}{a}\right)\frac{\mathrm{d}x}{|a|} \tag{6}$$

since $\gamma\{(x-b)/a\}$ is good. This also shows that equivalent $\{\gamma_n(x)\}$ give equivalent $\{\gamma_n(ax+b)\}$. Equation (6) can be stated as

Theorem 3.4. $\displaystyle\int_{-\infty}^{\infty} g(ax+b)\gamma(x)\mathrm{d}x = \int_{-\infty}^{\infty} g(x)\gamma\left(\frac{x-b}{a}\right)\frac{\mathrm{d}x}{|a|}.$

An immediate consequence of this theorem is that, if $g_1(x) = g_2(x)$, $g_1(ax+b) = g_2(ax+b)$.

Example 4.

$$\int_{-\infty}^{\infty} \delta(x-b)\gamma(x)\mathrm{d}x = \int_{-\infty}^{\infty} \delta(x)\gamma(x+b)\mathrm{d}x$$
$$= \gamma(b)$$

from Definition 3.3. More generally

$$\int_{-\infty}^{\infty} \delta(ax-b)\gamma(x)\mathrm{d}x = \frac{1}{|a|}\gamma\left(\frac{b}{a}\right).$$

Definition 3.6 defines addition for two generalised functions. Multiplication is much more difficult to handle and will be dealt with in detail in Chapter 6; for the moment we restrict ourselves to multiplication by a fairly good function.

Definition 3.8. *If ψ is fairly good the sequence $\{\psi\gamma_n\}$ is regular and defines a generalised function which will be denoted by $\psi(x)g(x)$.*

Regularity follows from

$$\lim_{n\to\infty}\int_{-\infty}^{\infty}\psi(x)\gamma_n(x)\gamma(x)\,dx = \lim_{n\to\infty}\int_{-\infty}^{\infty}\gamma_n(x)\{\psi(x)\gamma(x)\}\,dx$$
$$= \int_{-\infty}^{\infty} g(x)\{\psi(x)\gamma(x)\}\,dx$$

since $\psi\gamma$ is good by Theorem 2.1. This result may also be stated as

Theorem 3.5. $\displaystyle\int_{-\infty}^{\infty}\{\psi(x)g(x)\}\gamma(x)\,dx = \int_{-\infty}^{\infty} g(x)\{\psi(x)\gamma(x)\}\,dx.$

Example 5. Since x is fairly good $x\delta(x)$ is defined and

$$\int_{-\infty}^{\infty}\{x\delta(x)\}\gamma(x)\,dx = \int_{-\infty}^{\infty}\delta(x)\{x\gamma(x)\}\,dx$$
$$= [x\gamma(x)]_{x=0} = 0.$$

Thus

$$x\delta(x) = 0.$$

Similarly

$$e^{i\alpha x}\delta(x) = \delta(x).$$

Exercises

8. Evaluate

(i) $\displaystyle\int_{-\infty}^{\infty}\delta(x-b)e^{-x^2}\,dx,$ (ii) $\displaystyle\int_{-\infty}^{\infty}H(x)e^{-x^2}\,dx,$

(iii) $\displaystyle\int_{-\infty}^{\infty}\operatorname{sgn} x e^{-x^2}\,dx,$ (iv) $\displaystyle\int_{-\infty}^{\infty}H(x-b)\operatorname{sech} x\,dx.$

9. If ψ is fairly good and g_1, g_2 are generalised functions such that $\psi(x)g_1(x) = g_2(x)$ prove that $\psi(ax+b)g_1(ax+b) = g_2(ax+b)$.

10. If ψ is fairly good show that

$$\psi(x)\delta(x) = \psi(0)\delta(x).$$

11. Would Definition 3.8 remain valid if ψ were replaced by the moderately good function χ?

Definition 3.9. *The sequence* $\{d\gamma_n(x)/dx\}$ *is regular and defines a generalised function which will be denoted by* $g'(x)$ *and called the derivative of* g.

The sequence $\{d\gamma_n/dx\}$ is regular because

$$\int_{-\infty}^{\infty} \frac{d\gamma_n(x)}{dx}\gamma(x)\,dx = [\gamma_n(x)\gamma(x)]_{-\infty}^{\infty} - \int_{-\infty}^{\infty} \gamma_n(x)\frac{d\gamma(x)}{dx}dx$$

by an integration by parts. But $\gamma_n\gamma$, being good, vanishes at $\pm\infty$ and $d\gamma/dx$ is good so that

$$\lim_{n\to\infty}\int_{-\infty}^{\infty} \frac{d\gamma_n(x)}{dx}\gamma(x)\,dx = -\int_{-\infty}^{\infty} g(x)\frac{d\gamma(x)}{dx}dx.$$

This also demonstrates that equivalent $\{\gamma_n\}$ give equivalent $\{d\gamma_n/dx\}$; it may also be written as

Theorem 3.6. $$\int_{-\infty}^{\infty} g'(x)\gamma(x)\,dx = -\int_{-\infty}^{\infty} g(x)\frac{d\gamma(x)}{dx}dx.$$

According to Definition 3.9 every generalised function possesses a derivative. Since this derivative is a generalised function it will have a derivative which will be denoted by $g''(x)$. Applying Theorem 3.6 twice we see that

$$\int_{-\infty}^{\infty} g''(x)\gamma(x)\,dx = -\int_{-\infty}^{\infty} g'(x)\frac{d\gamma(x)}{dx}dx$$

$$= \int_{-\infty}^{\infty} g(x)\frac{d^2\gamma(x)}{dx^2}dx. \tag{7}$$

Obviously the process can be repeated so that a generalised function possesses a derivative of any order. It is now necessary to examine the relation between such a derivative and the conventional derivative, and especially to check consistency with Definition 3.5. The answer is provided by

Theorem 3.7. *If $f(x)$ is a continuous function with a derivative $\mathrm{d}f/\mathrm{d}x$ such that $f \in K_1$ and $\mathrm{d}f/\mathrm{d}x \in K_1$, then*

$$f' = \mathrm{d}f/\mathrm{d}x,$$

i.e. the generalised derivative of the generalised function f is equal to the generalised function $\mathrm{d}f/\mathrm{d}x$.

Proof. Since $f \in K_1$, it defines a generalised function, by Theorem 3.3, which is also denoted by f according to Definition 3.5. From Theorem 3.6

$$\begin{aligned} \int\!\!\!\!\!\!\rightarrow_{-\infty}^{\infty} f'(x)\gamma(x)\mathrm{d}x &= -\int\!\!\!\!\!\!\rightarrow_{-\infty}^{\infty} f(x)\frac{\mathrm{d}\gamma(x)}{\mathrm{d}x}\mathrm{d}x \\ &= -\int_{-\infty}^{\infty} f(x)\frac{\mathrm{d}\gamma(x)}{\mathrm{d}x}\mathrm{d}x \end{aligned}$$

by Theorem 3.3 because $\mathrm{d}\gamma/\mathrm{d}x$ is good. Now

$$\int_{-\infty}^{\infty} f(x)\frac{\mathrm{d}\gamma(x)}{\mathrm{d}x}\mathrm{d}x = [f(x)\gamma(x)]_{-\infty}^{\infty} - \int_{-\infty}^{\infty} \frac{\mathrm{d}f(x)}{\mathrm{d}x}\gamma(x)\mathrm{d}x,$$

the integral on the right-hand side existing because $\mathrm{d}f/\mathrm{d}x \in K_1$. Since the integrals exist so must the third term, i.e. $f(x)\gamma(x)$ tends to a definite limit as $|x| \to \infty$. But this limit must be zero because $\int_{-\infty}^{\infty} f(x)\gamma(x)\mathrm{d}x$ exists. Hence

$$\int_{-\infty}^{\infty} f(x)\frac{\mathrm{d}\gamma(x)}{\mathrm{d}x}\mathrm{d}x = -\int_{-\infty}^{\infty} \frac{\mathrm{d}f(x)}{\mathrm{d}x}\gamma(x)\mathrm{d}x.$$

Since $\mathrm{d}f/\mathrm{d}x \in K_1$, it defines a generalised function also denoted by $\mathrm{d}f/\mathrm{d}x$ such that

$$\int_{-\infty}^{\infty} \frac{\mathrm{d}f(x)}{\mathrm{d}x}\gamma(x)\mathrm{d}x = \int\!\!\!\!\!\!\rightarrow_{-\infty}^{\infty} \frac{\mathrm{d}f(x)}{\mathrm{d}x}\gamma(x)\mathrm{d}x.$$

Hence

$$\int\!\!\!\!\!\!\rightarrow_{-\infty}^{\infty} f'(x)\gamma(x)\mathrm{d}x = \int_{-\infty}^{\infty} \frac{\mathrm{d}f(x)}{\mathrm{d}x}\gamma(x)\mathrm{d}x$$

and the theorem is proved. □

Corollary 3.7. *If* $\mathrm{d}f/\mathrm{d}x \in K_1$ *then* $f' = \mathrm{d}f/\mathrm{d}x$.

For, if $\mathrm{d}f/\mathrm{d}x \in K_1$, $\mathrm{d}f/\mathrm{d}x$ is integrable over any finite interval $(-M, M)$ because

$$\int_{-M}^{M} \left|\frac{\mathrm{d}f}{\mathrm{d}x}\right| \mathrm{d}x \le (1+M^2)^N \int_{-M}^{M} \left|\frac{\mathrm{d}f}{\mathrm{d}x}\right| (1+x^2)^{-N} \mathrm{d}x < \infty$$

and so f can be defined by $f(x) = \int_0^x \{\mathrm{d}f(t)/\mathrm{d}t\}\,\mathrm{d}t + C$, C being a constant. Since $\mathrm{d}f/\mathrm{d}x \in K_1$, $\mathrm{d}f/\mathrm{d}x = o(x^{2N})$ as $|x| \to \infty$ and so $f(x) = o(x^{2N+1})$. Hence $f \in K_1$ and is continuous, by its definition; the result follows from Theorem 3.7. □

While this theorem covers a broad class of functions there are many generalised functions to which it does not apply. Here are one or two illustrations.

Example 6. By Theorem 3.6

$$\int\!\!\!\!\!\!\rightarrow_{-\infty}^{\infty} \mathrm{H}'(x)\gamma(x)\,\mathrm{d}x = -\int\!\!\!\!\!\!\rightarrow_{-\infty}^{\infty} \mathrm{H}(x)\frac{\mathrm{d}\gamma}{\mathrm{d}x}\mathrm{d}x.$$

But $\mathrm{H}(x) \in K_1$ so that, by Theorem 3.3,

$$\int\!\!\!\!\!\!\rightarrow_{-\infty}^{\infty} \mathrm{H}(x)\frac{\mathrm{d}\gamma}{\mathrm{d}x}\mathrm{d}x = \int_{-\infty}^{\infty} \mathrm{H}(x)\frac{\mathrm{d}\gamma}{\mathrm{d}x}\mathrm{d}x = \int_0^{\infty} \frac{\mathrm{d}\gamma}{\mathrm{d}x}\mathrm{d}x = -\gamma(0).$$

Hence

$$\int\!\!\!\!\!\!\rightarrow_{-\infty}^{\infty} \mathrm{H}'(x)\gamma(x)\,\mathrm{d}x = \gamma(0)$$

and therefore

$$\mathrm{H}'(x) = \delta(x). \tag{8}$$

Example 7. Since $\operatorname{sgn} x \in K_1$,

$$\begin{aligned}\int\!\!\!\!\!\!\rightarrow_{-\infty}^{\infty} \operatorname{sgn} x \frac{\mathrm{d}\gamma}{\mathrm{d}x}\mathrm{d}x &= \int_{-\infty}^{\infty} \operatorname{sgn} x \frac{\mathrm{d}\gamma}{\mathrm{d}x}\mathrm{d}x \\ &= \int_0^{\infty} \frac{\mathrm{d}\gamma}{\mathrm{d}x}\mathrm{d}x - \int_{-\infty}^{0} \frac{\mathrm{d}\gamma}{\mathrm{d}x}\mathrm{d}x \\ &= -2\gamma(0).\end{aligned}$$

Hence, as in the preceding example,

$$\operatorname{sgn}' x = 2\delta(x) \tag{9}$$

Example 8. By Theorem 3.6,

$$\int_{-\infty}^{\infty} \delta'(x)\gamma(x)\,dx = -\int_{-\infty}^{\infty} \delta(x)\frac{d\gamma(x)}{dx}dx = -\left[\frac{d\gamma(x)}{dx}\right]_{x=0}.$$

More generally,

$$\int_{-\infty}^{\infty} \delta'(x-a)\gamma(x)\,dx = -\int_{-\infty}^{\infty} \delta(x-a)\frac{d\gamma(x)}{dx}dx = -\left[\frac{d\gamma(x)}{dx}\right]_{x=a}.$$

Higher derivatives can be dealt with in the same way. For example, from (7),

$$\int_{-\infty}^{\infty} \delta''(x)\gamma(x)\,dx = \left[\frac{d^2\gamma(x)}{dx^2}\right]_{x=0}$$

and, in general,

$$\int_{-\infty}^{\infty} \delta^{(n)}(x)\gamma(x)\,dx = (-1)^n\left[\frac{d^n\gamma(x)}{dx^n}\right]_{x=0} = (-1)^n\gamma^{(n)}(0).$$

Exercises

12. If g_1 and g_2 are generalised functions show that

$$(g_1 + g_2)' = g_1' + g_2'.$$

13. Deduce (9) from (8) by proving, as a result for generalised functions, $H(x) = \frac{1}{2}(1 + \operatorname{sgn} x)$.

14. Evaluate $\int_{-\infty}^{\infty} H'(x-a)e^{-x^2}dx$ and $\int_{-\infty}^{\infty} \delta'(x-a)\operatorname{sech} x\,dx$.

15. Prove that $\{g(ax+b)\}' = ag'(ax+b)$. Show also that, if ψ is fairly good,

$$\{\psi(x)g(x)\}' = \psi'(x)g(x) + \psi(x)g'(x).$$

16. Prove that

$$x^n\delta^{(m)}(x) = \begin{cases} 0 & (m < n) \\ (-1)^n\dfrac{m!}{(m-n)!}\delta^{(m-n)}(x) & (m \geq n). \end{cases}$$

17. If ψ is fairly good show that

$$\psi(x)\delta'(x) = -\psi'(0)\delta(x) + \psi(0)\delta'(x).$$

Derive the corresponding result for $\psi(x)\delta^{(m)}(x)$.

18. If f is a continuous function with a derivative $\mathrm{d}f/\mathrm{d}x$ and both are in K_1 show that

$$[f(x)\mathrm{H}(x)]' = (\mathrm{d}f/\mathrm{d}x)\mathrm{H}(x) + f(0)\delta(x).$$

Use this result to show that, if f is continuous with derivative $\mathrm{d}f/\mathrm{d}x$ everywhere except at the points $x = a_1, a_2, \ldots, a_n$ where f has jumps of magnitude $b_1, b_2, \ldots, b_n$,

$$f'(x) = \mathrm{d}f/\mathrm{d}x + b_1\delta(x - a_1) + b_2\delta(x - a_2) + \ldots + b_n\delta(x - a_n).$$

19. Prove that $|x|' = \operatorname{sgn} x$.

A theorem in some ways more convenient than Theorem 3.7 is

Theorem 3.8. *If integrable f has a derivative* $\mathrm{d}f/\mathrm{d}x$ *in* (a, b) *such that* $\int_a^b f(x)\gamma(x)\mathrm{d}x$ *and* $\int_a^b \{\mathrm{d}f(x)/\mathrm{d}x\}\,\gamma(x)\mathrm{d}x$ *both exist for all good* γ *which vanish outside* (a, b), *and if the generalised function* $g(x) = f(x)$ *in* $a < x < b$ *then*

$$g'(x) = \mathrm{d}f(x)/\mathrm{d}x \qquad (a < x < b).$$

Proof. Let γ vanish outside (a, b). Then

$$⨍_{-\infty}^{\infty} g'(x)\gamma(x)\mathrm{d}x = -⨍_{-\infty}^{\infty} g(x)\frac{\mathrm{d}\gamma(x)}{\mathrm{d}x}\mathrm{d}x = -\int_a^b f(x)\frac{\mathrm{d}\gamma(x)}{\mathrm{d}x}\mathrm{d}x$$

by Definition 3.5. An integration by parts gives

$$\int_a^b f(x)\frac{\mathrm{d}\gamma(x)}{\mathrm{d}x}\mathrm{d}x = [f(x)\gamma(x)]_a^b - \int_a^b \frac{\mathrm{d}f(x)}{\mathrm{d}x}\gamma(x)\mathrm{d}x.$$

Suppose b is finite. Then $\gamma(x) = O(x - b)$ as $x \to b$ since $\mathrm{d}\gamma/\mathrm{d}x$ vanishes outside (a, b) and $f(x)$ cannot be as singular as $1/(x - b)$ since it is integrable. Hence the square bracket vanishes at upper limit. This is still true if b is infinite by the argument employed in Theorem 3.7. Similar conclusions apply at the lower limit and hence

$$⨍_{-\infty}^{\infty} g'(x)\gamma(x)\mathrm{d}x = \int_a^b \frac{\mathrm{d}f(x)}{\mathrm{d}x}\gamma(x)\mathrm{d}x.$$

The theorem follows from Definition 3.5. □

Example 9. Since $\delta(x)=0$ in $x>0$ we have, putting $f=0$ and $df/dx=0$ in Theorem 3.8, $\delta'(x)=0$ in $x>0$. Similarly $\delta'(x)=0$ in $x<0$. Repeating the process on $\delta'(x)$ we have $\delta''(x)=0$ in $x>0$ and in $x<0$. Obviously all derivatives of $\delta(x)$ have the same property. Thus $\delta(x)$ and its derivatives are examples of generalised functions which agree except in one neighbourhood and yet are different generalised functions.

Exercise

20. Show that $\delta''(ax+b)=0$ in $x>-b/a$ and $x<-b/a$.

3.4. The Fourier transform

Let Γ_n be the Fourier transform of the good function γ_n. Then

Definition 3.10. *If $\{\gamma_n\}$ is regular the sequence $\{\Gamma_n\}$ is regular and defines a generalised function which will be denoted by $G(\alpha)$ and called the Fourier transform of g. We write $G(\alpha)=\int_{-\infty}^{\infty} g(x)e^{-i\alpha x}dx$.*

To verify the statements made in the definition note that Γ_n is good by Theorem 2.4. If $\Gamma(\alpha)$ is any good function and $\gamma(x)$ is defined by $\gamma(x)=(1/2\pi)\int_{-\infty}^{\infty}\Gamma(\alpha)e^{i\alpha x}d\alpha$ then γ is good and Γ is the Fourier transform of γ by Theorems 2.4 and 2.5. Hence, by Theorem 2.6 with $y=0$ and f replaced by γ_n,

$$\int_{-\infty}^{\infty}\Gamma_n(\alpha)\Gamma(\alpha)d\alpha = 2\pi\int_{-\infty}^{\infty}\gamma_n(x)\gamma(-x)dx$$

$$\rightarrow 2\pi \int\!\!\!\!\!\!-_{-\infty}^{\infty} g(x)\gamma(-x)dx$$

as $n\rightarrow\infty$ because $\gamma(-x)$ is good. This shows that $\{\Gamma_n\}$ is regular and that equivalent $\{\gamma_n\}$ give equivalent $\{\Gamma_n\}$. Furthermore

Theorem 3.9. $$\int\!\!\!\!\!\!-_{-\infty}^{\infty} G(\alpha)\Gamma(\alpha)d\alpha = 2\pi\int\!\!\!\!\!\!-_{-\infty}^{\infty} g(x)\gamma(-x)dx.$$

Actually a more general result is true. For, if $y\in R_1$, $e^{i\alpha y}$ is fairly good and $e^{i\alpha y}G(\alpha)$ is a generalised function by Definition 3.8. Thus

$$\int_{-\infty}^{\infty} e^{i\alpha y}G(\alpha)\Gamma(\alpha)\,d\alpha = \lim_{n\to\infty}\int_{-\infty}^{\infty} e^{i\alpha y}\Gamma_n(\alpha)\Gamma(\alpha)\,d\alpha$$

$$= \lim_{n\to\infty} 2\pi\int_{-\infty}^{\infty} \gamma_n(x)\gamma(y-x)\,dx$$

$$= 2\pi\int_{-\infty}^{\infty} g(x)\gamma(y-x)\,dx$$

by Theorem 2.6. Hence

Theorem 3.9a. *If* $y\in R_1$,

$$\int_{-\infty}^{\infty} e^{i\alpha y}G(\alpha)\Gamma(\alpha)\,d\alpha = 2\pi\int_{-\infty}^{\infty} g(x)\gamma(y-x)\,dx.$$

It is often convenient to have this written in the alternative form

$$\int_{-\infty}^{\infty} G(\alpha)\Gamma(y-\alpha)\,d\alpha = 2\pi\int_{-\infty}^{\infty} e^{-iyx}g(x)\gamma(x)\,dx.$$

Example 10. According to Definition 3.3 the sequence $\{(n/\pi)^{1/2}e^{-nx^2}\}$ defines $\delta(x)$. In this case

$$\Gamma_n(\alpha) = \int_{-\infty}^{\infty}\left(\frac{n}{\pi}\right)^{1/2} e^{-nx^2-i\alpha x}\,dx$$

$$= (2\pi)^{-1/2}\int_{-\infty}^{\infty}\exp\left\{-\tfrac{1}{2}t^2 - \frac{i\alpha}{(2n)^{1/2}}t\right\}dt$$

$$= e^{-\alpha^2/4n}$$

by Theorem 2.2. But $\{e^{-\alpha^2/4n}\}$ defines the generalised function 1 (Example 1). Hence the Fourier transform of $\delta(x)$ is 1 or

$$\int_{-\infty}^{\infty}\delta(x)e^{-i\alpha x}\,dx = 1.$$

The Fourier transform has many properties which we now derive.

Theorem 3.10. *If* $G(\alpha)$ *is the Fourier transform of* $g(x)$, *the Fourier transform of* $g(ax+b)$ *is* $(1/|a|)\,G(\alpha/a)e^{ib\alpha/a}$.

Proof. By Definition 3.7, $g(ax+b)$ is defined by $\{\gamma_n(ax+b)\}$. The

Fourier transform of $\gamma_n(ax+b)$ is $(1/|a|)e^{ib\alpha/a}\Gamma_n(\alpha/a)$ (Exercise 14 of Chapter 2). Hence the Fourier transform of $g(ax+b)$ is defined by $\{(1/|a|)e^{ib\alpha/a}\Gamma_n(\alpha/a)\}$. Since $e^{ib\alpha/a}$ is fairly good the theorem follows from Definitions 3.8 and 3.7. □

Example 11. The Fourier transform of $\delta(x-b)$ is $e^{-ib\alpha}$, i.e.

$$\int_{-\infty}^{\infty} \delta(x-b)e^{-i\alpha x}\,dx = e^{-ib\alpha}. \tag{10}$$

More generally

$$\int_{-\infty}^{\infty} \delta(ax-b)e^{-i\alpha x}\,dx = \frac{e^{-ib\alpha/a}}{|a|}. \tag{11}$$

Perhaps the most important result is that the Fourier inversion theorem is always valid for generalised functions, i.e.

Theorem 3.11. *If $G(\alpha)$ is the Fourier transform of $g(x)$ then*

$$g(x) = \frac{1}{2\pi}\int_{-\infty}^{\infty} G(\alpha)e^{i\alpha x}\,d\alpha.$$

Proof. The sequence $\{\Gamma_n(\alpha)\}$ defines G. Therefore, by Definition 3.10, the sequence of the Fourier transforms of Γ_n defines the Fourier transform of G. Now

$$\int_{-\infty}^{\infty} \Gamma_n(\alpha)e^{-i\alpha x}\,d\alpha = 2\pi\gamma_n(-x)$$

by Theorem 2.5. Hence the Fourier transform of G is defined by $\{2\pi\gamma_n(-x)\}$ and must be $2\pi g(-x)$, i.e. $2\pi g(-x) = \int_{-\infty}^{\infty} G(\alpha)e^{-i\alpha x}\,d\alpha$. The theorem follows on changing the sign of x. □

Example 12. Since the Fourier transform of $\delta(x)$ is 1, Theorem 3.11 shows that

$$\frac{1}{2\pi}\int_{-\infty}^{\infty} e^{i\alpha x}\,d\alpha = \delta(x).$$

Similarly, from (10)

$$\frac{1}{2\pi}\int_{-\infty}^{\infty} e^{i\alpha x - ib\alpha}\,d\alpha = \delta(x-b).$$

Theorem 3.12. *The Fourier transform of $g'(x)$ is $i\alpha G(\alpha)$.*

Proof. Since g' is determined by $\{\gamma_n'\}$ its Fourier transform comes from

$$\int_{-\infty}^{\infty} \frac{d\gamma_n(x)}{dx} e^{-i\alpha x} dx = [\gamma_n(x)e^{-i\alpha x}]_{-\infty}^{\infty} + i\alpha \int_{-\infty}^{\infty} \gamma_n(x)e^{-i\alpha x} dx$$

$$= i\alpha\Gamma_n(\alpha)$$

because of the behaviour of a good function at infinity. Since α is fairly good $\{i\alpha\Gamma_n(\alpha)\}$ defines $i\alpha G(\alpha)$ by Definition 3.8 and the theorem is proved. □

Example 13. The Fourier transform of $\delta(x)$ is 1. Therefore the Fourier transform of $\delta'(x)$ is $i\alpha$ and of $\delta^{(n)}(x)$ is $(i\alpha)^n$. More generally,

$$\int_{-\infty}^{\infty} \delta'(x-b)e^{-i\alpha x} dx = i\alpha e^{-ib\alpha},$$

$$\int_{-\infty}^{\infty} \delta^{(n)}(x-b)e^{-i\alpha x} dx = (i\alpha)^n e^{-ib\alpha}.$$

Applying Theorem 3.11 to these two results we obtain

$$\frac{1}{2\pi}\int_{-\infty}^{\infty} i\alpha e^{i\alpha x - ib\alpha} d\alpha = \delta'(x-b), \tag{12}$$

$$\frac{1}{2\pi}\int_{-\infty}^{\infty} (i\alpha)^n e^{i\alpha x - ib\alpha} d\alpha = \delta^{(n)}(x-b). \tag{13}$$

The notation employed for the Fourier transform of a generalised function is justified by

Theorem 3.13. *If $f \in L_1$ it has an ordinary Fourier transform $F(\alpha) \in K_1$. The generalised function derived from F is the Fourier transform of the generalised function derived from f.*

Proof. $|F(\alpha)| = |\int_{-\infty}^{\infty} f(x)e^{-i\alpha x} dx| \le \int_{-\infty}^{\infty} |f(x)| dx \le M$ (say) since $f \in L_1$.

Therefore $\int_{-\infty}^{\infty} \{|F(\alpha)|/(1+\alpha^2)\} d\alpha \le \pi M$ and $F \in K_1$. By Theorem 3.3 f and F define generalised functions denoted by the same letters. Let F_1 denote the Fourier transform of the generalised function f.

Then

$$\int_{-\infty}^{\infty} F(\alpha)\Gamma(\alpha)d\alpha = \int_{-\infty}^{\infty} F(\alpha)\Gamma(\alpha)d\alpha$$

$$= 2\pi \int_{-\infty}^{\infty} f(x)\gamma(-x)dx$$

by Theorem 2.6. But

$$2\pi \int_{-\infty}^{\infty} f(x)\gamma(-x)dx = 2\pi \int_{-\infty}^{\infty} f(x)\gamma(-x)dx$$

$$= \int_{-\infty}^{\infty} F_1(\alpha)\Gamma(\alpha)d\alpha$$

by Theorem 3.9. Hence $F(\alpha) = F_1(\alpha)$ and the theorem is proved. □

According to this theorem the ordinary Fourier transform and the Fourier transform of Definition 3.10 agree when the former exists and so the notation introduced in the definition is justified.

Exercises

21. If the Fourier transform of $g(x)$ is $G(\alpha)$ what is the Fourier transform of $x^n g(x)$, n being a positive integer?

22. Find the Fourier transform of $x^m \delta^{(n)}(x)$, m and n being positive integers.

3.5. Limits

Definition 3.11. *If* $\{g_m\}$ *is a sequence of generalised functions then*

$$\operatorname*{Lim}_{m\to\infty} g_m = g$$

if, and only if, there is a generalised function g *such that*

$$\lim_{m\to\infty} \int_{-\infty}^{\infty} g_m(x)\gamma(x)dx = \int_{-\infty}^{\infty} g(x)\gamma(x)dx$$

for every good γ.

The capital letter is used in Lim to distinguish the generalised limit from one taken in the conventional sense.

More generally, we can consider a parameter μ which runs through the points of an interval of R_1 and suppose that $g_\mu(x)$ is a generalised function of x for each value of μ. Then

Definition 3.12 $\mathrm{Lim}_{\mu\to c}\, g_\mu = g$ *if, and only if,*

$$\lim_{\mu\to c} ⨍_{-\infty}^{\infty} g_\mu(x)\gamma(x)\,\mathrm{d}x = ⨍_{-\infty}^{\infty} g(x)\gamma(x)\,\mathrm{d}x$$

for every good γ.
(*Note*: Generalised functions which depend on more than one scalar parameter can be covered by this definition by allowing μ to run through the points of an interval of R_n.)

Example 14.

$$\mathrm{Lim}_{\mu\to 0} |x|^\mu = 1. \tag{14}$$

The proof of this is not a straightforward matter of saying that it is true for each fixed finite x because the question of what happens at infinity arises; the conditions of Definition 3.12 must be verified. The conventional function $|x|^\mu \in K_1$ for $\mu > -1$ and so is a generalised function by Theorem 3.3 for this interval of μ. Hence

$$⨍_{-\infty}^{\infty} \{|x|^\mu - 1\}\gamma(x)\,\mathrm{d}x = \int_{-\infty}^{\infty} \{|x|^\mu - 1\}\gamma(x)\,\mathrm{d}x.$$

Now, for $x \geq 0$,

$$\frac{x^\mu - 1}{x^\mu + 1} = \frac{e^{\mu \ln x} - 1}{e^{\mu \ln x} + 1} = \tanh(\tfrac{1}{2}\mu \ln x)$$

and $|\tanh y| \leq |y|$ so that

$$\left|\frac{x^\mu - 1}{x^\mu + 1}\right| \leq |\tfrac{1}{2}\mu \ln x|.$$

Therefore

$$\big||x|^\mu - 1\big| \leq \tfrac{1}{2}|\mu|(1 + |x|^\mu)|\ln |x||.$$

Consequently

$$\left|\int_{-\infty}^{\infty} \{|x|^\mu - 1\}\gamma(x)\,\mathrm{d}x\right| \leq \tfrac{1}{2}|\mu| \int_{-\infty}^{\infty} (1 + |x|^\mu)|\ln |x||\,|\gamma(x)|\,\mathrm{d}x.$$

The integral is finite because γ is good and so that the right-hand side tends to zero as $\mu \to 0$, i.e.

$$\lim_{\mu \to 0} \int_{-\infty}^{\infty} \{|x|^\mu - 1\}\gamma(x)dx = 0$$

and (14) is proved.

The limit has some remarkable properties as is demonstrated by

Theorem 3.14. *If* $\mathrm{Lim}_{\mu \to c}\, g_\mu = g$ *then*

(i) $\mathrm{Lim}_{\mu \to c}\, g'_\mu = g'$,
(ii) $\mathrm{Lim}_{\mu \to c}\, g_\mu(ax+b) = g(ax+b)$,
(iii) $\mathrm{Lim}_{\mu \to c}\, \psi g_\mu = \psi g$ *for any fairly good* ψ,
(iv) $\mathrm{Lim}_{\mu \to c}\, G_\mu = G$.

Proof

(i)
$$\lim_{\mu \to c} \int_{-\infty}^{\infty} g'_\mu(x)\gamma(x)dx = -\lim_{\mu \to c} \int_{-\infty}^{\infty} g_\mu(x)\gamma'(x)dx$$
$$= -\int_{-\infty}^{\infty} g(x)\gamma'(x)dx$$
$$= \int_{-\infty}^{\infty} g'(x)\gamma(x)dx$$

where Theorem 3.6 has been applied twice.

(ii) By Theorem 3.4

$$\lim_{\mu \to c} \int_{-\infty}^{\infty} g_\mu(ax+b)\gamma(x)dx = \lim_{\mu \to c} \int_{-\infty}^{\infty} g_\mu(x)\gamma\left(\frac{x-b}{a}\right)\frac{dx}{|a|}$$
$$= \int_{-\infty}^{\infty} g(x)\gamma\left(\frac{x-b}{a}\right)\frac{dx}{|a|}$$
$$= \int_{-\infty}^{\infty} g(ax+b)\gamma(x)dx.$$

(iii) Since ψ is fairly good $\psi\gamma$ is good and so

$$\lim_{\mu \to c} \int_{-\infty}^{\infty} \psi(x)g_\mu(x)\gamma(x)dx = \lim_{\mu \to c} \int_{-\infty}^{\infty} g_\mu(x)\{\psi(x)\gamma(x)\}dx$$
$$= \int_{-\infty}^{\infty} g(x)\psi(x)\gamma(x)dx$$

and this part is proved.

(iv) By Theorem 3.9

$$\lim_{\mu\to c} \int_{-\infty}^{\infty} G_\mu(\alpha)\Gamma(\alpha)\,\mathrm{d}\alpha = \lim_{\mu\to c} 2\pi \int_{-\infty}^{\infty} g_\mu(x)\gamma(-x)\,\mathrm{d}x$$

$$= 2\pi \int_{-\infty}^{\infty} g(x)\gamma(-x)\,\mathrm{d}x$$

$$= \int_{-\infty}^{\infty} G(\alpha)\Gamma(\alpha)\,\mathrm{d}\alpha$$

and the proof of the theorem is complete. □

Parts (i) and (iv) are particularly significant. The first shows that it is immaterial whether one follows a derivative by a limit or vice versa. The second shows that the Fourier transform of a limit is the limit of the Fourier transforms.

Example 15. By the same method that was used in proving (14) it may be shown that

$$\operatorname*{Lim}_{\mu\to 0} |x|^\mu \operatorname{sgn} x = \operatorname{sgn} x.$$

Hence by Theorem 3.14 (i),

$$\operatorname*{Lim}_{\mu\to 0} \{|x|^\mu \operatorname{sgn} x\}' = \operatorname{sgn}' x = 2\delta(x)$$

which may also be written as

$$\operatorname*{Lim}_{\mu\to 0} \mu|x|^{\mu-1} = 2\delta(x). \tag{15}$$

Since x^2 is fairly good Theorem 3.14 (iii) shows that

$$\operatorname*{Lim}_{\mu\to 0} \mu|x|^{\mu+1} = \operatorname*{Lim}_{\mu\to 0} x^2\mu|x|^{\mu-1} = 2x^2\delta(x) = 0.$$

On the other hand Theorem 3.14 (iv) applied to (15) gives

$$\operatorname*{Lim}_{\mu\to 0} \int_{-\infty}^{\infty} \mu|x|^{\mu-1} \mathrm{e}^{-\mathrm{i}\alpha x}\,\mathrm{d}x = \int_{-\infty}^{\infty} 2\delta(x)\mathrm{e}^{-\mathrm{i}\alpha x}\,\mathrm{d}x = 2.$$

Example 16. If $f \in L_1$ and $\int_{-\infty}^{\infty} f(t)\,\mathrm{d}t = 1$ then $\operatorname{Lim}_{m\to\infty} mf(mx) = \delta(x)$.

For $f \in K_1$ and so

$$\lim_{m\to\infty} \fint_{-\infty}^{\infty} mf(mx)\gamma(x)\,\mathrm{d}x = \lim_{m\to\infty} \int_{-\infty}^{\infty} mf(mx)\gamma(x)\,\mathrm{d}x$$

$$= \lim_{m\to\infty} \int_{-\infty}^{\infty} f(x)\left\{\gamma\left(\frac{x}{m}\right) - \gamma(0)\right\}\mathrm{d}x + \gamma(0).$$

A good function is bounded and therefore X can be chosen so that

$$\left|\int_X^{\infty} f(x)\left\{\gamma\left(\frac{x}{m}\right) - \gamma(0)\right\}\mathrm{d}x\right| \leq B \int_X^{\infty} |f(x)|\,\mathrm{d}x \leq \varepsilon$$

because $f \in L_1$. Similarly for $\int_{-\infty}^{-X}$. Having fixed X, m can be chosen so large that $|\gamma(x/m) - \gamma(0)| \leq \varepsilon/\{\int_{-X}^{X} |f(t)|\,\mathrm{d}t\}$ for $|x| \leq X$. Hence

$$\left|\int_{-X}^{X} f(x)\left\{\gamma\left(\frac{x}{m}\right) - \gamma(0)\right\}\mathrm{d}x\right| \leq \varepsilon.$$

Consequently $\lim_{m\to\infty} \fint_{-\infty}^{\infty} mf(mx)\gamma(x)\,\mathrm{d}x = \gamma(0)$, i.e.

$$\operatorname*{Lim}_{m\to\infty} mf(mx) = \delta(x).$$

Exercises

23. Verify the following.

(i) $\operatorname*{Lim}_{m\to\infty} \dfrac{x}{(x+m)^2+1} = 0,$ (ii) $\operatorname*{Lim}_{\mu\to 0} \mu|x|^{\mu-1}\mathrm{e}^{\mathrm{i}x} = 2\delta(x),$

(iii) $\operatorname*{Lim}_{m\to\infty} \cos mx = 0,$ (iv) $\operatorname*{Lim}_{m\to\infty} \dfrac{\sin mx}{\pi x} = \delta(x),$

(v) $\operatorname*{Lim}_{\mu\to 0} \dfrac{1}{\mu}(|x|^{\mu} - 1)\operatorname{sgn} x = \operatorname{sgn} x \ln|x|,$ (vi) $\operatorname*{Lim}_{m\to\infty} \dfrac{2}{\pi}\tan^{-1} mx = \operatorname{sgn} x.$

24. Is it true that $\operatorname{Lim}_{m\to\infty} m\mathrm{e}^{-mx}\mathrm{H}(x) = \delta(x)$?

25. Determine which of the following limits exist and find the resulting generalised function when the limit does exist.

(i) $\operatorname*{Lim}_{m\to\infty} \mathrm{i}m\mathrm{e}^{\mathrm{i}mx},$ (ii) $\operatorname*{Lim}_{m\to\infty} \dfrac{m}{\pi\{m^2(x+b)^2+1\}},$

(iii) $\operatorname*{Lim}_{m\to\infty} \{\mathrm{H}(m-x) - \mathrm{H}(-m-x)\}.$

26. If $\int_{-\infty}^{\infty} f(t)\,\mathrm{d}t = 1$ and $f \in K_1$ (but f does not necessarily belong to

$L_1(-\infty, \infty)$) show that $\mathrm{Lim}_{m\to\infty} \int_{-\infty}^{mx} f(t)\mathrm{d}t = \mathrm{H}(x)$. Deduce that $\mathrm{Lim}_{m\to\infty} mf(mx) = \delta(x)$. (This imposes less restriction on f than the example after Theorem 3.14.) Show that

$$\operatorname*{Lim}_{m\to\infty} \frac{m/\pi}{1+m^2x^2} = \delta(x), \qquad \operatorname*{Lim}_{m\to\infty} \tfrac{1}{2}me^{-m|x|} = \delta(x).$$

27. Prove Theorem 3.14 (iii) if ψ is replaced by a moderately good function.

28. Show that $\lim_{\mu\to\infty} \int_1^\mu - \cos tx\,\mathrm{d}t/t^2$ exists. Hence prove that $\int_0^\infty \cos 2\pi tx\,\mathrm{d}t$ exists and is $\frac{1}{2}\delta(x)$.

29. Prove that $\operatorname*{Lim}_{h\to 0} \dfrac{g(x+h)-g(x)}{h} = g'(x)$.

30. If $f_\mu(x) \in K_1$ and $f(x) \in K_1$ and f_μ tends to f uniformly in x as $\mu \to \mu_0$ show that $\mathrm{Lim}_{\mu\to\mu_0} f_\mu = f$ in the sense of generalised functions.

31. If $f(x)$ is continuous and $O(x^{-2})$ as $|x| \to \infty$ prove that, in the sequence $\{\gamma_n\}$ of Theorem 3.3, n can be chosen so large that $|f(x) - \gamma_n(x)| < \varepsilon$. Deduce that $\lim_{n\to\infty} \gamma_n(x) = f(x)$ as a pointwise limit and also in the generalised sense.

Definition 3.13. *If* $\mathrm{Lim}_{\mu_1\to\mu}\{g_{\mu_1}(x) - g_\mu(x)\}/(\mu_1 - \mu)$ *exists as a generalised function of x it is denoted by* $\partial g_\mu(x)/\partial\mu$.

Theorem 3.15. *If* $\partial g_\mu(x)/\partial\mu$ *exists then* $\{\partial g_\mu(x)/\partial\mu\}' = \partial g'_\mu(x)/\partial\mu$.

Proof. Since $\partial g_\mu(x)/\partial\mu$ is a generalised function it has a derivative and by Theorem 3.14 (i)

$$\left\{\frac{\partial g_\mu(x)}{\partial\mu}\right\}' = \operatorname*{Lim}_{\mu_1\to\mu} \frac{g'_{\mu_1}(x) - g'_\mu(x)}{\mu_1 - \mu}$$
$$= \partial g'_\mu(x)/\partial\mu$$

and the proof is complete. □

Example 17. If $\delta(x)$ were a conventional function one would expect that

$$\frac{\partial}{\partial\mu}\delta(x-\mu) = -\delta'(x-\mu). \tag{16}$$

That this is, in fact, true follows from

$$\lim_{\mu_1\to\mu} \int_{-\infty}^{\infty} \frac{\delta(x-\mu_1) - \delta(x-\mu)}{\mu_1 - \mu}\gamma(x)\mathrm{d}x = \lim_{\mu_1\to\mu} \frac{\gamma(\mu_1) - \gamma(\mu)}{\mu_1 - \mu} = \gamma'(\mu).$$

By Theorem 3.15

$$\delta''(x-\mu) = -\left\{\frac{\partial}{\partial\mu}\delta(x-\mu)\right\}'$$

$$= -\frac{\partial}{\partial\mu}\delta'(x-\mu).$$

The Fourier transform of (16) is, when account is taken of Theorem 3.14(iv),

$$\partial e^{-i\alpha\mu}/\partial\mu = -i\alpha e^{-i\alpha\mu}.$$

3.6. Classification of generalised functions

This section contains some useful information about the properties of generalised functions and, in particular, demonstrates that all generalised functions are of a particular type.

Theorem 3.16. *If $\{\gamma_m\}$ is a regular sequence there are finite integers k, r and a finite constant K such that*

$$\left|\int_{-\infty}^{\infty} \gamma_m(x)\gamma(x)\mathrm{d}x\right| \le K \max_{x\in R_1} |(1+x^2)^{k/2}\gamma^{(r)}(x)|$$

for every good γ and all m. In particular

$$\left|\int_{-\infty}^{\infty} g(x)\gamma(x)\mathrm{d}x\right| \le K \max_{x\in R_1} |(1+x^2)^{k/2}\gamma^{(r)}(x)|$$

for every good γ.

Proof. The sequence $\{\gamma_m\}$ is regular. Therefore there is a constant $K'(\gamma)$ which depends on γ, such that

$$\left|\int_{-\infty}^{\infty} \gamma_m(x)\gamma(x)\mathrm{d}x\right| \le K'(\gamma) \text{ for all } m \tag{17}$$

and, in particular, $|\int_{-\infty}^{\infty} g(x)\gamma(x)\mathrm{d}x| \le K'(\gamma)$.

Also

$$\left|\int_{-\infty}^{\infty} \gamma_m(x)\gamma(x)\mathrm{d}x\right| \le \left\{\int_{-\infty}^{\infty} |\gamma_m(x)|\,\mathrm{d}x\right\}\max|\gamma| \le K_{m0}\max|\gamma|$$

where K_{m0} is independent of γ. Now from Exercise 11 of Chapter 2,

$$\max(1+x^2)^{k/2}|\gamma^{(p)}(x)| \le (\tfrac{1}{2}\pi)^n \max(1+x^2)^{n+k/2}|\gamma^{(p+n)}(x)|. \qquad (18)$$

Hence, since $(1+x^2)^{k/2} \ge 1$ for $k \ge 0$, there is a finite K_{ms} such that

$$\left|\int_{-\infty}^{\infty} \gamma_m(x)\gamma(x)\mathrm{d}x\right| \le K_{ms} \max(1+x^2)^s|\gamma^{(s)}(x)| \qquad (19)$$

for every finite m. The theorem would therefore be proved if, for some s, K_{ms} were bounded as $m \to \infty$. We prove it to be true by assuming the opposite and showing that this leads to a contradiction. The method is to construct a special good function which shows that $\{\gamma_m\}$ cannot be regular.

Assume, therefore, that, for any given k and s, an m can be found such that $\int_{-\infty}^{\infty} \gamma_m(x)\gamma(x)\mathrm{d}x > L_m \max(1+x^2)^{k/2}|\gamma^{(s)}(x)|$ no matter how large L_m is. Then, having chosen a non-zero good function $\gamma_{10}(x)$ such that $\max|\gamma_{10}(x)| < \frac{1}{2}$, we can find an m_1 such that $\int_{-\infty}^{\infty} \gamma_{m_1}(x)\gamma_{10}(x)\mathrm{d}x > 2$. Then choose a non-zero good function $\gamma_{20}(x)$ such that $\max|(1+x^2)\gamma_{20}(x)| < 1/2^2$, $\max(1+x^2)|\gamma'_{20}(x)| < 1/2^2$, and $|\int_{-\infty}^{\infty} \gamma_{m_1}(x)\gamma_{20}(x)\mathrm{d}x| < 1/2^2$. Because of (18) and (19) this can be done by taking $\max(1+x^2)^2|\gamma'_{20}(x)|$ sufficiently small. Having fixed γ_{20} we can find an m_2 such that

$$\int_{-\infty}^{\infty} \gamma_{m_2}(x)\gamma_{20}(x)\mathrm{d}x > 3 + K'(\gamma_{10}).$$

Continue the process of forming a sequence of γ_{n0} and γ_{m_n} with the properties

(i) $\max(1+x^2)^n|\gamma_{n0}^{(p)}(x)| < 1/2^n$ for $p = 0, 1, \dots, n$,

(ii) $|\int_{-\infty}^{\infty} \gamma_{m_p}\gamma_{n0}(x)\mathrm{d}x| < 1/2^n$ for $p = 1, 2, \dots, n-1$,

(iii) $\int_{-\infty}^{\infty} \gamma_{m_n}\gamma_{n0}(x)\mathrm{d}x > 1 + n + \sum_{r=1}^{n-1} K'(\gamma_{r0})$.

For any n it is easily checked that this is possible once the preceding elements are known so that, by induction, the sequences exist.

Consider now $\sum_{n=1}^{\infty} \gamma_{n0}(x)$. Since $\max|\gamma_{n0}(x)| \le \max(1+x^2)^n$ $\gamma_{n0}(x)| \le 1/2^n$ by (i), the series is uniformly convergent (Weierstrass's M-test, Theorem 1.23) and so, by Theorem 1.24, represents a continuous function. The same argument applies to the series obtained by taking derivatives term by term any number of times. By Theorem 1.37 the series of derivatives is the derivative of the series. Hence

$\sum_{n=1}^{\infty} \gamma_{n0}(x)$ is a function which possesses all derivatives for every x. Also, for any fixed r,

$$\lim_{|x|\to\infty} x^r \sum_{n=1}^{\infty} \gamma_{n0}(x) = \lim_{|x|\to\infty} x^r \sum_{n=r}^{\infty} \gamma_{n0}(x)$$

$$= \lim_{|x|\to\infty} \frac{x^r}{(1+x^2)^r} \sum_{n=r}^{\infty} (1+x^2)^r \gamma_{n0}(x).$$

The series is bounded on account of (i) and hence the limit is zero. The same argument applies to any derivative and hence $\sum_{n=1}^{\infty} \gamma_{n0}(x)$ is a good function. Therefore $\int_{-\infty}^{\infty} \gamma_{m_n}(x) \sum_{p=1}^{\infty} \gamma_{p0}(x)\mathrm{d}x$ has a meaning. But

$$\left| \int_{-\infty}^{\infty} \gamma_{m_n}(x) \sum_{p=1}^{n-1} \gamma_{p0}(x)\mathrm{d}x \right| \le \sum_{p=1}^{n-1} K'(\gamma_{p0}),$$

$$\int_{-\infty}^{\infty} \gamma_{m_n}(x)\gamma_{n0}(x)\mathrm{d}x > 1 + n + \sum_{p=1}^{n-1} K'(\gamma_{p0})$$

and $\left|\int_{-\infty}^{\infty} \gamma_{m_n}(x) \sum_{p=n+1}^{\infty} \gamma_{p0}(x)\mathrm{d}x\right| < \sum_{p=n+1}^{\infty} 1/2^p < 1$. Hence

$$\int_{-\infty}^{\infty} \gamma_{m_n}(x) \sum_{p=1}^{\infty} \gamma_{p0}(x)\mathrm{d}x > n.$$

Allowing $n \to \infty$, we have a sequence $\{\gamma_{m_n}\}$ which is not regular. This is contrary to hypothesis and hence our assumption that K, k and r cannot be found must be incorrect. The theorem is proved. □

One consequence of Theorem 3.16 is that, if γ and all its derivatives tend to zero uniformly on R_1, $⨍_{-\infty}^{\infty} g(x)\gamma(x)\mathrm{d}x \to 0$. The connection with Schwartz's definition of $\mathscr{S}'$ is apparent.

A generalisation of Theorem 3.16 might be noted:

Theorem 3.17. *If a set of generalised functions g_μ is such that, for all μ, and every good γ, $\left|⨍_{-\infty}^{\infty} g_\mu(x)\gamma(x)\mathrm{d}x\right| \le K'(\gamma)$ there are finite fixed K, k and r such that*

$$\left| ⨍_{-\infty}^{\infty} g_\mu(x)\gamma(x)\mathrm{d}x \right| \le K \max |(1+x^2)^{k/2}\gamma^{(r)}(x)|$$

for every γ and all μ.

Proof. The hypothesis is the analogue of (17). The result of Theorem

3.16 is the analogue of (19). Thereafter, the proof is practically the same as that for Theorem 3.16, and so will not be given. □

Theorem 3.18. *In order that $g(x)$ be a generalised function it is necessary and sufficient that $g(x)$ be a generalised derivative*

$$g(x) = f^{(r)}(x)$$

where f is a continuous function such that, for some finite k,

$$(1 + x^2)^{-k/2} f(x)$$

is bounded on R_1.

Proof. If f is continuous, $f \in K_1$ and defines a generalised function by Theorem 3.3. Then, clearly, g is a generalised function. Thus the condition is sufficient.

To show that it is necessary for a given generalised function g to have this form, define a linear functional F over the space of good functions of the form $(1 + x^2)^{k/2}\gamma^{(r)}(x)$ (k, r being the integers of Theorem 3.16) by specifying that

$$F((1 + x^2)^{k/2}\gamma^{(r)}) = \int_{-\infty}^{\infty} g(x)\gamma(x)\mathrm{d}x.$$

Then, with $\| (1 + x^2)^{k/2}\gamma^{(r)} \| = \max|(1 + x^2)^{k/2}\gamma^{(r)}(x)|$, F is a bounded linear functional with $\|F\| = K$ because of Theorem 3.16. The functions $(1 + x^2)^{k/2}\gamma^{(r)}$ form a linear variety in the space of continuous functions. Hence, by the Hahn–Banach theorem (Theorem 1.11) F can be extended to a bounded linear functional over the space of continuous functions, the extension having the same norm as F. On account of the Riesz representation theorem (§1.12) this can be expressed as a Stieltjes integral and

$$F((1 + x^2)^{k/2}\gamma^{(r)}) = \int_{-\infty}^{\infty} (1 + x^2)^{k/2}\gamma^{(r)}(x)\mathrm{d}h(x)$$

where the total variation of h is K and, indeed, we can take $\max|h| \le K$. Now

$$\int_{-\infty}^{\infty} (1 + x^2)^{k/2}\gamma^{(r)}(x)\mathrm{d}h(x) = -\int_{-\infty}^{\infty} h(x)\frac{\mathrm{d}}{\mathrm{d}x}\{(1+x^2)^{k/2}\gamma^{(r)}(x)\}\,\mathrm{d}x$$

$$= -\int_{-\infty}^{\infty} h(x)\frac{\mathrm{d}}{\mathrm{d}x}\{(1 + x^2)^{k/2}\gamma^{(r)}(x)\}\,\mathrm{d}x$$

on account of Theorem 3.3 because $h \in K_1$. Hence

$$\int_{-\infty}^{\infty} g(x)\gamma(x)\mathrm{d}x = \int_{-\infty}^{\infty} [(-1)^{r+1}k\{h(x)x(1+x^2)^{(k/2)-1}\}^{(r)}$$
$$+(-1)^r\{h(x)(1+x^2)^{k/2}\}^{(r+1)}]\gamma(x)\mathrm{d}x.$$

Thus

$$g(x)=(-1)^{r+1}k\{h(x)x(1+x^2)^{(k/2)-1}\}^{(r)}+(-1)^r\{h(x)(1+x^2)^{k/2}\}^{(r+1)}$$

and this can be cast into the form $g(x)=f^{(r+2)}(x)$ where

$$f(x)=(-1)^r\int^x h(t)(1+t^2)^{k/2}\mathrm{d}t$$
$$+(-1)^{r+1}k\int^x\int^t h(u)u(1+u^2)^{(k/2)-1}\mathrm{d}u\,\mathrm{d}t$$

(Theorem 3.7), and obviously m can be chosen so that $f/(1+x^2)^{m/2}$ is bounded. The theorem is proved. □

Theorem 3.18 specifies precisely the class of generalised functions and shows that their growth at infinity is majorised by some power of x. Thus e^x *is not a generalised function.* Direct confirmation of this result can be obtained by noting that, if e^{ax} were a generalised function, $\int_{-\infty}^{\infty}\mathrm{e}^{ax}\gamma(x)\mathrm{d}x$ would be finite for every good γ. But, by choosing $\gamma(x)=\operatorname{sech} bx$ with b sufficiently small, we can see that there is a good γ such that the integral is not finite and so e^{ax} cannot be a generalised function.

It is immediately evident from Theorem 3.18 that, if g is a generalised function, there is a generalised function g_1 such that $g_1' = g$. Such a g_1 may be called the *indefinite integral* of g and the notation

$$g_1(x)=\int^x g(t)\mathrm{d}t$$

can be employed; it has the usual significance when $g \in K_1$ because of Corollary 3.7 (cf. §6.6, Ex. 64). Obviously g_1 is not unique since a constant can be added to it without affecting $g_1' = g$. It will be shown later (Corollary 4.2) that this is the only way in which g_1 is not unique.

There are some applications of the preceding theorems which are of some interest. The first shows that convergence of generalised functions is related to the convergence of the continuous functions of which they are derivatives.

Theorem 3.19. *If the sequence* $\{f_m\}$ *of continuous functions, bounded on* R_1, *converges uniformly to zero and the generalised function* $g_m(x) = [(1+x^2)^{k/2} f_m(x)]^{(r)}$ *then* $\text{Lim}_{m\to\infty} g_m = 0$.

Proof.

$$\int_{-\infty}^{\infty} g_m(x)\gamma(x)\mathrm{d}x = (-1)^r \int_{-\infty}^{\infty} (1+x^2)^{k/2} f_m(x)\gamma^{(r)}(x)\mathrm{d}x.$$

Since $\{f_m\}$ converges uniformly to zero there is an M independent of x such that $|f_m| < \varepsilon$ for $m \geq M$. Hence

$$\left| \int_{-\infty}^{\infty} g_m(x)\gamma(x)\mathrm{d}x \right| < \varepsilon \int_{-\infty}^{\infty} (1+x^2)^{k/2} |\gamma^{(r)}(x)| \mathrm{d}x$$

and the theorem is proved since the integral is independent of m. □

A converse to this theorem is provided by

Theorem 3.20. *If* $\text{Lim}_{m\to\infty} g_m = 0$ *there are continuous* f_m *such that* $g_m(x) = \{(1+x^2)^{k/2} f_m(x)\}^{(r)}$, *for fixed k and r, and* $\{f_m\}$ *converges uniformly to zero.*

Proof. The method consists in showing that the f_m can be taken to be equicontinuous and such that $\lim_{m\to\infty} \int_{-\infty}^{\infty} f_m(x)\gamma(x)\mathrm{d}x = 0$; then proving that this implies that $\{f_m\}$ converges uniformly to zero. Since $\text{Lim}_{m\to\infty} g_m = 0$, $|\int_{-\infty}^{\infty} g_m(x)\gamma(x)\mathrm{d}x| \leq K'(\gamma)$ and so, by Theorem 3.17,

$$\left| \int_{-\infty}^{\infty} g_m(x)\gamma(x)\mathrm{d}x \right| \leq K \max |(1+x^2)^{k/2} \gamma^{(r)}(x)|$$

for all m. As a result of Theorem 3.18 there exist continuous f_m, bounded on R_1, such that $g_m(x) = [(1+x^2)^{k/2} f_m(x)]^{(r)}$.

In the proof of Theorem 3.18 f is expressed as an integral involving a function h such that $\max|h| \leq K$ and, as a consequence, the f_m are bounded uniformly in m. Furthermore, because $\max|h_m| \leq K$, we have

$$|f_m(x)-f_m(x')| \le K\int_{x'}^{x} \lambda(t)\mathrm{d}t$$

where λ is a continuous function of t, independent of m. Hence there is δ, independent of m, such that $|f_m(x)-f_m(x')| < \varepsilon$ for $|x-x'|<\delta$, i.e. the f_m are *equicontinuous* on R_1.

Now

$$\int_{-\infty}^{\infty} g_m(x)\gamma(x)\mathrm{d}x = (-1)^r \int_{-\infty}^{\infty} (1+x^2)^{k/2} f_m(x)\gamma^{(r)}(x)\mathrm{d}x$$

so that

$$\lim_{m\to\infty} \int_{-\infty}^{\infty} (1+x^2)^{k/2} f_m(x)\gamma^{(r)}(x)\mathrm{d}x = 0 \tag{20}$$

for every good γ. Let γ_1 be any good function; then γ_2, defined by

$$\gamma_2(x) = \int_{-\infty}^{x} \left\{\gamma_1(x) - \frac{\mathrm{e}^{-x^2}}{\pi^{1/2}}\int_{-\infty}^{\infty} \gamma_1(t)\mathrm{d}t\right\}\mathrm{d}x,$$

is also a good function because it is the integral of a good function and tends to zero appropriately as $|x|\to\infty$. Hence

$$\int_{-\infty}^{\infty} (1+x^2)^{k/2} f_m(x)\left\{\gamma_1^{(r-1)}(x) - \pi^{-1/2}(\mathrm{e}^{-x^2})^{(r-1)}\int_{-\infty}^{\infty}\gamma_1(t)\mathrm{d}t\right\}\mathrm{d}x$$

$$= \int_{-\infty}^{\infty} (1+x^2)^{k/2} f_m(x)\gamma_2^{(r)}(x)\mathrm{d}x.$$

Denote $\int_{-\infty}^{\infty}(1+x^2)^{k/2}f_m(x)(\mathrm{e}^{-x^2})^{(r-1)}\mathrm{d}x$ by $\pi^{1/2}a_m$; it is a constant bounded for all m but independent of γ_1. Then, since

$$\int_{-\infty}^{\infty} \gamma_1(t)\mathrm{d}t = (-1)^{r-1}\int_{-\infty}^{\infty} \frac{t^{r-1}\gamma_1^{(r-1)}(t)\mathrm{d}t}{(r-1)!},$$

$$\int_{-\infty}^{\infty} \{(1+x^2)^{k/2}f_m(x) - (-1)^{r-1}a_m x^{r-1}/(r-1)!\}\gamma_1^{(r-1)}(x)\mathrm{d}x$$

$$= \int_{-\infty}^{\infty} (1+x^2)^{k/2} f_m(x)\gamma_2^{(r)}(x)\mathrm{d}x.$$

On account of (20) the right-hand side tends to zero as $m\to\infty$. Therefore the same is true of the left-hand side, which gives a result similar to (20) but with $r-1$ in place of r. Repeating the process $(r-1)$ times we obtain, for every good γ,

$$\lim_{m\to\infty}\int_{-\infty}^{\infty}\{(1+x^2)^{k/2}f_m(x)+P_m(x)\}\gamma(x)\mathrm{d}x=0 \tag{21}$$

where P_m is a polynomial of degree $r-1$ whose coefficients are bounded for all m. The P_m can be absorbed in the f_m without affecting g_m so that (21) can be written

$$\lim_{m\to\infty}\int_{-\infty}^{\infty}(1+x^2)^{k/2}f_m(x)\gamma(x)\mathrm{d}x=0.$$

Clearly these f_m are still equicontinuous and, by increasing k if necessary, tend to zero independently of m as $|x|\to\infty$. Then absorbing the $(1+x^2)^{k/2}$ in γ we have $\lim_{m\to\infty}\int_{-\infty}^{\infty}f_m(x)\gamma(x)\mathrm{d}x=0$ for every good γ.

Thus there is finite a such that $|f_m(x)|<\varepsilon$ for every m and $|x|\geq a$. Divide $(-a,a)$ into a finite number of intervals each of length less than δ. Then, for any pair x and x' from one of these sub-intervals, $|f_m(x)-f_m(x')|<\varepsilon$ for all m. Suppose now that in some sub-interval (δ_{n-1},δ_n) there is some p such that $f_p(x)\geq\varepsilon$ for all x in (δ_{n-1},δ_n). Then, if γ_0 is the good function defined by

$$\gamma_0(x)=\begin{cases}\exp\left\{-\left(\dfrac{1}{x-\delta_{n-1}}+\dfrac{1}{\delta_n-x}\right)\right\} & (\delta_{n-1}\leq x\leq\delta_n)\\[2ex] 0 & (x<\delta_{n-1}\text{ or }x>\delta_n),\end{cases}$$

$$\int_{-\infty}^{\infty}f_p(x)\gamma_0(x)\mathrm{d}x\geq\varepsilon\int_{\delta_{n-1}}^{\delta_n}\gamma_0(x)\mathrm{d}x.$$

Since $\lim_{m\to\infty}\int_{-\infty}^{\infty}f_m(x)\gamma_0(x)\mathrm{d}x=0$ it follows that there cannot be more than a finite number of p for which $f_p(x)\geq\varepsilon$ for all x in (δ_{n-1},δ_n). Hence there must be a finite m_n such that for $m\geq m_n$ there is an x_{mn} in (δ_{n-1},δ_n) for which $f_m(x_{mn})<\varepsilon$. Therefore

$$f_m(x)=f_m(x)-f_m(x_{mn})+f_m(x_{mn})$$

shows that $f_m(x)<2\varepsilon$ in (δ_{n-1},δ_n) for $m\geq m_n$. A similar argument demonstrates that $f_m(x)>-2\varepsilon$. Since there is only a finite number of sub-intervals there is a finite m_0 such that, in $(-a,a)$, $|f_m(x)|<2\varepsilon$

for $m \geq m_0$. Hence, for $m \geq m_0$ and all x, $|f_m(x)| < 2\varepsilon$, i.e. the f_m converge uniformly to zero as $m \to \infty$. The theorem is proved. □

Corollary 3.20*a*. *If* $\operatorname{Lim}_{m\to\infty} g_m = 0$ *there is an* m_0 *such that, for all good* γ,

$$\left|\int_{-\infty}^{\infty} g_m(x)\gamma(x)\mathrm{d}x\right| \leq \varepsilon \max|(1+x^2)^{k/2}\gamma^{(r)}(x)|$$

for $m \geq m_0$; *k and r are independent of* m_0.

Proof. By Theorem 3.20

$$\int_{-\infty}^{\infty} g_m(x)\gamma(x)\mathrm{d}x = \int_{-\infty}^{\infty} [(1+x^2)^{k/2} f_m(x)]^{(r)}\gamma(x)\mathrm{d}x$$

$$= (-1)^r \int_{-\infty}^{\infty} f_m(x)(1+x^2)^{k/2}\gamma^{(r)}(x)\mathrm{d}x.$$

By Theorem 3.20, m can be chosen sufficiently large for $|f_m| < \varepsilon/\pi$ and then the right-hand side does not exceed

$$\left(\frac{\varepsilon}{\pi}\right)\max|(1+x^2)^{(k/2)+1}\gamma^{(r)}(x)| \int_{-\infty}^{\infty} \frac{\mathrm{d}x}{1+x^2}$$

and the corollary is proved. □

By replacing g_m in Corollary 3.20*a* by $\gamma_m - g$ we have

Corollary 3.20*b*. *If* $\{\gamma_m\}$ *is a regular sequence of good functions defining the generalised function g, there is a finite* m_0 *such that*

$$\left|\int_{-\infty}^{\infty} \{\gamma_m(x) - g(x)\}\gamma(x)\mathrm{d}x\right| \leq \varepsilon \max|(1+x^2)^{k/2}\gamma^{(r)}(x)|$$

for all good γ *and* $m \geq m_0$.

There is one interesting consequence of Theorem 3.18 to note:

Theorem 3.21. *If the generalised function g is zero for* $x > 0$ *and for* $x < 0$ *then* $g(x) = \sum_{m=0}^{n} a_m \delta^{(m)}(x)$ *where n is finite.*

Proof. From Theorem 3.18

$$\int_{-\infty}^{\infty} g(x)\gamma(x)\mathrm{d}x = (-1)^r \int_{-\infty}^{\infty} f(x)(1+x^2)^{k/2}\gamma^{(r)}(x)\mathrm{d}x.$$

Since g vanishes for $x > 0$, $(1+x^2)^{k/2} f(x)$ must be a polynomial P_{r-1} of degree $(r-1)$ in $x > 0$. Now, by integration by parts,

$$\int_0^\infty P_{r-1}(x)\gamma^{(r)}(x)\mathrm{d}x = \sum_{m=1}^{r} (-1)^m P_{r-m}(0)\gamma^{(r-m)}(0).$$

On dealing with $x < 0$ similarly we see that

$$\int_{-\infty}^{\infty} g(x)\gamma(x)\mathrm{d}x = \sum_{m=0}^{r-1} a_m \gamma^{(m)}(0)$$

and the theorem follows. □

Exercises

32. Show that the Fourier transform of a generalised function $g(x)$ can be calculated by

$$G(\alpha) = (\mathrm{i}\alpha)^r D^N \int_{-\infty}^{\infty} \frac{f(x)\mathrm{e}^{-\mathrm{i}\alpha x}}{(1+x^2)^N}\mathrm{d}x$$

where r and N are finite integers and $DF(\alpha) = F(\alpha) - F''(\alpha)$. The function f is continuous and $O(|x|^k)$ as $|x| \to \infty$ so that only conventional integration is involved.

33. Show that

$$G(\alpha) = \operatorname*{Lim}_{\mu\to\infty} \int_{-\mu}^{\mu} g(x)\mathrm{e}^{-\mathrm{i}\alpha x}\mathrm{d}x$$

in the sense of generalised functions.

The results of Theorems 3.18 and 3.20 have been used as a starting point for the theory of generalised functions by several authors, for example Silva.† Essentially the idea consists in saying that, if $f(x, \mu)$ is a continuous function of x which converges uniformly as $\mu \to \mu_0$ to $f(x)$ and $f(x,\mu)$ has the ordinary derivative $\mathrm{d}f(x, \mu)/\mathrm{d}x$, then

$$\lim_{\mu\to\mu_0} \mathrm{d}f(x, \mu)/\mathrm{d}x$$

constitutes a representation of the generalised function $f'(x)$. The 'values' of the generalised function so defined can be found by forming integrals with good functions. In some connections this

† J. S. e Silva, *Pub. Cent. Estud. Mat. Lisboa* (2), 4, 79 (1954); *H. König, Math.-Phys. Semesterberichte*, 7, 26 (1960).

approach is simpler than the one we have adopted and in others it is less convenient. We shall use the original definition in terms of sequences or Theorem 3.18 according to the matter in hand.

The preceding theorems also provide another way of characterising generalised functions which is sometimes useful. Suppose that, when $\max|(1+x^2)^{k/2}\gamma_n(x)| \to 0$ as $n \to \infty$, $\int_{-\infty}^{\infty} g(x)\gamma_n(x)\mathrm{d}x \to 0$ for some k and all possible sequences $\{\gamma_n\}$ satisfying the given condition then we say that $g \in C_1$. If, when $\max|(1+x^2)^{k/2}\gamma_n(x)| \to 0$ and $\max|(1+x^2)^{k/2}\gamma_n'(x)| \to 0$ as $n \to \infty$ (the two conditions are related by (18)), $\int_{-\infty}^{\infty} g(x)\gamma_n(x)\mathrm{d}x \to 0$ then we say that $g \in C_2$. Clearly this process of classification can be continued indefinitely. If $g \in C_1$ then obviously $g \in C_2$ and, in general, $C_m \subset C_{m+1}$ so that as m increases the class of generalised functions widens.

If $g \in K_1$ then

$$\left|\int_{-\infty}^{\infty} g(x)\gamma_n(x)\mathrm{d}x\right|$$

$$= \left|\int_{-\infty}^{\infty} g(x)\gamma_n(x)\mathrm{d}x\right| \leq \max|(1+x^2)^N\gamma_n(x)| \int_{-\infty}^{\infty} \frac{|g(x)|}{(1+x^2)^N}\mathrm{d}x$$

and, since the integral is bounded independent of n, the right-hand side tends to zero as $\max|(1+x^2)^N\gamma_n(x)| \to 0$. Hence $g \in C_1$, i.e. if $g \in K_1$ then $g \in C_1$. But C_1 includes other generalised functions, for example $\delta(x)$. However, $\delta'(x)$ is not in C_1 because

$$\int_{-\infty}^{\infty} \delta'(x)\gamma_n(x)\mathrm{d}x = -\gamma_n'(0)$$

which does not necessarily tend to zero as $\max|\gamma_n| \to 0$. Consider, for instance, $\gamma_n(x) = n^{-1/2}\exp\{-n(x-n^{-1/2})^2\}$. Nevertheless $\delta' \in C_2$. Similarly $\delta'' \in C_3$ but is not in C_2.

On examining the proof of Theorem 3.18 it will be evident that, if $g \in C_m$, $g = f^{(m+2)}$ where f is a continuous function such that, for some k, $(1+x^2)^{-k/2}f(x)$ is bounded on R_1. Conversely, if g is of the form $f^{(m+2)}$ it can be shown, by using Stieltjes integrals, that $g \in C_m$.

A slightly different point of view is to observe that any generalised function $g = f^{(r)}$ where $f \in K_1$. Since

$$\int_{-\infty}^{\infty} g(x)\gamma(x)\mathrm{d}x = (-1)^r \int_{-\infty}^{\infty} f(x)\gamma^{(r)}(x)\mathrm{d}x \to 0$$

as $\max|(1+x^2)^{k/2}\gamma^{(r)}(x)| \to 0$, it follows that $g \in C_r$. If r_0 is the smallest value of r which can be used we see that $g \in C_{r_0}$ but is not in C_{r_0-1}.

A generalised function is called *positive*, and we write $g \geq 0$, if $\int_{-\infty}^{\infty} g(x)\gamma(x)\mathrm{d}x \geq 0$ for all good $\gamma \geq 0$. This is consistent with the normal usage for ordinary functions – we could use arguments along the lines of Theorem 3.1 to show that $\int_a^b f(x)\gamma(x)\mathrm{d}x \geq 0$ implied $f \geq 0$. What we now want to show is

Theorem 3.22. *A positive generalised function must be in* C_1.

Proof. Let $\max|(1+x^2)^{k_1/2}\gamma_n(x)| = \varepsilon_n \to 0$ as $n \to \infty$, where k_1 will be chosen later. Let $\{\phi_m\}$ be a sequence of real fine functions such that $0 \leq \phi_m(x) \leq 1$, $\mathrm{Lim}_{m\to\infty} \phi_m g = g$ and ϕ_m and its derivatives bounded independently of m (such fine functions exist by Theorem 5.15). Then, for each γ_n, a fine function can be selected – call it ϕ_n – such that

$$\left|\int_{-\infty}^{\infty} g(x)\gamma_n(x)\{1-\phi_n(x)\}\mathrm{d}x\right| < 1/2^n. \tag{22}$$

Now

$$|\gamma_n(x)\phi_n(x)| \leq \varepsilon_n\phi_n/(1+x^2)^{k_1/2}. \tag{23}$$

Hence, if γ_n is real,

$$\varepsilon_n\phi_n(1+x^2)^{-k_1/2} - \gamma_n(x)\phi_n(x) \geq 0,$$
$$\varepsilon_n\phi_n(1+x^2)^{-k_1/2} + \gamma_n(x)\phi_n(x) \geq 0.$$

Therefore, if g is positive,

$$-\varepsilon_n\int_{-\infty}^{\infty}\frac{g(x)\phi_n(x)}{(1+x^2)^{k_1/2}}\mathrm{d}x \leq \int_{-\infty}^{\infty} g(x)\gamma_n(x)\phi_n(x)\mathrm{d}x$$
$$\leq \varepsilon_n\int_{-\infty}^{\infty}\frac{g(x)\phi_n(x)}{(1+x^2)^{k_1/2}}\mathrm{d}x.$$

On account of Theorem 3.16 both sides of the inequality tend to zero as $n \to \infty$ provided that $k_1 \geq k$. Hence, from (22), $\int_{-\infty}^{\infty} g(x)\gamma_n(x)\mathrm{d}x \to 0$ and so $g \in C_1$.

If γ_n is complex we deal with the real and imaginary parts separately after (23), and the proof is complete. □

Exercise

34. The function $f(x)$ is defined in $(n, n+1]$, n being any integer, by $f(x) = \max_{n \le x \le n+1} |\gamma(x)|$ where γ is good. Prove that $\int_{-\infty}^{\infty} f(u)\rho\{m(u-x)\}\, m\,du$ is good and deduce that there is a good $\gamma_0 \ge 0$ such that $\gamma_0(x) \ge \gamma(x) \ge -\gamma_0(x)$ when γ is real.

If $g_m(x) \ge 0$ and $\lim_{m\to\infty} \int_{-\infty}^{\infty} g_m(x)\gamma(x)\,dx = 0$ for all good $\gamma \ge 0$ prove $\operatorname{Lim}_{m\to\infty} g_m(x) = 0$ in the generalised sense.

Show also that, if $g(x) \ge 0$, then $\operatorname{Im} g = 0$.

4

Powers of x

4.1. Change of notation

The operation $\fint_{-\infty}^{\infty}$ has many of the properties of integration. For example

$$\fint_{-\infty}^{\infty} \{g_1(x) + g_2(x)\}\gamma(x)\mathrm{d}x = \fint_{-\infty}^{\infty} g_1(x)\gamma(x)\mathrm{d}x + \fint_{-\infty}^{\infty} g_2(x)\gamma(x)\mathrm{d}x,$$

$$\fint_{-\infty}^{\infty} g(x)\{\gamma_1(x) + \gamma_2(x)\}\mathrm{d}x = \fint_{-\infty}^{\infty} g(x)\gamma_1(x)\mathrm{d}x + \fint_{-\infty}^{\infty} g(x)\gamma_2(x)\mathrm{d}x,$$

$$\fint_{-\infty}^{\infty} ag(x)\gamma(x)\mathrm{d}x = a\fint_{-\infty}^{\infty} g(x)\gamma(x)\mathrm{d}x.$$

Theorem 3.4 corresponds to the usual rule for a linear change of variable for an integral whereas Theorem 3.6 corresponds to the usual rule for integration by parts. Theorems 3.9 and 3.10 parallel corresponding results for conventional integrals and Fourier transforms. Moreover Theorem 3.3 shows that, if $g \in K_1$,

$$\fint_{-\infty}^{\infty} g(x)\gamma(x)\mathrm{d}x = \int_{-\infty}^{\infty} g(x)\gamma(x)\mathrm{d}x.$$

Accordingly, from now on *we shall change our notation and write* $\int_{-\infty}^{\infty}$ *for* $\fint_{-\infty}^{\infty}$, i.e. from now on $\int_{-\infty}^{\infty} g(x)\gamma(x)\mathrm{d}x$ will be understood to mean $\fint_{-\infty}^{\infty} g(x)\gamma(x)\mathrm{d}x$. The reader must beware of treating $\int_{-\infty}^{\infty} g(x)\gamma(x)\mathrm{d}x$ as an ordinary integral unless $g \in K_1$ and should not freely apply operations appropriate for conventional integrals unless these have been defined for generalised functions.

Occasionally, in the interests of clarity, it may be desirable to temporarily re-introduce the notation $\fint_{-\infty}^{\infty}$.

4.2. Powers of *x*

The function $1/|x|^{1/2}$ belongs to K_1 and therefore defines a generalised function. As a generalised function it possesses a derivative but as a conventional function it does not have a derivative at the origin although it has one elsewhere, the conventional and generalised derivative agreeing in $x > 0$ and $x < 0$ by Theorem 3.8. More generally, any fractional power greater than -1 of $|x|$ gives rise to a generalised function with similar properties. It therefore becomes necessary to state precisely what is meant by powers of $|x|$ regarded as generalised functions.

Now, if β is a complex number with $\operatorname{Re}\beta > 0$,

$$\frac{\mathrm{d}}{\mathrm{d}x}x^\beta = \beta x^{\beta-1}, \quad \frac{\mathrm{d}}{\mathrm{d}x}(-x)^\beta = -\beta(-x)^{\beta-1}$$

when $x > 0$. Therefore

$$\frac{\mathrm{d}}{\mathrm{d}x}|x|^\beta = \beta|x|^{\beta-1}\operatorname{sgn} x. \tag{1}$$

Similarly,

$$\frac{\mathrm{d}}{\mathrm{d}x}\{|x|^\beta \operatorname{sgn} x\} = \beta|x|^{\beta-1} \quad (\operatorname{Re}\beta > 0). \tag{2}$$

We intend to define the generalised functions $|x|^\beta$ and $|x|^\beta \operatorname{sgn} x$ so that these rules can be preserved as far as possible while identifying the generalised functions with the conventional when they are in K_1. It should, however, be noted that the generalised function $|x|^\beta \operatorname{sgn} x$ must be thought of as a single symbol when not in K_1, and not as a product.

Definition 4.1. *For all complex values of β except negative integers the generalised functions $|x|^\beta$ and $|x|^\beta \operatorname{sgn} x$ are defined by*

$$|x|^\beta = \frac{\{|x|^{\beta+n}(\operatorname{sgn} x)^n\}^{(n)}}{(\beta+1)(\beta+2)\dots(\beta+n)},$$

$$|x|^\beta \operatorname{sgn} x = \frac{\{|x|^{\beta+n}(\operatorname{sgn} x)^{n+1}\}^{(n)}}{(\beta+1)(\beta+2)\dots(\beta+n)}$$

where n is any non-negative integer such that $\operatorname{Re}\beta + n > -1$.

The definition clearly fails if β is a negative integer because then the denominator of the right-hand side vanishes for some n. Negative integer powers will consequently require a separate definition – a matter which will be discussed later.

Since Re $\beta+n>-1$, the generalised function $|x|^{\beta+n}$ $(\text{sgn } x)^n$ agrees with the conventional function denoted by the same symbols. Then it follows from (1) and (2) that

$$\{|x|^{\beta+n+1}(\text{sgn } x)^{n+1}\}' = \frac{\mathrm{d}}{\mathrm{d}x}\{|x|^{\beta+n+1}(\text{sgn } x)^{n+1}\}$$
$$= (\beta+n+1)|x|^{\beta+n}(\text{sgn } x)^n$$

which demonstrates that the definitions of $|x|^\beta$ and $|x|^\beta$ sgn x are independent of n. On account of Theorem 3.8 the two generalised functions agree with the conventional functions for $x>0$ and $x<0$; they also agree at the origin if Re $\beta \geq 0$. Furthermore,

$$\{|x|^\beta\}' = \frac{\{|x|^{\beta+n+1}(\text{sgn } x)^{n+1}\}^{(n+2)}}{(\beta+1)(\beta+2)\ldots(\beta+n+1)}$$
$$= \frac{\{|x|^{\beta+n}(\text{sgn } x)^n\}^{(n+1)}}{(\beta+1)(\beta+2)\ldots(\beta+n)}$$
$$= \beta|x|^{\beta-1}\text{ sgn } x \tag{3}$$

so that (1) holds as a relation between generalised functions. Similarly, (2) continues to hold in the form

$$\{|x|^\beta \text{ sgn } x\}' = \beta|x|^{\beta-1}. \tag{4}$$

Definition 4.1 is convenient for powers defined for all x but sometimes it is desirable to limit oneself to positive x. This necessitates a definition of $x^\beta \mathrm{H}(x)$, i.e. a generalised function which is zero for $x<0$. The notation x_+^β will also be employed for $x^\beta \mathrm{H}(x)$ which is, of course, to be regarded as a single symbol.

Definition 4.2. *For all complex values of* β *except negative integers, the generalised function* $x^\beta \mathrm{H}(x)$ *is defined by*

$$x^\beta \mathrm{H}(x) = \frac{\{x^{\beta+n}\mathrm{H}(x)\}^{(n)}}{(\beta+1)(\beta+2)\ldots(\beta+n)}$$

where n *is any non-negative integer such that* Re $\beta+n>-1$.

The reader will easily check that this definition is independent of n and that $\{x^\beta \mathrm{H}(x)\}' = \beta x^{\beta-1}\mathrm{H}(x)$. The meaning of $(-x)^\beta \mathrm{H}(-x)$ or x_-^β follows from Definition 3.7. Observe that $x^\beta \mathrm{H}(x-a)$ where $a>0$

does not require special treatment because it belongs to K_1 and so comes under Theorem 3.3.

Negative integer powers of x cannot be approached in exactly the same way because $1/x$ is not the derivative of a power of x. However, it is the derivative of a logarithm so we use

Definition 4.3. *The generalised function x^{-1} is defined by*

$$x^{-1} = \{\ln|x|\}'.$$

Since $\ln|x| \in K_1$ a generalised function is provided by this definition. Moreover, it coincides with the conventional function $1/x$ everywhere except possibly at the origin. Also, since x is fairly good,

$$\{x \ln|x|\}' = x\{\ln|x|\}' + \ln|x|$$

by Exercise 15 of Chapter 3. But, by Corollary 3.7,

$$\{x \ln|x|\}' = \frac{d}{dx}\{x \ln|x|\} = 1 + \ln|x|$$

since $1 + \ln|x| \in K_1$. Hence

$$x.x^{-1} = 1$$

Thus x^{-1} does have the desired properties.

The power x^{-m} can now be introduced via

Definition 4.4. *If m is a positive integer, the generalised function x^{-m} is defined by*

$$x^{-m} = \{(-1)^{m-1}/(m-1)!\}(x^{-1})^{(m-1)}.$$

Exercises

1. Verify that the definition of $x^\beta H(x)$ is independent of n.
2. Prove that (i) $\{x^\beta H(x)\}' = \beta x^{\beta-1} H(x)$,
   (ii) $\{(-x)^\beta H(-x)\}' = \beta(-x)^{\beta-1} H(-x)$.
3. Show that $x^m.x^{-m} = 1$.
4. Prove that $x.x^\beta H(x) = x^{\beta+1} H(x)$.
5. Prove that $x.x^{-m} = x^{1-m}$.
6. Show that $x^\beta H(x) = \frac{1}{2}(|x|^\beta + |x|^\beta \operatorname{sgn} x)$.

4.3. Even and odd generalised functions

Definition 4.5. *The generalised function $g(x)$ is said to be even if $g(x) = g(-x)$ and to be odd if $g(x) = -g(-x)$.*

This definition is consistent with the usual one for ordinary functions and so does not conflict with common usage when $g \in K_1$.

Since

$$\int_{-\infty}^{\infty} \delta(x)\gamma(x)\mathrm{d}x = \gamma(0) = \int_{-\infty}^{\infty} \delta(-x)\gamma(x)\mathrm{d}x$$

we have $\delta(x) = \delta(-x)$ so that $\delta(x)$ is even. Similarly it may be verified that $\delta'(x)$ is odd. This last result is a particular example of

Theorem 4.1. *If $g(x)$ is even (odd) then $g'(x)$ is odd (even) and $G(\alpha)$ is even (odd).*

Proof. If $g(x) = g(-x)$ then $g'(x) = -g'(-x)$ by Exercise 15 of Chapter 3, so that, if g is even, g' is odd. There is a similar proof if g is odd.

If $g(x) = g(-x)$ then, by Theorem 3.9, $G(\alpha) = G(-\alpha)$ so that the Fourier transform of an even g is even. There is a similar proof when g is odd, and the theorem is complete. □

Exercises

7. Show that $\delta^{(n)}(x)$ is even or odd according as n is even or odd.
8. If β is not a negative integer show that $|x|^\beta$ is even and that $|x|^\beta \operatorname{sgn} x$ is odd.
9. Show that x^{-m} (m a positive integer) is even or odd according as m is even or odd.
10. If $\ln x$ is defined by $\ln x = \ln|x| + i\pi \mathrm{H}(-x)$ show that $(\ln x)' = x^{-1} - i\pi\delta(x)$.
11. If $\ln z$, when z is complex, be defined by its principal value show that $\lim_{\varepsilon\to+0} \ln(x \pm i\varepsilon) = \ln|x| \pm i\pi \mathrm{H}(-x)$. Deduce that $\lim_{\varepsilon\to+0} 1/(x+i\varepsilon) = x^{-1} - i\pi\delta(x)$, $\lim_{\varepsilon\to+0} 1/(x-i\varepsilon) = x^{-1} + i\pi\delta(x)$ and $\lim_{\varepsilon\to0} x/(x^2+\varepsilon^2) = x^{-1}$.
12. If g is even prove that $\int_{-\infty}^{\infty} g(x)\gamma(x)\mathrm{d}x = 0$ for every odd γ. Prove also the converse of this statement.
13. If g is even and $h(x)$ is a conventional function such that $g(x) = h(x)$ in $a < x < b$ show that $g(x) = h(-x)$ in $-b < x < -a$. What is the corresponding statement when g is odd?

It is now convenient to ask what the solution of $xg(x) = 1$ is. One solution is x^{-1}, since $x.x^{-1} = 1$, and any other must differ from it only by solution of $xg(x) = 0$. The delineation of such solutions is provided by

Theorem 4.2. *If $xg(x) = 0$ then $g(x) = C\delta(x)$ where C is an arbitrary constant.*

Proof. Let γ be any good function and consider $\{\gamma(x) - \gamma(0)\mathrm{e}^{-x^2}\}/x$.

This behaves as a good function everywhere except possibly at the origin but there the Taylor expansions of $\gamma(x)$ and e^{-x^2} (Corollary 1.30*c*) show that any derivative of finite order exists. Hence $\{\gamma(x)-\gamma(0)e^{-x^2}\}/x$ is a good function. Therefore

$$\int_{-\infty}^{\infty} g(x)\gamma(x)dx = \int_{-\infty}^{\infty} g(x)\gamma(0)e^{-x^2}dx + \int_{-\infty}^{\infty} xg(x)\frac{\gamma(x)-\gamma(0)e^{-x^2}}{x}dx$$

$$= \gamma(0)\int_{-\infty}^{\infty} g(x)e^{-x^2}dx$$

since $xg(x)=0$. Since $\int_{-\infty}^{\infty} g(x)e^{-x^2}\,dx$ has a value C independent of γ the theorem follows. □

Corollary 4.2. *If $g'(x)=0$ then $g(x)=C$.*

Proof. By Theorem 3.12, $g'(x)=0$ implies that $i\alpha G(\alpha)=0$. Therefore $G(\alpha)=C\delta(\alpha)$ and the corollary follows because the Fourier transform of $\delta(\alpha)$ is 1. □

Turning now to the equation $xg(x)=1$ put $g(x)=x^{-1}+g_1(x)$ and then $xg_1(x)=0$ so that $g_1(x)=C\delta(x)$. Hence the general solution of $xg(x)=1$ is $g(x)=x^{-1}+C\delta(x)$. Since $\delta(x)$ is even and x^{-1} odd (Exercises 7 and 9 above) x^{-1} is the only odd solution of $xg(x)=1$.

Exercises

14. If $xg(x)=\delta^{(n)}(x)$ show that $g(x)=C\delta(x)-\delta^{(n+1)}(x)/(n+1)$.

15. If $x^m g(x)=0$ show that $g(x)=C_1\delta(x)+C_2\delta'(x)+\dots+C_m\delta^{(m-1)}(x)$ where $C_1, C_2, \dots, C_m$ are arbitrary constants.

16. Show that the general solution of $x^m g(x)=1$ is $g(x)=x^{-m}+\sum_{r=1}^{m} C_r\delta^{(r-1)}(x)$.

4.4. Singular integrals

The definitions of §4.2 permit the evaluation of integrals which, regarded as ordinary integrals, would be considered too singular for calculation.

Definition 4.6. *If $g(x)H(x-a)$ and $g(x)H(x-b)$ are generalised functions, by $\int_a^b g(x)\gamma(x)dx$ will be understood*

$$\int_{-\infty}^{\infty}\{g(x)H(x-a)-g(x)H(x-b)\}\,\gamma(x)dx.$$

The integral of g cannot always be obtained in this way because, for example, $\delta(x-a)\mathrm{H}(x-a)$ has not been defined. It might happen that $g(x)\mathrm{H}(a-x)$ and $g(x)\mathrm{H}(b-x)$ existed even if $g(x)\mathrm{H}(x-a)$ and $g(x)\mathrm{H}(x-b)$ did not; in that case an appropriate definition would be

$$\int_a^b g(x)\gamma(x)\mathrm{d}x = \int_{-\infty}^{\infty} \{g(x)\mathrm{H}(b-x) - g(x)\mathrm{H}(a-x)\}\gamma(x)\mathrm{d}x.$$

If both definitions existed they would agree provided that the generalised functions were such that

$$g(x)\mathrm{H}(x-a) + g(x)\mathrm{H}(a-x) = g(x),$$
$$g(x)\mathrm{H}(x-b) + g(x)\mathrm{H}(b-x) = g(x).$$

The definition is consistent with the meaning which would be attached to $\int_a^b$ if g were a conventional integrable function. There is an immediate consequence of this definition which should be noted. Let $\eta(x)$ be an infinitely differentiable function in the interval (c,d) where $c<a$ and $d>b$. Let $\gamma_1(x)$ be a good function which equals unity in (a,b) and vanishes outside (c,d). That there are such good functions can be seen from Exercise 10 of Chapter 2. Then $\eta(x)\gamma_1(x)$ is a good function and if $\int_{-\infty}^{\infty} g(x)\{\mathrm{H}(x-a)-\mathrm{H}(x-b)\}\eta(x)\gamma_1(x)\mathrm{d}x$ has a meaning it may be written $\int_a^b g(x)\eta(x)\gamma_1(x)\mathrm{d}x$. But, since $\eta(x)\gamma_1(x)=\eta(x)$ in (a,b), this last integral may equally well be written $\int_a^b g(x)\eta(x)\mathrm{d}x$. In this way a meaning is attached to integrals of a wider class of integrands. In particular, by choosing $\eta(x)=1$ a significance may be assigned to $\int_a^b g(x)\mathrm{d}x$. (See also §6.1.)

Suppose now that $g(x)=x^\beta$, where β is not a negative integer, and $a=0$, $b>0$. Then $x^\beta\mathrm{H}(x)$ has a meaning (Definition 4.2) and so does $x^\beta\mathrm{H}(x-b)$ with $b>0$. Now $x^\beta\mathrm{H}(x)$ is expressed as the nth derivative of $x^{\beta+n}\mathrm{H}(x)$. To obtain a similar form for $x^\beta\mathrm{H}(x-b)$ note that $(x^{\beta+n}-b^{\beta+n})\mathrm{H}(x-b)$ has no discontinuity at $x=b$. Hence

$$\begin{aligned}\{x^{\beta+n}\mathrm{H}(x-b)\}' &= \{(x^{\beta+n}-b^{\beta+n})\mathrm{H}(x-b)\}' + b^{\beta+n}\mathrm{H}'(x-b)\\ &= (\beta+n)x^{\beta+n-1}\mathrm{H}(x-b) + b^{\beta+n}\delta(x-b).\end{aligned}$$

Applying this method repeatedly we obtain

$$\begin{aligned}&\{x^{\beta+n}\mathrm{H}(x-b)\}^{(n)}\\ &= (\beta+n)(\beta+n-1)\ldots(\beta+1)x^\beta\mathrm{H}(x-b) + b^{\beta+n}\delta^{(n-1)}(x-b)\\ &\quad + (\beta+n)b^{\beta+n-1}\delta^{(n-2)}(x-b) + \ldots\\ &\quad + (\beta+n)\ldots(\beta+2)b^{\beta+1}\delta(x-b).\end{aligned}$$

Hence

$$\int_{-\infty}^{\infty}\{x^\beta H(x)-x^\beta H(x-b)\}\gamma(x)dx$$

$$=\int_{-\infty}^{\infty}\left[\frac{\{x^{\beta+n}H(x)-x^{\beta+n}H(x-b)\}^{(n)}}{(\beta+n)(\beta+n-1)\dots(\beta+1)}\right.$$

$$\left.+\frac{b^{\beta+n}\delta^{(n-1)}(x-b)}{(\beta+n)\dots(\beta+1)}+\dots+\frac{b^{\beta+1}\delta(x-b)}{\beta+1}\right]\gamma(x)dx$$

$$=\frac{(-1)^n}{(\beta+1)(\beta+2)\dots(\beta+n)}\int_{-\infty}^{\infty}x^{\beta+n}\{H(x)-H(x-b)\}\gamma^{(n)}(x)dx$$

$$+\frac{b^{\beta+1}}{\beta+1}\gamma(b)-\frac{b^{\beta+2}\gamma'(b)}{(\beta+1)(\beta+2)}+\dots+\frac{(-1)^{n-1}b^{\beta+n}\gamma^{(n-1)}(b)}{(\beta+1)(\beta+2)\dots(\beta+n)}.$$

But, with $\mathrm{Re}\,\beta+n>-1$ the integral on the right can be regarded as a conventional integral and therefore

$$\int_0^b x^\beta\gamma(x)dx=\frac{(-1)^n}{(\beta+1)(\beta+2)\dots(\beta+n)}\int_0^b x^{\beta+n}\gamma^{(n)}(x)dx$$

$$+\frac{b^{\beta+1}}{\beta+1}\gamma(b)-\frac{b^{\beta+2}\gamma'(b)}{(\beta+1)(\beta+2)}+\dots$$

$$+\frac{(-1)^{n-1}b^{\beta+n}\gamma^{(n-1)}(b)}{(\beta+1)(\beta+2)\dots(\beta+n)} \tag{5}$$

where the integral on the right is calculated in the usual way.

If one attempted to integrate the left-hand side of (5) by parts repeatedly one would obtain

$$\int_0^b x^\beta\gamma(x)dx=\frac{(-1)^n}{(\beta+1)(\beta+2)\dots(\beta+n)}\int_0^b x^{\beta+n}\gamma^{(n)}(x)dx$$

$$+\left[\frac{x^{\beta+1}}{\beta+1}\gamma(x)-\frac{x^{\beta+2}\gamma'(x)}{(\beta+1)(\beta+2)}+\dots\right.$$

$$\left.+\frac{(-1)^{n-1}x^{\beta+n}\gamma^{(n-1)}(x)}{(\beta+1)(\beta+2)\dots(\beta+n)}\right]_0^b.$$

This cannot be interpreted in the usual way, if n is the least integer with $\mathrm{Re}\,\beta+n>-1$, because the contribution of the lower limit is not finite. However, if only the *finite part* is retained, it can be shown, as was done by Hadamard, that many of the conventional rules of integration are valid for this finite part. Thus (5) gives the same interpretation to the left-hand side as the Hadamard finite part.

By replacing $\gamma(x)$ by $\eta(x)\gamma_1(x)$ we obtain an interpretation of $\int_0^b x^\beta \eta(x)\mathrm{d}x$ where $\eta(x)$ is infinitely differentiable. Since γ_1 is unity in (a,b) all its derivatives vanish there. Hence (5) remains valid if γ is replaced by η everywhere.

Let us turn now to integral powers and suppose that $g(x) = x^{-1}$, $a < 0$, $b > 0$. Then

$$\begin{aligned}
&[\{\mathrm{H}(x-a) - \mathrm{H}(x-b)\}\ln|x|]' \\
&\quad = \{\mathrm{H}(x-a)\ln(|x|/|a|) - \mathrm{H}(x-b)\ln(|x|/|b|) \\
&\qquad\quad + \mathrm{H}(x-a)\ln|a| - \mathrm{H}(x-b)\ln|b|\}' \\
&\quad = \{x^{-1}\mathrm{H}(x-a) - x^{-1}\mathrm{H}(x-b)\} + \delta(x-a)\ln|a| - \delta(x-b)\ln|b|.
\end{aligned} \tag{6}$$

Hence

$$\begin{aligned}
&\int_{-\infty}^{\infty} \{x^{-1}\mathrm{H}(x-a) - x^{-1}\mathrm{H}(x-b)\}\gamma(x)\mathrm{d}x \\
&\quad = \int_{-\infty}^{\infty} \{[\{\mathrm{H}(x-a) - \mathrm{H}(x-b)\}\ln|x|]' \\
&\qquad\quad + \delta(x-b)\ln|b| - \delta(x-a)\ln|a|\}\gamma(x)\mathrm{d}x \\
&\quad = \gamma(b)\ln|b| - \gamma(a)\ln|a| - \int_{-\infty}^{\infty} \{\mathrm{H}(x-a) - \mathrm{H}(x-b)\}\gamma'(x)\ln|x|\,\mathrm{d}x.
\end{aligned}$$

Consequently

$$\int_a^b x^{-1}\gamma(x)\mathrm{d}x = \gamma(b)\ln|b| - \gamma(a)\ln|a| - \int_a^b \gamma'(x)\ln|x|\,\mathrm{d}x \tag{7}$$

where the integral on the right-hand side is a conventional one.

This interpretation corresponds to the *Cauchy principal value* of ordinary integration. For the Cauchy principal value is defined as

$$\begin{aligned}
\lim_{\varepsilon\to 0}\left(\int_a^{-\varepsilon} + \int_\varepsilon^b\right)\frac{\gamma(x)}{x}\mathrm{d}x = \lim_{\varepsilon\to 0}\Bigg\{ &-\left(\int_a^{-\varepsilon} + \int_\varepsilon^b\right)\gamma'(x)\ln|x|\,\mathrm{d}x \\
&+ \gamma(b)\ln|b| - \gamma(\varepsilon)\ln\varepsilon + \gamma(-\varepsilon)\ln\varepsilon - \gamma(a)\ln|a|\Bigg\}
\end{aligned}$$

which is the same as the right-hand side of (7).

Again, $\gamma(x)$ can be replaced by $\eta(x)$, which is infinitely differentiable on an interval including (a,b), by the same argument as above. (For more general η see §6.1.)

By taking $m-1$ derivatives of (6) we obtain

$$\begin{aligned}
&[\{\mathrm{H}(x-a)-\mathrm{H}(x-b)\}\ln|x|]^{(m)} \\
&= (m-1)!(-1)^{m-1}x^{-m}\{\mathrm{H}(x-a)-\mathrm{H}(x-b)\} \\
&\quad + (m-2)!(-1)^{m-2}\{a^{-m+1}\delta(x-a)-b^{-m+1}\delta(x-b)\} \\
&\quad + (m-3)!(-1)^{m-3}\{a^{-m+2}\delta'(x-a)-b^{-m+2}\delta'(x-b)\} \\
&\quad + \ldots + \delta^{(m-1)}(x-a)\ln|a| - \delta^{(m-1)}(x-b)\ln|b|.
\end{aligned}$$

Consequently,

$$\begin{aligned}
&\int_a^b x^{-m}\gamma(x)\mathrm{d}x \\
&= \frac{-1}{(m-1)!}\int_a^b \gamma^{(m)}(x)\ln|x|\mathrm{d}x + \frac{a^{1-m}\gamma(a)-b^{1-m}\gamma(b)}{m-1} \\
&\quad + \frac{a^{2-m}\gamma'(a)-b^{2-m}\gamma'(b)}{(m-1)(m-2)} + \ldots + \frac{a^{-1}\gamma^{(m-2)}(a)-b^{-1}\gamma^{(m-2)}(b)}{(m-1)(m-2)\ldots 1} \\
&\quad + \frac{\gamma^{(m-1)}(b)\ln b - \gamma^{(m-1)}(a)\ln|a|}{(m-1)!}. \qquad (8)
\end{aligned}$$

Once again, $\gamma(x)$ can be replaced by the infinitely differentiable $\eta(x)$.

Exercises

17. Show that $\frac{1}{4}\int_{-\infty}^{\infty}|x|^{-3/2}\mathrm{e}^{-x^2}\mathrm{d}x = -(-\frac{1}{4})!$.

18. Prove that

(i) $$\int_a^b (b-x)^{-3/2}\mathrm{d}x = -2(b-a)^{-1/2},$$

(ii) $$\int_0^{1/4}\frac{x^{-3/2}}{x-1}\mathrm{d}x = 4-\ln 3,$$

(iii) $$\int_{-a}^{a}\frac{(b-x)^{-2}}{(a^2-x^2)^{1/2}}\mathrm{d}x = -\pi|b|(b^2-a^2)^{-3/2}\mathrm{H}(b^2-a^2).$$

19. If m is a positive integer prove that

(i) $$\int_0^a \frac{dx}{(a^2-x^2)^{m+1/2}} = -\int_{-a}^0 \frac{dx}{(a^2-x^2)^{m+1/2}}$$

(ii) $$\int_a^\infty \frac{dx}{(x^2-a^2)^{m+1/2}} = (m-1)!\frac{(-1)^m\pi^{1/2}}{(m-\frac{1}{2})!2a^{2m}}.$$

20. Show that $\int_{-\infty}^{\infty} x^{-2}e^{-x^2}dx = -2\pi^{1/2}$.

21. Show that $\int_{-\pi}^{\pi} x^{-3}\sin x\,dx = (1/\pi - \frac{1}{2})\int_{-\pi}^{\pi}(\sin x/x)dx$.

22. Prove that

(i) $\int_0^b x^\beta dx = b^{\beta+1}/(\beta+1)$, (ii) $\int_a^b x^{-1}dx = \ln|b/a|$ for $a<0<b$.

The meaning to be attached to the symbol $x^{-m}H(x)$ is less easy to decide. Suppose that we felt that the multiplication rule

$$x.x^{-1}H(x) = H(x)$$

must be obeyed. By taking a derivative we see, from Exercise 15 of Chapter 3, that

$$x.\{x^{-1}H(x)\}' + x^{-1}H(x) = \delta(x).$$

Therefore, if we said that $x^{-2}H(x) = -\{x^{-1}H(x)\}'$, the multiplication rule $x.x^{-2}H(x) = x^{-1}H(x)$ would fail because of the presence of the $\delta(x)$ on the right-hand side. One is therefore forced to abandon either the multiplication rule or the derivative rule. We shall retain the multiplication rule though it is perfectly possible to produce a self-consistent theory based on the derivative; the two will differ only at the origin.

Definition 4.7. *If m is a positive integer the generalised functions $x^{-m}H(x)$ and $(-x)^m H(-x)$ are defined by*

$$x^{-m}H(x) = \lim_{\mu\to 0}\left\{x^{\mu-m}H(x) - \frac{(-1)^{m-1}}{(m-1)!\mu}\delta^{(m-1)}(x)\right\},$$

$$(-x)^{-m}H(-x) = \lim_{\mu\to 0}\left\{(-x)^{\mu-m}H(-x) - \frac{\delta^{(m-1)}(x)}{(m-1)!\mu}\right\}.$$

The existence of the limits needs to be verified. We shall consider

$m = 1$ and leave the general case to the reader. By Theorem 3.14(i)

$$\operatorname*{Lim}_{\mu\to 0}\left\{x^{\mu-1}\mathrm{H}(x) - \frac{1}{\mu}\delta(x)\right\} = \left[\operatorname*{Lim}_{\mu\to 0}\frac{1}{\mu}\left\{x^{\mu}\mathrm{H}(x) - \mathrm{H}(x)\right\}\right]'$$

on account of Definition 4.2. But $x^{\mu}\mathrm{H}(x)$ is a conventional function and

$$x^{\mu}\mathrm{H}(x) = \mathrm{H}(x) + \mu\mathrm{H}(x)\ln|x| + o(\mu).$$

Hence

$$\operatorname*{Lim}_{\mu\to 0}\frac{1}{\mu}\left\{x^{\mu}\mathrm{H}(x) - \mathrm{H}(x)\right\} = \mathrm{H}(x)\ln|x|$$

and so Definition 4.7 implies that

$$x^{-1}\mathrm{H}(x) = \{\mathrm{H}(x)\ln|x|\}'. \tag{9}$$

Similarly

$$(-x)^{-1}\mathrm{H}(-x) = -\{\mathrm{H}(-x)\ln|x|\}'. \tag{10}$$

To check the multiplication property note that, by Theorem 3.14 (iii),

$$x.x^{-m}\mathrm{H}(x) = \operatorname*{Lim}_{\mu\to 0} x\left\{x^{\mu-m}\mathrm{H}(x) - \frac{(-1)^{m-1}}{(m-1)!\mu}\delta^{(m-1)}(x)\right\}.$$

Applying Exercise 4 and using the fact that $x\delta^{(m-1)}(x) = -(m-1)\delta^{(m-2)}(x)$ (Exercise 16 of Chapter 3) we deduce that

$$x.x^{-m}\mathrm{H}(x) = x^{1-m}\mathrm{H}(x).$$

Similarly

$$x.(-x)^{-m}\mathrm{H}(-x) = -(-x)^{1-m}\mathrm{H}(-x).$$

As far as the derivative is concerned, Theorem 3.14(i) gives

$$\begin{aligned}\{x^{-m}\mathrm{H}(x)\}' &= \operatorname*{Lim}_{\mu\to 0}\left\{(\mu-m)x^{\mu-m-1}\mathrm{H}(x) - \frac{(-1)^{m-1}}{(m-1)!\mu}\delta^{(m)}(x)\right\}\\ &= \operatorname*{Lim}_{\mu\to 0}\left[(\mu-m)\left\{x^{\mu-m-1}\mathrm{H}(x) - \frac{(-1)^{m}}{m!\mu}\delta^{(m)}(x)\right\} + \frac{(-1)^{m}}{m!}\delta^{(m)}(x)\right]\\ &= -mx^{-m-1}\mathrm{H}(x) + \frac{(-1)^{m}}{m!}\delta^{(m)}(x)\end{aligned} \tag{11}$$

from Definition 4.7. Similarly

$$\{(-x)^{-m}\mathrm{H}(-x)\}' = m(-x)^{-m-1}\mathrm{H}(-x) - \delta^{(m)}(x)/m!. \quad (12)$$

There is one further peculiarity to be remarked on. If a is positive

$$(ax)^{-1}\mathrm{H}(ax) = \{\mathrm{H}(ax)\ln|ax|\}'a^{-1}$$

according to (9). Since $\mathrm{H}(ax) = \mathrm{H}(x)$

$$\begin{aligned}(ax)^{-1}\mathrm{H}(ax) &= [\mathrm{H}(x)\{\ln|x| + \ln|a|\}]'a^{-1}\\ &= a^{-1}.x^{-1}\mathrm{H}(x) + a^{-1}\delta(x)\ln|a|\end{aligned}$$

so that $(ax)^{-1}\mathrm{H}(ax) \neq a^{-1}.x^{-1}\mathrm{H}(x)$ except when $a = 1$. Consequently the normal homogeneous property is not available. It could be recovered if one was prepared to redefine $x^{-1}\mathrm{H}(x)$ so that it contained an arbitrary multiple of $\delta(x)$; then $x^{-m}\mathrm{H}(x)$ would include an arbitrary multiple of $\delta^{(m-1)}(x)$ and (11) reveals that the usual sort of derivative rule would hold. Nevertheless, it has been deemed preferable to exclude an arbitrary element from the definition at the price of jettisoning homogeneity and the derivative. The reader should be warned, however, that not all writers adopt this stance.

The generalised function $x^{-m}\operatorname{sgn} x$ may now be defined by

$$x^{-m}\operatorname{sgn} x = x^{-m}\mathrm{H}(x) - (-1)^m(-x)^{-m}\mathrm{H}(-x).$$

If ε is a positive number, $x + \mathrm{i}\varepsilon$ never vanishes and $(x+\mathrm{i}\varepsilon)^\beta$ is a well-defined generalised function for all complex β being, in fact, identical with its conventional counterpart so long as the phase of $x + \mathrm{i}\varepsilon$ is prescribed to be between $-\pi$ and π. It satisfies the rule

$$\{(x+\mathrm{i}\varepsilon)^\beta\}' = \beta(x+\mathrm{i}\varepsilon)^{\beta-1} \quad (13)$$

for all β. Similar remarks apply to $(x - \mathrm{i}\varepsilon)^\beta$. Note that

$$(x \pm \mathrm{i}\varepsilon)^{-1} = \{\ln(x \pm \mathrm{i}\varepsilon)\}'. \quad (14)$$

We wish now to contemplate defining two generalised functions by

$$(x \pm \mathrm{i}0)^\beta = \operatorname*{Lim}_{\varepsilon\to+0}(x \pm \mathrm{i}\varepsilon)^\beta.$$

First, there is no problem if $\operatorname{Re}\beta > 0$ and

$$(x \pm \mathrm{i}0)^\beta = x_+^\beta + \mathrm{e}^{\pm\mathrm{i}\beta\pi}x_-^\beta \quad (15)$$

using the alternative notation. Secondly, (13) and Exercise 2 show that (15) is valid for all complex β apart from the negative integers. On the other hand, (14) gives

$$(x \pm \mathrm{i}0)^{-1} = \{\ln|x| \pm \mathrm{i}\pi \mathrm{H}(-x)\}' = x^{-1} \mp \mathrm{i}\pi\delta(x).$$

If follows from (13) that

$$(x \pm \mathrm{i}0)^{-m} = x^{-m} \mp \frac{(-1)^{m-1}\pi\mathrm{i}}{(m-1)!}\delta^{(m-1)}(x). \tag{16}$$

Exercises

23. Prove that

(i) $x^{-m}\mathrm{H}(x) = \dfrac{(-1)^{m-1}}{(m-1)!}\{\mathrm{H}(x)\ln|x|\}^{(m)} + \beta_m\delta^{(m-1)}(x)$,

(ii) $(-x)^{-m}\mathrm{H}(-x) = \dfrac{-1}{(m-1)!}\{\mathrm{H}(-x)\ln|x|\}^{(m)} + (-1)^{m-1}\beta_m\delta^{(m-1)}(x)$

where $\beta_1 = 0$ and

$$\beta_m = \frac{(-1)^{m-1}}{(m-1)!}\left(1 + \frac{1}{2} + \frac{1}{3} + \dots + \frac{1}{m-1}\right) \quad (m \geq 2).$$

24. Show that $x.x^{-m}\operatorname{sgn} x = x^{1-m}\operatorname{sgn} x$ and $(x^{-m}\operatorname{sgn} x)' = -mx^{-m-1}\operatorname{sgn} x + (-1)^m 2\delta^{(m)}(x)/m!$.

25. Prove that

(i) $\{(x \pm \mathrm{i}0)^\beta\}' = \beta(x \pm \mathrm{i}0)^{\beta-1}$,

(ii) $\{(x \pm \mathrm{i}0)^{-m}\}' = -m(x \pm \mathrm{i}0)^{-m-1}$,

(iii) $\lim\limits_{\mu\to 0}(x \pm \mathrm{i}0)^{\mu-m} = (x \pm \mathrm{i}0)^{-m}$,

(iv) $x.(x \pm \mathrm{i}0)^\beta = (x \pm \mathrm{i}0)^{\beta+1}$.

26. Show that

$$\begin{aligned} x^{-m} &= \lim_{\mu\to 0}\{x^{\mu-m}\mathrm{H}(x) + (-1)^m(-x)^{\mu-m}\mathrm{H}(-x)\} \\ &= x^{-m}\mathrm{H}(x) + (-1)^m(-x)^{-m}\mathrm{H}(-x). \end{aligned}$$

The factorial function $\nu!$ can be defined by $\nu! = \int_0^\infty x^\nu \mathrm{e}^{-x}\mathrm{d}x$ for $\mathrm{Re}\,\nu > -1$ and for other values by analytic continuation. Let us compare it with $f(\mu) = \int_0^\infty x^\mu \mathrm{e}^{-x}\,\mathrm{d}x$, the integral being interpreted in the sense of this section. Such an integral can be assigned a meaning, via Definitions 4.2 and 4.7, through

$$\int_{-\infty}^{\infty} x^{\mu} \mathrm{H}(x)\eta(x)\mathrm{e}^{-x}\mathrm{d}x$$

where η is any infinitely differentiable function which is unity for $x \geq 0$ and zero for $x \leq c < 0$ so that $\eta(x)\mathrm{e}^{-x}$ is good. If μ is not a negative integer, Exercise 2 gives

$$\begin{aligned}
\mu f(\mu - 1) &= \int_{-\infty}^{\infty} \{x^{\mu}\mathrm{H}(x)\}'\eta(x)\mathrm{e}^{-x}\mathrm{d}x \\
&= -\int_{-\infty}^{\infty} x^{\mu}\mathrm{H}(x)\{\eta(x)\mathrm{e}^{-x}\}'\mathrm{d}x \\
&= \int_{-\infty}^{\infty} x^{\mu}\mathrm{H}(x)\eta(x)\mathrm{e}^{-x}\mathrm{d}x = f(\mu)
\end{aligned}$$

since $\eta'(x) = 0$ for $x \geq 0$. Thus $f(\mu)$ satisfies the same recurrence relation as $\mu!$ and will therefore agree with $\mu!$ except when μ is a negative integer. Indeed, from (9),

$$f(-1) = \int_{0}^{\infty} \mathrm{e}^{-x}\ln x\,\mathrm{d}x$$

which is finite, but Definition 4.7 informs us that $f(\mu - 1)$ does not tend to $f(-1)$ as $\mu \to 0$. In fact

$$\begin{aligned}
\lim_{\mu\to 0} f(\mu - 1) &= \lim_{\mu \to 0} \int_{-\infty}^{\infty} \left\{x^{\mu - 1}\mathrm{H}(x) - \frac{\delta(x)}{\mu}\right\}\eta(x)\mathrm{e}^{-x}\,\mathrm{d}x + 1/\mu \\
&= f(-1) + \lim_{\mu \to 0}\frac{1}{\mu}.
\end{aligned}$$

Thus, as $\mu \to 0$, $f(\mu - 1)$ becomes infinite in the same way as a pole with residue 1. Because of the recurrence relation $f(\mu - m)$ behaves as a pole with residue $(-1)^{m-1}/(m-1)!$ as $\mu \to 0$. The factorial function has the same properties. Hence, it is feasible to write

$$\nu! = \int_{0}^{\infty} x^{\nu}\mathrm{e}^{-x}\mathrm{d}x$$

when ν has any complex value other than a negative integer.

Exercise

27. Show that, if ν is not a negative integer and $\operatorname{Re}\mu > 0$ ($|\operatorname{ph}\mu| < \frac{1}{2}\pi$),

$$\int_0^\infty x^\nu e^{-\mu x}dx = \nu!/\mu^{\nu+1}$$

where $\mu^\nu = \exp\{\nu(\ln|\mu| + i \operatorname{ph}\mu)\}$.

4.5. Fourier transforms

This section is concerned with deriving certain Fourier transforms connected with the powers of x which have been defined.

Theorem 4.3. *The Fourier transform of* $\operatorname{sgn} x$ *is* $-2i\alpha^{-1}$.

Proof. Let $G(\alpha)$ be the Fourier transform of $\operatorname{sgn} x$. By Theorem 3.12, the Fourier transform of $(\operatorname{sgn} x)'$ or $2\delta(x)$ is $i\alpha G(\alpha)$. Since the Fourier transform of $\delta(x)$ is 1 this implies that $i\alpha G(\alpha) = 2$. It follows from Theorem 4.2 that $G(\alpha) = -2i\alpha^{-1} + C\delta(\alpha)$. But, $\operatorname{sgn} x$ is odd so that G must be odd (Theorem 4.1) and therefore $C = 0$. The proof is complete.□

Corollary 4.3*a*. *The Fourier transform of* x^{-1} *is* $-\pi i \operatorname{sgn}\alpha$.

Proof. This follows immediately by applying the Fourier inversion Theorem 3.11 to Theorem 4.3.□

Corollary 4.3*b*. *The Fourier transform of* x^{-m} *is* $\dfrac{(-i)^m\pi}{(m-1)!}\alpha^{m-1}\operatorname{sgn}\alpha$.

Proof. $x^{-m} = (-1)^{m-1}(x^{-1})^{(m-1)}/(m-1)!$ so that, by Theorem 3.12, its Fourier transform is $(-1)^{m-1}(i\alpha)^{m-1}/(m-1)!$ times the Fourier transform of x^{-1} and the result follows from Corollary 4.3*a*.□

For powers on a half-line we have

Theorem 4.4. *If* β *is not an integer the Fourier transform of* $x^\beta H(x)$ *is*

$$\beta! e^{-\pi i(\beta+1)/2}(\alpha - i0)^{-\beta-1}$$

and of $(-x)^\beta H(-x)$ *is* $\beta! e^{\pi i(\beta+1)/2}(\alpha + i0)^{-\beta-1}$.

Proof. Consider firstly what happens when $\operatorname{Re}\beta > -1$. Then

$$\operatorname*{Lim}_{\mu\to+0} e^{-\mu x}x^\beta H(x) = x^\beta H(x)$$

because

$$\left|\int_0^\infty (1-e^{-\mu x})x^\beta \gamma(x)dx\right| \leq \int_0^\infty \mu x^{\beta+1}|\gamma(x)|dx \leq M\mu$$

where $M = \int_0^\infty x^{\beta+1}|\gamma(x)|dx$. By Theorem 3.14(iv) the Fourier transform of $x^\beta H(x)$ is

$$\operatorname*{Lim}_{\mu\to+0} \int_0^\infty e^{-\mu x} x^\beta e^{-i\alpha x} dx = \operatorname*{Lim}_{\mu\to+0} \beta!/(\mu+i\alpha)^{\beta+1}$$

since $\int_0^\infty t^z e^{-at} dt = z!/a^{z+1}$ when $\operatorname{Re} z > -1$, $|\operatorname{ph} a| < \frac{1}{2}\pi$ and the principal value of a^{z+1} is taken. Now

$$(\mu+i\alpha)^{-\beta-1} = e^{-\pi i(\beta+1)/2}(\alpha - i\mu)^{-\beta-1}$$

with the phase of $\alpha - i\mu$ between $-\pi$ and 0. The formula of the theorem now follows from the definition of the preceding section and the result has been demonstrated when $\operatorname{Re}\beta > -1$.

When $\operatorname{Re}\beta < -1$ the use of Definition 4.2 and Theorem 3.12 shows that the Fourier transform of $x^\beta H(x)$ is $(i\alpha)^n/(\beta+1)\dots(\beta+n)$ times the Fourier transform of $x^{\beta+n}H(x)$. From what has just been proved we obtain

$$\frac{(i\alpha)^n}{(\beta+1)\dots(\beta+n)}(\beta+n)!e^{-\pi i(\beta+n+1)/2}(\alpha - i0)^{-\beta-n-1}$$
$$= \beta! e^{-\pi i(\beta+1)/2}(\alpha - i0)^{-\beta-1}$$

from Exercise 25(iv). Thus the formula is valid for all β other than the integers. The Fourier transform of $(-x)^\beta H(-x)$ can now be deduced by Theorem 3.10 and the theorem is proved.□

Corollary 4.4. *The Fourier transform of* $|x|^\beta$ *is*

$$\beta!|\alpha|^{-\beta-1}2\cos\tfrac{1}{2}\pi(\beta+1)$$

and of $|x|^\beta \operatorname{sgn} x$ *is*

$$\beta!|\alpha|^{-\beta-1}\operatorname{sgn}\alpha(-2i)\sin\tfrac{1}{2}\pi(\beta+1)$$

when β *is not an integer.*

Proof. The corollary is an immediate consequence of Theorem 4.4 and

$$|x|^\beta = x^\beta H(x) + (-x)^\beta H(-x),$$
$$|x|^\beta \operatorname{sgn} x = x^\beta H(x) - (-x)^\beta H(-x).\square$$

It is worth remarking that, since $\text{Lim}_{\beta\to 0}|x|^{\beta} = 1$ ((14) of Chapter 3), we must have

$$\underset{\beta\to 0}{\text{Lim}}\ \beta!|\alpha|^{-\beta-1}2\cos\tfrac{1}{2}\pi(\beta+1) = 2\pi\delta(\alpha).$$

A direct confirmation is

$$\underset{\beta\to 0}{\text{Lim}}\ \beta!|\alpha|^{-\beta-1}2\cos\tfrac{1}{2}\pi(\beta+1) = \underset{\beta\to 0}{\text{Lim}}\ |\alpha|^{-\beta-1}(-\pi\beta) = 2\pi\delta(\alpha)$$

from (15) of Chapter 3. Similarly

$$\underset{\beta\to -1}{\text{Lim}}\ \beta!|\alpha|^{-\beta-1}(-2\mathrm{i})\sin\tfrac{1}{2}\pi(\beta+1)\operatorname{sgn}\alpha = -\pi\mathrm{i}\operatorname{sgn}\alpha$$

from which follows, by Corollary 4.3*a* and Theorem 3.14 (iv),

$$\underset{\beta\to -1}{\text{Lim}}\ |x|^{\beta}\operatorname{sgn} x = x^{-1}.$$

By taking *m* derivatives and using Theorem 3.14(i),

$$\underset{\beta\to -1}{\text{Lim}}\ |x|^{-m+\beta}\operatorname{sgn} x = x^{-m-1} \qquad (m\text{ even}), \tag{17}$$

$$\underset{\beta\to -1}{\text{Lim}}\ |x|^{-m+\beta} = x^{-m-1} \qquad (m\text{ odd}). \tag{18}$$

The Fourier transforms of quantities obtained from the preceding by replacing *x* by $ax + b$, with *a* and *b* real, can now be deduced by Theorem 3.10. One case of practical importance not covered by this is

Theorem 4.5. *The Fourier transform of* $(x+\mathrm{i}c)^{-m}$, *c real and non-zero, is* $-2\pi\mathrm{i}e^{-c\alpha}\mathrm{H}(\alpha c)(-\mathrm{i}\alpha)^{m-1}\operatorname{sgn} c/(m-1)!$.

Proof. Since $\int_0^\infty \alpha^{m-1}e^{-|c|\alpha+\mathrm{i}\alpha x}\,d\alpha = (m-1)!(-\mathrm{i})^{-m}(x+\mathrm{i}|c|)^{-m}$ the Fourier transform of $(x+\mathrm{i}|c|)^{-m}$ is $-2\pi\mathrm{i}(-\mathrm{i}\alpha)^{m-1}e^{-|c|\alpha}\mathrm{H}(\alpha)/(m-1)!$. Hence by Theorem 3.10, the Fourier transform of $(x+\mathrm{i}c)^{-m}$ or $(\operatorname{sgn} c)^m(x\operatorname{sgn} c+\mathrm{i}|c|)^{-m}$ is

$$(\operatorname{sgn} c)^m(-2\pi\mathrm{i})(-\mathrm{i}\alpha\operatorname{sgn} c)^{m-1}e^{-|c|\alpha\operatorname{sgn} c}\mathrm{H}(\alpha\operatorname{sgn} c)/(m-1)!$$

and the theorem is proved. □

It should be noted that, as $c\to 0$, the Fourier transform of $(x+\mathrm{i}c)^{-m}$ approaches

$$-2\pi\mathrm{i}\,\mathrm{H}(\alpha)(-\mathrm{i}\alpha)^{m-1}/(m-1)!$$

or

$$-2\pi i H(-\alpha)(-i\alpha)^{m-1}/(m-1)!$$

accordings as $c \to 0$ from above or below. Thus the Fourier transform of x^{-m} is not the limit of the Fourier transform of $(x+ic)^{-m}$ as $c \to 0$ either from above or below but is half the sum of the two since $H(\alpha)+H(-\alpha)=1$.

The meaning of an expression such as $(x^2-a^2)^{-1}$ does not follow immediately from our preceding definitions because the multiplication of arbitrary generalised functions has not been defined yet. However, a suitable interpretation is obtained by equating it to the partial fraction expansion that would be derived if it were treated as a conventional function. Thus

$$(x^2-a^2)^{-1} = (1/2a)\{(x-a)^{-1}-(x+a)^{-1}\},$$
$$x^3(x^2-a^2)^{-1} = x + \tfrac{1}{2}a^2\{(x-a)^{-1}+(x+a)^{-1}\}.$$

It follows that the Fourier transform of $x^3(x^2-a^2)^{-1}$ is

$$2\pi i\delta'(\alpha) - \pi i a^2 \cos a\alpha \operatorname{sgn}\alpha.$$

Exercises

28. If $a>0$ prove that

$$\int_0^\infty (x-a)^{-1}e^{-i\alpha x}dx = -\pi i e^{-i\alpha a} + \int_0^\infty \frac{e^{-t\alpha}}{t-ia}dt$$

if $\alpha>0$. What is the corresponding result if $\alpha<0$?

29. Verify Theorem 4.4 and Corollary 4.4 when β is complex.

30. Evaluate $\int_{-\infty}^{\infty}(x-a)^{-1}\sin\alpha x\,dx$.

31. Find the Fourier transform of $(1-x)^{-3/2}H(1-x)$.

32. Find the Fourier transforms of

(i) $(x^2-4)^{-1}$, (ii) $x^3(x^2+4x+3)^{-1}$, (iii) $(x^2-2x-3)^{-2}$.

33. Evaluate $\int_{-\infty}^{\infty}(x^2-a^2)^{-1}e^{-i\alpha x}dx$ for $\alpha>0$.

34. Show that the Fourier transform of $H(x)$ is $\pi\delta(\alpha)-i\alpha^{-1}$.

35. Prove that, if $\int_{-\infty}^{\infty}H(\alpha)e^{i\alpha x}d\alpha$ is written as $\int_0^\infty e^{i\alpha x}d\alpha$,

$$\int_0^\infty e^{i\alpha x}d\alpha = \lim_{R\to\infty}\int_0^R e^{i\alpha x}d\alpha = \lim_{\varepsilon\to 0}\int_0^\infty e^{-\alpha\varepsilon+i\alpha x}d\alpha = \pi\delta(x)+ix^{-1}.$$

36. Given that $\int_0^\infty e^{-\mu t}J_\nu(bt)t^\nu dt = (\nu-\frac{1}{2})!(2b)^\nu/(\mu^2+b^2)^{\nu+1/2}\pi^{1/2}$ when b is real, $\operatorname{Re}\nu > -\frac{1}{2}$ and $\operatorname{Re}\mu>0$ prove that

$$\int_0^\infty e^{-i\alpha t}J_\nu(bt)t^\nu dt = (\nu-\tfrac{1}{2})!(2b)^\nu(b+\alpha-i0)^{-\nu-1/2}(b-\alpha+i0)^{-\nu-1/2}/\pi^{1/2}$$

when Re $\nu > -\frac{1}{2}$. Deduce that

$$\int_0^\infty e^{-i\alpha t} J_\nu(bt) t^{\nu+1} \mathrm{d}t = (\nu+\tfrac{1}{2})!\,2i\alpha(2b)^\nu(b+\alpha-i0)^{-\nu-3/2}$$

$$\times(b-\alpha+i0)^{-\nu-3/2}/\pi^{1/2}.$$

(Hint: $\lim_{\mu\to+0} e^{-\mu x} J_\nu(bx)x^\nu = J_\nu(bx)x^\nu$.)

37. Verify that $\mathrm{Lim}_{a\to 0}(x^2-a^2)^{-1} = x^{-2}$.

The interpretation of an integral such as $\int_0^1 x^\mu(1-x)^\nu\,\mathrm{d}x$ is facilitated by using the good function $\tau(x)$ (defined in Exercise 10 of Chapter 2) which is zero for $|x| \geq 1$ and has the property $\tau(x)+\tau(x-1)=1$ when $0 \leq x \leq 1$. Then if we write

$$\int_0^1 x^\mu(1-x)^\nu \mathrm{d}x = \int_0^1 x^\mu(1-x)^\nu \tau(x)\mathrm{d}x + \int_0^1 x^\mu(1-x)^\nu \tau(x-1)\mathrm{d}x$$

the τ factors separate the two singularities. It is then easy to show that, if neither μ nor ν is a negative integer,

$$(\mu+\nu+1)\int_0^1 x^\mu(1-x)^\nu \mathrm{d}x = \nu\int_0^1 x^\mu(1-x)^{\nu-1}\mathrm{d}x$$

$$= \mu\int_0^1 x^{\mu-1}(1-x)^\nu \mathrm{d}x.$$

Since it is known that

$$\int_0^1 x^\mu(1-x)^\nu \mathrm{d}x = \frac{\mu!\,\nu!}{(\mu+\nu+1)!} \tag{19}$$

when Re $\mu > -1$ and Re $\nu > -1$, it follows without difficulty that the same formula holds when neither μ nor ν is a negative integer.

Expressed in terms of trigonometric functions this is

$$\int_0^{\pi/2} \sin^\mu\theta\cos^\nu\theta\,\mathrm{d}\theta = \frac{(\frac{1}{2}\mu-\frac{1}{2})!(\frac{1}{2}\nu-\frac{1}{2})!}{(\frac{1}{2}\mu+\frac{1}{2}\nu)!\,2}$$

when neither μ nor ν is an odd negative integer.

It may be shown similarly that

$$\int_0^\infty x^\mu(1+x)^{-\mu-\nu-2}\mathrm{d}x = \frac{\mu!\,\nu!}{(\mu+\nu+1)!}$$

when neither μ nor v is a negative integer.

In order to explain what happens when μ is a negative integer it is necessary to introduce the derivative $\psi(z)$ of the factorial function defined by

$$\psi(z) = \frac{1}{z!}\frac{\mathrm{d}}{\mathrm{d}z}(z!).$$

This function has the property that $\psi(0) = -\gamma$ where $\gamma = 0.5772\ldots$ is Euler's constant. Also $\psi(z) - \psi(z-1) = 1/z$ so that

$$\psi(n) = 1 + \frac{1}{2} + \frac{1}{3} + \ldots + \frac{1}{n} - \gamma.$$

It may also be noted that $z!(z-\frac{1}{2})! = (2z)!\pi^{1/2}/2^{2z}$ implies that

$$\psi(z) + \psi(z - \tfrac{1}{2}) = 2\psi(2z) - 2\ln 2.$$

When μ is the negative integer $-m$ we use Definition 4.7 to assert that

$$\begin{aligned}\int_0^1 x^{-m}(1-x)^v \mathrm{d}x &= \operatorname*{Lim}_{\mu\to 0} \int_0^1 \left\{ x^{\mu-m} - \frac{(-1)^{m-1}}{(m-1)!\mu}\delta^{(m-1)}(x)\right\}(1-x)^v \mathrm{d}x \\ &= \operatorname*{Lim}_{\mu\to 0}\left\{\frac{(\mu-m)!v!}{(\mu+v-m+1)!} - \frac{v!(-1)^{m-1}}{(m-1)!(v-m+1)!\mu}\right\}\end{aligned}$$

from (19) so long as v is not a negative integer. Since

$$(\mu - m)! = \frac{(-1)^{m-1}\pi}{(m-\mu-1)!\sin\mu\pi}$$

we have

$$\int_0^1 x^{-m}(1-x)^v \mathrm{d}x = \frac{v!(-1)^{m-1}}{(m-1)!(v-m+1)!}\{\psi(m-1) - \psi(v-m+1)\}$$

when v is not a negative integer.

Exercise

38. If neither μ nor v is a negative integer show that

$$\int_0^1 (1-x)^v x^\mu \ln x \, \mathrm{d}x = \frac{\mu!v!}{(\mu+v+1)!}\{\psi(\mu) - \psi(\mu+v+1)\}.$$

4.6. Generalised functions containing a logarithm

It is fairly easy to devise a rule for attaching a meaning to the product of a power of x and a logarithm. When $\beta > -1$, $\partial|x|^\beta/\partial\beta$ is a well-defined generalised function. Therefore $\{\partial|x|^\beta/\partial\beta\}^{(n)}$ is a generalised function which can be written, on account of Theorem 3.15, $\partial\{|x|^\beta\}^{(n)}/\partial\beta$. Similarly $\partial\{|x|^{\beta+n}\text{ sgn }x)^n\}^{(n)}/\partial\beta$, with $\beta+n > -1$, is a generalised function. It follows from Definition 4.1 that $\partial|x|^\beta/\partial\beta$ is a generalised function whether β be greater than -1 or not. Since $\partial|x|^\beta/\partial\beta = |x|^\beta \ln|x|$ when $\beta > -1$ this gives a method of defining $|x|^\beta \ln|x|$ when $\beta < -1$.

Definition 4.8. *The generalised functions* $|x|^\beta \ln|x|$, $|x|^\beta \ln|x| \text{ sgn } x$, $x^\beta \mathrm{H}(x)\ln x$ *are* ***defined by***

$$|x|^\beta \ln|x| = \frac{\partial}{\partial\beta}|x|^\beta, \qquad |x|^\beta \ln|x| \text{ sgn } x = \frac{\partial}{\partial\beta}(|x|^\beta \text{ sgn } x),$$

$$x^\beta \mathrm{H}(x)\ln x = \frac{\partial}{\partial\beta}\{x^\beta \mathrm{H}(x)\}.$$

The same definitions are valid when β is complex.

The standard rules for derivatives apply to these generalised functions because, from Theorem 3.15,

$$\{|x|^\beta \ln|x|\}' = \{\partial|x|^\beta/\partial\beta\}' = \partial\{|x|^\beta\}'/\partial\beta = \partial\{\beta|x|^{\beta-1}\text{ sgn }x\}/\partial\beta$$
$$= |x|^{\beta-1}\text{ sgn }x + \beta|x|^{\beta-1}\ln|x|\text{ sgn }x.$$

Definitions for integral powers of x follow immediately on account of (17) and (18).

Definition 4.9. *The generalised function* $x^{-m}\ln|x|$ *is defined by*

$$x^{-m}\ln|x| = \begin{cases} \operatorname*{Lim}_{\beta\to -m} |x|^\beta \ln|x| & (m \text{ even}) \\ \operatorname*{Lim}_{\beta\to -m} |x|^\beta \ln|x| \text{ sgn } x & (m \text{ odd}). \end{cases}$$

The definition of $x^{-m}\,\mathrm{H}(x)\ln x$ is similar to that of $x^{-m}\mathrm{H}(x)$, namely

Definition 4.10. *If m is a positive integer the generalised functions*

$x^{-m}\,\mathrm{H}(x)\ln x$ and $(-x)^{-m}\,\mathrm{H}(-x)\ln|x|$ are defined by

$$x^{-m}\,\mathrm{H}(x)\ln x = \operatorname*{Lim}_{\mu\to 0}\left\{x^{\mu-m}\,\mathrm{H}(x)\ln x + \frac{(-1)^{m-1}}{(m-1)!\,\mu^2}\,\delta^{(m-1)}(x)\right\},$$

$$(-x)^{-m}\,\mathrm{H}(-x)\ln|x| = \operatorname*{Lim}_{\mu\to 0}\left\{(-x)^{\mu-m}\,\mathrm{H}(-x)\ln|x| + \frac{\delta^{(m-1)}(x)}{(m-1)!\,\mu^2}\right\}.$$

As for Definition 4.7 it may be checked that

$$x^{-1}\,\mathrm{H}(x)\ln x = \{\tfrac{1}{2}\mathrm{H}(x)(\ln x)^2\}'.$$

One consequence of Definitions 4.7 and 4.10 is that

$$x^{\mu-m}\,\mathrm{H}(x) = \frac{(-1)^{m-1}}{(m-1)!\,\mu}\delta^{(m-1)}(x) + x^{-m}\,\mathrm{H}(x) + \mu x^{-m}\,\mathrm{H}(x)\ln x + o(\mu)$$

when μ is small. The reader should be able to confirm that

$$x.x^{-m}\,\mathrm{H}(x)\ln x = x^{1-m}\,\mathrm{H}(x)\ln x, \qquad (20)$$

$$\{x^{-m}\,\mathrm{H}(x)\ln x\}' = -mx^{-m-1}\,\mathrm{H}(x)\ln x + x^{-m-1}\,\mathrm{H}(x). \qquad (21)$$

As in §4.5 we define

$$x^{-m}\operatorname{sgn} x\ln|x| = x^{-m}\,\mathrm{H}(x)\ln x - (-1)^m(-x)^{-m}\,\mathrm{H}(-x)\ln|x|. \qquad (22)$$

Likewise

$$(x\pm \mathrm{i}0)^\beta\ln(x\pm \mathrm{i}0) = \operatorname*{Lim}_{\varepsilon\to +0}(x\pm \mathrm{i}\varepsilon)^\beta\ln(x\pm \mathrm{i}\varepsilon)$$

for any complex β so that

$$(x\pm \mathrm{i}0)^\beta\ln(x\pm \mathrm{i}0) = x_+^\beta\ln x + \mathrm{e}^{\pm \mathrm{i}\beta\pi}x_-^\beta\ln|x| \pm \mathrm{i}\pi\mathrm{e}^{\pm \mathrm{i}\beta\pi}x_-^\beta \quad (23)$$

if β is not a negative integer, $x_+^\beta\ln x$ and $x_-^\beta\ln|x|$ standing for $x^\beta\,\mathrm{H}(x)\ln x$ and $(-x)^\beta\,\mathrm{H}(-x)\ln|x|$ respectively. If m is a positive integer

$$(x\pm \mathrm{i}0)^{-m}\ln(x\pm \mathrm{i}0) = x^{-m}\ln|x| \pm \mathrm{i}\pi(-1)^m x_-^{-m} + \tfrac{1}{2}\pi^2\frac{(-1)^{m-1}}{(m-1)!}\,\delta^{(m-1)}(x). \qquad (24)$$

Fourier transforms may now be deduced by Theorem 3.14 (iv) which implies that the Fourier transform of $\partial g/\partial\beta$ is $\partial G/\partial\beta$.

Theorem 4.6. *If β is not an integer, the Fourier transform of $x^\beta H(x) \ln x$ is*

$$\beta!\,e^{-\pi i(\beta+1)/2}[\{\psi(\beta)-\tfrac{1}{2}\pi i\}(\alpha-i0)^{-\beta-1}-(\alpha-i0)^{-\beta-1}\ln(\alpha-i0)]$$

and of $(-x)^\beta H(-x)\ln|x|$ is

$$\beta!\,e^{\pi i(\beta+1)/2}[\{\psi(\beta)+\tfrac{1}{2}\pi i\}(\alpha+i0)^{-\beta-1}-(\alpha+i0)^{-\beta-1}\ln(\alpha+i0)].$$

Proof. By Theorem 4.4 the Fourier transform is

$$\frac{\partial}{\partial\beta}\{\beta!\,e^{-\pi i(\beta+1)/2}(\alpha-i0)^{-\beta-1}\}$$

and when this is evaluated the formula of the theorem is obtained. The second result is derived in a similar way. □

Corollary 4.6a. *If n is a non-negative integer, the Fourier transform of $x^n H(x)\ln x$ is*

$$n!\,e^{-\pi i(n+1)/2}[\pi i(-1)^{n+1}\alpha_-^{-n-1}+\{\psi(n)-\tfrac{1}{2}\pi i\}\alpha^{-n-1}$$
$$-\alpha^{-n-1}\ln|\alpha|+(-1)^n\pi i\psi(n)\delta^{(n)}(\alpha)/n!]$$

and of $(-x)^n H(-x)\ln|x|$ is

$$n!\,e^{\pi i(n+1)/2}[(-1)^n\pi i\alpha_-^{-n-1}+\{\psi(n)+\tfrac{1}{2}\pi i\}\alpha^{-n-1}$$
$$-\alpha^{-n-1}\ln|\alpha|-(-1)^n\pi i\psi(n)\delta^{(n)}(\alpha)/n!].$$

Proof. According to Exercise 25(iii) and (16)

$$\operatorname*{Lim}_{\beta\to n}(\alpha\pm i0)^{-\beta-1}=(\alpha\pm i0)^{-n-1}=\alpha^{-n-1}\mp(-1)^n\pi i\delta^{(n)}(\alpha)/n!.$$

Also, from (23) and Definitions 4.7, 4.10,

$$\operatorname*{Lim}_{\beta\to n}(\alpha\pm i0)^{-\beta-1}\ln(\alpha\pm i0)$$
$$=\mathrm{Lim}_{\beta\to n}[\alpha_+^{-\beta-1}\ln\alpha+(-1)^n\delta^{(n)}(\alpha)/n!(n-\beta)^2+$$
$$+e^{\mp i\pi(\beta+1)}\{\alpha_-^{-\beta-1}\ln|\alpha|+\delta^{(n)}(\alpha)/n!(n-\beta)^2\}$$
$$\pm i\pi e^{\mp i\pi(\beta+1)}\{\alpha_-^{-\beta-1}-\delta^{(n)}(\alpha)/n!(n-\beta)\}$$
$$+\{(\pm i\pi(n-\beta)-1)e^{\mp i\pi(\beta+1)}-(-1)^n\}\delta^{(n)}(\alpha)/n!(n-\beta)^2]$$
$$=\alpha_+^{-n-1}\ln\alpha+(-1)^{n+1}\alpha_-^{-n-1}\ln|\alpha|\pm i\pi(-1)^{n+1}\alpha_-^{-n-1}$$
$$+\tfrac{1}{2}\pi^2(-1)^n\delta^{(n)}(\alpha)/n!$$
$$=(\alpha\pm i0)^{-n-1}\ln(\alpha\pm i0)$$

from (24). Allowing $\beta \to n$ in Theorem 4.6 we obtain the corollary. □

By writing

$$x^n \ln|x| = x^n \mathrm{H}(x) \ln x + (-1)^n (-x)^n \mathrm{H}(-x) \ln|x|$$

we derive

Corollary 4.6*b*. *If n is a non-negative integer the Fourier transform of* $x^n \ln|x|$ *is*

$$\pi e^{-\pi i n/2}\{(-1)^n 2\psi(n)\delta^{(n)}(\alpha) - n!\alpha^{-n-1}\operatorname{sgn}\alpha\}$$

and, if m is a positive integer, the Fourier transform of x^{-m} sgn x *is*

$$\frac{2e^{-\pi i(m-1)/2}}{(m-1)!}\alpha^{m-1}\{\psi(m-1) - \ln|\alpha|\}.$$

The second part of the corollary follows from the first by Fourier inversion.

Corollary 4.6*c*. *If m is a positive integer, the Fourier transform of* $x^{-m}\mathrm{H}(x)$ *is*

$$\frac{(-i)^{m-1}}{(m-1)!}\alpha^{m-1}\{\psi(m-1) - \ln|\alpha| - \tfrac{1}{2}\pi i \operatorname{sgn}\alpha\}$$

and of $(-x)^{-m}\mathrm{H}(-x)$ *is*

$$\frac{(i\alpha)^{m-1}}{(m-1)!}\{\psi(m-1) - \ln|\alpha| + \tfrac{1}{2}\pi i \operatorname{sgn}\alpha\}.$$

Proof. This follows at once from (13), Exercise 26 and Corollaries 4.3*b*, 4.6*b*. □

Similarly the Fourier transform of $|x|^\beta \ln|x|$ is

$$\beta!|\alpha|^{-\beta-1} 2\cos\tfrac{1}{2}\pi(\beta+1)\{\psi(\beta) - \ln|\alpha| - \tfrac{1}{2}\pi\tan\tfrac{1}{2}\pi(\beta+1)\} \tag{25}$$

and of $|x|^\beta \ln|x| \operatorname{sgn} x$ is

$$\beta!|\alpha|^{-\beta-1}(-2i)\sin\tfrac{1}{2}\pi(\beta+1)\{\psi(\beta) - \ln|\alpha| + \tfrac{1}{2}\pi\cot\tfrac{1}{2}\pi(\beta+1)\}\operatorname{sgn}\alpha. \tag{26}$$

From Definition 4.9 and Theorem 3.14(iv) the Fourier transform of $x^{-m}\ln|x|$ is obtained by a limiting process from (25) and (26). Now a derivative of the relation

$$z!(-z)! = \pi z/\sin \pi z \tag{27}$$

gives

$$\psi(z) - \psi(-z) = 1/z - \pi \cot \pi z. \tag{28}$$

Hence (25) can be written as

$$\frac{\pi|\alpha|^{-\beta-1}}{(-\beta-1)!\cos\frac{1}{2}\pi\beta}\{\psi(-\beta-1) - \pi\cot\pi\beta - \ln|\alpha| + \tfrac{1}{2}\pi\cot\tfrac{1}{2}\pi\beta\}. \tag{29}$$

When $\beta \to -m$ (m even) this tends to

$$\frac{\pi(-1)^{m/2}|\alpha|^{m-1}}{(m-1)!}\{\psi(m-1) - \ln|\alpha|\}. \tag{30}$$

Similarly (26) can be rewritten in the form

$$\frac{\pi \mathrm{i}|\alpha|^{-\beta-1}}{(-\beta-1)!\sin\frac{1}{2}\pi\beta}\{\psi(-\beta-1) - \pi\cot\pi\beta - \ln|\alpha| - \tfrac{1}{2}\tan\tfrac{1}{2}\pi\beta)\operatorname{sgn}\alpha \tag{31}$$

and as $\beta \to -m$ (m odd) this tends to

$$\frac{\pi \mathrm{i}|\alpha|^{m-1}(-1)^{(m/2)+1/2}}{(m-1)!}\{\psi(m-1) - \ln|\alpha|\}\operatorname{sgn}\alpha. \tag{32}$$

Combining (30) and (32) into one formula we have

Theorem 4.7. *The Fourier transform of* $x^{-m}\ln|x|$ *is*

$$\frac{(-1)^m\pi\alpha^{m-1}}{(m-1)!}\mathrm{e}^{m\pi\mathrm{i}/2}\{\psi(m-1) - \ln|\alpha|\}\operatorname{sgn}\alpha.$$

Finally there is

Theorem 4.8. *If m is a positive integer, the Fourier transform of* $x^{-m}\mathrm{H}(x)\ln x$ *is*

$$\frac{(-\mathrm{i}\alpha)^{m-1}}{(m-1)!2}\left[\tfrac{1}{3}\pi^2 - \psi'(m-1) + \{\psi(m-1) - \ln|\alpha| - \tfrac{1}{2}\pi\mathrm{i}\operatorname{sgn}\alpha\}^2\right].$$

Proof. By Definition 4.10 the Fourier transform is

$$\lim_{\mu\to 0}\left[\frac{\partial}{\partial\mu}(\mu-m)!\,e^{-\pi i(\mu-m+1)/2}(\alpha-i0)^{m-\mu-1}+\frac{(-1)^{m-1}(i\alpha)^{m-1}}{(m-1)!\,\mu^2}\right]$$

$$=\lim_{\mu\to 0}\frac{\partial}{\partial\mu}\frac{(-i)^{m-1}(\alpha-i0)^{m-1}}{(m-1)!\,\mu}\left[1+\mu\{\psi(m-1)-\tfrac{1}{2}\pi i-\ln(\alpha-i0)\}\right.$$
$$+\tfrac{1}{6}\mu^2\pi^2+\tfrac{1}{2}\mu^2\{\psi(m-1)-\tfrac{1}{2}\pi i-\ln(\alpha-i0)\}^2$$
$$\left.-\tfrac{1}{2}\mu^2\psi'(m-1)+O(\mu^3)\right]+\frac{(-1)^{m-1}(i\alpha)^{m-1}}{(m-1)!\,\mu^2}$$

which supplies the result stated, on simplification.□

Corollary 4.8. *If m is a positive integer, the Fourier transform of $x^{-m}\operatorname{sgn} x\ln|x|$ is*

$$\frac{(-i\alpha)^{m-1}}{(m-1)!}\left[\{\ln|\alpha|-\psi(m-1)\}^2-\tfrac{1}{6}\pi^2-\psi'(m-1)\right]$$

and of $x^{m-1}\{\ln|x|-\psi(m-1)\}^2$ is

$$(m-1)!2\pi(-i)^{m-1}\alpha^{-m}\operatorname{sgn}\alpha\ln|\alpha|+2\pi i^{m-1}\{\tfrac{1}{6}\pi^2+\psi'(m-1)\}\delta^{(m-1)}(\alpha).$$

Proof. The first part stems from Theorem 4.8 and (22). The second part then follows by Fourier inversion.□

Exercises

39. Prove that (i) $x.(x\pm i0)^\beta\ln(x\pm i0)=(x\pm i0)^{\beta+1}\ln(x\pm i0)$,
(ii) $\partial(x\pm i0)^\beta/\partial\beta=(x\pm i0)^\beta\ln(x\pm i0)$.

40. If β is not an integer, show that the Fourier transform of $(x+i0)^\beta$ is $2\pi e^{\pi i\beta/2}\alpha_+^{-\beta-1}/(-\beta-1)!$ and of $(x-i0)^\beta$ is $2\pi e^{-\pi i\beta/2}\alpha_-^{-\beta-1}/(-\beta-1)!$. Verify that these are still true if β is replaced by the negative integer $-m$.

41. Show that, when m is a positive integer,

$$x^{-m}\ln|x|=\lim_{\mu\to 0}\{x^{\mu-m}(x)\ln x+(-1)^m(-x)^{\mu-m}H(-x)\ln|x|\}.$$

42. Prove that $x^{-m}\ln|x|=x_+^{-m}\ln x+(-1)^m x_-^{-m}\ln|x|$.

43. Prove Corollary 4.6c directly from Definition 4.7 by means of a limiting operation.

44. If n is a non-negative integer, show that the Fourier transform of $x^n\operatorname{sgn} x\ln|x|$ is $n!2(i\alpha)^{-n-1}\{\psi(n)-\ln|\alpha|\}$.

45. Find the Fourier transform of
(i) $(2-x)^{-5/2}H(2-x)\ln(2-x)$, (ii) $(x-1)^{-2}\ln|x-1|$,
(iii) $(x-1)^{-2}\operatorname{sgn}(x-1)\ln|x-1|$, (iv) $|x-2|^{-3/2}\ln|x-2|$.

46. Prove that $x^\beta H(x)\ln x = \frac{1}{2}\{|x|^\beta \ln|x| + |x|^\beta \operatorname{sgn} x \ln|x|\}$.

47. Show that the Fourier transform of $(-x)^{-m}H(-x)\ln|x|$, where m is a positive integer, is

$$\frac{(\mathrm{i}\alpha)^{m-1}}{(m-1)!2}\left[\tfrac{1}{3}\pi^2 - \psi'(m-1) + \{\psi(m-1) - \ln|\alpha| + \tfrac{1}{2}\pi\mathrm{i}\operatorname{sgn}\alpha\}^2\right].$$

48. Prove that $\psi(v)$ can be defined by $v!\psi(v) = \int_0^\infty x^v e^{-x}\ln x\,dx$ for all complex v except the negative integers for which a limiting process is involved.

49. Further logarithms can be introduced by extra derivatives with respect to β, for example $|x|^\beta \ln^2|x| = \partial^2|x|^\beta/\partial\beta^2$. Find the Fourier transform of $x^\beta H(x)\ln^2 x$.
How would you define $x^{-m}H(x)\ln^2 x$?

4.7. Integration

It was explained in §4.4 how definite integrals involving generalised functions could be defined in appropriate circumstances. This section is concerned with another definition which has some advantages in certain respects.

Definition 4.11. *If $\{\gamma_m^+\}$ is a sequence of infinitely differentiable functions defined on $x \geq 0$ and good at $+\infty$ and if $\lim_{m\to\infty}\int_0^\infty \gamma_m^+(x)\gamma(x)dx$ exists for any odd (even) good function γ then there is a unique odd (even) generalised function g such that*

$$\tfrac{1}{2}\int_{-\infty}^{\infty} g(x)\gamma(x)dx = \lim_{m\to\infty}\int_0^\infty \gamma_m^+(x)\gamma(x)dx \tag{33}$$

and we define $\int_0^\infty g(x)\gamma(x)dx$ by

$$\int_0^\infty g(x)\gamma(x)dx = \lim_{m\to\infty}\int_0^\infty \gamma_m^+(x)\gamma(x)dx. \tag{34}$$

There are several ways of justifying the definition; perhaps the most instructive is that in which the regular sequence defining g is constructed. It will be sufficient to limit our attention to the case when γ is odd since the analysis when γ is even is similar. In view of the assumptions concerning γ_m^+ there is an $\varepsilon > 0$ such that γ_m^+ and its derivatives are continuous for $x > -\varepsilon$. Therefore, if $\eta(x)$ is an infinitely differentiable function which equals 1 for $x \geq 0$ and vanishes identically for $x \leq -\frac{1}{2}\varepsilon$, $\eta(x)\gamma_m^+(x)$ is a good function; this good function can be split into its even and odd parts and each of them is of the same type as γ_m^+. Consequently there is no loss of generality in restricting the discussion to the two cases in which γ_m^+ is odd and even respectively.

Suppose firstly that γ_m^+ is odd. Then γ_m^+ is good and

$$\lim_{m\to\infty} \int_{-\infty}^{\infty} \gamma_m^+(x)\gamma(x)\mathrm{d}x = \lim_{m\to\infty} 2\int_0^{\infty} \gamma_m^+(x)\gamma(x)\mathrm{d}x \tag{35}$$

which shows that the limit on the left-hand side exists. If γ had been even the left-hand side would have been zero since γ_m^+ is odd. Therefore the limit of the left-hand side exists for all good γ and so defines a generalised function. This generalised function obviously satisfies (33) on account of (35) and is odd because γ_m^+ is odd.

Let us therefore turn to the case when γ_m^+ is even. It would be tempting to use $\gamma_m^+(x)$ sgn x to define the generalised function, but this cannot be done directly because $\gamma_m^+(x)$ sgn x is not good in general. Consider, however, the sequence of which a typical member is

$$\gamma_m(x) = \int_{-\infty}^{\infty} \gamma_m^+(u)\,\mathrm{sgn}\,u\,\rho\{n_m(u-x)\}n_m \mathrm{e}^{-u^2/n_m^2}\,\mathrm{d}u$$

where ρ is the fine function defined in §3.2, and $\{n_m\}$ is an increasing sequence of positive integers such that

$$\lim_{m\to\infty} \frac{1}{n_m}\int_{-\infty}^{\infty} |\gamma_m^+(u)|\,\mathrm{d}u = 0. \tag{36}$$

(Such a sequence can be constructed by choosing n_m to be the first integer larger than n_{m-1} and $m\int_{-\infty}^{\infty}|\gamma_m^+(u)|\,\mathrm{d}u$.) The function γ_m so defined is good (the proof is the same as for Theorem 3.3) and it is odd because $\rho(y)$ is an even function of y. Therefore, remembering that γ is odd,

$$\lim_{m\to\infty} \int_{-\infty}^{\infty} \gamma_m(x)\gamma(x)\mathrm{d}x = \lim_{m\to\infty} 2\int_0^{\infty} \gamma_m(x)\gamma(x)\mathrm{d}x.$$

Now

$$\left| \int_{-\infty}^{\infty} \gamma_m(x)\gamma(x)\mathrm{d}x - 2\int_0^{\infty} \gamma_m^+(x)\gamma(x)\mathrm{d}x \right|$$

$$= \left| \int_{-\infty}^{\infty} \gamma(x) \int_{-\infty}^{\infty} \gamma_m^+(u)\,\mathrm{sgn}\,u\,\rho\{(n_m(u-x)\}\,n_m \mathrm{e}^{-u^2/n_m{}^2}\mathrm{d}u\,\mathrm{d}x \right.$$

$$\left. - 2\int_0^{\infty} \gamma_m^+(x)\gamma(x)\mathrm{d}x \right|$$

$$= \left| \int_{-1}^{1} \rho(y) \int_{-\infty}^{\infty} \gamma_m^+(u)\,\mathrm{sgn}\,u\,\mathrm{e}^{-u^2/n_m{}^2}\gamma(u + y/n_m)\mathrm{d}u\,\mathrm{d}y \right.$$

$$\left. - 2\int_0^{\infty} \gamma_m^+(x)\gamma(x)\mathrm{d}x \right|$$

as in Theorem 3.3. In the integral with respect to u convert the range $(-\infty, 0)$ to $(0, \infty)$ by changing u to $-u$; then the fact that ρ and γ are even and odd functions respectively of their arguments gives

$$2\left| \int_{-1}^{1} \rho(y) \int_0^{\infty} \gamma_m^+(u)\mathrm{e}^{-u^2/n_m{}^2}\gamma\left(u + \frac{y}{n_m}\right)\mathrm{d}u\,\mathrm{d}y - \int_0^{\infty} \gamma_m^+(x)\gamma(x)\mathrm{d}x \right|$$

$$= 2\left| \int_{-1}^{1} \rho(y)\left[\int_0^{\infty} \gamma_m^+(u)\mathrm{e}^{-u^2/n_m{}^2}\left\{\gamma\left(u + \frac{y}{n_m}\right) - \gamma(u)\right\}\mathrm{d}u \right.\right.$$

$$\left.\left. + \int_0^{\infty} \gamma_m^+(u)\gamma(u)(\mathrm{e}^{-u^2/n_m{}^2} - 1)\mathrm{d}u \right]\mathrm{d}y \right|$$

$$\leq \left(\frac{B_1}{n_m} + \frac{B_2}{n_m}\right)\int_0^{\infty} |\gamma_m^+(u)|\mathrm{d}u$$

since

$$|\gamma(u + y/n_m) - \gamma(u)| \leq Ky/n_m,$$

$\int_{-1}^{1} \rho(y) y\,\mathrm{d}y$ is finite, $|\mathrm{e}^{-u^2/n_m{}^2} - 1| \leq u^2/n_m^2$ and $u^2\gamma(u)$ is bounded. On account of (27) the last expression tends to zero as $m \to \infty$. Hence

$$\lim_{m\to\infty} \int_{-\infty}^{\infty} \gamma_m(x)\gamma(x)\mathrm{d}x = \lim_{m\to\infty} \int_0^{\infty} \gamma_m^+(x)\gamma(x)\mathrm{d}x$$

so that $\{\gamma_m\}$ is a regular sequence of odd good functions, which defines an odd generalised function g satisfying (33).

That g is unique follows from the fact that, if g_1 and g_2 were two possibilities,

$$\int_{-\infty}^{\infty} \{g_1(x) - g_2(x)\}\gamma(x)\mathrm{d}x = 0$$

for every odd good γ. Therefore $g_1 - g_2$ must be even (Exercise 12) and so, since g_1 and g_2 are both odd, $g_1 = g_2$.

It is necessary to confirm that Definition 4.11 is not in conflict with Definition 4.6. Suppose that $\int_0^\infty g(x)\gamma(x)\mathrm{d}x$, where γ is odd, exists according to Definition 4.6. Then there is a generalised function $g(x)\mathrm{H}(x)$ such that $\int_{-\infty}^\infty g(x)\mathrm{H}(x)\gamma(x)\mathrm{d}x$ exists and $\int_0^\infty g(x)\gamma(x)\mathrm{d}x$ is defined by

$$\int_0^\infty g(x)\gamma(x)\mathrm{d}x = \int_{-\infty}^{\infty} g(x)\mathrm{H}(x)\gamma(x)\mathrm{d}x$$

$$= \tfrac{1}{2}\int_{-\infty}^{\infty} \{g(x)\mathrm{H}(x) - g(-x)\mathrm{H}(-x)\}\gamma(x)\mathrm{d}x$$

since γ is odd. If therefore $\{\gamma_m^\circ\}$ is a regular sequence of odd good functions defining the odd generalised function

$$g(x)\mathrm{H}(x) - g(-x)\mathrm{H}(-x)$$

the right-hand side may be written $\lim_{m\to\infty} \int_0^\infty \gamma_m^\circ(x)\gamma(x)\mathrm{d}x$. This is consistent with (34) and we see, therefore, that a semi-infinite integral obtained from Definition 4.6 will have the same value when derived from Definition 4.11. There may, however, be semi-infinite integrals which are covered by Definition 4.11 which do not come within the scope of Definition 4.6.

Example 1. If $\gamma_m^+(x) = (m/\pi)^{1/2}\mathrm{e}^{-mx^2}$,

$$\lim_{m\to\infty}\int_0^\infty \gamma_m^+(x)\gamma(x)\mathrm{d}x = \lim_{m\to\infty}\int_0^\infty \left(\frac{m}{\pi}\right)^{1/2}\mathrm{e}^{-mx^2}\{\gamma(x)-\gamma(0)\}\,\mathrm{d}x$$

$$+\tfrac{1}{2}\gamma(0) = \tfrac{1}{2}\gamma(0)$$

since

$$\left|\int_0^\infty \left(\frac{m}{\pi}\right)^{1/2}\mathrm{e}^{-mx^2}\{\gamma(x)-\gamma(0)\}\,\mathrm{d}x\right| \le \frac{B}{m^{1/2}}\int_0^\infty \mathrm{e}^{-x^2}x\,\mathrm{d}x \to 0.$$

If γ is odd the corresponding g of Definition 4.11 is 0. If γ is even the corresponding even g is $\delta(x)$. Thus (34) gives

$$\int_0^\infty \delta(x)\gamma(x)\,dx = \tfrac{1}{2}\gamma(0). \tag{37}$$

The sequence $\{\gamma_m^+\}$ can be regarded as defining the generalised function $g(x)$ on $x \geq 0$. In fact, if

$$\lim_{m\to\infty} \int_0^\infty \gamma_{1m}^+ \gamma(x)\,dx = \lim_{m\to\infty} \int_0^\infty \gamma_{2m}^+(x)\gamma(x)\,dx$$

for every odd (even) γ then we write

$$g_1(x) = g_2(x) \qquad (x \geq 0)$$

the equality being used to indicate that we have not restricted the good functions to those which, together with their derivatives, all vanish on $x \leq 0$.

One consequence of Definition 4.11 is the convention that the meaning to be assigned to $\int_0^\infty g(x)\gamma(x)\,dx$ is given by

$$\int_0^\infty g(x)\gamma(x)\,dx = \tfrac{1}{2}\int_{-\infty}^\infty g(x)\gamma(x)\,dx \tag{38}$$

when g and γ are either both even or both odd. Suppose now that g and γ' are both even (odd) and that $\int_0^\infty g(x)\gamma(x)\,dx$ exists. Then

$$\begin{aligned}\int_0^\infty g(x)\gamma'(x)\,dx &= \tfrac{1}{2}\int_{-\infty}^\infty g(x)\gamma'(x)\,dx \\ &= -\tfrac{1}{2}\int_{-\infty}^\infty g'(x)\gamma(x)\,dx \\ &= -\int_0^\infty g'(x)\gamma(x)\,dx \end{aligned} \tag{39}$$

from (38) since g' and γ are both odd (even).

It is absolutely vital when employing (39) not to forget the intermediate steps. For example, let g be equal in $x \geq 0$ to an ordinary continuous function $f(x)$ which possesses a derivative df/dx which is an ordinary function. Then, (39) would seem to be incorrect if $f(0) \neq 0$. However, $g(x) = f(x)$ or $g(x) = f(x)\operatorname{sgn} x$ according as g is even or odd if f is even and so $g' = df/dx$ when g is even and, in the other case, $g' = (\mathrm{d}f/\mathrm{d}x)\,\mathrm{sgn}\,x + 2f(x)\delta(x)$. Therefore (39) becomes

$$\int_0^\infty f(x)\gamma'(x)\mathrm{d}x = -\int_0^\infty (\mathrm{d}f/\mathrm{d}x)\gamma(x)\mathrm{d}x - f(0)\gamma(0)$$

from (37). This is the correct formula for integration by parts.

There is another point to be remarked. Suppose that

$$\lim_{m\to\infty}\int_0^\infty \gamma_m^+(x)\gamma'(x)\mathrm{d}x$$

exists. Then if γ is even (odd) there is an odd (even) generalised function $g(x)$ such that

$$\lim_{m\to\infty}\int_0^\infty \gamma_m^+(x)\gamma'(x)\mathrm{d}x = \int_0^\infty g(x)\gamma'(x)\mathrm{d}x.$$

Now, by integration by parts,

$$-\lim_{m\to\infty}\int_0^\infty \gamma_m^+(x)\gamma'(x)\mathrm{d}x = \lim_{m\to\infty}\left\{\gamma_m^+(0)\gamma(0) + \int_0^\infty \gamma_m^{+\prime}(x)\gamma(x)\mathrm{d}x\right\}. \tag{40}$$

If either γ or γ_m^+ is odd or $\lim_{m\to\infty}\gamma_m^+(0) = 0$ this shows that

$$\lim_{m\to\infty}\int_0^\infty \gamma_m^{+\prime}(x)\gamma(x)\mathrm{d}x$$

exists and there is an even (odd) generalised function g_1 when γ is even (odd) such that

$$\lim_{m\to\infty}\int_0^\infty \gamma_m^{+\prime}(x)\gamma(x)\mathrm{d}x = \int_0^\infty g_1(x)\gamma(x)\mathrm{d}x. \tag{41}$$

It follows from (39) that

$$\int_{-\infty}^\infty \{g_1(x) - g'(x)\}\gamma(x)\mathrm{d}x = 0.$$

Therefore $g_1 - g'$ is odd (even) when γ is even (odd) and since g_1, g' are both even (odd) the only possibility is $g_1 = g'$. If γ and γ_m^+ are both even and $\gamma_m^+(0)$ does not tend to zero as $m \to \infty$ this assertion cannot be made; in fact we cannot even be sure that $\lim_{m\to\infty}\int_0^\infty \gamma_m^{+\prime}(x)\gamma(x)\mathrm{d}x$ exists. Thus, if $\{\gamma_m^+\}$ determines g, $\{\gamma_m^{+\prime}\}$ may or may not determine g'.

If we know that $\lim_{m\to\infty}\int_0^\infty \gamma_m^+(x)\gamma'(x)dx$ and $\lim_{m\to\infty}\int_0^\infty \gamma_m^{+\prime}(x)\gamma(x)dx$ both exist then (40) shows that, when $\gamma(0) \neq 0$, $\lim_{m\to\infty}\gamma_m^+(0)$ exists; let its value be $\frac{1}{2}C$. Then, since (41) is still valid, (40) can be written

$$\tfrac{1}{2}\int_{-\infty}^{\infty}\{g'(x)-g_1(x)\}\gamma(x)dx = \tfrac{1}{2}C\gamma(0);$$

because γ must be even we deduce that $g' - g_1 - C\delta(x)$ must be odd. But g', g_1, δ are all even; therefore

$$g'(x) = g_1(x) + C\delta(x) \tag{42}$$

which demonstrates the relation between g' and the generalised function defined by $\{\gamma_m^{+\prime}\}$ when we know that the latter exists.

Example 2. If we apply (38) to (37) we obtain

$$\int_0^\infty \delta'(x)\gamma(x)dx = -\tfrac{1}{2}\gamma'(0). \tag{43}$$

Clearly, there is no new difficulty in defining $\int_{-\infty}^0$ and a simple change of origin gives $\int_a^\infty$. We have

$$\int_0^\infty g(x)\gamma(x)dx = \int_{-\infty}^0 g(-x)\gamma(-x)dx,$$

$$\int_0^\infty g(x)\gamma(x)dx = \int_a^\infty g(x-a)\gamma(x-a)dx$$

whenever one side exists.

Exercises

50. Prove that $\int_0^\infty x^{-1}\gamma(x)dx = -\int_0^\infty \gamma'(x)\ln x\,dx$ when γ is odd. What is the corresponding result when γ is even? Show that the same equalities are obtained when the left-hand side is regarded as $\int_{-\infty}^\infty x^{-1}\mathrm{H}(x)\gamma(x)dx$.

51. Show that $\gamma_m^+(x) = (m/\pi)^{1/2}e^{-m(x-a)^2}$ $(a>0)$ fulfils the conditions of Definition 4.11 and implies that

$$\int_0^\infty \{\delta(x-a)-\delta(x+a)\}\gamma(x)dx = \gamma(a)$$

when γ is odd. Find the corresponding result when γ is even and deduce that

$$\int_0^\infty \delta(x-a)\gamma(x)dx = \gamma(a), \qquad \int_0^\infty \delta^{(n)}(x-a)\gamma(x)dx = (-1)^n\gamma^{(n)}(a)$$

for any good γ.

5

Series

5.1. General properties

Series of generalised functions are defined in a similar way to that for conventional functions (§ 1.10) but have some different properties because the limit employed is that of Definition 3.11.

Definition 5.1. *The series* $\sum_{n=1}^{\infty} g_n(x)$ *is said to be generally convergent if, and only if,* $\mathrm{Lim}_{m\to\infty} \sum_{n=1}^{m} g_n(x)$ *exists as a generalised function, and we write* $\sum_{n=1}^{\infty} g_n(x) = \mathrm{Lim}_{m\to\infty} \sum_{n=1}^{m} g_n(x)$. *The series* $\sum_{n=-\infty}^{\infty} g_n(x)$ *is said to be generally convergent if, and only if,* $\mathrm{Lim}_{m\to\infty} \sum_{n=-m}^{m} g_n(x)$ *exists as a generalised function, and we write* $\sum_{n=-\infty}^{\infty} g_n(x) = \mathrm{Lim}_{m\to\infty} \sum_{n=-m}^{m} g_n(x)$.

In view of the similarity of definition the convergence properties of $\sum_{n=-\infty}^{\infty} g_n(x)$ are usually the same as those of $\sum_{n=1}^{\infty} g_n(x)$ and so, to avoid duplication, proofs will be given for the series $\sum_{n=1}^{\infty} g_n(x)$; the reader will easily supply any necessary modifications to cover the other type of series.

Series of generalised functions have remarkable properties which are very convenient for many purposes. For example,

Theorem 5.1. *If* $\sum_{n=1}^{\infty} g_n(x)$ *is generally convergent then*

(i) $\{\sum_{n=1}^{\infty} g_n(x)\}' = \sum_{n=1}^{\infty} g_n'(x)$,

(ii) $\sum_{n=1}^{\infty} G_n(\alpha)$ *is generally convergent and is the Fourier transform of* $\sum_{n=1}^{\infty} g_n(x)$,

(iii) *there is a generally convergent series* $\sum_{n=1}^{\infty} h_n(x)$ *such that*

$$\left\{\sum_{n=1}^{\infty} h_n(x)\right\}' = \sum_{n=1}^{\infty} g_n(x)$$

and $h_n'(x) = g_n(x)$.

Proof. According to Theorem 3.14(i)

$$\operatorname*{Lim}_{m\to\infty}\sum_{n=1}^{m} g_n'(x) = \left\{\operatorname*{Lim}_{m\to\infty}\sum_{n=1}^{m} g_n(x)\right\}'$$

and this is the statement contained in (i).

Similarly (ii) follows from Theorem 3.14(iv).

With regard to (iii) it was remarked in § 3.6 that to a generalised function $g_n(x)$ corresponds a generalised function $g_{n0}(x)$ such that $g_{n0}' = g_n$. However, it is not obvious that $\sum_{n=1}^{\infty} g_{n0}$ is generally convergent. To make sure that we have a generally convergent series define $h_n(x) = g_{n0}(x) + a_n$ where a_n is a constant such that

$$\int_{-\infty}^{\infty} \{g_{n0}(x) + a_n\}\, e^{-x^2}\, dx = 0. \tag{1}$$

Now, any good function γ can be written as

$$\frac{e^{-x^2}}{\pi^{1/2}}\int_{-\infty}^{\infty}\gamma(t)\,dt + \frac{d}{dx}\int_{-\infty}^{x}\left\{\gamma(t) - \frac{e^{-t^2}}{\pi^{1/2}}\int_{-\infty}^{\infty}\gamma(u)\,du\right\}dt.$$

The second term is the derivative of a good function since the integral tends to zero faster than any inverse power of x as $|x| \to \infty$. Hence

$$\int_{-\infty}^{\infty}\sum_{n=1}^{m} h_n(x)\gamma(x)\,dx$$

$$= \sum_{n=1}^{m}\int_{-\infty}^{\infty} h_n(x)\frac{e^{-x^2}}{\pi^{1/2}}dx\int_{-\infty}^{\infty}\gamma(t)\,dt$$

$$+ \sum_{n=1}^{m}\int_{-\infty}^{\infty} h_n(x)\frac{d}{dx}\int_{-\infty}^{x}\left\{\gamma(t) - \frac{e^{-t^2}}{\pi^{1/2}}\int_{-\infty}^{\infty}\gamma(u)\,du\right\}dt\,dx$$

$$= -\sum_{n=1}^{m}\int_{-\infty}^{\infty} g_n(x)\int_{-\infty}^{x}\left\{\gamma(t) - \frac{e^{-t^2}}{\pi^{1/2}}\int_{-\infty}^{\infty}\gamma(u)\,du\right\}dt\,dx$$

on account of (1) and $h_n' = g_n$. As $m \to \infty$ the right-hand side tends to a limit because $\sum_{n=1}^{\infty} g_n$ is generally convergent. Hence $\sum_{n=1}^{\infty} h_n$ is generally convergent and (iii) follows from (i). The proof is complete. □

This remarkable theorem shows that the derivative of a generally convergent series is obtained by taking derivatives term by term and that its Fourier transform is also supplied by taking Fourier transforms term by term. Part (iii) states, roughly speaking, that the integral of a generally convergent series is the sum of the integrals of the separate terms. However, the integrals are indefinite and not definite. In fact, h_n is not unique because $h_n(x)$ could be replaced by $h_n(x)+b_n$ without affecting (iii) provided that the series of constants $\sum_{n=1}^{\infty} b_n$ is convergent. This is supported by

Theorem 5.2. *The series of constants* $\sum_{n=1}^{\infty} b_n$ *is generally convergent or not according as it is convergent or not.*

Proof. The theorem follows at once from

$$\sum_{n=1}^{m} \int_{-\infty}^{\infty} b_n \gamma(x)\mathrm{d}x = \left(\sum_{n=1}^{m} b_n\right)\int_{-\infty}^{\infty} \gamma(x)\mathrm{d}x. \square$$

This theorem shows that general convergence is no better and no worse than ordinary convergence at summing series of constants. When we turn to quantities which are not constant the situation is different. Firstly we should note that there is no notion corresponding to pointwise convergence because the value of a generalised function is not defined at a point. (An exception to this is when a generalised function equals a conventional function in an interval; if this were true of the sum and all generalised functions in a generally convergent series we might think of it as converging pointwise at every interior point of the interval.) Secondly the convergence criterion is

Theorem 5.3. *A necessary and sufficient condition for* $\sum_{n=1}^{\infty} g_n(x)$ *to be generally convergent is that* $\mathrm{Lim}_{m\to\infty} \sum_{r=1}^{p} g_{m+r}(x) = 0$ *for every* $p \geq 1$.†

Proof. If $\sum_{n=1}^{\infty} g_n(x)$ is generally convergent, the series

$$\sum_{n=1}^{\infty} \int_{-\infty}^{\infty} g_n(x)\gamma(x)\mathrm{d}x$$

† An alternative statement of the theorem is: if $\{g_m\}$ is a sequence of generalised functions $\mathrm{Lim}_{m\to\infty} g_m = g$ if, and only if, the sequence of numbers $\{\int_{-\infty}^{\infty} g_m(x)\gamma(x)\mathrm{d}x\}$ converges, i.e. the space of generalised functions is complete under the generalised limit.

is convergent and hence, by Theorem 1.20,

$$\lim_{m\to\infty}\sum_{r=1}^{p}\int_{-\infty}^{\infty} g_{m+r}(x)\gamma(x)\,dx = 0$$

for every $p \geq 1$ and every good γ. Hence $\operatorname{Lim}_{m\to\infty}\sum_{r=1}^{p} g_{m+r}(x) = 0$.
Conversely, if $\operatorname{Lim}_{m\to\infty}\sum_{r=1}^{p} g_{m+r}(x) = 0$, Theorem 1.20 shows that

$$\lim_{m\to\infty}\sum_{n=1}^{m}\int_{-\infty}^{\infty} g_n(x)\gamma(x)\,dx$$

exists for every γ and has a finite value $F(\gamma)$. To show that $F(\gamma)$ determines a generalised function introduce the regular sequence $\{\gamma_{mn}\}$ defining the generalised function $\sum_{s=1}^{m} g_s(x)$. Then, by Corollary 3.20*b*, given $\varepsilon\,(0 < \varepsilon < 1)$, there is a γ_{mn_m} independent of γ such that

$$\left|\int_{-\infty}^{\infty}\left\{\gamma_{mn_m}(x) - \sum_{s=1}^{m} g_s(x)\right\}\gamma(x)\,dx\right| \leq \varepsilon^m \max\left|(1+x^2)^{k_m/2}\gamma^{(r_m)}(x)\right|$$

where k_m and r_m are bounded as $m \to \infty$ on account of Theorem 3.17. The sequence $\{\gamma_{mn_m}\}$ is regular because

$$\begin{aligned}\lim_{m\to\infty}\int_{-\infty}^{\infty}\gamma_{mn_m}(x)\gamma(x)\,dx &= \lim_{m\to\infty}\int_{-\infty}^{\infty}\left\{\gamma_{mn_m}(x) - \sum_{s=1}^{m} g_s(x)\right\}\gamma(x)\,dx \\ &\quad + \lim_{m\to\infty}\int_{-\infty}^{\infty}\sum_{s=1}^{m} g_s(x)\gamma(x)\,dx \\ &= F(\gamma).\end{aligned}$$

Thus $\{\gamma_{mn_m}\}$ defines a generalised function which takes the same values as $F(\gamma)$, i.e. $\sum_{s=1}^{\infty} g_s(x)$ is a generalised function. The proof is complete. □

One consequence of Theorem 5.3 is that, for a series to be generally convergent, it is necessary that $\operatorname{Lim}_{m\to\infty} g_m(x) = 0$. This is not so restrictive as it appears at first sight because the limit is in the sense of generalised functions. For example, $\operatorname{Lim}_{m\to\infty} \cos mx = 0$.

The connection between uniform convergence and general convergence is given by

Theorem 5.4. *If $g_n \in K_1$ for all n and, for some $N \geq 0$, $\sum_{n=1}^{\infty} g_n/(1+x^2)^N$ converges uniformly on R_1 then $\sum_{n=1}^{\infty} g_n$ is generally convergent.*

Proof. Because of the uniform convergence there is an n_0, independent of x, such that $|\sum_{r=1}^{p} g_{n+r}(x)| < \varepsilon(1+x^2)^N$ for $n > n_0, p \geq 1$. Hence (Theorem 3.3)

$$\left|\int_{-\infty}^{\infty} \sum_{r=1}^{p} g_{n+r}(x)\gamma(x)\,dx\right| \leq \varepsilon \int_{-\infty}^{\infty} |\gamma(x)|(1+x^2)^N\,dx$$

and the theorem follows from Theorem 5.3.□

In view of Theorems 5.2 and 5.4 the word 'generally' could be dropped from the description of the convergence of generalised functions. However in the interests of clarity it will be retained.

Exercises

1. Examine the general convergence of the following series.

(i) $\sum_{n=1}^{\infty} \frac{n^a}{n^b+1}$, (ii) $\sum_{n=1}^{\infty} \frac{(-1)^n}{(n+1)^{1/2}}$,

(iii) $\sum_{n=2}^{\infty} \frac{(-1)^n}{(-1)^n+n^{1/2}}$, (iv) $\sum_{n=1}^{\infty} \frac{\cos nx}{n^2}$,

(v) $\sum_{n=1}^{\infty} \frac{x \sin nx}{n}$, (vi) $\sum_{n=1}^{\infty} (x^2+n^2)^{-1}$,

(vii) $\sum_{n=0}^{\infty} g_n$ where $g_n = xe^{-nx}$ $(x \geq 0)$, $g_n = 0$ $(x > 0)$,

(viii) $\sum_{n=1}^{\infty} g_n$ where $g_n = 0$ $(x < 0)$, $g_n = \frac{x^{n+1}}{n^2(x^n+1)}$ $(x \geq 0)$,

(ix) $\sum_{n=1}^{\infty} \psi(nx)$, ψ being fairly good, (x) $\sum_{n=1}^{\infty} x^2 e^{inx}$.

2. If $g_{1m} \geq 0$, if $-g_{1m} \leq g_m \leq g_{1m}$ and $\sum_{m=1}^{\infty} g_{1m}$ is generally convergent prove that $\sum_{m=1}^{\infty} g_m$ is generally convergent. (Hint: Exercise 34 of Chapter 3.)

5.2. Hermite polynomials

In series of generalised functions the Hermite polynomials play a special role. To demonstrate this we shall derive firstly some of the properties of these polynomials.

The *Hermite polynomial* $H_n(x)$ is defined by

$$H_n(x) = (-1)^n e^{x^2} \frac{d^n}{dx^n} e^{-x^2}. \tag{2}$$

Since e^{-t^2+2tx}, regarded as a function of t, has no singularities in the complex t-plane it possesses a Taylor series which converges absolutely and uniformly for any finite t and x. Hence

$$e^{-t^2+2tx} = e^{x^2} \sum_{n=0}^{\infty} \frac{t^n}{n!} \left[\frac{d^n}{dt^n} e^{-(t-x)^2} \right]_{t=0}$$

$$= \sum_{n=0}^{\infty} \frac{t^n}{n!} H_n(x) \tag{3}$$

for any finite t and x. By putting $x = 0$ we see that

$$H_n(0) = 0 \ (n \text{ odd}), \qquad H_n(0) = \frac{n!(-1)^{n/2}}{(\frac{1}{2}n)!} \ (n \text{ even}). \tag{4}$$

There are various relations connecting the polynomials of different orders. Thus, from (2),

$$H_n'(x) = (-1)^n 2x e^{x^2} \frac{d^n}{dx^n} e^{-x^2} + (-1)^n e^{x^2} \frac{d^{n+1}}{dx^{n+1}} e^{-x^2}$$

$$= 2x H_n(x) - H_{n+1}(x). \tag{5}$$

On account of the uniform convergence of (3) and its derivatives

$$\sum_{n=0}^{\infty} \frac{t^n}{n!} H_n'(x) = 2t e^{-t^2+2tx} = 2t \sum_{n=0}^{\infty} \frac{t^n}{n!} H_n(x)$$

whence

$$H_n'(x) = 2n H_{n-1}(x). \tag{6}$$

Eliminating H_n' from (5) and (6) we obtain

$$H_{n+1}(x) - 2x H_n(x) + 2n H_{n-1}(x) = 0. \tag{7}$$

Further, by taking a derivative of (5) and then using (6) for H_{n+1}' we have

$$H_n''(x) - 2x H_n'(x) + 2n H_n(x) = 0 \tag{8}$$

which is the differential equation satisfied by H_n.

Also

$$\int_{-\infty}^{\infty} H_m(x)H_n(x)e^{-x^2}dx = \int_{-\infty}^{\infty} H_n(x)(-1)^m \frac{d^m}{dx^m}e^{-x^2}dx$$

$$= \left[(-1)^m H_n(x)\frac{d^{m-1}}{dx^{m-1}}e^{-x^2}\right]_{-\infty}^{\infty}$$

$$-\int_{-\infty}^{\infty} (-1)^m H_n'(x)\frac{d^{m-1}}{dx^{m-1}}e^{-x^2}dx$$

by integration by parts. Since H_n is a polynomial the first term vanishes and so

$$\int_{-\infty}^{\infty} H_m(x)H_n(x)e^{-x^2}dx = (-1)^{m+1}\int_{-\infty}^{\infty} H_n'(x)\frac{d^{m-1}}{dx^{m-1}}e^{-x^2}dx.$$

Repeating the process $m-1$ times we find

$$\int_{-\infty}^{\infty} H_m(x)H_n(x)e^{-x^2}dx = \int_{-\infty}^{\infty} e^{-x^2}H_n^{(m)}(x)dx. \tag{9}$$

For $m > n$, $H_n^{(m)}(x)$ is identically zero because H_n is a polynomial of degree n. Therefore the right-hand side of (9) is zero when $m > n$. Similarly, by interchanging m and n throughout the analysis, we see that it is zero when $m < n$. The remaining case is $m = n$ when, since the coefficient of x^n in H_n is 2^n,

$$\int_{-\infty}^{\infty} \{H_n(x)\}^2 e^{-x^2}dx = n!\,2^n\int_{-\infty}^{\infty} e^{-x^2}dx = n!\,2^n\pi^{1/2}. \tag{10}$$

These results are combined in the formula

$$\int_{-\infty}^{\infty} H_m(x)H_n(x)e^{-x^2}dx = n!\,2^n\pi^{1/2}\delta_{nm} \tag{11}$$

where $\delta_{nm} = 0\ (m \neq n)$, $\delta_{nm} = 1\ (m = n)$.

On account of (11) an orthonormal set $\{\mathscr{H}_n\}$ can be defined by putting

$$\mathscr{H}_n(x) = e^{-x^2/2}H_n(x)/\{n!\,2^n\pi^{1/2}\}^{1/2}. \tag{12}$$

For then (11) becomes

$$\int_{-\infty}^{\infty} \mathscr{H}_m(x)\mathscr{H}_n(x)dx = \delta_{nm}. \tag{13}$$

The properties of $\mathscr{H}_n$ can be deduced directly from (12) and (2) or from the corresponding ones for H_n. Thus

$$\mathscr{H}_n'(x) - x\mathscr{H}_n(x) = -\{2(n+1)\}^{1/2}\mathscr{H}_{n+1}(x), \tag{14}$$

$$\mathscr{H}_n'(x) + x\mathscr{H}_n(x) = (2n)^{1/2}\mathscr{H}_{n-1}(x), \tag{15}$$

$$\mathscr{H}_n''(x) + (2n+1-x^2)\mathscr{H}_n(x) = 0. \tag{16}$$

It is clear that $\mathscr{H}_n$, being the product of a polynomial and $e^{-x^2/2}$, is a good function. Conversely, any good function can be expressed in terms of the $\mathscr{H}_n$ as is demonstrated by

Theorem 5.5. *If γ is any good function then $\gamma(x) = \sum_{n=0}^{\infty} a_n \mathscr{H}_n(x)$ where $a_n = \int_{-\infty}^{\infty} \gamma(x)\mathscr{H}_n(x)\,dx$.*

Proof. The function $e^{-x^2/2}\gamma(x)$ is good, since γ is, and has a Fourier transform

$$\int_{-\infty}^{\infty} e^{-x^2/2}\gamma(x)e^{-i\alpha x}\,dx$$

$$= e^{-\alpha^2/4}\int_{-\infty}^{\infty}\gamma(x)\sum_{n=0}^{\infty}\left(\frac{\pi^{1/2}}{n!2^n}\right)^{1/2}(-i\alpha)^n\mathscr{H}_n(x)\,dx \tag{17}$$

from (12) and (3) with $t = -\frac{1}{2}i\alpha$. Now, by Theorem 1.27 and (13),

$$\int_{-\infty}^{\infty}|\gamma(x)\mathscr{H}_n(x)|\,dx \le \left\{\int_{-\infty}^{\infty}|\gamma(x)|^2\,dx\right\}^{1/2}$$

so that $\sum_{n=0}^{\infty}(\pi^{1/2}/n!2^n)^{1/2}\alpha^n\int_{-\infty}^{\infty}|\gamma(x)\mathscr{H}_n(x)|\,dx$ converges like the series $\sum_{n=0}^{\infty}\{\alpha^n/(n!2^n)^{1/2}\}$, i.e. for all finite α. It follows from Theorem 1.36 that the integration and summation in (17) can be interchanged. Consequently the Fourier transform of $e^{-x^2/2}\gamma(x)$ is $e^{-\alpha^2/4}\sum_{n=0}^{\infty}(\pi^{1/2}/n!2^n)^{1/2}(-i\alpha)^n a_n$. Hence, by Theorem 2.5(i),

$$e^{-x^2/2}\gamma(x) = \frac{1}{2\pi}\int_{-\infty}^{\infty}e^{-\alpha^2/4}\sum_{n=0}^{\infty}\left(\frac{\pi^{1/2}}{n!2^n}\right)^{1/2}(-i\alpha)^n a_n e^{i\alpha x}\,d\alpha. \tag{18}$$

Now, by Theorem 1.33,

$$\int_{-\infty}^{\infty}(-i\alpha)^n e^{i\alpha x-\alpha^2/4}\,d\alpha = (-1)^n\frac{d^n}{dx^n}\int_{-\infty}^{\infty}e^{i\alpha x-\alpha^2/4}\,d\alpha$$

$$= (-1)^n\frac{d^n}{dx^n}2\pi^{1/2}e^{-x^2}$$

$$= 2\pi^{1/2}e^{-x^2/2}\{n!2^n\pi^{1/2}\}^{1/2}\mathscr{H}_n(x) \tag{19}$$

by Theorem 2.2. Thus $\gamma(x)=\sum_{n=0}^{\infty}a_n\mathscr{H}_n(x)$ and the theorem is proved provided that the integration and summation in (18) can be interchanged.

That this is legitimate can be shown in the following way. By (15)

$$a_n=\frac{1}{\{2(n+1)\}^{1/2}}\int_{-\infty}^{\infty}\gamma(x)\{\mathscr{H}'_{n+1}(x)+x\mathscr{H}_{n+1}(x)\}\,dx$$

$$=\frac{-1}{\{2(n+1)\}^{1/2}}\int_{-\infty}^{\infty}\mathscr{H}_{n+1}(x)\{\gamma'(x)-x\gamma(x)\}\,dx \tag{20}$$

by integration by parts. Continuing the process we obtain

$$a_n=\frac{(-1)^r(n!)^{1/2}}{\{(n+r)!2^r\}^{1/2}}\int_{-\infty}^{\infty}\mathscr{H}_{n+r}(x)D^r_-\gamma\,dx$$

where $D_-\equiv(d/dx)-x$. The Schwarz inequality (Theorem 1.27) gives

$$\left|\int_{-\infty}^{\infty}\mathscr{H}_{n+r}(x)D^r_-\gamma\,dx\right|^2\le\int_{-\infty}^{\infty}|D^r_-\gamma|^2\,dx$$

which is independent of n so that $a_n=O(n^{-r})$ as $n\to\infty$, whatever the positive integer r. Also

$$\int_0^{\infty}\alpha^n e^{-\alpha^2/4}\,d\alpha=(\tfrac{1}{2}n-\tfrac{1}{2})!\,2^n \text{ so that } \sum_{n=0}^{\infty}\frac{|a_n|}{(n!2^n)^{1/2}}\int_{-\infty}^{\infty}\alpha^n e^{-\alpha^2/4}\,d\alpha$$

converges, on choosing r large enough. The application of Theorem 1.36 completes the proof. □

Notice that in the course of the proof we have shown

Corollary 5.5. *For every integer $r\ge 0$, $a_n=O(n^{-r})$ as $n\to\infty$.*

Theorem 5.6. *If $\gamma(x)=\sum_{n=0}^{\infty}\alpha_n\mathscr{H}_n(x)$ then $\gamma'(x)=\sum_{n=0}^{\infty}a_n\mathscr{H}'_n(x)$.*

Proof. Since $\gamma'(x)-x\gamma(x)$ is good, Theorem 5.5 gives

$$\gamma'(x)-x\gamma(x)=\sum_{n=0}^{\infty}b_n\mathscr{H}_n(x)$$

where

$$b_n=\int_{-\infty}^{\infty}\{\gamma'(x)-x\gamma(x)\}\,\mathscr{H}_n(x)\,dx=-(2n)^{1/2}a_{n-1}\quad(b_0=0)$$

by (20). Hence

$$\gamma'(x) = \sum_{n=0}^{\infty} a_n\{x\mathscr{H}_n(x) - (2n+2)^{1/2}\mathscr{H}_{n+1}(x)\}$$

$$= \sum_{n=0}^{\infty} a_n \mathscr{H}'_n(x)$$

from (14) and the proof is complete. □

Exercises

3. If, for every integer $r \geq 0$, $a_n = O(n^{-r})$ as $n \to \infty$ prove that $\sum_{n=0}^{\infty} a_n \mathscr{H}_n(x)$ is a good function.
4. Prove that the Fourier transform of $\mathscr{H}_n(x)$ is $(-\mathrm{i})^n(2\pi)^{1/2}\mathscr{H}_n(\alpha)$.
5. Prove that $\sum_{m=0}^{n} \frac{n!}{m!(n-m)!} H_m(2^{1/2}x)H_{n-m}(2^{1/2}y) = 2^{n/2}H_n(x+y)$.

$$\left(\text{Hint:} \left(\frac{\partial}{\partial x} + \frac{\partial}{\partial y}\right)^n g(x,y) = \frac{1}{f(x-y)}\left(\frac{\partial}{\partial x} + \frac{\partial}{\partial y}\right)^n \{g(x,y)f(x-y)\}.\right)$$

6. Can you prove Theorem 5.6 by means of Theorem 5.1 (i)?

5.3. The expansion of a generalised function in Hermite polynomials

In this section we consider the possibility of representing any generalised function in terms of Hermite polynomials. Several interesting results will come out of this investigation. The first is

Theorem 5.7. $\int_{-\infty}^{\infty} g(x)\mathscr{H}_n(x)\mathrm{d}x = 0$ $(n = 0, 1, 2, \ldots)$ *if, and only if,* $g = 0$.

Proof. If $g = 0$ then $\int_{-\infty}^{\infty} g(x)\mathscr{H}_n(x)\mathrm{d}x = 0$ since $\mathscr{H}_n$ is good.

Conversely, if $\int_{-\infty}^{\infty} g(x)\mathscr{H}_n(x)\mathrm{d}x = 0$ for every n, let γ be any good function and let $\gamma_M = \sum_{m=0}^{M} a_m \mathscr{H}_m$ where $a_m = \int_{-\infty}^{\infty} \gamma(x)\mathscr{H}_m(x)\mathrm{d}x$. Then $\int_{-\infty}^{\infty} g(x)\gamma_M(x)\mathrm{d}x = 0$ and therefore

$$\left|\int_{-\infty}^{\infty} g(x)\gamma(x)\mathrm{d}x\right| = \left|\int_{-\infty}^{\infty} g(x)\{\gamma(x) - \gamma_M(x)\}\mathrm{d}x\right|$$

$$\leq K \max_{x\in R_1} |(1+x^2)^{k/2}\{\gamma^{(r)}(x) - \gamma_M^{(r)}(x)\}|$$

from Theorem 3.16, K being independent of γ and γ_M. On account of

Theorem 5.6 we can, given $\varepsilon > 0$, choose M large enough for $|(1+x^2)^{k/2}\{\gamma^{(r)} - \gamma_M^{(r)}\}| < \varepsilon$ which implies

$$\int_{-\infty}^{\infty} g(x)\gamma(x)\mathrm{d}x = 0, \quad \text{i.e. } g = 0$$

and the proof is finished. □

Theorem 5.8. *If $\sum_{n=0}^{\infty} b_n \mathscr{H}_n(x)$ is a generalised function g then*

$$b_n = \int_{-\infty}^{\infty} g(x)\mathscr{H}_n(x)\mathrm{d}x.$$

Proof. Since $\mathscr{H}_m$ is good

$$\int_{-\infty}^{\infty} g(x)\mathscr{H}_m(x)\mathrm{d}x = \int_{-\infty}^{\infty} \mathscr{H}_m(x) \sum_{n=0}^{\infty} b_n \mathscr{H}_n(x)\mathrm{d}x$$

$$= \lim_{N\to\infty} \sum_{n=0}^{N} b_n \int_{-\infty}^{\infty} \mathscr{H}_m(x)\mathscr{H}_n(x)\mathrm{d}x$$

by Definition 5.1. The theorem follows immediately from (13). □

Theorem 5.7 indicates that if a generalised function does not contain an element of Hermite polynomial it vanishes and Theorem 5.8 gives a clue as to the form the coefficients must take. The validity of the representation is confirmed by

Theorem 5.9. *If g is a generalised function then $g(x) = \sum_{n=0}^{\infty} b_n \mathscr{H}_n(x)$ where $b_n = \int_{-\infty}^{\infty} g(x)\mathscr{H}_n(x)\mathrm{d}x$.*

Proof. By Theorem 3.18 $g = f^{(r)}$ where f is a function, continuous on R_1, which is $O(|x|^k)$ as $|x| \to \infty$. Hence

$$\int_{-\infty}^{\infty} g(x)\mathscr{H}_n(x)\mathrm{d}x = (-1)^r \int_{-\infty}^{\infty} f(x)\mathscr{H}_n^{(r)}(x)\mathrm{d}x.$$

We can write $f(x) = (1+x^2)^{k+1}h(x)$ where h is continuous on R_1 and $O(|x|^{-k-2})$ as $|x| \to \infty$. Consequently $h \in L_2$. Since, from (14) and (15),

$$2x\mathscr{H}_n(x) = (2n)^{1/2}\mathscr{H}_{n-1}(x) + (2n+2)^{1/2}\mathscr{H}_{n+1}(x)$$

it follows that b_n can be written as a finite number (which depends on k and r but not on n) of terms of the type

$$\{(n+1)(n+2)\dots(n+s)\}^{1/2} \int_{-\infty}^{\infty} h(x)\mathscr{H}_{n+s}(x)\mathrm{d}x.$$

But

$$\left|\int_{-\infty}^{\infty} h(x)\mathscr{H}_{n+s}(x)\,\mathrm{d}x\right|^2 \le \int_{-\infty}^{\infty} |h(x)|^2\,\mathrm{d}x$$

so that there is some N (independent of n) such that $b_n = O(n^N)$ as $n \to \infty$.

This information is sufficient to show that $\sum_{n=0}^{\infty} b_n \mathscr{H}_n(x)$ is a generalised function. For

$$\sum_{s=1}^{p} b_{m+s} \int_{-\infty}^{\infty} \mathscr{H}_{m+s}(x)\gamma(x)\,\mathrm{d}x = \sum_{s=1}^{p} b_{m+s} a_{m+s} \to 0$$

as $m \to \infty$ because a_m is certainly $O(m^{-N-2})$ (Corollary 5.5). Hence $\mathrm{Lim}_{m\to\infty} \sum_{s=1}^{p} b_{m+s} \mathscr{H}_{m+s}(x) = 0$ and $\sum_{n=0}^{\infty} b_n \mathscr{H}_n(x)$ is a generalised function, say g_1, by Theorem 5.3. On account of Theorem 5.8 $b_n = \int_{-\infty}^{\infty} g_1(x)\mathscr{H}_n(x)\,\mathrm{d}x$ and therefore

$$\int_{-\infty}^{\infty} \{g(x) - g_1(x)\}\,\mathscr{H}_n(x)\,\mathrm{d}x = 0 \qquad (n = 0, 1, 2, \dots).$$

It follows immediately from Theorem 5.7 that $g = g_1$ and the theorem is proved. □

Observe that we have also demonstrated

Corollary 5.9*a*. *The series $\sum_{n=0}^{\infty} b_n \mathscr{H}_n(x)$ is a generalised function if, and only if, $b_n = O(n^N)$ for some N as $n \to \infty$. If $\sum_{n=0}^{\infty} b_n \mathscr{H}_n(x) = 0$ then $b_n = 0$ $(n = 0, 1, 2, \dots)$.*

Another corollary devolves from Theorem 5.9 because

$$\int_{-\infty}^{\infty} g(x)\gamma(x)\,\mathrm{d}x = \lim_{M\to\infty} \sum_{m=0}^{M} b_m \int_{-\infty}^{\infty} \mathscr{H}_m(x)\gamma(x)\,\mathrm{d}x$$

$$= \lim_{M\to\infty} \sum_{m=0}^{M} a_m b_m = \sum_{m=0}^{\infty} a_m b_m.$$

Corollary 5.9*b*.

$$\int_{-\infty}^{\infty} g(x)\gamma(x)\,\mathrm{d}x = \sum_{m=0}^{\infty} a_m b_m, \qquad \int_{-\infty}^{\infty} \{\gamma(x)\}^2\,\mathrm{d}x = \sum_{m=0}^{\infty} a_m^2,$$

$$\int_{-\infty}^{\infty} |\gamma(x)|^2\,\mathrm{d}x = \sum_{m=0}^{\infty} |a_m|^2.$$

On account of Theorems 5.9, 3.14(iv) and Exercise 4, we have

Corollary 5.9c. *The Fourier transform G of g is given by*

$$G(\alpha) = (2\pi)^{1/2} \sum_{n=0}^{\infty} (-\mathrm{i})^n b_n \mathscr{H}_n(\alpha)$$

where

$$b_n = \int_{-\infty}^{\infty} g(x)\mathscr{H}_n(x)\,\mathrm{d}x = \frac{\mathrm{i}^n}{(2\pi)^{1/2}} \int_{-\infty}^{\infty} G(\alpha)\mathscr{H}_n(\alpha)\,\mathrm{d}\alpha.$$

It is evident from these results that another way of defining generalised functions would be as series of Hermite polynomials whose coefficients are $O(n^N)$ as $n \to \infty$. Such a process has no obvious advantage over the method adopted in this book but the interested reader might care to contemplate the rewriting of Chapter 3 from this point of view.

Example 1. For $g(x) = \delta(x-a)$, $b_n = \int_{-\infty}^{\infty} \delta(x-a)\mathscr{H}_n(x)\,\mathrm{d}x = \mathscr{H}_n(a)$ so that

$$\delta(x-a) = \sum_{n=0}^{\infty} \mathscr{H}_n(a)\mathscr{H}_n(x).$$

In particular

$$\delta(x) = \sum_{n=0}^{\infty} \mathscr{H}_n(0)\mathscr{H}_n(x).$$

Applying Corollary 5.9*c*

$$1 = (2\pi)^{1/2} \sum_{n=0}^{\infty} (-\mathrm{i})^n \mathscr{H}_n(0)\mathscr{H}_n(x).$$

Exercises

7. Prove that, if $|b| < 1$,

$$\frac{1}{\{\pi(1-b^2)\}^{1/2}} \exp\left\{\tfrac{1}{2}(x^2-a^2) - \frac{(x-ab)^2}{1-b^2}\right\} = \sum_{n=0}^{\infty} \mathscr{H}_n(x)\mathscr{H}_n(a) b^n.$$

Does this imply that

$$\lim_{b\to 1} \frac{1}{\{\pi(1-b^2)\}^{1/2}} \exp\left\{\tfrac{1}{2}(x^2-a^2) - \frac{(x-ab)^2}{1-b^2}\right\} = \delta(x-a)?$$

8. Assume that, if $\beta > -1$,

$$2\int_0^{\infty} x^{\beta} \frac{\mathrm{d}^{2r}}{\mathrm{d}x^{2r}} \mathrm{e}^{-x^2}\,\mathrm{d}x = \frac{\beta!(\frac{1}{2}\beta - r - \frac{1}{2})!}{(\beta - 2r)!}.$$

Show that, if $\beta > -1$,

$$\mathrm{e}^{-x^2/2}|x|^{\beta} = \frac{(\frac{1}{2}\beta - \frac{1}{2})!}{\pi^{1/4}} \sum_{n=0}^{\infty} \frac{\beta(\beta-2)\ldots(\beta-2n+2)}{\{(2n)!\}^{1/2} 2^n} \mathscr{H}_{2n}(x).$$

Deduce that

$$e^{-x^2/2}|x|^\beta \operatorname{sgn} x = \frac{(\frac{1}{2}\beta)!}{\pi^{1/4}} \sum_{n=0}^{\infty} \frac{(\beta-1)(\beta-3)\ldots(\beta-2n+1)}{\{(2n+1)!\}^{1/2}2^{n+1}} \mathscr{H}_{2n+1}(x).$$

Prove that these two expansions are still valid if β is any complex member other than a negative integer.

5.4. Functions of the class L_2

The functions which belong to L_2 occupy a special position in analysis. One reason is that many relations are symmetrical for such functions. Another is that they correspond to finite energy over a domain in physics. A useful theorem is

Theorem 5.10. *$g \in L_2$ if, and only if, $\sum_{n=0}^{\infty} |b_n|^2 < \infty$.*

Proof. If $g \in L_2$ then

$$0 \leq \int_{-\infty}^{\infty} \left| g(x) - \sum_{m=0}^{M} b_m \mathscr{H}_m(x) \right|^2 dx \leq \int_{-\infty}^{\infty} |g(x)|^2 dx - \sum_{m=0}^{M} |b_m|^2$$

on expanding the integrand and integrating term by term. Since $\sum_{m=0}^{M} |b_m|^2$ is monotone it follows that

$$\sum_{n=0}^{\infty} |b_n|^2 \leq \int_{-\infty}^{\infty} |g(x)|^2 dx < \infty.$$

The converse is more difficult to prove because, although $\sum_{n=0}^{\infty} |b_n|^2 < \infty$ shows that g is a generalised function on account of Corollary 5.9*a*, it must be shown that g is, in fact, a function in L_2. This depends upon Levi's theorem and Fatou's lemma though the following proof does not mention them explicitly.

Let $\gamma_n(x) = \sum_{m=0}^{n} b_m \mathscr{H}_m(x)$ and define $m_1 < m_2 < \ldots$ such that

$$\sum_{s=1}^{\infty} |b_{m_k+s}|^2 < 1/2^{2k}.$$

Then, for any (a, b),

$$\left\{ \int_a^b |\gamma_{m_{k+1}}(x) - \gamma_{m_k}(x)| dx \right\}^2 \leq \int_a^b dx \int_a^b |\gamma_{m_{k+1}}(x) - \gamma_{m_k}(x)|^2 dx$$

$$\leq (b-a) \sum_{s=m_k+1}^{m_{k+1}} |b_s|^2 \leq (b-a)/2^{2k}.$$

Therefore, for any finite a and b,

$$\sum_{k=1}^{\infty} \int_a^b |\gamma_{m_{k+1}}(x) - \gamma_{m_k}(x)| \mathrm{d}x < \infty.$$

It follows that

$$\sum_{k=1}^{\infty} \int_a^b \{\gamma_{m_{k+1}}(x) - \gamma_{m_k}(x)\} \mathrm{d}x < \infty.$$

Hence, if

$$h_k^+ = \tfrac{1}{2}\{|\gamma_{m_{k+1}} - \gamma_{m_k}| + \gamma_{m_{k+1}} - \gamma_{m_k}\}$$

and

$$h_k^- = \tfrac{1}{2}\{|\gamma_{m_{k+1}} - \gamma_{m_k}| - \gamma_{m_{k+1}} + \gamma_{m_k}\},$$

$\sum_{k=1}^{\infty} \int_a^b h_k^+(x)\mathrm{d}x < \infty$ and $\sum_{k=1}^{\infty} \int_a^b h_k^-(x)\mathrm{d}x < \infty$. Both h_k^+ and h_k^- are non-negative and members of the class $\mathscr{C}_1$ of §1.11. Let $\{f_{rn}^+\}$ be a sequence of step functions converging almost everywhere to $\sum_{k=1}^{r} h_k^+$. With $f_r^+ = \sup_{i \le r} f_{ir}^+$ we have $f_r^+ \le \sum_{k=1}^{r} h_k^+$ since $f_{ir}^+ \le \sum_{k=1}^{i} h_k^+ \le \sum_{k=1}^{r} h_k^+$ for $i \le r$. Therefore $\int_a^b f_r^+(x)\mathrm{d}x < \infty$ and, since $f_{r+1}^+ \ge f_r^+$, the sequence $\{f_r^+\}$ converges almost everywhere to a limit f in $\mathscr{C}_1$. On the other hand, $f_i^+ \ge f_{ri}^+$ for $r \le i$ so that, as $i \to \infty$, $f \ge \sum_{k=1}^{r} h_k^+$. Since $\sum_{k=1}^{r} h_k^+ \ge f_r^+$ it follows that $\sum_{k=1}^{r} h_k^+$ tends to f as $r \to \infty$. A similar argument can be applied to $\sum_{k=1}^{r} h_k^-$ and then $\sum_{k=1}^{r} (\gamma_{m_{k+1}} - \gamma_{m_k}) = \sum_{k=1}^{r} (h_k^+ - h_k^-)$ converges almost everywhere to a limit, i.e. $\gamma_{m_{r+1}}$ converges to a limit g almost everywhere on the arbitrary interval (a, b).

Furthermore

$$\begin{aligned}
&\left\{\int_{-\infty}^{\infty} |\gamma_{m_k}(x)|^2 \mathrm{d}x\right\}^{1/2} \\
&\le \left\{\int_{-\infty}^{\infty} |\gamma_{m_1}(x)|^2 \mathrm{d}x\right\}^{1/2} + \left\{\int_{-\infty}^{\infty} |\gamma_{m_k}(x) - \gamma_{m_1}(x)|^2 \mathrm{d}x\right\}^{1/2} \\
&\le \left\{\int_{-\infty}^{\infty} |\gamma_{m_1}(x)|^2 \mathrm{d}x\right\}^{1/2} + \left\{\sum_{s=m_1+1}^{m_k} |b_s|^2\right\}^{1/2} \\
&\le \left\{\int_{-\infty}^{\infty} |\gamma_{m_1}(x)|^2 \mathrm{d}x\right\}^{1/2} + \tfrac{1}{2} \le A \text{ (say)}
\end{aligned}$$

so that, if $h_n = \inf_{k \geq n} |\gamma_{m_k}|^2$,

$$\int_{-\infty}^{\infty} h_n(x)\,dx \leq \lim_{n\to\infty} \inf_{k \geq n} \int_{-\infty}^{\infty} |\gamma_{m_k}|^2\,dx \leq A^2.$$

Arguing now with h_r as we did with $\sum_{k=1}^{r} h_k^+$ we see that h_r converges almost everywhere to a limit, which must be $|g|^2$, such that

$$\int_{-\infty}^{\infty} |g|^2\,dx \leq A^2, \qquad \text{i.e. } g \in L_2.$$

To show that this is true for the whole sequence $\{\gamma_n\}$ we note that, if $n > m_r$,

$$\left\{\int_{-\infty}^{\infty} |\gamma_n - \gamma_{m_k}|^2\,dx\right\}^{1/2}$$

$$\leq \left\{\int_{-\infty}^{\infty} |\gamma_n - \gamma_{m_r}|^2\,dx\right\}^{1/2} + \left\{\int_{-\infty}^{\infty} |\gamma_{m_k} - \gamma_{m_r}|^2\,dx\right\}^{1/2} \leq 2^{1-r}$$

for $k > r$. Letting $k \to \infty$, we repeat the argument of the previous paragraph and

$$\int_{-\infty}^{\infty} |\gamma_n - g|^2\,dx \leq 2^{2-2r}.$$

Letting $r \to \infty$, and therefore $n \to \infty$, we have

$$\int_{-\infty}^{\infty} |\gamma_n - g|^2\,dx \to 0 \tag{21}$$

and the proof of the theorem is complete. □

As a consequence of (21) we have

Corollary 5.10. *If* $g \in L_2$, $\int_{-\infty}^{\infty} |g(x)|^2\,dx = \sum_{n=0}^{\infty} |b_n|^2$.

There are also two theorems concerning the Fourier transform which are worth consideration. One is

Theorem 5.11. *If* $g \in L_2$, *then* $G \in L_2$ *and*

$$\int_{-\infty}^{\infty} |g(x)|^2\,dx = \frac{1}{2\pi} \int_{-\infty}^{\infty} |G(\alpha)|^2\,d\alpha.$$

Proof. Since $g \in L_2$, Theorem 5.10 shows that $\sum_{n=0}^{\infty} |b_n|^2 < \infty$ and so $G \in L_2$ by Theorem 5.10 since $G = (2\pi)^{1/2} \sum_{n=0}^{\infty} (-\mathrm{i})^n b_n \mathscr{H}_n(\alpha)$. Applying Corollary 5.10 to both g and G we finish the proof. □

Evidently, if $g \in L_2$ it is immaterial whether we regard G as the Fourier transform in terms of generalised functions or in the L_2 sense.

If $g_1 \in L_2$ and $g_2 \in L_2$ then, from Theorem 5.11,

$$\int_{-\infty}^{\infty} |g_1(x) + g_2(x)|^2 \mathrm{d}x = \frac{1}{2\pi} \int_{-\infty}^{\infty} |G_1(\alpha) + G_2(\alpha)|^2 \mathrm{d}\alpha$$

or

$$\mathrm{Re}\left\{ \int_{-\infty}^{\infty} g_1(x)\bar{g}_2(x)\mathrm{d}x \right\} = \mathrm{Re}\left\{ \frac{1}{2\pi} \int_{-\infty}^{\infty} G_1(\alpha)\bar{G}_2(\alpha)\mathrm{d}\alpha \right\}.$$

The same result with Im in place of Re can be obtained by repeating the argument with $g_1 + \mathrm{i}g_2$ in place of $g_1 + g_2$. Hence

Theorem 5.12. *If $g_1 \in L_2$ and $g_2 \in L_2$ then*

$$\int_{-\infty}^{\infty} g_1(x)\bar{g}_2(x)\mathrm{d}x = \frac{1}{2\pi} \int_{-\infty}^{\infty} G_1(\alpha)\bar{G}_2(\alpha)\mathrm{d}\alpha.$$

5.5. Fourier series

A considerable body of analysis has been devoted to Fourier series, since their introduction by Fourier, in attempts to determine the circumstances under which it is permissible to expand a function in series of sines and cosines. Once the legitimacy of the process has been established it is easy to find the coefficients in the expansion, for, if $f(x) = \sum_{n=-\infty}^{\infty} c_n \mathrm{e}^{\mathrm{i}nx}$ in $(-\pi, \pi)$, multiplication by $\mathrm{e}^{-\mathrm{i}mx}$ and integration give

$$\int_{-\pi}^{\pi} f(x)\mathrm{e}^{-\mathrm{i}mx} \mathrm{d}x = \sum_{n=-\infty}^{\infty} \int_{-\pi}^{\pi} c_n \mathrm{e}^{\mathrm{i}(n-m)x} \mathrm{d}x = c_m. \tag{22}$$

However, the analysis necessary to prove the validity of the procedure is complicated even for relatively simple classes of functions.

When we turn to the corresponding problems in connection with generalised functions we find that many of the difficulties disappear and that Fourier series have many of the properties desirable for physical applications. Observe that (22) will not be satisfactory for generalised functions because the definite integral over a finite interval of a generalised function has not been defined in general. Nevertheless, a Fourier series is periodic so that the generalised function which it represents will be similarly periodic. This suggests that there must be some generalisation of the idea of integration over a period which is applicable to generalised functions. We shall see the form of this generalisation a little later.

The Fourier transform of e^{inx} is $2\pi\delta(\alpha - n)$, so that any discussion of a Fourier series involves the series $\sum c_n \delta(\alpha - n)$ and conversely. The series of delta functions is easier to handle because of the simple properties of the delta function. Therefore we start with

Theorem 5.13. *The series* $\sum_{n=-\infty}^{\infty} c_n \delta(\alpha - n)$ *is generally convergent if, and only if,* $c_n = O(|n|^N)$ *as* $|n| \to \infty$, *N being some fixed integer.*

Proof. If $\Gamma(\alpha)$ is any good function,

$$\sum_{r=1}^{p} \int_{-\infty}^{\infty} c_{m+r} \delta(\alpha - m - r)\Gamma(\alpha)\mathrm{d}\alpha = \sum_{r=1}^{p} c_{m+r}\Gamma(m+r). \tag{23}$$

If $c_n = O(|n|^N)$, this sum tends to zero as $m \to \infty$ because of the behaviour of a good function at infinity and so

$$\lim_{m\to\infty} \sum_{r=1}^{p} c_{m+r}\delta(\alpha - m - r) = 0.$$

It follows from Theorem 5.3 that $\sum_{n=-\infty}^{\infty} c_n \delta(\alpha - n)$ is generally convergent.

Conversely, if $\sum_{n=-\infty}^{\infty} c_n \delta(\alpha - n)$ is generally convergent, the right-hand side of (23) must be bounded as $p \to \infty$. We now have to show that this cannot be so unless c_n satisfies the condition stated in the theorem and we do this by constructing a special good function. If there is no N such that $c_n = O(|n|^N)$ as $|n| \to \infty$ there must be an increasing sequence $\{n_r\}$ such that $|n_1| < |n_2| < \dots$ and $|c_{n_r}| > |n_r|^r$ for $r = 1, 2, \dots$. Consider now $\sum_{r=1}^{\infty} \sigma(\alpha - n_r)/c_{n_r}$ where $\sigma(x) = 0$ $(|x| \geq 1)$, $\sigma(x) = e^{-x^2/(1-x^2)}$ $(|x| < 1)$. For any given α the series contains only a finite number of terms so that it represents a continuous function of α. The same is true of any derivative. Also, if $|n_t|+1<\alpha<|n_{t+1}|+1$,

$$\left|\alpha^s \sum_{r=1}^{\infty} \frac{\sigma(\alpha-n_r)}{c_{n_r}}\right| = \left|\alpha^s \sum_{r=t+1}^{\infty} \frac{\sigma(\alpha-n_r)}{c_{n_r}}\right| \leq \alpha^s \max_{|x|\leq 1} \sigma(x) \sum_{r=t+1}^{\infty} \frac{1}{|n_r|^r}$$

$$\leq \frac{\alpha^s}{|n_{t+1}|^{t-1}} \max_{|x|\leq 1} \sigma(x) \sum_{r=t+1}^{\infty} \frac{1}{|n_r|^2}$$

$$\to 0 \quad \text{as } t\to\infty.$$

The same argument applies to any derivative and therefore $\sum_{r=1}^{\infty} \sigma(\alpha-n_r)/c_{n_r}$ is a good function. Using it for $\Gamma(\alpha)$ in (23) we obtain for the right-hand side q where q is the number of n_r such that $m+1\leq n_r \leq m+p$. Since q increases without limit as $p\to\infty$ we have the contradiction sought and it is necessary that $c_n = O(|n|^N)$. The theorem is proved. □

Corollary 5.13. *The series* $\sum_{n=-\infty}^{\infty} c_n e^{inx}$ *is generally convergent if, and only if,* $c_n = O(|n|^N)$ *as* $|n|\to\infty$, *N being some fixed integer.*

Proof. This is an immediate consequence of Theorems 5.1 and 5.13. □

Our next object is to determine the coefficients c_n in terms of the generalised function which the series represents. To do this introduce the fine function $\tau(x)=0$ $(|x|\geq 1)$ and

$$\tau(x) = \int_{|x|}^{1} e^{-1/t(1-t)}\,dt \bigg/ \int_{0}^{1} e^{-1/t(1-t)}\,dt \qquad (|x|<1).$$

It has the property $\tau(x)+\tau(x-1)=1$ when $0\leq x\leq 1$. The function $\sum_{m=-n}^{n} \tau(x-m)$ is fine by Exercise 10 of Chapter 2 and equals 1 for $|x|\leq n$ and zero for $|x|\geq n+1$.

Theorem 5.14. *If* $\sum_{n=-\infty}^{\infty} c_n e^{inx}$ *is the generalised function g then*

$$c_n = \frac{1}{2\pi}\int_{-\infty}^{\infty} g(x)\tau(x/2\pi)e^{-inx}\,dx.$$

Proof. By Definition 5.1, τ being good,

$$\int_{-\infty}^{\infty} g(x)\tau(x/2\pi)e^{-inx}\,dx = \lim_{M\to\infty} \sum_{m=-M}^{M} \int_{-\infty}^{\infty} c_m e^{i(m-n)x}\tau(x/2\pi)\,dx.$$

Also, since $e^{i(m-n)x} \in K_1$ the integral on the right is a conventional one and so

$$\int_{-\infty}^{\infty} e^{i(m-n)x}\tau(x/2\pi)dx = \sum_{r=-\infty}^{\infty} \int_{(2r-1)\pi}^{(2r+1)\pi} e^{i(m-n)x}\tau(x/2\pi)dx$$

$$= \sum_{r=-\infty}^{\infty} 2\pi \int_{-1/2}^{1/2} e^{i(m-n)2\pi y}\tau(y+r)dy$$

$$= 2\pi \int_{-1/2}^{1/2} e^{i(m-n)2\pi y}\{\tau(y)+\tau(y-1)\}dy$$

$$= 2\pi \int_{-1/2}^{1/2} e^{i(m-n)2\pi y}dy$$

$$= \begin{cases} 0 & (m \neq n) \\ 2\pi & (m = n) \end{cases} \tag{24}$$

and the proof of the theorem is complete. □

An immediate consequence of Theorem 5.14 is

Corollary 5.14. *If* $\sum_{n=-\infty}^{\infty} c_n e^{inx} = 0$ *then* $c_n = 0$ *for all n.*

From this it follows that the Fourier expansion of a generalised function is unique. For, if a generalised function had two different Fourier expansions, their difference would represent zero without the coefficients being non-zero. Thus different Fourier series must represent different generalised functions.

Finally g is connected with $g\tau$ by

Theorem 5.15. $g(x) = \sum_{n=-\infty}^{\infty} g(x)\tau(x-n)$.

Proof. By Definition 5.1,

$$\int_{-\infty}^{\infty} \left\{ g(x) - \sum_{n=-\infty}^{\infty} g(x)\tau(x-n) \right\}\gamma(x)dx$$

$$= \lim_{n\to\infty} \int_{-\infty}^{\infty} g(x)\left\{1 - \sum_{m=-n}^{n} \tau(x-m)\right\}\gamma(x)dx. \tag{25}$$

Now $\{1-\sum_{m=-n}^{n}\tau(x-m)\}\gamma(x)$ is good and so, by Theorem 3.16,

$$\left|\int_{-\infty}^{\infty} g(x)\left\{1-\sum_{m=-n}^{n}\tau(x-m)\right\}\gamma(x)\mathrm{d}x\right|$$
$$\leq K\max\left|(1+x^2)^{k/2}\left[\left\{1-\sum_{m=-n}^{n}\tau(x-m)\right\}\gamma(x)\right]^{(r)}\right|$$

for some finite K, k and r independent of τ and γ. Since

$$1-\sum_{m=-n}^{n}\tau(x-m)=\begin{cases}0 & (|x|\leq n),\\ 1 & (|x|\geq n+1)\end{cases}$$

we see that the max comes only from $|x|>n$ and that the contribution from $|x|\geq n+1$ tends to zero as $n\to\infty$ since $x^k\gamma^{(r)}(x)\to 0$ as $|x|\to\infty$. With regard to $n<x<n+1$, the series reduces to $\tau(x-n)$ and so, since τ and its derivatives are bounded, the contribution of this interval vanishes in the limit because of the presence of γ. Similarly, the same is true for the portion $-n-1<x<-n$. Hence the limit on the right-hand side of (25) is zero and the theorem is proved. □

It should be remarked that the only properties of τ used in Theorems 5.14 and 5.15 are that it is zero for $|x|\geq 1$ and such that $\tau(x)+\tau(x-1)=1$ when $0\leq x\leq 1$. Therefore τ can be replaced by any fine function which has the same properties and Theorems 5.14 and 5.15 will remain true.

Theorem 5.14 does not tell us that any generalised function can be expanded in a Fourier series but only the value of the coefficients if the expansion is permissible. In fact, we have to restrict the class of generalised functions in order to make sure the series representation is valid. We do this in the next section.

Exercises

9. Show that $\sum_{n=-\infty}^{\infty}c_n\mathrm{e}^{in\pi x/l}$ is a generalised function if, and only if, $c_n=O(|n|^N)$ for some N as $|n|\to\infty$. If $\sum_{n=-\infty}^{\infty}c_n\mathrm{e}^{in\pi x/l}=g$ prove that

$$c_n=\frac{1}{2l}\int_{-\infty}^{\infty}g(x)\tau\left(\frac{x}{2l}\right)\mathrm{e}^{-in\pi x/l}\mathrm{d}x.$$

10. Which of the following are generalised functions?

(i) $\sum_{n=-\infty}^{\infty}\mathrm{e}^{inx}\sin n\beta$, (ii) $\sum_{n=-\infty}^{\infty}(n^2+2)^{1/2}\delta(n-\alpha)$,
(iii) $\sum_{n=-\infty}^{\infty}\mathrm{e}^{n+inx}$, (iv) $\sum_{n=-\infty}^{\infty}n^3\delta'(n-\alpha)$.

11. If $\sum_{n=-\infty}^{\infty} c_n e^{inx}$ is the generalised function 1 show that $c_n = 0$ $(n \neq 0)$, $c_n = 1$ $(n = 0)$ so that the series reduces to 1, as would be hoped.

5.6. Periodic generalised functions

Definition 5.2. *A generalised function $g(x)$ is said to be periodic with period $2l$ if $g(x+2l) = g(x)$.*

An obvious example is the generalised function 1. Another is $e^{in\pi x/l}$. More generally

Theorem 5.16. *If $c_n = O(|n|^N)$ as $|n| \to \infty$, $\sum_{n=-\infty}^{\infty} c_n e^{inx}$ is a periodic generalised function with period 2π.*

Proof. The series is a generalised function by Corollary 5.13. Hence by Definition 5.1,

$$\sum_{n=-\infty}^{\infty} c_n e^{in(x+2\pi)} = \lim_{M\to\infty} \sum_{m=-M}^{M} c_m e^{im(x+2\pi)}$$

$$= \lim_{M\to\infty} \sum_{m=-M}^{M} c_m e^{imx}$$

$$= \sum_{n=-\infty}^{\infty} c_n e^{inx}$$

and the proof is concluded. □

The application of Theorem 3.10 to Definition 5.2 gives

Theorem 5.17. *The generalised function g is periodic if, and only if, $G(\alpha) = e^{2li\alpha} G(\alpha)$.*

From this it is clear that $\delta(\alpha)$ is the Fourier transform of a periodic generalised function (in fact $1/2\pi$).

Exercises

12. If g has period $2l$ and if $\sum_{n=-\infty}^{\infty} g(x-2ln)$ is a generalised function show that it is periodic.

13. If $\sum_{n=-\infty}^{\infty} g(x-n)$ is a generalised function show that it has period 1.

14. If $c_n = O(|n|^N)$, is $\sum_{n=-\infty}^{\infty} c_n \delta(x-2ln)$ periodic?

Theorem 5.16 makes it transparently clear that a Fourier series cannot represent a generalised function which is not periodic. We shall now show that this is the only limitation which needs to be imposed. We commence by considering Fourier transforms.

Theorem 5.18. *If* $G(\alpha) = e^{2\pi i\alpha} G(\alpha)$ *and* $c'_n = \int_{-\infty}^{\infty} G(\alpha)\, T\{2\pi(n-\alpha)\} d\alpha$ *where* $T(\alpha)$ *is the Fourier transform of* $\tau(x)$ *then*

$$G(\alpha) = \sum_{n=-\infty}^{\infty} c'_n \delta(\alpha - n).$$

Proof. By hypothesis

$$\int_{-\infty}^{\infty} G(\alpha)(1 - e^{2\pi i\alpha})\Gamma(\alpha) d\alpha = 0$$

for any good function $\Gamma(\alpha)$. Choose Γ by

$$\Gamma(\alpha) = \frac{\Gamma_1(\alpha)\tau(\alpha - n) - \Gamma_1(n) T\{2\pi(n-\alpha)\}}{1 - e^{2\pi i\alpha}}$$

where Γ_1 is any good function. Γ, so chosen, is good because the numerator is good and the denominator can cause trouble only when it vanishes at say $\alpha = m$. But then the numerator is $\Gamma_1(m)\tau(m-n) - \Gamma_1(n)T\{2\pi(n-m)\}$; if $m \neq n$, $\tau(m-n) = 0$ and $T = 0$ by (24); if $m = n$, $\tau(m-n) = 1$ and $T\{2\pi(n-m)\} = 1$. In either case the numerator is zero and so Γ is good. Hence

$$\int_{-\infty}^{\infty} G(\alpha)\tau(\alpha - n)\Gamma_1(\alpha) d\alpha = \Gamma_1(n) \int_{-\infty}^{\infty} G(\alpha) T\{2\pi(n-\alpha)\} d\alpha$$

or

$$G(\alpha)\tau(\alpha - n) = c'_n \delta(\alpha - n)$$

since Γ_1 is arbitrary. Applying Theorem 5.15 we complete the proof. □

Corollary 5.18. *If* $G(\alpha) = e^{2li\alpha} G(\alpha)$ *and*

$$c'_n = \int_{-\infty}^{\infty} G(\alpha) T\{2(n\pi - \alpha l)\} d\alpha$$

then

$$G(\alpha) = \sum_{n=-\infty}^{\infty} c'_n \delta(\alpha - n\pi/l).$$

Proof. Let $\alpha=\alpha_1\pi/l$ and $G(\alpha_1\pi/l)=G_1(\alpha_1)$; then $G_1(\alpha_1)=e^{2\pi i\alpha_1}G_1(\alpha_1)$ and so, by Theorem 5.18, $G_1(\alpha_1)=\sum_{n=-\infty}^{\infty}c''_n\delta(\alpha_1-n)$ where

$$c''_n=\int_{-\infty}^{\infty}G_1(\alpha_1)T\{2\pi(n-\alpha_1)\}\,d\alpha_1.$$

Hence

$$G(\alpha)=\sum_{n=-\infty}^{\infty}c''_n\delta(\alpha l/\pi-n)=\sum_{n=-\infty}^{\infty}(\pi c''_n/l)\delta(\alpha-n\pi/l)$$

and the result follows on putting $\alpha_1=l\alpha/\pi$ in the integral for c''_n. □

Theorem 5.19. *If $g(x+2l)=g(x)$ and*

$$c_n=\frac{1}{2l}\int_{-\infty}^{\infty}g(x)\tau\left(\frac{x}{2l}\right)e^{-in\pi x/l}\,dx$$

then $g(x)=\sum_{n=-\infty}^{\infty}c_ne^{in\pi x/l}$ and $G(\alpha)=2\pi\sum_{n=-\infty}^{\infty}c_n\delta(\alpha-n\pi/l)$.

Proof. By Theorem 5.17 and Corollary 5.18

$$G(\alpha)=\sum_{n=-\infty}^{\infty}c'_n\delta(\alpha-n\pi/l).$$

The Fourier transform of this is

$$g(x)=\sum_{n=-\infty}^{\infty}(c'_n/2\pi)e^{in\pi x/l}.$$

Also, from Theorem 3.9*a*,

$$c_n=\frac{1}{2\pi}\int_{-\infty}^{\infty}G(-\alpha)T\{2(n\pi+\alpha l)\}\,d\alpha$$

$$=\frac{1}{2\pi}\int_{-\infty}^{\infty}G(\alpha)T\{2(n\pi-\alpha l)\}\,d\alpha=\frac{c'_n}{2\pi}$$

and the theorem is proved. □

There is an important corollary which connects the theory with that for Fourier series of conventional functions. It is

Corollary 5.19. *If $g(x+2l)=g(x)$ and if $g(x)=h(x)$ in $0<x<2l$ where $h\in L_1(0,2l)$ then $g(x)=\sum_{n=-\infty}^{\infty}c_ne^{in\pi x/l}$ where*

$$c_n=\frac{1}{2l}\int_0^{2l}h(x)e^{-in\pi x/l}\,dx,$$

with the last integral conventional, (g is to be regarded as the conventional function which is the periodic extension of h to R_1.)

Proof. By Theorem 5.19, and the fact that $g \in K_1$ because $h \in L_1$,

$$\begin{aligned} c_n &= \frac{1}{2l}\int_{-\infty}^{\infty} g(x)\tau\left(\frac{x}{2l}\right)e^{-in\pi x/l}\,dx \\ &= \frac{1}{2l}\sum_{m=-\infty}^{\infty}\int_{2ml}^{(2m+2)l} h(x)\tau\left(\frac{x}{2l}\right)e^{-in\pi x/l}\,dx \\ &= \frac{1}{2l}\sum_{m=-\infty}^{\infty}\int_{0}^{2l} h(x)\tau\left(\frac{x}{2l}+m\right)e^{-in\pi x/l}\,dx \\ &= \frac{1}{2l}\int_{0}^{2l} h(x)e^{-in\pi x/l}\,dx \end{aligned}$$

as in deriving (24) and the proof is complete. □

This corollary shows that, given any conventional integrable function in a finite interval we can make it a periodic generalised function by defining values outside the interval by periodicity. The generalised function has a Fourier expansion with coefficients calculated in the conventional way and therefore so does the conventional function in the given interval. However, it must not be forgotten that this is in the sense of generalised functions and therefore due caution must be taken when operating with the series.

Example 2. The generalised function

$$g(x) = x - 2m\pi \qquad \{2m\pi < x < (2m+2)\pi, \qquad m = 0, \pm 1, \pm 2, \ldots\}$$

satisfies the conditions of Corollary 5.19 with $h(x) = x$, $l = \pi$. Hence

$$c_n = \frac{1}{2\pi}\int_0^{2\pi} x e^{-inx}\,dx.$$

If $n \neq 0$, integration by parts gives

$$c_n = \frac{1}{2\pi}\left[\left(x + \frac{1}{in}\right)\frac{e^{-inx}}{-in}\right]_0^{2\pi} = \frac{i}{n}.$$

If $n = 0$, $c_0 = \pi$. Therefore

$$g(x) = \pi + \sum_{n=-\infty}^{\infty}{}' \frac{i}{n} e^{inx} \tag{26}$$

where the prime indicates that $n = 0$ is to be omitted. In particular,

$$x = \pi + \sum_{n=-\infty}^{\infty}{}' \frac{\mathrm{i}}{n} \mathrm{e}^{inx} \quad (0 < x < 2\pi). \tag{27}$$

At first sight, it seems that a derivative of this result gives

$$1 = -\sum_{n=-\infty}^{\infty}{}' \mathrm{e}^{inx}$$

which disagrees with Exercise 11; but we must remember that the correct formula is

$$g'(x) = -\sum_{n=-\infty}^{\infty}{}' \mathrm{e}^{inx} \tag{28}$$

because we can be sure of taking derivatives term by term only for generalised functions so that (26) must be employed. In fact,

$$g(x) = x - 2\pi \sum_{m=1}^{\infty} \mathrm{H}(x - 2m\pi) + 2\pi \sum_{m=0}^{-\infty} \mathrm{H}(2m\pi - x)$$

so that

$$g'(x) = 1 - 2\pi \sum_{m=-\infty}^{\infty} \delta(x - 2m\pi)$$

since $\delta(-x) = \delta(x)$. Therefore (28) is

$$2\pi \sum_{m=-\infty}^{\infty} \delta(x - 2m\pi) = \sum_{n=-\infty}^{\infty} \mathrm{e}^{inx}. \tag{29}$$

We now give an independent verification.

Example 3. $\sum_{m=-\infty}^{\infty} \delta(x - 2m\pi)$ is a generalised function by Theorem 5.13 and is periodic with period 2π, by Exercise 13. Therefore it has a Fourier expansion by Theorem 5.19 with coefficients given by

$$\begin{aligned} c_n &= \frac{1}{2\pi} \int_{-\infty}^{\infty} \sum_{m=-\infty}^{\infty} \delta(x - 2m\pi)\tau\left(\frac{x}{2\pi}\right)\mathrm{e}^{-inx}\,\mathrm{d}x \\ &= \frac{1}{2\pi} \lim_{M\to\infty} \sum_{m=-M}^{M} \int_{-\infty}^{\infty} \delta(x - 2m\pi)\tau\left(\frac{x}{2\pi}\right)\mathrm{e}^{-inx}\,\mathrm{d}x \\ &= \frac{1}{2\pi} \lim_{M\to\infty} \sum_{m=-M}^{M} \tau(m) = \frac{1}{2\pi}. \end{aligned}$$

Hence (29) is verified.

Example 4. The Fourier transform of the left-hand side of (29) is the same as that of the right-hand side. But this is $2\pi\sum_{n=-\infty}^{\infty}\delta(\alpha-n)$. Therefore the Fourier transform of $\sum_{m=-\infty}^{\infty}\delta(x-2m\pi)$ is $\sum_{n=-\infty}^{\infty}\delta(\alpha-n)$.

Exercises

15. Find the Fourier series of the generalised function with period 2π which equals $\frac{1}{2}(\pi-x)$ in $0<x<2\pi$.

16. Find the Fourier series of the generalised function with period 2π which equals -1 in $(0,\pi)$ and 1 in $(\pi,2\pi)$. What does the derivative of the series represent?

17. Is $\sum_{m=-\infty}^{\infty}\delta'(x-2ml)$ a periodic generalised function? If so, find its Fourier series.

18. Deduce from (27) that

$$1-\tfrac{1}{3}+\tfrac{1}{5}-\tfrac{1}{7}+\tfrac{1}{9}-\tfrac{1}{11}+\ldots=\tfrac{1}{4}\pi.$$

Why is it not straightforward to say that

$$0=\pi+\sum_{n=-\infty}^{\infty}{}'\,\frac{i}{n}?$$

19. Find the Fourier series of the generalised function with period 2π which equals e^x in $(-\pi,\pi)$. Deduce that

$$\sum_{n=0}^{\infty}\frac{(-1)^n}{1+n^2}=\tfrac{1}{2}(\pi\operatorname{cosech}\pi+1).$$

20. If g is an even periodic generalised function show that

$$g(x)=c_0+2\sum_{n=1}^{\infty}c_n\cos(n\pi x/l)$$

where $c_n=(1/2l)\int_{-\infty}^{\infty}g(x)\tau(x/2l)\cos(n\pi x/l)\,dx$. If, further, $g(x)=h(x)$ in $-l<x<l$, $h(x)=h(l-x)$ and $h\in L_1(-l,l)$ show that $c_{2n+1}=0$,

$$c_{2n}=\frac{2}{l}\int_0^{l/2}h(y)\cos\frac{2n\pi y}{l}\,dy \qquad (n=0,1,2,\ldots).$$

21. Show that the Fourier transform of $\sum_{m=-\infty}^{\infty}\delta(x-2ml)$ is

$$\frac{\pi}{l}\sum_{n=-\infty}^{\infty}\delta\left(\frac{n\pi}{l}-\alpha\right).$$

22. Prove that $(1/2\pi)\int_{-\infty}^{\infty}e^{-ia\cos x-i\alpha x}dx=\sum_{n=-\infty}^{\infty}e^{-in\pi/2}J_n(a)\delta(\alpha-n)$. (Hint: $(e^{in\pi/2}/2\pi)\int_0^{2\pi}e^{-ia\cos x-inx}dx=J_n(a)$.)

23. Prove that

$$\ln|2\sin\tfrac{1}{2}x| = -\sum_{m=1}^{\infty}\frac{1}{m}\cos mx.$$

Define $\cot x$, $\operatorname{cosec}^2 x$ as generalised functions by

$$\tfrac{1}{2}\cot\tfrac{1}{2}x = \{\ln|2\sin\tfrac{1}{2}x|\}', \qquad \tfrac{1}{2}\operatorname{cosec}^2\tfrac{1}{2}x = -(\cot\tfrac{1}{2}x)'$$

and deduce that

$$\tfrac{1}{2}\cot\tfrac{1}{2}x = \sum_{m=1}^{\infty}\sin mx,$$

$$-\tfrac{1}{4}\operatorname{cosec}^2\tfrac{1}{2}x = \sum_{m=1}^{\infty} m\cos mx.$$

24. Show that

$$\sum_{n=1}^{\infty} e^{inx} = \pi\sum_{m=-\infty}^{\infty}\delta(x-2m\pi) + \tfrac{1}{2}i\cot\tfrac{1}{2}x - \tfrac{1}{2}.$$

5.7. Poisson summation formula

A result which is sometimes useful in converting a poorly convergent series into one which is highly convergent is

Theorem 5.20. *For $\lambda > 0$ and any good γ,*

$$\sum_{m=-\infty}^{\infty}\gamma(\lambda m) = \frac{1}{\lambda}\sum_{n=-\infty}^{\infty}\Gamma\left(\frac{2\pi n}{\lambda}\right). \tag{30}$$

Proof. By Theorem 3.9

$$\int_{-\infty}^{\infty} G(\alpha)\Gamma(\alpha)\,d\alpha = 2\pi\int_{-\infty}^{\infty} g(x)\gamma(-x)\,dx.$$

Choose $g(x) = \sum_{m=-\infty}^{\infty}\delta(x-2ml)$. According to Exercise 21

$$G(\alpha) = \frac{\pi}{l}\sum_{n=-\infty}^{\infty}\delta\left(\frac{n\pi}{l}-\alpha\right).$$

Hence

$$\frac{\pi}{l}\sum_{n=-\infty}^{\infty}\Gamma\left(\frac{n\pi}{l}\right) = 2\pi\sum_{m=-\infty}^{\infty}\gamma(-2ml).$$

On putting $l = \tfrac{1}{2}\lambda$ and reversing the order of summation on the right-hand side we obtain the theorem. □

Example 5. If $\gamma(x) = e^{-x^2/2}$ then $\Gamma(\alpha) = (2\pi)^{1/2} e^{-\alpha^2/2}$ and Theorem 5.20 gives

$$\sum_{m=-\infty}^{\infty} e^{-m^2\lambda^2/2} = \frac{(2\pi)^{1/2}}{\lambda} \sum_{n=-\infty}^{\infty} e^{-2n^2\pi^2/\lambda^2}.$$

It will be observed that the larger λ is, the more rapidly the left-hand side converges whereas the smaller λ is, the fewer terms of the right-hand side are needed for a good approximation. Thus the two expressions are valuable in complementary regions. Neither has any advantage over the other when $\lambda = (2\pi)^{1/2}$. But if $\lambda > (2\pi)^{1/2}$ the magnitude of the term $m = 2$ on the left-hand side is less than $e^{-4\pi}$ or 0·000003, so that for most practical purposes the terms $m = 0, \pm 1$ would suffice. Similarly, for $\lambda < (2\pi)^{1/2}$ the terms $m = 0, \pm 1$ on the right-hand side would be sufficient for most calculations.

Formula (30) is valid for a wider class of functions than those which are good. Since values at particular points occur on both sides in (30) we cannot expect it to be true if γ is replaced by any generalised functions since there is no way of specifying the value of a generalised function at an arbitrary point. Therefore we shall restrict attention to those functions for which the quantities in (30) are sure to be defined. The reader might contemplate the possibility that (30) might hold for generalised functions regarded as functions of λ.

Theorem 5.21. (Poisson summation formula) *If f is continuous and of bounded variation on $(-\infty, \infty)$, tends to 0 as $|x| \to \infty$ and such that $\int_{-\infty}^{\infty} f(x)\mathrm{d}x$ converges, then*

$$\sum_{m=-\infty}^{\infty} f(\lambda m) = \frac{1}{\lambda} \sum_{n=-\infty}^{\infty} F\left(\frac{2\pi n}{\lambda}\right).$$

Proof. Since f is of bounded variation it is the difference between two bounded non-increasing functions. Since they are bounded and $f(x) \to 0$ as $x \to \infty$ they must tend to the same limit which can be taken to be zero by subtracting a suitable constant from each. Therefore f can be regarded as the difference between two bounded positive non-increasing functions each of which tends to zero as $x \to \infty$. If $\psi(x)$ be such a function and $x_2 \geq x_1 \geq x_0$ then, by the second mean value theorem (Theorem 1.38),

$$\left|\int_{x_1}^{x_2} \psi(t)e^{-i\alpha t}\,dt\right| = \left|\psi(x_1+0)\int_{x_1}^{\xi} e^{-i\alpha t}\,dt\right|$$

$$= \left|\frac{\psi(x_1+0)}{i\alpha}(e^{-i\alpha\xi} - e^{-i\alpha x_1})\right|$$

$$\leq (2/|\alpha|)\,|\psi(x_1+0)|.$$

For given $\alpha \neq 0$ we can choose x_0 so large that this is as small as we please and therefore $\int_a^\infty \psi(t)e^{-i\alpha t}\,dt$ exists for $\alpha \neq 0$ (p. 35). Therefore $\int_a^\infty f(t)e^{-i\alpha t}\,dt$ exists for $\alpha \neq 0$. Similarly, by regarding f as the difference between two non-decreasing functions which tend to 0 as $x \to -\infty$, we prove that $\int_{-\infty}^a f(t)e^{-i\alpha t}\,dt$ exists for $\alpha \neq 0$. Hence $\int_{-\infty}^\infty f(t)e^{-i\alpha t}\,dt$ exists for $\alpha \neq 0$. Since it exists for $\alpha = 0$ by assumption we have shown the existence of $F(\alpha)$ for all real α. Now

$$\sum_{n=-N}^{N} F\left(\frac{2\pi n}{\lambda}\right) = \int_{-\infty}^{\infty} f(t) \sum_{n=-N}^{N} e^{-2\pi i n t/\lambda}\,dt$$

$$= \int_{-\infty}^{\infty} f(t)\frac{\sin\{(N+\frac{1}{2})2\pi t/\lambda\}}{\sin(\pi t/\lambda)}\,dt$$

$$= \lim_{M\to\infty}\sum_{m=-M}^{M}\int_{(m-1/2)\lambda}^{(m+1/2)\lambda}\{f(t)-f(m\lambda)\}\frac{\sin\{(N+\frac{1}{2})2\pi t/\lambda\}}{\sin(\pi t/\lambda)}\,dt$$

$$+\lambda\sum_{m=-M}^{M} f(m\lambda).$$

Once again writing $f = \psi - \psi_1$ where ψ, ψ_1 are bounded, positive and non-increasing, we have

$$\int_{m\lambda}^{(m+1/2)\lambda}\{\psi(t)-\psi(m\lambda)\}\frac{\sin\{(N+\frac{1}{2})2\pi t/\lambda\}}{\sin(\pi t/\lambda)}\,dt$$

$$= \int_0^{\lambda/2}\{\psi(m\lambda+y)-\psi(m\lambda)\}\frac{\sin\{(N+\frac{1}{2})2\pi y/\lambda\}}{\sin(\pi y/\lambda)}\,dy$$

$$= [\psi\{(m+1)\lambda\} - \psi(m\lambda)]\int_{\xi}^{\lambda/2}\frac{\sin\{(N+\frac{1}{2})2\pi y/\lambda\}}{\sin(\pi y/\lambda)}\,dy$$

by the remark at the end of Theorem 1.38. But

$$\left|\int_{\xi}^{\lambda/2} \frac{\sin\{(N+\frac{1}{2})2\pi y/\lambda\}}{\sin(\pi y/\lambda)} dy\right|$$

$$\le \int_0^{\lambda/2} \left|\frac{1}{\sin(\pi y/\lambda)} - \frac{\lambda}{\pi y}\right| dy + \frac{\lambda}{\pi}\left|\int_{(2N+1)\pi\xi/\lambda}^{(N+1/2)\pi} \frac{\sin y}{y} dy\right|$$

which is bounded for all N and ξ. Since $\sum|\psi\{(m+1)\lambda\} - \psi(m\lambda)|$ converges the series of integrals converges as $M \to \infty$ and the convergence is uniform with respect to N. Clearly the same is true of the series of integrals involving f and, since

$$\lim_{N\to\infty} \int_{(m-1/2)\lambda}^{(m+1/2)\lambda} \{f(t) - f(m\lambda)\} \frac{\sin\{(N+\frac{1}{2})2\pi t/\lambda\}}{\sin(\pi t/\lambda)} dt = 0$$

by the Riemann–Lebesgue lemma (Theorem 9.1), it follows that the limit of the sum is 0. The theorem is proved. □

Example 6. If $f(x) = e^{-|x|}$ it satisfies the conditions of Theorem 5.21. Its Fourier transform is given by

$$F(\alpha) = \int_0^{\infty} e^{-(1+i\alpha)x} dx + \int_{-\infty}^{0} e^{(1-i\alpha)x} dx = \frac{2}{(1+\alpha^2)}.$$

Hence

$$\sum_{m=-\infty}^{\infty} e^{-|m\lambda|} = 2\sum_{n=-\infty}^{\infty} \frac{\lambda}{(\lambda^2 + 4n^2\pi^2)}$$

or

$$\tfrac{1}{2} + \sum_{m=1}^{\infty} e^{-m\lambda} = 2\left\{\frac{1}{2\lambda} + \sum_{n=1}^{\infty} \frac{\lambda}{\lambda^2 + 4n^2\pi^2}\right\}.$$

Once again the left-hand side converges rapidly for large λ whereas the right-hand side is more appropriate for small λ. However, in this case, the number of terms required for the same relative degree of accuracy is not the same on both sides. Of course, the left-hand side can be summed with the result

$$1 + 2\sum_{n=1}^{\infty} \frac{\lambda^2}{(\lambda^2 + 4n^2\pi^2)} = \lambda\left\{\frac{1}{2} + \frac{e^{-\lambda}}{1 - e^{-\lambda}}\right\} = \tfrac{1}{2}\lambda \coth \tfrac{1}{2}\lambda.$$

This is an example of how the Poisson summation formula sometimes leads to a comparatively simple expression for the sum of a series.

Exercises

25. By considering the good function $e^{-x^2-2\pi iax}$ prove that

$$1+2\sum_{m=1}^{\infty} e^{-m^2\lambda^2}\cos(2m\pi\lambda a)=\frac{\pi^{1/2}}{\lambda}e^{-\pi^2a^2}\left(1+2\sum_{n=1}^{\infty}e^{-n^2\pi^2/\lambda^2}\cosh\frac{2\pi^2na}{\lambda}\right).$$

If $a=\frac{1}{4}\lambda$ estimate the sum when (i) $\lambda=1$, (ii) $\lambda=10$.

26. Prove that

$$\sum_{m=-\infty}^{\infty}\mathscr{H}_n(\lambda m)=\frac{(2\pi)^{1/2}}{\lambda}(-\mathrm{i})^n\sum_{r=-\infty}^{\infty}\mathscr{H}_n\left(\frac{2\pi r}{\lambda}\right).$$

27. If γ is an even good function prove that

$$\tfrac{1}{2}\gamma(0)+\sum_{m=1}^{\infty}\gamma(m\lambda)=\frac{1}{\lambda}\left\{\tfrac{1}{2}\Gamma(0)+\sum_{n=1}^{\infty}\Gamma\left(\frac{2\pi n}{\lambda}\right)\right\}$$

where $\Gamma(\alpha)=2\int_0^\infty\gamma(x)\cos\alpha x\,\mathrm{d}x$.

28. Show that

$$\sum_{m=1}^{\infty}\frac{\cos m\lambda}{1+m^2}=\frac{\pi\cosh(\pi-\lambda)}{2\sinh\pi}-\frac{1}{2}$$

when $0<\lambda<2\pi$.

29. If f is continuous and of bounded variation on $0\le x<\infty$, tends to 0 as $x\to\infty$ and is such that $\int_0^\infty f(x)\mathrm{d}x$ converges prove that

$$\tfrac{1}{2}f(0)+\sum_{m=1}^{\infty}f(m\lambda)=\frac{1}{\lambda}\left\{\tfrac{1}{2}F(0)+\sum_{n=1}^{\infty}F\left(\frac{2\pi n}{\lambda}\right)\right\}$$

where $F(\alpha)=2\int_0^\infty f(x)\cos\alpha x\mathrm{d}x$.

30. By considering

$$f(x)=\begin{cases}\dfrac{2^{-\nu}(1-x^2)^{\nu-1/2}}{(\nu-\frac{1}{2})!\pi^{1/2}} & (|x|<1,\nu>\frac{1}{2})\\ 0 & (|x|\ge 1)\end{cases}$$

prove that

$$\frac{2^{1/2-\nu}}{(\nu-\frac{1}{2})!}\left\{\tfrac{1}{2}+\sum_{m\le 1/\lambda}(1-m^2\lambda^2)^{\nu-1/2}\right\}=\frac{(2\pi)^{1/2}}{\lambda}\left\{\frac{2^{-\nu-1}}{\nu!}+\sum_{n=1}^{\infty}\frac{J_\nu(2n\pi/\lambda)}{(2n\pi/\lambda)^\nu}\right\}$$

where J_ν is the Bessel function of order ν. (Hint: $F(\alpha)=\alpha^{-\nu}J_\nu(\alpha)(\alpha\ne 0)$.)

31. By considering $\frac{1}{2}\{1-\operatorname{sgn}(x+a\lambda)\}e^{-|x+a\lambda|}$ show that

$$\sum_{n=1}^{\infty}\frac{2\lambda\cos 2na\pi-4n\pi\sin 2na\pi}{\lambda^2+4n^2\pi^2}=\frac{e^{a\lambda}}{e^{\lambda}-1}$$

when $0<a<1$.

32. Obtain another representation of $\sum_{n=1}^{\infty}\cos na/(n^4+1)$.

5.8. An alternative definition of a periodic generalised function

Generalised functions have been defined by means of sequences of good functions. When the generalised functions are periodic one would expect that it might be possible to define them by sequences of periodic functions. Such functions would not, of course, be good. Nevertheless the sequences exist as is demonstrated by

Theorem 5.22. *The generalised function $g(x)$ is periodic, with period $2l$, if, and only if, there is a sequence $\{\sum_{n=-\infty}^{\infty}c_{mn}e^{in\pi x/l}\}$, with $\lim_{|n|\to\infty}|n|^r c_{mn}=0$ for any integer $r\geq 0$, such that*

$$\lim_{m\to\infty}\frac{1}{2l}\int_0^{2l}\sum_{n=-\infty}^{\infty}c_{mn}e^{in\pi x/l-ip\pi x/l}dx=c_p$$

with $|c_n|=O(|n|^N)$ for some fixed N and then

$$g(x)=\sum_{n=-\infty}^{\infty}c_ne^{in\pi x/l}=\operatorname{Lim}_{m\to\infty}\sum_{n=-\infty}^{\infty}c_{mn}e^{in\pi x/l}.$$

Proof. If g is periodic its Fourier expansion is unique (Corollary 5.14) and so it is determined completely by its Fourier coefficients c_n or, from Theorem 5.19, $\lim_{m\to\infty}(1/2l)\int_{-\infty}^{\infty}\gamma_m(x)\tau(x/2l)e^{-in\pi x/l}dx$ where $\{\gamma_m\}$ is a sequence of good functions defining g. Define

$$c_{mn}=\frac{1}{2l}\int_{-\infty}^{\infty}\gamma_m(x)\tau\left(\frac{x}{2l}\right)e^{-in\pi x/l}dx$$

so that $\lim_{m\to\infty}c_{mn}=c_n$. Considered in terms of n, c_{mn} is the Fourier transform of a good function and therefore $\lim_{|n|\to\infty}|n|^r c_{mn}=0$ for any integer $r\geq 0$. Therefore $\sum_{n=-\infty}^{\infty}c_{mn}e^{in\pi x/l}$ is an infinitely differentiable periodic function of x; it and its derivatives are also bounded and, consequently, it is fairly good. Hence, as in Corollary 5.19,

$$\lim_{m\to\infty}\frac{1}{2l}\int_0^{2l}\sum_{n=-\infty}^{\infty} c_{mn}\mathrm{e}^{\mathrm{i}(n-p)\pi x/l}\mathrm{d}x$$

$$=\lim_{m\to\infty}\frac{1}{2l}\int_{-\infty}^{\infty}\sum_{n=-\infty}^{\infty} c_{mn}\mathrm{e}^{\mathrm{i}(n-p)\pi x/l}\tau\left(\frac{x}{2l}\right)\mathrm{d}x$$

$$=\lim_{m\to\infty}\frac{1}{2l}\sum_{n=-\infty}^{\infty} c_{mn}\int_{-\infty}^{\infty}\tau\left(\frac{x}{2l}\right)\mathrm{e}^{\mathrm{i}(n-p)\pi x/l}\mathrm{d}x$$

$$=\lim_{m\to\infty} c_{mp}=c_p$$

where (24) has been used. This demonstrates half the theorem and shows that for each regular sequence $\{\gamma_m\}$ there is a corresponding sequence of periodic fairly good functions.

Conversely, if

$$\lim_{m\to\infty}\frac{1}{2l}\int_0^{2l}\sum_{n=-\infty}^{\infty} c_{mn}\mathrm{e}^{\mathrm{i}(n-p)\pi x/l}\mathrm{d}x=c_p$$

and $|c_p|=O(|p|^N)$ for some fixed N, $\sum_{p=-\infty}^{\infty} c_p\mathrm{e}^{\mathrm{i}p\pi x/l}$ is a periodic generalised function with period $2l$ by Theorem 5.16. Since the left-hand side can be written, as in Corollary 5.19, $\lim_{m\to\infty}(1/2l)\int_{-\infty}^{\infty}\sum_{n=-\infty}^{\infty} c_{mn}\,\mathrm{e}^{\mathrm{i}(n-p)\pi x/l}\,\tau(x/2l)\mathrm{d}x$ and a Fourier expansion is unique

$$\operatorname*{Lim}_{m\to\infty}\sum_{n=-\infty}^{\infty} c_{mn}\mathrm{e}^{\mathrm{i}n\pi x/l}=\sum_{p=-\infty}^{\infty} c_p\mathrm{e}^{\mathrm{i}p\pi x/l}$$

and the proof of the theorem is complete. □

6

Multiplication and the convolution product

6.1. Multiplication

It is desirable to define the product of two generalised functions g_1 and g_2 in such a way that it agrees with the conventional product $g_1 g_2$ when g_1 and g_2 are conventional functions. However, it is not possible to give a meaning to the product which is applicable to all generalised functions. One reason is that, if $g_1 \in K_1$ and $g_2 \in K_1$, it is not necessarily true that $g_1 g_2 \in K_1$ (e.g. $g_1(x) = 1/|x|^{1/4}, g_2(x) = 1/|x|^{3/4}$) and so the conventional product $g_1 g_2$ may not give rise to a generalised function. Another reason is that if $\{\gamma_{1m}\}$ and $\{\gamma_{2m}\}$ are regular sequences the sequence $\{\gamma_{1m}\gamma_{2m}\}$ need not be regular. Therefore restrictions must be placed on g_1 and g_2 in order that their product may be defined. We have already seen that, if g_1 is limited to the class of fairly good functions, the product $g_1 g_2$ can be satisfactorily specified for any generalised function g_2 (Definition 3.8). If we want g_1 to be less constrained then we must impose some conditions on g_2.

We start by defining multiplication in such a way as to include what is customarily meant by a product in so far as this is possible. Then the definition will be generalised so as to permit the products of a wider class of generalised functions.

Definition 6.1. *If $f_1 \in K_1, f_2 \in K_1$ and $f_1 f_2 \in K_1$ the generalised function determined by $f_1 f_2$ will also be denoted by $f_1 f_2$ (or $f_2 f_1$) and called the product of the generalised functions f_1 and f_2.*

If, temporarily, we signify by $(f_1 f_2)$ the generalised function obtained from $f_1 f_2$, Theorem 3.3 shows that

$$\int_{-\infty}^{\infty} (f_1 f_2)\gamma \mathrm{d}x = \int_{-\infty}^{\infty} f_1 f_2 \gamma \mathrm{d}x$$

where the product on the right-hand side is the usual one. Therefore the definition is reasonable and complies with standard usage for every pair of conventional functions which satisfy the conditions of the definition.

Theorem 6.1. *If* $f_1 \in K_p (p \geq 1)$ *and* $f_2 \in K_q \{q = p/(p-1)\}$ *the generalised function* $f_1 f_2$ *exists.*

Proof. If $p = 1, f_2 \in K_\infty$ and so is the product of a bounded function and a polynomial (Exercise 5, Chapter 3) so that $f_1 f_2 \in K_1$. Since $f_2 \in K_1$ (Theorem 3.2) the result is valid for $p = 1$. When $p > 1$, by Exercise 3 of Chapter 3, $f_1 f_2 \in K_1$ and, by Theorem 3.2, $f_1 \in K_1$ and $f_2 \in K_1$ so that the theorem follows immediately from Definition 6.1. □

It is evident that Definition 6.1 excludes the possibility of multiplying an arbitrary generalised function by a fairly good function because not all generalised functions are in K_1 and, indeed, only conventional functions are permitted so far. Therefore we extend the definition as follows:

Definition 6.2. *If* $g_1 g_3$ *and* $g_1 g_2$ *exist and if* $g_4 = g_1', g_2' = g_3$ *the product* $g_4 g_2$ *is defined by*

$$g_4 g_2 = (g_1 g_2)' - g_1 g_3. \tag{1}$$

The right-hand side of (1) has a definite meaning, because the derivative of a generalised function always exists.

Equation (1) can also be rewritten as

$$(g_1 g_2)' = g_1' g_2 + g_1 g_2'. \tag{2}$$

Since this is the customary rule for conventional functions we conclude that a product which could be defined either by Definition 6.1 or Definition 6.2 would turn out to be the same, i.e. Definition 6.2 is consistent with Definition 6.1. Note also that Definition 6.2 implies that the standard rule for the derivative of a product (2) holds if the product is defined and if either of the products on the right-hand side is known to exist.

The composite generalised function $x^\beta \mathrm{H}(x)$ is defined in Definition 4.2 when β is not a negative integer. If this were regarded as a product then the product $x^{\beta-1}\mathrm{H}(x)$ would also be defined and Definition 6.2 would give

$$\begin{aligned} x^\beta \delta(x) &= \{x^\beta \mathrm{H}(x)\}' - \beta x^{\beta-1}\mathrm{H}(x) \\ &= 0 \end{aligned}$$

on account of Definition 4.2. A similar conclusion would be obtained, when β is a negative integer, from Definition 4.7 and (4.10).

Theorem 6.2. *If $f_1 \in K_1, f_2 \in K_1$ and $f_1 f_2 \in K_1$ and if $g_1 = f_1', g_2' = f_2$ then the product $g_1 g_2$ exists.*

Proof. By hypothesis and Definition 6.1 the product $f_1 f_2$ or $f_1 g_2'$ exists. As in Corollary 3.7, $g_2 \in K_1$ and is continuous so that $f_1 g_2 \in K_1$. Hence the product $f_1 g_2$ exists as a generalised function and the theorem follows from Definition 6.2 with $g_1 g_2 = (f_1 g_2)' - f_1 g_2'$. □

Example 1.

$$\delta(x)(|x|^{3/2} \operatorname{sgn} x) = 0.$$

With $g_1(x) = \delta(x)$, $g_2(x) = |x|^{3/2} \operatorname{sgn} x$ we have $f_1 = \mathrm{H}(x)$, $f_2 = \frac{3}{2}|x|^{1/2}$ from which it is clear that $f_1 f_2 \in K_1$ and so $g_1 g_2$ exists by Theorem 6.2. In fact, by Definition 6.2,

$$\int_{-\infty}^{\infty} \delta(x)(|x|^{3/2} \operatorname{sgn} x)\gamma(x)\mathrm{d}x$$

$$= \int_{-\infty}^{\infty} (|x|^{3/2} \operatorname{sgn} x\, \mathrm{H}(x))'\gamma(x)\mathrm{d}x - \int_{-\infty}^{\infty} \tfrac{3}{2}|x|^{1/2}\mathrm{H}(x)\gamma(x)\mathrm{d}x$$

$$= -\int_{-\infty}^{\infty} |x|^{3/2} \operatorname{sgn} x\, \mathrm{H}(x)\gamma'(x)\mathrm{d}x - \tfrac{3}{2}\int_{0}^{\infty} x^{1/2}\gamma(x)\mathrm{d}x$$

$$= -[x^{3/2}\gamma(x)]_0^{\infty} = 0$$

which is the result stated.

Warning. By definition the product $g_1 g_2$ is commutative, i.e. $g_1 g_2 = g_2 g_1$, but it is not necessarily associative, i.e. the product $(g_1 g_2) g_3$ need not be the same as $g_1(g_2 g_3)$. This can be seen by considering the particular example in which $g_1(x) = \delta(x)$, $g_2(x) = x$, $g_3(x) = x^{-1}$. For $g_1 g_2 = 0$ and so $(g_1 g_2) g_3 = 0$. However $g_2 g_3 = 1$ and therefore $g_1(g_2 g_3) = \delta(x)$. Hence, in this case,

$$(g_1 g_2) g_3 \neq g_1(g_2 g_3).$$

Theorem 6.3. *If $g_1 = g^{(r)}$ and if $g g_2, g g_2', g g_2'', \dots g g_2^{(r)}$ all exist the product $g_1 g_2$ exists.*

Proof. The product $g' g_2$ exists, by Definition 6.2, since $g g_2$ and $g g_2'$

exist. So do the products $g'g_2', \ldots, g'g_2^{(r-1)}$. The existence of $g'g_2$ and $g'g_2'$ implies that of $g''g_2$. Similarly the products $g''g_2', \ldots, g''g_2^{(r-2)}$ exist. Continuing in this way we deduce that $g^{(r)}g_2$ exists and the theorem is proved. □

Corollary 6.3*a*. *If $g_2^{(r)} \in K_1$ for sufficiently large r the product gg_2 exists.*

Proof. By Theorem 3.18, $g = f^{(r)}$ where $f \in K_1$ and is continuous; in fact $|f(x)| \le M(1+x^2)^{k/2}$ for some finite k. Hence $fg_2^{(r)} \in K_1$ and $fg_2^{(r)}$ exists by Definition 6.1. As in Corollary 3.7 $g_2^{(r-1)} \in K_1$ and so, by continually repeating the argument, we see that $fg_2^{(r-1)}, \ldots, fg_2', fg_2$ are all in K_1 and exist as generalised products. The corollary now follows from Theorem 6.3. □

It demonstrates that any generalised function can be multiplied by a function provided that the latter is sufficiently differentiable. In particular

Corollary 6.3*b*. *If $\psi(x)$ is a fairly good function $\psi(x)g(x)$ exists for any generalised function g and*

$$\int_{-\infty}^{\infty} \psi(x)g(x)\gamma(x)\,dx = \int_{-\infty}^{\infty} g(x)\{\psi(x)\gamma(x)\}\,dx.$$

The purpose of this corollary is to show that multiplication as defined in this chapter is consistent with that of Definition 3.8 and Theorem 3.5.

Proof. Since all derivatives of finite order of ψ are certainly in K_1 the existence of $\psi(x)g(x)$ is an immediate consequence of Corollary 6.3*a*. Now, if $r = 1$, Definition 6.2 gives

$$\int_{-\infty}^{\infty} \psi(x)f'(x)\gamma(x)\,dx$$

$$= \int_{-\infty}^{\infty} \{\psi(x)f(x)\}'\gamma(x)\,dx - \int_{-\infty}^{\infty} \psi'(x)f(x)\gamma(x)\,dx$$

$$= -\int_{-\infty}^{\infty} f(x)\{\psi(x)\gamma'(x) + \psi'(x)\gamma(x)\}\,dx$$

$$= \int_{-\infty}^{\infty} f'(x)\{\psi(x)\gamma(x)\}\,dx$$

the rearrangement in the second line being permissible because it involves only conventional integrals. The corollary is thus proved for $g = f'$ and we prove it for $g = f^{(r)}$ by induction. Assume that it is true for $g = f^{(r-1)}$. Then, by Definition 6.2,

$$\int_{-\infty}^{\infty} \psi(x) f^{(r)}(x) \gamma(x) \mathrm{d}x$$

$$= \int_{-\infty}^{\infty} \{\psi(x) f^{(r-1)}(x)\}' \gamma(x) \mathrm{d}x - \int_{-\infty}^{\infty} \psi'(x) f^{(r-1)}(x) \gamma(x) \mathrm{d}x$$

$$= - \int_{-\infty}^{\infty} \psi(x) f^{(r-1)}(x) \gamma'(x) \mathrm{d}x - \int_{-\infty}^{\infty} \psi'(x) f^{(r-1)}(x) \gamma(x) \mathrm{d}x$$

$$= - \int_{-\infty}^{\infty} f^{(r-1)}(x) \{\psi(x)\gamma'(x) + \psi'(x)\gamma(x)\} \mathrm{d}x$$

by assumption since ψ' is also fairly good. The result now follows at once. □

Example 2. If $g' \in K_1$, $\delta(x)g(x)$ exists and equals $g(0)\delta(x)$. For, if $g' \in K_1$, we have $g \in K_1$, $g'(x)\mathrm{H}(x) \in K_1$, $g(x)\mathrm{H}(x) \in K_1$ so that $\mathrm{H}g$ and $\mathrm{H}g'$ exist. Since $\mathrm{H}' = \delta$ the existence of $\delta(x)g(x)$ follows from Theorem 6.3. Indeed

$$\int_{-\infty}^{\infty} \delta(x) g(x) \gamma(x) \mathrm{d}x$$

$$= \int_{-\infty}^{\infty} \{\mathrm{H}(x) g(x)\}' \gamma(x) \mathrm{d}x - \int_{-\infty}^{\infty} \mathrm{H}(x) g'(x) \gamma(x) \mathrm{d}x$$

$$= - \int_{-\infty}^{\infty} \mathrm{H}(x) g(x) \gamma'(x) \mathrm{d}x - \int_{0}^{\infty} g'(x) \gamma(x) \mathrm{d}x$$

$$= - [g(x)\gamma(x)]_0^\infty$$

$$= g(0)\gamma(0)$$

the contribution from the upper limit vanishing as in Theorem 3.7. Consequently

$$\delta(x) g(x) = g(0) \delta(x). \tag{3}$$

Example 3. If $g' \in K_1$ then $g' \ln|x| \in K_1$ and $g \ln|x| \in K_1$. Hence $x^{-1}g$ exists if $g \in K_1$. In fact

$$\int_{-\infty}^{\infty} x^{-1} g(x)\gamma(x)\,dx$$

$$= \int_{-\infty}^{\infty} \{g(x)\ln|x|\}'\gamma(x)\,dx - \int_{-\infty}^{\infty} g'(x)\ln|x|\gamma(x)\,dx$$

$$= -\int_{-\infty}^{\infty} \{g(x)\gamma(x)\}' \ln|x|\,dx.$$

In particular, when $g = 1$, this reduces to what would be expected from Definition 4.3. Here we have a generalisation of the types of integral involving x^{-1} which can be evaluated.

In Example 2, g' does not appear in the final formula. In fact it plays only an intermediary role in proving the result. When this observation is coupled with the knowledge that $\delta(x) = 0$ in $x > 0$ and $x < 0$ we arrive at the suggestion that (3) is valid if only g is suitably specified near the origin, without saying anything about g'. Therefore we introduce

Definition 6.3. *If, in some interval with the origin as an interior point, g is equal to a continuous function then $\delta(x)g(x)$ is defined by*

$$\delta(x)g(x) = g(0)\delta(x).$$

Notice that this is consistent with Example 2 but more general than it. Also nothing is specified about the behaviour of g outside the interval surrounding the origin.

Example 4. If, in some interval with the origin as an interior point, g' is equal to a continuous function then g is also continuous there. Therefore both $g'(x)\delta(x)$ and $g(x)\delta(x)$ are defined by Definition 6.3. Hence, by Definition 6.2,

$$\begin{aligned} g(x)\delta'(x) &= \{g(x)\delta(x)\}' - g'(x)\delta(x) \\ &= \{g(0)\delta(x)\}' - g'(0)\delta(x) \\ &= g(0)\delta'(x) - g'(0)\delta(x). \end{aligned}$$

The reader should check the consistency with Exercise 17 of Chapter 3.

It is evident that nothing we have said so far attaches a meaning to $\{\delta(x)\}^2$.

Exercises

1. Prove that if, in some interval with a as an interior point, g is equal to a continuous function, $\delta(x-a)g(x)$ is defined and

$$\delta(x-a)g(x) = g(a)\delta(x-a).$$

2. Is it true that $|x|^{-1/2}.|x|^{-1/2} = x^{-1}\operatorname{sgn} x$?
3. If $g^{(r)}(x)$ is equal to a continuous function in an interval surrounding the origin justify

$$g(x)\delta^{(r)}(x) = g(0)\delta^{(r)}(x) - rg'(0)\delta^{(r-1)}(x) + \frac{1}{2!}r(r-1)g''(0)\delta^{(r-2)}(x) - \dots$$
$$+(-1)^{r-1}rg^{(r-1)}(0)\delta'(x) + (-1)^r g^{(r)}(0)\delta(x).$$

4. Prove that

$$\int_a^b \delta(x)\mathrm{d}x = \begin{cases} 0 & (0 < a < b \text{ or } a < b < 0) \\ 1 & (a < 0 < b). \end{cases}$$

More generally, if g is continuous at the origin, $\int_a^b g(x)\delta(x)\mathrm{d}x$ is 0 or $g(0)$ in the two cases respectively.
5. The infinite integral of a generalised function can be defined by

$$\int_{-\infty}^{\infty} g(x)\mathrm{d}x = \lim_{n\to\infty} \int_{-n}^{n} g(x)\mathrm{d}x$$

when the right-hand side exists. Show that (i) $\int_{-\infty}^{\infty} \delta(x)\mathrm{d}x = 1$, (ii) $\int_0^\infty x^\beta\,\mathrm{d}x = 0$ if $\beta < -1$ but is not finite if $\beta > -1$, (iii) $\int_{-\infty}^{\infty} x^{-1}\,\mathrm{d}x = 0$.
6. If $2 > \beta > 1$ show that $|x|^\beta.|x|^{1-\beta}\operatorname{sgn} x = x$ and deduce that $|x|^\beta.|x|^{-\beta} = 1$. Prove also that $|x|^\beta.|x|^{-\beta} = 1$ when β is not a negative integer.
7. (i) Prove that $x^\beta.x^{-\beta}\mathrm{H}(x) = \mathrm{H}(x)$. (ii) Consider defining the product $|x|^\mu.|x|^\nu$ for all complex μ and ν by $|x|^\mu.|x|^\nu = |x|^{\mu+\nu}$.
8. If $h(x)$ is continuous in a neighbourhood which includes $x = a$ and also in one which includes $x = b$, show that $h(x)\{\mathrm{H}(x-a) - \mathrm{H}(x-b)\}$ is defined and that

$$[h(x)\{\mathrm{H}(x-a) - \mathrm{H}(x-b)\}]'$$
$$= h'(x)\{\mathrm{H}(x-a) - \mathrm{H}(x-b)\} + h(x)\{\delta(x-a) - \delta(x-b)\}.$$

Deduce that, if $g(x) = h'(x)$,

$$\int_a^b g(x)\mathrm{d}x = h(b) - h(a).$$

Hence show that, if h is continuous in $|x| > x_0$, $\int_{-\infty}^{\infty} g(x)\mathrm{d}x$ exists if $\lim_{n\to\infty} \{h(n) - h(-n)\}$ exists. (This is a generalisation of the usual formula for integration to integrals of generalised functions.)

9. If $\mathrm{Lim}_{\mu\to\mu_0} h_\mu(x) = h(x)$ in the sense of generalised functions and h_μ, h are continuous in intervals surrounding $x = a$ and $x = b$, and if $g_\mu = h'_\mu$ and $g = h'$ show that

$$\lim_{\mu\to\mu_0} \int_a^b g_\mu(x)\mathrm{d}x = h(b) - h(a).$$

10. If $g(x)$ is continuous in a neighbourhood of $x = a$ prove that

$$\int_0^\infty g(x)\delta(x-a)\mathrm{d}x = \begin{cases} g(a) & (a > 0), \\ \frac{1}{2}g(0) & (a = 0). \end{cases}$$

11. By starting from the formula

$$\{\mathrm{H}(x-a)\ln|x|\}' = x^{-1}\mathrm{H}(x-a) + \delta(x-a)\ln|a| \qquad (a \neq 0)$$

find conditions for the existence of $x^{-1}\mathrm{H}(x-a)g(x)$. Hence prove the generalisation of (4.7)

$$\int_a^b x^{-1} g(x)\gamma(x)\mathrm{d}x$$

$$= -\int_a^b \{g(x)\gamma(x)\}'\ln|x|\mathrm{d}x + g(b)\gamma(b)\ln|b| - g(a)\gamma(a)\ln|a|$$

for $a < 0 < b$ and g' continuous on an interval including $[a, b]$.

12. Is there a result corresponding to that in Exercise 11 for $\int_0^b x^\beta g(x)\gamma(x)\mathrm{d}x$?

13. Suppose that one attempts to define a product by requiring that $\{\gamma_{1m}\gamma_{2m}\}$ is regular for every regular sequence $\{\gamma_{1m}\}$ defining g_1 and every regular sequence $\{\gamma_{2m}\}$ defining g_2. Show that it is necessary and sufficient that the product shall be $\mathrm{Lim}_{m\to\infty} g_1\gamma_{2m}$ for every regular $\{\gamma_{2m}\}$. (Hint: Approximate g_1 by suitable γ_{1m} via Corollary 3.20*b*.)

14. With the multiplication of Exercise 13 show that $g(x)\delta(x)$ cannot be defined. (Hint: If $\{\gamma_{2m}\}$ is regular so is $\{\gamma_{2m} + m^{1/2}\mathrm{e}^{-x^2/m^2}\cos mx\}$.) This shows that the class of generalised functions which can be multiplied by the rule of Exercise 13 is not so wide as that provided by the rule adopted in the text. If we merely required that $\mathrm{Lim}_{m\to\infty} g_1\gamma_{2m}$ existed for *some* regular $\{\gamma_{2m}\}$ what would it be essential to prove?

15. If $g = f^{(r)}$ where $f \in K_1$ and is continuous, if f_1 is a continuous function with r continuous derivatives show, by using Definition 4.6, that a meaning may be attached to $\int_a^b g(x)f_1(x)\mathrm{d}x$ in suitable circumstances.

16. If $f_1 \in K_1$, $f_2 \in K_1$ and $f_1 f_2 \in K_1$ show that $\psi(f_1 f_2) - (\psi f_1)f_2 = f_1(\psi f_2)$ for a fairly good ψ. Deduce that if $g_1 g_2$ exists as in Theorem 6.3 $\psi(g_1 g_2) = (\psi g_1)g_2 = g_1(\psi g_2)$.

17. Suppose $f_{1\mu}, f_{2\mu}, f'_{2\mu} \in K_1$, are equicontinuous and $f_{1\mu}f_{2\mu} \in K_1$, $f_{1\mu}f'_{2\mu} \in K_1$.

Let $\lim_{\mu\to\mu_0} f_{1\mu} = f_1$, $\lim_{\mu\to\mu_0} f_{2\mu} = f_2$ where f_1 and f_2 have similar properties to $f_{1\mu}$ and $f_{2\mu}$. Prove that

$$\lim_{\mu\to\mu_0} f'_{1\mu} f_{2\mu} = f'_1 f_2.$$

Deduce that, if $g_\mu = f_{1\mu}^{(r)}$ and $f_{1\mu} f_{2\mu}, f_{1\mu} f'_{2\mu}, \ldots, f_{1\mu} f_{2\mu}^{(r)}$ all belong to K_1 and are equicontinuous, as well as their limits,

$$\operatorname*{Lim}_{\mu\to\mu_0} g_\mu f_{2\mu} = g f_2.$$

18. If f_1, f_2, f_3 all belong to K_∞, i.e. each is the product of a bounded function and a polynomial, show that $f_1(f_2 f_3) = (f_1 f_2) f_3$. More generally, show that this associativity holds if $f_1 \in K_1$ and $f_2, f_3 \in K_\infty$.

19. If $f_1 \in K_p$ $(p \geq 1)$, $f_2 \in K_q$ $(q = p/(p-1))$ and $f_3 \in K_\infty$ show that

$$f_1(f_2 f_3) = (f_1 f_2) f_3.$$

20. If g is infinitely differentiable for $x \geq b > 0$ and

$$\int_b^x g(t)\mathrm{d}t = \sum_{m=0}^{p} \sum_{n=0}^{q} a_{mn} x^m \ln^n x + o(1)$$

as $x \to \infty$, with p and q finite, a possible definition is $\int_b^\infty g(t)\mathrm{d}t = a_{00}$. Show that this is consistent with conventional integration when g is integrable.

Taking $\int_0^\infty = \int_0^b + \int_b^\infty$ with $\int_0^b$ calculated as in Chapter 4 show that $\int_0^\infty x^\beta \mathrm{d}x = 0$, $\int_0^\infty x^\beta \ln x \mathrm{d}x = 0$ for any complex β. Prove also that

(i) $\displaystyle\int_0^\infty \frac{x^{2m-1}}{1+x^2}\mathrm{d}x = 0,$ (ii) $\displaystyle\int_0^\infty \frac{x^{2m}}{1+x^2}\mathrm{d}x = (-1)^m \tfrac{1}{2}\pi,$

(iii) $\displaystyle\int_0^\infty \frac{x^3}{(1+x^2)^2}\mathrm{d}x = -\tfrac{1}{2},$

(iv) $\displaystyle\int_0^\infty (x^2+a^2)^\gamma \mathrm{d}x = \frac{(-\gamma-\frac{3}{2})!\,\pi^{1/2} a^{2\gamma+1}}{(-\gamma-1)!\,2}$ $\quad (a > 0, 2\gamma \neq -1, 1, 3 \ldots).$

This infinite integral covers more possibilities than that of Exercise 5.

Theorem 6.4. *If $g_1 = f_1^{(r)}$ where $f_1 \in K_p (p \geq 1)$, $g_2^{(r)} \in K_q (q = p/(p-1))$ and $g_3^{(r)} \in K_\infty$ then $g_1(g_2 g_3) = (g_1 g_2) g_3$.*

Proof. The theorem will be proved by induction, assuming it to be true for some $s \leq r-1$. We first prove that $g^{(s-1)} \in K_p$ if $g^{(s)} \in K_p (p \geq 1)$.

For

$$\left|\int_0^x g^{(s)}(t)\mathrm{d}t\right| \le (1+x^2)^N \int_0^{|x|} \frac{|g^{(s)}(t)|}{(1+t^2)^N}\mathrm{d}t \le (1+x^2)^N \int_{-\infty}^{\infty} \frac{|g^{(s)}(t)|}{(1+t^2)^N}\mathrm{d}t$$

and the integral is finite since $g^{(s)} \in K_1$ by Theorem 3.2. Thus the left-hand side belongs to K_∞ and, since $g^{(s-1)}$ differs from it by a constant at most, it follows that $g^{(s-1)} \in K_\infty \subset K_p$ by Theorem 3.2. Hence the given conditions are valid for $s-1$ if they are valid for s. One concludes that $g_2^{(s)} g_3^{(t)} \in K_q$ for $1 \le s \le r$, $1 \le t \le r$.

Assuming that associativity holds for $s-1$ $(1 \le s \le r)$, we have

$$\{f_1^{(s-1)}(g_2 g_3)\}' = \{(f_1^{(s-1)} g_2) g_3\}'$$

or, from Definition 6.2,

$$f_1^{(s)}(g_2 g_3) + f_1^{(s-1)}(g_2 g_3)' = (f_1^{(s-1)} g_2)' g_3 + (f_1^{(s-1)} g_2) g_3'$$

because (i) the second term on the left-hand side exists by Theorems 6.3 and 6.1 since $f_1 \in K_p$ and $(g_2 g_3)^{(t)} \in K_q$ by what has been proved above and (ii) the second term on the right-hand exists and equals $f_1^{(s-1)}(g_2 g_3')$ by assumption. Hence, applying Definition 6.2 again and noting that $g_2' g_3$ and $f_1^{(r)} g_2$ exist, we have

$$f_1^{(s)}(g_2 g_3) + f_1^{(s-1)}(g_2' g_3 + g_2 g_3') = (f_1^{(s)} g_2 + f_1^{(s-1)} g_2') g_3 + f_1^{(s-1)}(g_2 g_3').$$

By assumption $f_1^{(s-1)}(g_2' g_3) = (f_1^{(s-1)} g_2') g_3$ and consequently

$$f_1^{(s)}(g_2 g_3) = (f_1^{(s)} g_2) g_3.$$

Thus associativity, if it holds for $s-1$, is valid for $s(\le r)$ and, since it is known to be true for $s = 1$ (Exercise 19), the theorem is proved. □

The illustration of non-associativity in the warning (after Example 1), where two of the factors are first derivatives of functions in K_1 and the third belongs to K_∞, shows that there is little hope of weakening the conditions in Theorem 6.4 and still retaining associativity.

Finally we should note a definition related to Definition 6.3. Essentially this definition works because the points, where the two factors of the product are not continuous functions, are separated. With this idea in mind we formulate

Definition 6.4. *Let $\psi_1, \psi_2, \dots, \psi_M$ (where M may be infinite) be fairly good functions such that $\sum_{m=1}^M g_1(x)\psi_m(x) = g_1(x)$. Then, if*

$(g_1\psi_m)g_2$ exists for every m (and if, when M be infinite, $\sum_{m=1}^{\infty}$ $(g_1\psi_m)g_2$ exists) g_1g_2 is defined by $g_1g_2 = \sum_{m=1}^{M}\{(g_1\psi_m)g_2\}$.

We shall not attempt any verification that the product so obtained is independent of the particular choice of ψs since in most cases of interest it can be checked without difficulty.

Suppose that $g_1(x) = \eta_1(x)|x - a_1|^{\beta_1}$ and $g_2(x) = \eta_2(x)|x - a_2|^{\beta_2}$ where η_1 and η_2 are infinitely differentiable and $a_1 \neq a_2$. Then select ψ_1 as a fine function which equals 1 in $(a_1 - \varepsilon, a_1 + \varepsilon)$ and vanishes outside $(a_1 - 2\varepsilon, a_1 + 2\varepsilon)$ (ε is chosen so small that a_2 does not lie in this interval). Take $\psi_2 = 1 - \psi_1$; then ψ_2 is fairly good and vanishes in $(a_1 - \varepsilon, a_1 + \varepsilon)$. Since ψ_1 is non-zero only in $(a_1 - 2\varepsilon, a_1 + 2\varepsilon)$ and g_2 is infinitely differentiable in this interval it is clear that $(g_1\psi_1)g_2$ exists. Also $g_1\psi_2$ is infinitely differentiable since it possesses no singularity outside $(a_1 - \varepsilon, a_1 + \varepsilon)$ and so $(g_1\psi_2)g_2$ exists. Then

$$g_1g_2 = (g_1\psi_1)g_2 + (g_1\psi_2)g_2.$$

It is obvious that the product can be defined in this way when g_1 and g_2 have a finite number of distinct singularities. Further η_1 and η_2 need not be infinitely differentiable so long as they have sufficient derivatives (cf. Theorem 6.3).

It would be possible to define an infinite product by a limiting process as for series but, in view of the difficulties of multiplication, there seems little point in developing a general theory although particular infinite products may be useful.

6.2. Division

Division, the inverse of multiplication, has some peculiarities of its own where generalised functions are concerned. Part of the cause is the fact that multiplication is not in general associative. Consider the problem of finding g for given g_1 such that

$$xg(x) = g_1(x).$$

We know, from Theorem 4.2, that there is not a unique solution. In fact, if $g_0(x)$ is one solution, another is $g_0(x) + C\delta(x)$ where C is an arbitrary constant. But what can be said about g_0? Clearly if $x > 0$ or $x < 0$, $g_0(x) = g_1(x)/x$ since in these regions we are involved only with the conventional function $1/x$. However, it is not self-evident that we can write $g_0(x) = x^{-1}g_1(x)$ because the product $x^{-1}g_1(x)$ may not be defined under the rules of multiplication. For example, if $g_1(x) = \delta(x)$, $x^{-1}g_1(x)$ does not come within the scope of our definitions. Nevertheless g_0 exists, being $-\delta'(x)$; an attempt might be made to overcome this difficulty by defining $x^{-1}\delta(x) = -\delta'(x) + C_0\delta(x)$. It is not immediately obvious that this equation can be verified by multiplying by x, even though $xx^{-1} = 1$, because the example in the warning of §6.1 shows that if the δ is multiplied by x first a different answer is obtained. It would be necessary to show that the whole generalised function $x^{-1}\delta(x)$ when multiplied by x agreed with the corresponding result from the right-hand side.

On the other hand there are cases when multiplication by x^{-1} is permissible, for example $g_1(x) = 1$ when $g_0(x) = x^{-1}$, or $g_1(x) = \sin x$ when $g_0(x) = (\sin x)/x$.

Although it may not be possible to put $g_0(x) = x^{-1}g_1(x)$ always, yet g_0 always exists. This is due to

Theorem 6.5. *For any generalised function g_1 the equation $xg(x) = g_1(x)$ always has a solution; it is given by $g(x) = x^{-1}g_1(x) + C\delta(x)$ when $g'_1 \in K_1$.*

Proof. Evidently, Example 3 shows that a sufficient condition for $x^{-1}g_1(x)$ to exist is that $g'_1 \in K_1$ and then

$$\begin{aligned}
&\int_{-\infty}^{\infty} x\{x^{-1}g_1(x)\}\gamma(x)\,dx \\
&\qquad = -\int_{-\infty}^{\infty} \{g_1(x)x\gamma(x)\}' \ln|x|\,dx \\
&\qquad = -\int_{-\infty}^{\infty} [g_1(x)\gamma(x) + \{g_1(x)\gamma(x)\}'x] \ln|x|\,dx \\
&\qquad = -\int_{-\infty}^{\infty} g_1(x)\gamma(x)\ln|x|\,dx + \int_{-\infty}^{\infty} g_1(x)\gamma(x)\{x\ln|x|\}'\,dx \\
&\qquad = \int_{-\infty}^{\infty} g_1(x)\gamma(x)\,dx
\end{aligned}$$

since $\{x \ln|x|\}' = 1 + \ln|x|$. Hence $x^{-1} g_1(x)$ is a solution of the equation when $g_1' \in K_1$.

To show that g exists in general write, as in Theorem 4.2,

$$\int_{-\infty}^{\infty} g(x)\gamma(x)\mathrm{d}x = \int_{-\infty}^{\infty} g(x)\gamma(0)\mathrm{e}^{-x^2}\mathrm{d}x + \int_{-\infty}^{\infty} xg(x)\frac{\gamma(x)-\gamma(0)\mathrm{e}^{-x^2}}{x}\mathrm{d}x$$
$$= \gamma(0)\int_{-\infty}^{\infty} g(x)\mathrm{e}^{-x^2}\mathrm{d}x + \int_{-\infty}^{\infty} g_1(x)\frac{\gamma(x)-\gamma(0)\mathrm{e}^{-x^2}}{x}\mathrm{d}x.$$

The second term has a definite meaning since it involves the product of a generalised function and a good function. The first term can be definitely specified by ascribing a value, say C, to $\int_{-\infty}^{\infty} g(x)\mathrm{e}^{-x^2}\mathrm{d}x$. Hence the left-hand side has a meaning for every good γ, i.e. g exists as a generalised function. The proof is concluded. □

In the same way as we proved Theorem 6.5 we may show

Corollary 6.5. *If $f^{(m)} \in K_1$ the solution of $x^m g(x) = f(x)$ is*

$$g(x) = x^{-m} f(x) + \sum_{r=1}^{m} C_r \delta^{(r-1)}(x).$$

More general division can be carried out on the basis of Theorem 6.5. Let $\eta(x)$ be an infinitely differentiable function such that $|\eta(x)| \geq |x|^{-r}$ as $|x| \to \infty$, with r finite. If it has no zeros on the real axis $\eta(x)g(x) = g_1(x)$ is solved by $g(x) = g_1(x)/\eta(x)$. (Without the restriction on η at infinity, $1/\eta$ might grow too rapidly at infinity.) If it has one simple zero at $x = a$, then $\eta(x)/(x-a)$ and $(x-a)/\eta(x)$ are infinitely differentiable and without zeros. Hence an equivalent form to $\eta(x)g(x) = g_1(x)$ is

$$(x-a)g(x) = \frac{x-a}{\eta(x)} g_1(x)$$

which can now be solved by Theorem 6.5.

If $\eta(x)$ possessed more than one simple zero, say two, the problem would be reduced to one of solving

$$(x-a)(x-b)g(x) = \frac{(x-a)(x-b)}{\eta(x)} g_1(x)$$

which could be dealt with by two applications of Theorem 6.5. Thus, in suitable circumstances,

$$g(x) = (x-a)^{-1}(x-b)^{-1}\frac{(x-a)(x-b)}{\eta(x)}g_1(x)$$
$$+C_1\delta(x-a)+C_2\delta(x-b).$$

Of course, for suitable g_1 this could be expressed as

$$g(x) = \frac{g_1(x)}{\eta(x)} + C_1\delta(x-a) + C_2\delta(x-b)$$

with the convention that zeros of the denominator shall be understood as $(\ )^{-1}$ in the numerator.

In principle this method can be extended to any number of isolated simple zeros, though if there be an infinite number of them it may be more convenient to replace g_1 by $\sum_{n=-\infty}^{\infty} g_1(x)\tau(x-n)$ (Theorem 5.15) and solve the equations $\eta(x)g_n(x) = g_1(x)\tau(x-n)$. Each of these equations involves only a finite number of zeros where the right-hand side is non-vanishing and $g = \sum_{n=-\infty}^{\infty} g_n$.

There is no difficulty in extending these ideas to the case of multiple zeros; this extension is left as an exercise for the reader.

Exercises

21. If $x^k g(x) = f(x)$ and $f^{(k)} \in K_1$ show that

$$g(x) = x^{-k}f(x) + \sum_{r=0}^{k-1} C_r\delta^{(r-1)}(x).$$

22. If the zeros of $\eta(x)$ occur at $a_1, a_2, \ldots$ and are of multiplicity $n_1, n_2, \ldots$ show that any two solutions of $\eta(x)g(x) = g_1(x)$ differ by $\sum_{m=1}^{\infty}\sum_{r=0}^{n_m-1} C_{mr}\delta^{(r)}(x-a_m)$.

23. If $A(x) \in K_1$ and is an analytic function such that $|A(x)| \geq |x|^{-r}$, r finite, as $|x| \to \infty$ prove that $A(x)g(x) = 1$ always has a solution.

24. Give conditions (if any) under which $|x+a|^{1/2}g(x) = g_1(x)$ implies that $g(x) = |x+a|^{-1/2}g_1(x) + C\delta(x+a)$ when (i) a is not real, (ii) a is real.

25. Give a proof of Theorem 6.5 by using Fourier transforms and the remarks at the end of Theorem 3.18.

26. Show that the general solution of $(1-x^2)^k g(x) = 0$ is

$$g(x) = \sum_{r=0}^{k-1}\{C_r\delta^{(r)}(x-1) + C'_r\delta^{(r)}(x+1)\}.$$

27. Show that the solution of $(x-a)(x-b)g(x) = 1$ $(a \neq b)$ is

$$g(x) = \frac{1}{a-b}\{(x-a)^{-1} - (x-b)^{-1}\} + C_1\delta(x-a) + C_2\delta(x-b).$$

This justifies the use of partial fractions in §4.5 for expressions such as $(x-a)^{-1}(x-b)^{-1}$.

What is the solution when the right-hand side is $f(x)$ instead of 1 and $f' \in K_1$?

28. Show that the solution of $(x^2-a^2)g(x) = x^3$ is

$$g(x) = x + \tfrac{1}{2}a^2\{(x-a)^{-1} + (x+a)^{-1}\} + C_1\delta(x-a) + C_2\delta(x+a).$$

29. If $xg_1(x) = g_2(x)$ and $g_2(x)$ is the function $f(x)$ in $a < x < b$ where either $a > 0$ or $b < 0$ show that $g_1(x) = f(x)/x$ in $a < x < b$. (Hint: Definition 3.5.)

Enough has been said to indicate that division is a far from straightforward matter in general but that many cases which arise in practice can be dealt with although they may need individual treatment.

6.3. The convolution

We have defined multiplication of generalised functions only for certain classes. We turn now to a different kind of product which is a generalisation of that signified by $\int_{-\infty}^{\infty} g_1(t)g_2(x-t)\mathrm{d}t$. It is called the *convolution* (or *resultant* or *composition* or *Faltung*) of g_1 and g_2 and often written $g_1 * g_2$. There are problems similar to those of multiplication namely that the convolution cannot be defined for any pair of generalised functions: one or both have to be restricted. We start by applying sufficient restrictions on one member of the product for the other to be any generalised function.

Theorem 6.6. *If $\Psi(\alpha)$ is the Fourier transform of a fairly good function and $\Gamma(\alpha)$ is any good function then $\int_{-\infty}^{\infty} \Psi(\alpha)\Gamma(\beta-\alpha)\mathrm{d}\alpha$ is a good function of β.*

Proof. Since Ψ is a generalised function, Theorem 3.9*a* gives

$$\int_{-\infty}^{\infty} \Psi(\alpha)\Gamma(\beta-\alpha)\mathrm{d}\alpha = 2\pi \int_{-\infty}^{\infty} \mathrm{e}^{-i\beta t}\psi(t)\gamma(t)\mathrm{d}t. \quad (4)$$

But $\psi\gamma$ is good because ψ is fairly good (Theorem 2.1) and so the right-hand side is the Fourier transform of a good function and therefore good (Theorem 2.4). The theorem is proved. □

Definition 6.5. *If $\{\Gamma_m\}$ is a regular sequence defining the generalised* *function G and Ψ is the Fourier transform of a fairly good function then* $\{\int_{-\infty}^{\infty} \Psi(\alpha)\Gamma_m(x-\alpha)\mathrm{d}\alpha\}$ *is a regular sequence which defines a generalised function to be denoted by* $\int_{-\infty}^{\infty} \Psi(\alpha)G(x-\alpha)\mathrm{d}\alpha$ *or* $\int_{-\infty}^{\infty} \Psi(x-\alpha)G(\alpha)\mathrm{d}\alpha$ *or* $\Psi * G$.

In this way we define the convolution of any generalised function and the Fourier transform of a fairly good function. Before considering examples it is necessary to justify the statement that the sequence $\{\int_{-\infty}^{\infty} \Psi(\alpha)\Gamma_m(x-\alpha)\mathrm{d}\alpha\}$ is regular. Firstly, it is certainly a sequence of good functions by Theorem 6.6. Secondly, if $\Gamma(x)$ is any good function (4) shows that

$$\int_{-\infty}^{\infty} \Gamma(x)\left[\int_{-\infty}^{\infty} \Psi(\alpha)\Gamma_m(x-\alpha)\mathrm{d}\alpha\right]\mathrm{d}x$$
$$= \int_{-\infty}^{\infty} \Gamma(x)\left[2\pi\int_{-\infty}^{\infty} \mathrm{e}^{-\mathrm{i}xt}\psi(t)\gamma_m(t)\mathrm{d}t\right]\mathrm{d}x$$
$$= 2\pi\int_{-\infty}^{\infty} \psi(t)\gamma_m(t)\left[\int_{-\infty}^{\infty} \Gamma(x)\mathrm{e}^{-\mathrm{i}xt}\mathrm{d}x\right]\mathrm{d}t$$

the interchange being valid by Theorem 1.35 because only conventional integrals of good functions are involved. Hence

$$\int_{-\infty}^{\infty} \Gamma(x)\left[\int_{-\infty}^{\infty} \Psi(\alpha)\Gamma_m(x-\alpha)\mathrm{d}\alpha\right]\mathrm{d}x$$
$$= 4\pi^2\int_{-\infty}^{\infty} \psi(t)\gamma_m(t)\gamma(-t)\mathrm{d}t$$
$$\to 4\pi^2\int_{-\infty}^{\infty} g(t)\psi(t)\gamma(-t)\mathrm{d}t \qquad (5)$$

as $m \to \infty$ because $\psi(t)\gamma(-t)$ is good. Thus the sequence is regular. Furthermore, since the sequence can be written

$$\left\{\int_{-\infty}^{\infty} \Psi(x-\alpha)\Gamma_m(\alpha)\mathrm{d}\alpha\right\},$$

the second way of expressing the generalised function defined by the sequence is reasonable.

Another conclusion can be drawn by applying Theorem 3.9 to (5). It is

$$\int_{-\infty}^{\infty} \Gamma(x)\left[\int_{-\infty}^{\infty} \Psi(\alpha)G(x-\alpha)\mathrm{d}\alpha\right]\mathrm{d}x = 2\pi\int_{-\infty}^{\infty} G_0(x)\Gamma(x)\mathrm{d}x$$

where $G_0(x)$ is the Fourier transform of $g(t)\psi(t)$. Hence

Theorem 6.7. *If Ψ is the Fourier transform of a fairly good function*

$$\Psi * G = \int_{-\infty}^{\infty} \Psi(\alpha)G(x-\alpha)\mathrm{d}\alpha = 2\pi \int_{-\infty}^{\infty} g(t)\psi(t)\mathrm{e}^{-\mathrm{i}tx}\mathrm{d}t.$$

This theorem constitutes a generalisation of Theorem 3.9*a*.

Example 5. Since 1 is a fairly good function, one possibility for Ψ is $2\pi\delta(\alpha)$. Then, because $\int_{-\infty}^{\infty} g(t)\mathrm{e}^{-\mathrm{i}tx}\mathrm{d}t = G(x)$, Theorem 6.7 gives

$$\int_{-\infty}^{\infty} \delta(\alpha)G(x-\alpha)\mathrm{d}\alpha = \int_{-\infty}^{\infty} \delta(x-\alpha)G(\alpha)\mathrm{d}\alpha = G(x)$$

or, more briefly,

$$\delta * G = G. \tag{6}$$

Similarly, it may be proved that

$$\int_{-\infty}^{\infty} \delta'(x-\alpha)G(\alpha)\mathrm{d}\alpha = G'(x). \tag{7}$$

Note that (6) and (7) imply that

$$(\delta * G)' = \delta' * G = \delta * G' = G'$$

and, more generally,

$$(\delta * G)^{(n)} = \delta * G^{(n)} = G^{(n)} = \delta^{(n)} * G. \tag{8}$$

This property will be found to be of value in the discussion of differential equations.

Example 6. The function $(1+x^2)^{-\nu-1/2}$ is fairly good for $\nu + \frac{1}{2} \geq 0$. Its Fourier transform can be obtained by using Basset's formula† for the Bessel function K_ν. According to it, for $\mathrm{Re}(\nu+\frac{1}{2}) \geq 0$,

$$\int_{-\infty}^{\infty} \frac{\mathrm{e}^{-\mathrm{i}\alpha x}}{(1+x^2)^{\nu+1/2}}\mathrm{d}x = \frac{\pi^{1/2}|\alpha|^{\nu}}{(\nu-\frac{1}{2})!\,2^{\nu-1}} K_\nu(|\alpha|). \tag{9}$$

† See, for example, G. N. Watson, *Theory of Bessel Functions*, 2nd edition, Cambridge University Press (1944), p. 172.

Therefore Theorem 6.7 shows that for any generalised function G

$$\int_{-\infty}^{\infty} G(\alpha)|x-\alpha|^{\nu} K_{\nu}(|x-\alpha|)\mathrm{d}\alpha = (\nu-\tfrac{1}{2})!2^{\nu}\pi^{1/2}\int_{-\infty}^{\infty} \frac{g(t)\mathrm{e}^{-itx}}{(1+t^2)^{\nu+1/2}}\mathrm{d}t \tag{10}$$

when $\nu+\frac{1}{2}>0$.

There is a similar formula which stems from

$$\int_{1}^{\infty} \frac{\cos \alpha t}{(t^2-1)^{\nu+1/2}}\mathrm{d}t = (-\tfrac{1}{2}-\nu)!(-\tfrac{1}{2}\pi^{1/2})(\tfrac{1}{2}|\alpha|)^{\nu} Y_{\nu}(|\alpha|).$$

Exercises

30. Show, using the convolution, that

$$\tfrac{1}{2}\{G(x-y)+G(x+y)\} = \int_{-\infty}^{\infty} g(t)\cos yt\,\mathrm{e}^{-itx}\mathrm{d}t.$$

31. Prove that

$$\int_{-\infty}^{\infty} \mathrm{e}^{-|x-\alpha|}G(\alpha)\mathrm{d}\alpha = 2\int_{-\infty}^{\infty} \frac{g(t)\mathrm{e}^{-itx}}{1+t^2}\mathrm{d}t.$$

32. Show that, when $a>0$,

$$\int_{0}^{\infty} G(x-\alpha)\mathrm{e}^{-a\alpha}\mathrm{d}\alpha = \int_{-\infty}^{\infty} \frac{g(t)\mathrm{e}^{-itx}}{a-it}\mathrm{d}t.$$

33. Prove that

$$\int_{-\infty}^{\infty} \Psi(\alpha)G(x-\alpha)\mathrm{d}\alpha = 2\int_{-\infty}^{\infty} g(t)\frac{\sin at}{t}\mathrm{e}^{-itx}\mathrm{d}t$$

where $\Psi(\alpha)=\mathrm{H}(\alpha+a)-\mathrm{H}(\alpha-a)$. Adopting the convention of §4.6 we can write the left-hand side as $\int_{-a}^{a} G(x-\alpha)\mathrm{d}\alpha$.

34. Show that Definition 6.5 would remain legitimate if the phrase 'fairly good function' were replaced by 'moderately good function'. Deduce that

$$\int_{-\infty}^{\infty} \mathrm{e}^{(\pi/4)\mathrm{i}-\mathrm{i}\alpha^2/4\lambda}G(x-\alpha)\mathrm{d}\alpha = 2\pi^{1/2}\lambda^{1/2}\int_{-\infty}^{\infty} g(t)\mathrm{e}^{\mathrm{i}(\lambda t^2-tx)}\mathrm{d}t.$$

35. Prove that (10) is still valid if ν is complex but $\mathrm{Re}(\nu+\frac{1}{2})\geq 0$.

36. The Bessel function K_{ν} satisfies

$$\left(\frac{\mathrm{d}^2}{\mathrm{d}z^2}-1\right)z^{\nu}K_{\nu}(z) = (1-2\nu)z^{\nu-1}K_{\nu-1}(z)$$

for all z when Re $(\nu-1)>0$. Define $|\alpha|^\mu K_\mu(|\alpha|)$ for other μ by

$$|\alpha|^\mu K_\mu(|\alpha|)=\frac{D^k\{|\alpha|^{\mu+k}K_{\mu+k}(|\alpha|)\}}{2^k(\mu+\frac{1}{2})(\mu+\frac{3}{2})\dots(\mu+k-\frac{1}{2})}$$

where Re $\mu+k\geq 0$ and $Df=f-f''$. Show that the definition is independent of k and satisfies the equation of the first sentence. The definition fails when μ is a negative half-integer; this is because $K_{-1/2}(|\alpha|)=(\pi/2|\alpha|)^{1/2}$ $e^{-|\alpha|}$ so that there are difficulties similar to those for x^{-1} sgn x in §4.4.

Show that (9) and (10) remain valid for complex ν so long as ν is not a negative half-integer. What does the left-hand side of (9) equal when ν is a negative half-integer? Deduce that

$$\lim_{\mu\to-m+1/2}\frac{|\alpha|^\mu K_\mu(|\alpha|)}{(\mu-\frac{1}{2})!}=\frac{\pi^{1/2}}{2^{m-1/2}}D^{m-1}\delta(\alpha).$$

Explore the consequences of the definition

$$|\alpha|^{(1/2)-m}K_{(1/2)-m}(|\alpha|)=\lim_{\mu\to 0}\left\{|\alpha|^{\mu+(1/2)-m}K_{\mu+(1/2)-m}(|\alpha|)-\frac{(-1)^{m-1}\pi^{1/2}}{(m-1)!2^{m-1/2}\mu}D^{m-1}\delta(\alpha)\right\}.$$

6.4. Properties of the convolution

It has been shown in Theorem 3.3 that, if $f\in K_1$, the sequence $\{\int_{-\infty}^{\infty}f(u)\rho\{n(u-x)\}ne^{-u^2/n^2}du\}$ is regular and defines a generalised function, also denoted by f. Theorem 3.18 shows that any generalised function g can be expressed in the form $g=f^{(r)}$ where $f\in K_1$; indeed f is actually continuous. On account of Definition 3.9 the sequence

$$\left\{\frac{d^r}{dx^r}\int_{-\infty}^{\infty}f(u)\rho\{n(u-x)\}ne^{-u^2/n^2}du\right\}$$

or

$$\left\{\int_{-\infty}^{\infty}[f(u)e^{-u^2/n^2}]^{(r)}\rho\{n(u-x)\}n\,du\right\}$$

is regular and defines g. This shows how we can construct a regular sequence for any g.

Other sequences can of course be devised. For example, suppose that e^{-u^2/n^2} is replaced by the fine function $\phi_n(u)$ which is such that

$$\phi_n(u)=1\quad(|u|\leq n),\qquad \phi_n(u)=0\quad(|u|\geq n+1)$$

and $0 \leq \phi_n(u) \leq 1$ (as in §5.5, $\phi_n(u)$ might be $\sum_{m=-n}^{n} \tau(x-m)$). Then, if the analysis of Theorem 3.3 be repeated, it will be seen that we now have a sequence of fine functions which is regular and defines $f \in K_1$. (The only significant modification of proof required is that

$$\left| \int_{-\infty}^{\infty} f(u)\gamma(u)\{\phi_n(u)-1\}\,\mathrm{d}u \right| = \left| \int_{n}^{\infty} + \int_{-\infty}^{-n} f(u)\gamma(u)\{\phi_n(u)-1\}\,\mathrm{d}u \right|$$

$$\leq \max_{|u| \geq n}(1+u^2)^N |\gamma(u)| \int_{-\infty}^{\infty} \frac{|f(t)|}{(1+t^2)^N}\,\mathrm{d}t$$

$$\to 0$$

as $n \to \infty$ or, alternatively, apply Theorem 5.15.) Applying again the argument of the preceding paragraph we see that *any generalised function can be defined by a regular sequence of fine functions.*

In both cases the convolution of a generalised function and of a good function is involved. In one case the factor e^{-u^2/n^2} is added to ensure that a good function is obtained; in the other the factor $\phi_n(u)$ makes sure the function is fine. If one dispenses with such a requirement the sequence can be even simpler. Thus we can prove exactly as in Theorem 3.3 that, if $f \in K_1$,

$$\lim_{n \to \infty} \int_{-\infty}^{\infty} f(u)\rho\{n(u-x)\}\,n\,\mathrm{d}u = f(x)$$

and then, by applying the argument of the first paragraph,

$$\lim_{n \to \infty} \int_{-\infty}^{\infty} g(u)\rho\{n(u-x)\}\,n\,\mathrm{d}u = g(x).$$

Here we have a sequence of infinitely differentiable functions which converges to a generalised function.

Next we turn to the connection between the convolution and multiplication.

Theorem 6.8. *The Fourier transform of* $\Psi * G$ *is* $4\pi^2 g(-t)\psi(-t)$.

Proof. By Theorem 6.7 $\Psi * G/2\pi$ is the Fourier transform of $g(t)\psi(t)$. Hence, by Theorem 3.11,

$$g(t)\psi(t) = \frac{1}{4\pi^2} \int_{-\infty}^{\infty} \psi * G\mathrm{e}^{itx}\,\mathrm{d}x$$

and the theorem follows. □

One consequence of this theorem is that the Fourier transform of $\Psi_1 * (\Psi_2 * G)$ is

$$\begin{aligned} &2\pi\psi_1(-t) \times (\text{Fourier transform of } \Psi_2 * G) \\ &= 2\pi\psi_1(-t)\{4\pi^2\psi_2(-t)g(-t)\} \\ &= 8\pi^2\psi_2(-t)\{\psi_1(-t)g(-t)\} \\ &= 8\pi^2\{\psi_1(-t)\psi_2(-t)\}g(-t) \end{aligned}$$

because of the rules for multiplication by fairly good functions. The inverse Fourier transform of the last two results gives

$$\Psi_1 * (\Psi_2 * G) = \Psi_2 * (\Psi_1 * G) = (\Psi_1 * \Psi_2) * G, \tag{11}$$

i.e. the convolution is associative.

Example 7. By Theorem 6.8 and (9) the Fourier transform of

$$\int_{-\infty}^{\infty} \frac{\pi^{1/2}|\alpha|^{\mu}K_{\mu}(|\alpha|)}{(\mu-\frac{1}{2})!2^{\mu-1}} \cdot \frac{\pi^{1/2}|x-\alpha|^{\nu}K_{\nu}(|x-\alpha|)}{(\nu-\frac{1}{2})!2^{\nu-1}} \mathrm{d}\alpha$$

is $4\pi^2(1+t^2)^{-\mu-1/2}(1+t^2)^{-\nu-1/2}$ or $4\pi^2(1+t^2)^{-\mu-\nu-1}$. Using (9) again we have

$$\int_{-\infty}^{\infty} |\alpha|^{\mu}K_{\mu}(|\alpha|)|x-\alpha|^{\nu}K_{\nu}(|x-\alpha|)\mathrm{d}\alpha = \frac{(\mu-\frac{1}{2})!(\nu-\frac{1}{2})!}{(\mu+\nu)!2^{1/2}}\pi^{1/2}|x|^{\mu+\nu+1/2}K_{\mu+\nu+1/2}(|x|) \tag{12}$$

provided that none of μ, ν and $\mu+\nu+\frac{1}{2}$ is a negative half-integer. If μ and ν are not negative half-integers but $\mu+\nu+\frac{1}{2} = -\frac{1}{2}(2s+1)$ $(s = 0, 1, \dots)$ then

$$\int_{-\infty}^{\infty} |\alpha|^{\mu}K_{\mu}(|\alpha|)|x-\alpha|^{\nu}K_{\nu}(|x-\alpha|)\mathrm{d}\alpha = (\mu-\tfrac{1}{2})!(\nu-\tfrac{1}{2})!2^{\mu+\nu}\pi D^{s}\delta(x) \tag{13}$$

where $D\delta(x) = \delta(x) - \delta''(x)$.

Theorem 6.9.

$$\int_{-\infty}^{\infty} \Gamma(x)\left\{\int_{-\infty}^{\infty} \Psi(\alpha)G(x-\alpha)\mathrm{d}\alpha\right\}\mathrm{d}x = \int_{-\infty}^{\infty} G(x)\left\{\int_{-\infty}^{\infty} \Psi(\alpha)\Gamma(\alpha+x)\mathrm{d}\alpha\right\}\mathrm{d}x.$$

Proof. $\int_{-\infty}^{\infty} \Psi(\alpha)\Gamma(\alpha+x)\mathrm{d}\alpha$ is good, by Theorem 6.6, and is the Fourier transform of $2\pi\psi(-t)\gamma(t)$ by Theorem 6.8. Hence, by Theorem 3.9,

$$\int_{-\infty}^{\infty} G(x)\left\{\int_{-\infty}^{\infty} \Psi(\alpha)\Gamma(\alpha+x)\mathrm{d}\alpha\right\}\mathrm{d}x = 4\pi^2\int_{-\infty}^{\infty} g(t)\psi(t)\gamma(-t)\mathrm{d}t$$

and the theorem follows from (5).□

Finally we remark

Theorem 6.10. $(\Psi * G)' = \Psi' * G = \Psi * G'$.

Proof. The regular sequence defining $(\Psi * G)'$ is

$$\left\{\frac{\mathrm{d}}{\mathrm{d}x}\int_{-\infty}^{\infty} \Psi(\alpha)\Gamma_m(x-\alpha)\mathrm{d}\alpha\right\}$$

which may be written in the alternative forms $\{\int_{-\infty}^{\infty} \Psi(\alpha)\Gamma_m'(x-\alpha)\mathrm{d}\alpha\}$ and $\{\int_{-\infty}^{\infty} \Psi'(\alpha)\Gamma_m(x-\alpha)\mathrm{d}\alpha\}$ and the proof is complete.□

Exercises

37. If $\{\phi_n(x)\}$ is a sequence of fine functions such that $\operatorname{Lim}_{n\to\infty} \phi_n(x) = \delta(x)$ prove that

$$\operatorname*{Lim}_{n\to\infty}\int_{-\infty}^{\infty} \phi_n(\alpha)G(x-\alpha)\mathrm{d}\alpha = G(x).$$

38. Prove that

(i) $\int_{-\infty}^{\infty} \delta^{(m)}(\alpha)\delta^{(n)}(x-\alpha)\mathrm{d}\alpha = \delta^{(m+n)}(x)$,

(ii) $\int_{-\infty}^{\infty} \delta(\alpha-y)\delta(x-\alpha)\mathrm{d}\alpha = \delta(x-y)$,

(iii) $\int_{-\infty}^{\infty} \delta^{(m)}(\alpha-y)\delta^{(n)}(x-\alpha)\mathrm{d}\alpha = \delta^{(m+n)}(x-y)$.

6.5. A generalisation of the convolution

The reason that it is possible to form the convolution of an arbitrary generalised function with the Fourier transform of a fairly good function is that the latter tends to zero at infinity faster than any inverse power. This is obvious in particular examples and follows, as a general result, from the Riemann–Lebesgue lemma (§9.1) because fairly good functions are infinitely differentiable. In order to form the convolution when one of the members is not the Fourier transform of a fairly good function it will be necessary to apply restrictions so that the behaviour at infinity is appropriate. There are two ways of doing this. One is to prescribe integrability or some related condition so that the growth at infinity is not too great. The other is to limit consideration to those generalised functions which are zero from some point to infinity. We shall discuss the first possibility in this section and the second in the next section.

The spaces L_p have already been introduced, i.e. $f \in L_p$ if, and only if, $\int_{-\infty}^{\infty} |f(t)|^p \mathrm{d}t < \infty$. The space L_∞ is the space of bounded measurable functions on R_1. We now introduce

Definition 6.6. *The generalised function $g \in L'_p$ if, and only if, $g = \sum_{m=1}^{M} f_m^{(r_m)}$ where M and r_m are finite and $f_m \in L_p$ $(m = 1, \ldots, M)$.*

Thus L'_p consists of those generalised functions which are the generalised derivatives of functions in L_p. In particular L'_∞ is the space of derivatives of bounded functions. The spaces are related by

Theorem 6.11. *If $g \in L'_p$ then $g \in L'_{p'}$ for $p' \geq p \geq 1$.*

Proof. We prove firstly that, if $f \in L_p$, $f \in L'_{p'}$ for $p' \geq p \geq 1$. There is, of course, nothing to prove if $p' = p$ so that we can take $p' > p$. Let $q = p/(p-1)$, $q' = p'/(p'-1)$.

Consider $h(x)$ defined by

$$h(x) = \int_{-\infty}^{\infty} \mathrm{e}^{-|x-t|} f(t) \mathrm{d}t. \tag{14}$$

The cases $p = 1$ and $p > 1$ will be dealt with separately. For $p = 1$, and therefore $p' > 1$, $q' < \infty$

$$|h(x)| \leq \int_{-\infty}^{\infty} \mathrm{e}^{-|x-t|} |f(t)|^{1/p' + 1/q'} \mathrm{d}t$$

$$\leq \left\{ \int_{-\infty}^{\infty} \mathrm{e}^{-p'|x-t|} |f(t)| \mathrm{d}t \right\}^{1/p'} \left\{ \int_{-\infty}^{\infty} |f(t)| \mathrm{d}t \right\}^{1/q'}$$

by the Holder inequality (Theorem 1.28). Since $\int_{-\infty}^{\infty} |f(t)| \mathrm{d}t < M$, we have

$$\int_{-\infty}^{\infty} |h(x)|^{p'} \mathrm{d}x \leq M \int_{-\infty}^{\infty} \mathrm{e}^{-p'|x|} \mathrm{d}x \int_{-\infty}^{\infty} |f(t)| \mathrm{d}t < \infty.$$

Consequently $f \in L_1$ implies that $h \in L_{p'}$, $p' > 1$. Also

$$(e^{-|x|})'' = e^{-|x|} - 2\delta(x)$$

so that, from Theorem 6.10 and Exercise 31,

$$h''(x) - h(x) = -2f(x) \tag{15}$$

which shows that $f \in L'_{p'}$.

Suppose now that $p > 1$. Then $1/p_1 = 1/p' - 1/p + 1$ gives $p_1 > 1$ and, if $q_1 = p_1/(p_1 - 1)$,

$$|h(x)| \leq \int_{-\infty}^{\infty} e^{-(p_1/q)|x-t|} e^{-(1-p_1/q)|x-t|} |f(t)| \, dt$$

$$\leq \left\{ \int_{-\infty}^{\infty} e^{-p_1|x-t|} \, dt \right\}^{1/q} \left\{ \int_{-\infty}^{\infty} e^{-p(1-p_1/q)|x-t|} |f(t)|^p \, dt \right\}^{1/p}$$

by the Hölder inequality. Since $\int_{-\infty}^{\infty} e^{-p_1|t|} \, dt < M_1^q$ (say) we have

$$|h(x)| \leq M_1 \left\{ \int_{-\infty}^{\infty} e^{-p(1-p_1/q)|x-t|} |f(t)|^{p-(p^2/q_1)+p^2/q_1} \, dt \right\}^{1/p}$$

$$\leq M_1 \left[\left\{ \int_{-\infty}^{\infty} e^{-p_1|x-t|} |f(t)|^p \, dt \right\}^{(q_1-p)/q_1} \left\{ \int_{-\infty}^{\infty} |f(t)|^p \, dt \right\}^{p/q_1} \right]^{1/p}$$

on a further application of the Hölder inequality. With $\int_{-\infty}^{\infty} |f(t)|^p \, dt < M_2^{q_1}$ (say) we have

$$|h(x)| \leq M_2 M_1 \left\{ \int_{-\infty}^{\infty} e^{-p_1|x-t|} |f(t)|^p \, dt \right\}^{1/p'}.$$

Hence

$$\int_{-\infty}^{\infty} |h(x)|^{p'} \, dx \leq (M_2 M_1)^{p'} \int_{-\infty}^{\infty} e^{-p_1|x|} \, dx \int_{-\infty}^{\infty} |f(t)|^p \, dt < \infty \tag{16}$$

and so $h \in L_{p'}$. Since (15) is still valid we conclude that $f \in L_p$, $p > 1$, implies that $f \in L'_{p'}$ for $p' \geq p$.

Finally, $g \in L'_p$ means that $g = \sum_m f_m^{(r_m)}$ where $f_m \in L_p$ and so, by what has just been proved, $g \in L'_{p'}$, $p' \geq p$ and the proof is complete. □

An obvious corollary is

Corollary 6.11. *If $g \in L'_p$ then $g \in L'_\infty$.*

The spaces L'_p and K_q are associated. It can be shown† that $F \in L_q$ when $f \in L_p (1 < p \le 2)$. If $g \in L'_p$ it follows that $G = PF$ where P is some polynomial and consequently $G \in K_q$. Conversely, if $G \in K_p$ $(1 < p \le 2)$ then $g \in L'_q$. With regard to L'_1 it has been proved already in Theorem 3.13 that, if $f \in L_1$, F is bounded and so $F \in K_\infty$. Thus $g \in L'_1$ implies that $G \in K_\infty$. Also, $G \in K_1$ requires $g \in L'_\infty$. Because of Theorem 3.2 there results that $G \in K_q$ $(q \ge 1)$ necessitates $g \in L'_\infty$. The position of L'_∞ is different, for although $G \in K_1$ implies $g \in L'_\infty$, $g \in L'_\infty$ does not necessarily ensure that $G \in K_1$. For example, the function 1 belongs to L'_∞ but its Fourier transform $2\pi\delta$ does not lie in K_1. Similarly sgn x is an example of a function in K_∞ whose Fourier transform is not in L'_1.

In order to extend the range of the convolution we introduce

Definition 6.7. *If $g_1 * g_2$ exists $g'_1 * g_2$ and $g_1 * g'_2$ are defined by*

$$g'_1 * g_2 = g_1 * g'_2 = (g_1 * g_2)'.$$

This is consistent with Theorem 6.10 and does not, in fact, give any new information about products obtained via Definition 6.5. However

Theorem 6.12. *If $g_1 \in L'_p (p \ge 1)$ and $g_2 \in L'_q \{q = p/(p-1)\}$ the product $g_1 * g_2$ exists.*

Proof. By assumption $g_1 = \sum_m f_{1m}^{(r_m)}, g_2 = \sum_n f_{2n}^{(s_n)}$ where r_m and s_n are finite and $f_{1m} \in L_p, f_{2n} \in L_q$. The theorem will therefore follow from repeated application of Definition 6.7, with the result

$$g_1 * g_2 = \sum_m \sum_n (f_{1m} * f_{2n})^{(r_m + s_n)}$$

provided that $f_{1m} * f_{2n}$ exists. It will therefore be enough to consider a convolution of the type $f_1 * f_2$ where $f_1 \in L_p, f_2 \in L_q$. Now, by the Hölder inequality,

$$\left| \int_{-\infty}^{\infty} f_1(t) f_2(x-t) \mathrm{d}t \right| \le \left\{ \int_{-\infty}^{\infty} |f_1(t)|^p \mathrm{d}t \right\}^{1/p} \left\{ \int_{-\infty}^{\infty} |f_2(x-t)|^q \mathrm{d}t \right\}^{1/q}$$

$$\le \left\{ \int_{-\infty}^{\infty} |f_1(t)|^p \mathrm{d}t \right\}^{1/p} \left\{ \int_{-\infty}^{\infty} |f_2(t)|^q \mathrm{d}t \right\}^{1/q} \quad (17)$$

† See, for example, E. C. Titchmarsh, *The Theory of Fourier Integrals*, 2nd edition, Oxford University Press (1948).

which is finite on account of the assumed conditions and the theorem is proved for $p > 1$. If $p = 1$ use

$$\left|\int_{-\infty}^{\infty} f_1(t) f_2(x-t)\,\mathrm{d}t\right| \leq \sup |f_2(x)| \int_{-\infty}^{\infty} |f_1(t)|\,\mathrm{d}t.$$

In either case $f_1 * f_2 \in K_1$ and the result follows.

It is necessary to check that two different representations of a generalised function do not lead to different products. Now

$$\begin{aligned}
&\int_{-\infty}^{\infty} \gamma(x) \int_{-\infty}^{\infty} g_1(t) g_2(x-t)\,\mathrm{d}t\,\mathrm{d}x \\
&\quad = \sum_m \sum_n \int_{-\infty}^{\infty} \gamma(x) \left\{\int_{-\infty}^{\infty} f_{1m}(t) f_{2n}(x-t)\,\mathrm{d}t\right\}^{(r_m+s_n)} \mathrm{d}x \\
&\quad = \sum_m \sum_n (-1)^{r_m+s_n} \int_{-\infty}^{\infty} \gamma^{(r_m+s_n)}(x) \int_{-\infty}^{\infty} f_{1m}(t) f_{2n}(x-t)\,\mathrm{d}t\,\mathrm{d}x \\
&\quad = \sum_m \sum_n (-1)^{r_m+s_n} \int_{-\infty}^{\infty} f_{1m}(t) \int_{-\infty}^{\infty} \gamma^{(r_m+s_n)}(x) f_{2n}(x-t)\,\mathrm{d}x\,\mathrm{d}t \\
&\quad = \sum_m \sum_n (-1)^{r_m} \int_{-\infty}^{\infty} f_{1m}(t) \int_{-\infty}^{\infty} \gamma^{(r_m)}(x) f_{2n}^{(s_n)}(x-t)\,\mathrm{d}x\,\mathrm{d}t \\
&\quad = \sum_m (-1)^{r_m} \int_{-\infty}^{\infty} f_{1m}(t) \int_{-\infty}^{\infty} \gamma^{(r_m)}(x) g_2(x-t)\,\mathrm{d}x\,\mathrm{d}t.
\end{aligned}$$

Since the last quantity is independent of the particular representation of g_2 it follows that $g_1 * g_2$ is also independent of that representation. Similarly $g_1 * g_2$ is independent of the particular representation of g_1.□

Corollary 6.12. *If $g_1 \in L'_p (1 \leq p)$, $g_2 \in L'_{q'}$ $\{1 \leq q' \leq p/(p-1)\}$ then $g_1 * g_2$ exists.*

Proof. From Theorem 6.11 $g_2 \in L'_q$ and the corollary is a consequence of Theorem 6.12.□

Theorem 6.13. *If $g_1 \in L'_p$ $(1 \leq p \leq 2)$, $g_2 \in L'_{p'}$ $(1 \leq p' \leq 2)$ the Fourier transform of $g_1 * g_2$ is $G_1 G_2$.*

Proof. The convolution $g_1 * g_2$ exists on account of Corollary 6.12. Also, by Theorem 6.11, it is sufficient to consider $g_1 = f_1^{(r)}$, $g_2 = f_2^{(s)}$

where $f_1 \in L_2, f_2 \in L_2$. Since $f_2(t) \in L_2, \overline{f_2(x-t)} \in L_2$ and has Fourier transform $\overline{F_2(\alpha)}e^{-i\alpha x}$. Hence, from Theorem 5.12,

$$\int_{-\infty}^{\infty} f_1(t) f_2(x-t)\,dt = \frac{1}{2\pi}\int_{-\infty}^{\infty} F_1(\alpha)\overline{\overline{F_2(\alpha)}e^{-i\alpha x}}\,d\alpha$$

$$= \frac{1}{2\pi}\int_{-\infty}^{\infty} F_1(\alpha)F_2(\alpha)e^{i\alpha x}\,d\alpha$$

which shows that the Fourier transform of $f_1 * f_2$ is $F_1 F_2$. The product $F_1 F_2$ has the same meaning for generalised functions because of Theorem 6.1.

Now $g_1 * g_2 = (f_1 * f_2)^{(r+s)}$ so that the Fourier transform of $g_1 * g_2$ is $(i\alpha)^{r+s} F_1(\alpha) F_2(\alpha)$ (Theorem 3.12) which may be written

$$\{(i\alpha)^r F_1(\alpha)\}\{(i\alpha)^s F_2(\alpha)\}$$

(Theorem 6.1 and Exercise 16) or $G_1(\alpha) G_2(\alpha)$ and the proof is complete.□

A theorem, somewhat wider in some respects, is

Theorem 6.14. *If $g_1 \in L'_p (1 < p \le 2)$, $g_2 \in L'_q$, $G_1 \in K_q$ and $G_2 \in K_p$ then the Fourier transform of $g_1 * g_2$ is $G_1 G_2$.*

Proof. Note that both $g_1 * g_2$ and $G_1 G_2$ exist by Theorems 6.12 and 6.1. We can take a typical term of $g_1 * g_2$ to be $(f_1 * f_2)^{(r+s)}$ where $f_1 \in L_p, f_2 \in L_q, F_1 \in L_q$ and $F_2 \in L_p$ so that the first step is to prove the theorem for $f_1 * f_2$.

Consider $\int_{-\infty}^{\infty} e^{-t^2/4n} f_2(t) f_1(x-t)\,dt$. Now

$$\int_{-\infty}^{\infty} (e^{-t^2/4n} - 1) f_2(t) f_1(x-t)\,dt$$

$$= \left\{\int_{-n^{1/4}}^{n^{1/4}} + \int_{n^{1/4}}^{\infty} + \int_{-\infty}^{-n^{1/4}}\right\}(e^{-t^2/4n} - 1) f_2(t) f_1(x-t)\,dt$$

and

$$\left|\int_{-n^{1/4}}^{n^{1/4}} (e^{-t^2/4n} - 1) f_2(t) f_1(x-t)\,dt\right|$$

$$\le \frac{1}{4n^{1/2}}\int_{-n^{1/4}}^{n^{1/4}} |f_2(t) f_1(x-t)|\,dt$$

$$\le \frac{1}{4n^{1/2}}\left\{\int_{-\infty}^{\infty} |f_2(t)|^q\,dt\right\}^{1/q}\left\{\int_{-\infty}^{\infty} |f_1(t)|^p\,dt\right\}^{1/p} \qquad (18)$$

by the Hölder inequality. Also

$$\left|\int_{n^{1/4}}^{\infty}(e^{-t^2/4n}-1)f_2(t)f_1(x-t)\mathrm{d}t\right|$$

$$\leq 2\int_{n^{1/4}}^{\infty}|f_2(t)f_1(x-t)|\mathrm{d}t$$

$$\leq 2\left\{\int_{n^{1/4}}^{\infty}|f_2(t)|^q\mathrm{d}t\right\}^{1/q}\left\{\int_{-\infty}^{\infty}|f_1(t)|^p\mathrm{d}t\right\}^{1/p} \tag{19}$$

There is a similar result for the interval $(-\infty, -n^{1/4})$. As $n\to\infty$ the right-hand sides of (18) and (19) tend to zero. Hence

$$\lim_{n\to\infty}\int_{-\infty}^{\infty}e^{-t^2/4n}f_2(t)f_1(x-t)\mathrm{d}t=\int_{-\infty}^{\infty}f_2(t)f_1(x-t)\mathrm{d}t.$$

Now

$$\int_{-\infty}^{\infty}|e^{-t^2/4n}f_2(t)|\,\mathrm{d}t\leq\left\{\int_{-\infty}^{\infty}e^{-pt^2/4n}\mathrm{d}t\right\}^{1/p}\left\{\int_{-\infty}^{\infty}|f_2(t)|^q\mathrm{d}t\right\}^{1/q}$$

which demonstrates that $e^{-t^2/4n}f_2(t)\in L_1$. Since $f_1\in L_p$ Theorem 6.13 shows that the Fourier transform of $\int_{-\infty}^{\infty}e^{-t^2/4n}f_2(t)f_1(x-t)\,dt$ is $F_1(\alpha)\int_{-\infty}^{\infty}f_2(t)e^{-t^2/4n-i\alpha t}dt$ or $F_1(\alpha)\int_{-\infty}^{\infty}F_2(\alpha-\beta)(n/\pi)^{1/2}\,e^{-n\beta^2}d\beta$ from Theorem 6.7. Thus the Fourier transform of f_2*f_1 is

$$\lim_{n\to\infty}F_1(\alpha)\int_{-\infty}^{\infty}F_2(\alpha-\beta)\left(\frac{n}{\pi}\right)^{1/2}e^{-n\beta^2}\mathrm{d}\beta.$$

But

$$\int_{-\infty}^{\infty}\Gamma(\alpha)F_1(\alpha)\int_{-\infty}^{\infty}F_2(\alpha-\beta)\left(\frac{n}{\pi}\right)^{1/2}e^{-n\beta^2}\mathrm{d}\beta\mathrm{d}\alpha$$

$$=\int_{-\infty}^{\infty}\left(\frac{n}{\pi}\right)^{1/2}e^{-n\beta^2}\int_{-\infty}^{\infty}\Gamma(\alpha)F_1(\alpha)F_2(\alpha-\beta)\mathrm{d}\alpha\mathrm{d}\beta$$

the integral being absolutely convergent. The inner integral is a bounded function of β because

$$\left|\int_{-\infty}^{\infty}\Gamma(\alpha)F_1(\alpha)F_2(\alpha-\beta)\mathrm{d}\alpha\right|$$

$$\leq\left\{\int_{-\infty}^{\infty}|\Gamma(\alpha)F_1(\alpha)|^q\mathrm{d}\alpha\right\}^{1/q}\left\{\int_{-\infty}^{\infty}|F_2(\alpha)|^p\mathrm{d}\alpha\right\}^{1/p}<\infty$$

since $F_1 \in L_q$ and $F_2 \in L_p$. In fact it is continuous in β near $\beta = 0$ since

$$\left| \int_{-\infty}^{\infty} \Gamma(\alpha) F_1(\alpha) \{F_2(\alpha - \beta) - F_2(\alpha)\} \mathrm{d}\alpha \right|$$
$$\leq \left\{ \int_{-\infty}^{\infty} |\Gamma(\alpha) F_1(\alpha)|^q \mathrm{d}\alpha \right\}^{1/q} \left\{ \int_{-\infty}^{\infty} |F_2(\alpha - \beta) - F_2(\alpha)|^p \mathrm{d}\alpha \right\}^{1/p}$$

and the last { } tends to zero as $\beta \to 0$ by a well-known theorem of integration. Therefore, given $\varepsilon > 0$ choose $\eta > 0$ so that

$$\left| \int_{-\infty}^{\infty} \Gamma(\alpha) F_1(\alpha) \{F_2(\alpha - \beta) - F_2(\alpha)\} \mathrm{d}\alpha \right| < \varepsilon$$

for $|\beta| \leq \eta$. Then

$$\left| \int_{-\infty}^{\infty} \left(\frac{n}{\pi}\right)^{1/2} \mathrm{e}^{-n\beta^2} \int_{-\infty}^{\infty} \Gamma(\alpha) F_1(\alpha) \{F_2(\alpha - \beta) - F_2(\alpha)\} \mathrm{d}\alpha \, \mathrm{d}\beta \right|$$
$$\leq B\left(\int_{\eta}^{\infty} + \int_{-\infty}^{-\eta}\right)\left(\frac{n}{\pi}\right)^{1/2} \mathrm{e}^{-n\beta^2} \mathrm{d}\beta + \varepsilon \int_{-\eta}^{\eta} \left(\frac{n}{\pi}\right)^{1/2} \mathrm{e}^{-n\beta^2} \mathrm{d}\beta$$
$$\leq B n^{1/2} \mathrm{e}^{-n\eta^2/2} + \varepsilon$$

which tends to ε as $n \to \infty$. Since ε is arbitrarily small it follows that the Fourier transform of $f_2 * f_1$ is $F_1 F_2$.

The remainder of the proof is the same as in Theorem 6.13 and the theorem is complete. □

Corollary 6.14. *If* $g_1 \in L'_p$ $(1 < p \leq 2)$, $g_2 \in L'_{q'}$ $\{1 \leq q' \leq p/(p-1)\}$, $G_1 \in K_q$ *and* $G_2 \in K_{p'}$ $\{p' = q'/(q'-1)\}$ *then the Fourier transform of* $g_1 * g_2$ *is* $G_1 G_2$.

For $g_2 \in L'_q$ and $G_2 \in K_p$ and the corollary follows from Theorem 6.14.

Example 8. The Fourier transform of $|x|^\beta$ is $\beta! |\alpha|^{-\beta-1} 2 \cos \frac{1}{2}\pi(\beta + 1)$ (Corollary 4.4) if β is not a negative integer. By multiplying and dividing by $(1 + \alpha^2)^N$ we see that the transform is the product of a polynomial in α and a function in L_2 if $\beta < -\frac{1}{2}$. It follows from

Theorem 5.11 that $|x|^\beta \in L_2'$ if β is not a negative integer. The same argument applies if β is a negative integer (understanding $|x|^{-m}$ to mean $x^{-m}\operatorname{sgn} x$ when m is odd and x^{-m} when m is even) but the Fourier transform is different (Corollary 4.6*b*). Hence $|x|^\beta \in L_2'$ for $\beta < -\frac{1}{2}$. Therefore, if $\beta < -\frac{1}{2}$ and $\gamma < -\frac{1}{2}$, Theorem 6.13 gives, when neither β nor γ is a negative integer,

$$\int_{-\infty}^{\infty} e^{-i\alpha x}\int_{-\infty}^{\infty}|t|^\beta|x-t|^\gamma\,dt\,dx$$
$$= \beta!\gamma!|\alpha|^{-\beta-\gamma-2}\,4\cos\tfrac{1}{2}\pi(\beta+1)\cos\tfrac{1}{2}\pi(\gamma+1).$$

Alternatively,

$$\int_{-\infty}^{\infty}|t|^\beta|x-t|^\gamma\,dt$$
$$= \beta!\gamma!2\cos\tfrac{1}{2}\pi(\beta+1)\cos\tfrac{1}{2}\pi(\gamma+1)\int_{-\infty}^{\infty}|\alpha|^{-\beta-\gamma-2}e^{i\alpha x}\,d\alpha/\pi.$$

Thus, if $\beta < -\frac{1}{2}$, $\gamma < -\frac{1}{2}$ and $\beta+\gamma$ is not a negative integer

$$\int_{-\infty}^{\infty}|t|^\beta|x-t|^\gamma\,dt$$
$$= \beta!\gamma!(-\beta-\gamma-2)!\cos\tfrac{1}{2}\pi(\beta+1)\cos\tfrac{1}{2}\pi(\gamma+1)$$
$$\times\cos\tfrac{1}{2}\pi(\beta+\gamma+1)4|x|^{\beta+\gamma+1}/\pi \tag{20}$$
$$= \frac{\beta!\gamma!\cos\frac{1}{2}\pi(\beta+1)\cos\frac{1}{2}\pi(\gamma+1)}{(\beta+\gamma+1)!\cos\frac{1}{2}\pi(\beta+\gamma+2)}|x|^{\beta+\gamma+1} \tag{21}$$

on using the formula $z!(-z)! = \pi z/\sin\pi z$. On the other hand, if $\beta+\gamma = -2m-2$ $(m = 0, 1, 2, \ldots)$, (3.13) gives

$$\int_{-\infty}^{\infty}|t|^\beta|x-t|^\gamma\,dt = \beta!\gamma!4(-1)^m\cos\tfrac{1}{2}\pi(\beta+1)\cos\tfrac{1}{2}\pi(\gamma+1)\delta^{(2m)}(x) \tag{22}$$

whereas, if $\beta+\gamma = -2m-3$ $(m = 0, 1, 2, \ldots)$ it follows from Corollary 4.3*b* that (20) is still valid. If β is an even negative integer, but γ is not a negative integer, (20) and (21) are still true provided that $\beta!2\cos\frac{1}{2}\pi(\beta+1)$ is replaced by $\pi/\{(-\beta-1)!\cos\frac{1}{2}\pi\beta\}$.

Example 9. The Fourier transform of x^{-1} is $-\pi\mathrm{i}\,\operatorname{sgn}\alpha$ or

$$-\pi\mathrm{i}(1+\alpha^2)\operatorname{sgn}\alpha/(1+\alpha^2).$$

Hence

$$x^{-1} = -\tfrac{1}{2}\mathrm{i}\left(1-\frac{\mathrm{d}^2}{\mathrm{d}x^2}\right)\int_{-\infty}^{\infty}\frac{\operatorname{sgn}\alpha}{1+\alpha^2}\mathrm{e}^{\mathrm{i}\alpha x}\,\mathrm{d}\alpha.$$

Now

$$\int_{-\infty}^{\infty}\frac{\operatorname{sgn}\alpha}{1+\alpha^2}\mathrm{e}^{\mathrm{i}\alpha x}\,\mathrm{d}\alpha = 2\mathrm{i}\int_0^{\infty}\frac{\sin\alpha x}{1+\alpha^2}\mathrm{d}\alpha$$

and

$$\left|\int_0^{\infty}\frac{\sin\alpha x}{1+\alpha^2}\mathrm{d}\alpha\right| \le \int_0^{\infty}\frac{\mathrm{d}\alpha}{1+\alpha^2} \le \tfrac{1}{2}\pi. \tag{23}$$

Also, for $x \neq 0$, an integration by parts gives

$$\left|\int_0^{\infty}\frac{\sin\alpha x}{1+\alpha^2}\mathrm{d}\alpha\right| = \left|\frac{1}{x} - \int_0^{\infty}\frac{2\alpha\cos\alpha x}{x(1+\alpha^2)^2}\mathrm{d}\alpha\right| \le \frac{B}{|x|} \tag{24}$$

for some finite B. By using (23) for $0 \le |x| < X$ and (24) for $|x| \ge X$ we see that $\int_{-\infty}^{\infty}\{\operatorname{sgn}\alpha/(1+\alpha^2)\}\,\mathrm{e}^{\mathrm{i}\alpha x}\,\mathrm{d}\alpha \in L_p$ for $p > 1$. Hence $x^{-1} \in L'_p$ for $p > 1$. If follows from Definition 4.4 that $x^{-m} \in L'_p$ for $p > 1$. Consequently Corollary 6.12 shows that $\int_{-\infty}^{\infty}(x-t)^{-m}g(t)\,\mathrm{d}t$ exists for $g \in L'_{p'}$ $(1 \le p' < \infty)$. If $g \in L'_2$ we have, from Theorem 6.13,

$$\int_{-\infty}^{\infty}(x-t)^{-m}g(t)\,\mathrm{d}t = \frac{(-\mathrm{i})^m}{(m-1)!2}\int_{-\infty}^{\infty}\alpha^{m-1}\operatorname{sgn}\alpha\, G(\alpha)\mathrm{e}^{\mathrm{i}\alpha x}\,\mathrm{d}\alpha.$$

In particular

$$\int_{-\infty}^{\infty}(x-t)^{-1}g(t)\,\mathrm{d}t = -\tfrac{1}{2}\mathrm{i}\int_{-\infty}^{\infty}G(\alpha)\operatorname{sgn}\alpha\,\mathrm{e}^{\mathrm{i}\alpha x}\,\mathrm{d}\alpha. \tag{25}$$

$G(\alpha)$, being the transform of a generalised function in L'_2, is the product of a polynomial and a function in L_2. The same is therefore true of $G(\alpha)$ sgn α and so the right-hand side of (25) is in L_2'. Hence, from (25),

$$\begin{aligned}\int_{-\infty}^{\infty}(y-x)^{-1}\int_{-\infty}^{\infty}(x-t)^{-1}g(t)\,dt\,dx &= -\tfrac{1}{2}i\int_{-\infty}^{\infty}\{-\pi i G(\alpha)\operatorname{sgn}\alpha\}\operatorname{sgn}\alpha\, e^{i\alpha y}\,d\alpha \\ &= -\tfrac{1}{2}\pi\int_{-\infty}^{\infty}G(\alpha)e^{i\alpha y}\,d\alpha \\ &= -\pi^2 g(y). \qquad (26)\end{aligned}$$

Another way of writing (25) and (26) is that, if $h \in L_2'$ and

$$\int_{-\infty}^{\infty}(x-t)^{-1}g(t)\,dt = h(x),$$

then

$$g(t) = -\frac{1}{\pi^2}\int_{-\infty}^{\infty}(t-x)^{-1}h(x)\,dx.$$

This is an example of the *Hilbert transform*. The formulae hold whenever the Fourier transform of h is in K_p $(1 < p \leq 2)$.

Example 10. $(\sin x)/x \in L_p$ for $p > 1$ and $J_0(x) \in L_{p/(p-1)}$ $(p < 2)$ since $J_0(x) = O(|x|^{-1/2})$ as $|x| \to \infty$. The Fourier transform of $(\sin x)/x$ is $\pi\{H(\alpha+1) - H(\alpha-1)\}$ which $\in K_q$ for any $q \geq 1$. The Fourier transform of $J_0(x)$ is $2(1-\alpha^2)^{-1/2}$ for $|\alpha| < 1$ and zero elsewhere (Exercise 30 of Chapter 5); it is therefore in K_p for $1 \leq p < 2$. Hence Theorem 6.14 applies and

$$\int_{-\infty}^{\infty} J_0(t)\frac{\sin(x-t)}{x-t}\,dt = \frac{1}{2\pi}\int_{-1}^{1}\frac{2\pi}{(1-\alpha^2)^{1/2}}e^{i\alpha x}\,d\alpha = \pi J_0(x). \quad (27)$$

Changing x into $-x$ we have

$$\int_{-\infty}^{\infty} J_0(t)\frac{\sin(x+t)}{x+t}\,dt = \pi J_0(x).$$

A derivative of (27) gives

$$\int_{-\infty}^{\infty} J_1(t)\frac{\sin(x-t)}{x-t}\,dt = \pi J_1(x)$$

since $J_0'(x) = J_1(x)$. Similarly,

$$\int_{-\infty}^{\infty} J_n(t)\frac{\sin(x-t)}{x-t}\mathrm{d}t = \pi J_n(x).$$

Example 11. By (7) $\delta' * 1 = 0$ so that

$$(1 * \delta') * \mathrm{H} = 0.$$

However $\delta' * \mathrm{H} = \delta$ so that $1 * (\delta' * \mathrm{H}) = 1$ by (6). Thus

$$(1 * \delta') * \mathrm{H} \neq 1 * (\delta' * \mathrm{H})$$

so that the convolution, as now generalised, is not necessarily associative (cf. (11)).

Exercises

39. Prove that $\delta(x) \in L_1'$.

40. Show that $|x|^\beta \in L_1'$ for $\beta < -1$. Deduce, if $g \in L_2'$ and $\beta < -1$ but β is not a negative integer, that

$$\int_{-\infty}^{\infty} |x-t|^\beta g(t)\mathrm{d}t = \beta!\cos\tfrac{1}{2}\pi(\beta+1)\int_{-\infty}^{\infty} |\alpha|^{-\beta-1}G(\alpha)\mathrm{e}^{\mathrm{i}\alpha x}\mathrm{d}\alpha/\pi.$$

41. Show that, for all cases for which the convolution has been defined,

$$\int_{-\infty}^{\infty} \gamma(x)\left\{\int_{-\infty}^{\infty} g_1(t)g_2(x-t)\mathrm{d}t\right\}\mathrm{d}x = \int_{-\infty}^{\infty} g_1(x)\left\{\int_{-\infty}^{\infty} g_2(t)\gamma(t+x)\mathrm{d}t\right\}\mathrm{d}x.$$

42. If $G_1 \in K_2$ and $G_2 \in K_2$ prove that the Fourier transform of $G_1(\alpha)G_2(-\alpha)$ is $2\pi g_1 * g_2(-t)$.

43. If $g_1 \in L_1'$ and $g_2 \in L_2'$ prove that $g_1 * g_2 \in L_2'$.

44. If $g_1 \in L_p'$ $(1 \leq p \leq 2)$ and $g_2 \in L_{p'}'$ $(1 \leq p' \leq 2)$ prove that the Fourier transform of $g_1 * g_2$ is a conventional function in K_1.

45. If m is a positive integer and $m \geq 1$ find $g^{(m)}(t)$ when

$$\int_{-\infty}^{\infty} (x-t)^{-m} g(t)\mathrm{d}t = h(x),$$

given that $h \in L_2'$.

46. Show that $(\mathrm{H} * g)' = g$.

47. Prove that (26) is valid if $g \in L_q'$ $(1 \leq q < \infty)$ and $G \in K_p$.

48. Prove that, if $g_1 \in L_p'$ and $g_2 \in L_{p_1}'$ where $1 \leq p \leq \infty$, $1 \leq p_1 \leq \infty$ and $(1/p) + (1/p_1) - 1 \geq 0$, $g_1 * g_2 \in L_r'$ where $1/r = (1/p) + (1/p_1) - 1$. (Hint: A slight modification of the sort of argument used in deriving (16) from

(14) can be used to show that, if $f_1 \in L_p$ and $f_2 \in L_{p_1}$,

$$\left\{\int_{-\infty}^{\infty} |f_1 * f_2|^r \mathrm{d}x\right\}^{1/r} \leq \left\{\int_{-\infty}^{\infty} |f_1|^p \mathrm{d}x\right\}^{1/p} \left\{\int_{-\infty}^{\infty} |f_2|^{p_1} \mathrm{d}x\right\}^{1/p_1}.)$$

49. If ϕ is fine and $g \in L'_p$ prove that $g * \phi \in L_p$.

50. Show that

$$\int_0^\infty J_0(t) \frac{\sin(x-t)}{x-t} \mathrm{d}t = \tfrac{1}{2}\pi J_0(x).$$

(Hint: $\int_0^\infty J_0(x) \sin \alpha x \, \mathrm{d}x = 0$ $(|\alpha| < 1)$, $= \{\operatorname{sgn} \alpha/(\alpha^2 - 1)^{1/2}\}$ $(|\alpha| > 1)$.)

51. Prove that

$$\int_{-\infty}^{\infty} \frac{J_\nu(t)}{t^\nu} \frac{\sin(x+t)}{x+t} \mathrm{d}t = \pi \frac{J_\nu(x)}{x^\nu}.$$

52. Show that, under suitable restrictions on ν,

$$\int_{-\infty}^{\infty} |\alpha|^{2\nu} \operatorname{sgn} \alpha \frac{J_\nu(|\alpha|)}{|\alpha|^\nu} \mathrm{e}^{\mathrm{i}\alpha x} \mathrm{d}\alpha$$

$$= \frac{(2\nu)!\mathrm{i}2\cos\nu\pi}{(\nu - \frac{1}{2})!2^\nu \pi^{1/2}} \int_{-1}^{1} (1-t^2)^{\nu - 1/2} |x-t|^{-2\nu-1} \operatorname{sgn}(x-t) \, \mathrm{d}t.$$

Deduce that, if $-\frac{1}{2} < \nu < \frac{1}{2}$,

$$\int_0^\infty \alpha^\nu J_\nu(\alpha) \sin \alpha x \, \mathrm{d}\alpha = \begin{cases} 0 & (|x| < 1) \\ \dfrac{2^\nu \pi^{1/2} \operatorname{sgn} x}{(-\frac{1}{2} - \nu)!(x^2 - 1)^{\nu + 1/2}} & (|x| > 1). \end{cases}$$

Would you expect this to be true for $\nu \geq \frac{1}{2}$?

53. If $f \in K_1$ show that $f = f_1 f_2$ where $f_1 \in K_2$ and $f_2 \in K_2$. Deduce that for $f \in K_1$ it is necessary and sufficient that f should be the Fourier transform of $g_1 * g_2$ where $g_1 \in L'_2$ and $g_2 \in L'_2$.

54. Prove that, for a and b real,

$$\int_{-\infty}^{\infty} \{(t-b)(t-x)\}^{-1} \mathrm{e}^{\mathrm{i}at} \mathrm{d}t = \pi^2 \mathrm{e}^{\mathrm{i}ab} \delta(x-b) + \pi\mathrm{i}(x-b)^{-1} (\mathrm{e}^{\mathrm{i}ax} - \mathrm{e}^{\mathrm{i}ab}) \operatorname{sgn} a$$

the last term vanishing when $a = 0$.

55. If $g_1 \in L'_p (p \geq 1)$, $g_2 \in L'_q \{q = p/(p-1)\}$ and $g_3 \in L'_1$ prove that

$$g_1 * (g_2 * g_3) = (g_1 * g_2) * g_3.$$

56. If $f_\mu \in L_p$, $f_{\mu_0} \in L_p$ and $\lim_{\mu \to \mu_0} \int_{-\infty}^{\infty} |f_\mu - f_{\mu_0}|^p \mathrm{d}x = 0$ prove that, when $g_\mu = f_\mu^{(r)}$,

$$\operatorname*{Lim}_{\mu \to \mu_0} g_\mu * g_2 = g_{\mu_0} * g_2$$

when $g_2 \in L'_q$.

Somewhat different generalised functions are covered by

Theorem 6.15. *If, for some finite integer* $m \geq 0$, $(1+x^2)^{-m/2} g_1(x) \in L'_\infty$ *and* $(1+x^2)^{m/2} g_2(x) \in L'_1$ *the convolution* $g_1 * g_2$ *exists.*

Proof. It will be shown firstly that $(1+x^2)^{-m/2} g_1(x) \in L'_\infty$ implies that $g_1(x)$ is the sum of a finite number of terms of the type $\{(1+x^2)^{m/2} f_{10}(x)\}^{(r)}$ with r finite and $f_{10} \in L_\infty$. The method will be by induction since the statement is obviously valid if $g_1 \in L_\infty$. Now assume that both $(1+x^2)^{m/2} f_1^{(s-1)}(x)$ and $x(1+x^2)^{m/2-1} f_1^{(s-1)}(x)$ with $f_1 \in L_\infty$ can be written as the sum of terms of the type $\{(1+x^2)^{m/2} f_{10}\}^{(r)}$ with $r \leq s-1$. Then, from

$$(1+x^2)^{m/2} f_1^{(s)}(x) = \{(1+x^2)^{m/2} f_1^{(s-1)}(x)\}' - mx(1+x^2)^{m/2-1} f_1^{(s-1)}(x)$$

we see that $(1+x^2)^{m/2} f_1^{(s)}(x)$ is a sum of such terms but with $r \leq s$. A similar result for $x(1+x^2)^{(m/2)-1} f_1^{(s)}(x)$ can be deduced from

$$x(1+x^2)^{(m/2)-1} f_1^{(s)}(x) = \{x(1+x^2)^{m/2-1} f_1^{(s-1)}(x)\}'$$
$$- 2\{\tfrac{1}{2}m(1+x^2)^{m/2-1} - (\tfrac{1}{2}m-1)(1+x^2)^{m/2-2}\} f_1^{(s-1)}(x)$$

since $f_{10}/(1+x^2) \in L_\infty$ if $f_{10} \in L_\infty$. The statement at the beginning of the paragraph now follows by induction since $g_1(x) = (1+x^2)^{m/2} f_1^{(r)}(x)$.

We may show in a similar manner that $(1+x^2)^{m/2} g_2(x) \in L'_1$ implies that $g_2(x)$ is the sum of products of the type $\{(1+x^2)^{-m/2} f_{20}(x)\}^{(s)}$ where $f_{20} \in L_1$.

Thus the existence of $g_1 * g_2$ can be deduced via Definition 6.7 and from $g_1 = f_{11}^{(r)}, g_2 = f_{21}^{(s)}$ if we can prove that $f_{11} * f_{21}$ exists when

$$(1+x^2)^{-m/2} f_{11} = f_{10} \in L_\infty$$

and

$$(1+x^2)^{m/2} f_{21} = f_{20} \in L_1.$$

The use of the easily verified inequality

$$(1+\xi^2)(1+\eta^2) \geq \tfrac{1}{4}\{1+(\xi-\eta)^2\} \tag{28}$$

with $\xi = x$, $\eta = t$ gives

$$|f_{11} * f_{21}| \leq 2^m(1+x^2)^{m/2} \int_{-\infty}^{\infty} \{1+(x-t)^2\}^{-m/2} |f_{11}(x-t)|$$

$$\times (1+t^2)^{m/2} |f_{21}(t)| \, dt$$

$$\leq 2^m B(1+x^2)^{m/2} \tag{29}$$

the integral being bounded because of the properties of f_{11} and f_{21}. Hence $f_{11} * f_{21} \in K_\infty \subset K_1$ so that $f_{11} * f_{21}$ is a generalised function and the theorem is proved. □

Theorem 6.16. *If* $g_1 = f_1^{(r)}$, $(1+x^2)^{m/2} f_1 \in L_1$ *and* $g_2 = f_2^{(s)}$ *where* $(1+x^2)^{-m/2} f_2$ *is bounded and tends to zero as* $|x| \to \infty$, *if* $G_1^{(l)} \in K_\infty$, $G_2 = F_{22}^{(l)}$ *where* $F_{22} \in K_1$ *then the Fourier transform of* $g_1 * g_2$ *is* $G_1 G_2$.

Proof. The proof is that of Theorem 6.14 with some small modifications. The convolution $g_1 * g_2$ exists by Theorem 6.15 and $g_1 * g_2$ contains $(f_1 * f_2)^{(r+s)}$ as a typical term so that it will be sufficient to consider the product $f_1 * f_2$. The analogue of (19) is

$$\left| \int_{n^{1/4}}^{\infty} (e^{-t^2/4n} - 1) f_2(t) f_1(x-t) \, dt \right|$$

$$\leq 2 \int_{n^{1/4}}^{\infty} |f_2(t) f_1(x-t)| \, dt$$

$$\leq 2^m (1+x^2)^{m/2} \sup_{n^{1/4} \leq x} (1+x^2)^{-m/2} |f_2(x)| \int_{-\infty}^{\infty} (1+t^2)^{m/2} |f_1(t)| \, dt$$

by (28). The integral is finite and the sup tends to zero as $n \to \infty$ by assumption. The analogue of (18) has similar properties and so, since the product of $(1+x^2)^{m/2}$ and a good function is absolutely integrable,

$$\lim_{n \to \infty} \int_{-\infty}^{\infty} \gamma(x) \int_{-\infty}^{\infty} e^{-t^2/4n} f_2(t) f_1(x-t) \, dt \, dx$$

$$= \int_{-\infty}^{\infty} \gamma(x) \int_{-\infty}^{\infty} f_2(t) f_1(x-t) \, dt \, dx$$

i.e.

$$\lim_{n\to\infty}\int_{-\infty}^{\infty} e^{-t^2/4n} f_2(t) f_1(x-t)\,dt = f_1 * f_2$$

in the sense of generalised functions.

Now $e^{-t^2/4n} f_2(t) \in L_1$ since $e^{-t^2/4n}(1+t^2)^{m/2}$ is integrable; also $f_1 \in L_1$ because $(1+x^2)^{m/2} f_1 \in L_1$. Therefore, by Theorem 6.13, the Fourier transform of $f_1 * f_2$ is

$$\lim_{n\to\infty} F_1(\alpha)\int_{-\infty}^{\infty} F_2(\alpha-\beta)\left(\frac{n}{\pi}\right)^{1/2} e^{-n\beta^2}\,d\beta$$

after a use of Theorem 6.7. Since $F_2 = F_{22}^{(l)}$ this may be written as

$$\lim_{n\to\infty} F_1(\alpha)\int_{-\infty}^{\infty} F_{22}(\alpha-\beta)\left(\frac{n}{\pi}\right)^{1/2}\{e^{-n\beta^2}\}^{(l)}\,d\beta.$$

Therefore

$$\int_{-\infty}^{\infty}\Gamma(\alpha)F_1(\alpha)\int_{-\infty}^{\infty} F_2(\alpha-\beta)\left(\frac{n}{\pi}\right)^{1/2} e^{-n\beta^2}\,d\beta\,d\alpha$$

$$= \int_{-\infty}^{\infty}\left(\frac{n}{\pi}\right)^{1/2}\{e^{-n\beta^2}\}^{(l)}\int_{-\infty}^{\infty}\Gamma(\alpha)F_1(\alpha)F_{22}(\alpha-\beta)\,d\alpha\,d\beta$$

the integral being absolutely convergent. In view of the fact that the inner integral is a convolution this may be written as

$$\int_{-\infty}^{\infty}\left(\frac{n}{\pi}\right)^{1/2} e^{-n\beta^2}(-1)^l\int_{-\infty}^{\infty}\{\Gamma(\alpha)F_1(\alpha)\}^{(l)}F_{22}(\alpha-\beta)\,d\alpha\,d\beta.$$

The inequality (28) with $\xi = \alpha$, $\eta = \beta$ gives

$$\left|\int_{-\infty}^{\infty}\{\Gamma(\alpha)F_1(\alpha)\}^{(l)}F_{22}(\alpha-\beta)\,d\alpha\right|$$

$$\leq 4(1+\beta^2)^N \sup|\{\Gamma(\alpha)F_1(\alpha)\}^{(l)}|(1+\alpha^2)^N\int_{-\infty}^{\infty}\frac{F_{22}(t)\,dt}{(1+t^2)^N}$$

$$\leq B(1+\beta^2)^N$$

on account of the given properties of F_1 and F_{22}. As in Theorem 6.14

we may show that the integral is continuous in β near $\beta = 0$ and deduce that

$$\lim_{n\to\infty}\int_{-\infty}^{\infty}\Gamma(\alpha)F_1(\alpha)\int_{-\infty}^{\infty}F_2(\alpha-\beta)\left(\frac{n}{\pi}\right)^{1/2}e^{-n\beta^2}\,d\beta\,d\alpha$$

$$=(-1)^l\int_{-\infty}^{\infty}\{\Gamma(\alpha)F_1(\alpha)\}^{(l)}F_{22}(\alpha)\,d\alpha$$

$$=\int_{-\infty}^{\infty}\Gamma(\alpha)F_1(\alpha)F_2(\alpha)\,d\alpha.$$

Hence the Fourier transform of $f_1 * f_2$ is $F_1 F_2$ and the proof of the theorem is concluded. □

Theorem 6.17. *If, for some finite integer* $m \geq 0, (1+x^2)^{-m/2}g_1(x) \in L'_\infty, (1+x^2)^{m/2}g_2(x) \in L'_1$ *and* $(1+x^2)^{m/2}g_3(x) \in L'_1$ *then*

$$(g_1 * g_2) * g_3 = g_1 * (g_2 * g_3).$$

Proof. It is sufficient to consider the behaviour of $(f_1 * f_2) * f_3$ where $(1+x^2)^{-m/2}f_1 \in L_\infty, (1+x^2)^{m/2}f_2 \in L_1$ and $(1+x^2)^{m/2}f_3 \in L_1$. By (29)

$$|f_1 * f_2| \leq 2^m B(1+x^2)^{m/2}$$

and therefore

$$\int_{-\infty}^{\infty}\int_{-\infty}^{\infty}|f_3(y-x)|\,|f_1(t)|\,|f_2(x-t)|\,dt\,dx$$

$$\leq 2^m B\int_{-\infty}^{\infty}|f_3(y-x)|(1+x^2)^{m/2}\,dx$$

$$\leq 2^{2m}B(1+y^2)^{m/2}\int_{-\infty}^{\infty}|f_3(u)|(1+u^2)^{m/2}\,du$$

from (28) with $\xi = x - y, \eta = y$. Thus the integral is absolutely convergent for any finite y and the order of integration may be interchanged, i.e. $(f_1 * f_2) * f_3 = f_1 * (f_2 * f_3)$. Note that (28) can be employed to prove that $(1+x^2)^{m/2} f_2 * f_3 \in L_1$. The proof is complete. □

Example 11, in which two of the factors belong to L_∞ and one to L'_1, demonstrates that not much relaxation of the conditions in

Theorem 6.17 is permissible before the associativity of the convolution product disappears.

Exercise

57. Prove that $$\int_{-\infty}^{\infty} (x-t)^{-4} t \operatorname{sgn} t \, dt = \tfrac{1}{3}x^{-2}.$$

6.6. The convolution for the class K_+

At the beginning of §6.5 it was pointed out that the convolution could be extended to generalised functions which vanish from some point to infinity. Since the point can be converted to the origin by a translation, and the negative axis changed into the positive by changing the sign of the variable there is no loss of generality in restricting attention to those generalised functions which vanish on the negative real axis.

Definition 6.8. *$g \in K_+$ if, and only if, $g(x) = 0$ for $x < 0$.*
Both H(x) and $\delta(x)$ are in K_+.

Theorem 6.18. *If $g_1 \in K_+$ and $g_2 \in K_+$ then $g_1 * g_2$ exists; it vanishes for $x < 0$, i.e. $g_1 * g_2 \in K_+$.*

Proof. On account of Theorem 3.18 we can write $g_1 = f_1^{(r)}$, $g_2 = f_2^{(s)}$ where f_1 and f_2 are continuous and vanish for $x < 0$. Furthermore there is some finite k such that $(1+x^2)^{-k/2} f_1(x)$ and $(1+x^2)^{-k/2} f_2(x)$ are bounded on R_1. Because of Definition 6.7 it will be sufficient to prove that $f_1 * f_2$ exists and vanishes for $x < 0$.

Now, because $f_1(t)$ vanishes on $t < 0$,

$$\int_{-\infty}^{\infty} f_1(t) f_2(x-t)\,dt = \int_0^{\infty} f_1(t) f_2(x-t)\,dt$$

$$= \begin{cases} \displaystyle\int_0^{x} f_1(t) f_2(x-t)\,dt & (x \ge 0) \\ 0 & (x < 0) \end{cases}$$

since $f_2(x-t)=0$ if $t>x$. But

$$\left|\int_0^x f_1(t)f_2(x-t)\mathrm{d}t\right| \le \int_0^x M_1(1+t^2)^{k/2}M_2\{1+(x-t)^2\}^{k/2}\mathrm{d}t$$
$$\le M_1 M_2 x(1+x^2)^k$$

which shows that $f_1 * f_2 \in K_1$. It follows that $g_1 * g_2$ exists. Also $f_1 * f_2 = 0$ for $x<0$ implies that $g_1 * g_2 \in K_+$ and the theorem is complete. □

Corollary 6.18. *If g_1, g_2 and g_3 all belong to K_+ then*

$$g_1 *(g_2 * g_3) = (g_1 * g_2) * g_3.$$

Proof. With fs having the same properties as in Theorem 6.18

$$f_1 *(f_2 * f_3) = \int_0^x f_1(t)\int_0^{x-t} f_3(u)f_2(x-t-u)\mathrm{d}u\,\mathrm{d}t$$
$$= \int_0^x f_3(u)\int_0^{x-u} f_1(t)f_2(x-u-t)\mathrm{d}t\,\mathrm{d}u$$
$$=(f_1 * f_2) * f_3$$

the inversion of the order of integration being obviously justified for continuous functions. The corollary now follows by taking generalised derivatives. □

The convolution for the class K_+ permits generalised functions whose behaviour at infinity excludes them from the preceding section but imposes the requirement of being zero on the negative real axis. Two different classes of generalised function are therefore catered for though there is some overlap; for example, if g_1 and g_2 are both zero for $x<0$ in Theorem 6.12 the same conclusion can be drawn from Theorem 6.18. When $g_1 \in K_+, g_2 \in K_+$ and g_1, g_2 satisfy the conditions of Theorems 6.12, 6.13 or 6.14 or their corollaries $g_1 * g_2$ is the same whether it be determined from these theorems or from Theorem 6.18. For in these theorems it is only necessary to put $f_1 = \hat{f}_1^{(r_1)}, f_2 = \hat{f}_2^{(r_2)}$ and then they becomes particular cases of Theorem 6.18.

The form of $f_1 * f_2$ in Theorem 6.18 suggests that it will be convenient to write

$$g_1 * g_2 = \begin{cases} \int_0^x g_1(t) g_2(x-t)\,dt & (x \geq 0) \\ 0 & (x < 0) \end{cases}$$

when $g_1 \in K_+$ and $g_2 \in K_+$. But then it must be understood that

$$(g_1 * g_2)' = \int_0^x g_1'(t) g_2(x-t)\,dt = \int_0^x g_1(t) g_2'(x-t)\,dt$$

and not that

$$(g_1 * g_2)' = \int_0^x g_1(t) g_2'(x-t)\,dt + g_1(x) g_2(0).$$

Example 12. If $\operatorname{Re}\beta > -1$ and $\operatorname{Re}\gamma > -1$

$$\int_0^x t^\beta (x-t)^\gamma\,dt = x^{\beta+\gamma+1} \int_0^1 u^\beta (1-u)^\gamma\,du$$

by the substitution $t = xu$. The integral on the right is a Beta-function whose value is well known. Therefore

$$\int_0^x t^\beta (x-t)^\gamma\,dt = \frac{\beta!\,\gamma!\,x^{\beta+\gamma+1}}{(\beta+\gamma+1)!}. \tag{30}$$

An alternative way of writing this is

$$\int_{-\infty}^{\infty} t^\beta \mathrm{H}(t)(x-t)^\gamma \mathrm{H}(x-t)\,dt = \frac{\beta!\,\gamma!}{(\beta+\gamma+1)!} x^{\beta+\gamma+1} \mathrm{H}(x). \tag{31}$$

If β is not a negative integer we can find a non-negative integer such that $\operatorname{Re}\beta + n > -1$ so that, on account of (31),

$$\int_{-\infty}^{\infty} t^{\beta+n} \mathrm{H}(t)(x-t)^\gamma \mathrm{H}(x-t)\,dt = \frac{(\beta+n)!\,\gamma!}{(\beta+n+\gamma+1)!} x^{\beta+n+\gamma+1} \mathrm{H}(x).$$

It now follows from Definition 4.2 that

$$\int_{-\infty}^{\infty} t^\beta \mathrm{H}(t)(x-t)^\gamma \mathrm{H}(x-t)\,dt = \frac{\beta!\,\gamma!}{(\beta+n+\gamma+1)!} \{x^{\beta+n+\gamma+1} \mathrm{H}(x)\}^{(n)}$$

$$= \frac{\beta!\,\gamma!}{(\beta+\gamma+1)!} x^{\beta+\gamma+1} \mathrm{H}(x). \tag{32}$$

Thus (30) and (31) remain valid if β is a negative fraction (unless $\beta+\gamma+1$ is a negative integer in which case the right-hand side of (32) must be replaced by $\beta!\gamma!\delta^{(-\beta-\gamma-2)}(x)$). The same is true if γ is not a negative integer.

If $\beta=0$ (31) becomes

$$\int_{-\infty}^{\infty} H(t)(x-t)^{\gamma} H(x-t)\,dt = \gamma!\,x^{\gamma+1}H(x)/(\gamma+1)!.$$

Hence, $m+1$ generalised derivatives give

$$\int_{-\infty}^{\infty} \delta^{(m)}(t)\,(x-t)^{\gamma} H(x-t)\,dt = \gamma!\,x^{\gamma-m}H(x)/(\gamma-m)!$$

or

$$\int_{0}^{x} \delta^{(m)}(t)(x-t)^{\gamma}\,dt = \gamma!\,x^{\gamma-m}/(\gamma-m)!. \tag{33}$$

If $\beta=0$ and $\gamma=0$ then (31) is

$$\int_{-\infty}^{\infty} H(t)\,H(x-t)\,dt = xH(x)$$

whence

$$\int_{-\infty}^{\infty} \delta^{(m)}(t)H(x-t)\,dt = \int_{-\infty}^{\infty} \delta^{(m-r)}(t)\delta^{(r-1)}(x-t)\,dt = \delta^{(m-1)}(x)$$

or

$$\int_{0}^{x} \delta^{(m)}(t)\,dt = \int_{0}^{x} \delta^{(m-r)}(t)\delta^{(r-1)}(x-t)\,dt = \delta^{(m-1)}(x). \tag{34}$$

The formulae (30), (33) and (34) can be expressed in a compact form by introducing the generalised function x_β defined by

$$x_\beta = \begin{cases} \dfrac{x^{\beta-1}H(x)}{(\beta-1)!} & (\beta \neq 0, -1, -2, \dots) \\ \delta^{(-\beta)}(x) & (\beta = 0, -1, -2, \dots). \end{cases}$$

Then it may be easily verified that (30), (33) and (34) can all be expressed as

$$x_\beta * x_\gamma = x_{\beta+\gamma}. \tag{35}$$

Define $I^\beta g$, when $g \in K_+$, by

$$I^\beta g = x_\beta * g.$$

Then, on account of (35) and Corollary 6.18,

$$I^\beta(I^\gamma g) = (I^\beta I^\gamma)g = I^{\beta+\gamma} g. \tag{36}$$

The notation has been chosen so that $I^0 g = \delta * g = g$ from (6). Also if β is not 0 or a negative integer

$$I^\beta g = \int_0^x (x-t)^{\beta-1} g(t)\mathrm{d}t/(\beta-1)!.$$

For this reason $I^\beta g$ is often known as a *fractional integral of order* β. Furthermore, since $I^\beta(I^{-\beta} g) = I^0 g = g$, $I^{-\beta} g$ can be regarded as a *fractional derivative.*

If g_1 is the Fourier transform of a fairly good function as well as being in K_+ Theorem 6.8 shows that the Fourier transform of $g_1 * g_2$ is $G_1 G_2$. Similarly if g_1 and g_2 both belong to L_2' as well as K_+ the Fourier transform of $g_1 * g_2$ is $G_1 G_2$ by Theorem 6.13. However, it cannot be asserted that, in general, the Fourier transform of $g_1 * g_2$ is $G_1 G_2$. A counter-example is provided by taking $g_1(x) = g_2(x) = \mathrm{H}(x)$. If the Fourier transform were $G_1 G_2$ it would be $\{\pi\delta(\alpha) - \mathrm{i}\alpha^{-1}\}^2$, but such a product does not come within the definitions of multiplication given here.

There is one general result concerning the Fourier transform. It depends upon

Definition 6.9. *If* $g \in K_+$ *the Fourier transform of* $\mathrm{e}^{-ax} g(x) (a > 0)$ *is denoted by* $\hat{G}(\alpha)$.

An alternative notation is $G(\alpha - ia)$, thereby defining the Fourier transform in a complex domain. For the moment we shall use $\hat{G}$.

Theorem 6.19. *If* $g \in K_+$, $\hat{G}(\alpha)$ *is a regular function of* $\alpha - ia$ *in* $a > 0$ *and* $\mathrm{Lim}_{a \to +0} \hat{G}(\alpha) = G(\alpha)$.

Proof. $g = f^{(r)}$ where $f \in K_+$ is continuous and $O(x^k)$ for some finite k as $x \to \infty$. Hence

$$|\hat{F}(\alpha)| = \left|\int_0^\infty f(x)\mathrm{e}^{-ax - \mathrm{i}\alpha x}\mathrm{d}x\right| \le M \int_0^\infty (1+x^2)^{k/2}\mathrm{e}^{-ax}\mathrm{d}x < \infty \tag{37}$$

for all α and $a > 0$. Thus $\hat{F}$ is defined as a function of $\alpha - ia$ in $a > 0$. Denote $\alpha - ia$ by s. Then

$$\mathrm{d}\hat{F}(\alpha)/\mathrm{d}s = -\mathrm{i}\int_0^\infty xf(x)\mathrm{e}^{-\mathrm{i}sx}\mathrm{d}x$$

by Theorem 1.33 provided that the right-hand side is uniformly convergent. But this can be proved by the same method as for (37). Hence $\mathrm{d}\hat{F}/\mathrm{d}s$ exists and, similarly, so does any derivative of $\hat{F}$ with respect to s. Thus $\hat{F}$ is a regular function of $\alpha - ia$ in $a > 0$.

The Fourier transform of $\mathrm{e}^{-ax}f'(x)$ is the transform of

$$\{\mathrm{e}^{-ax}f(x)\}' + a\mathrm{e}^{-ax}f(x),$$

i.e. $\mathrm{i}\alpha\hat{F}(\alpha) + a\hat{F}(\alpha)$. Hence the Fourier transform of $\mathrm{e}^{-ax}g(x)$ is $(\mathrm{i}\alpha + a)^r\hat{F}(\alpha)$ and since the factor $(\mathrm{i}\alpha + a)^r$ has no effect on the regularity, $\hat{G}$ is regular in $a > 0$.

Finally

$$\lim_{a\to 0}\int_{-\infty}^\infty \{\hat{F}(\alpha) - F(\alpha)\}\Gamma(\alpha)\mathrm{d}\alpha = \lim_{a\to 0}\int_0^\infty 2\pi(\mathrm{e}^{-ax} - 1)f(x)\gamma(-x)\mathrm{d}x$$

by Theorem 3.9. Since $1 - \mathrm{e}^{-ax} \le ax$ for $x \ge 0$ the right-hand side tends to zero and $\mathrm{Lim}_{a\to 0}\hat{F} = F$. It follows that $\mathrm{Lim}_{a\to 0}\alpha^p\hat{F}(\alpha) = \alpha^p F(\alpha)$ and so

$$\mathrm{Lim}_{a\to 0}\hat{G}(\alpha) = \mathrm{Lim}_{a\to 0}(\mathrm{i}\alpha + a)^r\hat{F}(\alpha) = (\mathrm{i}\alpha)^r F(\alpha) = G(\alpha).$$

The proof is finished. □

As an example consider $g(x) = \mathrm{H}(x)$. Then $\hat{G} = 1/(\mathrm{i}\alpha + a)$ which is certainly regular for $a > 0$ though not for $a = 0$. The second part of the theorem shows that (cf. Exercise 11 of Chapter 4)

$$\mathrm{Lim}_{a\to +0}\frac{1}{\mathrm{i}\alpha + a} = \pi\delta(\alpha) - \mathrm{i}\alpha^{-1}.$$

This is consistent with the evaluation of $\lim_{a\to 0}\int_{-\infty}^\infty\{\Gamma(\alpha)/(\mathrm{i}\alpha + a)\}\mathrm{d}\alpha$ by contour integration.

Theorem 6.20. *If g_1 and g_2 belong to K_+ the Fourier transform of $g_1 * g_2$ is $\mathrm{Lim}_{a\to 0}\hat{G}_1\hat{G}_2$.*

Proof. By Theorem 6.19 the Fourier transform of $g_1 * g_2$ can be obtained by a limiting process from that of $e^{-ax}\int_0^x g_1(t)g_2(x-t)dt$ or of $\int_0^x e^{-at}g_1(t)e^{-a(x-t)}g_2(x-t)dt$. Since $e^{-at}g_1(t) = e^{-at}f_1^{(r)}(t)$ where f_1 is continuous and $O(x^k)$ as $x \to \infty$ it follows that $e^{-at}g_1(t) \in L_1'$. Similarly $e^{-at}g_2(t) \in L_1'$. Hence by Theorem 6.13, the Fourier transform of the convolution is $\hat{G}_1 \hat{G}_2$ and the theorem is proved. □

One consequence of Theorem 6.20 is obtained by putting $g_1 = g_2 = \mathrm{H}(x)$. Then the Fourier transform of $x\mathrm{H}(x)$ (Example 12) is

$$\lim_{a \to +0} (\mathrm{i}\alpha + a)^{-2} = -(\alpha - \mathrm{i}0)^{-2} = \pi \mathrm{i}\delta'(\alpha) - \alpha^{-2}$$

consistent with Chapter 4.

Theorem 6.21. *If g_1 and g_2 belong to K_+ and $g_1 * g_2 = 0$ then either $g_1 = 0$ or $g_2 = 0$.*

Proof. If $g_1 * g_2 = 0$ then

$$\int_0^x e^{-at}g_1(t)e^{-a(x-t)}g_2(x-t)dt = 0.$$

Hence, from the proof of Theorem 6.20, $\hat{G}_1 \hat{G}_2 = 0$. But $\hat{G}_1$ and $\hat{G}_2$ are both regular in $a > 0$ by Theorem 6.19 and therefore can have isolated zeros only. Hence either $\hat{G}_1 = 0$ or $\hat{G}_2 = 0$; consequently either $e^{-at}g_1(t) = 0$ or $e^{-at}g_2(t) = 0$ and the conclusion of the theorem follows. □

Exercises

58. If $g_1(x) = \mathrm{H}(x)\cos^2 x$, $g_2(x) = \mathrm{H}(x)\sin^2 x$, $g_3(x) = \mathrm{H}(x)$ show that

$$g_1 * g_3 + g_2 * g_3 = x\mathrm{H}(x).$$

59. If $g(x) = e^{-x}\mathrm{H}(x)$ show that $g * g = xg$. More generally, show that the convolution $g * g * g * \ldots$ containing n factors equals $x^{n-1}\, g/(n-1)!$.

60. If g_1 and g_2 belong to K_+ and

$$\int_0^x g_1(t)(x-t)^\beta dt = g_2(x),$$

β not being an integer, prove that

$$g_1(t) = \frac{1}{\beta!(-\beta-2)!}\int_0^t g_2(x)(t-x)^{-\beta-2}dx.$$

What is the corresponding result if β is a non-negative integer? (This is an example of an *Abel integral equation*. The factor $\{\beta!(-\beta-2)!\}^{-1}$ may also be written $\pi^{-1}(\beta+1)\sin\pi\beta$.)

61. If f is a conventional continuous function with conventional derivative $\mathrm{d}f/\mathrm{d}x$ prove that

$$\int_0^x f(t)(x-t)^{-1}\,\mathrm{d}t = f(0)\ln x + \int_0^x (\mathrm{d}f/\mathrm{d}t)\ln(x-t)\,\mathrm{d}t.$$

62. Prove that

$$\text{(i)}\quad \int_0^x \frac{J_1(t)}{t}J_0(x-t)\,\mathrm{d}t = J_1(x), \qquad \text{(ii)}\quad \int_0^x J_0(t)J_0(x-t)\,\mathrm{d}t = \sin x,$$

$$\text{(iii)}\quad \int_0^x \frac{J_\mu(t)J_\nu(x-t)}{t(x-t)}\,\mathrm{d}t = \frac{\mu+\nu}{\mu\nu}\cdot\frac{J_{\mu+\nu}(x)}{x}.$$

(Hint: The Fourier transforms of $J_0(x)\mathrm{H}(x)$, $J_1(x)\mathrm{H}(x)$, $J_\nu(x)\mathrm{H}(x)/x$ are $(1-\alpha^2)^{-1/2}$, $1-\mathrm{i}\alpha(1-\alpha^2)^{-1/2}$, $\{(1-\alpha^2)^{1/2}-\mathrm{i}\alpha\}^\nu/\nu$ respectively where $(1-\alpha^2)^{-1/2}$ means $-\mathrm{i}(\alpha^2-1)^{-1/2}\operatorname{sgn}\alpha$ when $|\alpha|>1$.)

63. If $g(x) = x^{1/2}\mathrm{e}^{-1/x}\mathrm{H}(x)$ show that the Fourier transform of $g*g$ is $\operatorname{Lim}_{a\to+0}\{\pi/(\mathrm{i}\alpha+a)\}\exp\{-4(\mathrm{i}\alpha+a)^{1/2}\}$. Is this the same as

$$\pi^2\delta(\alpha) - \pi\mathrm{i}\alpha^{-1}\exp\{-4(\mathrm{i}\alpha)^{1/2}\}\,?$$

64. If both g_1 and g_2 vanish for $x<-b$, $b>0$ show that g_1*g_2 exists. Show also that g_1*g_2 exists if g_1 and g_2 vanish for $x>b$. Hence prove that $g_1 = \mathrm{H}*\eta g - \underline{\mathrm{H}}*(1-\eta)g$ where $\underline{\mathrm{H}}(x) = \mathrm{H}(-x)$ and η is any infinitely differentiable function such that $\eta=0$ for $x\le -b$ and $\eta=1$ for $x\ge b$, satisfies $g_1' = g$. (This constitutes another proof that any generalised function possesses an indefinite integral; cf. §3.6.)

65. If g_1 and g_2 both belong to K_+ show that

$$\int_{-\infty}^{\infty}\gamma(x)g_1*g_2\,\mathrm{d}x = \int_{-\infty}^{\infty}g_1(x)\int_{-\infty}^{\infty}g_2(t)\gamma(x+t)\,\mathrm{d}t\,\mathrm{d}x.$$

66. If $g_\mu\in K_+$, $g\in K_+$ and $g_2\in K_+$ and if $\operatorname{Lim}_{\mu\to\mu_0}g_\mu = g$ show that

$$\operatorname*{Lim}_{\mu\to\mu_0} g_\mu * g_2 = g*g_2.$$

Deduce that, if $\partial g_\mu/\partial\mu$ exists, $\partial(g_\mu*g_2)/\partial\mu = (\partial g_\mu/\partial\mu)*g_2$.

67. Prove that

$$(\mu-\tfrac{1}{2})!\,2^\mu\pi^{1/2}x^{\mu/2}J_\mu(x^{1/2}) = \int_0^x (x-t)^{\mu-1/2}\frac{\cos t^{1/2}}{t^{1/2}}\,\mathrm{d}t.$$

Deduce that

$$I^{\nu+1}\{2^\mu\pi^{1/2}x^{\mu/2}J_\mu(x^{1/2})\} = 2^{\mu+\nu+1}\pi^{1/2}x^{(\mu+\nu+1)/2}J_{\mu+\nu+1}(x^{1/2})$$

and hence that

$$2^{\nu+1}x^{(\mu+\nu+1)/2}J_{\mu+\nu+1}(x^{1/2}) = \int_0^x t^{\mu/2}J_\mu(t^{1/2})\frac{(x-t)^\nu}{\nu!}\mathrm{d}t$$

which may also be written

$$J_{\mu+\nu+1}(x) = \frac{x^{\nu+1}}{\nu!2^\nu}\int_0^{\pi/2} J_\mu(x\sin\theta)\sin^{\mu+1}\theta\cos^{2\nu+1}\theta\,\mathrm{d}\theta.$$

(Sonine's first finite integral)

68. If $g \in K_+$ and $g(x) = O(x^\nu)(\mathrm{Re}\,\nu > -1)$ for $x \geq 0$ prove that

$$\operatorname*{Lim}_{a\to+0} \mathrm{e}^{-ax}g(x) = g(x)$$

and that

$$\operatorname*{Lim}_{a\to+0} \hat{G}(\alpha) = G(\alpha).$$

One further result should also be noted. Namely, that if g_2 is doubly restricted so as to vanish outside a finite interval then no restriction need be imposed on g_1. For

Theorem 6.22. *If g_2 vanishes outside a finite interval then $g_1 * g_2$ exists for any generalised function g_1 and $g_1 * g_2 = g_2 * g_1$.*

Proof. As in Theorem 6.14 it is only necessary to prove this result for continuous functions with similar properties. Now

$$\int_{-\infty}^{\infty} f_1(t)f_2(x-t)\mathrm{d}t = \int_{x-b}^{x+b} f_1(t)f_2(x-t)\mathrm{d}t$$

where b is finite. The modulus of the right-hand side does not exceed $M_1M_2(1+x^2)^k$ and so the right-hand side defines a function in K_1. Obviously $f_1 * f_2 = f_2 * f_1$ and the theorem follows. □

Again, by considering functions we can show

Corollary 6.22. *If g_2 and g_3 both vanish outside a finite interval*

$$g_1 * (g_2 * g_3) = (g_1 * g_2) * g_3.$$

Clearly this can be extended to any number of factors provided that all, except possibly one, vanish outside a finite interval.

Exercises

69. If g is a generalised function which vanishes outside a finite interval show that it must be the Fourier transform of a fairly good function. Hence Theorem 6.22 can be deduced from §§6.3 and 6.4.

70. If g vanishes outside a finite interval prove that

$$\left\{\int_{-\infty}^{\infty} g(t)|x-t|\,\mathrm{d}t\right\}'' = 2g(x).$$

Prove also that this is true if $(1+x^2)^{1/2}g(x)\in L_1'$. (Hint: $|x|(1+x^2)^{-1/2}\in L_\infty$.)

7
Several variables

The notion of a generalised function of a single variable can be extended to the case of several variables and this chapter will be concerned with the extension. Many of the definitions and theorems are straightforward generalisations of those for a single variable and may therefore be dealt with briefly. Certain new features will however be treated in detail.

The complex-valued functions considered in this chapter are defined on R_n. The rectangular coordinates in R_n will be denoted by $x_1, x_2, \dots, x_n$ or occasionally by $x, y, z, \dots$. The letter r will be used to signify the radial distance so that $r^2 = x_1^2 + x_2^2 + \dots + x_n^2$. Often we shall use $\boldsymbol{x}$ as an abbreviation for $x_1, x_2, \dots, x_n$; thus $f(\boldsymbol{x})$ will mean $f(x_1, x_2, \dots, x_n)$ and $\int_a^b f(\boldsymbol{x})\mathrm{d}\boldsymbol{x}$ will mean

$$\int_a^b \int_a^b \cdots \int_a^b f(x_1, x_2, \dots, x_n)\mathrm{d}x_1\,\mathrm{d}x_2 \dots \mathrm{d}x_n.$$

In this notation it is sometimes convenient to write $r = |\boldsymbol{x}|$. The partial derivative $\partial^{p_1+p_2+\dots+p_n} f/\partial x_1^{p_1}\partial x_2^{p_2}\dots\partial x_n^{p_n}$ will be abridged to $\partial^p f$ on occasions.

There is one result concerning the transformation of integrals which is often useful. Let C_X denote $|x_1| \le X, |x_2| \le X, \dots, |x_n| \le X$ and S_X denote $r \le X$. Then

$$\int_{-\infty}^{\infty} f(\boldsymbol{x})\mathrm{d}\boldsymbol{x} = \lim_{X\to\infty} \int_{C_X} f(\boldsymbol{x})\mathrm{d}\boldsymbol{x}.$$

Now

$$\left|\int_{C_X} f(\boldsymbol{x})\mathrm{d}\boldsymbol{x} - \int_{S_X} f(\boldsymbol{x})\mathrm{d}\boldsymbol{x}\right| = \left|\int_{C_X - S_X} f(\boldsymbol{x})\mathrm{d}\boldsymbol{x}\right|$$

$$\le \int_{C_X - S_X} |f(\boldsymbol{x})|\mathrm{d}\boldsymbol{x}$$

$$\leq \int_{\Omega - S_X} |f(\boldsymbol{x})| \mathrm{d}\boldsymbol{x} \tag{1}$$

$$\leq \int_{\Omega - C_{X/2}} |f(\boldsymbol{x})| \mathrm{d}\boldsymbol{x} \tag{2}$$

where $\Omega - S_X$, $\Omega - C_{X/2}$ are the regions outside S_X and $C_{X/2}$ respectively. If $\int_{-\infty}^{\infty} |f(\boldsymbol{x})| \mathrm{d}\boldsymbol{x}$ exists we can choose X so large that the right-hand side of (2) is as small as we please so that $\lim_{X\to\infty} \int_{S_X} f(\boldsymbol{x})\mathrm{d}\boldsymbol{x} = \int_{-\infty}^{\infty} f(\boldsymbol{x})\mathrm{d}\boldsymbol{x}$. On the other hand if $\lim_{X\to\infty} \int_{S_X} f(\boldsymbol{x})\mathrm{d}\boldsymbol{x}$ exists the right-hand side of (1) can be chosen as small as we please and then

$$\lim_{X\to\infty} \int_{C_X} f(\boldsymbol{x})\mathrm{d}\boldsymbol{x} = \lim_{X\to\infty} \int_{S_X} f(\boldsymbol{x})\mathrm{d}\boldsymbol{x}.$$

Since C_X corresponds to rectangular coordinates and S_X to spherical polars, we have

Theorem 7.1. *If f is absolutely integrable its infinite integral can be calculated by means of rectangular or spherical polar coordinates. If f is absolutely integrable in spherical polars its infinite integral can be calculated either in rectangular or spherical polar coordinates.*

Exercise

1. Show that, if f is absolutely integrable, its infinite integral can be calculated by means of cylindrical polar coordinates, i.e. coordinates in which x_n is unaltered but $x_1, \dots, x_{n-1}$ are replaced by the spherical polar coordinates of R_{n-1}.

7.1. Good functions

Definition 7.1. *A function γ is said to be good if it possesses partial derivatives of all orders everywhere on R_n and if*

$$\lim_{r\to\infty} |r^k \partial^p \gamma| = 0$$

for all radial directions, every integer $k \geq 0$ and all integers $p_1 \geq 0$, $p_2 \geq 0, \dots, p_n \geq 0$ with $p = p_1 + p_2 + \dots + p_n$.

Obviously e^{-r^2}, $x_1^2 \mathrm{e}^{-r^2}$ are good functions whereas $\mathrm{e}^{-x_2^2}$ is not. Evidently, the sum and product of two good functions are good. If γ is good so are $\gamma(a_1 x_1 + b_1, a_2 x_2 + b_2, \dots, a_n x_n + b_n)$ and $\partial^p \gamma$. It is easy to see that a function which is good on R_n is also good on $R_1, R_2, \dots, R_{n-1}$. Theorem 7.1 shows that $\int_{-\infty}^{\infty} |\gamma(\boldsymbol{x})| \mathrm{d}\boldsymbol{x} < \infty$.

Definition 7.2. *A function ψ is said to be fairly good if it possesses partial derivatives of all orders everywhere on R_n and if there is some fixed N such that*

$$\lim_{r\to\infty} |r^{-N}\partial^p\gamma| = 0$$

for all radial directions and all integers $p_1 \geq 0, p_2 \geq 0, \ldots, p_n \geq 0$ with $p = p_1 + p_2 + \cdots + p_n$.

Any polynomial is fairly good; so is e^{ix_1}. The product and sum of fairly good functions are fairly good. If ψ is fairly good so are $\partial^p \psi$ and $\psi(a_1x_1 + b_1, a_2x_2 + b_2, \ldots, a_nx_n + b_n)$. As in Theorem 2.1 it may be shown that $\psi\gamma$ is good when ψ is fairly good and γ good.

Definition 7.3. *A function ϕ is said to be fine if it possesses partial derivatives of all orders everywhere on R_n and if it and all its partial derivatives vanish identically outside some finite sphere.*

The function which equals $e^{-r^2/(1-r^2)}$ for $r < 1$ and vanishes for $r \geq 1$ is fine. The sum and product of two fine functions are fine. A fine function is good. The product of a fairly good and a fine function is fine. Any partial derivative of a fine function is fine. A function which is fine on R_n is also fine on $R_1, R_2, \ldots, R_{n-1}$.

Definition 7.4. *If $\int_{-\infty}^{\infty} f(\boldsymbol{x})e^{-i(\alpha_1x_1+\alpha_2x_2+\ldots+\alpha_nx_n)}d\boldsymbol{x}$ exists for every $\boldsymbol{\alpha}$ in R_n it is called the Fourier transform of f.*

By introducing the notation

$$\boldsymbol{\alpha}.\boldsymbol{x} = \alpha_1x_1 + \alpha_2x_2 + \ldots + \alpha_nx_n$$

we can write the Fourier transform compactly as

$$F(\boldsymbol{\alpha}) = \int_{-\infty}^{\infty} f(\boldsymbol{x})e^{-i\boldsymbol{\alpha}.\boldsymbol{x}}d\boldsymbol{x}.$$

Theorem 7.2. *The Fourier transform of $e^{-r^2/2}$ is $(2\pi)^{n/2} e^{-\alpha^2/2}$ where $\alpha^2 = \alpha_1^2 + \alpha_2^2 + \ldots + \alpha_n^2$.*

Proof.

$$F(\boldsymbol{\alpha}) = \int_{-\infty}^{\infty} e^{-(x_1^2/2)-i\alpha_1x_1}dx_1 \int_{-\infty}^{\infty} e^{-(x_2^2/2)-i\alpha_2x_2}dx_2 \ldots$$

$$\times \int_{-\infty}^{\infty} e^{-(x_n^2/2)-i\alpha_nx_n}dx_n$$

and the theorem is an immediate consequence of Theorem 2.2.□

More generally, the Fourier transform of $\exp\{-\frac{1}{2}\sum_{j=1}^{n}\sum_{k=1}^{n} a_{jk}x_jx_k\}$ $(a_{jk}=a_{kj})$ can be considered. If $\boldsymbol{x}$ be regarded as a column matrix, $\boldsymbol{x}^{\mathrm{T}}$ as its transpose and A as a matrix with typical element a_{jk} we can write

$$\sum_{j=1}^{n}\sum_{k=1}^{n} a_{jk}x_jx_k = \boldsymbol{x}^{\mathrm{T}}A\boldsymbol{x}.$$

Suppose that $\boldsymbol{x}^{\mathrm{T}}A\boldsymbol{x}>0$ for all $\boldsymbol{x}$ except $\boldsymbol{x}=\boldsymbol{0}$. Then there is an orthogonal mapping $\boldsymbol{x}=L\boldsymbol{X}$ which converts the quadratic form $\boldsymbol{x}^{\mathrm{T}}A\boldsymbol{x}$ into

$$\lambda_1X_1^2+\lambda_2X_2^2+\ldots+\lambda_nX_n^2$$

or $\boldsymbol{X}^{\mathrm{T}}\Lambda\boldsymbol{X}$ where $\lambda_1,\lambda_2,\ldots,\lambda_n$ are all positive. If we put $\boldsymbol{\alpha}=L\boldsymbol{\beta}$, $\boldsymbol{\beta}=L^{\mathrm{T}}\boldsymbol{\alpha}$, $\boldsymbol{\beta}^{\mathrm{T}}=\boldsymbol{\alpha}L$ because $L^{\mathrm{T}}L=I$ for an orthogonal matrix. Hence

$$\boldsymbol{\alpha}.\boldsymbol{x}=\boldsymbol{\alpha}^{\mathrm{T}}\boldsymbol{x}=\boldsymbol{\alpha}^{\mathrm{T}}L\boldsymbol{X}=\boldsymbol{\beta}^{\mathrm{T}}\boldsymbol{X}$$

so that, as the element of volume is unaltered by an orthogonal mapping, the Fourier transform becomes

$$\int_{-\infty}^{\infty}\exp(-\tfrac{1}{2}\lambda_1X_1^2-\mathrm{i}\beta_1X_1)\,\mathrm{d}X_1\int_{-\infty}^{\infty}\exp(-\tfrac{1}{2}\lambda_2X_2^2-\mathrm{i}\beta_2X_2)\mathrm{d}X_2\ldots$$
$$\times\int_{-\infty}^{\infty}\exp(-\tfrac{1}{2}\lambda_nX_n^2-\mathrm{i}\beta_nX_n)\mathrm{d}X_n$$
$$=\left(\frac{2^n\pi^n}{\lambda_1\lambda_2\ldots\lambda_n}\right)^{1/2}\exp\left\{-\frac{1}{2}\left(\frac{\beta_1^2}{\lambda_1}+\frac{\beta_2^2}{\lambda_2}+\ldots+\frac{\beta_n^2}{\lambda_n}\right)\right\}$$
$$=\left(\frac{2^n\pi^n}{\det\Lambda}\right)^{1/2}\exp(-\tfrac{1}{2}\boldsymbol{\beta}^{\mathrm{T}}\Lambda^{-1}\boldsymbol{\beta}).$$

Since $\Lambda=L^{\mathrm{T}}AL$, $\det\Lambda=\det A$ because $\det L^{\mathrm{T}}L=1$, and $\Lambda^{-1}=L^{-1}A^{-1}(L^{\mathrm{T}})^{-1}$. Hence

Theorem 7.3. *The Fourier transform of* $\exp(-\frac{1}{2}\boldsymbol{x}^{\mathrm{T}}A\boldsymbol{x})$, *$A$ being positive definite, is* $(2^n\pi^n/\det A)^{1/2}\exp(-\frac{1}{2}\boldsymbol{\alpha}^{\mathrm{T}}A^{-1}\boldsymbol{\alpha})$.

Theorem 7.4. *If $\gamma(\boldsymbol{x})$ is good so is its Fourier transform $\Gamma(\boldsymbol{\alpha})$.*

Proof. The proof is along the lines of Theorem 2.4. As there, we reach

$$\partial^q \Gamma(\boldsymbol{\alpha}) = \int_{-\infty}^{\infty} \gamma_1(\boldsymbol{x}) e^{-i\boldsymbol{\alpha}.\boldsymbol{x}} d\boldsymbol{x}$$

where $\gamma_1(\boldsymbol{x})$ is the good function $(-ix_1)^{q_1}(-ix_2)^{q_2} \dots (-ix_n)^{q_n} \gamma(\boldsymbol{x})$, with $q_1 + q_2 + \dots + q_n = q$. Integration by parts with respect to x_n gives

$$\partial^q \Gamma(\boldsymbol{\alpha}) = \int_{-\infty}^{\infty} \dots \int_{-\infty}^{\infty} \left[\frac{\gamma_1(\boldsymbol{x}) e^{-i\boldsymbol{\alpha}.\boldsymbol{x}}}{-i\alpha_n} \right]_{x_n=-\infty}^{x_n=\infty} dx_1 \dots dx_{n-1}$$

$$+ \frac{1}{i\alpha_n} \int_{-\infty}^{\infty} e^{-i\boldsymbol{\alpha}.\boldsymbol{x}} \frac{\partial \gamma_1}{\partial x_n} d\boldsymbol{x}$$

$$= \frac{1}{i\alpha_n} \int_{-\infty}^{\infty} e^{-i\boldsymbol{\alpha}.\boldsymbol{x}} \frac{\partial \gamma_1}{\partial x_n} d\boldsymbol{x}$$

since the integral in the first term converges uniformly with respect to x_n and $\gamma_1 \to 0$ as $|x_n| \to \infty$. Repetition of this process leads to

$$\partial^q \Gamma(\boldsymbol{\alpha}) = \frac{1}{(i\alpha_1)^{m_1}(i\alpha_2)^{m_2} \dots (i\alpha_n)^{m_n}} \int_{-\infty}^{\infty} e^{-i\boldsymbol{\alpha}.\boldsymbol{x}} \partial^m \gamma_1 d\boldsymbol{x}$$

from which we deduce that $\lim_{|\boldsymbol{\alpha}|\to\infty} \left| |\boldsymbol{\alpha}|^k \partial^q \Gamma(\boldsymbol{\alpha}) \right| = 0$ and the proof is finished.□

Theorem 7.5. *If $\gamma(\boldsymbol{x})$ is a good function*

(i) $\gamma(\boldsymbol{x}) = \{1/(2\pi)^n\} \int_{-\infty}^{\infty} \Gamma(\boldsymbol{\alpha}) e^{i\boldsymbol{\alpha}.\boldsymbol{x}} d\boldsymbol{\alpha}$,

(ii) *the Fourier transform of $\partial\gamma/\partial x_p$ is $i\alpha_p \Gamma(\boldsymbol{\alpha})$.*

Proof. $\int_{-\infty}^{\infty} \dots \int_{-\infty}^{\infty} \gamma(\boldsymbol{x}) e^{-i(\alpha_1 x_1 + \dots + \alpha_{n-1} x_{n-1})} dx_1 \dots dx_{n-1}$ is a good function of x_n for

$$\left| \int_{-\infty}^{\infty} \dots \int_{-\infty}^{\infty} \frac{\partial^q \gamma}{\partial x_n^q} e^{-i(\alpha_1 x_1 + \dots + \alpha_{n-1} x_{n-1})} dx_1 \dots dx_{n-1} \right|$$

$$\leq B \int_{-\infty}^{\infty} \dots \int_{-\infty}^{\infty} \frac{dx_1 \dots dx_{n-1}}{(1+x_1^2)^2 \dots (1+x_{n-1}^2)^2}$$

$$< \infty$$

so that the integral on the left converges uniformly with respect to x_n and the result follows at once. Hence, by Theorem 2.5(i),

$$\int_{-\infty}^{\infty} \cdots \int_{-\infty}^{\infty} \gamma(\boldsymbol{x}) e^{-i(\alpha_1 x_1 + \ldots + \alpha_{n-1} x_{n-1})} dx_1 \ldots dx_{n-1}$$

$$= \frac{1}{2\pi} \int_{-\infty}^{\infty} \Gamma(\boldsymbol{\alpha}) e^{i\alpha_n x_n} d\alpha_n.$$

Repeating the process with n replaced by $n - 1$ we obtain

$$\int_{-\infty}^{\infty} \cdots \int_{-\infty}^{\infty} \gamma(\boldsymbol{x}) e^{-i(\alpha_1 x_1 + \ldots + \alpha_{n-2} x_{n-2})} dx_1 \ldots dx_{n-2}$$

$$= \frac{1}{2\pi} \int_{-\infty}^{\infty} e^{i\alpha_{n-1} x_{n-1}} \left\{ \int_{-\infty}^{\infty} \cdots \int_{-\infty}^{\infty} \gamma(\boldsymbol{x}) e^{-i(\alpha_1 x_1 + \ldots + \alpha_{n-1} x_{n-1})} \times dx_1 \ldots dx_{n-1} \right\} d\alpha_{n-1}$$

$$= \frac{1}{(2\pi)^2} \int_{-\infty}^{\infty} \int_{-\infty}^{\infty} \Gamma(\boldsymbol{x}) e^{i(\alpha_{n-1} x_{n-1} + \alpha_n x_n)} d\alpha_{n-1} d\alpha_n.$$

Continuing in this way we derive the result stated in (i) of the theorem.

As for (ii) the proof is the same as for Theorem 2.5 modified as in Theorem 7.4.□

Theorem 7.6. *If γ is good and $\int_{-\infty}^{\infty} |f(\boldsymbol{x})| d\boldsymbol{x} < \infty$ then*

$$\int_{-\infty}^{\infty} \Gamma(\boldsymbol{\alpha}) F(\boldsymbol{\alpha}) e^{i\boldsymbol{\alpha}.\boldsymbol{y}} d\boldsymbol{\alpha} = (2\pi)^n \int_{-\infty}^{\infty} f(\boldsymbol{x}) \gamma(\boldsymbol{y} - \boldsymbol{x}) d\boldsymbol{x}.$$

Proof. The same as for Theorem 2.6.□

Exercises

2. Which of the following are good on R_n?

$$r^2 e^{-r^2 + \pi i/4}, \sin r, 1/(r^4 + 1), e^{-r^2} \sin 4x_1, 0 \, (r > 1) \text{ and } e^{-1/(1-r^2)} \, (r \le 1).$$

3. Which of the following are fairly good on R_n?

$$\sin r, \sin x_2, \tan 2x_3, (1 + r^2)^m.$$

4. Show that $\sigma(\boldsymbol{x})$ and $\tau(\boldsymbol{x})$ defined by

$$\sigma(\boldsymbol{x}) = e^{-r^2/(1-r^2)}, \quad \tau(\boldsymbol{x}) = \int_r^1 e^{-1/t(1-t)} dt \Big/ \int_0^1 e^{-1/t(1-t)} dt \qquad (r < 1),$$

$$\sigma(\boldsymbol{x}) = 0, \quad \tau(\boldsymbol{x}) = 0 \qquad (r \ge 1)$$

are fine. Show also that

$$\sum_{m=1}^{M} \sigma(x_1 - m^2, x_2 - m^2, \dots, x_n - m^2)/m^4 \text{ and } \sum_{l=-m}^{m} \tau(x_1 - l, x_2 - l, \dots, x_n - l)$$

are fine.

5. If γ is good prove that

$$\max_{x \in R_n}(1 + r^2)^{k/2}|\partial^p \gamma(x)| \leq (\tfrac{1}{2}\pi)^{nm} \max_{x \in R_n}(1 + r^2)^{(k/2)+m}|\partial^{p+m}\gamma(x)|.$$

6. Determine the Fourier transform of $r^2 e^{-r^2}$.

7. If γ is good, with Fourier transform Γ, prove that the Fourier transform of $\gamma(a_1x_1 + b_1, a_2x_2 + b_2, \dots, a_nx_n + b_n)$ is

$$\left[\exp\left\{i\left(\frac{b_1\alpha_1}{a_1} + \frac{b_2\alpha_2}{a_2} + \dots + \frac{b_n\alpha_n}{a_n}\right)\right\} \Big/ |a_1 a_2 \dots a_n|\right] \Gamma\left(\frac{\alpha_1}{a_1}, \frac{\alpha_2}{a_2}, \dots, \frac{\alpha_n}{a_n}\right).$$

7.2. Generalised functions

Definition 7.5. *A sequence $\{\gamma_m\}$ of good functions is said to be regular if, for every good γ, $\lim_{m\to\infty} \int_{-\infty}^{\infty} \gamma_m(x)\gamma(x)dx$ exists and is finite. Two regular sequences which give the same limit are said to be equivalent. An equivalence class of regular sequences is a generalised function.*

If g is the generalised function associated with the equivalence class of which $\{\gamma_m\}$ is a typical member we shall write

$$\lim_{m\to\infty} \int_{-\infty}^{\infty} \gamma_m(x)\gamma(x)dx = \int_{-\infty}^{\infty} g(x)\gamma(x)dx.$$

The proofs that $\{e^{-r^2/m}\}$ and $\{(e^{-r^2/m})/m\}$ are regular sequences defining the generalised functions 1 and 0 respectively are unaltered from the single variable case.

Definition 7.6. *The sequence $\{(m/\pi)^{n/2} e^{-mr^2}\}$ is regular and defines a generalised function, denoted by $\delta(x)$, such that*

$$\int_{-\infty}^{\infty} \delta(x)\gamma(x)dx = \gamma(0).$$

For, from Theorem 7.2,

$$\int_{-\infty}^{\infty} \left(\frac{m}{\pi}\right)^{n/2} e^{-mr^2}\gamma(x)dx = \gamma(0) + \int_{-\infty}^{\infty} \left(\frac{m}{\pi}\right)^{n/2} \{\gamma(x) - \gamma(0)\}e^{-mr^2}dx.$$

Since

$$|\gamma(\boldsymbol{x})-\gamma(\mathbf{0})|$$
$$=\left|\int_0^{x_1}\frac{\partial\gamma(t_1,0,0,\dots)}{\partial t_1}\mathrm{d}t_1+\int_0^{x_2}\frac{\partial\gamma(x_1,t_2,0,\dots)}{\partial t_2}\mathrm{d}t_2+\dots\right.$$
$$\left.+\int_0^{x_n}\frac{\partial\gamma(x_1,x_2,\dots,t_n)}{\partial t_n}\mathrm{d}t_n\right| \tag{3}$$
$$\leq M(|x_1|+|x_2|+\dots+|x_n|)$$

the integral does not exceed $Mn/m^{1/2}$ which tends to zero as $m\to\infty$ and the definition is justified.

Definition 7.7. *$f\in\boldsymbol{K}_p$ if, and only if, $\int_{-\infty}^{\infty}\{|f(\boldsymbol{x})|^p/(1+r^2)^N\}\,\mathrm{d}\boldsymbol{x}<\infty$ for some finite N.*

As in the case of the single variable, $\boldsymbol{K}_p\subset\boldsymbol{K}_{p_1}$ for $p\geq p_1$. We also use $\boldsymbol{L}_p$ to denote the analogue of L_p.

Theorem 7.7. *If $f(x_1)\in K_p$ then $f\in\boldsymbol{K}_p$.*

Proof.

$$\int_{-\infty}^{\infty}\frac{|f(x_1)|^p}{(1+r^2)^{N+m}}\mathrm{d}\boldsymbol{x}\leq\int_{-\infty}^{\infty}\frac{|f(x_1)|^p}{(1+x_1^2)^N}\mathrm{d}x_1\int_{-\infty}^{\infty}\cdots\int_{-\infty}^{\infty}\frac{\mathrm{d}x_2\dots\mathrm{d}x_n}{(1+r^2-x_1^2)^m}$$
$$<\infty$$

since the first integral is finite because $f\in K_p$ and m can be chosen so large that the $(n-1)$-dimensional integral is finite. The proof is complete. □

This shows that the space K_1, which has proved so important in the case of a single variable, is embedded in $\boldsymbol{K}_1$. Since we expect $\boldsymbol{K}_1$ to play the same role in R_n that K_1 played in R_1 we anticipate that generalised functions of a single variable will have the same properties whether considered on R_1 or on R_n. This will be proved in full generality later. Here we note that support is provided by

Theorem 7.8. *If $f\in\boldsymbol{K}_1$ the sequence $\{\gamma_m\}$, where*

$$\gamma_m(\boldsymbol{x})=\int_{-\infty}^{\infty}f(\boldsymbol{t})\rho\{m(t_1-x_1)\}\rho\{m(t_2-x_2)\}\dots$$
$$\rho\{m(t_n-x_n)\}m^n\mathrm{e}^{-(t_1^2+t_2^2+\dots+t_n^2)/m^2}\,\mathrm{d}\boldsymbol{t},$$

is regular and defines a generalised function g such that

$$\int_{-\infty}^{\infty} g(x)\gamma(x)\mathrm{d}x = \int_{-\infty}^{\infty} f(x)\gamma(x)\mathrm{d}x.$$

The function ρ is the same as that defined in Chapter 3. It is possible to find, as in §6.4, a sequence of fine functions which define g and satisfy the theorem; replace $\mathrm{e}^{-(t_1^2+\ldots+t_n^2)/m^2}$ by $\phi_m(t_1)\phi_m(t_2)\ldots\phi_m(t_n)$.

Proof. We proceed as in Theorem 3.3 (note that this is possible by the remarks at the end of Chapter 1). The principal differences are that, in the inequality for $\partial^q\gamma_m$, $\{1+(|x|+1)^2\}^N\mathrm{e}^{-(|x|-1)^2/m^2}$ is replaced by

$$\{1+(r+1)^2\}^N \exp\left\{-\sum_{j=1}^{n}(|x_j|-1)^2/m^2\right\}$$

and that $\gamma(u-y/m)-\gamma(u)$ is replaced by

$$\gamma\left(\boldsymbol{u}-\frac{\boldsymbol{y}}{m}\right)-\gamma(\boldsymbol{u}) = \sum_{j=1}^{n}\frac{y_j}{m}\frac{\partial\gamma}{\partial\xi_j}$$

from (3) and the first mean value theorem (Theorem 1.30). □

Definition 7.8. *If f is a function and g a generalised function such that $\int_{-\infty}^{\infty} g(x)\gamma(x)\mathrm{d}x = \int_{-\infty}^{\infty} f(x)\gamma(x)\mathrm{d}x$ for every good γ we say that $g(x)=f(x)$ and conversely. If $\int_{-\infty}^{\infty} g(x)\gamma(x)\mathrm{d}x = \int_{-\infty}^{\infty} f(x)\gamma(x)\mathrm{d}x$ for every good γ which vanishes outside $(\boldsymbol{a},\boldsymbol{b})$ we say that*

$$g(x)=f(x) \qquad (a_1<x_1<b_1, a_2<x_2<b_2,\ldots,a_n<x_n<b_n)$$

and conversely.

This definition is consistent with Definition 3.5. It allows for a possibility such as

$$g(x_1,x_2)=f(x_1,x_2) \qquad (a_1<x_1<b_1)$$

which means that there is no restriction on x_2. One consequence is that we could say $\delta(x)=0$ for $r>0$.

Definition 7.9. *The sum g_1+g_2 of two generalised functions is defined by the regular sequence $\{\gamma_{1m}+\gamma_{2m}\}$. The generalised function*

$$g(a_1x_1+b_1, a_2x_2+b_2,\ldots,a_nx_n+b_n)$$

is defined by the regular sequence

$$\{\gamma_m(a_1x_1+b_1, a_2x_2+b_2, \ldots, a_nx_n+b_n)\}.$$

The generalised function ψg, ψ being fairly good, is defined by $\{\psi\gamma_m\}$.

Corresponding to Theorem 3.4 we have

Theorem 7.9.

$$\int_{-\infty}^{\infty} g(a_1x_1+b_1, \ldots, a_nx_n+b_n)\gamma(\boldsymbol{x})\mathrm{d}\boldsymbol{x}$$
$$= \int_{-\infty}^{\infty} g(\boldsymbol{x})\gamma\left(\frac{x_1-b_1}{a_1}, \ldots, \frac{x_n-b_n}{a_n}\right)\frac{\mathrm{d}\boldsymbol{x}}{|a_1a_2\ldots a_n|}.$$

One particular example of this is that

$$\int_{-\infty}^{\infty} \delta(a_1x_1-b_1, \ldots, a_nx_n-b_n)\gamma(\boldsymbol{x})\mathrm{d}\boldsymbol{x}$$
$$= \frac{1}{|a_1a_2\ldots a_n|}\gamma\left(\frac{b_1}{a_1}, \frac{b_2}{a_2}, \ldots, \frac{b_n}{a_n}\right). \quad (4)$$

The product ψg has properties similar to those for the single variable. In particular

$$x_j\delta(\boldsymbol{x})=0, \qquad \mathrm{e}^{\mathrm{i}\boldsymbol{\alpha}.\boldsymbol{x}}\delta(\boldsymbol{x})=\delta(\boldsymbol{x}).$$

Exercises

8. Are the following sequences regular and, if so, what generalised functions do they determine? (i) $\{\mathrm{e}^{-r/m}\}$, (ii) $\{m\mathrm{e}^{-mr^2}\}$, (iii) $m^{n/4}\mathrm{e}^{-mr^4}$

9. If $f_1 \in \boldsymbol{K}_p$ and $f_2 \in \boldsymbol{K}_p$ prove that $f_1+f_2 \in \boldsymbol{K}_p$.

10. If $f_1 \in \boldsymbol{K}_p (p>1)$ and $f_2 \in \boldsymbol{K}_q \{q=p/(p-1)\}$ prove that $f_1f_2 \in \boldsymbol{K}_1$.

11. Which of the following functions belong to $\boldsymbol{K}_1$?

$1+x_1^2$, $1+r^2$, $\cos(x_1^2+x_2^2)$, $\ln(x_1x_2x_3)$, $\mathrm{e}^{x_1x_2}$, $\mathrm{e}^{\mathrm{i}x_1x_2}$, $\mathrm{e}^{\mathrm{i}r^2}/r^{n-1/2}$, $|x_1^2-1|^{-1/2}|x_2^2-2|^{-1/3}$, $1/x_1$.

12. Evaluate

(i) $\displaystyle\int_{-\infty}^{\infty} \delta(\boldsymbol{x}-\boldsymbol{b})\mathrm{e}^{-r^2}\mathrm{d}\boldsymbol{x}$, (ii) $\displaystyle\int_{-\infty}^{\infty} \mathrm{H}(x_1)\mathrm{e}^{-r^2}\mathrm{d}\boldsymbol{x}$,

(iii) $\displaystyle\int_{-\infty}^{\infty} \mathrm{H}(x_1)\,\mathrm{sgn}\,(x_2)\mathrm{e}^{-r^2}\mathrm{d}\boldsymbol{x}$, (iv) $\displaystyle\int_{-\infty}^{\infty} x_2^2\,\mathrm{sgn}\,(x_1)\mathrm{e}^{-r^2}\mathrm{d}\boldsymbol{x}$.

13. If ψ is fairly good show that

$$\psi(\mathbf{x})\delta(\mathbf{x}) = \psi(\mathbf{0})\delta(\mathbf{x}).$$

7.3. The derivative

The partial derivative of a generalised function is defined in a similar manner to that for a single variable. However, the notation is necessarily more complicated. To prevent any possible confusion with the conventional partial derivative we shall add an extra suffix. Thus

Definition 7.10. *The sequence $\{\partial\gamma_m/\partial x_p\}$ is regular and defines a generalised function which will be denoted by $\partial_p g$ and called the partial derivative of g with respect to x_p.*

The sequence $\{\partial\gamma_m/\partial x_p\}$ is regular because

$$\int_{-\infty}^{\infty} \frac{\partial\gamma_m}{\partial x_p}\gamma(\mathbf{x})\mathrm{d}\mathbf{x}$$

$$= \int_{-\infty}^{\infty} \dots \int_{-\infty}^{\infty} [\gamma_m\gamma]_{x_p=-\infty}^{x_p=\infty} \mathrm{d}x_1 \dots \mathrm{d}x_{p-1}\mathrm{d}x_{p+1} \dots \mathrm{d}x_n$$

$$- \int_{-\infty}^{\infty} \gamma_m(\mathbf{x})\frac{\partial\gamma}{\partial x_p}\mathrm{d}\mathbf{x}$$

by an integration by parts. The first integral on the right-hand side converges uniformly in x_p (Theorem 1.31 in $n-1$ dimensions) and so, by Theorem 1.32, vanishes since $\gamma_m\gamma$ is zero at $\pm\infty$. Hence

Theorem 7.10.

$$\int_{-\infty}^{\infty} \{\partial_p g(\mathbf{x})\}\gamma(\mathbf{x})\mathrm{d}\mathbf{x} = -\int_{-\infty}^{\infty} g(\mathbf{x})\frac{\partial\gamma(\mathbf{x})}{\partial x_p}\mathrm{d}\mathbf{x}.$$

Consequently, the first partial derivatives of any generalised function are defined. Since these are generalised functions they will possess partial derivatives and so the partial derivative of any order can be obtained. In general

$$\int_{-\infty}^{\infty} \{\partial_1^{q_1} \dots \partial_n^{q_n} g(\mathbf{x})\}\gamma(\mathbf{x})\mathrm{d}\mathbf{x} = (-1)^q \int_{-\infty}^{\infty} g(\mathbf{x})\partial^q\gamma(\mathbf{x})\mathrm{d}\mathbf{x}$$

where $q = q_1 + q_2 + \ldots + q_n$. Obviously the order in which the derivatives are taken is unimportant since it is unimportant for γ. Thus it is always true for generalised functions that

$$\partial_1 \partial_2 g = \partial_2 \partial_1 g.$$

The connection between the conventional and generalised partial derivatives is provided by

Theorem 7.11. *If f is a continuous function such that $\partial f/\partial x_p \in \boldsymbol{K}_1$ and $f \in \boldsymbol{K}_1$ then*

$$\partial_p f = \partial f/\partial x_p.$$

Proof. This runs parallel to that of Theorem 3.7.□

Example 1. If $g(\boldsymbol{x}) = \mathrm{H}(x_1)\mathrm{H}(x_2)\ldots\mathrm{H}(x_n)$ we have

$$\begin{aligned}\int_{-\infty}^{\infty} \{\partial_1 g(\boldsymbol{x})\}\gamma(\boldsymbol{x})\mathrm{d}\boldsymbol{x} &= -\int_{-\infty}^{\infty} g(\boldsymbol{x})\frac{\partial\gamma(\boldsymbol{x})}{\partial x_1}\mathrm{d}\boldsymbol{x} \\ &= -\int_0^{\infty}\cdots\int_0^{\infty}\frac{\partial\gamma}{\partial x_1}\mathrm{d}x_1\ldots\mathrm{d}x_n \\ &= \int_0^{\infty}\cdots\int_0^{\infty}\gamma(0,x_2,\ldots,x_n)\mathrm{d}x_2\ldots\mathrm{d}x_n\end{aligned}$$

so that

$$\partial_1\{\mathrm{H}(x_1)\mathrm{H}(x_2)\ldots\mathrm{H}(x_n)\} = \delta(x_1)\mathrm{H}(x_2)\ldots\mathrm{H}(x_n).$$

Example 2. If $g(\boldsymbol{x}) = 1 \quad (r \leq 1),\ g(\boldsymbol{x}) = 0 \quad (r > 1)$,

$$\int_{-\infty}^{\infty} \{\partial_1 g(\boldsymbol{x})\}\gamma(\boldsymbol{x})\mathrm{d}\boldsymbol{x} = -\int_{-\infty}^{\infty} g(\boldsymbol{x})\frac{\partial\gamma(\boldsymbol{x})}{\partial x_1}\mathrm{d}\boldsymbol{x}.$$

In view of the value of g this is actually the integral of $\partial\gamma/\partial x_1$ over the interior of the unit sphere. Integrating by parts with respect to x_1 we obtain $-\int_S \gamma(\boldsymbol{x})\mathrm{d}x_2\ldots\mathrm{d}x_n$ or $-\int_S\gamma(\boldsymbol{x})\cos\theta_1\,\mathrm{d}S$ where S is the surface of the unit sphere, $\mathrm{d}S$ an element of surface and θ_1 the angle between the radius vector from the origin and the x_1-axis.

One way of interpreting this result is to think of g as $\mathrm{H}(1-r)$ and $\partial_1 g$ as $-\delta(r-1)\cos\theta_1$ which is what one would obtain formally via

$$\frac{\partial}{\partial x_1}\mathrm{H}(1-r) = \frac{x_1}{r}\frac{\partial}{\partial r}\mathrm{H}(1-r) = -\cos\theta_1\,\delta(r-1).$$

Example 3. More generally, let $g(\boldsymbol{x}) = f(\boldsymbol{x})$ inside the n-dimensional volume T, bounded by the hypersurface S, and vanish outside. Suppose f is continuous and possesses a continuous partial derivative on $T + S$. Then, if S is sufficiently smooth for integration by parts to be valid,

$$\int_{-\infty}^{\infty} \{\partial_1 g(\boldsymbol{x})\} \gamma(\boldsymbol{x}) \mathrm{d}\boldsymbol{x} = - \int_T f(\boldsymbol{x}) \frac{\partial \gamma(\boldsymbol{x})}{\partial x_1} \mathrm{d}\boldsymbol{x}$$

$$= \int_T \frac{\partial f}{\partial x_1} \gamma(\boldsymbol{x}) \mathrm{d}\boldsymbol{x} - \int_S f(\boldsymbol{x}) \gamma(\boldsymbol{x}) \cos \theta_1 \,\mathrm{d}S \tag{5}$$

where now θ_1 is the angle between the outward normal to S and the x_1-axis. Thus the generalised partial derivative consists of the conventional partial derivative $\partial f/\partial x_1$ together with a generalised function which corresponds to a surface distribution on S of density $-f \cos \theta_1$.

An alternative interpretation is that, if S can be represented by $h(\boldsymbol{x}) = 0$, and $h > 0$ in T,

$$g(\boldsymbol{x}) = f(\boldsymbol{x}) \mathrm{H}\{h(\boldsymbol{x})\}$$

and

$$\partial_1 g(\boldsymbol{x}) = \frac{\partial f}{\partial x_1} \mathrm{H}\{h(\boldsymbol{x})\} + f(\boldsymbol{x}) \frac{\partial h}{\partial x_1} \delta\{h(\boldsymbol{x})\}. \tag{6}$$

This can be identified with (5) provided that a suitable meaning is attached to $\delta(h)$; this point will be considered later in §8.4.

It may be shown in a similar manner to (5) that

$$\int_{-\infty}^{\infty} \{\partial_1^2 g(\boldsymbol{x})\} \gamma(\boldsymbol{x}) \mathrm{d}\boldsymbol{x} = \int_T \frac{\partial^2 f}{\partial x_1^2} \gamma(\boldsymbol{x}) \mathrm{d}\boldsymbol{x} - \int_S \frac{\partial f}{\partial x_1} \gamma(\boldsymbol{x}) \cos \theta_1 \,\mathrm{d}S$$

$$+ \int_S f \frac{\partial \gamma}{\partial x_1} \cos \theta_1 \,\mathrm{d}S \tag{7}$$

when f possesses a second conventional partial derivative. In this case, as well as a surface distribution of density $-(\partial f/\partial x_1) \cos \theta_1$, there is a distribution of doublets or dipoles all parallel to the x_1-axis and of density $f \cos \theta_1$. From this result we derive the generalised Laplacian of g, namely

$$\int_{-\infty}^{\infty} \{(\partial_1^2 + \partial_2^2 + \dots + \partial_n^2)g(\boldsymbol{x})\}\gamma(\boldsymbol{x})\mathrm{d}\boldsymbol{x}$$

$$= \int_T (\nabla^2 f)\gamma(\boldsymbol{x})\mathrm{d}\boldsymbol{x} + \int_S \left(f\frac{\partial\gamma}{\partial\nu} - \frac{\partial f}{\partial\nu}\gamma\right)\mathrm{d}S \tag{8}$$

where

$$\nabla^2 f = \frac{\partial^2 f}{\partial x_1^2} + \frac{\partial^2 f}{\partial x_2^2} + \dots + \frac{\partial^2 f}{\partial x_n^2},$$

and $\partial/\partial\nu$ is a derivative in the direction of the outward normal to S.

Exercises

14. In two-dimensional space evaluate

(i) $\displaystyle\int_{-\infty}^{\infty} \{\partial_1 g(\boldsymbol{x})\}\mathrm{e}^{-r^2}\mathrm{d}\boldsymbol{x}$ when $g(\boldsymbol{x}) = \begin{cases} x_1 & (r \le 1), \\ 0 & (r > 1), \end{cases}$

(ii) $\displaystyle\int_{-\infty}^{\infty} (\partial_1^2 g + \partial_2^2 g)\mathrm{e}^{-r^2}\mathrm{d}\boldsymbol{x}$ when $g(\boldsymbol{x}) = \begin{cases} r^2 & (r \le 1), \\ 0 & (r > 1). \end{cases}$

15. Show that, under the conditions of Example 3,

$$\int_{-\infty}^{\infty} \left(\sum_{j=1}^{n}\sum_{k=1}^{n} a_{jk}\partial_j\partial_k g(\boldsymbol{x})\right)\gamma(\boldsymbol{x})\mathrm{d}\boldsymbol{x} = \int_T \gamma(\boldsymbol{x}) \sum_{j=1}^{n}\sum_{k=1}^{n} a_{jk}\frac{\partial^2 f}{\partial x_j \partial x_k}\mathrm{d}\boldsymbol{x}$$

$$+ \sum_{j=1}^{n}\sum_{k=1}^{n} \int_S a_{jk}\left(f\frac{\partial\gamma}{\partial x_k} - \gamma\frac{\partial f}{\partial x_k}\right)\cos\theta_j \mathrm{d}S.$$

Theorem 7.12. *If continuous integrable f has a partial derivative $\partial f/\partial x_1$ in $a_1 < x_1 < b_1, a_2 < x_2 < b_2, \dots, a_n < x_n < b_n$ such that the integrals over this region of $f(\boldsymbol{x})\gamma(\boldsymbol{x})$ and $\gamma(\boldsymbol{x})\,\partial f/\partial x_1$ both exist for all good γ which vanish outside and if $g(\boldsymbol{x}) = f(\boldsymbol{x})$ in this region then*

$$\partial_1 g(\boldsymbol{x}) = \partial f(\boldsymbol{x})/\partial x_1 \qquad (a_1 < x_1 < b_1, \dots, a_n < x_n < b_n).$$

Proof. The proof of Theorem 3.8 can be taken over virtually without modification.□

7.4. The Fourier transform

Definition 7.11. *If the sequence* $\{\gamma_m\}$ *defines* g *then the sequence* $\{\Gamma_m\}$ *is regular and defines a generalised function* $G(\boldsymbol{\alpha})$ *called the Fourier transform of* g. *We write*

$$G(\boldsymbol{\alpha}) = \int_{-\infty}^{\infty} g(\boldsymbol{x})\mathrm{e}^{-\mathrm{i}\boldsymbol{\alpha}.\boldsymbol{x}}\,\mathrm{d}\boldsymbol{x}.$$

The verification of the regularity proceeds as for the single-variable but using Theorems 7.4, 7.5 and 7.6. As a consequence

Theorem 7.13. *If* $\boldsymbol{y} \in R_n$,

$$\int_{-\infty}^{\infty} \mathrm{e}^{\mathrm{i}\boldsymbol{\alpha}.\boldsymbol{y}}\, G(\boldsymbol{\alpha})\Gamma(\boldsymbol{\alpha})\mathrm{d}\boldsymbol{\alpha} = (2\pi)^n \int_{-\infty}^{\infty} g(\boldsymbol{x})\gamma(\boldsymbol{y}-\boldsymbol{x})\mathrm{d}\boldsymbol{x}.$$

An alternative way of writing this is

$$\int_{-\infty}^{\infty} G(\boldsymbol{\alpha})\Gamma(\boldsymbol{y}-\boldsymbol{\alpha})\mathrm{d}\boldsymbol{\alpha} = (2\pi)^n \int_{-\infty}^{\infty} \mathrm{e}^{-\mathrm{i}\boldsymbol{y}.\boldsymbol{x}}\, g(\boldsymbol{x})\gamma(\boldsymbol{x})\mathrm{d}\boldsymbol{x}.$$

Example 4. From Definition 7.6 the Fourier transform of $\delta(\boldsymbol{x})$ is defined by

$$\Gamma_m(\boldsymbol{\alpha}) = \int_{-\infty}^{\infty} \left(\frac{m}{\pi}\right)^{n/2} \exp(-mr^2 - \mathrm{i}\boldsymbol{\alpha}.\boldsymbol{x})\mathrm{d}\boldsymbol{x}$$

$$= \frac{1}{(2\pi)^{n/2}} \int_{-\infty}^{\infty} \exp\{-\tfrac{1}{2}r^2 - \mathrm{i}\boldsymbol{\alpha}.\boldsymbol{x}/(2m)^{1/2}\}\,\mathrm{d}\boldsymbol{x}$$

$$= \mathrm{e}^{-\alpha^2/4m}$$

from Theorem 7.2. It follows immediately that

$$\int_{-\infty}^{\infty} \delta(\boldsymbol{x})\mathrm{e}^{-\mathrm{i}\boldsymbol{\alpha}.\boldsymbol{x}}\,\mathrm{d}\boldsymbol{x} = 1.$$

Theorem 7.14. *If* G *is the Fourier transform of* g

(i) $g(\boldsymbol{x}) = \{1/(2\pi)^n\} \int_{-\infty}^{\infty} G(\boldsymbol{\alpha})\mathrm{e}^{\mathrm{i}\boldsymbol{\alpha}.\boldsymbol{x}}\,\mathrm{d}\boldsymbol{\alpha}$,

(ii) *the Fourier transform of* $g(a_1x_1+b_1,\dots,a_nx_n+b_n)$ is

$$\left[\exp\left\{\mathrm{i}\left(\frac{b_1\alpha_1}{a_1}+\dots+\frac{b_n\alpha_n}{a_n}\right)\right\}\Big/|a_1\dots a_n|\right]G\left(\frac{\alpha_1}{a_1},\dots,\frac{\alpha_n}{a_n}\right),$$

(iii) *the Fourier transform of* $\partial_p g$ *is* $\mathrm{i}\alpha_p G(\boldsymbol{\alpha})$.

Proof. As for Theorem 3.10, 3.11 and 3.12 but using the definitions and theorems of this chapter. □

Example 5. From Example 4 and Theorem 7.14(i)

$$\delta(\boldsymbol{x})=\frac{1}{(2\pi)^n}\int_{-\infty}^{\infty}\mathrm{e}^{\mathrm{i}\boldsymbol{\alpha}.\boldsymbol{x}}\,\mathrm{d}\boldsymbol{\alpha}$$

and

$$\delta(\boldsymbol{x}-\boldsymbol{b})=\frac{1}{(2\pi)^n}\int_{-\infty}^{\infty}\mathrm{e}^{\mathrm{i}\boldsymbol{\alpha}.\boldsymbol{x}\,-\mathrm{i}\boldsymbol{\alpha}.\boldsymbol{b}}\,\mathrm{d}\boldsymbol{\alpha}$$

by Theorem 7.14(ii). By Theorem 7.14 (iii)

$$\int_{-\infty}^{\infty}\{\partial_1^{q_1}\dots\partial_n^{q_n}\delta(\boldsymbol{x})\}\,\mathrm{e}^{-\mathrm{i}\boldsymbol{\alpha}.\boldsymbol{x}}\,\mathrm{d}\boldsymbol{x}=(\mathrm{i}\alpha_1)^{q_1}\dots(\mathrm{i}\alpha_n)^{q_n}$$

and so

$$\partial_1^{q_1}\dots\partial_n^{q_n}\delta(\boldsymbol{x})=\frac{1}{(2\pi)^n}\int_{-\infty}^{\infty}(\mathrm{i}\alpha_1)^{q_1}\dots(\mathrm{i}\alpha_n)^{q_n}\mathrm{e}^{\mathrm{i}\boldsymbol{\alpha}.\boldsymbol{x}}\,\mathrm{d}\boldsymbol{\alpha}.$$

More generally

$$\partial_1^{q_1}\dots\partial_n^{q_n}\delta(\boldsymbol{x}-\boldsymbol{b})=\frac{1}{(2\pi)^n}\int_{-\infty}^{\infty}(\mathrm{i}\alpha_1)^{q_1}\dots(\mathrm{i}\alpha_n)^{q_n}\mathrm{e}^{\mathrm{i}\boldsymbol{\alpha}.(\boldsymbol{x}-\boldsymbol{b})}\,\mathrm{d}\boldsymbol{\alpha}.$$

Theorem 7.15. *If* $f\in \boldsymbol{L}_1$ *it has a conventional Fourier transform* $F\in\boldsymbol{K}_1$. *The generalised function F is the Fourier transform of the generalised function f.*

Proof. As for Theorem 3.13. □

This theorem shows that we can identify the Fourier transform defined in the conventional way or as a generalised function when $f\in\boldsymbol{L}_1$. The terminology is therefore justified.

Exercises

16. Find the Fourier transform of $x_1^{q_1}x_2^{q_2}\dots x_n^{q_n}g(\boldsymbol{x})$ knowing that the Fourier transform of g is G.
17. Find the Fourier transform of $x_1^{q_n}\dots x_n^{q_n}\partial_1^{q_n}\dots\partial_n^{q_n}\delta(\boldsymbol{x})$.

Definition 7.12. $\mathrm{Lim}_{\mu\to c}\, g_\mu = g$ *if, and only if, there is a generalised function g such that*

$$\lim_{\mu\to c}\int_{-\infty}^{\infty} g_\mu(\boldsymbol{x})\gamma(\boldsymbol{x})\mathrm{d}\boldsymbol{x} = \int_{-\infty}^{\infty} g(\boldsymbol{x})\gamma(\boldsymbol{x})\mathrm{d}\boldsymbol{x}$$

for every good γ.

It is not necessary that μ be a scalar parameter; it could be a point of R_m where m need not be the same as n.

Theorem 7.16. *If* $\mathrm{Lim}_{\mu\to c}\, g_\mu = g$ *then*

(i) $\mathrm{Lim}_{\mu\to c}\,\partial_p g_\mu = \partial_p g$,
(ii) $\mathrm{Lim}_{\mu\to c}\, g_\mu(a_1x_1+b_1,\dots,a_nx_n+b_n) = g(a_1x_1+b_1,\dots,a_nx_n+b_n)$,
(iii) $\mathrm{Lim}_{\mu\to c}\,\psi g_\mu = \psi g$ for fairly good ψ,
(iv) $\mathrm{Lim}_{\mu\to c}\, G_\mu = G$.

Proof. The proof runs parallel to that for Theorem 3.14. □

Example 6. Consider $m^n\mathrm{e}^{-mr}$ as $m\to\infty$. We have, because $m^n\mathrm{e}^{-mr}\in \boldsymbol{K}_1$,

$$\int_{-\infty}^{\infty} m^n\mathrm{e}^{-mr}\gamma(\boldsymbol{x})\mathrm{d}\boldsymbol{x}$$

$$= \int_{-\infty}^{\infty} m^n\mathrm{e}^{-mr}\{\gamma(\boldsymbol{x})-\gamma(\boldsymbol{0})\}\mathrm{d}\boldsymbol{x} + \gamma(\boldsymbol{0})\int_{-\infty}^{\infty} m^n\mathrm{e}^{-mr}\mathrm{d}\boldsymbol{x}. \qquad (9)$$

On account of Theorem 7.1 the last integral can be calculated by means of the spherical polar coordinates $r,\theta_1,\theta_2,\dots,\theta_{n-1}$ where

$$x_1 = r\cos\theta_1,\quad x_2 = r\sin\theta_1\cos\theta_2,$$
$$x_3 = r\sin\theta_1\sin\theta_2\cos\theta_3,\dots,x_{n-1} = r\sin\theta_1\dots\sin\theta_{n-2}\cos\theta_{n-1},$$
$$x_n = r\sin\theta_1\dots\sin\theta_{n-2}\sin\theta_{n-1}$$
$$(0\le\theta_1\le\pi,\ 0\le\theta_2\le\pi,\dots,0\le\theta_{n-2}\le\pi,\ 0\le\theta_{n-1}\le 2\pi).$$

Thus

$$\int_{-\infty}^{\infty} m^n e^{-mr}\, dx = \int_0^\infty \int_0^\pi \cdots \int_0^\pi \int_0^{2\pi} m^n e^{-mr} r^{n-1} \sin^{n-2}\theta_1 \sin^{n-3}\theta_2 \cdots$$
$$\times \sin\theta_{n-2}\, dr\, d\theta_1 \ldots d\theta_{n-2}\, d\theta_{n-1}$$
$$= \frac{2\pi^{n/2}}{(\tfrac{1}{2}n-1)!}\int_0^\infty m^n e^{-mr} r^{n-1}\, dr$$

since $\int_0^\pi \sin^\nu\theta\, d\theta = (\tfrac{1}{2}\nu - \tfrac{1}{2})!\pi^{1/2}/(\tfrac{1}{2}\nu)!$. Hence

$$\int_{-\infty}^{\infty} m^n e^{-mr}\, dx = \frac{(n-1)!2\pi^{n/2}}{(\tfrac{1}{2}n-1)!}.$$

The first integral of (9) can be written as $\int_{-\infty}^{\infty} e^{-r}\{\gamma(x/m) - \gamma(0)\}\, dx$. Now, if T_r and $\Omega - T_r$ are the interior and exterior respectively of the sphere of radius r and centre the origin, we can choose R so large that

$$\left|\int_{\Omega - T_R} e^{-r}\{\gamma(x/m) - \gamma(0)\}\, dx\right| \le B\int_{\Omega - T_R} e^{-r}\, dx \le \varepsilon$$

and R is independent of m since a good function is bounded. We then choose m so large that

$$|\gamma(x/m) - \gamma(0)| \le \varepsilon \Big/ \int_{T_R} e^{-r}\, dx$$

for x in T_R; then

$$\left|\int_{T_R} e^{-r}\{\gamma(x/m) - \gamma(0)\}\, dx\right| \le \varepsilon$$

and consequently

$$\lim_{m\to\infty}\int_{-\infty}^{\infty} m^n e^{-mr}\gamma(x)\, dx = \frac{(n-1)!2\pi^{n/2}}{(\tfrac{1}{2}n-1)!}\gamma(0)$$

or

$$\operatorname*{Lim}_{m\to\infty} m^n e^{-mr} = \frac{(n-1)!2\pi^{n/2}}{(\tfrac{1}{2}n-1)!}\delta(x).$$

It follows from Theorem 7.16 (ii) that

$$\operatorname*{Lim}_{m\to\infty} m^n \exp[-m\{(a_1x_1+b_1)^2+\ldots+(a_nx_n+b_n)^2\}^{1/2}]$$

$$=\frac{(n-1)!2\pi^{n/2}}{(\frac{1}{2}n-1)!}\delta(a_1x_1+b_1,\ldots,a_nx_n+b_n).$$

Exercises

18. Verify that (i) $\mathrm{Lim}_{m\to\infty} r\{(r+m)^2+1\}^{-1}=0$,

(ii) $\mathrm{Lim}_{\mu\to 0}\mu|x_1|^{\mu-1}=2\delta(x_1)$,

(iii) $\mathrm{Lim}_{\mu\to+0}\mu r^{\mu-n}=\{2\pi^{n/2}/(\frac{1}{2}n-1)!\}\delta(\boldsymbol{x})$.

19. Prove that

$$\operatorname*{Lim}_{\mu\to 0}\frac{1}{\mu}\{g(x_1+\mu,x_2,\ldots,x_n)-g(\boldsymbol{x})\}=\partial_1 g.$$

7.5. The general form of a generalised function

This section is concerned with extending the ideas of §3.6 to several variables.

Theorem 7.17. *If a set of generalised functions g_μ is such that, for all μ and every good γ, $|\int_{-\infty}^{\infty} g_\mu(\boldsymbol{x})\gamma(\boldsymbol{x})\mathrm{d}\boldsymbol{x}|\le K'(\gamma)$ there are finite fixed K, k and p such that*

$$\left|\int_{-\infty}^{\infty} g_\mu(\boldsymbol{x})\gamma(\boldsymbol{x})\mathrm{d}\boldsymbol{x}\right|\le K\max_{\boldsymbol{x}\in R_n}|(1+r^2)^{k/2}\partial^p\gamma(\boldsymbol{x})|$$

for every γ and all μ.

Proof. The proof proceeds by the method of Theorems 3.16 and 3.17. The analogue of (19) of §3.6 is

$$\left|\int_{-\infty}^{\infty} g_m(\boldsymbol{x})\gamma(\boldsymbol{x})\mathrm{d}\boldsymbol{x}\right|\le K_{ms}\max(1+r^2)^s|\partial^s\gamma(\boldsymbol{x})|.$$

The proof requires an inequality analogous to (18) of §3.6. This is provided by noting that

$$\gamma(\boldsymbol{x})=\begin{cases}\displaystyle\int_{-\infty}^{x_1}\frac{\partial}{\partial t_1}\gamma(t_1,x_2,\ldots,x_n)\mathrm{d}t_1 & (x_1\le 0)\\[2ex] \displaystyle-\int_{x_1}^{\infty}\frac{\partial}{\partial t_1}\gamma(t_1,x_2,\ldots,x_n)\mathrm{d}t_1 & (x_1\ge 0).\end{cases}$$

Consequently

$$\max|\gamma| \le \tfrac{1}{2}\pi \max(1+x_1^2)|\partial\gamma/\partial x_1| \le \tfrac{1}{2}\pi \max(1+r^2)|\partial\gamma/\partial x_1|$$

and one deduces that

$$\max(1+r^2)^{k/2}|\partial^p\gamma| \le (\tfrac{1}{2}\pi)^s \max(1+r^2)^{(k/2)+s}|\partial^{p+s}\gamma|,$$

s being a non-negative integer. This is the required analogue.

Thereafter the proof is virtually unchanged.□

Theorem 7.18. *In order that g be a generalised function it is necessary and sufficient that g be a generalised partial derivative*

$$g(\boldsymbol{x}) = \partial_1^{p_1}\partial_2^{p_2}\dots\partial_n^{p_n} f(\boldsymbol{x})$$

where $f \in \boldsymbol{K}_1$ and is continuous on R_n.

Proof. As in Theorem 3.18 we form a linear functional F over the space of good functions of the form $(1+r^2)^{k/2}\partial^p\gamma(\boldsymbol{x})$ and then extend it to a bounded linear functional over the space of continuous functions by the Hahn–Banach theorem. The Riesz representation theorem then applies and

$$F\{(1+r^2)^{k/2}\partial^p\gamma\} = \int_{-\infty}^{\infty} (1+r^2)^{(k/2)}\partial^p\gamma(\boldsymbol{x})\,\mathrm{d}h(\boldsymbol{x})$$

where the n-dimensional total variation of h is K and $\max|h| \le K$. An n-dimensional integration by parts gives

$$\int_{-\infty}^{\infty} g(\boldsymbol{x})\gamma(\boldsymbol{x})\,\mathrm{d}\boldsymbol{x} = (-1)^n \int_{-\infty}^{\infty} h(\boldsymbol{x})\frac{\partial^n}{\partial x_1\dots\partial x_n}\{(1+r^2)^{k/2}\partial^p\gamma(\boldsymbol{x})\}\,\mathrm{d}\boldsymbol{x}$$

$$= (-1)^p \int_{-\infty}^{\infty} \gamma(\boldsymbol{x})\partial_1^{p_1}\dots\partial_n^{p_n}\{(1+r^2)^{k/2}\partial_1\dots\partial_n h(\boldsymbol{x})\}\mathrm{d}\boldsymbol{x}.$$

Thus

$$g(\boldsymbol{x}) = (-1)^p\partial_1^{p_1}\dots\partial_n^{p_n}\{(1+r^2)^{k/2}\partial_1\dots\partial_n h(\boldsymbol{x})\}.$$

Since

$$(1+r^2)^{k/2}\partial_1 h = \partial_1\{(1+r^2)^{k/2}h\} - kx_1(1+r^2)^{(k/2)-1}h$$

it follows that

$$g(\boldsymbol{x}) = \partial_1^{p_1+1}\dots\partial_n^{p_n+1}h_1$$

where h_1 is $O(r^k)$ as $r \to \infty$. Consequently

$$g(x) = \partial_1^{p_1+2} \dots \partial_n^{p_n+2} f$$

where

$$f(x) = \int^{x_1} \int^{x_2} \dots \int^{x_n} h_1(t)\mathrm{d}t$$

and the theorem is proved, since f is continuous and $\in K_1$ on account of the behaviour of h_1 at infinity. □

One consequence of this theorem is that, if g is any generalised function, there is a generalised function g_1 such that $\partial_p g_1 = g$.

Corollary 7.18 *If $g(x_1)$ is a generalised function on R_1 then it is one on R_n.*

Proof. By Theorems 3.18 and 7.7, $g(x_1)$ can be cast into the form required by Theorem 7.18. □

Theorem 7.19. *If $\operatorname{Lim}_{m\to\infty} g_m = 0$ there are continuous f_m such that $g_m(x) = \partial_1^{p_1} \dots \partial_n^{p_n}\{(1+r^2)^{k/2} f_m(x)\}$ for fixed k and p and $\{f_m\}$ converges uniformly to zero.*

Proof. The method of Theorem 3.20 is used. On account of Theorem 7.18 there are f_m bounded uniformly in m and equicontinuous on R_n such that

$$\int_{-\infty}^{\infty} g_m(x)\gamma(x)\mathrm{d}x = (-1)^p \int_{-\infty}^{\infty} (1+r^2)^{k/2} f_m(x)\partial^p \gamma(x)\mathrm{d}x$$

so that

$$\lim_{m\to\infty} \int_{-\infty}^{\infty} (1+r^2)^{k/2} f_m(x)\partial^p \gamma(x)\mathrm{d}x = 0.$$

By introducing, for any good γ_1, the good function

$$\gamma_2(x) = \int_{-\infty}^{x_1} \left\{\gamma_1(u_1, x_2, \dots, x_n) - \frac{\mathrm{e}^{-u_1^2}}{\pi^{1/2}} \int_{-\infty}^{\infty} \gamma_1(t_1, x_2, \dots, x_n)\mathrm{d}t_1\right\}\mathrm{d}u_1,$$

we prove eventually that, for every good γ,

$$\lim_{m\to\infty} \int_{-\infty}^{\infty} \{(1+r^2)^{k/2} f_m(x) + P_m(x)\}\gamma(x)\mathrm{d}x = 0$$

where

$$P_m(x) = \sum_{k=1}^{n} \sum_{j=0}^{p_k - 1} p(x_1, \dots, x_{k-1}, x_{k+1}, \dots, x_n) x_k^j.$$

Absorbing the P_m in f_m and the $(1 + r^2)^{k/2}$ in γ, increasing k if necessary, we have

$$\lim_{m \to \infty} \int_{-\infty}^{\infty} f_m(x)\gamma(x)\mathrm{d}x = 0$$

with the f_m equicontinuous and tending to zero independently of m as $r \to \infty$.

The argument now proceeds as in Theorem 3.20 with the intervals of length δ replaced by n-dimensional cubes of side δ and γ_0 replaced by a product of such good functions, one for each variable. The proof is complete. □

Corollary 7.19. *If* $\mathrm{Lim}_{m \to \infty} g_m = 0$ *there is an* m_0 *such that, for all good* γ,

$$\left| \int_{-\infty}^{\infty} g_m(x)\gamma(x)\mathrm{d}x \right| \le \varepsilon \max \left| (1 + r^2)^{k/2} \partial^p \gamma(x) \right|$$

for $m \ge m_0$; *k and p are independent of* m_0.

Proof. As for Corollary 3.20*a*. □

Exercises

20. If the sequence $\{f_m\}$ of continuous functions, bounded on R_n, converges uniformly to zero and

$$g_m(x) = \partial_1^{p_1} \dots \partial_n^{p_n} \{(1 + r^2)^{k/2} f_m(x)\}$$

prove that $\mathrm{Lim}_{m \to \infty} g_m = 0$.

21. If g vanishes for $r > 0$ show that

$$g(x) = \sum_{p=0}^{m} a_p \partial_1^{p_1} \dots \partial_n^{p_n} \delta(x)$$

where m is finite and $p_1 + p_2 + \dots + p_n = p$.

Theorem 7.20. *If* γ *is any good function then*

$$\gamma(x) = \sum_{m_1=0}^{\infty} \dots \sum_{m_n=0}^{\infty} a_{m_1 \dots m_n} \mathscr{H}_{m_1}(x_1) \dots \mathscr{H}_{m_n}(x_n)$$

where $a_{m_1 \dots m_n} = \int_{-\infty}^{\infty} \gamma(x) \mathscr{H}_{m_1}(x_1) \dots \mathscr{H}_{m_n}(x_n) \mathrm{d}x$.

Proof. This theorem may be proved in the same way as Theorem 5.5 by starting from $e^{-r^2/2}\gamma(\boldsymbol{x})$. □

Theorem 7.21. $\int_{-\infty}^{\infty} g(\boldsymbol{x})\mathscr{H}_{m_1}(x_1)\ldots\mathscr{H}_{m_n}(x_n)\mathrm{d}\boldsymbol{x} = 0$ *for all non-negative* $m_1, m_2, \ldots, m_n$ *if, and only if,* $g = 0$.

Proof. Use the method of Theorem 5.7. □

Corollary 7.21*a* $\int_{-\infty}^{\infty} g(\boldsymbol{x})\gamma_1(x_1)\ldots\gamma_n(x_n)\mathrm{d}\boldsymbol{x} = 0$ *for all good* $\gamma_1, \ldots, \gamma_n$ *on* R_1 *if, and only if,* $g = 0$.

Proof. If $g = 0$ the conclusion is immediate since $\gamma_1\gamma_2\ldots\gamma_n$ is good on R_n. Conversely, if the integral vanishes we take $\gamma_1 = \mathscr{H}_{m_1}, \ldots, \gamma_n = \mathscr{H}_{m_n}$ and then $g = 0$ by Theorem 7.21. □

The importance of Corollary 7.21*a* is that it shows that the most general good functions of several variables need not be used; it is sufficient to work with good functions which are the product of n good functions of a single variable.

Corollary 7.21*b*. *If* $\int_{-\infty}^{\infty} f(\boldsymbol{x})\gamma(\boldsymbol{x})\mathrm{d}\boldsymbol{x} = 0$ *for all good functions* γ *which vanish outside the domain of continuity* $(\boldsymbol{a}, \boldsymbol{b})$ *of the function* f *then* $f \equiv 0$ *in* $(\boldsymbol{a}, \boldsymbol{b})$.

Proof. The proof is the same as Theorem 3.1 but with γ_1 replaced by a product of such functions. □

This corollary indicates that the method of definition by good functions does provide a way of distinguishing between different continuous functions.

Theorem 7.22. *If* $\{\gamma_m\}$ *defines* g,

$$\lim_{m\to\infty}\int_{-\infty}^{\infty} \gamma_m(\boldsymbol{x})\gamma_{(n)}(x_n)\mathrm{d}x_n$$

is a generalised function on R_{n-1}, *which will be denoted by*

$$\int_{-\infty}^{\infty} g(\boldsymbol{x})\gamma_{(n)}(x_n)\mathrm{d}x_n$$

and

$$\int_{-\infty}^{\infty} g(\boldsymbol{x})\gamma_{(1)}(x_1)\ldots\gamma_{(n)}(x_n)\,\mathrm{d}\boldsymbol{x}$$

$$= \int_{-\infty}^{\infty} \gamma_{(1)}(x_1)\int_{-\infty}^{\infty} \gamma_{(2)}(x_2)\ldots\int_{-\infty}^{\infty} g(\boldsymbol{x})\gamma_{(n)}(x_n)\,\mathrm{d}x_n\ldots\mathrm{d}x_2\,\mathrm{d}x_1.$$

Proof. Let

$$\gamma_m^{\circ}(x_1,\ldots,x_{n-1}) = \int_{-\infty}^{\infty} \gamma_m(\boldsymbol{x})\gamma_{(n)}(x_n)\,\mathrm{d}x_n.$$

We first show that γ_m° is good on R_{n-1}. Since γ_m is good there is a constant M independent of $\boldsymbol{x}$ such that $|\partial^q\gamma_m| < M$, the ∂ referring only to R_{n-1}. Hence $\int_{-\infty}^{\infty} \partial^q\gamma_m\gamma_{(n)}(x_n)\mathrm{d}x_n$ converges uniformly on R_{n-1} and

$$\partial^q\gamma_m^{\circ} = \int_{-\infty}^{\infty} \partial^q\gamma_m(\boldsymbol{x})\gamma_{(n)}(x_n)\,\mathrm{d}x_n.$$

Also, for large enough $x_1,\ldots,x_{n-1}$, $|(x_1^2+\ldots+x_{n-1}^2)^s\partial^q\gamma_m| < \varepsilon$ so that

$$|(x_1^2+\ldots+x_{n-1}^2)^s\partial^q\gamma_m^{\circ}| < \varepsilon\int_{-\infty}^{\infty} |\gamma_{(n)}(x_n)|\,\mathrm{d}x_n$$

and consequently γ_m° is good on R_{n-1}. Now

$$\lim_{m\to\infty}\int_{-\infty}^{\infty} \gamma_m^{\circ}\gamma_{(1)}(x_1)\ldots\gamma_{(n-1)}(x_{n-1})\,\mathrm{d}x_1\ldots\mathrm{d}x_{n-1}$$

$$= \lim_{m\to\infty}\int_{-\infty}^{\infty} \gamma_m(\boldsymbol{x})\gamma_{(1)}(x_1)\ldots\gamma_{(n)}(x_n)\,\mathrm{d}\boldsymbol{x}$$

$$= \int_{-\infty}^{\infty} g(\boldsymbol{x})\gamma_1(x_1)\ldots\gamma_{(n)}(x_n)\,\mathrm{d}\boldsymbol{x}. \qquad (10)$$

Thus $\mathrm{Lim}_{m\to\infty}\,\gamma_m^{\circ}$ defines a generalised function on R_{n-1} on account of Corollary 7.21*a*.

The final part of the theorem follows from (10) and the proof is complete. □

Corollary 7.22. $\partial_1\int_{-\infty}^{\infty} g(\boldsymbol{x})\gamma_{(n)}(x_n)\mathrm{d}x_n = \int_{-\infty}^{\infty}\{\partial_1 g(\boldsymbol{x})\}\,\gamma_{(n)}(x_n)\mathrm{d}x_n.$

Proof. By Theorem 7.16(i)

$$\partial_1 \int_{-\infty}^{\infty} g(\boldsymbol{x})\gamma_{(n)}(x_n)\mathrm{d}x_n = \operatorname*{Lim}_{m\to\infty} \partial_1 \int_{-\infty}^{\infty} \gamma_m(\boldsymbol{x})\gamma_{(n)}(x_n)\mathrm{d}x_n$$
$$= \operatorname*{Lim}_{m\to\infty} \int_{-\infty}^{\infty} (\partial_1 \gamma_m)\gamma_{(n)}(x_n)\mathrm{d}x_n$$
$$= \int_{-\infty}^{\infty} \{\partial_1 g(\boldsymbol{x})\}\gamma_{(n)}(x_n)\mathrm{d}x_n$$

from Theorem 7.22.□

Quite often it is desirable to have available the single-variable Fourier transform of a generalised function on R_n instead of the n-dimensional Fourier transform. This possibility is allowed for by

Theorem 7.23 *If* $\{\gamma_m\}$ *defines* g *then* $\{\int_{-\infty}^{\infty} \gamma_m(\boldsymbol{x})\mathrm{e}^{-\mathrm{i}\alpha_n x_n}\mathrm{d}x_n\}$ *is regular and defines on* $x_1, x_2, \dots, x_{n-1}, \alpha_n$ *a generalised function which will be denoted by* $\int_{-\infty}^{\infty} g(\boldsymbol{x})\mathrm{e}^{-\mathrm{i}\alpha_n x_n}\mathrm{d}x_n$, *and*

$$\partial_1 \int_{-\infty}^{\infty} g(\boldsymbol{x})\mathrm{e}^{-\mathrm{i}\alpha_n x_n}\mathrm{d}x_n = \int_{-\infty}^{\infty} \{\partial_1 g(\boldsymbol{x})\}\mathrm{e}^{-\mathrm{i}\alpha_n x_n}\mathrm{d}x_n.$$

Proof. There is no difficulty in verifying that $\int_{-\infty}^{\infty} \gamma_m(\boldsymbol{x})\mathrm{e}^{-\mathrm{i}\alpha_n x_n}\mathrm{d}x_n$ is good on $x_1, \dots, x_{n-1}, \alpha_n$. Also

$$\int_{-\infty}^{\infty} \cdots \int_{-\infty}^{\infty} \gamma_{(1)}(x_1)\dots\gamma_{(n-1)}(x_{n-1})\Gamma_{(n)}(\alpha_n)$$
$$\times\left\{\int_{-\infty}^{\infty} \gamma_m(\boldsymbol{x})\mathrm{e}^{-\mathrm{i}\alpha_n x_n}\mathrm{d}x_n\right\}\mathrm{d}x_1\dots\mathrm{d}x_{n-1}\,\mathrm{d}\alpha_n$$
$$= \int_{-\infty}^{\infty} \cdots \int_{-\infty}^{\infty} \gamma_{(1)}(x_1)\dots\gamma_{(n-1)}(x_{n-1})$$
$$\times\left\{\int_{-\infty}^{\infty} \Gamma_{(n)}(\alpha_n)\int_{-\infty}^{\infty} \gamma_m(\boldsymbol{x})\mathrm{e}^{-\mathrm{i}\alpha_n x_n}\mathrm{d}x_n\,\mathrm{d}\alpha_n\right\}\mathrm{d}x_1\dots\mathrm{d}x_{n-1}$$
$$= 2\pi \int_{-\infty}^{\infty} \gamma_{(1)}(x_1)\dots\gamma_{(n-1)}(x_{n-1})\gamma_{(n)}(-x_n)\gamma_m(\boldsymbol{x})\mathrm{d}\boldsymbol{x}$$
$$\to 2\pi \int_{-\infty}^{\infty} \gamma_{(1)}(x_1)\dots\gamma_{(n)}(-x_n)g(\boldsymbol{x})\mathrm{d}\boldsymbol{x} \tag{11}$$

as $m \to \infty$. Thus the first part of the theorem is confirmed.

With regard to the second part note that (11) is valid with g replaced by $\partial_1 g$ if, in the first line, either $\gamma_{(1)}$ is replaced by $-\partial\gamma_{(1)}/\partial x_1$ or γ_m is replaced by $\partial\gamma_m/\partial x_1$ and the proof is complete. □

Obviously it is possible, by repeated applications of Theorem 7.23, to obtain the Fourier transform

$$\int_{-\infty}^{\infty} e^{-i\alpha_r x_r} \int_{-\infty}^{\infty} e^{-i\alpha_{r+1}x_{r+1}} \ldots \int_{-\infty}^{\infty} g(\boldsymbol{x}) e^{-i\alpha_n x_n}\, dx_n \ldots dx_{r+1}\, dx_r.$$

By continually using results of the form (11) and Theorem 7.13 we find

Corollary 7.23.

$$G(\boldsymbol{\alpha}) = \int_{-\infty}^{\infty} g(\boldsymbol{x}) e^{-i\boldsymbol{\alpha}.\boldsymbol{x}}\, d\boldsymbol{x}$$

$$= \int_{-\infty}^{\infty} e^{-i\alpha_1 x_1} \ldots \int_{-\infty}^{\infty} g(\boldsymbol{x}) e^{-i\alpha_n x_n}\, dx_n \ldots dx_1,$$

i.e. calculation of G as a repeated integral is permissible.

7.6. The direct product

We now turn to a product which has no analogue in the theory of a single variable, namely the product of a generalised function in one space with a generalised function in another space.

Definition 7.13. *If $g_1(\boldsymbol{x})$ is a generalised function on R_n and $g_2(\boldsymbol{y})$ is a generalised function on R_m the direct product $g_1(\boldsymbol{x})g_2(\boldsymbol{y})$ is a generalised function on R_{n+m} defined by*

$$\int_{-\infty}^{\infty} g_1(\boldsymbol{x}) g_2(\boldsymbol{y}) \gamma_1(\boldsymbol{x}) \gamma_2(\boldsymbol{y})\, d\tau$$

$$= \int_{-\infty}^{\infty} g_1(\boldsymbol{x}) \gamma_1(\boldsymbol{x})\, d\boldsymbol{x} \int_{-\infty}^{\infty} g_2(\boldsymbol{y}) \gamma_2(\boldsymbol{y})\, d\boldsymbol{y}$$

where $d\tau$ is the volume element of R_{n+m} and γ_1, γ_2 are any good functions in R_n and R_m respectively.

Observe that, since both quantities on the right-hand side have a meaning, the left-hand side has one. Also, by Corollary 7.21*a*, giving a meaning to the left-hand side does attach a significance to $g_1 g_2$ on R_{n+m}. The definition is also consistent with the usual properties when g_1 and g_2 are conventional functions.

It is obvious that $g_1(\boldsymbol{x})g_2(\boldsymbol{y}) = g_2(\boldsymbol{y})g_1(\boldsymbol{x})$ and that

$$g_1(\boldsymbol{x})\{g_2(\boldsymbol{y})g_3(\boldsymbol{z})\} = \{g_1(\boldsymbol{x})g_2(\boldsymbol{y})\}g_3(\boldsymbol{z}).$$

The definition can also be interpreted as saying that integration can be regarded as repeated integration and that the order of integration can be inverted at will.

Example 7. By Definition 7.13 $\delta(x_1)\delta(x_2)\ldots\delta(x_n)$ is a generalised function on R_n and is such that

$$\int_{-\infty}^{\infty} \delta(x_1)\ldots\delta(x_n)\gamma_1(x_1)\ldots\gamma_n(x_n)\mathrm{d}\boldsymbol{x}$$
$$= \int_{-\infty}^{\infty} \delta(x_1)\gamma_1(x_1)\mathrm{d}x_1 \ldots \int_{-\infty}^{\infty} \delta(x_n)\gamma_n(x_n)\mathrm{d}x_n$$
$$= \gamma_1(0)\gamma_2(0)\ldots\gamma_n(0).$$

But

$$\int_{-\infty}^{\infty} \delta(\boldsymbol{x})\gamma_1(x_1)\ldots\gamma_n(x_n)\mathrm{d}\boldsymbol{x} = \gamma_1(0)\gamma_2(0)\ldots\gamma_n(0).$$

Hence

$$\delta(\boldsymbol{x}) = \delta(x_1)\delta(x_2)\ldots\delta(x_n).$$

More generally, it can be shown that

$$\delta(a_1x_1+b_1,\ldots,a_nx_n+b_n) = \delta(a_1x_1+b_1)\delta(a_2x_2+b_2)\ldots\delta(a_nx_n+b_n). \tag{12}$$

The proof is left as an exercise for the reader.

Theorem 7.24. $\partial_1\{g_1(\boldsymbol{x})g_2(\boldsymbol{y})\} = \{\partial_1 g_1(\boldsymbol{x})\}g_2(\boldsymbol{y}).$

Proof.

$$\int_{-\infty}^{\infty} \partial_1 \{g_1(x)g_2(y)\}\gamma_1(x)\gamma_2(y)\mathrm{d}\tau$$

$$= -\int_{-\infty}^{\infty} g_1(x)g_2(y)\frac{\partial}{\partial x_1}\gamma_1(x)\gamma_2(y)\mathrm{d}\tau$$

$$= -\int_{-\infty}^{\infty} g_1(x)\frac{\partial}{\partial x_1}\gamma_1(x)\mathrm{d}x \int_{-\infty}^{\infty} g_2(y)\gamma_2(y)\mathrm{d}y$$

by Definition 7.13. The product on the right can be written

$$\int_{-\infty}^{\infty} \{\partial_1 g(x)\}\gamma_1(x)\mathrm{d}x \int_{-\infty}^{\infty} g_2(y)\gamma_2(y)\mathrm{d}y$$

$$= \int_{-\infty}^{\infty} \{\partial_1 g_1(x)\}g_2(y)\gamma_1(x)\gamma_2(y)\mathrm{d}\tau$$

by a further application of Definition 7.13. The theorem is proved. □

Theorem 7.25. *The Fourier transform of $g_1(x)g_2(y)$ is $G_1(\alpha)G_2(\beta)$.*

Proof. By Definition 7.13

$$\int_{-\infty}^{\infty} g_1(x)g_2(y)\gamma_1(-x)\gamma_2(-y)\mathrm{d}\tau$$

$$= \int_{-\infty}^{\infty} g_1(x)\gamma_1(-x)\mathrm{d}x \int_{-\infty}^{\infty} g_2(y)\gamma_2(-y)\mathrm{d}y$$

$$= \frac{1}{(2\pi)^{n+m}}\int_{-\infty}^{\infty} G_1(\alpha)\Gamma_1(\alpha)\mathrm{d}\alpha \int_{-\infty}^{\infty} G_2(\beta)\Gamma_2(\beta)\mathrm{d}\beta$$

from Theorem 7.13. Applying Definition 7.13 we obtain for the right-hand side

$$\frac{1}{(2\pi)^{n+m}}\int_{-\infty}^{\infty} G_1(\alpha)G_2(\beta)\Gamma_1(\alpha)\Gamma_2(\beta)\mathrm{d}\alpha\,\mathrm{d}\beta$$

and the theorem follows from Theorem 7.13. □

Example 8. $\mathrm{H}(x_1)\mathrm{H}(x_2)\ldots\mathrm{H}(x_k)(1 \le k \le n)$ is a generalised function on R_k or R_n. By Theorem 7.25 and Exercise 34 of Chapter 4 its

Fourier transform is

$$\{\pi\delta(\alpha_1) - \mathrm{i}\alpha_1^{-1}\} \dots \{\pi\delta(\alpha_k) - \mathrm{i}\alpha_k^{-1}\}(2\pi)^{n-k}\,\delta(\alpha_{k+1}) \dots \delta(\alpha_n).$$

Also, by Theorem 7.24,

$$\partial_1 \mathrm{H}(x_1) \dots \mathrm{H}(x_k) = \delta(x_1)\mathrm{H}(x_2) \dots \mathrm{H}(x_k)$$

with Fourier transform

$$\{\pi\delta(\alpha_2) - \mathrm{i}\alpha_2^{-1}\} \dots \{\pi\delta(\alpha_k) - \mathrm{i}\alpha_k^{-1}\}(2\pi)^{n-k}\,\delta(\alpha_{k+1}) \dots \delta(\alpha_n).$$

Clearly

$$\partial_1 \partial_2 \dots \partial_k \mathrm{H}(x_1) \dots \mathrm{H}(x_k) = \delta(x_1)\delta(x_2) \dots \delta(x_k)$$

and

$$\partial_1 \partial_2 \dots \partial_n \mathrm{H}(x_1) \dots \mathrm{H}(x_n) = \delta(\boldsymbol{x})$$

from (12).

Similarly $\mathrm{H}(x_1) \dots \mathrm{H}(x_k)\delta(x_{k+1}) \dots \delta(x_n)$ satisfies

$$\partial_1 \partial_2 \dots \partial_k \mathrm{H}(x_1) \dots \mathrm{H}(x_k)\delta(x_{k+1}) \dots \delta(x_n) = \delta(\boldsymbol{x})$$

and has Fourier transform

$$\{\pi\delta(\alpha_1) - \mathrm{i}\alpha_1^{-1}\} \dots \{\pi\delta(\alpha_k) - \mathrm{i}\alpha_k^{-1}\}.$$

Theorem 7.24 corresponds to the customary rule for taking partial derivatives and prompts the introduction of

Definition 7.14. *g is said to be independent of x_1 if, and only if, $\partial_1 g = 0$.*

If $g(\boldsymbol{x}) = g(x_1, \dots, x_{n-1})$ then $\partial_n g = 0$ by Theorem 7.13 so that the definition is justified if the converse holds. The converse is contained in

Theorem 7.26. *If $\partial_n g = 0$ then $g(\boldsymbol{x}) = g(x_1, \dots, x_{n-1})$.*

Proof. If $\gamma_1(x_1) \dots \gamma_n(x_n)$ is a good function on R_n then so is

$$\gamma_1(x_1) \dots \gamma_{n-1}(x_{n-1}) \left[\int_{-\infty}^{x_n} \left\{ \gamma_n(t_n) - \frac{\mathrm{e}^{-t_n^2}}{\pi^{1/2}} \int_{-\infty}^{\infty} \gamma_n(u)\,\mathrm{d}u \right\} \mathrm{d}t_n \right]$$

because the quantity in square brackets is a good function of x_n.

Therefore

$$\int_{-\infty}^{\infty} g(\boldsymbol{x})\gamma_1(x_1)\ldots\gamma_{n-1}(x_{n-1})\left\{\gamma_n(x_n) - \frac{e^{-x_n^2}}{\pi^{1/2}}\int_{-\infty}^{\infty}\gamma_n(u)\,du\right\}d\boldsymbol{x}$$

$$= \int_{-\infty}^{\infty} g(\boldsymbol{x})\partial_n\gamma_1(x_1)\ldots\gamma_{n-1}(x_{n-1})$$

$$\times\left[\int_{-\infty}^{x_n}\left\{\gamma_n(t_n) - \frac{e^{-t_n^2}}{\pi^{1/2}}\int_{-\infty}^{\infty}\gamma_n(u)\,du\right\}dt_n\right]d\boldsymbol{x}$$

$$= -\int_{-\infty}^{\infty}\{\partial_n g(\boldsymbol{x})\}\gamma_1(x_1)\ldots\gamma_{n-1}(x_{n-1})$$

$$\times\left[\int_{-\infty}^{x_n}\left\{\gamma_n(t_n) - \frac{e^{-t_n^2}}{\pi^{1/2}}\int_{-\infty}^{\infty}\gamma_n(u)\,du\right\}dt_n\right]d\boldsymbol{x} = 0$$

since $\partial_n g = 0$. Hence

$$\int_{-\infty}^{\infty} g(\boldsymbol{x})\gamma_1(x_1)\ldots\gamma_n(x_n)\,d\boldsymbol{x}$$

$$= \int_{-\infty}^{\infty}\gamma_n(u)\,du\int_{-\infty}^{\infty} g(\boldsymbol{x})\gamma_1(x_1)\ldots\gamma_{n-1}(x_{n-1})\frac{e^{-x_n^2}}{\pi^{1/2}}\,d\boldsymbol{x}$$

$$= \int_{-\infty}^{\infty}\gamma_1(x_1)\ldots\gamma_{n-1}(x_{n-1})\gamma_n(u)$$

$$\times\int_{-\infty}^{\infty} g(\boldsymbol{x})\frac{e^{-x_n^2}}{\pi^{1/2}}\,dx_n\,dx_1\ldots dx_{n-1}\,du$$

from Theorem 7.22. Consequently

$$g(\boldsymbol{x}) = \int_{-\infty}^{\infty} g(\boldsymbol{x})\frac{e^{-x_n^2}}{\pi^{1/2}}\,dx_n$$

and the right-hand side is a generalised function on R_{n-1} by Theorem 7.22. The proof is finished. □

Corollary 7.26 *If $x_n g(\boldsymbol{x}) = 0$ then $g(\boldsymbol{x}) = \delta(x_n)g_1(x_1,\ldots,x_{n-1})$ where g_1 is an arbitrary generalised function on R_{n-1}.*

Proof. On account of Theorem 7.14 (iii) $x_n g(\boldsymbol{x}) = 0$ implies that $\partial_n G(\boldsymbol{\alpha}) = 0$. It follows from Theorem 7.26 that $G(\boldsymbol{\alpha}) = G(\alpha_1,\ldots,\alpha_{n-1})$. Theorem 7.25 completes the proof. □

Exercises

22. Prove that, on R_2,

$$(x_1^2+x_2^2)\sum_{m=0}^{M}(-1)^m\frac{(2M)!}{(2m)!(2M-2m)!}\partial_1^{2m}\partial_2^{2M-2m}\delta(x)=0$$

for any positive integer M.

23. If $x_n g(x)=\delta^{(m)}(x_n)$ show that
$g(x)=\delta(x_n)g_1(x_1,\ldots,x_{n-1})-\delta^{(m+1)}(x_n)/(m+1)$
where g_1 is arbitrary.

24. If $x_n^m g(x)=0$ show that

$$g(x)=g_1(x_1,\ldots,x_{n-1})\delta(x_n)$$
$$+g_2(x_1,\ldots,x_{n-1})\delta'(x_n)+\ldots+g_m(x_1,\ldots,x_{n-1})\delta^{(m-1)}(x_n)$$

where $g_1,g_2,\ldots,g_m$ are arbitrary generalised functions on R_{n-1}.

25. If, on R_2, $\partial_1\partial_2 g(x_1,x_2)=0$ prove that

$$g(x_1,x_2)=g_1(x_1)+g_2(x_2)$$

where g_1 and g_2 are arbitrary. Deduce that, if $x_1x_2g(x_1,x_2)=0$ then

$$g(x_1,x_2)=g_1(x_1)\delta(x_2)+g_2(x_2)\delta(x_1).$$

26. If $\partial_{n-1}\partial_n g=0$ prove that

$$g(x)=g_1(x_1,\ldots,x_{n-1})+g_2(x_1,\ldots,x_{n-2},x_n),$$

and deduce that, if $x_{n-1}x_n g(x)=0$,

$$g(x)=g_1(x_1,\ldots,x_{n-1})\delta(x_n)+g_2(x_1,\ldots,x_{n-2},x_n)\delta(x_{n-1}).$$

27. If $x_{n-1}x_n^2g(x)=0$ prove that

$$g(x)=g_1(x_1,\ldots,x_{n-2},x_n)\delta(x_{n-1})$$
$$+g_2(x_1,\ldots,x_{n-1})\delta(x_n)+g_3(x_1,\ldots,x_{n-1})\delta'(x_n).$$

Theorem 7.27. *If $g(x)=0$ for $x_n>0$ and for $x_n<0$ then*

$$g(x)=\sum_{p=0}^{m}g_p(x_1,\ldots,x_{n-1})\delta^{(p)}(x_n)$$

where m is finite and g_p is independent of x_n.

Proof. Since g is of the form $\partial_1^{p_1}\ldots\partial_n^{p_n}f(x)$ and vanishes for $x_n>0$ it follows (cf. Exercise 24) that

$$\partial_1^{p_1}\ldots\partial_{n-1}^{p_{n-1}}f(x)=\sum_{p=0}^{p_n-1}g_p(x_1,\ldots,x_{n-1})x_n^p.$$

Now

$$\int_{-\infty}^{\infty} g(\boldsymbol{x})\gamma(x_1,\dots,x_{n-1})\gamma_n(x_n)\mathrm{d}\boldsymbol{x}$$

$$=(-1)^{p_n}\int_{-\infty}^{\infty}(\partial_1^{p_1}\dots\partial_{n-1}^{p_{n-1}}f)\gamma\gamma_n^{(p_n)}(x_n)\mathrm{d}\boldsymbol{x}$$

and

$$\int_0^{\infty} x_n^p\gamma_n^{(p_n)}(x_n)\mathrm{d}x_n = C_p\gamma_n^{(p_n-p-1)}(0).$$

Thus

$$\int_{-\infty}^{\infty} g(\boldsymbol{x})\gamma(x_1,\dots,x_{n-1})\gamma_n(x_n)\mathrm{d}\boldsymbol{x}$$

$$=\sum_{p=0}^{p_n-1} C_p\gamma_n^{(p_n-p-1)}(0)\int_{-\infty}^{\infty} g_p\gamma\mathrm{d}x_1\dots\mathrm{d}x_{n-1}$$

and the result follows. □

Corollary 7.27. *If* $g(\boldsymbol{x})=0$ *for* $x_k, x_{k+1},\dots, x_n \gtrless 0$ *then*

$$g(\boldsymbol{x})=\sum_{p=0}^{m} g_p(x_1,\dots,x_{k-1})\delta^{(p_k)}(x_k)\delta^{(p_{k+1})}(x_{k+1})\dots\delta^{(p_n)}(x_n)$$

where $p=p_k+p_{k+1}+\dots+p_n$.

This is a consequence of repeated application of Theorem 7.27.

Theorem 7.28. *In order that the equations*

$$\partial_1 g=g_1,\quad \partial_2 g=g_2,\quad \dots,\quad \partial_k g=g_k$$

possess a solution it is necessary and sufficient that $\partial_i g_j=\partial_j g_i$ *for* $i\le k$ *and* $j\le k$. *Then any two solutions differ by a generalised function independent of* $x_1, x_2,\dots,x_k$.

Proof. The last statement follows at once from Definition 7.14.

Also, if the equations have a solution,

$$\partial_i g_j=\partial_i\partial_j g=\partial_j\partial_i g=\partial_j g_i$$

and necessity is proved.

With regard to sufficiency, the general solution of $\partial_1 g = g_1$ is

$$g(x) = g_0(x) + h(x_2, \dots, x_n)$$

where g_0 is any generalised function such that $\partial_1 g_0 = g_1$ and h is an arbitrary generalised function of $x_2, \dots, x_n$. Then $g_2 = \partial_2 g$ becomes

$$g_2 = \partial_2 g_0 + \partial_2 h.$$

This equation is consistent because

$$\partial_1(g_2 - \partial_2 g_0) = \partial_1 g_2 - \partial_2 \partial_1 g_0 = \partial_2 g_1 - \partial_2 g_1 = 0$$

since $\partial_1 g_2 = \partial_2 g_1$ by assumption. Thus

$$\partial_2 h = g_2 - \partial_2 g_0$$

involves only $x_2, \dots, x_n$ and provides h. The general form of h contains an arbitrary element of $x_3, \dots, x_n$ which satisfies a partial differential equation determined by the third equation on g. The process continues until all k equations have been used. The proof is complete. □

Exercises

28. If $\partial_1 g = 0, \partial_2 g = 0, \dots, \partial_n g = 0$ show that g is a constant.

29. Show that $\alpha_1 G(\alpha) = G_1(\alpha)$, $\alpha_2 G(\alpha) = G_2(\alpha), \dots, \alpha_k G(\alpha) = G_k(\alpha)$ are consistent if, and only if, $\alpha_i G_j = \alpha_j G_i$ for $i \le k, j \le k$. What do any two solutions differ by?

30. If $\partial_1^{m_1} \dots \partial_n^{m_n} g = 0$ for all $m_1, \dots, m_n$ such that $m_1 + m_2 + \dots + m_n = m$ show that g is a polynomial of degree not exceeding $m - 1$.

7.7. Some special generalised functions

In Chapter 4 meanings have been attached to quantities such as

$$|x|^\beta, \quad |x|^\beta \operatorname{sgn} x, \quad x^\beta \mathrm{H}(x), \quad x^{-m}, \quad x^{-m} \operatorname{sgn} x,$$

$$|x|^\beta \ln |x|, \quad x^\beta \mathrm{H}(x) \ln x, \quad x^{-m} \ln |x|, \quad x^{-m} \operatorname{sgn} x \ln |x|$$

so that, by Definition 7.13, the significance of, for example,

$$|x_1|^{\beta_1} |x_2|^{\beta_2} \operatorname{sgn} x_2, \quad x_1^{-m} x_2^{\beta_2} \mathrm{H}(x_2) x_3^{-1} \operatorname{sgn} x_3,$$

$$|x_1|^{\beta_1} |x_2|^{\beta_2} \ln |x_2| \quad \text{and} \quad x_1^{-m_1} \operatorname{sgn} x_1\, x_2^{-m_2} \operatorname{sgn} x_2 \ln |x_2|$$

as generalised functions is known. By Theorem 7.25 and Theorems 4.3, 4.4, 4.6–4.8, their Fourier transform are

$$-\mathrm{i}2^n\pi^{n-2}\beta_1!\beta_2!|\alpha_1|^{-\beta_1-1}\cos\tfrac{1}{2}\pi(\beta_1+1)|\alpha_2|^{-\beta_2-1}$$
$$\times\sin\tfrac{1}{2}\pi(\beta_2+1)\,\mathrm{sgn}\,\alpha\,\delta(\alpha_3)\dots\delta(\alpha_n),$$

$$\frac{(-\mathrm{i})^m 2^{n-3}\pi^{n-2}}{(m-1)!}\beta_2!\alpha_1^{m-1}\,\mathrm{sgn}\,\alpha_1|\alpha_2|^{-\beta_2-1}\mathrm{e}^{-(\pi/2)\mathrm{i}(\beta_2+1)\mathrm{sgn}\alpha_2}$$
$$\times\{2\psi(0)-2\ln|\alpha_3|\}\delta(\alpha_4)\dots\delta(\alpha_n),$$

$$2^n\pi^{n-2}\beta_1!\beta_2!|\alpha_1|^{-\beta_1-1}\cos\tfrac{1}{2}\pi(\beta_1+1)|\alpha_2|^{-\beta_2-1}\cos\tfrac{1}{2}\pi(\beta_2+1)$$
$$\times\{\psi(\beta_2)-\ln|\alpha_2|-\tfrac{1}{2}\pi\tan\tfrac{1}{2}\pi(\beta_2+1)\}\delta(\alpha_3)\dots\delta(\alpha_n),$$

$$\frac{-(-\mathrm{i})^{m_1+m_2}2^{n-2}\pi^{n-2}}{(m_1-1)!(m_2-1)!}\alpha_1^{m_1-1}\{2\psi(m_1-1)-2\ln|\alpha_1|\}\alpha_2^{m_2-1}$$
$$\times[\{\ln|\alpha_1|-\psi(m_2-1)\}^2-\tfrac{1}{6}\pi^2-\psi'(m_2-1)]\delta(\alpha_3)\dots\delta(\alpha_n)$$

respectively. The partial Fourier transform of Theorem 7.23 can also be calculated. For example,

$$\int_{-\infty}^{\infty}|x_1|^{\beta_1}|x_2|^{\beta_2}\,\mathrm{sgn}\,x_2\,\mathrm{e}^{-\mathrm{i}\alpha_1x_1}\,\mathrm{d}x_1$$
$$=\beta_1!|\alpha_1|^{-\beta_1-1}2\cos\tfrac{1}{2}\pi(\beta_1+1)|x_2|^{\beta_2}\,\mathrm{sgn}\,x_2.$$

The most direct generalisation of the definitions of Chapter 4 occurs when one considers the possibility of replacing $|x|$ by r. Let us therefore investigate the meaning to be assigned to r^β. When $\mathrm{Re}\,\beta > 2-n$ the standard interpretation can be applied because $r^\beta \in \boldsymbol{K}_1$ for $\mathrm{Re}\,\beta + n > 0$ and we have

$$\partial r^\beta/\partial x_1 = \beta x_1 r^{\beta-2}, \quad \partial^2 r^\beta/\partial x_1^2 = \beta r^{\beta-2} + \beta(\beta-2)x_1^2 r^{\beta-4}$$

and so

$$\nabla^2 r^\beta = n\beta r^{\beta-2} + \beta(\beta-2)r^{\beta-2} = \beta(\beta+n-2)r^{\beta-2}. \tag{13}$$

We now define r^β when $\mathrm{Re}\,\beta < -n$ so that this relation is preserved apart from in certain exceptional cases. One exception is $\beta = 2-n$. To examine this it is convenient to prove first a general theorem which is often useful.

Theorem 7.29. *If $f \in \boldsymbol{K}_1$ and depends only on r then*

$$(\partial_1^2 + \ldots + \partial_n^2)f = \frac{1}{r^{n-1}}\partial_r(r^{n-1}\partial_r f)$$

in the sense that

$$\int_{-\infty}^{\infty} \{(\partial_1^2 + \ldots + \partial_n^2)f\}\gamma\,\mathrm{d}\boldsymbol{x} = \int_{-\infty}^{\infty} f\frac{1}{r^{n-1}}\frac{\partial}{\partial r}\left(r^{n-1}\frac{\partial\gamma}{\partial r}\right)\mathrm{d}\boldsymbol{x}.$$

Proof.

$$\int_{-\infty}^{\infty} \{(\partial_1^2 + \ldots + \partial_n^2)f\}\gamma\,\mathrm{d}\boldsymbol{x} = \textstyle\int_{-\infty}^{\infty} f\nabla^2\gamma\,\mathrm{d}\boldsymbol{x}.$$

The integral on the right is a conventional absolutely convergent integral since $f \in \boldsymbol{K}_1$ (Theorem 7.8) and so can be calculated in spherical polar coordinates by Theorem 7.1. Hence the right-hand side is

$$\int_0^\infty r^{n-1}f(r)\int_0^\pi \ldots \int_0^\pi \int_0^{2\pi} (\nabla^2\gamma)\sin^{n-2}\theta_1 \sin^{n-3}\theta_2 \ldots$$

$$\times \sin\theta_{n-2}\,\mathrm{d}\theta_1 \ldots \mathrm{d}\theta_{n-2}\,\mathrm{d}\theta_{n-1}\,\mathrm{d}r \tag{14}$$

where

$$\nabla^2\gamma = \frac{1}{r^{n-1}}\frac{\partial}{\partial r}\left(r^{n-1}\frac{\partial\gamma}{\partial r}\right) + \frac{1}{r^2\sin^{n-2}\theta_1}\frac{\partial}{\partial\theta_1}\left(\sin^{n-2}\theta_1\frac{\partial\gamma}{\partial\theta_1}\right)$$

$$+\frac{1}{r^2\sin^2\theta_1\sin^{n-3}\theta_2}\frac{\partial}{\partial\theta_2}\left(\sin^{n-3}\theta_2\frac{\partial\gamma}{\partial\theta_2}\right) + \ldots$$

$$+\frac{1}{r^2\sin^2\theta_1\sin^2\theta_2\ldots\sin^2\theta_{n-3}\sin\theta_{n-2}}\frac{\partial}{\partial\theta_{n-2}}\left(\sin\theta_{n-2}\frac{\partial\gamma}{\partial\theta_{n-2}}\right)$$

$$+\frac{1}{r^2\sin^2\theta_1\sin^2\theta_2\ldots\sin^2\theta_{n-2}}\frac{\partial^2\gamma}{\partial\theta_{n-1}^2} \tag{15}$$

and γ is expressed in terms of $r, \theta_1, \ldots, \theta_{n-1}$. Now

$$\int_0^{2\pi}\frac{\partial^2\gamma}{\partial\theta_{n-1}^2}\mathrm{d}\theta_{n-1} = \left[\frac{\partial\gamma}{\partial\theta_{n-1}}\right]_0^{2\pi} = 0$$

since γ must have period 2π in θ_{n-1} because it is good on R_n. Also

$$\int_0^\pi \frac{\partial}{\partial\theta_{n-2}}\left(\sin\theta_{n-2}\frac{\partial\gamma}{\partial\theta_{n-2}}\right)\mathrm{d}\theta_{n-2} = \left[\sin\theta_{n-2}\frac{\partial\gamma}{\partial\theta_{n-2}}\right]_0^\pi = 0$$

since $\partial\gamma/\partial\theta_{n-2}$ must be bounded and continuous because γ is good.

Hence all the terms except the first on the right-hand side of (15) give a zero contribution to (14). The proof is complete. □

Example 9. In this example it is shown that

$$(\partial_1^2+\ldots+\partial_n^2)r^{2-n}=\begin{cases}-\dfrac{4\pi^{n/2}\delta(\boldsymbol{x})}{(\frac{1}{2}n-2)!} & (n\neq 2)\\[2ex] 0 & (n=2)\end{cases}\tag{16}$$

For $r^{2-n}\in \boldsymbol{K}_1$ and so, by Theorem 7.29,

$$\int_{-\infty}^{\infty}\{(\partial_1^2+\ldots+\partial_n^2)r^{2-n}\}\gamma\,\mathrm{d}\boldsymbol{x}$$

$$=\int_0^\infty r^{2-n}\frac{\partial}{\partial r}r^{n-1}\frac{\partial}{\partial r}$$

$$\int_0^\pi\ldots\int_0^{2\pi}\gamma\sin^{n-2}\theta_1\ldots\sin\theta_{n-2}\,\mathrm{d}\theta_1\ldots\mathrm{d}\theta_{n-1}\,\mathrm{d}r$$

$$=\left[r\frac{\partial}{\partial r}\int_0^\pi\ldots\int_0^{2\pi}\gamma\sin^{n-2}\theta_1\ldots\mathrm{d}\theta_1\ldots\mathrm{d}\theta_{n-1}\right]_0^\infty$$

$$-(2-n)\int_0^\infty\frac{\partial}{\partial r}\int_0^\pi\ldots\int_0^{2\pi}\gamma\sin^{n-2}\theta_1\ldots\mathrm{d}\theta_1\ldots\mathrm{d}\theta_{n-1}\,\mathrm{d}r$$

by integration by parts. The first term vanishes at the upper limit because γ is good and at the lower because $r=0$ and the other factor is bounded. If $n=2$ the second term vanishes and the result stated is confirmed for this case. In the remaining cases another integration gives

$$\int_{-\infty}^{\infty}\{(\partial_1^2+\ldots+\partial_n^2)r^{2-n}\}\gamma\,\mathrm{d}\boldsymbol{x}$$

$$=(2-n)\gamma(\mathbf{0})\int_0^\pi\ldots\int_0^{2\pi}\sin^{n-2}\theta_1\ldots\mathrm{d}\theta_1\ldots\mathrm{d}\theta_{n-1}$$

$$=(2-n)\gamma(\mathbf{0})\pi^{(n-2)/2}\frac{\{\frac{1}{2}(n-3)\}!\,\{\frac{1}{2}(n-4)\}!}{\{\frac{1}{2}(n-2)\}!\,\{\frac{1}{2}(n-3)\}!}\cdots\frac{0!}{\frac{1}{2}!}2\pi$$

$$=\frac{2(2-n)\pi^{n/2}\gamma(\mathbf{0})}{(\frac{1}{2}n-1)!}$$

since

$$\int_0^{\pi} \sin^{\nu}\theta \, \mathrm{d}\theta = (\tfrac{1}{2}\nu - \tfrac{1}{2})!\,\pi^{1/2}/(\tfrac{1}{2}\nu)! \tag{17}$$

The given result now follows immediately.

The conclusion in (16) when $n = 2$ is only to be expected since then r^{2-n} is, in fact, unity. A δ-function is obtained on the right-hand side when $n = 2$ from a logarithmic term as is demonstrated in the following example.

Example 10. If $n = 2$, $\ln r \in \boldsymbol{K}_1$ and so, by Theorem 7.29,

$$\int_{-\infty}^{\infty} \{(\partial_1^2 + \partial_2^2)\ln r\}\gamma \, \mathrm{d}\boldsymbol{x} = \int_0^{\infty} \ln r \frac{\partial}{\partial r} r \frac{\partial}{\partial r} \int_0^{2\pi} \gamma \, \mathrm{d}\theta_1 \, \mathrm{d}r$$

$$= -\int_0^{\infty} \frac{\partial}{\partial r} \int_0^{2\pi} \gamma \, \mathrm{d}\theta_1 \, \mathrm{d}r$$

$$= 2\pi\gamma(\mathbf{0}).$$

Hence, when $n = 2$,

$$(\partial_1^2 + \partial_2^2)\ln r = 2\pi\delta(\boldsymbol{x}). \tag{18}$$

Example 9 indicates that (13) cannot be expected to hold in general. Nevertheless there are cases when it will as may be seen from

Definition 7.15. *For any complex* $\beta(\mathrm{Re}\,\beta < -n)$ *except* $\beta = -n-2k$ *(k a non-negative integer)* r^{β} *is defined by*

$$r^{\beta} = \frac{(\partial_1^2 + \partial_2^2 + \ldots + \partial_n^2)^m r^{\beta+2m}}{(\beta+2m)(\beta+2m-2)\ldots(\beta+2)(\beta+2m+n-2)(\beta+2m+n-4)\ldots(\beta+n)}$$

where m is a non-negative integer such that $n > \mathrm{Re}\,\beta + 2m + n > 0$.

On account of Theorem 7.12 the generalised function r^{β} agrees with the conventional function denoted by r^{β} when $r > 0$. Furthermore

$$(\partial_1^2 + \partial_2^2 + \ldots + \partial_n^2)r^{\beta} = \frac{(\partial_1^2 + \ldots + \partial_n^2)^{m+1} r^{\beta+2m}}{(\beta+2m)\ldots(\beta+2)(\beta+2m+n-2)\ldots(\beta+n)}$$

$$= \frac{\beta(\beta+n-2)(\partial_1^2 + \ldots + \partial_n^2)^{m+1} r^{\beta-2+2m+2}}{(\beta+2m)\ldots\beta(\beta+2m+n-2)\ldots(\beta+n-2)}$$

since neither β nor $\beta+n-2$ is zero when Re $\beta \le -n$. Applying Definition 7.15 with β and m replaced by $\beta-2$ and $m+1$ respectively, we obtain

$$(\partial_1^2+\partial_2^2+\ldots+\partial_n^2)r^\beta = \beta(\beta+n-2)r^{\beta-2}. \tag{19}$$

Because of (13) and Theorem 7.12 this relation holds for all β except $\beta = 2-n$ (see Example 9) and $\beta = -n-2k$.

To deal with the exceptional values of β we introduce

Definition 7.16. *The generalised function r^{-n-2k} where k is a non-negative integer is defined by*

$$r^{-n-2k} = \operatorname*{Lim}_{\mu\to 0}\left\{r^{\mu-n-2k} - \frac{\pi^{n/2}(\partial_1^2+\ldots+\partial_n^2)^k\,\delta(\boldsymbol{x})}{(\frac{1}{2}n+k-1)!k!2^{2k-1}\mu}\right\}.$$

To demonstrate that the definition has a meaning we consider r^{-n}. Now, by (19),

$$r^{\mu-n} = \frac{(\partial_1^2+\ldots+\partial_n^2)r^{\mu-n+2}}{(\mu-n+2)\mu}.$$

Since $r^{\mu-n+2}$ is a conventional function we can write

$$r^{\mu-n+2} = r^{2-n} + \mu r^{2-n}\ln r + \tfrac{1}{2}\mu^2 r^{2-n}\ln^2 r + O(\mu^3).$$

The expansion to so many terms is necessary because of the different behaviour for $n=2$ and $n\neq 2$. If $n=2$

$$\begin{aligned} r^{\mu-n} &= (1/\mu)(\partial_1^2+\partial_2^2)(\ln r + \tfrac{1}{2}\mu\ln^2 r) + O(\mu) \\ &= (2\pi/\mu)\delta(\boldsymbol{x}) + (\partial_1^2+\partial_2^2)\tfrac{1}{2}\ln^2 r + O(\mu) \end{aligned}$$

from (18). It follows from Definition 7.16 that

$$r^{-2} = (\partial_1^2+\partial_2^2)\tfrac{1}{2}\ln^2 r \quad (n=2). \tag{20}$$

In like manner we obtain from (16) when $n\neq 2$

$$r^{-n} = \{1/(2-n)\}(\partial_1^2+\ldots+\partial_n^2)r^{2-n}\ln r \quad (n\neq 2). \tag{21}$$

By virtue of (19)

$$(\partial_1^2+\ldots+\partial_n^2)r^{-n-2k} = \operatorname*{Lim}_{\mu\to 0}\left\{(\mu-n-2k)(\mu-2k-2)r^{\mu-n-2k-2} - \frac{\pi^{n/2}(\partial_1^2+\ldots+\partial_n^2)^{k+1}\,\delta(\boldsymbol{x})}{(\frac{1}{2}n+k-1)!k!2^{2k-1}\mu}\right\}$$

$$= \operatorname*{Lim}_{\mu\to 0}\left[(\mu-n-2k)(\mu-2k-2)\right.$$
$$\times\left\{r^{\mu-n-2k-2}-\frac{\pi^{n/2}(\partial_1^2+\dots+\partial_n^2)^{k+1}\delta(\boldsymbol{x})}{(\frac{1}{2}n+k)!(k+1)!2^{2k+1}\mu}\right\}$$
$$\left.+\frac{\pi^{n/2}\{\mu-(n+4k+2)\}}{(\frac{1}{2}n+k)!(k+1)!2^{2k+1}}(\partial_1^2+\dots+\partial_n^2)^{k+1}\delta(\boldsymbol{x})\right].$$

Hence, by Definition 7.16,

$$(\partial_1^2+\dots+\partial_n^2)r^{-n-2k}=(n+2k)(2k+2)r^{-n-2k-2}-\frac{(n+4k+2)\pi^{n/2}(\partial_1^2+\dots+\partial_n^2)^{k+1}\delta(\boldsymbol{x})}{(\frac{1}{2}n+k)!(k+1)!2^{2k+1}} \tag{22}$$

which is the analogue of (19). By means of (20)–(22) expressions for r^{-n-2k} as derivatives of logarithms and $\delta(\boldsymbol{x})$ can be provided.

As far as multiplication is concerned there is the slight difficulty that r is not a fairly good function. However, r^2 is fairly good and

$$r^2.r^{-n-2k}=\operatorname*{Lim}_{\mu\to 0}\left\{r^{\mu-n-2k+2}-\frac{\pi^{n/2}r^2(\partial_1^2+\dots+\partial_n^2)^k\delta(\boldsymbol{x})}{(\frac{1}{2}n+k-1)!k!2^{2k-1}\mu}\right\}.$$

Since $r^2\delta(\boldsymbol{x})=0$(§7.2) the application of $(\partial_1^2+\dots+\partial_n^2)^k$ gives

$$r^2(\partial_1^2+\dots+\partial_n^2)^k\delta(\boldsymbol{x})=2k(n+2k-2)(\partial_1^2+\dots+\partial_n^2)^{k-1}\delta(\boldsymbol{x}) \tag{23}$$

which, on substitution, supplies

$$r^2.r^{-n-2k}=r^{2-n-2k}. \tag{24}$$

It should be remarked that r^{-n-2k} does not have the homogeneous property. For, if $a>0$, (20) supplies

$$(ar)^{-2}=(r^{-2}+2\pi\delta(\boldsymbol{x})\ln a)/a^2 \quad (n=2)$$

while for $n\neq 2$

$$(ar)^{-n}=\left(r^{-n}+\frac{2\pi^{n/2}\delta(\boldsymbol{x})}{(\frac{1}{2}n-1)!}\ln a\right)\Big/a^n$$

from (21).

Exercises

31. Evaluate $(\partial_1^2 + \dots + \partial_n^2)(a_1^2 x_1^2 + a_2^2 x_2^2 + \dots + a_n^2 x_n^2)^{\beta/2}$ when (i) $\beta = 2 - n$, (ii) $\beta \neq 2 - n$.

32. If k is a positive integer, prove that

$$(\partial_1^2 + \dots + \partial_n^2)^k r^{k-n} = \frac{(k-1)!2^k \pi^{n/2}}{(\frac{1}{2}n - 1)!}(2k-n)(2k-n-2)\dots(2-n)\delta(\boldsymbol{x}),$$

the right-hand side being zero when n is even and $2k - n \geq 0$.

33. What is the difference between $(ar)^{-n-2k}$ and r^{-n-2k}/a^{n+2k} when $n = 2$ and $n \neq 2$?

34. Prove that $r^2 . r^\beta = r^{\beta+2}$.

35. Show that

$$\sum_{i=1}^{n} x_i \partial_i r^{-n-2k} = -(n+2k)r^{-n-2k} + \frac{\pi^{n/2}(\partial_1^2 + \dots + \partial_n^2)^k \delta(\boldsymbol{x})}{(\frac{1}{2}n + k - 1)! k! 2^{2k-1}}.$$

The calculation of the Fourier transform of a function of r is assisted by

Theorem 7.30. *If $f \in \boldsymbol{L}_1$ and f is a function of r only*

$$F(\boldsymbol{\alpha}) = (2\pi)^{n/2} \int_0^\infty f(r) r^{n-1} \frac{J_{(n-2)/2}(\alpha r)}{(\alpha r)^{(n/2)-1}} \mathrm{d}r$$

where $\alpha = |\boldsymbol{\alpha}|$.

Proof. Since $\int_{-\infty}^{\infty} f(r) \mathrm{e}^{-i\boldsymbol{\alpha}.\boldsymbol{x}} \,\mathrm{d}\boldsymbol{x}$ is absolutely convergent because $f \in \boldsymbol{L}_1$ it may be evaluated by using spherical polar coordinates. There are two ways of doing this.

In the first we make the direct substitution $x_1 = r\cos\theta_1$, etc.; the resulting integrals can then be calculated by means of the formulae

$$J_m(z) = \frac{1}{2\pi} \int_{\theta_0}^{2\pi + \theta_0} \mathrm{e}^{\mathrm{i}(m\theta_{n-1} - z\sin\theta_{n-1})} \mathrm{d}\theta_{n-1}, \tag{25}$$

$$\int_0^\pi \mathrm{e}^{iz\cos\theta\cos\psi} J_\nu(z\sin\theta\sin\psi)\sin^{\nu+1}\theta \,\mathrm{d}\theta = \left(\frac{2\pi}{z}\right)^{1/2} \sin^\nu\psi\, J_{\nu+1/2}(z). \tag{26}$$

In the second way choose the polar axis to lie along the direction of $\boldsymbol{\alpha}$; then $\boldsymbol{\alpha}.\boldsymbol{x} = \alpha r \cos\theta_1$ and

$$F(\boldsymbol{\alpha}) = \int_0^\infty f(r) r^{n-1} \int_0^\pi e^{-i\alpha r\cos\theta_1} \sin^{n-2}\theta_1 \, d\theta_1$$

$$\times \int_0^\pi \dots \int_0^{2\pi} \sin^{n-3}\theta_2 \dots \sin\theta_{n-2} \, d\theta_2 \dots d\theta_{n-1} \, dr$$

$$= \int_0^\infty f(r) r^{n-1} \frac{2\pi^{(n-1)/2}}{(\frac{1}{2}n - \frac{3}{2})!} \int_0^\pi e^{-i\alpha r\cos\theta_1} \sin^{n-2}\theta_1 \, d\theta_1 \, dr$$

from (17). When $\operatorname{Re}\nu > -\frac{1}{2}$

$$J_\nu(z) = \frac{(\frac{1}{2}z)^\nu}{(\nu - \frac{1}{2})!\pi^{1/2}} \int_0^\pi e^{-iz\cos\theta} \sin^{2\nu}\theta \, d\theta \tag{27}$$

so that

$$F(\boldsymbol{\alpha}) = (2\pi)^{n/2} \int_0^\infty f(r) r^{n-1} \frac{J_{(n-2)/2}(\alpha r)}{(\alpha r)^{(n/2)-1}} dr. \tag{28}$$

If $\alpha = 0$, (17) replaces (27) but (28) converges uniformly with respect to α, because $J_{(n-2)/2}(x)/x^{(n/2)-1}$ is bounded for all x and $f \in \boldsymbol{L}_1$, so that (28) is continuous at $\alpha = 0$ by Theorem 1.32. Hence $F(\mathbf{0})$ can be obtained by continuity from (28). □

Theorem 7.30 can be extended in various ways which are useful for handling certain types of generalised function. For example, suppose that f is a function of r only and $f \in \boldsymbol{K}_1$. Then Theorem 7.30 may not be directly applicable. However, we can choose N large enough for $f/(1 + r^2)^N \in \boldsymbol{L}_1$ and then Theorem 7.30 gives

$$F(\boldsymbol{\alpha}) = (1 - \partial_1^2 - \partial_2^2 - \dots - \partial_n^2)^N (2\pi)^{n/2} \int_0^\infty \frac{f(r) r^{n-1} J_{(n-2)/2}(\alpha r)}{(1 + r^2)^N (\alpha r)^{(n/2)-1}} dr \tag{29}$$

by Theorem 7.14(iii). Here $\partial_1, \partial_2, \dots$ are generalised derivatives with respect to $\alpha_1, \alpha_2, \dots$. A further application of Theorem 7.14 (iii) shows that the Fourier transform of the generalised function $g(\boldsymbol{x}) = \partial_1^{p_1} \dots \partial_n^{p_n} f(r)$ where $f \in \boldsymbol{K}_1$ is of the form

$$G(\boldsymbol{\alpha}) = (i\alpha_1)^{p_1} \dots (i\alpha_n)^{p_n} (1 - \partial_1^2 \dots - \partial_n^2)^N (2\pi)^{n/2}$$

$$\times \int_0^\infty \frac{f(r) r^{n-1} J_{(n-2)/2}(\alpha r)}{(1 + r^2)^N (\alpha r)^{(n/2)-1}} dr. \tag{30}$$

In this formula only a conventional integral has to be evaluated.

For the special generalised function r^β we have

Theorem 7.31. *If* $\beta \neq 2k$ *and* $\beta \neq -n-2k\,(k=0,1,2,\ldots)$ *then*

$$\int_{-\infty}^{\infty} r^\beta \mathrm{e}^{-\mathrm{i}\boldsymbol{\alpha}.\boldsymbol{x}}\,\mathrm{d}\boldsymbol{x} = \frac{(\frac{1}{2}\beta+\frac{1}{2}n-1)!}{(-\frac{1}{2}\beta-1)!}2^{\beta+n}\pi^{n/2}\alpha^{-\beta-n}.$$

Proof. Consider firstly the case in which $\frac{1}{2}-\frac{1}{2}n > \operatorname{Re}\beta > -n$. Then the conventional Fourier transform of r^β exists and is a continuous function for $\boldsymbol{\alpha} \neq \mathbf{0}$. Now, if $\mu > 0$ a transformation to spherical polar coordinates gives, for $\operatorname{Re}\beta > -n$,

$$\int_{-\infty}^{\infty} r^\beta \mathrm{e}^{-r^2/\mu}\,\mathrm{d}\boldsymbol{x} = \Omega_{n-1}\int_0^\infty r^{\beta+n-1}\mathrm{e}^{-r^2/\mu}\,\mathrm{d}r$$

where Ω_n is the area of the unit sphere in n-dimensional space, i.e. $\Omega_n = 2\pi^{n/2}/(\frac{1}{2}n-1)!$. Hence

$$\int_{-\infty}^{\infty} r^\beta \mathrm{e}^{-r^2/\mu}\,\mathrm{d}\boldsymbol{x} = (\tfrac{1}{2}\beta+\tfrac{1}{2}n-1)!\tfrac{1}{2}\Omega_{n-1}\mu^{\beta/2+n/2}.$$

This implies, for $\operatorname{Re}\beta < 0$,

$$\int_{-\infty}^{\infty} \alpha^{-\beta-n}\mathrm{e}^{-\mu\alpha^2/4}\,\mathrm{d}\boldsymbol{\alpha} = (-\tfrac{1}{2}\beta-1)!\tfrac{1}{2}\Omega_{n-1}(\tfrac{1}{4}\mu)^{\beta/2}.$$

On the other hand, from Theorems 7.13 and 7.3,

$$\int_{-\infty}^{\infty} r^\beta \mathrm{e}^{-r^2/\mu}\,\mathrm{d}\boldsymbol{x} = \left(\frac{\mu}{4\pi}\right)^{n/2}\int_{-\infty}^{\infty} F(\boldsymbol{\alpha})\mathrm{e}^{-\mu\alpha^2/4}\,\mathrm{d}\boldsymbol{\alpha}$$

where $F(\boldsymbol{\alpha})$ is the Fourier transform of r^β. Accordingly,

$$\int_{-\infty}^{\infty} F(\boldsymbol{\alpha})\mathrm{e}^{-\mu\alpha^2/4}\,\mathrm{d}\boldsymbol{\alpha} = \frac{(\frac{1}{2}\beta+\frac{1}{2}n-1)!}{(-\frac{1}{2}\beta-1)!}2^{\beta+n}\pi^{n/2}\int_{-\infty}^{\infty}\alpha^{-\beta-n}\mathrm{e}^{-\mu\alpha^2/4}\,\mathrm{d}\boldsymbol{\alpha}.$$

By Lerch's theorem (*Acta Mathematica*, **27**, 339 (1903)) which says that, if $f(\boldsymbol{\alpha})$ is continuous for $\alpha > 0$, then $\int_{-\infty}^{\infty} f(\boldsymbol{\alpha})\mathrm{e}^{-\mu\alpha^2}\,\mathrm{d}\boldsymbol{\alpha} = 0$ for all sufficiently large positive μ implies $f(\boldsymbol{\alpha})$ is zero, we deduce that

$$F(\boldsymbol{\alpha}) = \frac{(\frac{1}{2}\beta+\frac{1}{2}n-1)!}{(-\frac{1}{2}\beta-1)!}2^{\beta+n}\pi^{n/2}\alpha^{-\beta-n}$$

and the Theorem is proved for $\frac{1}{2}-\frac{1}{2}n > \operatorname{Re}\beta > -n$. By applying (19) and Theorem 7.14 (iii) we conclude that it is true for $\mathrm{Re}\,\beta < \frac{1}{2} - \frac{1}{2}n$. Taking advantage of the Fourier inversion theorem we confirm that it holds for $\mathrm{Re}\,\beta > -\frac{1}{2} - \frac{1}{2}n$. The proof is complete. □

We turn now to the cases excluded from Theorem 7.31.

Theorem 7.32. *For $k = 0, 1, \dots$*

$$\int_{-\infty}^{\infty} r^{2k} e^{-i\boldsymbol{\alpha}.\boldsymbol{x}} \, d\boldsymbol{x} = (2\pi)^n (-1)^k (\partial_1^2 + \dots + \partial_n^2)^k \delta(\boldsymbol{\alpha})$$

where the partial derivatives are with respect to $\boldsymbol{\alpha}$.

Proof. The Fourier transform of $\delta(\boldsymbol{x})$ is 1. Therefore, by Theorem 7.14 (iii), the Fourier transform of $(\partial_1^2 + \dots + \partial_n^2)^k \delta(\boldsymbol{x})$ is $(-\alpha^2)^k$. The Fourier inversion theorem (Theorem 7.14 (i)) now gives

$$(\partial_1^2 + \dots + \partial_n^2)^k \delta(\boldsymbol{x}) = \frac{1}{(2\pi)^n} \int_{-\infty}^{\infty} (-1)^k \alpha^{2k} e^{i\boldsymbol{\alpha}.\boldsymbol{x}} \, d\boldsymbol{\alpha}.$$

Interchanging the variables supplies the result stated. □

Theorem 7.33. *The Fourier transform of* $\ln(1/r)$ *is*

$$(\tfrac{1}{2}n - 1)!\,2^{n-1} \pi^{n/2} \alpha^{-n} - (2\pi)^n \{\tfrac{1}{2}\psi(\tfrac{1}{2}n - 1) + \tfrac{1}{2}\psi(0) + \ln 2\} \delta(\boldsymbol{\alpha})$$

and of r^{-n-2k} $(k = 0, 1, \dots)$ *is*

$$\frac{(-1)^k \pi^{n/2} \alpha^{2k}}{(\frac{1}{2}n + k - 1)!\,k!\,2^{2k-1}} \{\tfrac{1}{2}\psi(\tfrac{1}{2}n + k - 1) + \tfrac{1}{2}\psi(k) + \ln 2 - \ln \alpha\}.$$

Proof. Since $\ln r = \mathrm{Lim}_{\mu \to 0} (1 - r^{-\mu})/\mu$ the Fourier transform of $\ln(1/r)$ is

$$\underset{\mu \to 0}{\mathrm{Lim}} \frac{1}{\mu} \left\{ \frac{(-\frac{1}{2}\mu + \frac{1}{2}n - 1)!}{(\frac{1}{2}\mu - 1)!} 2^{n-\mu} \pi^{n/2} \alpha^{\mu - n} - (2\pi)^n \delta(\boldsymbol{\alpha}) \right\}$$

by Theorems 7.16 (iv), 7.31 and 7.32. Rewriting this as

$$\underset{\mu \to 0}{\mathrm{Lim}} \left[\frac{(-\frac{1}{2}\mu + \frac{1}{2}n - 1)!}{(\frac{1}{2}\mu)!} 2^{n-\mu-1} \pi^{n/2} \left\{ \alpha^{\mu-n} - \frac{2\pi^{n/2} \delta(\boldsymbol{\alpha})}{(\frac{1}{2}n - 1)!\,\mu} \right\} + \frac{(2\pi)^n}{\mu} \left\{ \frac{(-\frac{1}{2}\mu + \frac{1}{2}n - 1)!\,2^{-\mu}}{(\frac{1}{2}n - 1)!\,(\frac{1}{2}\mu)!} - 1 \right\} \right]$$

we see that the first result stated follows from Definition 7.16.

By the Fourier inversion theorem

$$\frac{1}{(2\pi)^n}\int_{-\infty}^{\infty}\left[(\tfrac{1}{2}n-1)!2^{n-1}\pi^{n/2}\alpha^{-n}-(2\pi)^n\{\tfrac{1}{2}\psi(\tfrac{1}{2}n-1)\right.$$
$$\left.+\tfrac{1}{2}\psi(0)+\ln 2\}\delta(\boldsymbol{\alpha})\right]e^{i\boldsymbol{\alpha}.\boldsymbol{x}}\,d\boldsymbol{\alpha}=\ln(1/r)$$

which, on a change of notation, leads to

$$\int_{-\infty}^{\infty} r^{-n}e^{-i\boldsymbol{\alpha}.\boldsymbol{x}}\,d\boldsymbol{x}=\frac{2\pi^{n/2}}{(\tfrac{1}{2}n-1)!}\{\tfrac{1}{2}\psi(\tfrac{1}{2}n-1)+\tfrac{1}{2}\psi(0)+\ln 2-\ln\alpha\}$$

so that the second part of the theorem is verified for $k=0$.

Suppose now that it holds for some k. Then, by (22),

$$\int_{-\infty}^{\infty}\left\{(n+2k)(2k+2)r^{-n-2k-2}\right.$$
$$\left.-\frac{(n+4k+2)\pi^{n/2}(\partial_1^2+\dots+\partial_n^2)^{k+1}\delta(\boldsymbol{x})}{(\tfrac{1}{2}n+k)!(k+1)!2^{2k+1}}\right\}e^{-i\boldsymbol{\alpha}.\boldsymbol{x}}\,d\boldsymbol{x}$$
$$=\frac{(-1)^{k+1}\pi^{n/2}\alpha^{2k+2}}{(\tfrac{1}{2}n+k-1)!k!2^{2k-1}}\{\tfrac{1}{2}\psi(\tfrac{1}{2}n+k-1)+\tfrac{1}{2}\psi(k)+\ln 2-\ln\alpha\}$$

on account of Theorem 7.14 (iii). Since

$$\frac{n+4k+2}{(\tfrac{1}{2}n+k)(k+1)}=\frac{2}{\tfrac{1}{2}n+k}+\frac{2}{k+1}$$

and $\psi(k)+1/(k+1)=\psi(k+1)$ we find that the Fourier transform of $r^{-n-2k-2}$ is the same as that for r^{-n-2k} with $k+1$ for k. Since it is known to be true for $k=0$ induction ensures that it is universally valid. The proof is complete.□

Products such as $\alpha^{2m}\ln\alpha$ are well defined because α^{2m} is fairly good. A product such as $r^{\beta}\ln r$ has not so far been defined and we now turn our attention to this matter. The procedure is similar to that of §4.6.

Definition 7.17. *When* $\beta\neq -n-2m\,(m=0,1,\dots)$ *the generalised function* $r^{\beta}\ln r$ *is defined by*

$$r^{\beta}\ln r=\partial r^{\beta}/\partial\beta.$$

Definition 7.18. *The generalised function* $r^{-n-2k}\ln r$ $(k=0,1,\dots)$ *is defined by*

$$r^{-n-2k}\ln r = \operatorname*{Lim}_{\mu\to 0}\left\{r^{\mu-n-2k}\ln r + \frac{\pi^{n/2}(\partial_1^2+\dots+\partial_n^2)^k\delta(\boldsymbol{x})}{(\frac{1}{2}n+k-1)!k!2^{2k-1}\mu^2}\right\}.$$

Since

$$r^{\mu-n}\ln r = \frac{\partial}{\partial\mu}r^{\mu-n} = \frac{\partial}{\partial\mu}\frac{(\partial_1^2+\dots+\partial_n^2)r^{\mu-n+2}}{(\mu-n+2)\mu}$$

and

$$r^{\mu-n+2} = r^{2-n} + \mu r^{2-n}\ln r + \tfrac{1}{2}\mu^2 r^{2-n}\ln^2 r + \tfrac{1}{6}\mu^3 r^{2-n}\ln^3 r + O(\mu^4)$$

we can say when $n=2$ that

$$\begin{aligned} r^{\mu-n}\ln r &= \frac{\partial}{\partial\mu}(\partial_1^2+\partial_2^2)\left\{\frac{\ln r}{\mu} + \tfrac{1}{6}\mu\ln^3 r\right\} + O(\mu) \\ &= -(2\pi/\mu^2)\delta(\boldsymbol{x}) + \tfrac{1}{6}(\partial_1^2+\partial_2^2)\ln^3 r + O(\mu) \end{aligned}$$

from (18). Thus it can be inferred from Definition 7.18 that

$$r^{-2}\ln r = \tfrac{1}{6}(\partial_1^2+\partial_2^2)\ln^3 r \quad (n=2). \tag{31}$$

Similarly

$$r^{-n}\ln r = \frac{1}{2-n}\left\{(\partial_1^2+\dots+\partial_n^2)\tfrac{1}{2}r^{2-n}\ln^2 r - r^{-n} + \frac{2\pi^{n/2}\delta(\boldsymbol{x})}{(\frac{1}{2}n-1)!(2-n)}\right\} \quad (n\neq 2). \tag{32}$$

With regard to derivatives we note from (19) that

$$\begin{aligned} (\partial_1^2+\dots+\partial_n^2)r^\beta\ln r &= (\partial_1^2+\dots+\partial_n^2)\frac{\partial r^\beta}{\partial\beta} = \frac{\partial}{\partial\beta}(\partial_1^2+\dots+\partial_n^2)r^\beta \\ &= \frac{\partial}{\partial\beta}\beta(\beta+n-2)r^{\beta-2} \\ &= (2\beta+n-2)r^{\beta-2} + \beta(\beta+n-2)r^{\beta-2}\ln r \end{aligned} \tag{33}$$

provided that $\beta\neq 2-n$ and $\beta\neq -n-2k$. Taking advantage of (33) in Definition 7.18 we have

$$(\partial_1^2 + \ldots + \partial_n^2) r^{-n-2k} \ln r = \operatorname*{Lim}_{\mu \to 0} \Big\{ (2\mu - n - 4k - 2) r^{\mu - n - 2k - 2}$$

$$+ (\mu - n - 2k)(\mu - 2k - 2) r^{\mu - n - 2k - 2} \ln r$$

$$+ \frac{\pi^{n/2} (\partial_1^2 + \ldots + \partial_n^2)^{k+1}}{(\tfrac{1}{2}n + k - 1)!\, k!\, 2^{2k-1} \mu^2} \delta(\boldsymbol{x}) \Big\}.$$

Replace $r^{\mu - n - 2k - 2}$ by

$$r^{\mu - n - 2k - 2} - \frac{\pi^{n/2} (\partial_1^2 + \ldots + \partial_n^2)^{k+1} \delta(\boldsymbol{x})}{(\tfrac{1}{2}n + k)!\,(k+1)!\, 2^{2k+1} \mu}$$

and replace $r^{\mu - n - 2k - 2} \ln r$ by

$$r^{\mu - n - 2k - 2} \ln r + \frac{\pi^{n/2} (\partial_1^2 + \ldots + \partial_n^2)^{k+1} \delta(\boldsymbol{x})}{(\tfrac{1}{2}n + k)!\,(k+1)!\, 2^{2k+1} \mu^2}$$

and add in the necessary correction terms. As a consequence of Definitions 7.16 and 7.18 we obtain

$$(\partial_1^2 + \ldots + \partial_n^2) r^{-n-2k} \ln r$$

$$= (n + 2k)(2k + 2) r^{-n-2k-2} \ln r$$

$$- (n + 4k + 2) r^{-n-2k-2} + \frac{\pi^{n/2} (\partial_1^2 + \ldots + \partial_n^2)^{k+1} \delta(\boldsymbol{x})}{(\tfrac{1}{2}n + k)!\,(k+1)!\, 2^{2k+1}}. \qquad (34)$$

Theorem 7.34. *The Fourier transform of $r^\beta \ln r$ when $\beta \neq 2k$ and $\beta \neq -n - 2k (k = 0, 1, \ldots)$ is*

$$\frac{(\tfrac{1}{2}\beta + \tfrac{1}{2}n - 1)!}{(-\tfrac{1}{2}\beta - 1)!} 2^{\beta + n} \pi^{n/2} \alpha^{-\beta - n} \{ \tfrac{1}{2}\psi(\tfrac{1}{2}\beta + \tfrac{1}{2}n - 1)$$

$$+ \tfrac{1}{2}\psi(-\tfrac{1}{2}\beta - 1) - \ln \tfrac{1}{2}\alpha \}$$

and of $r^{2k} \ln r$ *is*

$$(\tfrac{1}{2}n + k - 1)!\, k!\, 2^{n + 2k - 1} \pi^{n/2} (-1)^{k+1} \alpha^{-n-2k}$$

$$+ (-1)^k (2\pi)^n \{ \tfrac{1}{2}\psi(\tfrac{1}{2}n + k - 1) + \tfrac{1}{2}\psi(k)$$

$$+ \ln 2 \} (\partial_1^2 + \ldots + \partial_n^2)^k \delta(\boldsymbol{\alpha}).$$

Proof. The first part is derived by taking a derivative with respect to β of the transform in Theorem 7.31. The second part can be inferred from Theorem 7.33 and the proof is concluded. □

For the Fourier transform of $r^{-n-2k} \ln r$ we have

Theorem 7.35. *The Fourier transform of $r^{-n-2k}\ln r$ is*

$$\frac{(-1)^k\pi^{n/2}\alpha^{2k}}{(\frac{1}{2}n+k-1)!k!2^{2k}}\Big[\{\tfrac{1}{2}\psi(\tfrac{1}{2}n+k-1)+\tfrac{1}{2}\psi(k)-\ln\tfrac{1}{2}\alpha\}^2$$
$$-\tfrac{1}{4}\{\psi'(\tfrac{1}{2}n+k-1)+\psi'(k)\}+\frac{\pi^2}{12}\Big].$$

Proof. From Definition 7.18, Theorems 7.32 and 7.31 the Fourier transform is

$$\lim_{\mu\to 0}\frac{\partial}{\partial\mu}\left\{\frac{(\frac{1}{2}\mu-k-1)!2^{\mu-2k}}{(-\frac{1}{2}\mu+\frac{1}{2}n+k-1)!}\pi^{n/2}\alpha^{2k-\mu}-\frac{\pi^{n/2}(-\alpha^2)^k}{(\frac{1}{2}n+k-1)!k!2^{2k-1}\mu}\right\}$$
$$=\lim_{\mu\to 0}\tfrac{1}{2}\frac{\partial^2}{\partial\mu^2}\frac{2^{\mu-2k}\pi^{n/2}(-1)^k\alpha^{2k-\mu}}{(-\frac{1}{2}\mu+\frac{1}{2}n+k-1)!(k-\frac{1}{2}\mu)!}\frac{\mu\pi}{\sin\frac{1}{2}\mu\pi}$$

which, on evaluation, provides the formula of the theorem. □

By invoking the Fourier inversion theorem we derive

Corollary 7.35. *The Fourier transform of*

$$\{\tfrac{1}{2}\psi(\tfrac{1}{2}n+k-1)+\tfrac{1}{2}\psi(k)-\ln\tfrac{1}{2}r\}r^{2k}$$

is

$$(\tfrac{1}{2}n+k-1)!k!(-1)^k\pi^{n/2}2^{n+2k}\alpha^{-n-2k}\ln\alpha$$
$$-(-1)^k(2\pi)^n\{\tfrac{1}{12}\pi^2-\tfrac{1}{4}\psi'(\tfrac{1}{2}n+k-1)-\tfrac{1}{4}\psi'(k)\}(\partial_1^2+\ldots+\partial_n^2)^k\delta(\boldsymbol{\alpha}).$$

There is no difficulty in seeing how definitions along the lines of this section could be provided for generalised functions such as $r^\beta(\ln r)^2$ and the like but we shall not go into detail here.

Exercises

36. If $\beta>-n$ prove that the Fourier transform of r^β is

$$\lim_{\mu\to+0}\frac{(2\pi)^{n/2}}{\alpha^{(n/2)-1}}\int_0^\infty e^{-\mu r}r^{\beta+n/2}J_{(n-2)/2}(\alpha r)dr$$
$$=\lim_{\mu\to+0}\frac{(\beta+n-1)!2\pi^{n/2}}{(\frac{1}{2}n-1)!(\mu^2+\alpha^2)^{(\beta+n)/2}}\,{}_2F_1\left(\tfrac{1}{2}\beta+\tfrac{1}{2}n,-\tfrac{1}{2}\beta-\tfrac{1}{2};\tfrac{1}{2}n;\frac{\alpha^2}{\mu^2+\alpha^2}\right).$$

Deduce that

$$\lim_{\mu\to+0}\frac{{}_2F_1\left(\frac{1}{2}\beta+\frac{1}{2}n,-\frac{1}{2}\beta-\frac{1}{2};\frac{1}{2}n;\frac{\alpha^2}{\mu^2+\alpha^2}\right)}{(\mu^2+\alpha^2)^{(\beta+n)/2}}=\frac{(\frac{1}{2}n-1)!\pi^{1/2}\alpha^{-\beta-n}}{(-\frac{1}{2}\beta-1)!(\frac{1}{2}n+\frac{1}{2}\beta-\frac{1}{2})!}.$$

This should be compared with the statement

$$ {}_2F_1(a,b;c;1) = \frac{(c-1)!(c-a-b-1)!}{(c-a-1)!(c-b-1)!} $$

when Re $(c-a-b)>0$.

37. Assume that the Fourier transform of r^β ($\beta \neq 2k$ or $-n-2k$) is $a\alpha^{-\beta-n}$ where a is a constant. By calculating $\int_{-\infty}^{\infty} r^\beta e^{-r^2} dx$ and using Theorem 7.13, determine a and check that your result is consistent with Theorem 7.31.

38. Prove that

$$ \operatorname*{Lim}_{\mu\to+0} \mu\alpha^{\mu-n} = \frac{2\pi^{n/2}\delta(\alpha)}{(\frac{1}{2}n-1)!}. $$

39. If β and ν are real show that $\operatorname{Lim}_{\nu\to 0} r^{\beta+i\nu} = r^\beta$ when $\beta \neq -n-2k$.

40. Repeat Exercise 39 for $r^{\beta+i\nu}\ln r$ with β and ν real.

41. Prove that, when Re $\beta > 2-n$,

$$ \partial_1 r^\beta = \beta x_1 r^{\beta-2}. $$

Deduce, by Fourier transforms, that this is generally true if $\beta \neq 2-n-2k$. What equation is valid when $\beta = 2-n-2k$?

42. Prove, stating any restrictions on β, that

$$ \partial_1(r^\beta \ln r) = x_1 r^{\beta-2} + \beta x_1 r^{\beta-2}\ln r. $$

43. Show that (i) $r^2(r^\beta \ln r) = r^{\beta+2}\ln r$, (ii) $r^2.r^{-n-2k}\ln r = r^{2-n-2k}\ln r$.

44. If n is even and $2k-n \geq 0$ prove that

$$ (\partial_1^2+\dots+\partial_n^2)^k(r^{2k-n}\ln r) = (k-1)!(k-\tfrac{1}{2}n)!(-1)^{(n/2)-1}2^{2k-1}\pi^{n/2}\delta(\boldsymbol{x}). $$

Compare Exercise 32.

45. Obtain the Fourier transform of $r^{-2}\ln r$ when $n=2$ from (31) by using the result.

$$ \tfrac{1}{2}(\ln r)^2 = \operatorname*{Lim}_{\mu\to 0}(r^{-\mu}-1+\mu\ln r)/\mu^2. $$

7.8. Multiplication

The operation of multiplication for generalised functions of several variables is analogous to that for those of one variable.

Definition 7.19. *If $g_1 \in \boldsymbol{K}_1, g_2 \in \boldsymbol{K}_1$ and $g_1 g_2 \in \boldsymbol{K}_1$ the generalised function determined by $g_1 g_2$ will be called the product of g_1 and g_2 and written $g_1 g_2$ or $g_2 g_1$. If $g_1 g_2$ and $(\partial_1 g_1) g_2$ exist then $g_1 \partial_1 g_2$ is defined by*

$$ g_1 \partial_1 g_2 = \partial_1(g_1 g_2) - g_2(\partial_1 g_1). $$

As in Theorem 6.1 we can prove without difficulty that $g_1 g_2$ exists when $g_1 \in \boldsymbol{K}_p (p \geq 1)$ and $g_2 \in \boldsymbol{K}_q \{q = p/(p-1)\}$. The theorem analogous to Theorem 6.3 is

Theorem 7.36. *If $g_1 = \partial_1^{p_1} \dots \partial_n^{p_n} g$ and if $g \partial_1^{r_1} \dots \partial_n^{r_n} g_2$ exists for all $r_1, \dots, r_n$ satisfying $0 \leq r_1 \leq p_1, \dots, 0 \leq r_n \leq p_n$ then $g_1 g_2$ exists. In particular, if $g \in \boldsymbol{K}_1$ and is continuous and $\partial_1^{r_1} \dots \partial_n^{r_n} g_2 \in \boldsymbol{K}_1$ for the same range of $r_1, \dots, r_n$ then $g_1 g_2$ exists.*

The proof of this theorem runs parallel to that of Theorem 6.3. One particular consequence is that the product of a fairly good function and any generalised function is defined and agrees with Definition 7.9.

As in Definition 6.3 we may define $g(\boldsymbol{x})\delta(\boldsymbol{x})$ when g is continuous in a neighbourhood surrounding the origin by

$$g(\boldsymbol{x})\delta(\boldsymbol{x}) = g(\boldsymbol{0})\delta(\boldsymbol{x}).$$

By a proof similar to that of Theorem 6.4 we have

Theorem 7.37. *If $g_1 = \partial_1^{p_1} \dots \partial_n^{p_n} f_1$ where $f_1 \in \boldsymbol{K}_p (p \geq 1)$, if*

$$\partial_1^{r_1} \dots \partial_n^{r_n} g_2 \in \boldsymbol{K}_q \quad \{q = p/(p-1)\}$$

and $\partial_1^{r_1} \dots \partial_n^{r_n} g_3 \in \boldsymbol{K}_\infty$ for any $r_1, \dots, r_n$ satisfying $0 \leq r_1 \leq p_1, \dots, 0 \leq r_n \leq p_n$ then

$$g_1(g_2 g_3) = (g_1 g_2) g_3 .$$

Exercises

46. If $g(\boldsymbol{x})$ is continuous in a neighbourhood of the origin prove that $\int_T g(\boldsymbol{x})\delta(\boldsymbol{x})\mathrm{d}\boldsymbol{x}$ is 0 or $g(\boldsymbol{0})$ according as the origin is outside or inside T.

47. Can you find a result corresponding to Exercise 17 of Chapter 6 for several variables?

48. If $g_1 g_2$ and $g_3 g_4$ exist show that the product of $g_1(\boldsymbol{x})g_3(\boldsymbol{y})$ and $g_2(\boldsymbol{x})g_4(\boldsymbol{y})$ is $g_1(\boldsymbol{x})g_2(\boldsymbol{x})g_3(\boldsymbol{y})g_4(\boldsymbol{y})$.

Consider now the convolution product in several variables. Proceeding as in Definition 6.5 and Theorems 6.6–6.9 we may show

Theorem 7.38. *If Ψ is the Fourier transform of a fairly good function the sequence $\{\int_{-\infty}^{\infty} \Psi(\boldsymbol{\alpha})\Gamma_m(\boldsymbol{x}-\boldsymbol{\alpha})\mathrm{d}\boldsymbol{\alpha}\}$ defines*

$$\Psi * G = \int_{-\infty}^{\infty} \Psi(\boldsymbol{x}-\boldsymbol{\alpha})G(\boldsymbol{\alpha})\mathrm{d}\boldsymbol{\alpha} = \int_{-\infty}^{\infty} \Psi(\boldsymbol{\alpha})G(\boldsymbol{x}-\boldsymbol{\alpha})\mathrm{d}\boldsymbol{\alpha}.$$

Also $\Psi * G = (2\pi)^n \int_{-\infty}^{\infty} g(\mathbf{t})\psi(\mathbf{t})\mathrm{e}^{-i\mathbf{t}.\mathbf{x}}\,\mathrm{d}\mathbf{t}$ *and*

$$\partial_1(\Psi * G) = (\partial_1 \Psi) * G = \Psi * \partial_1 G.$$

The Fourier transform of $\Psi * G$ *is* $(2\pi)^{2n} g(-\boldsymbol{\alpha})\psi(-\boldsymbol{\alpha})$.

It follows, as in the case of the single variable, that

$$\delta * G = G \tag{35}$$

and

$$\partial_1(\delta * G) = (\partial_1 \delta) * G = \partial_1 G. \tag{36}$$

By starting from the fairly good function $(1 + r^2)^{-\mu}$ and employing the formula

$$\int_0^\infty \frac{x^{\nu+1} J_\nu(ax)}{(x^2+1)^\mu}\mathrm{d}x$$

$$= \frac{a^{\mu-1} K_{\nu-\mu+1}(a)}{(\mu-1)!2^{\mu-1}} \qquad (a > 0,\ -1 < \operatorname{Re}\nu < 2\operatorname{Re}\mu - \tfrac{1}{2}) \tag{37}$$

which is due to Sonine† we see from Theorem 7.30 that

$$2^{(n/2)-\mu+1}\pi^{n/2} K_{\mu-n/2}(\alpha)/(\mu-1)!$$

with $\alpha = |\boldsymbol{\alpha}|$ is the Fourier transform of $(1 + r^2)^{-\mu}$ when $\operatorname{Re}\mu > \frac{1}{4}(n-1)$. Hence

$$\int_{-\infty}^{\infty} |\mathbf{x}-\boldsymbol{\alpha}|^{\mu-n/2} K_{\mu-n/2}(|\mathbf{x}-\boldsymbol{\alpha}|)G(\boldsymbol{\alpha})\mathrm{d}\boldsymbol{\alpha}$$

$$= (\mu-1)!2^{(n/2)+\mu-1}\pi^{n/2}\int_{-\infty}^{\infty}\frac{g(\mathbf{t})\mathrm{e}^{-i\mathbf{t}.\mathbf{x}}}{(1+t^2)^\mu}\mathrm{d}\mathbf{t} \tag{38}$$

when $\operatorname{Re}\mu > \frac{1}{4}(n-1)$. In particular

$$\int_{-\infty}^{\infty} |\mathbf{x}-\boldsymbol{\alpha}|^{\mu-n/2} K_{\mu-n/2}(|\mathbf{x}-\boldsymbol{\alpha}|)|\boldsymbol{\alpha}|^{\nu-n/2} K_{\nu-n/2}(|\boldsymbol{\alpha}|)\mathrm{d}\boldsymbol{\alpha}$$

$$= (\mu-1)!(\nu-1)!2^{\mu+\nu-2}\int_{-\infty}^{\infty}\frac{\mathrm{e}^{-i\mathbf{t}.\mathbf{x}}}{(1+t^2)^{\mu+\nu}}\mathrm{d}\mathbf{t}$$

$$= \frac{(\mu-1)!(\nu-1)!}{(\mu+\nu-1)!}2^{(n/2)-1}\pi^{n/2}|\mathbf{x}|^{\mu+\nu-n/2}K_{\mu+\nu-n/2}(|\mathbf{x}|) \tag{39}$$

† See, for example, G. N. Watson, *Theory of Bessel Functions*, 2nd edition, Cambridge University Press (1944), p. 434.

from Theorem 7.30 and (38) provided that Reμ and Reν are both greater than $\frac{1}{4}(n-1)$.

Exercises

49. Prove that

$$\int_{-\infty}^{\infty} e^{-|x-\alpha|}G(\alpha)d\alpha = (\tfrac{1}{2}n-\tfrac{1}{2})!\pi^{(n/2)-1/2}2^n\int_{-\infty}^{\infty}\frac{g(t)e^{-it.x}}{(1+t^2)^{(n/2)+1/2}}dt.$$

50. Show that $\int_{-\infty}^{\infty}\delta(\alpha)\partial_1\delta(x-\alpha)d\alpha = \partial_1\delta(x)$.

51. Prove that

$$(\partial_1^2+\dots+\partial_n^2-1)^n\{|x|^{n/2}K_{n/2}(|x|)\} = (n-1)!(-1)^n2^{n/2}\pi^{1-n/2}\delta(x).$$

Definition 7.20. *The generalised function $g \in L'_p$ if, and only if, $g=\sum_m^M \partial_1^{m_1}\dots\partial_n^{m_n}f_m$ where $M, m_1, \dots, m_n$ are finite and $f_m \in L_p$.*

We can show that, if $g \in L'_p$, then $g \in L'_{p_1}$ for $p_1 \geq p \geq 1$. The proof is similar to that of Theorem 6.11 with $e^{-|x|}$ replaced by $|x|^{n/2}K_{n/2}(|x|)$ and the result of Exercise 51 being noted.

The spaces L'_p and K'_q are connected in a similar way to L'_p and K_q. However, for our purposes it will usually be sufficient to know that if $g \in L'_2$ then $G \in K_2$ and conversely. This follows from the analogue of Theorem 5.11 which may be proved either directly or as follows.

Theorem 7.39. *If $f \in L_2$, then $F \in L_2$ and*

$$\int_{-\infty}^{\infty}|f|^2dx = \frac{1}{(2\pi)^n}\int_{-\infty}^{\infty}|F|^2d\alpha.$$

If, in addition, $f_1 \in L_2$ then

$$\int_{-\infty}^{\infty}f(x)f_1(x)dx = \frac{1}{(2\pi)^n}\int_{-\infty}^{\infty}F(\alpha)\bar{F}_1(\alpha)d\alpha.$$

Proof. The proof will be given when f is a function of only two variables, the extension to general n then being obvious. Since $f(x_1,x_2) \in L_2$, it follows from Fubini's theorem that $f \in L_2$ as a function of x_2 for almost all x_1. Hence, from Theorem 5.11,

$$\int_{-\infty}^{\infty}|f(x_1,x_2)|^2dx_2 = \frac{1}{2\pi}\int_{-\infty}^{\infty}|\hat{f}(x_1,\alpha_2)|^2d\alpha_2$$

for almost all x_1 where $\hat{f}(x_1, \alpha_2) = \int_{-\infty}^{\infty} f(x_1, x_2) e^{-i\alpha_2 x_2} dx_2$. Hence

$$\int_{-\infty}^{\infty}\int_{-\infty}^{\infty} |f(x_1,x_2)|^2 dx_2 dx_1 = \frac{1}{2\pi}\int_{-\infty}^{\infty}\int_{-\infty}^{\infty} |\hat{f}(x_1,\alpha_2)|^2 d\alpha_2 dx_1$$

and a further application of Fubini's theorem shows that $\hat{f}(x_1,\alpha_2) \in L_2$ as a function of x_1 for almost all α_2. Therefore Theorem 5.11 gives

$$\int_{-\infty}^{\infty} |\hat{f}(x_1,\alpha_2)|^2 dx_1 = \frac{1}{2\pi}\int_{-\infty}^{\infty} |\hat{\hat{f}}(\alpha_1,\alpha_2)|^2 d\alpha_1$$

where

$$\hat{\hat{f}}(\alpha_1,\alpha_2) = \int_{-\infty}^{\infty} \hat{f}(x_1,\alpha_2) e^{-i\alpha_1 x_1} dx_1 = F(\alpha_1,\alpha_2)$$

and the first half of the theorem is proved. The second half follows as in the proof of Theorem 5.12. □

There is now no difficulty in confirming that Theorems 6.12–6.17 are valid for several variables. The proofs need little modification other than those to be expected naturally. For instance, in Theorem 6.14, the interval $|x| \le n^{1/4}$ is replaced by the sphere $r \le n^{1/4}$. Similarly (6.28) is correct in the form

$$(1 + |\xi|^2)(1 + |\eta|^2) \ge \tfrac{1}{4}\{1 - |\xi - \eta|^2\}. \tag{40}$$

In view of these remarks the corresponding theorems will be given as exercises for the reader.

Exercises

52. If $g_1 \in L_p'(p \ge 1)$ and $g_2 \in L_q'\{q = p/(p-1)\}$ show that $g_1 * g_2$ exists. If, in particular, $p = 2$ prove that the Fourier transform of $g_1 * g_2$ is $G_1 G_2$.

53. If $g_1 \in L_p'(1 < p \le 2)$, $g_2 \in L_q'$, $G_1 \in K_q$ and $G_2 \in K_p$ prove that the Fourier transform of $g_1 * g_2$ is $G_1 G_2$.

54. If, for some finite $m \ge 0$, $(1 + r^2)^{-m/2} g_1(x) \in L_\infty'$ and $(1 + r^2)^{m/2} g_2(x) \in L_1'$ prove that $g_1 * g_2$ exists.

55. If $g_1 = \partial_1^{m_1} \ldots \partial_n^{m_n} f_1$ where $(1 + r^2)^{m/2} f_1 \in L_1$, if $g_2 = \partial_1^{l_1} \ldots \partial_n^{l_n} f_2$ where $(1 + r^2)^{-m/2} f_2$ is bounded and tends to zero as $r \to \infty$, if $\partial_1^{k_1} \ldots \partial_n^{k_n} G_1 \in K_\infty$ and $G_2 = \partial_1^{k_1} \ldots \partial_n^{k_n} F_{22}$ where $F_{22} \in K_1$ prove that the Fourier transform of $g_1 * g_2$ is $G_1 G_2$.

56. If, for some finite $m \ge 0$, $(1 + r^2)^{-m/2} g_1(x) \in L_\infty'$, $(1 + r^2)^{m/2} g_2(x) \in L_1'$ and $(1 + r^2)^{m/2} g_3(x) \in L_1'$ prove that $(g_1 * g_2) * g_3 = g_1 * (g_2 * g_3)$.

57. If $\mathrm{Re}\,\beta < -\frac{1}{2}n$, $\mathrm{Re}\,\gamma < -\frac{1}{2}n$, if $\beta + n$ and $\gamma + n$ are not non-positive even integers prove that

$$\int_{-\infty}^{\infty} |t|^{\beta}|x-t|^{\gamma}\,\mathrm{d}t$$

$$= \frac{(\frac{1}{2}\beta+\frac{1}{2}n-1)!(\frac{1}{2}\gamma+\frac{1}{2}n-1)!(-\frac{1}{2}\beta-\frac{1}{2}\gamma-\frac{1}{2}n-1)!}{(-\frac{1}{2}\beta-1)!(-\frac{1}{2}\gamma-1)!(\frac{1}{2}\beta+\frac{1}{2}\gamma+n-1)!}\pi^{n/2}r^{\beta+\gamma+n}$$

when neither $\beta+\gamma+2n$ nor $-\beta-\gamma-n$ is a non-positive even integer. If $\beta+\gamma = -2m-2n$ $(m = 0, 1, \dots)$ prove that

$$\int_{-\infty}^{\infty} |t|^{\beta}|x-t|^{\gamma}\,\mathrm{d}t$$

$$= \frac{(\frac{1}{2}\beta+\frac{1}{2}n-1)!(\frac{1}{2}\gamma+\frac{1}{2}n-1)!}{(-\frac{1}{2}\beta-1)!(-\frac{1}{2}\gamma-1)!}2^{\beta+\gamma+2n}\pi^{n}(-1)^{n}(\partial_1^2+\dots+\partial_n^2)^{m}\delta(x).$$

Example 11. It follows from Theorem 7.31 that

$$(1-\partial_1^2-\dots-\partial_n^2)^{n}\int_{-\infty}^{\infty}\frac{\alpha^{-2}\mathrm{e}^{\mathrm{i}\alpha.x}}{(1+\alpha^2)^{n}}\mathrm{d}\alpha = \int_{-\infty}^{\infty}\alpha^{-2}\mathrm{e}^{\mathrm{i}\alpha.x}\,\mathrm{d}\alpha$$

$$= (\tfrac{1}{2}n-2)!\,2^{n-2}\pi^{n/2}r^{2-n} \quad (41)$$

when $n \neq 2$. The modulus of the integral is bounded by $\int_{-\infty}^{\infty}\alpha^{-2}(1+\alpha^2)^{-n}\mathrm{d}\alpha$ and therefore $r^{2-n}\in L'_{\infty}$. Hence the convolution of r^{2-n} and any generalised function in L'_1 can be formed. Consequently, Example 9 shows that, if $g\in L'_1$,

$$(\partial_1^2+\dots+\partial_n^2)\int_{-\infty}^{\infty} g(t)|x-t|^{2-n}\mathrm{d}t = -\frac{4\pi^{n/2}g(x)}{(\frac{1}{2}n-2)!} \quad (n\neq 2). \quad (42)$$

More generally, the integral on the left-hand side of (41) is obviously bounded for any finite $|x|$ and behaves like a constant multiple of r^{2-n} as $r\to\infty$ (§9.4). Consequently the integral belongs to $L_p\{p>n/(n-2)\}$ and so $r^{2-n}\in L'_p$. Therefore (42) is also valid for $g\in L'_q(1\leq q<\frac{1}{2}n)$.

Let T be the interior of the closed surface S whose equation is $f(x)=0\,(f>0$ in $T)$ with grad $f\neq 0$ on S. Suppose $g = g_1\mathrm{H}(f)$ where

$$(\partial_1^2+\dots+\partial_n^2)g_1 = 0$$

in $T\cup S$. Let it be permissible to regard $g_1\mathrm{H}$ as a product (e.g. g_1 continuous with continuous first derivatives in the neighbourhood of S), and assume that (42) holds. Then

$$-\frac{4\pi^{n/2}g_1\mathrm{H}(f)}{(\frac{1}{2}n-2)!}=\int_{-\infty}^{\infty}|\boldsymbol{x}-\boldsymbol{t}|^{2-n}(\partial_1^2+\ldots+\partial_n^2)g\,\mathrm{d}\boldsymbol{t}$$
$$=\int_S\left\{g_1(\boldsymbol{t})\frac{\partial}{\partial v}|\boldsymbol{x}-\boldsymbol{t}|^{2-n}-|\boldsymbol{x}-\boldsymbol{t}|^{2-n}\frac{\partial g_1}{\partial v}\right\}\mathrm{d}S \quad (43)$$

where $\partial/\partial v$ is a derivative along the outward normal and $\boldsymbol{t}$ is a point of S (cf. Example 3 and (8)). In this way is obtained a representation of a solution of Laplace's equation in terms of sources with density $-\partial g_1/\partial v$ and doublets with density g_1.

Exercises

58. If $n \neq 2$ and $(1+r^2)^{1-n/2}g \in \boldsymbol{L}'_1$ prove that (42) is valid.

59. If $n=2$ prove that

$$(\partial_1^2+\partial_2^2)\int_{-\infty}^{\infty}g(\boldsymbol{t})\ln|\boldsymbol{x}-\boldsymbol{t}|\,\mathrm{d}\boldsymbol{t}=2\pi g(\boldsymbol{x})$$

if $(1+r^2)^{1+a}g \in \boldsymbol{L}'_\infty$ $(a>0)$. Show that this is also true if $g\ln(1+r^2) \in \boldsymbol{L}'_1$.

7.9. Integration with respect to a parameter

Let $g(\mu, \boldsymbol{x})$ be a generalised function of $\boldsymbol{x}$ for all μ in some domain M, which may be of one or more dimensions. If

$$\int_M\left\{\int_{-\infty}^{\infty}g(\mu,\boldsymbol{x})\gamma(\boldsymbol{x})\,\mathrm{d}\boldsymbol{x}\right\}\mathrm{d}\mu$$

exists for every good γ we say that $\int_M g(\mu, \boldsymbol{x})\,\mathrm{d}\mu$ exists as a generalised function and is defined by

$$\int_{-\infty}^{\infty}\gamma(\boldsymbol{x})\left\{\int_M g(\mu,\boldsymbol{x})\,\mathrm{d}\mu\right\}\mathrm{d}\boldsymbol{x}=\int_M\left\{\int_{-\infty}^{\infty}g(\mu,\boldsymbol{x})\gamma(\boldsymbol{x})\,\mathrm{d}\boldsymbol{x}\right\}\mathrm{d}\mu.$$

In this way an integral with respect to a parameter is defined.

Theorem 7.40. *If $\int_M g(\mu, \boldsymbol{x})\,\mathrm{d}\mu$ exists then so does $\int_M \partial_1 g(\mu, \boldsymbol{x})\,\mathrm{d}\mu$ and $\partial_1 \int_M g(\mu, \boldsymbol{x})\,\mathrm{d}\mu = \int_M \partial_1 g(\mu, \boldsymbol{x})\,\mathrm{d}\mu$.*

Proof. If $\int_M\{\int_{-\infty}^{\infty} g(\mu, \boldsymbol{x})\gamma(\boldsymbol{x})\,\mathrm{d}\boldsymbol{x}\}\,\mathrm{d}\mu$ exists for every good γ then so does $\int_M\{\int_{-\infty}^{\infty} g(\mu,\boldsymbol{x})\partial\gamma(\boldsymbol{x})/\partial x_1\,\mathrm{d}\boldsymbol{x}\}\,\mathrm{d}\mu$ since $\partial\gamma/\partial x_1$ is good and this proves the first part of the theorem.

Also

$$\int_{-\infty}^{\infty} \gamma(\boldsymbol{x})\left\{\partial_1 \int_M g(\mu, \boldsymbol{x})\,\mathrm{d}\mu\right\}\mathrm{d}\boldsymbol{x} = -\int_{-\infty}^{\infty} (\partial_1 \gamma)\left\{\int_M g(\mu, \boldsymbol{x})\,\mathrm{d}\mu\right\}\mathrm{d}\boldsymbol{x}$$

$$= -\int_M \left\{\int_{-\infty}^{\infty} g(\mu, \boldsymbol{x})\partial_1 \gamma\,\mathrm{d}\boldsymbol{x}\right\}\mathrm{d}\mu$$

$$= \int_M \left\{\int_{-\infty}^{\infty} \gamma\partial_1 g(\mu, \boldsymbol{x})\,\mathrm{d}\boldsymbol{x}\right\}\mathrm{d}\mu$$

and the proof is complete. □

Theorem 7.41. *If M is a bounded domain and if, for every $\mu_0 \in M$, $\operatorname{Lim}_{\mu\to\mu_0} g(\mu, \boldsymbol{x}) = g(\mu_0, \boldsymbol{x})$ then $\int_M g(\mu, \boldsymbol{x})\,\mathrm{d}\mu$ exists.*

Proof. On account of the given condition $\int_{-\infty}^{\infty} g(\mu, \boldsymbol{x})\gamma(\boldsymbol{x})\,\mathrm{d}\boldsymbol{x}$ is a continuous function of μ on M and therefore its integral with respect to μ is finite since M is bounded and the theorem is proved. □

Theorem 7.42. *If $\int_M g(\mu, \boldsymbol{x})\,\mathrm{d}\mu$, $\int_M g_\nu(\mu, \boldsymbol{x})\,\mathrm{d}\mu$ exist and $\operatorname{Lim}_{\nu\to c} g_\nu(\mu, \boldsymbol{x}) = g(\mu, \boldsymbol{x})$ uniformly in μ then*

$$\operatorname*{Lim}_{\nu\to c} \int_M g_\nu(\mu, \boldsymbol{x})\,\mathrm{d}\mu = \int_M g(\mu, \boldsymbol{x})\,\mathrm{d}\mu$$

when M is bounded.

Proof. Under the hypotheses

$$\lim_{\nu\to c} \int_M \left[\int_{-\infty}^{\infty} \{g_\nu(\mu, \boldsymbol{x}) - g(\mu, \boldsymbol{x})\}\gamma(\boldsymbol{x})\,\mathrm{d}\boldsymbol{x}\right]\mathrm{d}\mu = 0$$

and the theorem follows immediately from the definition. □

Example 12. Let $\boldsymbol{\omega}$ be a unit vector from the origin to the unit sphere. Then, by Definition 8.3, $|\boldsymbol{\omega}.\boldsymbol{x}|^\nu$ is a generalised function and

$$\lim_{\boldsymbol{\omega}\to\boldsymbol{\omega}_0} |\boldsymbol{\omega}.\boldsymbol{x}|^\nu = |\boldsymbol{\omega}_0.\boldsymbol{x}|^\nu.$$

Hence $\int_\Omega |\boldsymbol{\omega}.\boldsymbol{x}|^\nu\,\mathrm{d}\Omega$, Ω being the unit sphere, exists by Theorem 7.41.

When Re $\nu > -1$,

$$\int_\Omega |\omega.x|^\nu \, d\Omega = Ar^\nu$$

where A is some constant. Putting $x_1 = 1, x_2 = x_3 = \ldots = x_n = 0$, we obtain

$$\int_0^\pi \cdots \int_0^\pi \int_0^{2\pi} |\cos\theta_1|^\nu \sin^{n-2}\theta_1 \sin^{n-3}\theta_2 \ldots \sin\theta_{n-2} \, d\theta_1 \ldots d\theta_{n-1} = A,$$

or

$$A = \frac{(\frac{1}{2}\nu - \frac{1}{2})! 2\pi^{(n-1)/2}}{(\frac{1}{2}\nu + \frac{1}{2}n - 1)!}.$$

Hence, for Re $\nu > -1$,

$$\int_\Omega |\omega.x|^\nu \, d\Omega = \frac{(\frac{1}{2}\nu - \frac{1}{2})! 2\pi^{(n/2)-1/2} r^\nu}{(\frac{1}{2}\nu + \frac{1}{2}n - 1)!}. \tag{44}$$

Apply the operator $\partial_1^2 + \ldots + \partial_n^2$ to both sides. On account of Theorems 7.40, 8.6 and (3) and (4) of Chapter 4

$$\int_\Omega |\omega.x|^{\nu-2} \, d\Omega = \frac{(\frac{1}{2}\nu - \frac{3}{2})! \pi^{(n/2)-1/2}}{(\frac{1}{2}\nu + \frac{1}{2}n - 1)!} (\partial_1^2 + \ldots + \partial_n^2) r^\nu$$

provided that $\nu - 2$ is not a negative integer. It follows from (19) that the right-hand side is the same as that of (44) with $\nu - 2$ in place of ν. By repeated applications of this process (44) is established for all complex ν except negative integral values.

If m is a non-negative integer (18) of Chapter 4 shows that

$$\lim_{\nu \to -2m} |\omega.x|^\nu = (\omega.x)^{-2m}$$

and therefore, from Theorem 7.42,

$$\int_\Omega (\omega.x)^{-2m} \, d\Omega = \frac{(-m - \frac{1}{2})! 2\pi^{(n/2)-1/2}}{(-m + \frac{1}{2}n - 1)!} r^{-2m} \quad (2m < n). \tag{45}$$

On the other hand $\lim_{\mu \to 0} \mu |x|^{\mu-1} = 2\delta(x)$ gives

$$\lim_{\nu \to -2m-1} (\nu + 2m + 1)|\omega.x|^\nu = \frac{2\delta^{(2m)}(\omega.x)}{(2m)!}.$$

Hence

$$\int_\Omega \delta^{(2m)}(\omega.x)\,d\Omega = \frac{(2m)!(-1)^m \pi^{(n/2)-1/2} 2r^{-2m-1}}{m!(-m+\frac{1}{2}n-\frac{3}{2})!} \qquad (2m+1<n)$$

$$= \frac{(m-\frac{1}{2})!(-1)^m 2^{2m+1} \pi^{(n/2)-1} r^{-2m-1}}{(-m+\frac{1}{2}n-\frac{3}{2})!}. \qquad (46)$$

If n is even and $n \neq 2$, the application of $\partial_1^2 + \ldots + \partial_n^2$ to (45) with $m = \frac{1}{2}(n-2)$ gives, on account of Definition 4.4,

$$\int_\Omega (\omega.x)^{-n}\,d\Omega = \frac{(-\frac{1}{2}n-\frac{1}{2})!}{(\frac{1}{2}n-1)!} 2\pi^{n-1/2}\delta(x)$$

from (16). The right-hand side can also be written as $(2\pi)^n(-1)^{n/2} \times \delta(x)/(n-1)!$ by using $(2z)! = z!(z-\frac{1}{2})!2^{2z}/\pi^{1/2}$ and $z!(-z)! = \pi z/\sin \pi z$. Similarly, when n is odd (46) gives

$$\int_\Omega \delta^{(n-1)}(\omega.x)\,d\Omega = (-1)^{(n/2)-1/2} 2^n \pi^{n-1} \delta(x).$$

Exercises

60. Prove that

$$\int_\Omega |\omega.x|^\nu \ln|\omega.x|\,d\Omega$$

$$= \frac{(\frac{1}{2}\nu-\frac{1}{2})!\pi^{(n/2)-1/2} r^\nu}{(\frac{1}{2}\nu+\frac{1}{2}n-1)!}\{\psi(\tfrac{1}{2}\nu-\tfrac{1}{2}) - \psi(\tfrac{1}{2}\nu+\tfrac{1}{2}n-1) + \ln r\}$$

if ν is not a negative integer.

61. Prove that, if $n = 2$, $\int_\Omega (\omega.x)^{-2}\,d\Omega = -4\pi^2\delta(x)$.

62. Show that the Fourier transform of $\int_M g(\mu, x)\,d\mu$ is $\int_M G(\mu, \alpha)\,d\mu$.

63. Show that $\int_\Omega \omega.x(\ln|\omega.x| - 1)\,d\Omega = 0$ and deduce that, if n is odd, $\int_\Omega (\omega.x)^{-n}\,d\Omega = 0$.

64. Find the value of $\int_\Omega H(\omega.x)\,d\Omega$ and show that, if n is even, $\int_\Omega \delta^{(n-1)}(\omega.x)\,d\Omega = 0$.

65. Prove that $\int_\Omega (\omega.x - i0)^{-n}\,d\Omega = \{(2\pi i)^n/(n-1)!\}\,\delta(x)$.

8

Change of variables and related topics

8.1. Rotation of axes

One significant way in which calculations in R_n differ from those in R_1 is that the axes can be chosen fairly freely. Furthermore one often wishes to calculate multiple integrals by means of spherical polars or cylindrical polars instead of Cartesians. In order to provide similar facilities for generalised functions it is necessary to see what effect a change of variable has.

We commence by examining the effect of choosing a different set of Cartesian axes with the same origin. (A change of origin without alteration of the directions of the axes is covered by Definition 7.9.) Regarding $\boldsymbol{x}$ as a column matrix we can obtain any other Cartesian set with the same origin by a linear transformation $\boldsymbol{y} = L\boldsymbol{x}$ where L is an orthogonal matrix, i.e. $L^{\mathrm{T}}L = I$ where L^{T} is the transpose of L. The determinant of L, det L, is either 1 or -1. If det $L = 1$ the new axes are derived from the old by a proper rotation; if det $L = -1$ an improper rotation, i.e. a proper rotation together with a reflection, is involved.

Definition 8.1. *If* $\{\gamma_m\}$ *is a regular sequence defining* g, *the sequence* $\{\gamma_m(L\boldsymbol{x})\}$ *is regular and defines a generalised function which is denoted by* $g(L\boldsymbol{x})$.

If $\gamma_m(\boldsymbol{x})$ is good, so is $\gamma_m(L\boldsymbol{x})$ because a change in the directions of of the axes does not affect the property of being good. Also

$$\int_{-\infty}^{\infty} \gamma_m(L\boldsymbol{x})\gamma(\boldsymbol{x})\,\mathrm{d}\boldsymbol{x} = \int_{-\infty}^{\infty} \gamma_m(\boldsymbol{y})\gamma(L^{\mathrm{T}}\boldsymbol{y})\,\mathrm{d}\boldsymbol{y}$$

since the integrals are absolutely convergent and can be evaluated by any system of Cartesian axes. The integral on the right tends to a limit as $m \to \infty$ since $\gamma(L^{\mathrm{T}}\boldsymbol{y})$ is a good function of $\boldsymbol{y}$. Thus $\gamma_m(L\boldsymbol{x})$ is regular and

$$\int_{-\infty}^{\infty} g(L\boldsymbol{x})\gamma(\boldsymbol{x})\,\mathrm{d}\boldsymbol{x} = \int_{-\infty}^{\infty} g(\boldsymbol{x})\gamma(L^{\mathrm{T}}\boldsymbol{x})\,\mathrm{d}\boldsymbol{x}. \tag{1}$$

The definition is therefore justified and (1) corresponds to the standard rule for rotation of axes in conventional integrals.

Example 1. By (1)

$$\int_{-\infty}^{\infty} \delta(L\boldsymbol{x})\gamma(\boldsymbol{x})\,\mathrm{d}\boldsymbol{x} = \int_{-\infty}^{\infty} \delta(\boldsymbol{x})\gamma(L^{\mathrm{T}}\boldsymbol{x})\,\mathrm{d}\boldsymbol{x} = \gamma(\mathbf{0}).$$

Hence

$$\delta(L\boldsymbol{x}) = \delta(\boldsymbol{x}).$$

Definition 8.2. *If $g(L\boldsymbol{x}) = g(\boldsymbol{x})$ for any orthogonal L, g is said to be invariant under a rotation, and conversely. If*

$$g(-x_1, x_2, \ldots, x_n) = g(x_1, x_2, \ldots, x_n)$$

then g is said to be symmetric about the x_1-plane, and conversely. If $g(-x_1, x_2, \ldots, x_n) = -g(x_1, x_2, \ldots, x_n), g$ is said to be anti-symmetric about the x_1-plane.

If g is invariant under a rotation it is symmetric about the x_j-plane ($j = 1, 2, \ldots, n$) because a possible choice for L is one which leaves $x_i (i \neq j)$ unaltered and changes x_j to $-x_j$.

Example 2. $\delta(\boldsymbol{x})$ is invariant under a rotation by Example 1 but $\delta(x_1)$ is not. However $\delta(x_1)$ is symmetric about the x_1-plane; $\delta'(x_1)$ is antisymmetric about the x_1-plane.

Theorem 8.1. *If g is invariant under a rotation so is its Fourier transform G. If g is symmetric (anti-symmetric) about the x_j-plane G is symmetric (anti-symmetric) about the α_j-plane.*

Proof. By (1)

$$\int_{-\infty}^{\infty} G(L\boldsymbol{\alpha})\Gamma(\boldsymbol{\alpha})\,\mathrm{d}\boldsymbol{\alpha} = \int_{-\infty}^{\infty} G(\boldsymbol{\alpha})\Gamma(L^{\mathrm{T}}\boldsymbol{\alpha})\,\mathrm{d}\boldsymbol{\alpha}$$

$$= (2\pi)^n \int_{-\infty}^{\infty} g(\boldsymbol{x})\gamma(-L^{\mathrm{T}}\boldsymbol{x})\,\mathrm{d}\boldsymbol{x}$$

from Theorem 7.13, recalling that the Fourier transform of $\gamma(L\boldsymbol{x})$ is $\Gamma(L\boldsymbol{\alpha})$. The last integral can be written

$$(2\pi)^n \int_{-\infty}^{\infty} g(L\boldsymbol{x})\gamma(-\boldsymbol{x})\mathrm{d}\boldsymbol{x} = (2\pi)^n \int_{-\infty}^{\infty} g(\boldsymbol{x})\gamma(-\boldsymbol{x})\mathrm{d}\boldsymbol{x}$$

because g is invariant under a rotation. Applying Theorem 7.13 again we see that $G(L\boldsymbol{\alpha}) = G(\boldsymbol{\alpha})$, i.e. G is invariant under a rotation.

On choosing L to be the special transformation which changes x_j to $-x_j$ and leaves the other coordinates alone we see that the same method of proof proves the second part of the theorem. The theorem is complete. □

Although, in a practical situation, it will probably be readily apparent whether or not a generalised function is invariant under a rotation it is desirable for some purposes to have available a criterion which can be checked easily. This is provided by

Theorem 8.2. *g is invariant under a rotation if, and only if,*

$$x_j\partial_k g - x_k\partial_j g = 0 \quad (j = 1, \dots, n; k = 1, \dots, n).$$

Proof. Take

$$L = \begin{pmatrix} \cos\theta & \sin\theta & 0 & 0\dots0 \\ -\sin\theta & \cos\theta & 0 & 0\dots0 \\ 0 & 0 & 1 & 0\dots0 \\ 0 & 0 & 0 & 1 \\ \vdots & \vdots & \vdots & & \ddots \\ 0 & 0 & 0 & & & 1 \end{pmatrix}$$

Then L corresponds to a simple rotation in which the axes x_1, x_2 are turned through an angle θ and the other axes are unchanged. If g is invariant under a rotation,

$$g(x_1 \cos\theta + x_2 \sin\theta, -x_1 \sin\theta + x_2 \cos\theta, x_3, \dots) = g(x_1, x_2, \dots)$$

for all real θ. Therefore a partial derivative of the left-hand side with respect to θ must vanish, i.e.

$$(-x_1 \sin\theta + x_2 \cos\theta)\partial_1 g + (-x_1 \cos\theta - x_2 \sin\theta)\partial_2 g = 0.$$

Putting $\theta = 0$ we obtain $x_2 \partial_1 g - x_1 \partial_2 g = 0$. Similarly, by choosing L as a simple rotation in which x_j and x_k are rotated through an angle θ and the other axes fixed, we prove that $x_j \partial_k g - x_k \partial_j g = 0$. It has thus been proved that the conditions stated are necessary when g is invariant.

Conversely, if $x_j \partial_k g - x_k \partial_j g = 0$, let

$$y_1 = x_1 \cos\theta + x_2 \sin\theta,$$
$$y_2 = -x_1 \sin\theta + x_2 \cos\theta.$$

Then

$$\frac{\partial}{\partial\theta} g(y_1, y_2, x_3, \ldots, x_n) = \left(y_2 \frac{\partial}{\partial y_1} - y_1 \frac{\partial}{\partial y_2} \right) g(y_1, y_2, x_3, \ldots, x_n) = 0$$

by assumption. Therefore $g(y_1, y_2, \ldots, x_n)$ is independent of θ. In other words, a simple rotation affecting only x_1 and x_2 does not change g. Similarly, a simple rotation involving x_j and x_k does not alter g. But any rotation can be built up from at most $\frac{1}{2}n$ simple rotations together with (in the improper case) the reversal of one axis. Hence g must be invariant under a rotation. The theorem is proved.□

Exercises

1. If $\boldsymbol{y} = L\boldsymbol{x}$ prove that

$$\partial_k g(\boldsymbol{y}) = \sum_{j=1}^{n} l_{jk} \frac{\partial}{\partial y_j} g(\boldsymbol{y})$$

where $L \equiv (l_{jk})$.

2. Prove that the Fourier transform of $g(L\boldsymbol{x})$ is $G(L\boldsymbol{\alpha})$.
3. If $f \in K_1$ and depends only on r prove that f is invariant under a rotation.
4. If g is invariant under a rotation prove that $(\partial_1^2 + \ldots + \partial_n^2)g$ is also invariant.
5. Show that r^β is invariant under a rotation.
6. Prove the first half of Theorem 8.1 by means of Theorem 8.2.

More generally, we can consider a linear mapping which leaves the origin unaltered (since this is covered by Definition 7.9) given by the matrix A provided that the mapping is non-singular, i.e. $\det A \neq 0$. A then possesses an inverse A^{-1}. Then

Definition 8.3. *If $\{\gamma_m\}$ is a regular sequence defining g, the sequence $\{\gamma_m(A\boldsymbol{x})\}$ is regular and defines a generalised function which is denoted by $g(A\boldsymbol{x})$. If $g(A\boldsymbol{x}) = g(\boldsymbol{x})$, g is said to be invariant under A.*

The justification is similar to that for Definition 8.1 and we note only that

$$\int_{-\infty}^{\infty} g(A\boldsymbol{x})\gamma(\boldsymbol{x})\mathrm{d}\boldsymbol{x} = \int_{-\infty}^{\infty} g(\boldsymbol{y})\gamma(A^{-1}\boldsymbol{y})\frac{\mathrm{d}\boldsymbol{y}}{|\det A|}. \tag{2}$$

In particular

$$\int_{-\infty}^{\infty} \delta(A\boldsymbol{x})\gamma(\boldsymbol{x})\mathrm{d}\boldsymbol{x} = \frac{\gamma(\mathbf{0})}{|\det A|}$$

or

$$\delta(A\boldsymbol{x}) = \delta(\boldsymbol{x})/|\det A|. \tag{3}$$

Definition 8.3*a*. *$g(\boldsymbol{x})$ is said to be homogeneous and of degree μ if, and only if, $g(a\boldsymbol{x}) = a^{\mu}g(\boldsymbol{x})$ for every real positive a.*

From either §7.4 or (3) we see that $\delta(a\boldsymbol{x}) = a^{-n}\delta(\boldsymbol{x})$ when $a > 0$. Thus $\delta(\boldsymbol{x})$ is homogeneous and of degree $-n$. Similarly $\partial_1\delta(\boldsymbol{x})$ is of degree $-n-1$. It is obvious that the sum of two generalised functions, which are both homogeneous and of degree μ, is also homogeneous and of degree μ. If the fairly good function ψ is homogeneous and of degree ν then ψg is homogeneous and of degree $\mu + \nu$ when g is homogeneous and of degree μ.

Theorem 8.3. *If $g(\boldsymbol{x})$ is homogeneous and of degree μ then $\partial_j g$ is homogeneous and of degree $\mu - 1$.*

Proof. If $g_1(\boldsymbol{x}) = \partial_j g(\boldsymbol{x})$ then, as in Exercise 1,

$$g_1(a\boldsymbol{x}) = a^{-1}\partial_j g(a\boldsymbol{x}) = a^{\mu-1}\partial_j g(\boldsymbol{x}) = a^{\mu-1}g_1(\boldsymbol{x})$$

and the proof is complete. □

Theorem 8.4. *If $g_j(\boldsymbol{x})$ is homogeneous and of degree μ_j with $\mu_j \neq \mu_k (j \neq k)$ then $\sum_{j=1}^{m} c_j g_j(\boldsymbol{x}) = 0$ implies that $c_j = 0$ $(j = 1, \ldots, m)$.*

Proof. Since the given condition implies that $\sum_{j=1}^{m} c_j a^{\mu_j} g_j(\boldsymbol{x}) = 0$ for any $a > 0$ we can, by giving m different values to a, obtain m equations from which we can eliminate all gs except one, i.e. we can reduce to $c_j g_j = 0$. Since $g_j \neq 0$ we must have $c_j = 0$ and the proof is complete. □

Theorem 8.5. *g is homogeneous and of degree μ if and only if*

$$\sum_{j=1}^{n} x_j \partial_j g(\boldsymbol{x}) = \mu g.$$

Proof. If g is homogeneous and of degree μ, a derivative of $g(a\boldsymbol{x}) = a^{\mu} g(\boldsymbol{x})$ with respect to a gives $a^{-1}\sum_{j=1}^{n} x_j \partial_j g(a\boldsymbol{x}) = \mu a^{\mu-1} g$ and the necessity of the given condition follows at once.

Conversely, if the given condition holds, the derivative of $a^{-\mu} g(a\boldsymbol{x})$ with respect to a is $a^{-\mu-1}\{\sum_{j=1}^{n} y_j \partial_j g(\boldsymbol{y}) - \mu g(\boldsymbol{y})\} = 0$ where $\boldsymbol{y} = a\boldsymbol{x}$. Hence $a^{-\mu} g(a\boldsymbol{x})$ is independent of a and therefore equal to $g(\boldsymbol{x})$ so that g is homogeneous and of degree μ. □

Exercises

7. Prove that the Fourier transform of $g(A\boldsymbol{x})$ is $G\{(A^{-1})^T \boldsymbol{\alpha}\}/|\det A|$. Use this result to verify Theorem 7.3 from Theorem 7.2.
8. If $\sum_{j=1}^{n} a_j^2 = 1$ prove that $\int_{-\infty}^{\infty} \delta(\sum_{j=1}^{n} a_j x_j)\gamma(\boldsymbol{x})\mathrm{d}\boldsymbol{x} = \int_S \gamma \mathrm{d}S$ and that
$$\int_{-\infty}^{\infty} \delta^{(m)}\left(\sum_{j=1}^{n} a_j x_j\right)\gamma(\boldsymbol{x})\mathrm{d}\boldsymbol{x} = \int_S \frac{\partial^m \gamma}{\partial v^m}\mathrm{d}S$$
where S is the hyperplane $\sum_{j=1}^{n} a_j x_j = 0$ and $\partial/\partial v$ is a normal derivative on the side of S on which $\sum_{j=1}^{n} a_j x_j < 0$.
9. Prove that
$$\int_{-\infty}^{\infty} \delta\left(\sum_{j=1}^{n} a_j x_j, \sum_{j=1}^{n} b_j x_j\right)\gamma(\boldsymbol{x})\mathrm{d}\boldsymbol{x}$$
$$= \int_{-\infty}^{\infty} \cdots \int_{-\infty}^{\infty} \frac{[\gamma]}{|a_n b_{n-1} - b_n a_{n-1}|}\mathrm{d}x_1 \ldots \mathrm{d}x_{n-2}$$
where $[\gamma]$ is γ with $\sum_{j=1}^{n} a_j x_j = 0$ and $\sum_{j=1}^{n} b_j x_j = 0$ and it is supposed that $a_n b_{n-1} \neq b_n a_{n-1}$. The integral can be regarded as one over a hypersurface of $n-2$ dimensions.
10. If $\sum_{j=1}^{n} a_j^2 \neq 0$ prove that $(\sum_{j=1}^{n} a_j x_j)\delta(\sum_{k=1}^{n} a_k x_k) = 0$ and that
$$\left(\sum_{j=1}^{n} a_j x_j\right)\delta^{(m)}\left(\sum_{k=1}^{n} a_k x_k\right) + m\delta^{(m-1)}\left(\sum_{k=1}^{n} a_k x_k\right) = 0.$$
11. Show that, if $a_1 \neq 0$, the Fourier transform of $\delta(\sum_{j=1}^{n} a_j x_j)$ can be written as
$$\frac{(2\pi)^{n-1}}{a_1}\delta\left(\alpha_2 - \frac{a_2\alpha_1}{a_1}\right)\delta\left(\alpha_3 - \frac{a_3\alpha_1}{a_1}\right)\ldots\delta\left(\alpha_n - \frac{a_n\alpha_1}{a_1}\right).$$
12. Show that $\mathrm{x}^{\beta}\mathrm{H}(\mathrm{x})$ and $|\mathrm{x}|^{\beta}$ are homogeneous and of degree β when β is not a negative integer.

13. Show that r^{β} is homogeneous and of degree $\beta(\beta \neq -n-2k)$ but that r^{-n-2k} is not homogeneous. (Hint: Exercise 35 of Chapter 7.)
14. If $g(\boldsymbol{x})$ is homogeneous and of degree μ prove that $G(\boldsymbol{\alpha})$ is homogeneous and of degree $-\mu-n$.
15. If $\psi(\boldsymbol{x})$ does not vanish anywhere show that

$$\delta^{(m)}\left(\sum_{j=1}^{m} a_j x_j \psi\right) = |\psi|^{-m-1}\delta^{(m)}\left(\sum_{j=1}^{n} a_j x_j\right).$$

16. Prove that $2\delta(x_1 - x_2)\delta(x_1 + x_2) = \delta(x_1)\delta(x_2)$.

8.2. Change of variable: one-dimensional case

When one wishes to consider general changes of variables and especially non-linear transformations the situation becomes much more complicated. One reason is that one may not want each of the new variables to range from $-\infty$ to ∞ as is required at present. Another reason is that, in general, a good function will not remain good under an arbitrary transformation. Therefore we shall start by discussing the one-dimensional case, where the problems are somewhat simpler, and discover how some of the difficulties may be resolved. The problems of several variables are deferred until the next section.

We commence by restricting the class of transformations so that we can be sure that the change of variables is straightforward.

Definition 8.4. *Let $f(x)$ be a real-valued fairly good function, with $f'(x) \neq 0$ for any x, which increases steadily from $-\infty$ to ∞ as x goes from $-\infty$ to ∞ in such a way that $|f'(x)| \geq K|x|^{\nu-1}$ $(\nu > 0)$ for sufficiently large $|x|$. Then $\{\gamma_m(f(x))\}$ is a regular sequence and defines a generalised function which will be denoted by $g\{f(x)\}$.*

To justify the definition it is necessary to show that the sequence $\{\gamma_m(f)\}$ is regular. Firstly observe that $[\gamma_m(f)]^{(n)}$ consists of a finite sum of products of $f^{(n)}, f^{(n-1)}, \ldots, f', \gamma_m', \ldots, \gamma_m^{(n)}$ and so $\gamma_m(f)$ possesses derivatives of all orders. Confirmation that $\gamma_m(f)$ is good will be obtained if we can demonstrate that $x^r[\gamma_m(f)]^{(n)}$ tends to zero as $|x| \to \infty$. Since f is fairly good a typical term in the sum will be less than $|x|^{r+M}|\gamma_m^{(k)}(f)|$ for some finite M. But, when $|f|$ is sufficiently large, as it is when $|x|$ is big enough, $|\gamma_m^{(k)}(f)| < |f|^{-s}$ where s can be taken arbitrarily large since $|f'| \geq K|x|^{\nu-1}$ and $|f| \geq \frac{1}{2}K|x|^{\nu}$ as $|x|$

increases. Hence a typical term is less than $2^s|x|^{r+M-sv}/K^s$ which tends to zero on choosing s large enough. Therefore $\gamma_m(f)$ is good.

The relation $y=f(x)$ is one-to-one since f increases steadily with x. Therefore the inverse relation $x=f^{-1}(y)$ is also one-to-one. It is now convenient to show that, if $\gamma(x)$ is a good function of x, $\gamma(f^{-1}(y))$ is a good function of y. Now $\mathrm{d}^n\gamma(f^{-1}(y))/\mathrm{d}y^n$ is a combination of products of $\mathrm{d}^n f^{-1}/\mathrm{d}y^n, \dots, \mathrm{d}f^{-1}/\mathrm{d}y, \gamma', \dots, \gamma^{(n)}$. Also $\mathrm{d}^m f^{-1}/\mathrm{d}y^m$, expressed in terms of x, is a sum of products of $f^{(m)}(x), \dots, f''(x)$, $1/f'(x)$ since $\mathrm{d}f^{-1}/\mathrm{d}y = 1/f'(x)$ which exists since $f' \neq 0$. Hence there is a finite M such that $|\mathrm{d}^m f^{-1}/\mathrm{d}y^m| = O(|x|^M)$. Therefore

$$\left| y^r \frac{\mathrm{d}^n\gamma(f^{-1}(y))}{\mathrm{d}y^n} \right| = O\{|f(x)|^r |x|^M |\gamma^{(s)}(x)|\}$$
$$\to 0$$

as $|x| \to \infty$ because f is fairly good. Since $|x| \to \infty$ as $|y| \to \infty$ it follows that $\gamma(f^{-1}(y))$ is a good function of y.

With this information at our disposal we can verify the statement of Definition 8.4. For

$$\int_{-\infty}^{\infty} \gamma_m\{f(x)\}\gamma(x)\,\mathrm{d}x = \int_{-\infty}^{\infty} \gamma_m(y)\frac{\gamma(f^{-1}(y))}{f'(x)}\mathrm{d}y.$$

Since $1/f'(x) = \mathrm{d}f^{-1}(y)/\mathrm{d}y$ the preceding paragraph makes it clear that $\gamma(f^{-1}(y))/f'(x)$ is a good function of y. Hence the right-hand side tends to a limit as $m \to \infty$ and the sequence is regular.

In the course of the investigation it has been shown that

$$\int_{-\infty}^{\infty} g\{f(x)\}\gamma(x)\,\mathrm{d}x = \int_{-\infty}^{\infty} g(y)\frac{\gamma(f^{-1}(y))}{f'(x)}\mathrm{d}y$$
$$= \int_{-\infty}^{\infty} g(y)\gamma(f^{-1})\frac{\mathrm{d}f^{-1}}{\mathrm{d}y}\mathrm{d}y. \qquad (4)$$

Note: If f, instead of increasing, decreased steadily from ∞ to $-\infty$ there would be little modification to the foregoing analysis. The main difference would be a change of sign in the right-hand side of (4). Thus f either increasing or decreasing is covered by

$$\int_{-\infty}^{\infty} g\{f(x)\}\gamma(x)\,\mathrm{d}x = \int_{-\infty}^{\infty} g(y)\frac{\gamma(f^{-1}(y))}{|f'(x)|}\mathrm{d}y. \qquad (5)$$

Example 3. The function x^3+3x has derivative $3(x^2+1)$ and therefore satisfies all the conditions of Definition 8.4. Hence, from (4)

$$\int_{-\infty}^{\infty} \delta(x^3+3x)\gamma(x)\mathrm{d}x = \int_{-\infty}^{\infty} \delta(y)\frac{\gamma(f^{-1}(y))}{3(x^2+1)}\mathrm{d}y$$
$$= \left[\frac{\gamma(f^{-1}(y))}{3(x^2+1)}\right]_{y=0} = \tfrac{1}{3}\gamma(0)$$

since $x=0$ when $y=0$. Thus

$$\delta(x^3+3x) = \tfrac{1}{3}\delta(x).$$

More generally, if f satisfies the conditions of Definition 8.4,

$$\int_{-\infty}^{\infty} \delta\{f(x)\}\gamma(x)\mathrm{d}x = \int_{-\infty}^{\infty} \delta(y)\frac{\gamma(f^{-1}(y))}{f'(x)}\mathrm{d}y$$
$$= \gamma(x_0)/f'(x_0)$$

where $f(x_0)=0$. Therefore

$$\delta\{f(x)\} = \frac{\delta(x-x_0)}{f'(x_0)}. \tag{6}$$

Theorem 8.6. *If f satisfies the conditions of Definition* 8.4.

$$[g\{f(x)\}]' = g'\{f(x)\}f'(x).$$

This is formally the same as the customary rule for the derivative of a function of a function.

Proof. By definition

$$[g\{f(x)\}]' = \lim_{m\to\infty}[\gamma_m\{f(x)\}]' = \lim_{m\to\infty}\gamma_m'\{f(x)\}f'(x)$$
$$= g'\{f(x)\}f'(x)$$

since f' is fairly good. This demonstration is valid whether f is steadily increasing or steadily decreasing and the proof is finished. □

Exercises

17. If f satisfies the conditions of Definition 8.4, prove that $f^{-1}(y)$ is a fairly good function of y.
18. Prove that $\sinh^{-1} x$ does not satisfy the conditions of Definition 8.4. Do you see any difficulty about defining $g(\sinh^{-1} x)$? Show that $g\{\sinh^{-1} x + x(1+x^2)^{1/2}\}$ can be defined.

19. Does $f(x) = \tanh x$ satisfy Definition 8.4?

20. Prove that

(i) $\delta'(x^3 + 3x) = \frac{1}{9}\delta'(x)$,

(ii) $\delta''(\frac{1}{3}x^3 + x) = \delta''(x) - 2\delta(x)$,

(iii) $\delta(3x - \frac{1}{2}\sin 2x) = \frac{1}{2}\delta(x)$,

(iv) $\delta'(3x^5 + 5x^3 + 15x - 23) = \frac{1}{2025}\{\delta'(x-1) + 2\delta(x-1)\}$.

21. Show that $[H\{f(x)\}]' = \delta(x - x_0)$ where $f(x_0) = 0$.

22. Prove that

$$\delta'\{f(x)\} = \frac{1}{\{f'(x_0)\}^2}\left\{\delta'(x - x_0) + \frac{f''(x_0)}{f'(x_0)}\delta(x - x_0)\right\}.$$

23. Prove that, under the conditions of Definition 8.4,

$$[g\{f(x)\}]'' = g''\{f(x)\}\{f'(x)\}^2 + g'\{f(x)\}f''(x).$$

24. Prove that Definition 8.4 is still satisfactory if the condition that f be fairly good be replaced by the requirement that f be moderately good.

It is not an easy matter to relax the conditions in Definition 8.4 and retain a change of variable which is valid for all generalised functions; of course, for particular generalised functions the use of a much wider class of f may be permissible. Exercise 24 gives one slight relaxation of the conditions imposed on f but we cannot dispense with the restrictions of infinite differentiability and of $f'(x) \neq 0$. Nor can we dispose of the constraint that f be bounded by a power of x at infinity. For, if $f(x) = e^x$, the generalised function y would be transformed to e^x which is not a generalised function. On the other hand, there is trouble also if f grows too slowly at infinity. Suppose $f(x) = \ln x$ for large x then the good function sech y would be transformed to $2x/(1 + x^2)$ which is not good. Thus regular sequences would not remain regular in general. Since $f'(x) = 1/x$ this example shows that the condition $|f'(x)| \geq K|x|^{\nu-1}$ must be kept. Little improvement in the restrictions of Definition 8.4 can be expected therefore while we are concerned with transformations from $(-\infty, \infty)$ to $(-\infty, \infty)$. Consequently, we turn our attention to changes of variable in which one interval is not $(-\infty, \infty)$.

Let us first recall that the behaviour of $g(x)$ in $a < x < b$, a and b being finite, is entirely determined by $\int_{-\infty}^{\infty} g(x)\gamma(x)\,dx$ where γ is any good function which vanishes identically outside (a, b). This implies that it is sufficient to consider $\lim_{m\to\infty}\int_a^b \gamma_m(x)\gamma(x)\,dx$ where $\{\gamma_m\}$ is a regular sequence defining g. We shall now show that $\{\gamma_m\}$ can be replaced by a sequence of fine functions which are zero outside (a, b), without loss of generality. Let $\phi_n(x)$ be a fine function which equals 1 in $(a + 1/n, b - 1/n)$ and is identically zero in $(-\infty, a + 1/(n+1))$ and $(b - 1/(n+1), \infty)$. Such a fine function can be obtained by defining it to be zero in $(-\infty, a)$, (b, ∞) and to be, in (a, b),

$$\sum_{m=-n}^{n} \tau(\tan y - m)$$

where $x = a + (y + \frac{1}{2}\pi)(b - a)/\pi$ and τ is the fine function defined in §5.5. Then, on account of Theorem 3.16,

$$\left|\int_a^b \gamma_m(x)\{\phi_m(x) - 1\}\gamma(x)\,dx\right| \leq K \sup_{a\leq x\leq b} [\gamma(x)\{\phi_m(x) - 1\}]^{(r)}$$

for some finite r. Now

$$\sup_{a\leq x\leq b} |\gamma^{(r_1)}(x)\{\phi_m(x) - 1\}^{(r_2)}| \leq \sup|\gamma^{(r_1)}(x)\{\phi_m(x) - 1\}^{(r_2)}| \quad (7)$$

with $a \leq x \leq a + 1/m$ and $b - 1/m \leq x \leq b$ because of the properties of ϕ_m. If $r_2 = 0$, $|\phi_m(x) - 1| \leq 1$ and the right-hand side tends to zero as $m \to \infty$ because γ and all its derivatives vanish at a and b. If $r_2 \neq 0$, the facts that $d/dx = \{\pi \sec^2 y/(b - a)\}\{d/d(\tan y)\}$ and that any derivative of τ with respect to the argument is bounded show that the right-hand side of (7) does not exceed

$$\sup_{a\leq x\leq a+1/m} (x - a)^{-M_1}|\gamma^{(r_1)}(x)| + \sup_{b-1/m\leq x\leq b} (b - x)^{-M_2}|\gamma^{(r_2)}(x)|$$

for some finite M_1, M_2. Since γ and its derivatives vanish faster than any power of $(x - a)$ at a and faster than any power of $(b - x)$ at b this quantity tends to zero as $m \to \infty$. Hence

$$\lim_{m\to\infty}\int_a^b \gamma_m(x)\gamma(x)\,dx = \lim_{m\to\infty}\int_a^b \phi_m(x)\gamma_m(x)\gamma(x)\,dx$$

for any good γ which is zero outside (a, b). Therefore, for each regular sequence $\{\gamma_m\}$ there is an equivalent regular sequence of fine functions which are identically zero outside (a, b). Consequently, as far as the behaviour of $g(x)$ in $a < x < b$ is concerned, it is no restriction to consider only defining sequences which vanish outside (a, b).

Definition 8.5. *Let $f(x)$ be a real-valued infinitely differentiable function, with $f'(x) \neq 0$, which increases steadily from $-\infty$ to ∞ as x goes from a to b. Suppose that there are finite positive constants N_m, N'_m, v, v' such that*

$$|f^{(m)}(x)| < K_m(x-a)^{-N_m}(m=0,1,\dots),|f'(x)| > K(x-a)^{-1-v}$$

in a neighbourhood $x \geq a$ and such that

$$|f^{(m)}(x)| < K'_m(b-x)^{-N'_m}(m=0,1,\dots),|f'(x)| > K'(b-x)^{-1-v'}$$

in a neighbourhood $x \leq b$. Then, if $\{\gamma_m(y)\}$ is a regular sequence defining $g(y)$ on R_1, $\{\gamma_m(f(x))\}$ defines a generalised function denoted by $g\{f(x)\}$ on $a < x < b$.

To justify the definition we have to show that $\{\gamma_m(f)\}$ is a regular sequence of fine functions which vanish at a and b. Now $[\gamma_m(f)]^{(n)}$ consists of a finite sum of products of $f^{(n)}, f^{(n-1)}, \dots, f', \gamma'_m, \dots, \gamma_m^{(n)}$ so that $\gamma_m(f)$ possesses derivatives of all orders with respect to x away from a and b. To complete the verification we have to prove that $(x-a)^{-r}[\gamma_m(f)]^{(n)}$ tends to zero as $x \to a+0$ and that

$$(b-x)^{-r}[\gamma_m(f)]^{(n)} \to 0$$

as $x \to b-0$. A typical term in $(x-a)^{-r}[\gamma_m(f)]^{(n)}$ does not exceed $K(x-a)^{-r-M}|\gamma_m^{(k)}(f)|$ where M is finite since $|f^{(m)}(x)| < (x-a)^{-N_m}$. As $x \to a+0, f \to -\infty$ and $|f| > (x-a)^{-v}$ because $|f'| > (x-a)^{-1-v}$. Therefore $|\gamma_m^{(k)}(f)| < |f|^{-s} < (x-a)^{vs}$ where s can be chosen arbitrarily large. Thus $\gamma_m(f)$ has appropriate behaviour as $x \to a+0$. Similarly its conduct as $x \to b-0$ is satisfactory and $\gamma_m(f)$ is fine.

It will now be shown that, if $\gamma(x)$ is a fine function which vanishes identically outside (a, b), $\gamma(f^{-1}(y))$ is a good function of y on R_1. (There is no difficulty in defining $x = f^{-1}(y)$ since the relation $y = f(x)$ is one-to-one.) Now $\mathrm{d}^m f^{-1}/\mathrm{d}y^m$, expressed in terms of x, is a sum of products of $f^{(m)}(x), \dots, f''(x)$, $1/f'(x)$ and hence as $x \to a$ or $y \to -\infty$ there is a finite M such that $|\mathrm{d}^m f^{-1}/\mathrm{d}y^m| = O\{(x-a)^{-M}\}$. Therefore

$$\left| y^r \frac{\mathrm{d}^n \gamma(f^{-1}(y))}{\mathrm{d}y^n} \right| = O\{|f(x)|^r (x-a)^{-M} |\gamma^{(s)}(x)|\}$$

$$= O\{(x-a)^{-M-rN_0}|\gamma^{(s)}(x)|\} \to 0$$

as $x \to a$ and $y \to -\infty$ because γ vanishes identically outside (a, b). There is a similar result as $y \to \infty$ and so $\gamma(f^{-1})$ is good. It is also clear that $\gamma(f^{-1})\mathrm{d}f^{-1}/\mathrm{d}y$ is good.

The fact that $\{\gamma_m(f)\}$ is regular now follows from

$$\int_a^b \gamma_m\{f(x)\}\gamma(x)\mathrm{d}x = \int_{-\infty}^{\infty} \gamma_m(y)\gamma(f^{-1}(y))\frac{\mathrm{d}f^{-1}(y)}{\mathrm{d}y}\mathrm{d}y$$

because the right-hand side tends to a limit as $m \to \infty$. It has thus been shown that

$$\int_a^b g\{f(x)\}\gamma(x)\mathrm{d}x = \int_{-\infty}^{\infty} g(y)\gamma(f^{-1})\frac{\mathrm{d}f^{-1}}{\mathrm{d}y}\mathrm{d}y \tag{8}$$

for any fine γ which vanishes outside (a, b). Once again if f decreases steadily, we have

$$\int_a^b g\{f(x)\}\gamma(x)\,\mathrm{d}x = \int_{-\infty}^{\infty} g(y)\gamma(f^{-1})\left|\frac{\mathrm{d}f^{-1}}{\mathrm{d}y}\right|\mathrm{d}y. \tag{9}$$

Example 4. The function $f(x) = \tan x$ satisfies the conditions of Definition 8.5 with $a = -\frac{1}{2}\pi$, $b = \frac{1}{2}\pi$. Hence, from (8),

$$\int_{-\pi/2}^{\pi/2} g(\tan x)\gamma(x)\,\mathrm{d}x = \int_{-\infty}^{\infty} g(y)\gamma(\tan^{-1} y)(1 + y^2)^{-1}\,\mathrm{d}y$$

for any fine γ vanishing outside $(-\frac{1}{2}\pi, \frac{1}{2}\pi)$. In particular

$$\delta(\tan x) = \delta(x) \quad (-\tfrac{1}{2}\pi < x < \tfrac{1}{2}\pi)$$

and, more generally,

$$\delta\{f(x)\} = \delta(x - x_0)/f'(x_0) \quad (-\tfrac{1}{2}\pi < x < \tfrac{1}{2}\pi)$$

where $f(x_0) = 0$.

It may be shown that Theorem 8.6 is valid under the conditions of Definition 8.5; the proof will be left as an exercise for the reader.

Once again the restrictions on f cannot be lightened to any great extent where transformations for all generalised functions are concerned. Thus if $f(x)$ behaves as $\ln(x - a)$ as $x \to a$ the good function sech y becomes $2(x - a)/\{1 + (x - a)^2\}$ which is not fine so that regularity of sequences is not preserved; for that reason $v > 0$ cannot be dispensed with. However, if $f(x)$ has the form $\mathrm{e}^{1/(x-a)}$ as $x \to a$ the fine function $\mathrm{e}^{1/(a-x)}$ becomes $1/y$ which is not good so that (8) could not be asserted; therefore f and its derivatives must not approach infinity more rapidly than some inverse power of $x-a$.

If $b=\infty$ replace the conditions

$$|f^{(m)}(x)|<K'_m(b-x)^{-N'_m},\quad |f'(x)|>K'(b-x)^{-1-\nu'}$$

in Definition 8.5 by

$$|f^{(m)}(x)|<K'_m x^{N'_m},\quad f'(x)>Kx^{\nu'-1} \tag{10}$$

for sufficiently large x. Then we have a change of variable relating (a,∞) and $(-\infty,\infty)$ whose validity may be checked by applying the method of justification of Definition 8.5 at one end and that of Definition 8.4 at the other end. Similar considerations apply to the connection between $(-\infty,b)$ and $(-\infty,\infty)$.

If we wish to proceed from generalised functions defined on $a<x<b$ to those on $(-\infty,\infty)$ we can use the transformation $f^{-1}(y)$. It is sometimes convenient to have the conditions stated on the function of y rather than x. Thus if $f_1(y)$ increases steadily from a to b as y goes from $-\infty$ to ∞ we require that f_1 be moderately good and that there be finite positive constants ν_1 and ν_2 such that

$$b-f_1(y)<K_0 y^{-\nu_1},\quad f_1'(y)>Ky^{-\nu_2} \tag{11}$$

as $y\to\infty$ and similarly, with a in place of b, as $y\to-\infty$. Then given $g(x)$ on $a<x<b$ we can define $g\{f_1(y)\}$ on $(-\infty,\infty)$ in an analogous way to that in Definition 8.5.

It remains to consider what can be done when neither interval occupies $(-\infty,\infty)$. Let (a,b), (c,∞) be the two intervals where a, b and c are finite. Then

Definition 8.6. *Let $f(x)$ be a real-valued infinitely differentiable function, with $f'(x)\neq 0$, which increases steadily from c to ∞ as x goes from a to b. Suppose that there are finite positive constants N_m, N'_m, ν, ν', N such that*

$$|f^{(m)}(x)|<K_m(x-a)^{-N_m}\qquad (m=1,2,\ldots),$$
$$|f(x)-c|<K_0(x-a)^{\nu},\quad |f'(x)|>K(x-a)^N$$

in a neighbourhood $x\geq a$ and

$$|f^{(m)}(x)|<K'_m(b-x)^{-N'_m}\quad (m=0,1,\ldots),$$
$$|f'(x)|>K'(b-x)^{-1-\nu'}$$

in a neighbourhood $x \le b$. Then, if $\{\gamma_m(y)\}$ is a regular sequence of good functions vanishing identically outside (c, ∞) and defining $g(y)$ on $y > c$, $\{\gamma_m(f(x))\}$ defines a generalised function denoted by $g\{f(x)\}$ on $a < x < b$.

The justification of this definition is not greatly different from that for Definition 8.5 and will not be given here.

If $b = \infty$ we alter the conditions near $x = b$ to (10). If $a = -\infty$ we employ inequalities of the type (11) instead of those in $x \ge a$.

Finally, if we are concerned with the mapping from (a, b) to (c, d) where d is finite we alter the conditions near $x = b$ in Definition 8.6 to be of the same type as those near $x = a$. Consequently we have now obtained classes of transformations for which it is permissible to go from generalised functions on one interval to those on another. In all cases we have

$$\int_a^b g\{f(x)\}\gamma(x)\,dx = \int_c^d g(y)\gamma(f^{-1})\left|\frac{df^{-1}}{dy}\right| dy \tag{12}$$

for any good γ which vanishes identically outside (a, b) and

$$[g\{f(x)\}]' = g'\{f(x)\}f'(x). \tag{13}$$

Given $g(y)$ on $c < y < d$, (12) determines $g\{f(x)\}$ on $a < x < b$ whether any of a, b, c and d be finite or infinite.

Example 5. The function $f(x) = x^2$ with $a = 0$, $b = \infty$, $c = 0$, $d = \infty$ satisfies Definition 8.6 at a and (10) at b. Hence

$$\int_0^\infty g(x^2)\gamma(x)\,dx = \int_0^\infty g(y)\gamma(y^{1/2})(\tfrac{1}{2}y^{-1/2})\,dy \tag{14}$$

for any good γ which is zero outside $(0, \infty)$. Taking $g(y) = \delta(y - a^2)$ with $a > 0$ we see that

$$\delta(x^2 - a^2) = \frac{1}{2a}\delta(x - a) \quad (x > 0). \tag{15}$$

On the other hand $g(y) = \delta(y)$ gives

$$\delta(x^2) = 0 \quad (x > 0).$$

By choosing $a = -\infty, b = 0$ we find

$$\int_{-\infty}^{0} g(x^2)\gamma(x)\,dx = \int_{0}^{\infty} g(y)\gamma(-y^{1/2})(\tfrac{1}{2}y^{-1/2})\,dy$$

for any good γ which vanishes outside $(-\infty, 0)$. Put $\gamma(x) = \gamma_1(-x)$ so that γ_1 is zero outside $(0, \infty)$. Then changing x to $-x$ on the left-hand side supplies

$$\int_{0}^{\infty} g\{(-x)^2\}\gamma_1(x)\,dx = \int_{0}^{\infty} g(y)\gamma_1(y^{1/2})(\tfrac{1}{2}y^{-1/2})\,dy.$$

Comparison with (14) reveals that

$$g\{(-x)^2\} = g(x^2) \quad (x > 0) \tag{16}$$

which is the same property as for conventional functions. In particular, if $g(x) = 0$ for $x > 0$ then $g(x^2) = 0$ for $x > 0$ and $x < 0$. If it is known that $g(x^2)$ is a generalised function for x on $(-\infty, \infty)$ and that $g(x^2) = g_0(x^2)$ for $x > 0$, it follows from Theorem 3.21 that

$$g(x^2) = g_0(x^2) + \sum_{m=0}^{n} c_m \delta^{(m)}(x).$$

Example 6. The transformation $f(x) = \cos x$ with $a = 0$, $b = \pi$, $c = -1, d = 1$ satisfies our conditions and so

$$\int_{0}^{\pi} g(\cos x)\gamma(x)\,dx = \int_{-1}^{1} g(y)\gamma(\cos^{-1} y)(1 - y^2)^{-1/2}\,dy$$

for any good γ which vanishes outside $(0, \pi)$. Here $\cos^{-1} y$ lies in $[0, \pi]$. Consequently

$$\delta(\cos x) = \delta(x - \tfrac{1}{2}\pi) \quad (0 < x < \pi).$$

One extension of the range of transformations is provided by the following rule. *If, in any finite closed or open interval $g(x)$ is equal to a conventional integrable function $h(x)$ we understand by $g\{f(x)\}$ the function $h\{f(x)\}$ for any conventional function f for which this expression is a conventional integrable function and the new interval is finite.* The restriction to finite intervals can be removed provided that one confines one's attention to those h and f which do not violate the constraints on the behaviour of generalised functions at infinity. (Theorem 3.18.) Equation (12) will be valid if it is valid for conventional functions.

One particular application of this rule is that, if $g(x)$ is zero for x in the interval (x_0, x_1) or $[x_0, x_1]$ then $g\{f(x)\}$ is zero for the interval traced by $f(x)$ as x goes from x_0 to x_1 provided that f is a permissible operation.

Example 7. Since $\delta(y - a^2)\,(a > 0)$ is zero for $0 \le y \le \frac{1}{4}a^2$ we have

$$\delta(x^2 - a^2) = 0 \quad (-\tfrac{1}{2}a \le x \le \tfrac{1}{2}a). \tag{17}$$

From (16) and (15)

$$\delta(x^2 - a^2) = \frac{1}{2a}\delta(-x - a) = \frac{1}{2a}\delta(x + a) \quad (x < 0)$$

and this may be combined with (15) and (17) to give

$$\delta(x^2 - a^2) = \frac{1}{2a}\{\delta(x - a) + \delta(x + a)\} \tag{18}$$

without restriction on x.

Example 8. $\mathrm{H}\{f(x)\}$ can be defined for a continuous f whose only zeros are isolated by giving it the value 1 when $f(x) > 0$ and the value 0 when $f(x) < 0$. Let $x_1, x_2, \ldots$ be the zeros of $f(x)$ and let f be positive in the intervals $(x_{m_1}, x_{m_2}), (x_{m_3}, x_{m_4}), \ldots$ and negative elsewhere. Then, for any good γ,

$$\begin{aligned}\int_{-\infty}^{\infty} [\mathrm{H}\{f(x)\}]'\gamma(x)\,dx &= -\int_{-\infty}^{\infty} \mathrm{H}\{f(x)\}\gamma'(x)\,dx \\ &= -\sum_{s=0} \int_{x_{m_{2s+1}}}^{x_{m_{2s+2}}} \gamma'(x)\,dx \\ &= \sum_{s=0} \{\gamma(x_{m_{2s+1}}) - \gamma(x_{m_{2s+2}})\}.\end{aligned}$$

Hence

$$[\mathrm{H}\{f(x)\}]' = \sum_{s=0} \{\delta(x - x_{m_{2s+1}}) - \delta(x - x_{m_{2s+2}})\}. \tag{19}$$

Thus there is a contribution from each zero when f goes from negative to positive as x increases, a contribution of opposite sign when f goes from positive to negative and no contribution where f does not change sign. Therefore another way of writing (19) is

$$[\mathrm{H}\{f(x)\}]' = \sum_{m=1} \tfrac{1}{2}\delta(x - x_m)\{\operatorname{sgn} f(x_m + \varepsilon_m) - \operatorname{sgn} f(x_m - \varepsilon_m)\}, \tag{20}$$

ε_m being chosen so small that $(x_m - \varepsilon_m, x_m + \varepsilon_m)$ contains no zero other than x_m.

Exercises

25. Verify that the conditions in the sentence containing (11) are sufficient for defining a generalised function on $(-\infty, \infty)$ given one on (a,b).

26. Prove that, for $x > 0$,

$$\delta(a - 1/x) = \begin{cases} (1/a^2)\delta(x - 1/a) & (a > 0) \\ 0 & (a \leq 0). \end{cases}$$

27. If $g(y) = 0$ for $y > 0$ show that $g(1/x) = 0$ for $x > 0$.

28. Prove that, if $a > 0$,

$$\delta(x^4 - a^4) = (1/3a^3)\delta(x - a) \qquad (x > 0)$$

and deduce that

$$\delta(x^4 - a^4) = (1/3a^3)\{\delta(x - a) + \delta(x + a)\}$$

without restriction on x.

29. Prove that $x\delta'(x^2 - a^2) = (1/2a)\{\delta'(x - a) + \delta'(x + a)\}$ when $a > 0$. Does this imply that

$$\delta'(x^2 - a^2) = (1/2a^3)\{\delta'(x - a) - \delta'(x + a) + a\delta(x - a) + a\delta(x + a)\}\ ?$$

30. Prove that

$$(x^2 - 1)\delta(x^2 - 1) = 0$$

$$(x^2 - 1)\delta^{(m)}(x^2 - 1) + m\delta^{(m-1)}(x^2 - 1) = 0.$$

Show that the Fourier transform of $\delta(x^2 - 1)$ is $\cos\alpha$. Show also that, if $G_m(\alpha)$ is the Fourier transform of $\delta^{(m)}(x^2 - 1)$, $G'_{m+1} = \tfrac{1}{2}\alpha G_m$ and

$$2G_{m+1} = (2m + 1)G_m - \alpha G'_m.$$

Deduce that G_m can be written as

$$(-1)^m \pi^{1/2}(\tfrac{1}{2}\alpha)^{m+1/2} J_{-m-1/2}(\alpha) \quad \text{or} \quad (-1)^m \frac{\alpha^{2m+1}}{2^m}\left(\frac{\mathrm{d}}{\alpha\,\mathrm{d}\alpha}\right)^m \frac{\cos\alpha}{\alpha}.$$

31. Prove that

$$\delta(\sin x) = \frac{1}{\pi} \sum_{m=-\infty}^{\infty} e^{2imx}.$$

32. Suppose that f satisfies the conditions of either Definition 8.5 or 8.6 in $[a_1, b_1]$. Then $\delta\{f(x)\} = \delta(x - x_0)/f'(x_0)$ in $a_1 < x < b_1$, where x_0 is the zero (if there is one) of f in $[a_1, b_1]$. Suppose f decreases steadily in $[b_1, a_2]$, increases steadily in $[a_2, b_2]$, decreases steadily in $[b_2, a_3]$ and so on, and that R_1 can be covered by a countable number of such intervals. If $f'(x) \neq 0$ at any zero $x = x_m$ of f, prove that

$$\delta\{f(x)\} = \sum_m \frac{\delta(x - x_m)}{|f'(x_m)|}$$

the summation being over all the zeros of f. Deduce that

$$f'(x)\delta'\{f(x)\} = \sum_m \frac{\delta'(x - x_m)}{|f'(x_m)|}.$$

33. If $g(x) = g_1(x) g_2(x)$ and $f(x)$ satisfies any of the conditions of Definitions 8.4–8.6, show that $g\{f(x)\} = g_1\{f(x)\} g_2\{f(x)\}$ for $a < x < b$.

34. Is the answer to Exercise 32 consistent with (20) and (13)?

35. Define sgn $\{f(x)\}$ in a similar way to $\mathrm{H}\{f(x)\}$ and deduce that

$$\operatorname{sgn}(|x|^{1/2}) = \mathrm{H}(|x|^{1/3}).$$

Determine $[\operatorname{sgn}\{f(x)\}]'$.

Finally, it is necessary to consider the effect of change of variable on integrals. By definition (§4.4) we mean by $\int_{c_1}^{d_1} g(y)\,dy$ the value (if it exists) of

$$\int_c^d \{g(y)\,\mathrm{H}(y - c_1) - g(y)\,\mathrm{H}(y - d_1)\}\gamma_1(y)\,dy$$

where $c < c_1 < d_1 < d$ and γ_1 is a fine function which equals 1 in $[c_1, d_1]$ and vanishes identically outside (c, d). If (c, d) is an interval for which (12) is valid, the integral can be expressed as

$$\int_a^b [g\{f(x)\}\,\mathrm{H}\{f(x) - c_1\} - g\{f(x)\}\,\mathrm{H}\{f(x) - d_1\}]\gamma_1\{f(x)\} f'(x)\,dx$$

$$= \int_{a_1}^{b_1} g\{f(x)\} f'(x)\,dx$$

where $f(a) = c$, $f(a_1) = c_1$, $f(b) = d$, $f(b_1) = d_1$. Hence, if $a < a_1 < b_1 < b$ and (12) is true,

$$\int_{a_1}^{b_1} g\{f(x)\}f'(x)\,dx = \int_{f(a_1)}^{f(b_1)} g(y)\,dy \tag{21}$$

provided that either side has a meaning.

With regard to change of variable in a convolution we note that

$$\int_{-\infty}^{\infty} \gamma(x) \int_{-\infty}^{\infty} g_1(t) g_2(x-t)\,dt\,dx = \int_{-\infty}^{\infty} g_2(x) \int_{-\infty}^{\infty} g_1(t)\gamma(t+x)\,dt\,dx. \tag{22}$$

If f satisfies the conditions of Definition 8.5, (9) demonstrates that

$$\int_{-\infty}^{\infty} g_1(t)\gamma(t+x)\,dt = \int_a^b g_1\{f(u)\}\gamma\{f(u)+x\}f'(u)\,du$$

where $f(b) = \infty, f(a) = -\infty$. Making the change of variable $x + f(u) = y$ on the right-hand side of (22) we obtain

$$\int_{-\infty}^{\infty} \gamma(x) \int_{-\infty}^{\infty} g_1(t) g_2(x-t)\,dt\,dx = \int_{-\infty}^{\infty} \gamma(y) \int_a^b g_1\{f(u)\} g_2\{y - f(u)\} f'(u)\,du\,dy.$$

Hence, under the conditions of Definition 8.5,

$$\int_{-\infty}^{\infty} g_1(t) g_2(x-t)\,dt = \int_a^b g_1\{f(u)\} g_2\{x - f(u)\} f'(u)\,du. \tag{23}$$

Exercises

36. The substitution $y = x/a (a > 0)$ in $\int_0^1 y^{-1}dy$ gives formally $\int_0^a x^{-1}\,dx$. In what sense (if any) are the two integrals equal? What is the connection with

$$\int_1^a x^{-1}\,dx = \ln a\,?$$

37. Prove that

$$\int_1^{\infty} g_1(t)(x-t)^{-1}\,dt = \int_0^1 g(1/u)\{u(ux-1)\}^{-1}\,du.$$

8.3. Change of variables: several dimensions

In the consideration of several variables we turn firstly to the natural generalisation of the changes of variables which have been discussed in one dimension. Let $y_1, y_2, \ldots, y_n$ be variables connected to the variables $x_1, \ldots, x_n$ by relations which may be denoted by $\boldsymbol{y} = \boldsymbol{f}(\boldsymbol{x})$ or $y_i = f_i(\boldsymbol{x})$ or $y_i = f_i(x_1, \ldots, x_n)$ according to convenience. The Jacobian J is defined by

$$J = \partial(y_1, \ldots, y_n)/\partial(x_1, \ldots, x_n)$$

i.e. it is the determinant whose ijth element is $\partial y_i/\partial x_j$. Then

Definition 8.7. *Let $f(\boldsymbol{x})$ be a real-valued moderately good function such that* (i) *the correspondence between $\boldsymbol{x}$ and $\boldsymbol{y}$ is one-to-one,*† (ii) *as $|\boldsymbol{x}| \to \infty$ along any direction, $|\boldsymbol{y}| \to \infty$ and conversely,* (iii) *for sufficiently large $|\boldsymbol{x}|$, $|J| \geq K|\boldsymbol{x}|^{\nu-n}$ $(\nu > 0)$. Then, if $\{\gamma_m(\boldsymbol{x})\}$ is a regular sequence defining $g(\boldsymbol{x})$, $\{\gamma_m(\boldsymbol{f}(\boldsymbol{x}))\}$ is a regular sequence defining a generalised function which will be denoted by $g\{\boldsymbol{f}(\boldsymbol{x})\}$.*

In one respect the justification is a little more complicated than that of Definition 8.4. Let S denote $|\boldsymbol{y}| = R$. Then, for large enough R,

$$CR^n = \int_{|\boldsymbol{y}| \leq R} \mathrm{d}\boldsymbol{y} = \int_{|\boldsymbol{y}| \leq R} |J|\,\mathrm{d}\boldsymbol{x} \geq \int_{|\boldsymbol{y}| \leq R - T_0} K|\boldsymbol{x}|^{\nu-n}\,\mathrm{d}\boldsymbol{x} + \int_{T_0} |J|\,\mathrm{d}\boldsymbol{x}$$

if $|J| \geq K|\boldsymbol{x}|^{\nu-n}$ outside T_0. Hence

$$CR^n \geq K \min_{\boldsymbol{x} \in S} |\boldsymbol{x}|^{\nu} + \text{constant}.$$

This information enables one to show, in a manner similar to that for Definition 8.4. that $\gamma_m(\boldsymbol{f})$ is good. Similarly, and noting $|J\partial x_j/\partial y_i| = |\text{ cofactor } \partial y_i/\partial x_j| = O(|\boldsymbol{x}|^m)$ since $\boldsymbol{f}$ is fairly good, we prove that $\gamma(\boldsymbol{f}^{-1})$ is good when γ is. The analogue of (5) is

$$\int_{-\infty}^{\infty} g\{\boldsymbol{f}(\boldsymbol{x})\}\gamma(\boldsymbol{x})\mathrm{d}\boldsymbol{x} = \int_{-\infty}^{\infty} g(\boldsymbol{y})\gamma(\boldsymbol{f}^{-1}(\boldsymbol{y}))\frac{\mathrm{d}\boldsymbol{y}}{|J|}. \tag{24}$$

We may also prove by exactly the same method as was used in Theorem 8.6 that

$$\partial_1 g\{\boldsymbol{f}(\boldsymbol{x})\} = \left[\sum_{j=1}^{n} \{\partial_{yj} g(\boldsymbol{y})\}\partial_1 y_j\right]_{\boldsymbol{y}=\boldsymbol{f}} \tag{25}$$

† A mapping $\boldsymbol{y} = \boldsymbol{f}(\boldsymbol{x})$ which is one-to-one and such that $\boldsymbol{f}$ and $\boldsymbol{f}^{-1}$ are both infinitely differentiable is sometimes called a *diffeomorphism.*

where ∂_{yj} is the generalised partial derivative with respect to y_j. This is in agreement with the standard rule for the partial derivative of a function of several variables.

When we turn to domains of other sizes the complication is one of wording more than anything else. Let T be a finite simply-connected domain bounded by an infinitely differentiable hypersurface. Let d denote the distance between a typical interior point and a typical boundary point. Then

Definition 8.8. *Let $f(x)$ be a real-valued infinitely differentiable function such that the correspondence between T and R_n is one-to-one, $|y| \to \infty$ as x tends to any boundary point of T. Let there be finite positive constants N_p, ν such that $|\partial_1^{p_1} \dots \partial_n^{p_n} f(x)| < K_p d^{-N_p}$ ($p_1, \dots, p_n$ any non-negative integers) and $|J| > Kd^{-n-\nu}$ as any boundary point is approached from the interior. Then, if $\{\gamma_m(y)\}$ is a regular sequence defining $g(y)$ on R_n, $\{\gamma_m(f(x))\}$ defines a generalised function denoted by $g\{f(x)\}$ on T.*

As in Definition 8.5, this definition does not determine $g(f)$ on the boundary of T. The justification, which runs parallel to that for Definition 8.5, will be left to the reader.

Definitions 8.7 and 8.8 cover the cases when a point at infinity and point at a finite distance from the origin respectively in x-space correspond to a point at infinity in y-space. The two other possibilities are (i) a point at infinite in x-space corresponds to a point at a finite distance in y-space and (ii) a point at a finite distance corresponds to a point at a finite distance. We shall now state the conditions appropriate to these two cases, leaving the reader to convince himself of their validity:

(i) $f(x)$ to be moderately good and, as $|x|$ tends to the relevant point at infinity

$$|b - f(x)| < K_0 |x|^{-\nu_1}, \qquad |J| > K|x|^{-\nu_2} \tag{26}$$

where b is the boundary point and ν_1, ν_2 are positive.

(ii) f infinitely differentiable and

$$\left.\begin{aligned} &|b - f(x)| < Kd^{\nu}, \qquad |J| > Kd^{N}, \\ &|\partial_1^{p_1} \dots \partial_n^{p_n} f(x)| < K_p d^{-N_p} \qquad (p_1, \dots, p_n \text{ positive integers}) \end{aligned}\right\} \tag{27}$$

in a neighbourhood of the boundary point corresponding to b.

In all of these cases, if the domain T becomes U under the transformation,

$$\int_T g\{\boldsymbol{f}(\boldsymbol{x})\}\gamma(\boldsymbol{x})\,\mathrm{d}\boldsymbol{x} = \int_U g(\boldsymbol{y})\gamma(\boldsymbol{f}^{-1}(\boldsymbol{y}))\frac{\mathrm{d}\boldsymbol{y}}{|J|} \tag{28}$$

for a good γ vanishing identically outside T. Also

$$\partial_1 g\{\boldsymbol{f}(\boldsymbol{x})\} = \left[\sum_{j=1}^{n} \partial_{yj} g(\boldsymbol{y})\partial_1 y_j\right]_{\boldsymbol{y}=\boldsymbol{f}} \quad (\boldsymbol{x}\in T). \tag{29}$$

Further

$$\partial_2\partial_1 g\{\boldsymbol{f}(\boldsymbol{x})\} = \left[\sum_{k=1}^{n}\sum_{j=1}^{n} \partial_{yk}\partial_{yj} g(\boldsymbol{y})\partial_2 y_k \partial_1 y_j + \sum_{j=1}^{n} \partial_{yj} g(\boldsymbol{y})\partial_2\partial_1 y_j\right]_{\boldsymbol{y}=\boldsymbol{f}}$$

$$= \partial_1\partial_2 g\{\boldsymbol{f}(\boldsymbol{x})\}. \tag{30}$$

Example 9. The transformation $y_1 = x_1^2$, $y_2 = \cos x_2$ relates the strip $-\infty < x_1 < \infty, 0 < x_2 < \pi$ in R_2 with $0 < y_1 < \infty, -1 < y_2 < 1$ and satisfies our conditions. Hence

$$\int_{-\infty}^{\infty}\int_0^{\pi} g(x_1^2, \cos x_2)\gamma(x_1, x_2)\,\mathrm{d}x_1\,\mathrm{d}x_2$$

$$= \int_0^{\infty}\int_{-1}^{1} g(y_1, y_2)\frac{\gamma(y_1^{1/2}, \cos^{-1} y_2)}{2y_1^{1/2}(1-y_2^2)^{1/2}}\,\mathrm{d}y_1\,\mathrm{d}y_2$$

and so, if $g(y_1, y_2) = \delta(y_1 - a^2, y_2)$ with $a > 0$, we obtain

$$\delta(x_1^2 - a^2, \cos x_2) = (1/2a)\delta(x_1 - a, x_2 - \tfrac{1}{2}\pi) \quad (x_1 > 0, 0 < x_2 < \pi).$$

Example 10. Let $\boldsymbol{y} = \boldsymbol{f}(\boldsymbol{x})$ be a transformation of the type covered by the preceding theory and let $g(\boldsymbol{y})$ be zero when $y_k, y_{k+1}, \ldots, y_n$ are non-zero. Then Corollary 7.27 shows that

$$g(\boldsymbol{y}) = \sum_{p=0}^{m} g_p(y_1, \ldots, y_{k-1})\partial_{yk}^{p_k} \ldots \partial_{yn}^{p_n}\delta(y_k)\ldots\delta(y_n).$$

Expressing this in terms of $\boldsymbol{x}$ we obtain the form of a generalised function which is zero except on the manifold $f_k(\boldsymbol{x}) = 0, \ldots, f_n(\boldsymbol{x}) = 0$.

Example 11. If $f(x)$ is a transformation of the preceding type

$$\int_{-\infty}^{\infty} \frac{\partial f_n}{\partial x_n}\delta\{f_n(x)\}\gamma(x)\,\mathrm{d}x$$
$$= \int_{-\infty}^{\infty} \frac{\partial f_n}{\partial x_n}\delta(y_n)\gamma(f^{-1}(y))\frac{\mathrm{d}y}{|J|}$$
$$= \int_{-\infty}^{\infty} \cdots \int_{-\infty}^{\infty} \left[\frac{\gamma(f^{-1}(y))}{|J|}\frac{\partial f_n}{\partial x_n}\right]_{y_n=0} \mathrm{d}y_1 \ldots \mathrm{d}y_{n-1}.$$

The relation $y_n = 0$ implies that x_n can be expressed in terms of $x_1, \ldots, x_{n-1}$ (and uniquely because f is one-to-one) so that $y_1, \ldots, y_{n-1}$ can be regarded as functions $\psi_1, \ldots, \psi_{n-1}$ of $x_1, \ldots, x_{n-1}$ only in the last integral. Therefore

$$\int_{-\infty}^{\infty} \frac{\partial f_n}{\partial x_n}\delta\{f_n(x)\}\gamma(x)\,\mathrm{d}x$$
$$= \int_{-\infty}^{\infty} \cdots \int_{-\infty}^{\infty} \left[\frac{\gamma(x)}{|J|}\frac{\partial f_n}{\partial x_n}\right]_{y_n=0} |J_0|\,\mathrm{d}x_1 \ldots \mathrm{d}x_{n-1}$$

where

$$J_0 = \partial(\psi_1, \ldots, \psi_{n-1})/\partial(x_1, \ldots, x_{n-1}).$$

Now

$$J = \begin{vmatrix} \frac{\partial f_1}{\partial x_1} & \cdots & \frac{\partial f_1}{\partial x_n} \\ \vdots & & \vdots \\ \frac{\partial f_n}{\partial x_1} & \cdots & \frac{\partial f_n}{\partial x_n} \end{vmatrix} = \begin{vmatrix} \frac{\partial f_1}{\partial x_1} - \frac{\partial f_n}{\partial x_1}\frac{\partial f_1}{\partial x_n}\Big/\frac{\partial f_n}{\partial x_n} & \cdots & 0 \\ \vdots & & \vdots \\ \frac{\partial f_{n-1}}{\partial x_1} - \frac{\partial f_n}{\partial x_1}\frac{\partial f_{n-1}}{\partial x_n}\Big/\frac{\partial f_n}{\partial x_n} & \cdots & 0 \\ \frac{\partial f_n}{\partial x_1} & \cdots & \frac{\partial f_n}{\partial x_n} \end{vmatrix}$$

on taking multiples of the last row from the others, and

$$\frac{\partial \psi_i}{\partial x_j} = \left[\frac{\partial f_i}{\partial x_j} - \frac{\partial f_i}{\partial x_n}\frac{\partial f_n}{\partial x_j}\Big/\frac{\partial f_n}{\partial x_n}\right]_{y_n=0} \qquad (i, j = 1, \ldots, n-1).$$

Hence

$$[J]_{y_n=0} = J_0[\partial f_n/\partial x_n]_{y_n=0}$$

and therefore

$$\int_{-\infty}^{\infty} \frac{\partial f_n}{\partial x_n} \delta\{f_n(\boldsymbol{x})\}\gamma(\boldsymbol{x})\mathrm{d}\boldsymbol{x} = \pm \int_{-\infty}^{\infty} \dots \int_{-\infty}^{\infty} [\gamma(\boldsymbol{x})]_{y_n=0} \mathrm{d}x_1 \dots \mathrm{d}x_{n-1}$$

the upper or lower sign being used according as $[\partial f_n/\partial x_n]_{y_n=0} \gtrless 0$. This result can be expressed more conveniently by introducing the hypersurface S_0 defined by $f_n = 0$. Then

$$\int_{-\infty}^{\infty} \frac{\partial f_n}{\partial x_n} \delta\{f_n(\boldsymbol{x})\}\gamma(\boldsymbol{x})\mathrm{d}\boldsymbol{x} = -\int_{S_0} \gamma(\boldsymbol{x})\cos\theta_n \mathrm{d}S \tag{31}$$

where θ_n is the angle between the positive direction of the x_n-axis and the normal to S_0 drawn on the side on which $f_n < 0$. An alternative way of writing (31) is

$$\int_{-\infty}^{\infty} [\partial_n \mathrm{H}\{f_n(\boldsymbol{x})\}]\gamma(\boldsymbol{x})\mathrm{d}\boldsymbol{x} = -\int_{S_0} \gamma(\boldsymbol{x}) \cos\theta_n \mathrm{d}S. \tag{32}$$

The result of the last two examples are subject to certain restrictions on f. If we wanted to extend them to more general f we can see that although they might be valid locally they may not be true globally; for example, $f_n = 0$ might not determine x_n uniquely in terms of $x_1, \dots, x_{n-1}$. However, local behaviour is often sufficient provided that we can apply the rule of the preceding section to determine global behaviour. Thus, if f_n is sufficiently smooth and $f_n = 0$ is a hypersurface S with interior T in which $f_n > 0$, we could say

$$\int_T [\partial_n \mathrm{H}\{f_n(\boldsymbol{x})\}]\gamma(\boldsymbol{x})\mathrm{d}\boldsymbol{x} = -\int_S \gamma(\boldsymbol{x}) \cos\theta_n \mathrm{d}S \tag{33}$$

and similarly for (31). Now θ_n can be described as the angle between the x_n-axis and the outward normal to S.

In a different context the application of the rule gives, when $a > 0$,

$$\delta(x_1^2 - a^2, \cos x_2) = \frac{1}{2a} \sum_{m=-\infty}^{\infty} [\delta\{x_1 - a, x_2 - (m + \tfrac{1}{2})\pi\} + \delta\{x_1 + a, x_2 - (m + \tfrac{1}{2})\pi\}]$$

without restriction on x_1 or x_2.

Exercises

38. Prove that, if $\boldsymbol{x}_0 \neq \mathbf{0}$,

(i) $\delta(\boldsymbol{x}-\boldsymbol{x}_0) = r^{-1}\delta(r-r_0)\delta(\phi-\phi_0) \qquad (r>0, 0<\phi<2\pi)$

where $x_1 = r\cos\phi$, $x_2 = r\sin\phi$,

(ii) $\delta(\boldsymbol{x}-\boldsymbol{x}_0) = \delta(r-r_0)\delta(\theta-\theta_0)\delta(\phi-\phi_0)/r^2\sin\theta$
$(r>0, 0<\theta<\pi, 0<\phi<2\pi)$

where $x_1 = r\sin\theta\cos\phi$, $x_2 = r\sin\theta\sin\phi$, $x_3 = r\cos\theta$,

(iii) $$\delta(\boldsymbol{x}-\boldsymbol{x}_0) = \frac{\delta(r-r_0)\prod_{m=1}^{n-1}\delta(\theta_m-\theta_{m0})}{r^{n-1}\sin^{n-2}\theta_1\sin^{n-3}\theta_2\ldots\sin\theta_{n-2}}$$

for $r>0, 0<\theta_1<\pi, \ldots, 0<\theta_{n-2}<\pi, 0<\theta_{n-1}<2\pi$ and

$$x_1 = r\cos\theta_1, x_2 = r\sin\theta_1\cos\theta_2, \ldots, x_n = r\sin\theta_1\ldots\sin\theta_{n-1}.$$

39. If $x_1 = l\cosh u\cos v$, $x_2 = l\sinh u\sin v$ show that

$$\delta(\boldsymbol{x}-\boldsymbol{x}_0) = \delta(u-u_0)\delta(v-v_0)/l^2(\cosh^2 u - \cos^2 v) \qquad (u>0, 0<v<2\pi).$$

40. If $\boldsymbol{f}$ satisfies Definition 8.7 prove that

$$\delta\{\boldsymbol{f}(\boldsymbol{x})\} = \delta(\boldsymbol{x}-\boldsymbol{x}_0)/|J_0|$$

where $\boldsymbol{f}(\boldsymbol{x}_0) = \mathbf{0}$ and J_0 is the value of J at $\boldsymbol{x} = \boldsymbol{x}_0$.
Is there a result analogous to Exercise 32?

8.4. Delta functions

It will be noted that (31) involves only f_n and not all of $\boldsymbol{f}$. Suppose, therefore, that we are given an infinitely differentiable $f(\boldsymbol{x})$ such that grad $f \neq 0$ anywhere on $f=0$; then the hypersurface $f=0$ has no singular points and we can find (at any rate, locally) an $\boldsymbol{f}$ whose f_n coincides with f. Then (31) will be valid with f_n replaced by this f; it can be written in the alternative form

$$\int_{-\infty}^{\infty}\delta(f)\gamma(\boldsymbol{x})\,\mathrm{d}\boldsymbol{x} = \int_S \frac{\gamma(\boldsymbol{x})}{|\text{grad}\, f|}\mathrm{d}S \tag{34}$$

where S is the hypersurface $f=0$. It can be deduced at once that

$$f\delta(f) = 0. \tag{35}$$

A derivative with respect to f gives

$$f\delta'(f) + \delta(f) = 0 \tag{36}$$

and, repeating the process, we obtain

$$f\delta^{(k)}(f)+k\delta^{(k-1)}(f)=0. \tag{37}$$

If f_1 and f_2 are both of the same type as f, i.e. they are both infinitely differentiable and grad $f_1 \neq 0$ on $f_1=0$, grad $f_2 \neq 0$ on $f_2=0$ then $f_1 f_2$ can be used for f provided that the hypersurfaces $f_1=0$ and $f_2=0$ have no point in common. With this proviso (34) gives

$$\begin{aligned}\int_{-\infty}^{\infty}\delta(f_1 f_2)\gamma(\boldsymbol{x})\,\mathrm{d}\boldsymbol{x} &= \int_{S_1+S_2}\frac{\gamma(\boldsymbol{x})}{|\mathrm{grad}\, f_1 f_2|}\mathrm{d}S \\ &= \int_{S_1}\frac{\gamma(\boldsymbol{x})}{|f_2\,\mathrm{grad}\, f_1|}\mathrm{d}S+\int_{S_2}\frac{\gamma(\boldsymbol{x})}{|f_1\,\mathrm{grad}\, f_2|}\mathrm{d}S\end{aligned}$$

where S_1 and S_2 are the hypersurfaces $f_1=0$ and $f_2=0$ respectively. Thus, if $f_1=0$ and $f_2=0$ have no points in common,

$$\delta(f_1 f_2)=|f_2|^{-1}\delta(f_1)+|f_1|^{-1}\delta(f_2). \tag{38}$$

In particular, if f_2 does not vanish anywhere,

$$\delta(f_1 f_2)=|f_2|^{-1}\delta(f_1). \tag{39}$$

Repeated derivatives with respect to f_1 then supply

$$\delta^{(k)}(f_1 f_2)=f_2^{-k}|f_2|^{-1}\delta^{(k)}(f_1) \tag{40}$$

when f_2 does not vanish anywhere.

Example 12. If $a>0$, $r-a$ is a suitable f and (34) becomes

$$\begin{aligned}\int_{-\infty}^{\infty}\delta(r-a)\gamma(\boldsymbol{x})\,\mathrm{d}\boldsymbol{x} &= \int_{\Omega_a}\frac{\gamma(\boldsymbol{x})}{|\mathrm{grad}\,(r-a)|}\mathrm{d}\Omega \\ &= \int_{\Omega_a}\gamma(\boldsymbol{x})\,\mathrm{d}\Omega,\end{aligned}$$

Ω_a being the sphere $r=a$.

Somewhat more complicated formulae arise for δ-functions with more than one argument. Let $f_1, f_2, \ldots, f_m$ be infinitely differentiable functions such that at each point of the hypersurface $f_1=0, f_2=0, \ldots, f_m=0$ a local system of coordinates $\boldsymbol{y}$ can be chosen so that $\boldsymbol{y}=\boldsymbol{f}(\boldsymbol{x})$

has the properties described at the beginning of the section. Then

$$\int_{-\infty}^{\infty} \delta(f_1, \dots, f_m)\gamma(\boldsymbol{x})\mathrm{d}\boldsymbol{x} = \int_{-\infty}^{\infty} \delta(y_1, \dots, y_m)\gamma(f^{-1})\frac{\mathrm{d}\boldsymbol{y}}{|J|}$$

$$= \int_{-\infty}^{\infty} \cdots \int_{-\infty}^{\infty} \left[\frac{\gamma(f^{-1})}{|J|}\right]_{y_1=0,\dots,y_m=0} \times \mathrm{d}y_{m+1} \dots \mathrm{d}y_n.$$

The integral on the right is essentially an integral over the surface $f_1 = 0, \dots, f_m = 0$ and so is independent of the particular choice of $y_{m+1}, \dots, y_n$, though some choices may be more convenient than others for calculation in certain cases.

It is clear that

$$f_i \delta(f_1, \dots, f_m) = 0 \qquad (i = 1, \dots, m) \tag{41}$$

and derivatives with respect to f_i lead to

$$f_i \frac{\partial^k}{\partial f_i^k} \delta(f_1, \dots, f_m) + k \frac{\partial^{k-1}}{\partial f_i^{k-1}} \delta(f_1, \dots, f_m) = 0. \tag{42}$$

Exercises

41. Prove that, if $a > 0$,

(i) $$\int_{-\infty}^{\infty} \delta^{(k)}(r-a)\gamma(\boldsymbol{x})\mathrm{d}\boldsymbol{x} = \frac{(-1)^k}{a^{n-1}} \int_{\Omega_a} \left[\frac{\partial^k}{\partial r^k}(r^{n-1}\gamma)\right]_{r=a} \mathrm{d}\Omega,$$

(ii) $$\int_{-\infty}^{\infty} \delta^{(k)}(r^2-a^2)\gamma(\boldsymbol{x})\mathrm{d}\boldsymbol{x} = \frac{(-1)^k}{2a^{n-1}} \int_{\Omega_a} \left[\left(\frac{\partial}{2r\partial r}\right)^k (r^{n-2}\gamma)\right]_{r=a} \mathrm{d}\Omega.$$

42. Show that, when $t > 0$,

$$(\partial_1^2 + \dots + \partial_n^2 - \partial_t^2)\delta(r^2 - t^2) = (2n-6)\delta'(r^2 - t^2)$$

so that, when $n = 3$,

$$(\partial_1^2 + \dots + \partial_n^2 - \partial_t^2)\delta(r^2 - t^2) = 0.$$

More generally, when n is odd,

$$(\partial_1^2 + \dots + \partial_n^2 - \partial_t^2)\delta^{(n-3)/2}(r^2 - t^2) = 0.$$

(Hint: Equation (37).)

43. Prove that, if $a > 0$,

$$\delta(r^2 - a^2) = \frac{1}{2a}\{\delta(r-a) + \delta(r+a)\}$$

and that

$$\delta^{(k)}(r^2 - a^2) = (r+a)^{-k-1}\delta^{(k)}(r-a).$$

44. If $a > 0$ and $n = 2$ show that

$$\int_{-\infty}^{\infty} \delta(x_1 x_2 - a)\gamma(\boldsymbol{x})\mathrm{d}\boldsymbol{x} = \int_{-\infty}^{\infty} \gamma\left(x_1, \frac{a}{x_1}\right)\frac{\mathrm{d}x_1}{|x_1|}.$$

45. Prove that

$$\int_{-\infty}^{\infty} \delta^{(k)}(f)\gamma(\boldsymbol{x})\mathrm{d}\boldsymbol{x} = \int_S \left(\frac{1}{|\operatorname{grad} f|}\frac{\partial}{\partial \nu}\right)^k \frac{\gamma(\boldsymbol{x})}{|\operatorname{grad} f|}\mathrm{d}S.$$

(Hint: Equation (8) of Chapter 7.)

46. If $i \neq j$ and $\delta(f_1, \ldots, f_m)$ is abbreviated by δ show that

$$\delta + f_i \frac{\partial \delta}{\partial f_i} + f_j \frac{\partial \delta}{\partial f_j} + f_i f_j \frac{\partial^2 \delta}{\partial f_i \partial f_j} = 0.$$

When we turn to a generalised function such as $\delta(x_1 x_2)$, (38) can no longer be applied since $x_1 = 0$ and $x_2 = 0$ intersect and have points in common. In fact it is necessary to specify a meaning for $\delta(x_1 x_2)$ because it does not come within the scope of the preceding theory. We shall define it so that it complies with (38).

Definition 8.9. *The generalised function* $\delta^{(k)}(x_1 x_2)$ $(k = 0, 1, 2, \ldots)$ *is defined by*

$$\delta^{(k)}(x_1 x_2) = \delta^{(k)}(x_2)x_1^{-k-1}\operatorname{sgn} x_1 + \delta^{(k)}(x_1)x_2^{-k-1}\operatorname{sgn} x_2.$$

Since $x_1^{-k-1}\operatorname{sgn} x_1$ and $\delta^{(k)}(x_2)$ are generalised functions their direct product is well defined (Definition 7.3) and so the right-hand sides do give generalised functions.

There is no difficulty in verifying that

$$\delta^{(k)}(-x_1 x_2) = (-1)^k \delta^{(k)}(x_1 x_2)$$

and that

$$x_1 x_2 \delta(x_1 x_2) = 0.$$

By means of a rotation of axes (§8.1) we can deduce that

$$\delta(x_1^2 - x_2^2) = |x_1 - x_2|^{-1}\delta(x_1 + x_2) + |x_1 + x_2|^{-1}\delta(x_1 - x_2) \quad (43)$$

with a corresponding formula for $\delta^{(k)}(x_1^2 - x_2^2)$.

Exercises

47. Prove that $\partial_1 \mathrm{H}(x_1 x_2) = x_2 \delta(x_1 x_2)$ but that $\partial_1 \partial_2 \mathrm{H}(x_1 x_2) = 2\delta(x_1)\delta(x_2)$.

48. Prove that, when $n = 2$, $(\partial_1^2 - \partial_2^2)\mathrm{H}(x_1^2 - x_2^2) = 4\delta(\boldsymbol{x})$.

49. Show that $x_1 x_2 \delta^{(k)}(x_1 x_2) + k\delta^{(k-1)}(x_1 x_2) = 0$.

50. By means of Exercise 44 and the corresponding result for $\delta(x_1 x_2)$ show that

$$\operatorname*{Lim}_{a\to+0} \{\delta(x_1 x_2 - a) + 2\delta(x_1, x_2)\ln a\} = \delta(x_1 x_2).$$

51. Prove that

$$(x_1\partial_1 + x_2\partial_2)\delta^{(k-1)}(x_1 x_2) = -2k\delta^{(k-1)}(x_1 x_2) - \frac{(-1)^k}{(k-1)!}4\delta^{(k-1)}(x_1)\delta^{(k-1)}(x_2)$$

showing that $\delta^{(k)}(x_1 x_2)$ is not a homogeneous generalised function.

52. When $n = 2$ prove that the Fourier transform of $\delta(x_1 x_2)$ is $2\{2\psi(0) - \ln|\alpha_1||\alpha_2|\}$ and deduce that the Fourier transform of $\delta(x_1^2 - x_2^2)$ is $2\psi(0) - \ln|\alpha_1 + \alpha_2||\alpha_1 - \alpha_2|$. The Fourier transform of $\delta^{(k)}(x_1^2 - x_2^2)$ is $\{(\alpha_1^2 - \alpha_2^2)^k/k!2^{2k}\}\{2\psi(k) - \ln|\alpha_1^2 - \alpha_2^2|\}$.

The corresponding formulae for spaces of a higher number of dimensions are more complicated. Proceed formally in order to determine the appropriate definition. Consider $\delta(x_1^2 - x_2^2 - \ldots - x_n^2)$ with $n \geq 3$. By changing to spherical polar coordinates in the $x_2, \ldots, x_n$-space we can write this as $\delta(x_1^2 - r_{n-1}^2)$ where r_{n-1} is the radial distance in a space of $n-1$ dimensions. This entity could be interpreted by means of (43) provided that an appropriate significance can be attached to the terms. Now

$$\int_{-\infty}^{\infty} \delta(x_1^2 - x_2^2 - \ldots - x_n^2)\gamma(\boldsymbol{x})\,d\boldsymbol{x}$$

$$= \int_0^{\infty}\int_0^{\infty} \delta(x_1^2 - r_{n-1}^2) r_{n-1}^{n-2}\{\gamma_1(x_1, r_{n-1}) + \gamma_1(-x_1, r_{n-1})\} \times dx_1\,dr_{n-1}$$

where

$$\gamma_1(x_1, r_{n-1}) = \int_{\Omega_{n-1}} \gamma(x_1, r_{n-1}\hat{x}_2, \ldots, r_{n-1}\hat{x}_n)\,d\Omega_{n-1} \tag{44}$$

and $(\hat{x}_2, \ldots, \hat{x}_n)$ is a point on the unit sphere Ω_{n-1} in $n-1$ dimensions. Clearly, γ_1 is a good function in a space with Cartesian coordinates x_1 and r_{n-1}. Putting $x_1 = \frac{1}{2}(x+y)$, $r_{n-1} = \frac{1}{2}(y-x)$ we obtain

$$2^{1-n}\int_0^{\infty}\int_{-y}^{y} \delta(xy)(y-x)^{n-2}\gamma_2(x, y)\,dx\,dy$$

where

$$\gamma_2(x, y) = \gamma_1(x_1, r_{n-1}) + \gamma_1(-x_1, r_{n-1}). \tag{45}$$

From Definition 8.9 $\delta(xy)$ is zero except on the x-and y-axes. Of these only the positive y-axis is involved in the integration and this suggests

$$\int_{-\infty}^{\infty} \delta(x_1^2 - x_2^2 - \dots - x_n^2)\gamma(\boldsymbol{x})\,d\boldsymbol{x}$$

$$= 2^{1-n} \int_0^{\infty} \int_{-y}^{y} y^{-1}\delta(x)(y - x)^{n-2}\gamma_2(x, y)\,dx\,dy$$

$$= 2^{1-n} \int_0^{\infty} y^{n-3}\gamma_2(0, y)\,dy$$

with the integration understood in the sense of §4.4. The substitution $x = t\xi, y = \xi$ also gives the formula

$$\int_{-\infty}^{\infty} \delta(x_1^2 - x_2^2 - \dots - x_n^2)\gamma(\boldsymbol{x})\,d\boldsymbol{x}$$

$$= 2^{1-n} \int_0^{\infty} \int_{-1}^{1} \xi^{n-3}\delta(t)(1 - t)^{n-2}\gamma_2(t\xi, \xi)\,dt\,d\xi.$$

This forms the basis of

Definition 8.10. *The generalised function* $\delta^{(k)}(x_1^2 - x_2^2 - \dots - x_n^2)$ *is defined so that*

$$\int_{-\infty}^{\infty} \delta^{(k)}(x_1^2 - x_2^2 - \dots - x_n^2)\gamma(\boldsymbol{x})\,d\boldsymbol{x}$$

$$= 2^{1-n} \int_0^{\infty} \int_{-1}^{1} \xi^{n-2k-3}\delta^{(k)}(t)(1 - t)^{n-2}\gamma_2(t\xi, \xi)\,dt\,d\xi.$$

A similar technique may be employed for the generalised function $\delta(x_1^2 + \dots + x_p^2 - x_{p+1}^2 - \dots - x_n^2)$. In this case transform to spherical polar coordinates in $x_1, \dots, x_p$-space and in $x_{p+1}, \dots, x_n$-space so as to arrive at $\delta(r_p^2 - r_{n-p}^2)$ where r_p and r_{n-p} are the radial distances in the two spaces respectively. We then reach

$$\int_0^{\infty} \int_0^{\infty} \delta(r_p^2 - r_{n-p}^2)r_p^{p-1} r_{n-p}^{n-p-1} \gamma_3(r_p, r_{n-p})\,dr_p\,dr_{n-p}$$

where

$$\gamma_3(r_p, r_{n-p}) = \int_{\Omega_p} \int_{\Omega_{n-p}} \gamma(r_p\hat{x}_1, \dots, r_p\hat{x}_p, r_{n-p}\hat{x}_{p+1}, \dots, r_{n-p}\hat{x}_n)\mathrm{d}\Omega_p\mathrm{d}\Omega_{n-p} \tag{46}$$

and $(\hat{x}_1, \dots, \hat{x}_p)$, $(\hat{x}_{p+1}, \dots, \hat{x}_n)$ are points on Ω_p, Ω_{n-p} respectively, the unit spheres of p and $n-p$ dimensions. Thereafter, the argument runs parallel and if we write

$$\gamma_2(x, y) = \gamma_3\{\tfrac{1}{2}(x+y), \tfrac{1}{2}(y-x)\} \tag{47}$$

we have

Definition 8.11. *The generalised function* $\delta^{(k)}(x_1^2 + \dots + x_p^2 - \dots - x_n^2)$ *with* $n-1 \geq p \geq 1$ *is defined so that*

$$\int_{-\infty}^{\infty} \delta^{(k)}(x_1^2 + \dots + x_p^2 - \dots - x_n^2)\gamma(\boldsymbol{x})\mathrm{d}\boldsymbol{x}$$

$$= 2^{1-n} \int_0^{\infty} \int_{-1}^{1} \xi^{n-2k-3}\delta^{(k)}(t)(1+t)^{p-1}(1-t)^{n-p-1}\gamma_2(t\xi, \xi)\mathrm{d}t\,\mathrm{d}\xi.$$

The case $p = 1$ is included in Definition 8.11 and the same symbol γ_2 used as in Definition 8.10 because the latter can be subsumed in the former provided that a certain convention be adopted about integration over Ω_1, namely that

$$\int_{\Omega_1} \gamma(x_1, x_2, \dots, x_n)\mathrm{d}\Omega_1 = \gamma(x_1, x_2, \dots, x_n) + \gamma(-x_1, x_2, \dots, x_n)$$

so that (45) and (47) agree. With this convention

$$\int_{\Omega_n} \mathrm{d}\Omega_n = \Omega_n = \frac{2\pi^{n/2}}{(\frac{1}{2}n - 1)!} \tag{48}$$

for $n \geq 1$.

For the calculation of derivatives it is convenient to know that if $\gamma(\boldsymbol{x})$ is replaced by $(\partial_1^2 + \dots + \partial_p^2 - \dots - \partial_n^2)\gamma_0(\boldsymbol{x})$ then in Definition 8.11, $\gamma_2(t\xi, \xi)$ is replaced by $l\gamma_2$ where the operator l is specified by

$$l \equiv \frac{4}{\xi}\frac{\partial^2}{\partial t \partial \xi} - \frac{4t}{\xi^2}\frac{\partial^2}{\partial t^2} + \left\{(p-1)\frac{1-t}{1+t} + (n-p-1)\frac{1+t}{1-t} - 2\right\}\frac{2}{\xi^2}\frac{\partial}{\partial t}$$
$$+ \left(\frac{p-1}{1+t} - \frac{n-p-1}{1-t}\right)\frac{2}{\xi}\frac{\partial}{\partial \xi}. \tag{49}$$

Note also that

$$2\left(\frac{\partial}{\partial \xi} - \frac{t}{\xi}\frac{\partial}{\partial t}\right)\left\{\xi^{n-3}(1+t)^{p-1}(1-t)^{n-p-1}\frac{\partial \gamma_2}{\partial t}\right\}$$
$$+ 2\xi^{n-3}\frac{\partial}{\partial t}\left[(1+t)^{p-1}(1-t)^{n-p-1}\left\{\frac{\partial}{\partial \xi} - \frac{t}{\xi}\frac{\partial}{\partial t}\right\}\gamma_2\right]$$
$$= \xi^{n-2}(1+t)^{p-1}(1-t)^{n-p-1} l\gamma_2. \tag{50}$$

If $\sum_{j=1}^n \sum_{k=1}^n a_{jk}x_j x_k$ is a quadratic form with real coefficients a_{jk} and $a_{jk} = a_{kj}$ there is an orthogonal mapping (rotation of axes) which converts it to $\lambda_1 y_1^2 + \ldots + \lambda_p y_p^2 - \lambda_{p+1} y_{p+1}^2 - \ldots - \lambda_n y_n^2$ where $\lambda_1, \ldots, \lambda_n$ are all positive. It may happen that $p = n$ or $p = 0$. A change of scale $y_j\sqrt{\lambda_j} = z_j$ now leads to $z_1^2 + \ldots + z_p^2 - z_{p+1}^2 - \ldots - z_n^2$. Proceeding in this way and taking advantage of the theory of §8.1 we can assign a meaning to $\delta(\sum_{j=1}^n \sum_{k=1}^n a_{jk}x_j x_k)$ by means of the definitions that have just been given, provided that both $+1$ and -1 occur as coefficients in the quadratic form in z.

Exercises

53. Show that

(i) $(x_1^2 - \ldots - x_n^2)\delta^{(k)}(x_1^2 - \ldots - x_n^2) + k\delta^{(k-1)}(x_1^2 - \ldots - x_n^2) = 0,$

(ii) $(x_1^2 + \ldots + x_p^2 - \ldots - x_n^2)\delta^{(k)}(x_1^2 + \ldots + x_p^2 - \ldots - x_n^2)$
$+ k\delta^{(k-1)}(x_1^2 + \ldots + x_p^2 - \ldots - x_n^2) = 0.$

54. Prove that $\partial_1 \mathrm{H}(v) = 2x_1\delta(v)$ for $v = x_1^2 - \ldots - x_n^2$. Show also that for $1 \le p < n-1$

(i) $(\partial_1^2 + \ldots + \partial_p^2 - \ldots - \partial_n^2)\mathrm{H}(x_1^2 + \ldots + x_p^2 - \ldots - x_n^2)$
$= 2(n-2)\delta(x_1^2 + \ldots + x_p^2 - \ldots - x_n^2) \quad (n > 2),$

(ii) $(\partial_1^2 + \ldots + \partial_p^2 - \ldots - \partial_n^2)\delta^{(k)}(x_1^2 + \ldots + x_p^2 - \ldots - x_n^2)$
$= 2(n-2k-4)\delta^{(k+1)}(x_1^2 + \ldots + x_p^2 - \ldots - x_n^2)$

if n is odd or if n is even and $k < \frac{1}{2}n - 2$. (Hint: Equation (11) of Chapter 4; the odd derivatives of γ_3 vanish at $\xi = 0$.)

(iii) $(\partial_1^2 + \ldots + \partial_p^2 - \ldots - \partial_n^2)\,\delta^{((n/2)-2)}(x_1^2 + \ldots + x_p^2 - \ldots - x_n^2)$

$$= 4\pi^{(n/2)-1} \sin \tfrac{1}{2}p\pi\ \delta(\boldsymbol{x})$$

when n is even. $\left(\text{Hint: } \gamma_2(0,0) = \Omega_p \Omega_{n-p} \gamma(\mathbf{0}) \text{ and, at } t = 0,\right.$

$$\left.\frac{\mathrm{d}^{(n/2)-1}}{\mathrm{d}t^{(n/2)-1}}(1+t)^{p-1}(1-t)^{n-p-1} = \frac{(n-p-1)!(-1)^{(n/2)-1}2^{p-1}\pi^{1/2}}{(\frac{1}{2}n - \frac{1}{2}p - \frac{1}{2})!(-\frac{1}{2}p)!}.\right)$$

55. Generalise Exercises 53 and 54 to $\sum_{j=1}^{n} \sum_{k=1}^{n} a_{jk} x_j x_k$ when the quadratic form is neither positive-definite nor negative-definite. Show, in particular, that if $\det A \neq 0$ and $n > 2$

$$\sum_{r=1}^{n} \sum_{s=1}^{n} a^{rs} \partial_r \partial_s \mathrm{H}\left(\sum_{j=1}^{n} \sum_{k=1}^{n} a_{jk} x_j x_k\right) = 2(n-2)\delta\left(\sum_{j=1}^{n} \sum_{k=1}^{n} a_{jk} x_j x_k\right)$$

where a^{rs} is such that $\sum_{s=1}^{n} a^{rs} a_{sk}$ equals 1 if $r = k$ and is zero if $r \neq k$.

8.5. Periodic and other changes of variables

The changes of variable delineated in §8.3 are not completely satisfactory for all practical purposes. The reason why this is so can be seen in the use of polar coordinates in two dimensions $x_1 = r\cos\phi$, $x_2 = r\sin\phi$. In order to maintain the one-to-one character of the mapping it is necessary to exclude the origin $r = 0$ and some line, say $\phi = 0$. Then all our results for generalised functions would be subject to $r > 0, 0 < \phi < 2\pi$ (cf. Exercise 38) which is an undesirable constraint in applications. The problem involving ϕ can be overcome by considering periodic generalised functions and will be dealt with now; the problem at $r = 0$ is rather more difficult.

Definition 8.12. *The generalised function $g(\boldsymbol{x})$ is said to be periodic with period $2l_1$ in $x_1, \ldots, 2l_k$ in x_k $(k \leq n)$ if, and only if,*

$$g(x_1 + 2l_1, \ldots, x_k + 2l_k, x_{k+1}, \ldots, x_n) = g(\boldsymbol{x}).$$

Theorem 8.7. *If $g(\boldsymbol{x})$ is periodic as in Definition* 8.12 *then*

$$g(\boldsymbol{x}) = \sum_{m_1=-\infty}^{\infty} \cdots \sum_{m_k=-\infty}^{\infty} g_{m_1 \ldots m_k}(x_{k+1}, \ldots, x_n) e^{im_1\pi x_1/l_1 + \ldots + im_k \pi x_k/l_k}$$

where

$$g_{m_1\dots m_k} = \frac{1}{2^k l_1 \dots l_k}\int_{-\infty}^{\infty}\dots\int_{-\infty}^{\infty} g(\boldsymbol{x})\tau\left(\frac{x_1}{2l_1}\right)\dots\tau\left(\frac{x_k}{2l_k}\right)$$
$$\times\, \mathrm{e}^{-\mathrm{i}m_1\pi x_1/l_1 - \dots \mathrm{i}m_k\pi x_k/l_k}\,\mathrm{d}x_1\dots\mathrm{d}x_k$$

and τ *is as defined just after Corollary* 5.13.

Proof. Let $\gamma_{(1)}(x_1)$ be a good function of x_1 and $\gamma_{(2)}(x_2,\dots,x_n)$ a good function of $x_2,\dots,x_n$. By Theorem 7.22 $\int_{-\infty}^{\infty}\dots\int_{-\infty}^{\infty} g\gamma_{(2)}\,\mathrm{d}x_2\dots\mathrm{d}x_n$ is a generalised function on R_1 and is periodic with period $2l_1$ in x_1. Hence, from Theorem 5.19,

$$\int_{-\infty}^{\infty}\dots\int_{-\infty}^{\infty} g(\boldsymbol{x})\gamma_{(2)}\,\mathrm{d}x_2\dots\mathrm{d}x_n = \sum_{m=-\infty}^{\infty} c_m\,\mathrm{e}^{\mathrm{i}m\pi x_1/l_1}$$

where

$$c_m = \frac{1}{2l_1}\int_{-\infty}^{\infty}\tau\left(\frac{x_1}{2l_1}\right)\mathrm{e}^{-\mathrm{i}m\pi x_1/l_1}\int_{-\infty}^{\infty}\dots\int_{-\infty}^{\infty} g(\boldsymbol{x})\gamma_{(2)}\,\mathrm{d}x_2\dots\mathrm{d}x_n\,\mathrm{d}x_1$$
$$= \frac{1}{2l_1}\int_{-\infty}^{\infty}\dots\int_{-\infty}^{\infty}\gamma_{(2)}\int_{-\infty}^{\infty} g(\boldsymbol{x})\tau\left(\frac{x_1}{2l_1}\right)\mathrm{e}^{-\mathrm{i}m\pi x_1/l_1}\,\mathrm{d}x_1\,\mathrm{d}x_2\dots\mathrm{d}x_n$$

by Theorem 7.22. Consequently

$$g(\boldsymbol{x}) = \sum_{m=-\infty}^{\infty} g_m(x_2,\dots,x_n)\mathrm{e}^{\mathrm{i}m\pi x_1/l_1} \tag{51}$$

where

$$g_m = \frac{1}{2l_1}\int_{-\infty}^{\infty} g(\boldsymbol{x})\tau\left(\frac{x_1}{2l_1}\right)\mathrm{e}^{-\mathrm{i}m\pi x_1/l_1}\,\mathrm{d}x_1.$$

This proves the theorem for $k = 1$. Using the fact that g_m is periodic in x_2 we obtain an expansion of the type (51) for g_m. The theorem is then proved for $k = 2$ and the proof for general k follows from repetition of the process. □

Corollary 8.7. *The Fourier transform of a periodic* g *is*

$$G(\alpha) = (2\pi)^k \sum_{m_1=-\infty}^{\infty}\dots\sum_{m_k=-\infty}^{\infty} G_{m_1\dots m_k}(\alpha_{k+1},\dots,\alpha_n)$$
$$\times\,\delta(\alpha_1 - m_1\pi/l_1)\dots\delta(\alpha_k - m_k\pi/l_k).$$

This is an immediate consequence of Theorem 8.7 and Corollary 7.23.

Now suppose that $f(\boldsymbol{x})$ is periodic with period $2l_n$ in x_n. We also suppose that, for fixed x_n, a domain T of R_{n-1} is mapped onto U in such a way that, as $\boldsymbol{x}$ approaches a boundary point of T, $\boldsymbol{y}$ approaches a boundary point of U consistent with the conditions of §8.3, and conversely. We are interested only in cases where the domain in $\boldsymbol{y}$-space does not impose any restriction on x_n since the other cases have been covered already. With this understanding we have

Theorem 8.8. *If* $\{\gamma_m(\boldsymbol{y})\}$ *is a regular sequence defining* $g(\boldsymbol{y})$ *then* $\{\gamma_m(f(\boldsymbol{x}))\}$ *determines a periodic generalised function which will be denoted by* $g\{f(\boldsymbol{x})\}$.

Proof. Since f is periodic $\gamma_m(f)$ is periodic with period $2l_n$ in x_n. Therefore there is an expansion

$$\gamma_m(f) = \sum_{p=-\infty}^{\infty} \gamma_{mp}(x_1, \dots, x_{n-1}) e^{ip\pi x_n/l_n}$$

and

$$\int_T \gamma_m(f)\gamma(x_1, \dots, x_{n-1}) dx_1 \dots dx_{n-1} = \sum_{p=-\infty}^{\infty} c_{mp} e^{ip\pi x_n/l_n}$$

where γ is a good function which vanishes identically outside T and c_{mp} is a constant. Now

$$\int_0^{2l_n} e^{-ip\pi x_n/l_n} \int_T \gamma_m(f)\gamma(x_1, \dots, x_{n-1}) d\boldsymbol{x}$$
$$= \int_0^{2l_n} \int_U \gamma_m(\boldsymbol{y})\gamma(x_1, \dots, x_{n-1}) e^{-ip\pi x_n/l_n} \frac{d\boldsymbol{y}}{|J|}.$$

But, by the assumptions made on f, $\gamma e^{-ip\pi x_n/l_n}$ is a good function on $\boldsymbol{y}$-space and so the right-hand side tends to a limit as $m \to \infty$, γ_m being replaced by g in the limit. It follows, from Theorem 5.22, that

$$\lim_{m\to\infty} \int_T \gamma_m(f)\gamma(x_1, \dots, x_{n-1}) dx_1 \dots dx_{n-1} = g_n(x_n)$$

where g_n is periodic. Consequently, if $\gamma_{(n)}(x_n)$ is any good function of x_n,

$$\lim_{m\to\infty}\int_{-\infty}^{\infty}\gamma_{(n)}(x_n)\int_T \gamma_m(f)\gamma(x_1,\dots,x_{n-1})\,\mathrm{d}\boldsymbol{x}=\int_{-\infty}^{\infty}g_n(x_n)\gamma_{(n)}(x_n)\,\mathrm{d}x_n$$

which implies, on account of Corollary 7.21*a*, that a generalised function is defined without restriction on x_n (other than that of periodicity) for $x_1,\dots,x_{n-1}\in T$. The theorem is therefore proved. □

In the course of the proof it has been shown that

$$\int_{-\infty}^{\infty}\tau\left(\frac{x_n}{2l_n}\right)\mathrm{e}^{-\mathrm{i}p\pi x_n/l_n}\int_T g\{f(\boldsymbol{x})\}\gamma(x_1,\dots,x_{n-1})\,\mathrm{d}\boldsymbol{x}$$

$$=\lim_{m\to\infty}\int_{-\infty}^{\infty}\tau\left(\frac{x_n}{2l_n}\right)\mathrm{e}^{-\mathrm{i}p\pi x_n/l_n}\int_T \gamma_m\{f(\boldsymbol{x})\}\gamma(x_1,\dots,x_{n-1})\,\mathrm{d}\boldsymbol{x}$$

$$=\lim_{m\to\infty}\int_0^{2l_n}\mathrm{e}^{-\mathrm{i}p\pi x_n/l_n}\int_T \gamma_m\{f(\boldsymbol{x})\}\gamma(x_1,\dots,x_{n-1})\,\mathrm{d}\boldsymbol{x}$$

$$=\int_0^{2l_n}\int_U g(\boldsymbol{y})\gamma(x_1,\dots,x_{n-1})\mathrm{e}^{-\mathrm{i}p\pi x_n/l_n}\frac{\mathrm{d}\boldsymbol{y}}{|J|}. \tag{52}$$

An alternative way of expressing this is to say that, because of Theorem 8.7,

$$g\{f(\boldsymbol{x})\}=\sum_{p=-\infty}^{\infty}g_p(x_1,\dots,x_{n-1})\mathrm{e}^{\mathrm{i}p\pi x_n/l_n} \tag{53}$$

where

$$\int_T g_p(x_1,\dots,x_{n-1})\gamma(x_1,\dots,x_{n-1})\,\mathrm{d}x_1\dots\mathrm{d}x_{n-1}$$

$$=\frac{1}{2l_n}\int_0^{2l_n}\int_U g(\boldsymbol{y})\gamma(x_1,\dots,x_{n-1})\mathrm{e}^{-\mathrm{i}p\pi x_n/l_n}\frac{\mathrm{d}\boldsymbol{y}}{|J|}. \tag{54}$$

Example 13. In two dimensions $y_1=r\cos\phi$, $y_2=r\sin\phi$ satisfies the conditions on the transformation provided that $r>0$, $x_n=\phi$, $l_n=\pi$. Hence, from (53) and (54),

$$g(y_1,y_2)=\sum_{p=-\infty}^{\infty}g_p(r)\mathrm{e}^{\mathrm{i}p\phi}\quad(r>0) \tag{55}$$

where

$$\int_0^\infty g_p(r)\gamma(r)\,\mathrm{d}r = \frac{1}{2\pi}\int_{-\infty}^\infty\int_{-\infty}^\infty g(y_1, y_2)\gamma(r)\,\mathrm{e}^{-\mathrm{i}p\phi}(r^{-1})\,\mathrm{d}y_1\,\mathrm{d}y_2, \tag{56}$$

$\gamma(r)$ being any good function which, together with all its derivatives, vanishes at the origin and r, ϕ being understood to be $(y_1^2 + y_2^2)^{1/2}$ and $\tan^{-1}(y_2/y_1)$ with $0 \le \phi \le 2\pi$ respectively.

Suppose $g(y_1, y_2) = \delta(\boldsymbol{y} - \boldsymbol{y}_0)$ where $\boldsymbol{y}_0 \ne 0$. Then the right-hand side of (56) is $\gamma(r_0)\mathrm{e}^{-\mathrm{i}p\phi_0}/2\pi r_0$ where r_0, ϕ_0 are the values of r, ϕ which correspond to $\boldsymbol{y}_0$. Hence $g_p(r) = \mathrm{e}^{-\mathrm{i}p\phi_0}\delta(r - r_0)/2\pi r_0$ and

$$\delta(\boldsymbol{y} - \boldsymbol{y}_0) = \sum_{p=-\infty}^{\infty} \mathrm{e}^{\mathrm{i}p(\phi - \phi_0)}\delta(r - r_0)/2\pi r_0 \quad (r > 0, r_0 > 0).$$

It now follows from (5.29) that

$$\delta(\boldsymbol{y} - \boldsymbol{y}_0) = \sum_{p=-\infty}^{\infty} \frac{1}{r_0}\delta(r - r_0)\delta(\phi - \phi_0 - 2p\pi) \quad (r > 0, r_0 > 0). \tag{57}$$

If $0 < \phi_0 < 2\pi$ the application of our rule concerning local behaviour enables us to say that

$$\delta(\boldsymbol{y} - \boldsymbol{y}_0) = (1/r_0)\delta(r - r_0)\delta(\phi - \phi_0) \quad (r > 0, r_0 > 0, 0 \le \phi \le 2\pi). \tag{58}$$

It is necessary to omit the origin in the example we have just considered because the Jacobian vanishes there. Consequently the one-to-one property of the transformation is lost there; in fact when $r = 0$ any value can be given to ϕ. It is most desirable not to exclude the origin from polar coordinates and we shall now indicate how this can be done. We shall confine our attention to polar coordinates in two dimensions but the reader will be able to see how to generalise the ideas to other transformations which have similar problems.

Suppose that $g(y_1, y_2)$ is a generalised function and we wish to write it in terms of the polar coordinates r, ϕ. Let $\{\gamma_m(\boldsymbol{y})\}$ be a regular sequence defining $g(\boldsymbol{y})$ and let $\gamma(\boldsymbol{y})$ be any good function.

Then

$$\int_{-\infty}^{\infty} g(y)\gamma(y)\,dy = \lim_{m\to\infty}\int_{-\infty}^{\infty}\gamma_m(y)\gamma(y)\,dy$$

$$= \lim_{m\to\infty}\int_0^{\infty}\int_0^{\infty}\gamma_m\gamma r\,dr\,d\phi$$

where now γ_m and γ are expressed in terms of r and ϕ, the change of variable being permissible since the integral on the right is a conventional one. Since γ and γ_m are periodic in ϕ,

$$\gamma = \sum_{n=-\infty}^{\infty} c_n(r)e^{in\phi}, \qquad \gamma_m = \sum_{n=-\infty}^{\infty}\gamma_{mn}(r)e^{in\phi}$$

where

$$c_n(r) = \frac{1}{2\pi}\int_0^{2\pi}\gamma e^{-in\phi}\,d\phi, \qquad \gamma_{mn} = \frac{1}{2\pi}\int_0^{2\pi}\gamma_m e^{-in\phi}\,d\phi \tag{59}$$

Hence

$$\int_{-\infty}^{\infty} g(y)\gamma(y)\,dy = \lim_{m\to\infty}\sum_{n=-\infty}^{\infty} 2\pi\int_0^{\infty}\gamma_{mn}(r)c_{-n}(r)r\,dr$$

the interchange of summation and integration being obviously valid. Now a permissible choice for γ is $c_p e^{ip\phi}$ (p any integer), and so the limit of each of the separate terms on the right must exist and

$$\int_{-\infty}^{\infty} g(y)\gamma(y)dy = \sum_{n=-\infty}^{\infty}\lim_{m\to\infty} 2\pi\int_0^{\infty}\gamma_{mn}(r)c_{-n}(r)r\,dr.$$

It is clear, from (59), that $|c_n| \leq |\gamma|$ and that $|c_n^{(p)}| \leq |\gamma^{(p)}|$; therefore c_n (and similarly γ_{mn}) is infinitely differentiable and a good function of r as $r \to \infty$. Further

$$c_n(-r) = \int_0^{2\pi}\gamma(-r\cos\phi,\ -r\sin\phi)e^{-in\phi}\,d\phi$$

$$= \int_{-\pi}^{\pi}\gamma(r\cos\theta, r\sin\theta)e^{-in(\pi+\theta)}\,d\theta \qquad (\phi = \pi + \theta)$$

$$= (-1)^n c_n(r)$$

on changing θ in $(-\pi, 0)$ to $\theta - 2\pi$. Hence c_n (and similarly γ_{mn}) is even or odd in r according as n is even or odd. Therefore rc_{-n} is an odd or even good function on $-\infty < r < \infty$ and so, by Definition 4.11, γ_{mn} defines an odd or even generalised function g_n according as n is even or odd. Hence we can write

$$g(y) = \sum_{n=-\infty}^{\infty} g_n(r)e^{in\phi} \qquad (r \geq 0) \tag{60}$$

where

$$\int_{-\infty}^{\infty} g(y)\gamma(y)\,dy = \sum_{n=-\infty}^{\infty} 2\pi \int_0^{\infty} g_n(r)c_{-n}(r)r\,dr. \tag{61}$$

In particular, by choosing $\gamma(y) = c_{-n}(r)e^{-in\phi}$, where c_{-n} is infinitely differentiable on $r \geq 0$ and good as $r \to \infty$,

$$\int_{-\infty}^{\infty} g(y)c_{-n}(r)e^{-in\phi}\,dy = 2\pi \int_0^{\infty} g_n(r)c_{-n}(r)r\,dr. \tag{62}$$

This method enables one to go from Cartesian to polar coordinates without any restrictions on r and ϕ.

Example 14. Let $g(y) = \delta(y)$. Clearly $g_n = 0$ $(n \neq 0)$ and, from (62),

$$c_0(0) = 2\pi \int_0^{\infty} g_0(r)rc_0(r)\,dr$$

where g_0 is odd. It follows from (4.34) that $g_0(r) = -\delta'(r)/\pi$. Hence

$$\delta(y) = -\delta'(r)/\pi \qquad (r \geq 0). \tag{63}$$

Since $r\delta'(r) = -\delta(r)$ this is sometimes written as $\delta(r)/\pi r$ but, if so, its significance is purely symbolic since no meaning has been attached to $\delta(r)r^{-1}$.

Exercises

56. If $g(x)$ is periodic as in Definition 8.12 and equal, for $0 < x_1 < 2l_1, \ldots, 0 < x_k < 2l_k$, to $f(x_1, \ldots, x_k)g_0(x_{k+1}, \ldots, x_n)$ where f is an integrable function show that

$$g(x) = g_0 \sum_{m_1=-\infty}^{\infty} \cdots \sum_{m_k=-\infty}^{\infty} c_{m_1 \ldots m_k} e^{im_1\pi x_1/l_1 + \ldots + im_k\pi x_k/l_k}$$

where

$$c_{m_1 \ldots m_k} = \frac{1}{2^k l_1 \ldots l_k} \int_0^{2l_1} \cdots \int_0^{2l_k} f(x_1, \ldots, x_k)e^{-im_1\pi x_1/l_1 - \ldots - im_k\pi x_k/l_k}\,dx_1 \ldots dx_k.$$

57. Prove that

$$\sum_{m_1=-\infty}^{\infty} \cdots \sum_{m_n=-\infty}^{\infty} \delta(x_1 - 2m_1 l_1) \ldots \delta(x_n - 2m_n l_n)$$
$$= \frac{1}{2^n l_1 \ldots l_n} \sum_{m_1=-\infty}^{\infty} \cdots \sum_{m_n=-\infty}^{\infty} e^{im_1\pi x_1/l_1 + \ldots + im_n\pi x_n/l_n}$$

and deduce that the Fourier transform of the left-hand side is

$$\frac{\pi^n}{l_1 \ldots l_n} \sum_{m_1=-\infty}^{\infty} \cdots \sum_{m_n=-\infty}^{\infty} \delta\left(\frac{m_1 \pi}{l_1} - \alpha_1\right) \ldots \delta\left(\frac{m_n \pi}{l_n} - \alpha_n\right).$$

58. Prove that, if $\lambda_1, \lambda_2, \ldots, \lambda_n$ are all positive,

$$\sum_{m_1=-\infty}^{\infty} \cdots \sum_{m_n=-\infty}^{\infty} \gamma(m_1 \lambda_1, \ldots, m_n \lambda_n)$$
$$= \frac{1}{\lambda_1 \ldots \lambda_n} \sum_{m_1=-\infty}^{\infty} \cdots \sum_{m_n=-\infty}^{\infty} \Gamma\left(\frac{2m_1 \pi}{\lambda_1}, \ldots, \frac{2m_n \pi}{\lambda_n}\right).$$

59. Prove that

$$1 + 2^n \sum_{m_1=1}^{\infty} \cdots \sum_{m_n=1}^{\infty} e^{-m_1^2\lambda_1^2 - \cdots - m_n^2\lambda_n^2} \cos(2m_1 \pi\lambda_1 a_1) \ldots \cos(2m_n \pi\lambda_n a_n)$$
$$= \frac{\pi^{n/2}}{\lambda_1 \ldots \lambda_n} e^{-\pi^2(a_1^2 + \cdots + a_n^2)} \left(1 + 2^n \sum_{m_1=1}^{\infty} \cdots \sum_{m_n=1}^{\infty} e^{-m_1^2\pi^2/\lambda_1^2 - \cdots - m_n^2\pi^2/\lambda_n^2}\right.$$
$$\left. \times \cosh \frac{2\pi^2 m_1 a_1}{\lambda_1} \ldots \cosh \frac{2\pi^2 m_n a_n}{\lambda_n}\right).$$

60. If $y_1 = r\cos\phi$, $y_2 = r\sin\phi$ prove that

$$r\partial_{y_1} g(y_1, y_2) = \tfrac{1}{2} \sum_{n=-\infty}^{\infty} [r\{g'_{n-1}(r) + g'_{n+1}(r)\} - (n-1)g_{n-1}$$
$$+ (n+1)g_{n+1}]e^{in\phi}$$

in $r \geq 0$. Is this the same as

$$r\partial_{y_1} g(y_1, y_2) = r\cos\phi\, \partial_r g - \sin\phi\, \partial_\phi g\,?$$

61. In three-dimensional space, prove that

$$\delta(y) = \delta''(r)/4\pi \quad (r \geq 0, 0 < \theta < \pi).$$

Prove also that, under analogous conditions in n-dimensional space,

$$\delta(y) = \frac{(-1)^{n-1}\delta^{(n-1)}(r)}{(\frac{1}{2}n - \frac{1}{2})!(2\pi^{1/2})^{n-1}}.$$

62. Prove that

$$r^{1-n}\partial_r(r^{n-1}\partial_r r^{\beta+2}) = r^\beta.$$

8.6. Singular integrals

Let T be a volume bounded by the hypersurface S with equation $f(\mathbf{x})=0$ and such that $f>0$ in T and $f<0$ outside T. Then, in analogy with the single variable case, the meaning to be attached to $\int_T g(\mathbf{x})\gamma(\mathbf{x})\,\mathrm{d}\mathbf{x}$ is $\int_{-\infty}^{\infty} g(\mathbf{x})\mathrm{H}\{f(\mathbf{x})\}\gamma(\mathbf{x})\,\mathrm{d}\mathbf{x}$ whenever this has a significance. By considering those good functions which equal unity in T it may be possible to attach a meaning to $\int_T g(\mathbf{x})\,\mathrm{d}\mathbf{x}$. Thus, if integrals involving g_0 are defined and g_0 is continuous in a neighbourhood of S,

$$\int_T (\partial_1 g_0)\gamma\,\mathrm{d}\mathbf{x} = \int_{-\infty}^{\infty} (\partial_1 g_0)\mathrm{H}(f)\gamma\,\mathrm{d}\mathbf{x}$$

$$= -\int_T g_0 \frac{\partial\gamma}{\partial x_1}\mathrm{d}\mathbf{x} + \int_S g_0\gamma\cos\theta_1\,\mathrm{d}S \qquad (64)$$

as in deriving (33). In particular, by taking $\gamma=1$ on $T+S$,

$$\int_T (\partial_1 g_0)\,\mathrm{d}\mathbf{x} = \int_S g_0\cos\theta_1\,\mathrm{d}S \qquad (65)$$

when g_0 is continuous in a neighbourhood of S. Similarly, if both $\partial_1 g_0$ and g_0 are continuous in a neighbourhood of S,

$$\int_T (\partial_1^2 g_0)\gamma\,\mathrm{d}\mathbf{x} = \int_T g_0\frac{\partial^2\gamma}{\partial x_1^2}\mathrm{d}\mathbf{x} + \int_S\left(\gamma\partial_1 g_0 - g_0\frac{\partial\gamma}{\partial x_1}\right)\cos\theta_1\,\mathrm{d}S. \qquad (66)$$

When $\beta<-n$ and $\beta\neq -n-2,\,-n-4,\ldots,$

$$r^\beta = \frac{(\partial_1^2+\ldots+\partial_n^2)r^{\beta+2}}{(\beta+2)(\beta+n)}$$

so that (66) gives

$$\int_T r^\beta\gamma(\mathbf{x})\,\mathrm{d}\mathbf{x} = \frac{1}{(\beta+2)(\beta+n)}\int_T r^{\beta+2}\nabla^2\gamma\,\mathrm{d}\mathbf{x}$$

$$+\frac{1}{(\beta+2)(\beta+n)}\int_S\left(\gamma\frac{\partial r^{\beta+2}}{\partial\nu} - r^{\beta+2}\frac{\partial\gamma}{\partial\nu}\right)\mathrm{d}S$$

so long as S does not pass through the origin and the integral over T on the right has a meaning. As usual, $\partial/\partial\nu$ is a derivative along the outward normal. That a meaning can be attached to the integral over T can be seen by continual repetition of the result until a power of r greater than $-n$ is obtained. By employing (20) and (21) of Chapter 7 a value can be attributed to the integral over T of r^{-n} and thence of r^{-n-2k}.

If $\operatorname{Re}\beta + n > 0$

$$\int_{-\infty}^{\infty} r^{\beta}\gamma(\boldsymbol{x})\,\mathrm{d}\boldsymbol{x} = \int_{0}^{\infty} r^{\beta+n-1}\gamma_0(r)\,\mathrm{d}r$$

where

$$\gamma_0(r) = \int_{\Omega_n} \gamma(r\hat{x}_1, \ldots, r\hat{x}_n)\,\mathrm{d}\Omega$$

and $(\hat{x}_1, \ldots, \hat{x}_n)$ is a point on the unit sphere in n dimensions. Since the right-hand side is a one-dimensional integral the equality may be written as

$$\int_{-\infty}^{\infty} r^{\beta}\gamma(\boldsymbol{x})\,\mathrm{d}\boldsymbol{x} = \int_{-\infty}^{\infty} x_+^{\beta+n-1}\gamma_0(x)\,\mathrm{d}x. \tag{67}$$

If $\operatorname{Re}(\beta+n) > -2$, express r^{β} as a derivative of $r^{\beta+2}$ and then transfer the derivative to γ as above. It follows from Theorem 7.29 that γ_0 is replaced by $x^{1-n}\mathrm{d}\{x^{n-1}(\mathrm{d}\gamma_0/\mathrm{d}x)\}/\mathrm{d}x$. After moving the derivatives from γ_0 to x_+ we recover (67). In this way (67) can be established for any complex β except $\beta = -n - 2k$. On the other hand

$$\int_{-\infty}^{\infty} \frac{2\pi^{n/2}\,\delta(\boldsymbol{x})}{(\frac{1}{2}n-1)!}\gamma(\boldsymbol{x})\,\mathrm{d}\boldsymbol{x} = \Omega_n\gamma(\boldsymbol{0}) = \int_{-\infty}^{\infty} \delta(x)\gamma_0(x)\,\mathrm{d}x$$

from (48). Thus, using (67) with $\beta = \mu - n$ and remembering Definitions 4.7 and 7.16, we see that (67) is confirmed when $\beta = -n$. More generally, it is true for $\beta = -n - 2k$ and so (67) holds for any complex β. Thus the definitions adopted for r^{β} are such that integrals may be evaluated in spherical polar coordinates provided that the interpretation in (67) is understood. Obviously, (67) will also be valid if r^{β} and $x_+^{\beta+n-1}$ are replaced by $r^{\beta}\ln r$ and $x_+^{\beta+n-1}\ln x$ respectively.

Exercises

63. Prove that $\int_T \delta(\boldsymbol{x})\,\mathrm{d}\boldsymbol{x} = 1$ if the origin is inside T but is equal to zero if the origin is outside T.

64. Show that, if $\beta \neq -n$,

$$\int_{r\le 1} r^\beta \,\mathrm{d}\boldsymbol{x} = \frac{2\pi^{n/2}}{(\frac{1}{2}n-1)!(\beta+n)}$$

and determine $\int_{r\le a} r^{-n}\mathrm{d}\boldsymbol{x}$.

8.7. The hyperbolic and ultrahyperbolic distances

The function $x_1^2 - x_2^2 - \dots - x_n^2$ is well defined. Formally, it differs from r^2 in having $\mathrm{i}x_2, \dots, \mathrm{i}x_n$ in place of $x_2, \dots, x_n$. One is therefore tempted to define fractional powers by means of those for r^β with the substitution of $\mathrm{i}x_2, \dots, \mathrm{i}x_n$ for $x_2, \dots, x_n$. But the differences are, in fact, profound: r^2 vanishes only at the origin whereas $x_1^2 - \dots - x_n^2$ vanishes on a hypersurface, namely the circular cone with equation

$$x_1^2 = x_2^2 + \dots + x_n^2.$$

This cone has its vertex at the origin and its axis occupies the entire x_1-axis. Outside this cone $x_1^2 - x_2^2 - \dots - x_n^2$ is negative whereas r^2 is never negative. These differences are sufficient to require consideration of $x_1^2 - \dots - x_n^2$ separate from that of r.

A convenient notation is provided by writing $q = x_1^2 - x_2^2 - \dots - x_n^2$ and introducing q_+ and q_- defined by

$$q_+^\beta = q^\beta \mathrm{H}(q), \qquad q_-^\beta = (-q)^\beta \mathrm{H}(-q). \tag{68}$$

The meaning to be attached to these symbols has still to be specified but it is clear that q_+ is zero outside the cone $q \ge 0$ while q_- is zero inside the cone. We may also consider $q_+^\beta\,\mathrm{H}(x_1)$ which is zero outside the cone $q \ge 0$ and in $x_1 < 0$ whereas $q_+^\beta\,\mathrm{H}(-x_1)$ is zero outside the cone and in $x_1 > 0$.

The case $n = 2$ is easier in some respects than general n (the cone being replaced by a pair of straight lines). Formally

$$\int_{-\infty}^{\infty} \frac{q_+^\beta}{(1+r^2)^N}\mathrm{d}\boldsymbol{x} = \int_{-\infty}^{\infty}\int_{-|x_1|}^{|x_1|} \frac{(x_1^2 - x_2^2)^\beta}{(1+r^2)^N}\mathrm{d}x_2\,\mathrm{d}x_1.$$

N can be chosen large enough to make the integral converge at infinity. The substitution $x_2 = x_1 t$ shows that convergence in the finite part of the plane exists if $\operatorname{Re}\beta > -1$. Therefore $q_+^\beta \in \boldsymbol{K}_1$ for $\operatorname{Re}\beta > -1$ and thereby defines a generalised function. Similar considerations apply to q_-, $q_+^\beta \mathrm{H}(x_1)$, $q_+^\beta \mathrm{H}(-x_1)$ when $\mathrm{Re}\,\beta > -1$ and so only the case $\mathrm{Re}\,\beta \leq -1$ requires special attention.

For general values of n it may be shown in a similar way that q_+^β, etc. are generalised functions for $\mathrm{Re}\,\beta > -1$.

The quadratic form may be made more widely applicable by taking $q = x_1^2 + \dots + x_p^2 - \dots - x_n^2$ and continuing to use the specification (68). Generalised functions will still be obtained when $\mathrm{Re}\,\beta > -1$. Since $\partial_1^2 - \dots - \partial_n^2$ is a hyperbolic operator and $\partial_1^2 + \dots + \partial_p^2 - \dots - \partial_n^2$ an ultrahyperbolic operator $|q|^{1/2}$ can be called the *hyperbolic distance* when $p = 1$ and the *ultrahyperbolic distance* when $1 < p < n - 1$. It will be convenient to write

$$L \equiv \partial_1^2 + \dots + \partial_p^2 - \dots - \partial_n^2.$$

Definition 8.13. *If $\beta \neq -m$ (m a positive integer) and $\beta \neq -\frac{1}{2}n - k$ (k a non-negative integer) we define*

$$q_+^\beta = \frac{Lq_+^{\beta+1}}{(2\beta+2)(2\beta+n)}, \qquad q_-^\beta = \frac{-Lq_-^{\beta+1}}{(2\beta+2)(2\beta+n)}.$$

Since q_+^β and q_-^β are generalised functions for $\mathrm{Re}\,\beta > -1$ the definition does supply generalised functions for all complex β other than the exceptional values stipulated in the definition. Clearly they have the properties

$$Lq_+^\beta = 2\beta(2\beta+n-2)q_+^{\beta-1} \quad (\beta \neq 2-m, \beta \neq 2-\tfrac{1}{2}n-k), \tag{69}$$
$$Lq_-^\beta = -2\beta(2\beta+n-2)q_-^{\beta-1} \quad (\beta \neq 2-m, \beta \neq 2-\tfrac{1}{2}n-k), \tag{70}$$
$$q.q_+^\beta = q_+^{\beta+1}, \quad q.q_-^\beta = -q_-^{\beta+1}. \tag{71}$$

Turning now to the exception values we introduce

Definition 8.14. *The generalised functions q_+^{-m}, q_-^{-m} are defined by*

$$q_+^{-m} = \lim_{\mu\to 0}\left\{q_+^{\mu-m} - \frac{(-1)^{m-1}}{(m-1)!\,\mu}\delta^{(m-1)}(q)\right\},$$

$$q_-^{-m} = \lim_{\mu\to 0}\left\{q_-^{\mu-m} - \frac{\delta^{(m-1)}(q)}{(m-1)!\,\mu}\right\}$$

provided that $m \leq \frac{1}{2}n - 1$ when n is even.

To justify the definition note that, by Exercise 54 and Definition 8.13,

$$q_+^{-m} = \operatorname*{Lim}_{\mu\to 0} L^m \left\{ \frac{(-\mu-1)!(\frac{1}{2}n-m-1+\mu)!(-1)^m q_+^\mu}{(m-1-\mu)!(\frac{1}{2}n-1+\mu)!2^{2m}} - \frac{(\frac{1}{2}n-m-1)!(-1)^{m-1}\mathrm{H}(q)}{(m-1)!(\frac{1}{2}n-1)!2^{2m}\mu} \right\}.$$

But $q_+^\mu = \mathrm{H}(q) + \mu\mathrm{H}(q)\ln q + O(\mu^2)$ and so

$$\begin{aligned} q_+^{-m} &= L^m \frac{(\frac{1}{2}n-m-1)!\mathrm{H}(q)(-1)^{m-1}}{(m-1)!(\frac{1}{2}n-1)!2^{2m}} \\ &\quad \times \{\ln q + \psi(m-1) - \psi(0) + \psi(\tfrac{1}{2}n-m-1) - \psi(\tfrac{1}{2}n-1)\} \\ &= \frac{(\frac{1}{2}n-m-1)!(-1)^{m-1}L^m\mathrm{H}(q)\ln q}{(m-1)!(\frac{1}{2}n-1)!2^{2m}} \\ &\quad + \frac{(-1)^{m-1}}{(m-1)!}\{\psi(m-1) - \psi(0) + \psi(\tfrac{1}{2}n-m-1) \\ &\qquad - \psi(\tfrac{1}{2}n-1)\}\delta^{(m-1)}(q) \end{aligned} \tag{72}$$

from Exercise 54. Similarly

$$\begin{aligned} q_-^{-m} &= \frac{(\frac{1}{2}n-m-1)!(-1)L^m\mathrm{H}(-q)\ln|q|}{(m-1)!(\frac{1}{2}n-1)!2^{2m}} \\ &\quad + \frac{1}{(m-1)!}\{\psi(m-1) - \psi(0) \\ &\qquad + \psi(\tfrac{1}{2}n-m-1) - \psi(\tfrac{1}{2}n-1)\}\delta^{(m-1)}(q). \end{aligned} \tag{73}$$

From (72)

$$Lq_+^{-m} = -2m(n-2m-2)q_+^{-m-1} + \frac{(-1)^m}{m!}2(n-4m-2)\delta^{(m)}(q) \tag{74}$$

and from (73)

$$Lq_-^{-m} = 2m(n-2m-2)q_-^{-m-1} - \frac{2}{m!}(n-4m-2)\delta^{(m)}(q) \tag{75}$$

without restriction on m when n is odd and subject to $m < \frac{1}{2}n - 1$ when n is even.

Definition 8.14 implies that

$$q.q_{+}^{-m} = \operatorname*{Lim}_{\mu\to 0}\left\{q_{+}^{\mu-m+1} - \frac{(-1)^{m-1}q}{(m-1)!\,\mu}\delta^{(m-1)}(q)\right\}$$

$$= \operatorname*{Lim}_{\mu\to 0}\left\{q_{+}^{\mu-m+1} - \frac{(-1)^{m-2}}{(m-2)!\,\mu}\delta^{(m-2)}(q)\right\} = q_{+}^{1-m} \qquad (76)$$

when advantage is taken of Exercise 53. Similarly

$$q.q_{-}^{-m} = -q_{-}^{1-m}. \qquad (77)$$

Results may be derived in an alternative way by remarking that, when $\operatorname{Re}\beta > -1$,

$$\int_{-\infty}^{\infty} q_{+}^{\beta}\,\gamma(\boldsymbol{x})\,\mathrm{d}\boldsymbol{x}$$

$$= \int_{0}^{\infty}\int_{0}^{r_p} (r_p^2 - r_{n-p}^2)^{\beta}\, r_p^{p-1}\, r_{n-p}^{n-p-1}\,\gamma_3(r_p, r_{n-p})\,\mathrm{d}r_p\,\mathrm{d}r_{n-p}$$

where γ_3 is as specified in (46). The substitution $r_1 - r_2 = x = t\xi$ and $r_1 + r_2 = y = \xi$ then leads to

$$\int_{-\infty}^{\infty} q_{+}^{\beta}\,\gamma(\boldsymbol{x})\,\mathrm{d}\boldsymbol{x}$$

$$= 2^{1-n}\int_{0}^{\infty}\int_{0}^{1} \xi^{2\beta+n-1}\,t^{\beta}(1+t)^{p-1}(1-t)^{n-p-1}\gamma_2(t\xi,\xi)\,\mathrm{d}t\,\mathrm{d}\xi \qquad (78)$$

where γ_2 is as defined in (47) with the convention described after Definition 8.11. It may be asserted that (78) holds for all complex β apart from the exceptional values in Definition 8.13 by using the definition and then applying (49) and (50).

One consequence of (78) is that

$$\lim_{\mu\to 0}\mu\int_{-\infty}^{\infty} q_{+}^{\mu-n/2}\gamma(\boldsymbol{x})\,\mathrm{d}\boldsymbol{x}$$

$$= \lim_{\mu\to 0}\mu 2^{1-n}\int_{0}^{\infty}\int_{0}^{1} \xi^{2\mu-1}\,t^{\mu-n/2}(1+t)^{p-1}(1-t)^{n-p-1}\gamma_2(t\xi,\xi)\,\mathrm{d}t\,\mathrm{d}\xi.$$

Now, it has been observed in Chapter 4 that $\operatorname{Lim}_{\mu\to 0}\mu\xi^{2\mu-1} = \frac{1}{2}\delta(\xi)$

and, if n is odd, $\mathrm{Lim}_{\mu\to 0} t^{\mu-n/2} = t^{-n/2}$. Hence, if n is odd,

$$\lim_{\mu\to 0} \mu \int_{-\infty}^{\infty} q_+^{\mu-n/2}\gamma(\boldsymbol{x})\,\mathrm{d}\boldsymbol{x}$$

$$= 2^{-n}\Omega_p\Omega_{n-p}\gamma(\mathbf{0}) \lim_{\mu\to 0} \int_0^1 t^{\mu-n/2}(1+t)^{p-1}(1-t)^{n-p-1}\,\mathrm{d}t.$$

The change of variable

$$t = \{1-(1-u)^{1/2}\}/\{1+(1-u)^{1/2}\}$$

gives

$$\int_0^1 t^{\mu-n/2}(1+t)^{p-1}(1-t)^{n-p-1}\,\mathrm{d}t$$

$$= 2^{n-2}\int_0^1 \frac{u^{\mu-n/2}(1-u)^{((n-p)/2)-1}}{\{1+(1-u)^{1/2}\}^{2\mu}}\,\mathrm{d}u.$$

Allowing $\mu \to 0$ we obtain

$$\lim_{\mu\to 0} \mu \int_{-\infty}^{\infty} q_+^{\mu-n/2}\gamma(\boldsymbol{x})\mathrm{d}\boldsymbol{x} = \tfrac{1}{4}\Omega_p\Omega_{n-p}\gamma(\mathbf{0})\int_0^1 u^{-n/2}(1-u)^{(n/2)-(p/2)-1}\,\mathrm{d}u$$

$$= (-\tfrac{1}{2}n)!\,\pi^{(n/2)-1}\gamma(\mathbf{0})\sin\tfrac{1}{2}p\pi$$

from (19) of Chapter 4 and (48). Thus

$$\mathrm{Lim}_{\mu\to 0}\, \mu q_+^{\mu-n/2} = (-\tfrac{1}{2}n)!\,\pi^{(n/2)-1}\sin\tfrac{1}{2}p\pi\,\delta(\boldsymbol{x}) \tag{79}$$

when n is odd. Therefore

$$\begin{aligned} Lq_+^{1-n/2} &= \mathrm{Lim}_{\mu\to 0}\, Lq_+^{1-(n/2)+\mu} \\ &= \mathrm{Lim}_{\mu\to 0}\,(2-n+2\mu)2\mu q_+^{\mu-n/2} \\ &= (1-\tfrac{1}{2}n)!\,4\pi^{(n/2)-1}\sin\tfrac{1}{2}p\pi\,\delta(\boldsymbol{x}) \end{aligned} \tag{80}$$

when n is odd.

When n is even the above process fails because $q_+^{1-(n/2)+\mu}$ does not approach $q_+^{1-n/2}$ as $\mu \to 0$ on account of Definition 8.14; this is supported by the fact that $t^{\mu-n/2}$ becomes unbounded as $\mu \to 0$.

Indeed, according to Definition 4.7,

$$2\mu\xi^{2\mu-1} = \delta(\xi) + 2\mu\xi^{-1} + o(\mu),$$

$$t^{\mu-n/2} = \frac{(-1)^{(n/2)-1}}{(\frac{1}{2}n-1)!\mu}\delta^{((n/2)-1)}(t) + t^{-n/2} + o(1) \tag{81}$$

in view of the interpretation to be adopted for the integrals we are concerned with. Hence

$$\begin{aligned}(2-n+2\mu)2\mu\,\xi^{2\mu-1}t^{\mu-n/2} \\ &= \frac{(-1)^{n/2}2\delta^{((n/2)-1)}(t)}{(\frac{1}{2}n-2)!}\left\{\frac{\delta(\xi)}{\mu} - \frac{\delta(\xi)}{\frac{1}{2}n-1} + 2\xi^{-1}\right\} \\ &\quad + (2-n)\delta(\xi)t^{-n/2} + o(1).\end{aligned}$$

Therefore

$$\begin{aligned}\lim_{\mu\to 0}\int_{-\infty}^{\infty}&\left\{(2-n+2\mu)2\mu\,q_+^{\mu-n/2}\right.\\ &\left. - \frac{(-1)^{n/2}4\pi^{(n/2)-1}}{(\frac{1}{2}n-2)!\mu}\sin\tfrac{1}{2}p\pi\delta(\boldsymbol{x})\right\}\gamma(\boldsymbol{x})\,\mathrm{d}\boldsymbol{x}\\
&= \lim_{\mu\to 0} - \frac{2^{2-n}}{(\frac{1}{2}n-2)!}\left(\frac{1}{\mu} - \frac{1}{\frac{1}{2}n-1}\right)\\
&\quad\times\left[\frac{\mathrm{d}^{(n/2)-1}}{\mathrm{d}t^{(n/2)-1}}(1+t)^{p-1}(1-t)^{n-p-1}\right]_{t=0}\Omega_p\Omega_{n-p}\gamma(\boldsymbol{0})\\
&\quad + \frac{2^{3-n}(-1)^{n/2}}{(\frac{1}{2}n-2)!}\int_0^{\infty}\int_0^1 \xi^{-1}\delta^{((n/2)-1)}(t)\\
&\qquad\qquad \times(1+t)^{p-1}(1-t)^{n-p-1}\gamma_2(t\xi,\xi)\,\mathrm{d}t\,\mathrm{d}\xi\\
&\quad + 2^{1-n}(2-n)\Omega_p\Omega_{n-p}\gamma(\boldsymbol{0})\int_0^1 t^{-n/2}(1+t)^{p-1}(1-t)^{n-p-1}\,\mathrm{d}t\\
&\quad - \frac{(-1)^{n/2}4\pi^{(n/2)-1}}{(\frac{1}{2}n-2)!\mu}\sin\tfrac{1}{2}p\pi\,\gamma(\boldsymbol{0})\\
&= \frac{(-1)^{(n/2)-1}4\pi^{(n/2)-1}}{(\frac{1}{2}n-1)!}\sin\tfrac{1}{2}p\pi\,\gamma(\boldsymbol{0}) + \frac{(-1)^{n/2}4}{(\frac{1}{2}n-2)!}\int_{-\infty}^{\infty}\delta^{((n/2)-1)}(q)\gamma(\boldsymbol{x})\,\mathrm{d}\boldsymbol{x}\\
&\quad + 2^{1-n}(2-n)\Omega_p\Omega_{n-p}\gamma(\boldsymbol{0})\int_0^1 t^{-n/2}(1+t)^{p-1}(1-t)^{n-p-1}\,\mathrm{d}t\end{aligned}$$

from Definition 8.11 and Exercise 54 (iii). Consequently

$$\lim_{\mu\to 0}\left\{(2-n+2\mu)2\mu q_+^{\mu-n/2}-\frac{(-1)^{n/2}4\pi^{(n/2)-1}}{(\frac{1}{2}n-2)!\,\mu}\sin\tfrac{1}{2}p\pi\,\delta(\boldsymbol{x})\right\}$$

$$=\left\{\frac{(-1)^{(n/2)-1}4\pi^{(n/2)-1}}{(\frac{1}{2}n-1)!}\sin\tfrac{1}{2}p\pi+2^{1-n}(2-n)\Omega_p\Omega_{n-p}I_{n,p}\right\}\delta(\boldsymbol{x})$$

$$+\frac{(-1)^{n/2}4}{(\frac{1}{2}n-2)!}\delta^{((n/2)-1)}(q)$$

where $I_{n,p}=\int_0^1 t^{-n/2}(1+t)^{p-1}(1-t)^{n-p-1}\,dt$. Therefore

$$Lq_+^{1-n/2}=\left\{\frac{(-1)^{(n/2)-1}4\pi^{(n/2)-1}}{(\frac{1}{2}n-1)!}\sin\tfrac{1}{2}p\pi+2^{1-n}(2-n)\Omega_p\Omega_{n-p}I_{n,p}\right\}\delta(\boldsymbol{x})$$

$$+\frac{(-1)^{n/2}4}{(\frac{1}{2}n-2)!}\delta^{((n/2)-1)}(q) \tag{82}$$

when n is even.

It may be shown in a similar way that

$$Lq_-^{1-n/2}=(1-\tfrac{1}{2}n)!(-4\pi^{(n/2)-1})\sin\tfrac{1}{2}(n-p)\pi\,\delta(\boldsymbol{x}) \tag{83}$$

when n is odd and

$$Lq_-^{1-n/2}$$

$$=\left\{\frac{(-1)^{n/2}4\pi^{(n/2)-1}}{(\frac{1}{2}n-1)!}\sin\tfrac{1}{2}(n-p)\pi+2^{1-n}(n-2)\Omega_p\Omega_{n-p}I_{n,n-p}\right\}\delta(\boldsymbol{x})$$

$$+\frac{4\delta^{((n/2)-1)}(q)}{(\frac{1}{2}n-2)!} \tag{84}$$

when n is even.

To determine an expression for $I_{n,p}$ when n is even remark firstly that

$$I_{n,p+2}=\int_0^1 t^{-n/2}(1+t)^{p-1}\{(1-t)^2+4t\}(1-t)^{n-p-3}\,dt$$

$$=I_{n,p}+4I_{n-2,p} \tag{85}$$

which is a recurrence relation enabling values for higher p to be deduced from those for lower p. Secondly, since $t^{-n/2} = \operatorname{Lim}_{\mu\to 0}\{\partial(\mu t^{\mu - n/2})/\partial\mu\}$, (cf. (81)),

$$I_{n,p} = \lim_{\mu\to 0}\frac{\partial}{\partial\mu}\int_0^1 \mu t^{\mu - n/2}(1+t)^{p-1}(1-t)^{n-p-1}\,\mathrm{d}t. \tag{86}$$

Thus

$$\begin{aligned} I_{n,1} &= \lim_{\mu\to 0}\frac{\partial}{\partial\mu}\frac{(\mu - \frac{1}{2}n)!(n-2)!}{(\mu + \frac{1}{2}n - 1)!}\mu \\ &= \lim_{\mu\to 0}\frac{(n-2)!\pi(-1)^{(n/2)-1}\mu}{(\mu + \frac{1}{2}n - 1)!(\frac{1}{2}n - 1 - \mu)!\sin\mu\pi} = 0 \end{aligned}$$

because the derivative of an even function of μ vanishes at the origin. The implication of this and (85) is that

$$I_{n,p} = 0 \qquad (p \text{ odd}). \tag{87}$$

When p is even use the change of variable employed to demonstrate (79); then

$$\begin{aligned} I_{n,p} &= \lim_{\mu\to 0}\frac{\partial}{\partial\mu}2^{n-2}\mu\int_0^1 \frac{u^{\mu - n/2}(1-u)^{(n/2)-(p/2)-1}}{\{1+(1-u)^{1/2}\}^{2\mu}}\mathrm{d}u \\ &= \lim_{\mu\to 0}\frac{\partial}{\partial\mu}2^{n-2}\mu\int_0^0 u^{\mu - n/2}(1-u)^{(n/2)-(p/2)-1}\,\mathrm{d}u - \frac{(-1)^{(n/2)-1}2^{n-1}}{(\frac{1}{2}n-1)!} \\ &\quad\times\int_0^1 \delta^{((n/2)-1)}(u)(1-u)^{(n/2)-(p/2)-1}\ln\{1+(1-u)^{1/2}\}\,\mathrm{d}u \end{aligned}$$

when (81) is borne in mind. Hence

$$\begin{aligned} I_{n,p} &= \lim_{\mu\to 0}\frac{\partial}{\partial\mu}\frac{(\frac{1}{2}n - \frac{1}{2}p - 1)!(\frac{1}{2}p - 1 - \mu)!(-1)^{(n/2)-p/2}2^{n-2}\mu}{(\frac{1}{2}n - 1 - \mu)!} \\ &\quad - \frac{2^{n-1}}{(\frac{1}{2}n-1)!}\left[\frac{\mathrm{d}^{(n/2)-1}}{\mathrm{d}u^{(n/2)-1}}(1-u)^{(n/2)-(p/2)-1}\ln\{1+(1-u)^{1/2}\}\right]_{u=0} \\ &= \frac{(\frac{1}{2}p-1)!}{(\frac{1}{2}n-1)!}(\tfrac{1}{2}n - \tfrac{1}{2}p - 1)!(-1)^{(n/2)-p/2}2^{n-2} \\ &\quad - \frac{(-1)^{(n/2)-1}2^{n-1}}{(\frac{1}{2}n-1)!}\left[\frac{\mathrm{d}^{(n/2)-1}}{\mathrm{d}v^{(n/2)-1}}v^{(n/2)-(p/2)-1}\ln(1+v^{1/2})\right]_{v=1} \end{aligned}$$

when it is recalled that p is even. Now, if

$$J_{n,p}=\left[\frac{\mathrm{d}^{(n/2)-1}}{\mathrm{d}v^{(n/2)-1}}v^{(n/2)-(p/2)-1}\ln(1+v^{1/2})\right]_{v=1},$$

$$J_{n,p}=\left[\frac{\mathrm{d}^{(n/2)-2}}{\mathrm{d}v^{(n/2)-2}}\left\{(\tfrac{1}{2}n-\tfrac{1}{2}p-1)v^{(n/2)-(p/2)-2}\ln(1+v^{1/2})\right.\right.$$

$$\left.\left.+\tfrac{1}{2}\frac{v^{(n/2)-(p/2)-1}(v^{1/2}-1)}{v^{1/2}(v-1)}\right\}\right]_{v=1}$$

$$=(\tfrac{1}{2}n-\tfrac{1}{2}p-1)J_{n-2,p}$$

$$+\frac{(\tfrac{1}{2}n-2)!}{(\tfrac{1}{2}n-1)!2}\left[\frac{\mathrm{d}^{(n/2)-1}}{\mathrm{d}v^{(n/2)-1}}(v^{(n/2)-(p/2)-1}-v^{(n/2)-(p/2)-3/2})\right]_{v=1}$$

$$=(\tfrac{1}{2}n-\tfrac{1}{2}p-1)J_{n-2,p}-\frac{(\tfrac{1}{2}n-\tfrac{1}{2}p-\tfrac{3}{2})!}{(-\tfrac{1}{2}p-\tfrac{1}{2})!(n-2)}$$

since $\tfrac{1}{2}n-\tfrac{1}{2}p-1$ is a positive integer less than $\tfrac{1}{2}n-1$. Thus

$$J_{n,p}=(\tfrac{1}{2}n-\tfrac{1}{2}p-1)(\tfrac{1}{2}n-\tfrac{1}{2}p-2)J_{n-4,p}$$

$$-\frac{(\tfrac{1}{2}n-\tfrac{1}{2}p-\tfrac{5}{2})!}{(-\tfrac{1}{2}p-\tfrac{1}{2})!(n-4)}(\tfrac{1}{2}n-\tfrac{1}{2}p-1)-\frac{(\tfrac{1}{2}n-\tfrac{1}{2}p-\tfrac{3}{2})!}{(-\tfrac{1}{2}p-\tfrac{1}{2})!(n-2)}.$$

But

$$J_{p+2,p}=\left[\frac{\mathrm{d}^{p/2}}{\mathrm{d}v^{p/2}}\ln(1+v^{1/2})\right]_{v=1}$$

$$=\left[\frac{\mathrm{d}^{(p/2)-1}}{\mathrm{d}v^{(p/2)-1}}\frac{v^{1/2}-1}{2v^{1/2}(v-1)}\right]_{v=1}$$

$$=\frac{(-\tfrac{1}{2})!(-1)}{(-\tfrac{1}{2}p-\tfrac{1}{2})!p}$$

and so

$$J_{n,p}=-\frac{(\tfrac{1}{2}n-\tfrac{1}{2}p-1)!}{(-\tfrac{1}{2}p-\tfrac{1}{2})!}\sum_{r=0}^{(n/2)-(p/2)-1}\frac{(r-\tfrac{1}{2})!}{r!(2r+p)}.$$

Consequently

$$I_{n,p} = \frac{(\frac{1}{2}n - \frac{1}{2}p - 1)!}{(\frac{1}{2}n - 1)!} 2^{n-2} \Bigg\{ (\tfrac{1}{2}p - 1)!(-1)^{(n/2)-p/2}$$

$$+ \frac{(-1)^{(n/2)-1} 2}{(-\frac{1}{2}p - \frac{1}{2})!} \sum_{r=0}^{(n/2)-(p/2)-1} \frac{(r - \frac{1}{2})!}{r!(2r+p)} \Bigg\}. \tag{88}$$

By putting $p = n - 2$ in (88) we obtain

$$I_{n,n-2} = \frac{2^{n-1}}{2-n} \left\{ 1 + \frac{(-\frac{1}{2})!(-1)^{n/2}}{(\frac{1}{2} - \frac{1}{2}n)!(\frac{1}{2}n - 1)!} \right\}. \tag{89}$$

On the other hand

$$I_{n,2} = \frac{(-1)^{(n/2)-1} 2^{n-1}}{n-2} \left\{ 1 + \frac{1}{(-\frac{3}{2})!} \sum_{r=0}^{(n/2)-2} \frac{(r - \frac{1}{2})!}{(r+1)!} \right\}$$

$$= \frac{2^{n-1}(-1)^{(n/2)-1}}{n-2} \Bigg[1 + \frac{(\frac{1}{2}n - \frac{3}{2})!}{(\frac{1}{2}n - 1)!(-\frac{1}{2})!} \times \left\{ 1 - \frac{(\frac{1}{2}n - 1)!(-\frac{1}{2})!}{(\frac{1}{2}n - \frac{3}{2})!} \right\} \Bigg]$$

since

$$\sum_{s=0}^{m-1} \frac{(u+s)!}{(v+s)!} = \frac{(u+m)!}{(v+m-1)!(u-v+1)} \left\{ 1 - \frac{(v+m-1)!u!}{(v-1)!(u+m)!} \right\}.$$

Therefore

$$I_{n,2} = \frac{(\frac{1}{2}n - \frac{3}{2})!(-1)^{(n/2)-1} 2^{n-1}}{(\frac{1}{2}n - 1)!(-\frac{1}{2})!(n-2)}. \tag{90}$$

Hence

$$I_{n,2} + (-1)^{n/2} I_{n,n-2} = (-1)^{(n/2)-1} 2^{n-1}/(n-2). \tag{91}$$

Now suppose that

$$I_{n,r} + (-1)^{n/2} I_{n,n-r} = \frac{(\frac{1}{2}r - 1)!}{(\frac{1}{2}n - 1)!} (\tfrac{1}{2}n - \tfrac{1}{2}r - 1)!(-1)^{(n/2)-r/2} 2^{n-2}$$

for $r = 2, 4, \ldots, p$. Then, from (85)

$$I_{n,p+2} = \frac{(\frac{1}{2}p-1)!}{(\frac{1}{2}n-1)!}(\tfrac{1}{2}n-\tfrac{1}{2}p-1)!(-1)^{(n/2)-p/2}2^{n-2} - (-1)^{n/2}I_{n,n-p}$$
$$+\frac{(\frac{1}{2}p-1)!}{(\frac{1}{2}n-2)!}(\tfrac{1}{2}n-\tfrac{1}{2}p-2)!(-1)^{(n/2)-(p/2)-1}2^{n-2}$$
$$+(-1)^{n/2}4I_{n-2,n-2-p}.$$

However, (85) implies that $4I_{n-2,n-2-p} - I_{n,n-p} = -I_{n,n-p-2}$ and so

$$I_{n,p+2} + (-1)^{n/2}I_{n,n-p+2}$$
$$= \frac{(\frac{1}{2}p)!}{(\frac{1}{2}n-1)!}(\tfrac{1}{2}n-\tfrac{1}{2}p-2)!(-1)^{(n/2)-(p/2)+1}2^{n-2}$$

which is the same as was assumed but with $p+2$ for p. Therefore induction will establish

$$I_{n,p} + (-1)^{n/2}I_{n,n-p}$$
$$= (\tfrac{1}{2}p-1)!(\tfrac{1}{2}n-\tfrac{1}{2}p-1)!(-1)^{(n/2)-p/2}2^{n-2}/(\tfrac{1}{2}n-1)! \qquad (92)$$

so long as it is true for $p = 2$. But, when $p = 2$, this is the same as (91). The equation (92) therefore holds for all even p.

The formulae take a somewhat simpler form by specifying

$$\lambda_{n,p} = \frac{(\frac{1}{2}n-1)!\,\pi^{n/2}(-1)^{(n/2)-1}I_{n,p}}{(\frac{1}{2}p-1)!(\frac{1}{2}n-\frac{1}{2}p-1)!\,2^{n-2}}.$$

Then

$$\lambda_{n,p} = \begin{cases} 0 & (p \text{ odd}) \quad (93) \\ (-1)^{(p/2)-1}\pi^{n/2} + \dfrac{2\pi^{n/2}}{(\frac{1}{2}p-1)!(-\frac{1}{2}p-\frac{1}{2})!} & \\ \quad \times \displaystyle\sum_{r=0}^{(n/2)-(p/2)-1} \frac{(r-\frac{1}{2})!}{r!(2r+p)} & (p \text{ even}) \quad (94) \end{cases}$$

from (88). Also

$$\lambda_{n,p} + (-1)^{n/2}\lambda_{n,n-p} = (-1)^{(p/2)-1}\pi^{n/2} \qquad (95)$$

from (92) when p is even.

In terms of $\lambda_{n,p}$ (82) and (84) become

$$Lq_{+}^{1-n/2}=\frac{(-1)^{n/2}4}{(\frac{1}{2}n-2)!}\left\{\left(\lambda_{n,p}-\frac{\pi^{(n/2)-1}}{\frac{1}{2}n-1}\sin\tfrac{1}{2}p\pi\right)\delta(\boldsymbol{x})+\delta^{((n/2)-1)}(q)\right\}, \tag{96}$$

$$Lq_{-}^{1-n/2}=\frac{(-1)^{(n/2)-1}4}{(\frac{1}{2}n-2)!}\left[\left\{\lambda_{n,n-p}-\frac{\pi^{(n/2)-1}}{\frac{1}{2}n-1}\sin\tfrac{1}{2}(n-p)\pi\right\}\delta(\boldsymbol{x})\right.$$
$$\left.+(-1)^{(n/2)-1}\delta^{((n/2)-1)}(q)\right] \tag{97}$$

when n is even.

Turning now to the other exceptional values we have

Definition 8.15. *When n is odd the generalised functions $q_{+}^{-(n/2)-k}$, $q_{-}^{-(n/2)-k}$ $(k=0,1,2,\ldots)$ are defined by*

$$q_{+}^{-(n/2)-k}=\lim_{\mu\to 0}\left\{q_{+}^{\mu-(n/2)-k}-\frac{(-1)^{(n/2)-1/2}\pi^{n/2}\sin\frac{1}{2}p\pi}{(\frac{1}{2}n+k-1)!k!2^{2k}\mu}L^{k}\delta(\boldsymbol{x})\right\},$$

$$q_{-}^{-(n/2)-k}=\lim_{\mu\to 0}\left\{q_{-}^{\mu-(n/2)-k}-\frac{(-1)^{(n/2)-1/2+k}\pi^{n/2}\sin\frac{1}{2}(n-p)\pi}{(\frac{1}{2}n+k-1)!k!2^{2k}\mu}L^{k}\delta(\boldsymbol{x})\right\}$$

and, when n is even, by

$$q_{+}^{-(n/2)-k}=\lim_{\mu\to 0}\left\{q_{+}^{\mu-(n/2)-k}+\frac{(-1)^{n/2}L^{k}}{(\frac{1}{2}n+k-1)!k!2^{2k}\mu}\left[\left\{\frac{1}{\mu}+\psi(k)-\psi(0)\right.\right.\right.$$
$$\left.+\psi(\tfrac{1}{2}n+k-1)-\psi(\tfrac{1}{2}n-1)\right\}\pi^{(n/2)-1}\sin\tfrac{1}{2}p\pi\,\delta(\boldsymbol{x})$$
$$\left.\left.+\lambda_{n,p}\delta(\boldsymbol{x})+\delta^{((n/2)-1)}(q)\right]\right\},$$

$$q_{-}^{-(n/2)-k}=\lim_{\mu\to 0}\left\{q_{-}^{\mu-(n/2)-k}+\frac{(-1)^{(n/2)+k}L^{k}}{(\frac{1}{2}n+k-1)!k!2^{2k}\mu}\left[\left\{\frac{1}{\mu}+\psi(k)-\psi(0)\right.\right.\right.$$
$$\left.+\psi(\tfrac{1}{2}n+k-1)-\psi(\tfrac{1}{2}n-1)\right\}\pi^{(n/2)-1}\sin\tfrac{1}{2}(n-p)\pi\,\delta(\boldsymbol{x})$$
$$\left.\left.+\lambda_{n,n-p}\delta(\boldsymbol{x})+(-1)^{(n/2)-1}\delta^{((n/2)-1)}(q)\right]\right\}.$$

In checking that the definitions exist, consideration will be limited to $k = 0$, the general case being left as an exercise for the reader. Now, when n is odd,

$$\begin{aligned} q_+^{\mu - n/2} &= \frac{Lq_+^{\mu+1-n/2}}{(2\mu + 2 - n)2\mu} \\ &= \frac{1}{(2\mu + 2 - n)2\mu} L\{q_+^{1-n/2} + \mu q_+^{1-n/2} \ln q + O(\mu^2)\} \\ &= \frac{(1 - \frac{1}{2}n)!\,4\pi^{(n/2)-1}}{(2\mu + 2 - n)2\mu} \sin \tfrac{1}{2}p\pi\, \delta(\boldsymbol{x}) + \frac{1}{2(2-n)} Lq_+^{1-n/2} \ln q + o(1) \end{aligned}$$

$$= \frac{(-1)^{(n/2)-1/2}\pi^{n/2}(2-n)}{(\frac{1}{2}n - 1)!\,\mu(2\mu + 2 - n)} \sin \tfrac{1}{2}p\pi\, \delta(\boldsymbol{x}) + \frac{1}{2(2-n)} Lq_+^{1-n/2} \ln q + o(1)$$

from (80). Hence, when n is odd,

$$q_+^{-n/2} = \frac{1}{2(2-n)} Lq_+^{1-n/2} \ln q + \frac{(-1)^{(n/2)-1/2}\pi^{n/2}}{(\frac{1}{2}n - 1)!(\frac{1}{2}n - 1)} \sin \tfrac{1}{2}p\pi\, \delta(\boldsymbol{x}). \quad (98)$$

Similarly

$$q_-^{-n/2} = \frac{1}{2(n-2)} Lq_-^{1-n/2} \ln |q| + \frac{(-1)^{(n/2)-1/2}\pi^{n/2}}{(\frac{1}{2}n - 1)!(\frac{1}{2}n - 1)} \sin \tfrac{1}{2}(n-p)\pi\, \delta(\boldsymbol{x}) \quad (99)$$

when n is odd.

A parallel technique based on Definition 8.14, Exercise 54(iii) and (96) leads to

$$\begin{aligned} q_+^{-n/2} &= \frac{1}{2(2-n)} Lq_+^{1-n/2} \ln q \\ &\quad + \frac{(-1)^{(n/2)-1}}{(\frac{1}{2}n - 1)!(\frac{1}{2}n - 1)} \{\lambda_{n,p}\delta(\boldsymbol{x}) + \delta^{((n/2)-1)}(q)\}, \end{aligned} \quad (100)$$

$$\begin{aligned} q_-^{-n/2} &= \frac{1}{2(n-2)} Lq^{1-n/2} \ln |q| \\ &\quad + \frac{1}{(\frac{1}{2}n - 1)!(\frac{1}{2}n - 1)} \{(-1)^{(n/2)-1}\lambda_{n,n-p}\delta(\boldsymbol{x}) + \delta^{((n/2)-1)}(q)\} \end{aligned} \quad (101)$$

when n is even ($n \neq 2$) and

$$q_+^{-1} = \tfrac{1}{8} LH(q) \ln^2 q, \quad (102)$$

$$q_-^{-1} = -\tfrac{1}{8} LH(-q) \ln^2 |q| \quad (103)$$

when $n=2$.

The reader should have no difficulty in showing that

$$q.q_+^{-(n/2)-k} = q_+^{1-(n/2)-k}, \qquad q.q_-^{-(n/2)-k} = -q_-^{1-(n/2)-k}. \quad (104)$$

Further

$$Lq_+^{-(n/2)-k} = 2(n+2k)(k+1)q_+^{-(n/2)-k-1} - \frac{(-1)^{(n/2)-1/2}(n+4k+2)\pi^{n/2}\sin\frac{1}{2}p\pi}{(\frac{1}{2}n+k)!(k+1)!2^{2k+1}}L^{k+1}\delta(\boldsymbol{x}), \quad (105)$$

$$Lq_-^{-(n/2)-k} = -2(n+2k)(k+1)q_-^{-(n/2)-k-1} - \frac{(-1)^{(n/2)-1/2+k}(n+4k+2)\pi^{n/2}\sin\frac{1}{2}(n-p)\pi}{(\frac{1}{2}n+k)!(k+1)!2^{2k+1}}L^{k+1}\delta(\boldsymbol{x}) \quad (106)$$

when n is odd and

$$Lq_+^{-(n/2)-k} = 2(n+2k)(k+1)q_+^{-(n/2)-k-1} - \frac{(-1)^{n/2}L^{k+1}}{(\frac{1}{2}n+k)!(k+1)!2^{2k+2}}\{[4-2(n+4k+2)\{\psi(k+1) - \psi(0)+\psi(\tfrac{1}{2}n+k)-\psi(\tfrac{1}{2}n-1)\}]\pi^{(n/2)-1}\sin\tfrac{1}{2}p\pi\,\delta(\boldsymbol{x}) - 2(n+4k+2)[\lambda_{n,p}\delta(\boldsymbol{x})+\delta^{((n/2)-1)}(q)]\}, \quad (107)$$

$$Lq_-^{-(n/2)-k} = -2(n+2k)(k+1)q_-^{-(n/2)-k-1} - \frac{(-1)^{(n/2)+k}L^{k+1}}{(\frac{1}{2}n+k)!(k+1)!2^{2k+2}}\{[4-2(n+4k+2)\{\psi(k+1) - \psi(0)+\psi(\tfrac{1}{2}n+k)-\psi(\tfrac{1}{2}n-1)\}]\pi^{(n/2)-1}\sin\tfrac{1}{2}(n-p)\pi\,\delta(\boldsymbol{x}) - 2(n+4k+2)[\lambda_{n,n-p}\delta(\boldsymbol{x})+(-1)^{(n/2)-1}\delta^{((n/2)-1)}(q)]\} \quad (108)$$

when n is even.

Exercises

65. Prove that $\partial_1 q_+^\beta = 2\beta x_1 q_+^{\beta-1}$ when $\beta \neq -k$ and $\beta \neq 1-\frac{1}{2}n-k$ ($k=0,1,2,\ldots$).

66. If $q = x_1^2 - x_2^2$ show that $(\partial_1^2 - \partial_2^2)\{\mathrm{H}(q)\ln q\} = 4\delta(q)$.

67. Prove that

(i) $(x_1\partial_1 + x_2\partial_2 + \ldots + x_n\partial_n)q_+^\beta = 2\beta q_+^\beta$ with β restricted as in Exercise 65,

(ii) $(x_1\partial_1 + x_2\partial_2 + \dots + x_n\partial_n)\mathrm{H}(q)\ln q = 2\mathrm{H}(q)$,

(iii) $(x_1\partial_1 + x_2\partial_2 + \dots + x_n\partial_n)\{\mathrm{H}(q)(\ln q)^2\} = 4\mathrm{H}(q)\ln q$,

(iv) $(x_1\partial_1 + x_2\partial_2 + \dots + x_n\partial_n)\mathrm{H}(q) = 0$,

(v) $Lq_+ = 2n\,\mathrm{H}(q)$.

68. If $a > 0$ does $(aq)_+^{-m}$ equal $a^{-m}q_+^{-m}$? If note, what is the difference? How do $(aq)_+^{-(n/2)-k}$ and $a^{-(n/2)-k}\,q_+^{-(n/2)-k}$ compare?

69. If n is odd and m is a positive integer show that

$$L^m\,q_+^{m-n/2} = (m - \tfrac{1}{2}n)!(m-1)!2^{2m}\pi^{(n/2)-1}\sin\tfrac{1}{2}p\pi\,\delta(\boldsymbol{x}).$$

70. Demonstrate (84).

71. Prove that

$$L^{(n/2)-1/2}\,q_+^{\beta} = \frac{\beta!(\beta+\frac{1}{2}n-1)!2^{n-1}}{(\beta-\frac{1}{2})!(\beta+\frac{1}{2}-\frac{1}{2}n)!}q_+^{\beta+1/2-n/2}$$

when n is odd and deduce that

$$L^{(n/2)+1/2}\,q_+^{1/2} = (\tfrac{1}{2}n-\tfrac{1}{2})!2^n\pi^{(n/2)-1/2}\sin\tfrac{1}{2}p\pi\,\delta(\boldsymbol{x}).$$

72. When n is even show that

$$L^{(n/2)-1}\,q_+^{\beta} = (\beta+\tfrac{1}{2}n-1)!2^{n-2}\,q_+^{\beta+1-n/2}/(\beta+1-\tfrac{1}{2}n)!$$

and deduce that

$$L^{n/2}\,\mathrm{H}(q) = (\tfrac{1}{2}n-1)!2^n\pi^{(n/2)-1}\sin\tfrac{1}{2}p\pi\,\delta(\boldsymbol{x}).$$

73. Prove that

$$q^{-(n/2)-k} = \frac{(-1)^{(n/2)-1/2}}{(\frac{1}{2}n+k-1)!k!2^{2k}}\left[\frac{\pi^{1/2}2^{-n}}{(\frac{1}{2}n-\frac{1}{2})!}L^{k+(n/2)+1/2}\,q_+^{1/2}\ln|q| + \{\psi(\tfrac{1}{2}n+k-1) - \psi(\tfrac{1}{2}n-\tfrac{1}{2}) + \psi(k) - \psi(-\tfrac{3}{2})\}\pi^{n/2}\sin\tfrac{1}{2}p\pi\,L^k\delta(\boldsymbol{x})\right]$$

when n is odd and

$$q_+^{-(n/2)-k} = \frac{(-1)^{(n/2)+1}}{(\frac{1}{2}n+k-1)!(\frac{1}{2}n-1)!k!2^{2k+n}}\Big[\tfrac{1}{2}L^{k+n/2}\,\mathrm{H}(q)\ln^2 q + \{\psi(k)-\psi(0) + \psi(\tfrac{1}{2}n+k-1) - \psi(\tfrac{1}{2}n-1)\}\,L^{k+n/2}\,\mathrm{H}(q)\ln q$$
$$+ \{[\psi(k)-\psi(0)+\psi(\tfrac{1}{2}n+k-1)-\psi(\tfrac{1}{2}n-1)]^2 + 3\psi'(0) - \psi'(k) - \psi'(\tfrac{1}{2}n+k-1) - \psi'(\tfrac{1}{2}n-1)\}(\tfrac{1}{2}n-1)!2^{n-1}\pi^{(n/2)-1}\sin\tfrac{1}{2}p\pi\,L^k\delta(\boldsymbol{x})\Big]$$

when n is even. (Hint: $\psi'(0) = \pi^2/6$.)

Other generalised functions can be defined in terms of q_+^β. For example, we have

Definition 8.16. *If $\beta \neq -m$ (m a positive integer) and $\beta \neq -\frac{1}{2}n - k$ (k a non-negative integer)*

$$|q|^\beta = q_+^\beta + q_-^\beta,$$
$$|q|^\beta \operatorname{sgn} q = q_+^\beta - q_-^\beta,$$
$$(q \pm \mathrm{i}0)^\beta = q_+^\beta + \mathrm{e}^{\pm \mathrm{i}\beta\pi} q_-^\beta$$

The properties of these generalised functions can be inferred easily from those already derived for q_+, q_- and so will be left as exercises.

For the exceptional values we introduce

Definition 8.17. *If m is a positive integer and n is odd or n is even and $m \leq \frac{1}{2}n - 1$*

$$q^{-m} = q_+^{-m} + (-1)^m q_-^{-m},$$
$$q^{-m} \operatorname{sgn} q = q_+^{-m} - (-1)^m q_-^{-m},$$
$$(q \pm \mathrm{i}0)^{-m} = \operatorname*{Lim}_{\mu \to 0} (q \pm \mathrm{i}0)^{\mu - m}$$

while, if k is a non-negative integer,

$$(q \pm \mathrm{i}0)^{-(n/2)-k} = \operatorname*{Lim}_{\mu \to 0} \left\{ (q \pm \mathrm{i}0)^{\mu - (n/2) - k} - \frac{\pi^{n/2} \mathrm{e}^{\pm \pi \mathrm{i}(p-n)/2}}{(\frac{1}{2}n + k - 1)!k!2^{2k}\mu} L^k \delta(\boldsymbol{x}) \right\}.$$

Again the properties will be found in exercises.

Logarithmic factors can be inserted in the same way as in Chapter 4. For brevity, only definitions for q_+ will be given since those for other generalised functions run along similar lines. Also β, m and k will be understood to have the same significance as in previous definitions.

Definition 8.18. *The generalised function $q_+^\beta \ln q$ is defined by*

$$q_+^\beta \ln q = \partial q_+^\beta / \partial \beta,$$

$$q_+^{-m} \ln q = \operatorname*{Lim}_{\mu \to 0} \left\{ q_+^{\mu - m} \ln q + \frac{(-1)^{m-1}}{(m-1)!\mu^2} \delta^{(m-1)}(q) \right\},$$

$$q_+^{-(n/2)-k} \ln q$$
$$= \operatorname*{Lim}_{\mu \to 0} \left\{ q_+^{\mu - (n/2) - k} \ln q + \frac{(-1)^{(n/2) - 1/2} \pi^{n/2} \sin \frac{1}{2} p\pi}{(\frac{1}{2}n + k - 1)!k!2^{2k}\mu^2} L^k \delta(\boldsymbol{x}) \right\}$$

when n is odd and

$$q_+^{-(n/2)-k}\ln q = \operatorname*{Lim}_{\mu\to 0}\left\{ q_+^{\mu-(n/2)-k}\ln q - \frac{(-1)^{n/2}L^k}{(\frac{1}{2}n+k-1)!k!2^{2k}\mu^2}\right.$$

$$\times\left[\left\{\frac{2}{\mu}+\psi(k)-\psi(0)+\psi(\tfrac{1}{2}n+k-1)\right.\right.$$

$$\left.-\psi(\tfrac{1}{2}n-1)\right\}\pi^{(n/2)-1}\sin\tfrac{1}{2}p\pi\,\delta(\boldsymbol{x})$$

$$\left.\left.+\lambda_{n,p}\delta(\boldsymbol{x})+\delta^{((n/2)-1)}(q)\right]\right\}$$

when n is even.

Observe that

$$Lq_+^\beta \ln q = 2\beta(2\beta+n-2)q_+^{\beta-1}\ln q + 2(4\beta+n-2)q_+^{\beta-1} \tag{109}$$

for the non-exceptional values of β.

Exercises

74. Prove that $q.|q|^\beta = |q|^{\beta+1}\operatorname{sgn} q$, $q.|q|^\beta\operatorname{sgn} q = |q|^{\beta+1}$, $q.(q\pm \mathrm{i}0)^\beta = (q\pm \mathrm{i}0)^{\beta+1}$.

75. Show that

(i) $L|q|^\beta = 2\beta(2\beta+n-2)|q|^{\beta-1}\operatorname{sgn} q$,

(ii) $L|q|^\beta\operatorname{sgn} q = 2\beta(2\beta+n-2)|q|^{\beta-1}$,

(iii) $L(q\pm \mathrm{i}0)^\beta = 2\beta(2\beta+n-2)(q\pm \mathrm{i}0)^{\beta-1}$.

76. Show that $q.q^{-m} = q^{1-m}$, $q.q^{-m}\operatorname{sgn} q = q^{1-m}\operatorname{sgn} q$, $q.(q\pm \mathrm{i}0)^{-m} = (q\pm \mathrm{i}0)^{1-m}, q.(q\pm \mathrm{i}0)^{-(n/2)-k} = (q\pm \mathrm{i}0)^{1-(n/2)-k}$.

77. Prove that $\operatorname{Lim}_{\mu\to 0}|q|^{\mu-m} = q^{-m}$ when m is even and $\operatorname{Lim}_{\mu\to 0}|q|^{\mu-m}\operatorname{sgn} q = q^{-m}$ when m is odd.

78. Prove that $\quad Lq^{-m} = -2m(n-2m-2)q^{-m-1}$,

$$Lq^{-m}\operatorname{sgn} q = -2m(n-2m-2)q^{-m-1}\operatorname{sgn} q + \frac{(-1)^m}{m!}4(n-4m-2)\delta^{(m)}(q)$$

and that, when n is even,

$$Lq^{1-n/2} = \frac{(-1)^{(n/2)-(p/2)-1}}{(\frac{1}{2}n-2)!}4\pi^{n/2}\delta(\boldsymbol{x})$$

when p is even and $Lq^{1-n/2} = 0$ when p is odd.

79. Show that

(i) $L(q \pm \mathrm{i}0)^{-m} = -2m(n-2m-2)(q \pm \mathrm{i}0)^{-m-1}$,

(ii) $L(q \pm \mathrm{i}0)^{1-n/2} = -\dfrac{4\pi^{n/2}}{(\frac{1}{2}n-2)!}\mathrm{e}^{\pm\pi\mathrm{i}(p-n)/2}\delta(\boldsymbol{x})$.

80. Show that

$$(q \pm \mathrm{i}0)^{-m} = q^{-m} \mp \frac{(-1)^{m-1}\pi\mathrm{i}}{(m-1)!}\delta^{(m-1)}(q).$$

81. Show that

$$L(q \pm \mathrm{i}0)^{-(n/2)-k} = 2(k+1)(n+2k)(q \pm \mathrm{i}0)^{-(n/2)-k-1} - \frac{(n+4k+2)\pi^{n/2}\mathrm{e}^{\pm\pi\mathrm{i}(p-n)/2}}{(\frac{1}{2}n+k)!(k+1)!2^{2k+1}}L^{k+1}\delta(\boldsymbol{x}).$$

82. Prove that $\mathrm{Lim}_{\varepsilon\to+0}\,(q \pm \mathrm{i}\varepsilon r^2)^\beta = (q \pm \mathrm{i}0)^\beta$ for all complex β with the understanding that $0 < \mathrm{ph}(q + \mathrm{i}\varepsilon r^2) < \pi$ and $0 > \mathrm{ph}(q - \mathrm{i}\varepsilon r^2) > -\pi$.

83. Prove that

$$(q \pm \mathrm{i}0)^{-n/2} = \frac{1}{2(2-n)}L(q \pm \mathrm{i}0)^{1-n/2}\ln(q \pm \mathrm{i}0) + \frac{\pi^{n/2}\mathrm{e}^{\pm\pi\mathrm{i}(p-n)/2}}{(\frac{1}{2}n-1)!(\frac{1}{2}n-1)}\delta(\boldsymbol{x})$$

if $n \neq 2$ and that, when $n = 2$,

$$(q \pm \mathrm{i}0)^{-1} = \tfrac{1}{8}L\ln^2(q \pm \mathrm{i}0).$$

84. Find the formula corresponding to (78) for $q_+^\beta \ln q$.

85. Show that, when n is odd,

$$L^{(n/2)-1/2}q_+^\beta \ln q = \frac{\beta!(\beta+\frac{1}{2}n-1)!}{(\beta+\frac{1}{2}-\frac{1}{2}n)!(\beta-\frac{1}{2})!}2^{n-1} \times [\,q_+^{\beta+1/2-n/2}\ln q + \{\psi(\beta)+\psi(\beta+\tfrac{1}{2}n-1) - \psi(\beta+\tfrac{1}{2}-\tfrac{1}{2}n)-\psi(\beta-\tfrac{1}{2})\}q_+^{\beta+1/2-n/2}\,]$$

and derive the analogous expression for $L^{(n/2)-1}q_+^\beta \ln q$ when n is even.

86. Prove that

$$Lq_+^{-m}\ln q = -2m(n-2m-2)q_+^{-m-1}\ln q + 2(n-2-4m)q_+^{-m-1} + (-1)^m 4\delta^{(m)}(q)/m!.$$

87. Demonstrate that, when n is odd,

$$Lq_+^{1-n/2}\ln q = 2(2-n)q_+^{-n/2} + (-1)^{(n/2)-1/2}4\pi^{n/2}\sin\tfrac{1}{2}p\pi\,\delta(\boldsymbol{x})/(\tfrac{1}{2}n-1)!.$$

88. When n is odd prove that

$$Lq^{-(n/2)-k}\ln q = 2(k+1)(n+2k)q_+^{-(n/2)-k-1}\ln q - 2(n+4k+2)q_+^{-(n/2)-k-1} + \frac{(-1)^{(n/2)-1/2}\pi^{n/2}\sin\frac{1}{2}p\pi}{(\frac{1}{2}n+k)!(k+1)!2^{2k}}L^{k+1}\delta(x)$$

and verify that, when n is even,

$$Lq^{-(n/2)-k}\ln q = 2(k+1)(n+2k)q_+^{-(n/2)-k-1}\ln q - 2(n+4k+2)q_+^{-(n/2)-k-1} + \frac{(-1)^{(n/2)+1}\pi^{(n/2)-1}}{(\frac{1}{2}n+k)!(k+1)!2^{2k}}L^{k+1}\Big[\{\psi(k+1)-\psi(0)+\psi(\tfrac{1}{2}n+k) - \psi(\tfrac{1}{2}n-1)\}\sin\tfrac{1}{2}p\pi\delta(x) + \pi^{1-n/2}\{\lambda_{n,p}\delta(x)+\delta^{((n/2)-1)}(q)\}\Big].$$

89. Show that

$$(q\pm \mathrm{i}0)^{-(n/2)-k} = q_+^{-(n/2)-k} + \mathrm{e}^{\mp\mathrm{i}((n/2)+k)\pi}q_-^{-(n/2)-k} + \frac{\pi^{(n/2)+1}\sin\frac{1}{2}(n-p)\pi}{(\frac{1}{2}n+k-1)!k!2^{2k}}L^k\delta(x)$$

when n is odd and

$$(q\pm \mathrm{i}0)^{-(n/2)-k} = q_+^{-(n/2)-k} + (-1)^{(n/2)+k}q_-^{-(n/2)-k} \mp \frac{\mathrm{i}\pi}{(\frac{1}{2}n+k-1)!k!2^{2k}}L^k\{\lambda_{n,n-p}\delta(x)+(-1)^{(n/2)-1}\delta^{((n/2)-1)}(q)\}$$

when n is even and p is even. Show also that

$$(q\pm \mathrm{i}0)^{-(n/2)-k} = q_+^{-(n/2)-k} + (-1)^{(n/2)+k}q_-^{-(n/2)-k} \mp \frac{\mathrm{i}\pi}{(\frac{1}{2}n+k-1)!k!2^{2k}}L^k\Big[(-1)^{n/2-1}\delta^{((n/2)-1)}(q) \pm \pi^{(n/2)-1}\mathrm{i}\mathrm{e}^{\pm\pi\mathrm{i}(p-n)/2}\{\psi(k)-\psi(0) + \psi(\tfrac{1}{2}n+k-1)-\psi(\tfrac{1}{2}n-1)\pm\tfrac{1}{2}\pi\mathrm{i}\}\delta(x)\Big]$$

when n is even and p is odd.

When $p=1$ the generalised functions $q_+^\beta\,\mathrm{H}(x_1)$ and $q_+^\beta\,\mathrm{H}(-x_1)$ can also be considered. However, before doing so, it is appropriate to discuss $\delta^{(k)}(q)\mathrm{H}(x_1)$. For this, Definition 8.10 may be taken over except that now $\gamma_2(x,y)=\gamma_1(x_1,r_{n-1})$ instead of being given by (45). Properties of $\delta^{(k)}(q)\mathrm{H}(x_1)$ now follow immediately and those of

$\delta^{(k)}(q)\mathrm{H}(-x_1)$ can be deduced via

$$\delta^{(k)}(q)\mathrm{H}(-x_1) = \delta^{(k)}(q) - \delta^{(k)}(q)\mathrm{H}(x_1).$$

For $q_+^\beta \mathrm{H}(x_1)$ use Definition 8.13 with q_+^β replaced by $q_+^\beta \mathrm{H}(x_1)$ throughout. A similar replacement in Definition 8.14 provides $q_+^{-m}\mathrm{H}(x_1)$ so long as $\delta^{(m-1)}(q)\mathrm{H}(x_1)$ is inserted for $\delta^{(m-1)}(q)$. In Definition 8.15, as well as putting $q_+^{-(n/2)-k}\mathrm{H}(x_1)$ for $q_+^{-(n/2)-k}$, change $\delta(\boldsymbol{x})$ to $\frac{1}{2}\delta(\boldsymbol{x})$ and replace $\delta^{((n/2)-1)}(q)$ *by* $\delta^{((n/2)-1)}(q)\mathrm{H}(x_1)$. Finally $q_+^\beta \mathrm{H}(-x_1)$ is obtained from

$$q_+^\beta \mathrm{H}(-x_1) = q_+^\beta - q_+^\beta \mathrm{H}(x_1).$$

Exercises

90. Show that

$$L\delta^{(k)}(q)\mathrm{H}(x_1) = 2(n-2k-4)\delta^{(k+1)}(q)\mathrm{H}(x_1)$$

if n is odd or if n is even and $k < \frac{1}{2}n - 2$.

91. Prove that

$$L\mathrm{H}(q)\mathrm{H}(x_1) = 2(n-2)\delta(q)\mathrm{H}(x_1) \quad (n > 2),$$

and that

$$L\delta^{((n/2)-2)}(q)\mathrm{H}(x_1) = 2\pi^{n/2-1}\sin\tfrac{1}{2}p\pi\,\delta(\boldsymbol{x})$$

when n is even.

92. Prove that

$$Lq_+^{-m}\mathrm{H}(x_1) = -2m(n-2m-2)q_+^{-m-1}\mathrm{H}(x_1) + (-1)^m 2(n-4m-2)\delta^{(m)}(q)\mathrm{H}(x_1)/m!.$$

93. Obtain the formula for $Lq_+^{-(n/2)-k}\mathrm{H}(x_1)$.

8.8. Fourier transforms of the hyperbolic and ultrahyperbolic distances

In determining the Fourier transforms of the generalised functions studied in the preceding section it is convenient to employ the notation

$$\tilde{q} = \alpha_1^2 + \dots + \alpha_p^2 - \dots - \alpha_n^2$$

whereas $\tilde{L}$ is the same as L except that the variables are $\alpha_1, \dots, \alpha_n$ instead of $x_1, \dots, x_n$.

Theorem 8.9. *If β is neither a non-negative integer nor of the form $-\frac{1}{2}n-k$ the Fourier transform of $(q \pm \mathrm{i}0)^\beta$ is*

$$(\beta + \tfrac{1}{2}n - 1)!\,\pi^{n/2}2^{2\beta+n}e^{\pm\pi\mathrm{i}(p-n)/2}(\tilde{q} \mp \mathrm{i}0)^{-\beta-n/2}/(-\beta-1)!.$$

Proof. If $0 > \operatorname{Re}\beta > -1$ and $\varepsilon > 0$

$$\int_{-\infty}^{\infty} (q + i\varepsilon r^2)^{\beta} e^{-i\boldsymbol{\alpha}.\boldsymbol{x}} d\boldsymbol{x}$$

$$= \int_{-\infty}^{\infty} \int_{0}^{\infty} \frac{t^{-\beta-1}}{(-\beta-1)!} \exp[i\{\tfrac{1}{2}\pi\beta + t(q + i\varepsilon r^2) - \boldsymbol{\alpha}.\boldsymbol{x}\}] dt\, d\boldsymbol{x}$$

$$= \int_{0}^{\infty} \frac{t^{-\beta-1}}{(-\beta-1)!} \int_{-\infty}^{\infty} \exp[i\{\tfrac{1}{2}\pi\beta + t(q + i\varepsilon r^2) - \boldsymbol{\alpha}.\boldsymbol{x}\}] d\boldsymbol{x}\, dt.$$

Applying Theorem 2.2. repeatedly or Theorem 7.3 we obtain

$$\int_{-\infty}^{\infty} (q + i\varepsilon r^2)^{\beta} e^{-i\boldsymbol{\alpha}.\boldsymbol{x}} d\boldsymbol{x}$$

$$= \frac{\pi^{n/2} e^{\pi i\beta/2}}{(-\beta-1)!(\varepsilon - i)^{p/2}(\varepsilon + i)^{(n/2)-p/2}}$$

$$\times \int_{0}^{\infty} t^{-\beta-1-n/2} \exp\left\{-\frac{1}{4t}\left(\frac{\alpha_1^2}{\varepsilon - i} + \dots + \frac{\alpha_p^2}{\varepsilon - i} - \dots - \frac{\alpha_n^2}{\varepsilon + i}\right)\right\} dt$$

with the understanding that $-\tfrac{1}{2}\pi < \operatorname{ph}(\varepsilon \pm i) < \tfrac{1}{2}\pi$. Making the transformation $t = 1/u$ we are led to

$$\int_{-\infty}^{\infty} (q + i\varepsilon r^2)^{\beta} e^{-i\boldsymbol{\alpha}.\boldsymbol{x}} d\boldsymbol{x}$$

$$= \frac{(\beta + \tfrac{1}{2}n - 1)!\, \pi^{n/2} e^{\pi i\beta/2} 2^{2\beta+n}}{(-\beta-1)!(\varepsilon - i)^{p/2}(\varepsilon + i)^{(n-p)/2}}$$

$$\times \left(\frac{\alpha_1^2}{\varepsilon - i} + \dots + \frac{\alpha_p^2}{\varepsilon - i} - \dots - \frac{\alpha_n^2}{\varepsilon + i}\right)^{-\beta-(n/2)}$$

Now allow $\varepsilon \to 0$. Remembering the rule for the determination of phase $\varepsilon \pm i \to e^{\pm\pi i/2}$ and Exercise 82 we obtain the result stated in the theorem for $(q + i0)^{\beta}$ when $0 > \operatorname{Re}\beta > -1$.

On the other hand the Fourier transform of $L(q + i0)^{\beta}$ is $-\tilde{q}$ times the Fourier transform of $(q + i0)^{\beta}$. Invoking Exercises 74 and 75 we see that the formula of theorem can be extended to $\operatorname{Re}\beta < 0$ apart from the exceptional values and $(q + i0)^{-m}$. However, Definition 8.17 indicates that $(q + i0)^{-m}$ can be dealt with by placing $\beta = -m$ in the transform. Thus the theorem is established for

$(q+\mathrm{i}0)^\beta$ when $\operatorname{Re}\beta<0$ and $\beta\neq-\frac{1}{2}n-k$. The result for $(q-\mathrm{i}0)^\beta$ follows on changing the sign of i.

By virtue of the Fourier inversion theorem we can confirm the formula for $\operatorname{Re}\beta>-\frac{1}{2}n$ and β not a non-negative integer. The proof is complete. □

Corollary 8.9a. *The Fourier transform of* $(q\pm\mathrm{i}0)^k$ *is* $(2\pi)^n(-\tilde{L})^k\delta(\boldsymbol{\alpha})$ *and of* $(q\pm\mathrm{i}0)^{-(n/2)-k}$ *is*

$$\frac{(-1)^k\pi^{n/2}\mathrm{e}^{\pm\pi\mathrm{i}(p-n)/2}}{(\frac{1}{2}n+k-1)!k!2^{2k}}\left[\tilde{q}^k\{2\ln 2+\psi(\tfrac{1}{2}n+k-1)+\psi(k)\}-\tilde{q}^k\ln(\tilde{q}\mp\mathrm{i}0)\right]\quad(k=0,1,\ldots).$$

Proof. Since $(q\pm\mathrm{i}0)^k=q^k$ there is no problem with both having the same Fourier transform. Also the Fourier transform of q^k is $(-L)^k$ times the Fourier transform of 1 which is $(2\pi)^n\delta(\boldsymbol{\alpha})$ by Example 5 of Chapter 7, and the first part of the corollary is proved.

By virtue of Definition 8.17 and Theorem 8.9 the Fourier transform of $(q\pm\mathrm{i}0)^{-(n/2)-k}$ is

$$\lim_{\mu\to 0}2^{-2k}\pi^{n/2}\mathrm{e}^{\pm\pi\mathrm{i}(p-n)/2}\left\{\frac{(\mu-k-1)!2^{2\mu}}{(\frac{1}{2}n+k-1-\mu)!}(\tilde{q}\mp\mathrm{i}0)^{k-\mu}-\frac{(-\tilde{q})^k}{(\frac{1}{2}n+k-1)!k!\mu}\right\}.$$

Write $(\mu-k-1)!=(-1)^k\pi/(k-\mu)!\sin\mu\pi$ and then proceed to the limit to obtain the second part of the corollary. The proof is complete. □

Corollary 8.9b. *If* β *is not a non-negative integer and* $\beta\neq-m$, $\beta\neq-\frac{1}{2}n-k$ *the Fourier transform of* q_+^β *is*

$$\beta!(\beta+\tfrac{1}{2}n-1)!\mathrm{i}\pi^{(n/2)-1}2^{2\beta+n-1}\{\mathrm{e}^{\mathrm{i}\pi(\beta-(p/2)+n/2)}(\tilde{q}+\mathrm{i}0)^{-\beta-n/2}-\mathrm{e}^{-\mathrm{i}\pi(\beta-(p/2)+n/2)}(\tilde{q}-\mathrm{i}0)^{-\beta-n/2}\}$$

and of q_-^β *is*

$$\beta!(\beta+\tfrac{1}{2}n-1)!\mathrm{i}\pi^{(n/2)-1}2^{2\beta+n-1}\{\mathrm{e}^{\pi\mathrm{i}(p-n)/2}(\tilde{q}-\mathrm{i}0)^{-\beta-n/2}-\mathrm{e}^{-\pi\mathrm{i}(p-n)/2}(\tilde{q}+\mathrm{i}0)^{-\beta-n/2}\}.$$

Proof. Under the stated conditions on β

$$2\mathrm{i}q_+^\beta \sin\beta\pi = (q-\mathrm{i}0)^\beta \mathrm{e}^{\mathrm{i}\beta\pi} - (q+\mathrm{i}0)^\beta \mathrm{e}^{-\mathrm{i}\beta\pi}, \tag{110}$$

$$2\mathrm{i}q_-^\beta \sin\beta\pi = (q+\mathrm{i}0)^\beta - (q-\mathrm{i}0)^\beta, \tag{111}$$

according to Definition 8.16. Thus Theorem 8.9 gives the corollary on the insertion of $-\pi/(-\beta-1)!\sin\beta\pi = \beta!$ and the proof is finished. □

The transforms in Corollary 8.9*b* can be written alternatively as

$$q_+^\beta,\quad \beta!(\beta+\tfrac{1}{2}n-1)!\pi^{(n/2)-1}2^{2\beta+n}\{\tilde{q}_-^{-\beta-n/2}\sin\tfrac{1}{2}p\pi - \tilde{q}_+^{-\beta-n/2}\sin(\beta-\tfrac{1}{2}p+\tfrac{1}{2}n)\pi\},$$

$$q_-^\beta,\quad \beta!(\beta+\tfrac{1}{2}n-1)!\pi^{(n/2)-1}2^{2\beta+n}\{\tilde{q}_+^{-\beta-n/2}\sin\tfrac{1}{2}(n-p)\pi - \tilde{q}_-^{-\beta-n/2}\sin(\beta+\tfrac{1}{2}p)\pi\}.$$

Let $\beta\to 0$ in Corollary 8.9*b* and take note of Definition 8.17. Then the Fourier transform of $\mathrm{H}(q)$ is

$$(\tfrac{1}{2}n-1)!\,\pi^{(n/2)-1}\mathrm{i}2^{n-1}\{\mathrm{e}^{-\pi\mathrm{i}(p-n)/2}(\tilde{q}+\mathrm{i}0)^{-n/2} - \mathrm{e}^{\pi\mathrm{i}(p-n)/2}(\tilde{q}-\mathrm{i}0)^{-n/2}\} + (2\pi)^n\delta(\boldsymbol{\alpha})$$

and that for $\mathrm{H}(-q)$ may be obtained in a like manner. The Fourier transform can be expressed in terms of $\tilde{q}_+$ and $\tilde{q}_-$ by means of Exercise 89. By multiplying $\mathrm{H}(q)$ by q^k and thereby applying the operator $(-\tilde{L})^k$ to the transform the result for q_+^k can be derived via Exercise 81. In contrast, the application of L^m to $\mathrm{H}(q)$ enables us to find the transform of $\delta^{(m-1)}(q)$. These results are summarised in

Theorem 8.10. *The Fourier transform of* q_+^k *is*

$$(\tfrac{1}{2}n+k-1)!k!(-1)^k\pi^{(n/2)-1}\mathrm{i}2^{n+2k-1}\{\mathrm{e}^{-\pi\mathrm{i}(p-n)/2}(\tilde{q}+\mathrm{i}0)^{-(n/2)-k} - \mathrm{e}^{\pi\mathrm{i}(p-n)/2}(\tilde{q}-\mathrm{i}0)^{-(n/2)-k}\} + (-1)^k2^n[\pi^n - \mathrm{i}\pi^{n-1}\{\psi(\tfrac{1}{2}n+k-1)+\psi(k) - \psi(\tfrac{1}{2}n-1)-\psi(0)\}]\tilde{L}^k\delta(\boldsymbol{\alpha})$$

and of $\delta^{(k)}(q)$ *is*

$$(\tfrac{1}{2}n-k-2)!\pi^{(n/2)-1}\mathrm{i}2^{n-2k-3}(-1)^{k+1}\{\mathrm{e}^{-\pi\mathrm{i}(p-n)/2}(\tilde{q}+\mathrm{i}0)^{k+1-n/2} - \mathrm{e}^{\pi\mathrm{i}(p-n)/2}(\tilde{q}-\mathrm{i}0)^{k+1-n/2}\}$$

if n is odd or if n is even and $k<\tfrac{1}{2}n-1$.

It is also possible to allow $\beta \to -m$ in Corollary 8.9*b* and take note of Definitions 8.14, 8.16, 8.17 as well as Theorem 8.10. This gives

Corollary 8.10. *The Fourier transform of* q_+^{-m} *where* $m = 1, 2, \dots$ *if* n *is odd and* $m = 1, 2, \dots, \frac{1}{2}n - 1$ *if* n *is even is*

$$\frac{(\frac{1}{2}n - m - 1)!}{(m-1)!} \mathrm{i}\pi^{(n/2)-1} 2^{n-2m-1}$$
$$\times [\mathrm{e}^{-\pi\mathrm{i}(p-n)/2}(\tilde{q} + \mathrm{i}0)^{m-n/2} \ln(\tilde{q} + \mathrm{i}0) - \mathrm{e}^{\pi\mathrm{i}(p-n)/2}(\tilde{q} - \mathrm{i}0)^{m-n/2}$$
$$\times \ln(\tilde{q} - \mathrm{i}0) - \{\psi(\tfrac{1}{2}n - m - 1) + \psi(m-1) + 2\ln 2\}$$
$$\times \{\mathrm{e}^{-\pi\mathrm{i}(p-n)/2}(\tilde{q} + \mathrm{i}0)^{m-n/2} - \mathrm{e}^{\pi\mathrm{i}(p-n)/2}(\tilde{q} - \mathrm{i}0)^{m-n/2}\}$$
$$- \pi\mathrm{i}\{\mathrm{e}^{-\pi\mathrm{i}(p-n)/2}(\tilde{q} + \mathrm{i}0)^{m-n/2} + \mathrm{e}^{\pi\mathrm{i}(p-n)/2}(\tilde{q} - \mathrm{i}0)^{m-n/2}\}].$$

The combination of Definition 8.18 and Corollary 8.9*b* leads to

Theorem 8.11. *If* β *is not a non-negative integer and* $\beta \neq -m$, $\beta \neq -\frac{1}{2}n - k$ *the Fourier transform of* $q_+^{\beta} \ln q$ *is*

$$\beta!(\beta + \tfrac{1}{2}n - 1)!\,\mathrm{i}\pi^{(n/2)-1} 2^{2\beta+n-1}$$
$$\times [\mathrm{e}^{-\mathrm{i}\pi(\beta-(p/2)+n/2)}(\tilde{q} - \mathrm{i}0)^{-\beta-n/2} \ln(\tilde{q} - \mathrm{i}0)$$
$$- \mathrm{e}^{\mathrm{i}\pi(\beta-(p/2)+n/2)}(\tilde{q} + \mathrm{i}0)^{-\beta-n/2} \ln(\tilde{q} + \mathrm{i}0)$$
$$+ \mathrm{i}\pi\{\mathrm{e}^{\mathrm{i}\pi(\beta-(p/2)+n/2)}(\tilde{q} + \mathrm{i}0)^{-\beta-n/2}$$
$$+ \mathrm{e}^{-\mathrm{i}\pi(\beta-(p/2)+n/2)}(\tilde{q} - \mathrm{i}0)^{-\beta-n/2}\} + \{\psi(\beta) + \psi(\beta + \tfrac{1}{2}n - 1) + 2\ln 2\}$$
$$\times \{\mathrm{e}^{\mathrm{i}\pi(\beta-(p/2)+n/2)}(\tilde{q} + \mathrm{i}0)^{-\beta-n/2}$$
$$- \mathrm{e}^{-\mathrm{i}\pi(\beta-(p/2)+n/2)}(\tilde{q} - \mathrm{i}0)^{-\beta-n/2}\}].$$

Corollary 8.11. *The Fourier transform of* $q_+^k \ln q$ *is*

$$k!(\tfrac{1}{2}n + k - 1)!\,\mathrm{i}\pi^{(n/2)-1}(-1)^k 2^{2k+n-1}$$
$$\times [\mathrm{e}^{\pi\mathrm{i}(p-n)/2}(\tilde{q} - \mathrm{i}0)^{-(n/2)-k} \ln(\tilde{q} - \mathrm{i}0)$$
$$- \mathrm{e}^{-\pi\mathrm{i}(p-n)/2}(\tilde{q} + \mathrm{i}0)^{-(n/2)-k} \ln(\tilde{q} + \mathrm{i}0)$$
$$+ \mathrm{i}\pi\{\mathrm{e}^{-\pi\mathrm{i}(p-n)/2}(\tilde{q} + \mathrm{i}0)^{-(n/2)-k}$$
$$+ \mathrm{e}^{\pi\mathrm{i}(p-n)/2}(\tilde{q} - \mathrm{i}0)^{-(n/2)-k}\} + \{\psi(k) + \psi(k + \tfrac{1}{2}n - 1) + 2\ln 2\}$$
$$\times \{\mathrm{e}^{-\pi\mathrm{i}(p-n)/2}(\tilde{q} + \mathrm{i}0)^{-(n/2)-k} - \mathrm{e}^{\pi\mathrm{i}(p-n)/2}(\tilde{q} - \mathrm{i}0)^{-(n/2)-k}\}]$$
$$+ (2\pi)^n(-1)^k\{\psi(k) + \psi(\tfrac{1}{2}n + k - 1) + 2\ln 2\}\, \tilde{L}^k \delta(\boldsymbol{\alpha})$$

and of $\lambda_{n,p}\delta(x) + \delta^{((n/2)-1)}(q)$ *when n is even is*

$$(-1)^{n/2}\pi^{(n/2)-1}[\{\ln|\tilde{q}| - \psi(0) - \psi(\tfrac{1}{2}n - 1) \\ - 2\ln 2\}\sin\tfrac{1}{2}(n-p)\pi - \pi \mathrm{H}(\tilde{q})\cos\tfrac{1}{2}(n-p)\pi].$$

Proof. The first part is a consequence of allowing $\beta \to k$ in Theorem 8.11 and recalling Definition 8.17. As to the second part remark that, from (72) and (96),

$$\lambda_{n,p}\delta(x) + \delta^{((n/2)-1)}(q) = \frac{L^{n/2}\mathrm{H}(q)\ln q}{(\frac{1}{2}n-1)!2^n}. \tag{112}$$

The Fourier transform of $\mathrm{H}(q)\ln q$ is furnished by the first part of the corollary when $k = 0$. Multiplying that Fourier transform by $(-\tilde{q})^{n/2}$ then provides the transform of $L^{n/2}\mathrm{H}(q)\ln q$ and hence the required transform by (112). The proof is finished. □

The result in the second part of Corollary 8.11, coupled with Corollary 8.9*b*, permits us to state

Theorem 8.12. *The Fourier transform of* $q_+^{-(n/2)-k}$ *is*

$$\frac{\pi^{n/2}(-1)^{k+(n/2)-1/2}\tilde{q}^k}{(\frac{1}{2}n+k-1)!k!2^{2k}}[\{2\ln 2 + \psi(\tfrac{1}{2}n+k-1) + \psi(k) \\ - \ln|\tilde{q}|\}\sin\tfrac{1}{2}p\pi - \pi\mathrm{H}(\tilde{q})\cos\tfrac{1}{2}p\pi]$$

when n is odd and

$$\frac{(-1)^{k+(n/2)+1}\pi^{(n/2)-1}\tilde{q}^k}{(\frac{1}{2}n+k-1)!k!2^{2k}}\{[\tfrac{1}{6}\pi^2 - \tfrac{1}{2}\pi^2\mathrm{H}(\tilde{q}) + \tfrac{1}{2}\{2\ln 2 \\ + \psi(\tfrac{1}{2}n+k-1) + \psi(k)\}^2 - \tfrac{1}{2}\{\psi'(\tfrac{1}{2}n+k-1) + \psi'(k)\} \\ - \{2\ln 2 + \psi(\tfrac{1}{2}n+k-1) + \psi(k)\}\ln|\tilde{q}| + \tfrac{1}{2}\ln^2|\tilde{q}|]\sin\tfrac{1}{2}p\pi \\ + \pi\mathrm{H}(\tilde{q})[\ln|\tilde{q}| - 2\ln 2 - \psi(\tfrac{1}{2}n+k-1) - \psi(k)]\cos\tfrac{1}{2}p\pi\}$$

when n is even.

It is now possible to deduce the Fourier transforms of other generalised functions which have been considered with the exception of $q_+^\beta\mathrm{H}(x_1)$ when $p = 1$. For this we may proceed directly as in

Theorem 8.13. *If* β *is not a non-negative integer and* $\beta \neq -m$,

$\beta \neq -\frac{1}{2}n - k$ the Fourier transform of $q_+^\beta \mathrm{H}(x_1)$ is

$$\beta!(\beta + \tfrac{1}{2}n - 1)!\pi^{(n/2)-1}2^{2\beta+n-1}\{\tilde{q}_-^{-\beta-n/2} + \mathrm{e}^{\mathrm{i}\pi(\beta+(n/2))}\tilde{q}_+^{-\beta-n/2} - \tilde{q}_+^{-\beta-n/2}\mathrm{H}(\alpha_1)2\mathrm{i}\sin(\beta + \tfrac{1}{2}n)\pi\}$$

when $p = 1$.

Proof. The case where $\mathrm{Re}\,\beta > -1$ will be tackled first. Then $\mathrm{Lim}_{\mu\to+0}\mathrm{e}^{-\mu x_1}q_+^\beta \mathrm{H}(x_1) = q_+^\beta \mathrm{H}(x_1)$ and so the Fourier transform can be inferred from that of $\mathrm{e}^{-\mu x_1}q_+^\beta \mathrm{H}(x_1)$. For this the integration over Ω_{n-1} can be carried out as in Theorem 7.30 and brings one to

$$(2\pi)^{(n/2)-1/2}\int_0^\infty\int_{r_{n-1}}^\infty \mathrm{e}^{-\mu x_1 - \mathrm{i}\alpha_1 x_1}(x_1^2 - r_{n-1}^2)^\beta J_{(n/2)-3/2}(\alpha_v r_{n-1}) \times (\alpha_v r_{n-1})^{(3/2)-n/2} r_{n-1}^{n-2}\,\mathrm{d}x_1\,\mathrm{d}r_{n-1}$$

where $\alpha_v^2 = \alpha_2^2 + \dots + \alpha_n^2$. Make the change of variables $x_1 = r_{n-1}t$, $r_{n-1} = r$ to obtain

$$\begin{aligned}\frac{(2\pi)^{(n/2)-1/2}}{\alpha_v^{(n/2)-3/2}}\int_0^\infty r^{2\beta+(n/2)+1/2}J_{(n/2)-3/2}(\alpha_v r)\int_1^\infty \mathrm{e}^{-(\mu+\mathrm{i}\alpha_1)rt}(t^2-1)^\beta\mathrm{d}t\,\mathrm{d}r \\ = \frac{\beta!2^{\beta+n/2}\pi^{(n/2)-1}}{\alpha_v^{(n/2)-3/2}(\mu+\mathrm{i}\alpha_1)^{\beta+1/2}}\int_0^\infty r^{\beta+n/2}J_{(n/2)-3/2}(\alpha_v r) \\ \times K_{\beta+1/2}\{(\mu+\mathrm{i}\alpha_1)r\}\,\mathrm{d}r\end{aligned} \tag{113}$$

on employing the formula

$$K_v(z) = \frac{\pi^{1/2}(\frac{1}{2}z)^v}{(v-\frac{1}{2})!}\int_1^\infty \mathrm{e}^{-zt}(t^2-1)^{v-1/2}\,\mathrm{d}t \tag{114}$$

for the Bessel function K_v. It is valid for $\mathrm{Re}\,v > -\frac{1}{2}$ and $|\mathrm{ph}\,z| < \frac{1}{2}\pi$. Now

$$\int_0^\infty K_\lambda(at)J_v(bt)t^{\lambda+v+1}\,\mathrm{d}t = \frac{(\lambda+v)!2^{\lambda+v}a^\lambda b^v}{(a^2+b^2)^{\lambda+v+1}} \tag{115}$$

provided that $\mathrm{Re}\ a > |\mathrm{Im}\,b|$ and $\mathrm{Re}\,v > -1$ if $\mathrm{Re}\,\lambda > 0$ but $\mathrm{Re}\,(\lambda + v) > -1$ if $\mathrm{Re}\,\lambda < 0$. These conditions are complied with in (113) and so we arrive at

$$\frac{\beta!(\beta+\frac{1}{2}n-1)!2^{2\beta+n-1}\pi^{(n/2)-1}}{\{(\mu+\mathrm{i}\alpha_1)^2+\alpha_v^2\}^{\beta+n/2}}.$$

However

$$\{(\mu+\mathrm{i}\alpha_1)^2+\alpha_\nu^2\}^{-\beta-n/2}$$
$$=\frac{(-\beta-\frac{1}{2}n)!(-\beta-1)!(-1)^s\tilde{L}^s\{(\mu+\mathrm{i}\alpha_1)^2+\alpha_\nu^2\}^{s-\beta-n/2}}{(s-\beta-\frac{1}{2}n)!(s-\beta-1)!2^{2s}}$$

Choose s so large that $s-\frac{1}{2}n>\operatorname{Re}\beta$ and then

$$\{(\mu+\mathrm{i}\alpha_1)^2+\alpha_\nu^2\}^{s-\beta-n/2}\to\tilde{q}_-^{s-\beta-n/2}+\tilde{q}_+^{s-\beta-n/2}\mathrm{H}(\alpha_1)\mathrm{e}^{\mathrm{i}\pi(s-\beta-n/2)}$$
$$+\tilde{q}_+^{s-\beta-n/2}\mathrm{H}(-\alpha_1)\mathrm{e}^{-\mathrm{i}\pi(s-\beta-n/2)}$$

as $\mu\to+0$. Applying the operator $\tilde{L}^s$ we come to the expression set out in the theorem. Thus the theorem has been demonstrated for $\operatorname{Re}\beta>-1$.

The transform of $L^m q_+^\beta\mathrm{H}(x_1)$ is $(-\tilde{q})^m$ times that of $q_+^\beta\mathrm{H}(x_1)$ and so the result of the theorem can be checked for all β other than those excluded by the conditions of the theorem. There is nothing further to prove. □

Exercises

94. Show that the Fourier transform of $\mathrm{H}(-q)$ is

$$(\tfrac{1}{2}n-1)!\pi^{(n/2)-1}\mathrm{i}2^{n-1}\{\mathrm{e}^{\pi\mathrm{i}(p-n)/2}(\tilde{q}-\mathrm{i}0)^{-n/2}-\mathrm{e}^{-\pi\mathrm{i}(p-n)/2}(\tilde{q}+\mathrm{i}0)^{-n/2}\}.$$

95. Prove that, when n is odd, the Fourier transform of $\delta^{(m)}(q)$ is

$$(\tfrac{1}{2}n-m-2)!\pi^{(n/2)-1}2^{n-2m-2}\{\tilde{q}_-^{m+1-n/2}\sin\tfrac{1}{2}p\pi$$
$$+(-1)^m\tilde{q}_+^{m+1-n/2}\sin\tfrac{1}{2}(n-p)\pi\}.$$

96. Show that the Fourier transform of q_-^{-m} is

$$\frac{(\frac{1}{2}n-m-1)!}{(m-1)!}(-1)^{m-1}\mathrm{i}\pi^{(n/2)-1}2^{n-2m-1}$$
$$\times[\mathrm{e}^{-\pi\mathrm{i}(p-n)/2}(\tilde{q}+\mathrm{i}0)^{m-n/2}\ln(\tilde{q}+\mathrm{i}0)-\mathrm{e}^{\pi\mathrm{i}(p-n)/2}(\tilde{q}-\mathrm{i}0)^{m-n/2}\ln(\tilde{q}-\mathrm{i}0)$$
$$+\{\psi(\tfrac{1}{2}n-m-1)+\psi(m-1)+2\ln 2\}\{\mathrm{e}^{\pi\mathrm{i}(p-n)/2}(\tilde{q}-\mathrm{i}0)^{m-n/2}$$
$$-\mathrm{e}^{-\pi\mathrm{i}(p-n)/2}(\tilde{q}+\mathrm{i}0)^{m-n/2}\}].$$

97. Prove that the Fourier transform of q^{-m} is

$$\frac{(\frac{1}{2}n-m-1)!}{(m-1)!}\pi^{n/2}2^{n-2m}\{(-1)^{(n/2)-1/2}\tilde{q}_+^{m-n/2}\sin\tfrac{1}{2}p\pi$$
$$+(-1)^m\tilde{q}_-^{m-n/2}\cos\tfrac{1}{2}p\pi\}$$

when n is odd.

98. Prove that when n is even and $m<\frac{1}{2}n-1$ the Fourier transform of q^{-m} is

$$\frac{(\frac{1}{2}n-m-1)!}{(m-1)!}\pi^{n/2}2^{n-2m}(-1)^{n/2}\tilde{q}^{m-n/2}\cos\tfrac{1}{2}p\pi$$

$$+\frac{2^{n-2m}\pi^{(n/2)+1}}{(m-1)!}(-1)^{m}\delta^{((n/2)-m-1)}(\tilde{q})\sin\tfrac{1}{2}p\pi.$$

If $n=2$ and $p=1$ prove that the Fourier transform of q^{-1} is $-\frac{1}{2}\pi^2\operatorname{sgn}\tilde{q}$.

99. When $p=1$ show that the Fourier transform of $\mathrm{H}(q)\,\mathrm{H}(x_1)$ is

$$(\tfrac{1}{2}n-1)!\,\pi^{(n/2)-1}2^{n-1}\{\tilde{q}_-^{-n/2}+\mathrm{e}^{\mathrm{i}n\pi/2}\tilde{q}_+^{-n/2}-\tilde{q}_+^{-n/2}\mathrm{H}(\alpha_1)2\mathrm{i}\sin\tfrac{1}{2}n\pi\}$$
$$+\tfrac{1}{2}(2\pi)^n\delta(\boldsymbol{\alpha})$$

when n is odd and

$$(\tfrac{1}{2}n-1)!\,\pi^{(n/2)-1}2^{n-1}\{\tilde{q}_-^{-n/2}+(-1)^{n/2}\tilde{q}_+^{-n/2}\}$$
$$+2^{n-2}\pi^n\delta(\boldsymbol{\alpha})+2^{n-1}\pi^{n/2}\mathrm{i}\{\delta^{((n/2)-1)}(\tilde{q})-2\delta^{((n/2)-1)}(\tilde{q})\mathrm{H}(\alpha_1)\}$$

when n is even.

100. If $p=1$ prove that the Fourier transform of $\delta^{(m-1)}(q)\mathrm{H}(x_1)$ is

$$(\tfrac{1}{2}n-m-1)!\,\pi^{(n/2)-1}2^{n-2m-1}(-1)^m\{(-1)^m\tilde{q}_-^{m-n/2}$$
$$+\mathrm{e}^{\mathrm{i}n\pi/2}\tilde{q}_+^{m-n/2}-\tilde{q}_+^{m-n/2}\mathrm{H}(\alpha_1)2\mathrm{i}\sin\tfrac{1}{2}n\pi\}$$

when n is odd and

$$(\tfrac{1}{2}n-m-1)!\,\pi^{(n/2)-1}2^{n-2m-1}(-1)^{m-n/2}(\tilde{q}+\mathrm{i}0)^{m-n/2}$$
$$-2^{n-2m}\pi^{n/2}\mathrm{i}\delta^{((n/2)-m-1)}(\tilde{q})\,\mathrm{H}(\alpha_1)$$

if n is even and $1\le m\le\frac{1}{2}n-1$.

8.9. The general quadratic form

Consider now the quadratic form $Q=\sum_{j=1}^{n}\sum_{k=1}^{n}q_{jk}x_jx_k$ where q_{jk} is real and $q_{jk}=q_{kj}$. It will be assumed that all the eigenvalues associated with Q are non-zero. Then there is an orthogonal mapping $\boldsymbol{x}=L\boldsymbol{y}$ such that

$$Q=\sum_{j=1}^{n}\lambda_j y_j^2$$

where the eigenvalues λ_j are real and non-zero. The substitution $|\lambda_j|^{1/2}y_j=z_j$ or $\boldsymbol{y}=B\boldsymbol{z}$ gives $Q=\boldsymbol{z}^{\mathrm{T}}Z\boldsymbol{z}$ where Z is a matrix all of whose elements are zero except the diagonal ones which are either 1 or -1 depending on the sign of λ_j. In this representation Q has the same form as either $\pm r^2$ or $\pm q$ and since the matrix LB is non-singular, properties of Q can be deduced from those of r^2 and q by means of §8.1.

In particular, if ∂_{jz} stands for $\partial/\partial z_j$,

$$(\partial_{1z}^2 + \dots + \partial_{pz}^2 - \dots - \partial_{nz}^2)Q = \frac{\partial^{\mathrm{T}}}{\partial \boldsymbol{z}} Z \frac{\partial}{\partial \boldsymbol{z}} Q.$$

Since $\boldsymbol{z} = A\boldsymbol{x}$ where $A^{-1} = LB$,

$$\frac{\partial}{\partial z_k} = \sum_{j=1}^{n} a_{jk}^{-1} \frac{\partial}{\partial x_j}$$

or $\partial/\partial\boldsymbol{z} = (A^{-1})^{\mathrm{T}}\partial/\partial\boldsymbol{x}$. Hence

$$\frac{\partial^{\mathrm{T}}}{\partial \boldsymbol{z}} Z \frac{\partial}{\partial \boldsymbol{z}} Q = \frac{\partial^{\mathrm{T}}}{\partial \boldsymbol{x}} A^{-1} Z (A^{-1})^{\mathrm{T}} \frac{\partial}{\partial \boldsymbol{z}} Q.$$

However, $Q = \boldsymbol{x}^{\mathrm{T}} A^{\mathrm{T}} Z A \boldsymbol{x}$ and the inverse of $A^{\mathrm{T}} Z A$ is $A^{-1} Z (A^{-1})^{\mathrm{T}}$ because $Z = Z^{-1}$. Hence $A^{-1} Z (A^{-1})^{\mathrm{T}}$ is the inverse of the matrix (q_{jk}). Therefore

$$(\partial_{1z}^2 + \dots + \partial_{pz}^2 - \dots - \partial_{nz}^2)Q = \mathscr{L}Q$$

with $\mathscr{L} \equiv \sum_{j=1}^{n} \sum_{k=1}^{n} Q_{jk} \partial_j \partial_k$ where

$$\sum_{j=1}^{n} q_{ij} Q_{jk} = \begin{cases} 1 & (i = k) \\ 0 & (i \neq k). \end{cases}$$

As a consequence, a formula such as (69) becomes

$$\mathscr{L} Q_+^{\beta} = 2\beta(2\beta + n - 2) Q_+^{\beta - 1}. \tag{116}$$

On the other hand, (80) will transform to

$$\mathscr{L} Q_+^{1-n/2} = (1 - \tfrac{1}{2}n)! 4\pi^{(n/2)-1} \sin \tfrac{1}{2}p\pi\, \delta(A\boldsymbol{x})$$

where p is the number of positive units in Z. Observe that p is an invariant of Q and so is well defined. Bearing in mind (3) we have

$$\mathscr{L} Q_+^{1-n/2} = (1 - \tfrac{1}{2}n)! 4\pi^{(n/2)-1} \sin \tfrac{1}{2}p\pi\, \delta(\boldsymbol{x})/|\det A|.$$

Now

$$\begin{aligned}|\det A| &= |\det L^{-1}||\det B^{-1}| = |\det B^{-1}| = |\lambda_1 \lambda_2 \dots \lambda_n|^{1/2} \\ &= |\det Q|^{1/2}\end{aligned}$$

where det Q is the determinant formed by the coefficients q_{jk}. Hence

$$\mathscr{L}Q_+^{1-n/2} = (1-\tfrac{1}{2}n)!4\pi^{(n/2)-1}\sin\tfrac{1}{2}p\pi\,\delta(\boldsymbol{x})/|\det Q|^{1/2} \quad (117)$$

when n is odd. Remark that this is still true if $p = n$ by Example 9 of Chapter 7 or if $p = 0$ (when Q_+ is zero everywhere).

In this connection we should recognise that $(r^2 + \mathrm{i}0)^\beta = r^{2\beta}$ and $(-r + \mathrm{i}0)^\beta = \mathrm{e}^{\mathrm{i}\beta\pi}r^{2\beta}$.

With regard to Fourier transforms Exercise 7 indicates that the whole transform must be divided by $|\det Q|^{1/2}$ and a quantity such as $\tilde{q}$ replaced by $\tilde{Q}$ where $\tilde{Q} = \sum_{j=1}^{n}\sum_{k=1}^{n} Q_{jk}\alpha_j\alpha_k$.

Exercises

101. Show that

$$\mathscr{L}(Q \pm \mathrm{i}0)^{1-n/2} = -\frac{4\pi^{n/2}}{(\frac{1}{2}n-2)!}\frac{\mathrm{e}^{\pm\pi\mathrm{i}(p-n)/2}}{|\det Q|^{1/2}}\delta(\boldsymbol{x}).$$

102. The generalised function $\delta(Q)$ can be obtained, when $n > 2$, from $\mathscr{L}\mathrm{H}(Q) = 2(n-2)\delta(Q)$.

103. Show that $\mathscr{L}\delta^{(k)}(Q) = 2\,(n-2k-4)\delta^{(k+1)}(Q)$ when n is odd or if n is even and $k < \frac{1}{2}n - 2$. When n is even show that $\mathscr{L}\delta^{((n/2)-2)}(Q) = 4\pi^{(n/2)-1}\sin\frac{1}{2}p\pi\,\delta(\boldsymbol{x})/|\det Q|^{1/2}$.

104. Prove that when n is even

$$\mathscr{L}\mathrm{H}(Q)\ln Q = (\tfrac{1}{2}n-1)!2^n\{\delta^{((n/2)-1)}(\tilde{Q}) + \lambda_{n,p}\delta(\boldsymbol{x})/|\det Q|^{1/2}\}.$$

105. Prove that the Fourier transform of $(Q \pm \mathrm{i}0)^\beta$ is

$$(\beta + \tfrac{1}{2}n - 1)!\pi^{n/2}2^{2\beta+n}\mathrm{e}^{\pm\pi\mathrm{i}(p-n)/2}(\tilde{Q} \mp \mathrm{i}0)^{-\beta-n/2}/(-\beta-1)!|\det Q|^{1/2}$$

when β is neither a non-negative integer nor of the form $-\frac{1}{2}n - k$.

106. Show that the Fourier transform of $(r^2 + a^2)^\beta$ $(a > 0)$ is

$$\beta!\mathrm{i}\pi^{(n/2)-1}(2a)^{\beta+n/2}\mathrm{e}^{-\pi\mathrm{i}\beta}\alpha^{-\beta-n/2}K_{\beta+n/2}(a\alpha)$$

when β is neither an integer nor of the form $-\frac{1}{2}n - k$. Show also that the Fourier transform of $(Q + a^2)^\beta$ is

$$\frac{\beta!\mathrm{i}\pi^{(n/2)-1}}{|\det Q|^{1/2}}(2a)^{\beta+n/2}\mathrm{e}^{\pi\mathrm{i}(p-n-2\beta)/2}(\tilde{Q}-\mathrm{i}0)^{-(\beta/2)-n/4}K_{\beta+n/2}\{a(\tilde{Q}-\mathrm{i}0)^{1/2}\}.$$

(Hint: §6.3 and the results after Theorem 7.38.)

8.10. The class K_+

This section is concerned with the generalisation of the class K_+ introduced in §6.6. Let K be a cone consisting of the union of semi-infinite straight lines which start at the origin and go to infinity. Thus K might be a single semi-infinite straight line, or a quadrant, or a conical volume. It will be assumed that K is simply-connected. The further restriction will be imposed that there is at least one hyperplane through the vertex which does not intersect the cone in a generator. By a rotation of axes it can then be arranged that all points of K except the vertex lie entirely in $x_1 > 0$; it will be supposed that this has been done. If $n = 2$, K consists of a single semi-infinite straight line or two such lines and the area between them.

Definition 8.19. *$g \in \boldsymbol{K}_+$ if, and only if, $g = 0$ outside K.*

Clearly $\boldsymbol{K}_+$ will vary with the choice of K but, in all cases, $\delta(\boldsymbol{x}) \in \boldsymbol{K}_+$. In what follows, a fixed choice of K will be assumed to have been made.

Theorem 8.14. *If $g_1 \in \boldsymbol{K}_+$ and $g_2 \in \boldsymbol{K}_+$ then $g_1 * g_2 \in \boldsymbol{K}_+$.*

Proof. As in Theorem 6.18 it is sufficient to prove the theorem for the continuous functions f_1 and f_2 which vanish outside K and are such that $(1 + r^2)^{-k/2} f_1$ and $(1 + r^2)^{-k/2} f_2$ are bounded on R_n. Now

$$f_1 * f_2 = \int_K f_1(\boldsymbol{t}) f_2(\boldsymbol{x} - \boldsymbol{t}) \, \mathrm{d}\boldsymbol{t}$$

$$= \begin{cases} \displaystyle\int_{K_\cap(\boldsymbol{x})} f_1(\boldsymbol{t}) f_2(\boldsymbol{x} - \boldsymbol{t}) \, \mathrm{d}\boldsymbol{t} & (\boldsymbol{x} \in K) \\ 0 & (\boldsymbol{x} \notin K) \end{cases}$$

where $K_\cap(\boldsymbol{x}) = K \cap K_-(\boldsymbol{x})$, $K_-(\boldsymbol{0})$ is the cone obtained from K by reflecting the generators in the origin and $K_-(\boldsymbol{x})$ is the cone $K_-(\boldsymbol{0})$ translated so that its vertex is at $\boldsymbol{x}$. Also

$$\left| \int_{K \cap (\boldsymbol{x})} f_1(\boldsymbol{t}) f_2(\boldsymbol{x} - \boldsymbol{t}) \, \mathrm{d}\boldsymbol{t} \right| \le M_1 M_2 \int_{K \cap (\boldsymbol{x})} (1 + |\boldsymbol{t}|^2)^{k/2} (1 + |\boldsymbol{x} - \boldsymbol{t}|^2)^{k/2} \, \mathrm{d}\boldsymbol{t}$$

$$\le M x_1^n (1 + a x_1^2)^k$$

because $K_\cap(\boldsymbol{x})$ certainly lies within a sphere, centre the origin, passing through the intersection of $t_1 = x_1$ and the generator(s) making the largest angle with the t_1-axis. Therefore $f_1 * f_2 \in K_1$ and is a generalised function. Thus $f_1 * f_2 \in K_+$ and the proof is complete. □

Corollary 8.14. *If g_1, g_2 and g_3 all belong to K_+ then*

$$g_1 * (g_2 * g_3) = (g_1 * g_2) * g_3.$$

Proof. Along the lines of Corollary 6.18. □

It is sometimes convenient to write

$$g_1 * g_2 = \begin{cases} \displaystyle\int_{K \cap (x)} g_1(t) g_2(x - t) \mathrm{d}t & (x \in K) \\ 0 & (x \notin K). \end{cases}$$

Definition 8.20. *If $g \in K_+$, the Fourier transform of $\mathrm{e}^{-ax_1} g(x)$ $(a > 0)$ is denoted by $\hat{G}(\alpha)$.*

Theorem 8.15. *If $g \in K_+$, $\hat{G}(\alpha)$ is a regular function of $\alpha_1 - \mathrm{i}a$ in $a > 0$ and* $\mathrm{Lim}_{a \to 0} \hat{G}(\alpha) = G(\alpha)$.

Proof. $g = \partial_1^{p_1} \ldots \partial_n^{p_n} f$ where $f \in K_+$, is continuous and $O(r^k)$ for some finite k as $r \to \infty$. Hence

$$|\hat{F}(\alpha)| \leq M \int_K (1 + r^2)^{k/2} \mathrm{e}^{-ax_1} \mathrm{d}x$$

$$\leq M \int_0^\infty \mathrm{e}^{-ax_1} \int_X (1 + r^2)^{k/2} \mathrm{d}x_2 \ldots \mathrm{d}x_n \mathrm{d}x_1$$

where X is the intersection of K and the hyperplane perpendicular to the x_1-axis through x_1. If θ $(< \frac{1}{2}\pi)$ is the largest angle made by the generators with the x_1-axis the maximum distance of a point of X from the x_1-axis is $x_1 \tan \theta$ and the maximum value of r is $x_1 \sec \theta$. Hence the integral over X is $O(x_1^{k+n-1})$ as $x_1 \to \infty$. Thus $\hat{F}(\alpha)$ is defined as a conventional function for $a > 0$.

The remainder of the proof is along the lines of Theorem 6.19 and will be omitted here. □

Theorem 8.16. *If g_1 and g_2 belong to K_+ and $g_1 * g_2 = 0$ then either $g_1 = 0$ or $g_2 = 0$.*

Proof. This is similar to Theorem 6.21. □

Example 15. Let K be the cone $x_1^2 \geq x_2^2 + \ldots + x_n^2 (x_1 \geq 0)$. Write

$$u_+^\beta(\boldsymbol{x}) = q_+^\beta \mathrm{H}(x_1)$$

the slight difference of notation from §8.8 being in order to exhibit the explicit dependence on the variables.

Consider $\int_{-\infty}^{\infty} u_+^\beta(\boldsymbol{t}) u_+^\gamma(\boldsymbol{x}-\boldsymbol{t})\,\mathrm{d}\boldsymbol{t}$. According to the analogue of Theorem 6.20 its Fourier transform is $\mathrm{Lim}_{a\to 0} \hat{G}_1 \hat{G}_2$ where g_1 is u_+^β. In the proof of Theorem 8.13 it is shown that, when $\mathrm{Re}\,\beta > -1$,

$$\hat{G}_1 = \frac{\beta!(\beta+\frac{1}{2}n-1)!2^{2\beta+n-1}\pi^{n/2-1}}{\{(a+\mathrm{i}\alpha_1)^2+\alpha_v^2\}^{\beta+n/2}}$$

where $\alpha_v^2 = \alpha_2^2 + \ldots + \alpha_n^2$. Thus, when $\mathrm{Re}\,\beta > -1$ and $\mathrm{Re}\,\gamma > -1$,

$$\hat{G}_1 \hat{G}_2 = \frac{\beta!(\beta+\frac{1}{2}n-1)!\gamma!(\gamma+\frac{1}{2}n-1)!2^{2\beta+2\gamma+2n-2}\pi^{n-2}}{\{(a+i\alpha_1)^2+\alpha v^2\}^{\beta+\gamma+n}}$$

from which it is evident that

$$\int_{-\infty}^{\infty} u_+^\beta(\boldsymbol{t}) u_+^\gamma(\boldsymbol{x}-\boldsymbol{t})\,\mathrm{d}\boldsymbol{t}$$

$$= \frac{\beta!(\beta+\frac{1}{2}n-1)!\gamma!(\gamma+\frac{1}{2}n-1)!}{(\beta+\gamma+\frac{1}{2}n)!(\beta+\gamma+n-1)!2}\pi^{(n/2)-1}u_+^{\beta+\gamma+n/2}(\boldsymbol{x}) \quad (118)$$

when $\mathrm{Re}\,\beta > -1$, $\mathrm{Re}\,\gamma > -1$. By applying the operator L a sufficient number of times we deduce that (118) is valid for any complex β and γ such that none of β, γ, $\beta+\gamma+\frac{1}{2}n$ is of the form $-m$ or $-\frac{1}{2}n-k$.

If neither β nor γ is of the form $-m$ or $-\frac{1}{2}n-k$ but $\beta+\gamma+\frac{1}{2}\mathrm{n}=0$ (118) still holds. Applying the operator L then leads to the formula

$$u_+^\beta * u_+^\gamma = \frac{\beta!(\beta+\frac{1}{2}n-1)!\gamma!(\gamma+\frac{1}{2}n-1)!}{(\frac{1}{2}n-m-1)!2}\pi^{(n/2)-1}\delta^{(m-1)}(q)\mathrm{H}(x_1) \quad (119)$$

for $\beta+\gamma+\frac{1}{2}n = -m$ when n is odd or when n is even and $m < \frac{1}{2}n-1$ (Exercise 90). On the other hand, if $\beta+\gamma+\frac{1}{2}n = -\frac{1}{2}n-k$,

$$u_+^\beta * u_+^\gamma = \beta!(\beta+\tfrac{1}{2}n-1)!\gamma!(\gamma+\tfrac{1}{2}n-1)!\pi^{n-2}2^{-2-2k}L^k\delta(\boldsymbol{x}). \quad (120)$$

If we begin with (118) for $\beta = 0$ it may be shown similarly that

$$\int_{-\infty}^{\infty} \delta^{(m-1)}(t_1^2 - \ldots - t_n^2)\mathrm{H}(t_1)u_+^\gamma(\boldsymbol{x}-\boldsymbol{t})\mathrm{d}\boldsymbol{t}$$

$$= \frac{(\frac{1}{2}n - m - 1)!\gamma!(\gamma + \frac{1}{2}n - 1)!\pi^{(n/2)-1}}{(\gamma + \frac{1}{2}n - m)!(\gamma + n - m - 1)!2} u_+^{\gamma + (n/2) - m} \tag{121}$$

when n is odd or when n is even and $m \leq \frac{1}{2}n - 1$ provided that neither γ nor $\gamma + \frac{1}{2}n - m$ belongs to the two exceptional sets described above. It will also be discovered, as must be true from other considerations, that

$$\int_{-\infty}^{\infty} \delta(\boldsymbol{t})\, u_+^\gamma(\boldsymbol{x}-\boldsymbol{t})\mathrm{d}\boldsymbol{t} = u_+^\gamma(\boldsymbol{x}). \tag{122}$$

Define the generalised function u_β by

$$u_\beta = u_+^{\beta - n/2}/(\beta - \tfrac{1}{2}n)!(\beta - 1)!\pi^{(n/2)-1}2^{2\beta-1}$$
$$(\beta \neq 0, -1, -2, \ldots; \tfrac{1}{2}n - 1, \tfrac{1}{2}n - 2, \ldots)$$
$$= \delta^{((n/2)-\beta-1)}(q)\mathrm{H}(x_1)/(\beta - 1)!\pi^{(n/2)-1}2^{2\beta-1}$$
$$(\beta = \tfrac{1}{2}n - 1, \tfrac{1}{2}n - 2, \ldots \text{ and, if } n \text{ is even, } \beta \geq 1)$$
$$= L^{-\beta}\delta(\boldsymbol{x}) \qquad (\beta = 0, -1, -2, \ldots).$$

Then, (118)–(122) are all included in

$$u_\beta * u_\gamma = u_{\beta+\gamma} \tag{123}$$

and there is no difficulty in verifying this relation for all complex β and γ.

When $g \in \boldsymbol{K}_+$ define $I^\beta g$ by

$$I^\beta g = u_\beta * g.$$

Then

$$I^\beta(I^\gamma g) = (I^\beta I^\gamma)g = I^{\beta+\gamma}g, \quad I^0 g = g. \tag{124}$$

Exercises

107. If g_1 and g_2 belong to $\boldsymbol{K}_+$ prove that the Fourier transform of $g_1 * g_2$ is $\operatorname{Lim}_{a \to 0} \hat{G}_1 \hat{G}_2$.

108. If K is defined as in Example 15, if $g_1 \in \boldsymbol{K}_+$ and

$$\int_K g_1(\boldsymbol{t})u_+^{\beta - n/2}(\boldsymbol{x}-\boldsymbol{t})\mathrm{d}\boldsymbol{t} = g_2(\boldsymbol{x})$$

with $\beta \neq 0, -1, -2, \ldots; \frac{1}{2}n-1, \frac{1}{2}n-2, \ldots$, prove that

$$g_1(t) = \frac{-4\beta \sin \pi\beta \int_K g_2(x) u_+^{-\beta-n/2}(t-x)\mathrm{d}x}{(\beta-\frac{1}{2}n)!(-\beta-\frac{1}{2}n)!\pi^{n-1}}.$$

109. If g_1 and g_2 both belong to K_+ show that

$$\int_{-\infty}^{\infty} \gamma(x) g_1 * g_2 \,\mathrm{d}x = \int_{-\infty}^{\infty} g_1(x) \int_{-\infty}^{\infty} g_2(t)\gamma(x+t)\mathrm{d}t\mathrm{d}x.$$

110. If $g_\mu \in K_+, g \in K_+$ and $g_2 \in K_+$ and if $\operatorname{Lim}_{\mu\to\mu_0} g_\mu = g$ show that

$$\operatorname*{Lim}_{\mu\to\mu_0} g_\mu * g_2 = g * g_2.$$

Deduce that, if $\partial g_\mu/\partial\mu$ exists, $\partial(g_\mu * g_2)/\partial\mu = (\partial g_\mu/\partial\mu) * g_2$.

111. If g_1 vanishes outside a finite sphere show that it must be the Fourier transform of a fairly good function. Deduce that $g_1 * g$ exists for any generalised function g. (Hint: consider the Fourier transform of $g_1\phi$, ϕ being fine and equal to unity on the sphere, and use Theorem 7.17.)
If g_2 vanishes outside a finite sphere prove that

$$g_2 * (g_1 * g) = (g_2 * g_1) * g.$$

112. Is it possible to define K as $x_1^2 + \ldots + x_p^2 \geq x_{p+1}^2 + \ldots + x_n^2$, $x_1 \geq 0$ and repeat the analysis of Example 15?

9

Asymptotic behaviour of Fourier integrals

9.1. The Riemann–Lebesgue lemma

There are many problems in which solutions are obtained as Fourier integrals which cannot be evaluated exactly. Nevertheless these solutions are important because it is often possible to derive asymptotic developments for the integrals, and so determine the significant behaviour of a solution for some range of parameters. An *asymptotic development* consists of two parts, one of which is dominant and one of which becomes less and less significant relatively as the parameters approach some particular set of values. The term *asymptotic series* will usually be reserved for an expansion of the function $f(x)$ in the form $\sum_{m=0}^{\infty} a_m/x^m$ as $x \to \infty$ where $\lim_{x\to\infty} x^M\{f(x) - \sum_{m=0}^{M} a_m/x^m\} = 0$ although the series itself may be divergent; a similar series in powers of $x-a$ may arise when x tends to the finite number a.

We shall not be concerned with the general theory of asymptotic developments. Attention will be restricted to certain types of Fourier integral which occur frequently in practice. One of the most important results is the Riemann–Lebesgue lemma which will now be proved in R_n.

Theorem 9.1. (Riemann–Lebesgue lemma) *If* $f \in L_1$ *then* $F(\boldsymbol{\alpha}) \to 0$ *as* $|\boldsymbol{\alpha}| \to \infty$ *along a radius.*

Proof. As $|\boldsymbol{\alpha}| \to \infty$ along a radius one at least of $|\alpha_1|, |\alpha_2|, \ldots, |\alpha_n|$ must tend to infinity; let it be $|\alpha_j|$. Then

$$F(\boldsymbol{\alpha}) = \int_{-\infty}^{\infty} f(\mathbf{x}) e^{-i\boldsymbol{\alpha}.\mathbf{x}} \, d\mathbf{x}$$

$$= -\int_{-\infty}^{\infty} f\left(x_1, \ldots, x_j + \frac{\pi}{\alpha_j}, \ldots, x_n\right) e^{-i\alpha \cdot x} \, d\mathbf{x}$$

by changing the variable x_j to $x_j + \pi/\alpha_j$. Hence

$$F(\boldsymbol{\alpha}) = \frac{1}{2}\int_{-\infty}^{\infty}\left\{f(\mathbf{x}) - f\left(\ldots, x_j + \frac{\pi}{\alpha_j}, \ldots\right)\right\}e^{-i\boldsymbol{\alpha}.\mathbf{x}}d\mathbf{x}$$

and so

$$|F(\boldsymbol{\alpha})| \le \frac{1}{2}\int_{-\infty}^{\infty}\left|f(\mathbf{x}) - f\left(\ldots, x_j + \frac{\pi}{\alpha_j}, \ldots\right)\right|d\mathbf{x}$$

which tends to zero as $|\alpha_j| \to \infty$ by a fundamental theorem of integration. The proof is complete. □

It is clear from this that, if f is absolutely integrable over any finite domain and $f = O(r^{-n-\varepsilon})(\varepsilon > 0)$ as $r \to \infty$, $F \to 0$ as $|\boldsymbol{\alpha}| \to \infty$.

For subsequent analysis it is convenient to introduce the following notation.

Definition 9.1. $g(\mathbf{x}) = O\{g_1(\mathbf{x})\}$ *if, and only if, there is a positive generalised function* g_1 *and positive* a *such that* $-ag_1(\mathbf{x}) < g(\mathbf{x}) < ag_1(\mathbf{x})$. *The statement* $g(\mathbf{x}) \to 0$ *as* $\mathbf{x} \to \mathbf{c}$ *(which may be infinite) means that, in some neighbourhood which includes* $\mathbf{c}$, g *equals the conventional function* f_1 *and* $f_1 \to 0$ *as* $\mathbf{x} \to \mathbf{c}$. *Similarly* $g(\mathbf{x}) = o\{f(\mathbf{x})\}$ *means* $f_1/f \to 0$ *as* $\mathbf{x} \to \mathbf{c}$.

The definition of inequality for generalised functions on R_1 is given at the end of §3.6; the definition for other intervals is obtained by considering only those γ which vanish outside these intervals. We may proceed similarly in R_n and so the order relation in Definition 9.1 agrees with that for conventional functions when g and g_1 are conventional functions.

Example 1. $\delta(x) + x^{-1} \to 0$ as $|x| \to \infty$. Also $\delta(\mathbf{x}) + r^{-1} \to 0$ as $r \to \infty$. Further $\delta(x) + \sin x \to 0$ as $x \to \pi$. Other illustrations are

$$\delta'(x) + \cos x = O(1) \qquad (x > \tfrac{1}{2}\pi),$$

$$\delta^{(m)}(x - \pi) + \sin x = o(x^2)$$

as $x \to \infty$.

Example 2. If $f \in L_1$, the Fourier transform of $f + |x|^{\beta}e^{iax}(\mathrm{Re}\,\beta > -1, a$ real) tends to zero as $|\alpha| \to \infty$. For $F \to 0$ by Theorem 9.1, and the Fourier transform of $|x|^\beta e^{iax}$ tends to zero by Corollaries 4.4 and 4.3*b* when β is not an even integer, and by (3.13) and Example 1 when β is a non-negative even integer.

The corresponding result in n dimensions is: if $f \in \boldsymbol{L}_1$ the Fourier transform of $f + r^\beta e^{i\mathbf{a}.\mathbf{x}}$ tends to zero as $|\boldsymbol{\alpha}| \to \infty$ when $\operatorname{Re}\beta > -n$. This is easily confirmed from Theorems 9.1, 7.31 and 7.32.

Example 2 demonstrates that it is not necessary for a function to be integrable at infinity in order that its Fourier transform tends to zero at infinity. On the other hand the Fourier transform of r^β ($\operatorname{Re}\beta \le -n$) certainly does not tend to zero as $|\boldsymbol{\alpha}| \to \infty$. This suggests that the behaviour of a Fourier transform at infinity is dictated by the behaviour of a generalised function in the finite part of space. Applying a Fourier inverse to this idea we come to the suggestion that the behaviour of a Fourier transform in the finite part of space is determined by the behaviour of the generalised function at infinity. Neither of these suggestions is strictly true – there are exceptions (see Exercise 1)–but there is a theorem of sufficiently wide validity to justify thinking in these terms in many cases.

Theorem 9.2. *If $f \in L_1$ the Fourier transform of*

$$f + \sum_{m=1}^{M} \{c_{1m}|x|^{\beta_{1m}} e^{ia_{1m}x} + c_{2m}|x|^{\beta_{2m}} \operatorname{sgn} x \, e^{ia_{2m}x}$$
$$+ c_{3m}|x|^{\beta_{3m}} \ln|x| \, e^{ia_{3m}x} + c_{4m}|x|^{\beta_{4m}} \ln|x| \operatorname{sgn} x \, e^{ia_{4m}x}\}$$

tends to zero as $|\alpha| \to \infty$. Here c_{jm} is complex, a_{jm} is real and $\operatorname{Re}\beta_{jm} > -1$.

Proof. The theorem is an immediate consequence of the Riemann–Lebesgue lemma and the Fourier transforms derived in Chapter 4. □

Corollary 9.2. *Theorem* 9.2 *is still true if any $|x|^{\beta_{jm}}$ is replaced by* $|x|^{-1} \mathrm{H}(x - X_m)$ *or* $|x|^{-1} \mathrm{H}(X_m - x)(X_m > 0)$.

Proof. Let $\eta(x)$ be an infinitely differentiable function which is unity for $x \ge X_m$ and vanishes for $x \le \frac{1}{2}X_m$. Then $|x|^{-1}\eta(x) - |x|^{-1} \times \mathrm{H}(x - X_m)$, being zero for $x > X_m$ and $x \le \frac{1}{2}X_m$ and finite elsewhere, is absolutely integrable and so its Fourier transform tends to zero.

Therefore we can limit our consideration to terms of the type $e^{iax}|x|^{-1}\eta(x)$ (terms which include the factors $\ln|x|$ and $\operatorname{sgn} x$ are dealt with similarly). Now

$$\{e^{iax}|x|^{-1}\eta(x)\}' - ia\, e^{iax}|x|^{-1}\eta(x) = \{-|x|^{-2}\eta(x) + |x|^{-1}\eta'(x)\}\, e^{iax}.$$

Clearly the right-hand side is in L_1 and so its Fourier transform tends to zero as $|\alpha| \to \infty$. Hence the Fourier transform of the left-hand side, or $i(\alpha - a)$ times the Fourier transform of $e^{iax}|x|^{-1}\eta(x)$, tends to zero which implies that the Fourier transform of $e^{iax}|x|^{-1}\eta(x)$ tends to zero as $|\alpha| \to \infty$.

Proceed similarly for $|x|^{-1}H(X_m - x)$ and the corollary is proved. □

The corresponding results in n dimensions are

Theorem 9.3. *If $f \in L_1$ the Fourier transform of*

$$f + \sum_{m=1}^{M} \{c_{1m} r^{\beta_{1m}} e^{i\boldsymbol{a}_{1m}\cdot\boldsymbol{x}} + c_{2m} r^{\beta_{2m}} \ln r\, e^{i\boldsymbol{a}_{2m}\cdot\boldsymbol{x}}$$
$$+ c_{3m} r^{-n} H(r - r_{3m}) e^{i\boldsymbol{a}_{3m}\cdot\boldsymbol{x}} + c_{4m} r^{-n} \ln r\, e^{i\boldsymbol{a}_{4m}\cdot\boldsymbol{x}} H(r - r_{4m})\}$$

tends to zero as $|\boldsymbol{\alpha}| \to \infty$ along a radius. Here c_{jm} is complex, $\boldsymbol{a}_{jm}$ is real, $r_{jm} > 0$ and $\operatorname{Re}\beta_{jm} > -n$.

Definition 9.2. *A generalised function which can be written in one of the forms displayed in Theorem 9.2, Corollary 9.2 or Theorem 9.3 is called well behaved at infinity.*

Example 3. $x^{-\nu}J_\nu(x)$ is continuous everywhere and therefore absolutely integrable over every finite interval. At infinity the asymptotic development of J_ν gives

$$x^{-\nu}J_\nu(x)$$
$$= (1/2\pi)^{1/2} x^{-\nu-1/2}\{e^{i\{x-(\nu/2)\pi-\pi/4\}} + e^{-i\{x-(\nu/2)\pi-\pi/4\}}\}$$
$$+ O(x^{-\nu-3/2})$$

as $x \to \infty$. Thus $x^{-\nu}J_\nu$ is absolutely integrable for $\operatorname{Re}\nu > \frac{1}{2}$. On the other hand the order term is absolutely integrable for $\frac{1}{2} \geq \operatorname{Re}\nu > -\frac{1}{2}$ and the first term is well behaved so that $x^{-\nu}J_\nu(x)$ is well behaved for $\operatorname{Re}\nu > -\frac{1}{2}$ and so its Fourier transform tends to zero.

In fact the Fourier transform of $x^{-\nu}J_\nu(x)$ is identically zero in $|\alpha| > 1$ when $\mathrm{Re}\,\nu > \frac{1}{2}$ (Exercise 30 of Chapter 5). By taking more terms in the asymptotic development of J_ν we can extend the range of ν for which the Fourier transform tends to zero.

Example 4. In contrast to Example 3, $|x|^\nu J_\nu(|x|)$ is absolutely integrable in every finite interval only when $\mathrm{Re}\,\nu > -\frac{1}{2}$. As in Example 3 it may be shown that $|x|^\nu J_\nu(|x|)$ is well behaved at infinity and so its Fourier transform tends to zero as $|\alpha| \to \infty$ when $\mathrm{Re}\,\nu > -\frac{1}{2}$.

Exercises

1. Prove that the Fourier transform of e^{ix^2} is $e^{-i\pi^2\alpha^2/2}(1+i)\sqrt{(\frac{1}{2}\pi)}$.
2. Prove that $\mathrm{Lim}_{\varepsilon\to+0}\, e^{-\varepsilon|x|}|x|^\nu J_\nu(|x|) = |x|^\nu J_\nu(|x|)$ when $\mathrm{Re}\,\nu > -1$ and deduce that the Fourier transform of $|x|^\nu J_\nu(|x|)$ is

$$\frac{(\nu-\frac{1}{2})!\,2^{\nu+1}}{\pi^{1/2}(1-\alpha^2)^{\nu+1/2}} \qquad (|\alpha| < 1),$$

$$-\frac{(\nu-\frac{1}{2})!\,2^{\nu+1}\sin\nu\pi}{\pi^{1/2}(\alpha^2-1)^{\nu+1/2}} \qquad (|\alpha| > 1).$$

Check that the Fourier transform does not tend to zero for $-1 < \mathrm{Re}\,\nu \le -\frac{1}{2}$.
3. Prove that the Fourier transform of $(a^2+r^2)^\beta$ tends to zero as $|\boldsymbol{\alpha}| \to \infty$.

9.2. Generalised functions with a finite number of singularities in one dimension

Having established a class of generalised functions whose Fourier transforms behave at infinity in a manner determined by what the generalised function does in the finite part of space we must now investigate what the behaviour at infinity is. In this section attention will be restricted to generalised functions in one dimension.

Definition 9.3. *A generalised function is said to have a finite number of singularities if in each of the $M+1$ intervals (M finite) $-\infty < x < x_1$, $x_1 < x < x_2, \dots, x_{M-1} < x < x_M$, $x_M < x < \infty$ the generalised function is equal to an infinitely differentiable function.*

Thus, a singularity is a point where a generalised function is not infinitely differentiable in the conventional sense.

Example 5. $\delta(x)$ has a finite number of singularities, being equal to the infinitely differentiable function 0 in $x > 0$ and $x < 0$. Thus it has only the single singularity $x = 0$.

$\delta^{(m)}(x) + |x^2 - 3x + 2|^{-1}$ has a finite number of singularities, located at $x = 0$, 1, and 2.

Let

$$g_m(x) = \sum_{p=1}^{P_m} \{c_{1p}|x - x_m|^{\beta_{1p}} + c_{2p}|x - x_m|^{\beta_{2p}} \operatorname{sgn}(x - x_m) + c_{3p}|x - x_m|^{\beta_{3p}} \ln|x - x_m| + c_{4p}|x - x_m|^{\beta_{4p}} \ln|x - x_m| \operatorname{sgn}(x - x_m) + c_{5p}\delta^{(p)}(x - x_m)\}$$

where the c_{jp} are complex and need not be the same for every m. Then

Theorem 9.4. *If the generalised function $g(x)$ has a finite number of singularities at $x = x_1, \ldots, x_M$ and if $g^{(N)}(x) - g_m^{(N)}(x)$ is absolutely integrable in some interval including $x_m (m = 1, \ldots, M)$, and if $g^{(N)}(x)$ is well behaved at infinity then*

$$G(\alpha) = \sum_{m=1}^{M} G_m(\alpha) + o(|\alpha|^{-N})$$

as $|\alpha| \to \infty$.

Proof. Since g has only a finite number of singularities, $g^{(N)}$ is absolutely integrable in any finite interval which excludes the singularities. Also $g_m^{(N)}$ is absolutely integrable in any finite interval which excludes x_m whereas $g^{(N)} - g_m^{(N)}$ is absolutely integrable in some interval including x_m. Therefore $g^{(N)} - g_m^{(N)}$ is absolutely integrable in any finite interval which excludes $x_1, \ldots, x_{m-1}, x_{m+1}, \ldots, x_M$. Now $g_\mu^{(N)} (\mu \neq m)$ is absolutely integrable in an interval including x_m. Hence $g^{(N)} - \sum_{m=1}^{M} g_m^{(N)}$ is absolutely integrable in any finite interval. Since $g^{(N)}$ and $g_m^{(N)}$ are well behaved at infinity it follows from Theorem 9.2 and Corollary 9.2 that the Fourier transform of $g^{(N)} - \sum_{m=1}^{M} g_m^{(N)}$ tends to zero as $|\alpha| \to \infty$, i.e.

$$(\mathrm{i}\alpha)^N \left\{ G(\alpha) - \sum_{m=1}^{M} G_m(\alpha) \right\} = o(1)$$

and the theorem is proved. □

(*Note*: The proof of this theorem does not require g to be infinitely differentiable between singularities but only that $g^{(N)}$ is absolutely integrable in any finite interval excluding singularities (and, of course, well behaved at infinity). To put it another way, if g has singularities $|x - x_1|^\beta$ and $|x - x_2|^{\beta+N}$ ($\beta > -1$) the second can be ignored in Theorem 9.4.)

Example 6. The only singularity of $|x|^\nu J_\nu(|x|)$ is $x = 0$ near which

$$g(x) = |x|^\nu \sum_{p=0}^{\infty} \frac{(-1)^p(\frac{1}{2}|x|)^{\nu+2p}}{p!(\nu+p)!}.$$

Therefore

$$g(x) - \frac{|x|^{2\nu}}{\nu!2^\nu} = O(|x|^{2\nu+2}) \tag{1}$$

near $x = 0$ and the Nth derivative is absolutely integrable in an interval including $x = 0$ if N is the largest integer less than $2\nu + 3$. Now g and its derivatives are well behaved at infinity by Example 4; hence Theorem 9.4 and Corollary 4.4 show that

$$G(\alpha) = \frac{(2\nu)! \cos \frac{1}{2}\pi(2\nu+1)}{\nu!2^{\nu-1}|\alpha|^{2\nu+1}} + o(|\alpha|^{-N}) \tag{2}$$

as $|\alpha| \to \infty$ where N is the integer satisfying $2\nu + 2 \leq N < 2\nu + 3$.

The reader should check that the same result is obtained from Exercise 2 (remember $(2\nu)! = \nu!(\nu - \frac{1}{2})!2^{2\nu}/\pi^{1/2}$).

If a smaller error term in G is required more terms of the series for g near $x = 0$ must be retained.

From the form of the order term in (1) we expect its Fourier transform to be $O(|\alpha|^{-2\nu-3})$ and therefore $o(|\alpha|^{-N})$ in (2) could be replaced by $O(|\alpha|^{-2\nu-3})$.

Example 7. $|x + 1|^{1/2}|x - 1|^{-1/2}$ is well behaved at infinity and has singularities $x_1 = -1$, $x_2 = 1$. Near $x = -1$,

$$g(x) = 2^{-1/2}|x+1|^{1/2} + O(|x+1|^{3/2})$$

and, near $x = 1$,

$$g(x) = 2^{1/2}|x-1|^{-1/2} + 2^{-3/2}|x-1|^{1/2}\,\mathrm{sgn}\,(x-1) + O(|x-1|^{3/2}).$$

Therefore, if

$$g_1(x) = 2^{-1/2}|x+1|^{1/2}$$

and

$$g_2(x) = 2^{1/2}|x-1|^{-1/2} + 2^{-3/2}|x-1|^{1/2}\,\mathrm{sgn}\,(x-1),$$

the conditions of Theorem 9.4 are satisfied with $N = 2$. Hence

$$G(\alpha) = \tfrac{1}{2}!2^{1/2}\cos\tfrac{3}{4}\pi\frac{\mathrm{e}^{\mathrm{i}\alpha}}{|\alpha|^{3/2}}$$

$$+\mathrm{e}^{-\mathrm{i}\alpha}\left\{\frac{(-\frac{1}{2})!2^{3/2}\cos\frac{1}{4}\pi}{|\alpha|^{1/2}} - \frac{\frac{1}{2}!2^{1/2}\mathrm{i}\sin\frac{3}{4}\pi\,\mathrm{sgn}\,\alpha}{|\alpha|^{3/2}}\right\} + o(|\alpha|^{-2})$$

$$= \frac{2\pi^{1/2}}{|\alpha|^{1/2}}\mathrm{e}^{-\mathrm{i}\alpha} - \frac{\pi^{1/2}}{2|\alpha|^{3/2}}(\mathrm{e}^{\mathrm{i}\alpha} + \mathrm{i}\mathrm{e}^{-\mathrm{i}\alpha}\,\mathrm{sgn}\,\alpha) + o(|\alpha|^{-2}).$$

As in the previous example the error term $o(|\alpha|^{-2})$ can be replaced by $O(|\alpha|^{-5/2})$.

The reader will observe that the dominant term in the asymptotic development comes from the most singular point of g; that this is generally true is evident from Theorem 9.4.

Example 8. Consider $\int_a^b (x-\alpha)^{\mu-1}(b-x)^{\nu-1}f(x)\mathrm{e}^{-\mathrm{i}\alpha x}\,\mathrm{d}x$ where $0 < \mu \le 1$, $0 < \nu \le 1$ and f is a function which is N times continuously differentiable in $a \le x \le b$. Here we are concerned with the Fourier transform of

$$(x-a)^{\mu-1}(b-x)^{\nu-1}f(x)\,\mathrm{H}(x-a)\,\mathrm{H}(b-x)$$

with singularities at $x = a$ and b. Near $x = a$, put

$$g_1(x) = (x-a)^{\mu-1}\,\mathrm{H}(x-a)\sum_{p=0}^{N-1}\frac{(x-a)^p}{p!}\frac{\mathrm{d}^p}{\mathrm{d}a^p}\{(b-a)^{\nu-1}f(a)\}$$

and, near $x = b$,

$$g_2(x) = (b-x)^{\nu-1}\,\mathrm{H}(b-x)\sum_{p=0}^{N-1}\frac{(x-b)^p}{p!}\frac{\mathrm{d}^p}{\mathrm{d}b^p}\{(b-a)^{\mu-1}f(b)\}.$$

Hence, by Theorem 9.4 (see Note) and Theorem 4.4,

$$\int_a^b (x-a)^{\mu-1}(b-x)^{\nu-1}f(x)\mathrm{e}^{-\mathrm{i}\alpha x}\,\mathrm{d}x$$

$$= \mathrm{e}^{-\mathrm{i}\alpha a}\sum_{p=0}^{N-1}\frac{(\mu+p-1)!\,\mathrm{e}^{-(\pi/2)\mathrm{i}(\mu+p)\operatorname{sgn}\alpha}}{p!\,|\alpha|^{\mu+p}}\frac{\mathrm{d}^p}{\mathrm{d}a^p}\{(b-a)^{\nu-1}f(a)\}$$

$$+\,\mathrm{e}^{-\mathrm{i}b\alpha}\sum_{p=0}^{N-1}\frac{(\nu+p-1)!\,\mathrm{e}^{(\pi/2)\mathrm{i}(\nu+p)\operatorname{sgn}\alpha}}{p!\,|\alpha|^{\nu+p}}(-1)^p$$

$$\times\frac{\mathrm{d}^p}{\mathrm{d}b^p}\{(b-a)^{\mu-1}f(b)\}+o(|\alpha|^{-N})$$

as $|\alpha|\to\infty$.

Exercises

4. Prove that the Fourier transform of $|x|\,|x+a|^{1/2}|x-a|^{-1/2}(a>0)$ is

$$\frac{2\pi^{1/2}a^{3/2}}{|\alpha|^{1/2}}\mathrm{e}^{-\mathrm{i}a\alpha}-\frac{\pi^{1/2}a^{1/2}}{2|\alpha|^{3/2}}(\mathrm{e}^{\mathrm{i}a\alpha}+\tfrac{5}{2}\mathrm{i}\,\mathrm{e}^{-\mathrm{i}a\alpha}\operatorname{sgn}\alpha)-\frac{2}{\alpha^2}+o(|\alpha|^{-2})$$

as $|\alpha|\to\infty$.

5. Examine the differences between the behaviour, as $|\alpha|\to\infty$, of the Fourier transforms of e^{-x^2} and $\mathrm{e}^{-x^2}\operatorname{sgn} x$.

6. Derive an asymptotic development of $\int_{-\infty}^{\infty}\mathrm{e}^{-\mathrm{i}\alpha x}x^{-1}\ln|x-1|\,\mathrm{d}x$ with an error $o(|\alpha|^{-2})$.

7. If $f(x)$ is continuously differentiable in $x\geq 0$ and if f and its derivatives are well behaved at infinity show that

$$\int_{-\infty}^{\infty} f(|x|)\operatorname{sgn} x\,\mathrm{e}^{-\mathrm{i}\alpha x}\,\mathrm{d}x=\sum_{p=0}^{N-1}\frac{2f^{(2p)}(0)}{(\mathrm{i}\alpha)^{2p+1}}+o(|\alpha|^{-2N})$$

for any N. Deduce that

$$\int_0^{\infty} f(x)\sin\alpha x\,\mathrm{d}x=\sum_{p=0}^{N-1}\frac{(-1)^p f^{(2p)}(0)}{\alpha^{2p+1}}+o(|\alpha|^{-2N}).$$

Show also that

$$\int_0^{\infty} f(x)\cos\alpha x\,\mathrm{d}x=\sum_{p=1}^{N}\frac{(-1)^p f^{(2p-1)}(0)}{\alpha^{2p}}+o(|\alpha|^{-2N-1}).$$

8. If $0 < \mu \leq 1$, $0 < \nu \leq 1$ and f is a function which is N times continuously differentiable in $a \leq x \leq b$ show that

$$\int_a^b e^{-i\alpha x}(x-a)^{\mu-1}(b-x)^{\nu-1}f(x)\ln(x-a)\,dx$$

$$= \sum_{p=0}^{N-1} \frac{(\mu+p-1)!\,e^{-ia\alpha-\pi i(\mu+p)/2}}{p!\,\alpha^{\mu+p}}\{-\tfrac{1}{2}\pi i - \ln\alpha + \psi(\mu+p-1)\}$$

$$\times \frac{d^p}{da^p}\{(b-a)^{\nu-1}f(a)\} + \sum_{p=0}^{N-1}\frac{(\nu+p-1)!\,e^{-ib\alpha+\pi i(\nu-p)/2}}{p!\,\alpha^{\nu+p}}$$

$$\times \frac{d^p}{db^p}\{(b-a)^{\mu-1}f(b)\ln(b-a)\} + o(\alpha^{-N})$$

as $\alpha \to \infty$.

9. Derive an asymptotic development of

$$\int_1^2 e^{-i\alpha x}\frac{\ln(2-x)}{(x-1)^{4/3}}\,dx$$

with an error $o(|\alpha|^{-2})$.

10. Obtain an asymptotic development of

$$\int_0^2 e^{-i\alpha x}x\ln|x-1|\,dx$$

with an error $o(|\alpha|^{-3})$. What difference does it make if, in $0 \leq x < 1$, $\ln|x-1|$ is replaced by $i\pi + \ln|x-1|$?

11. If $g(x) \in K_+$ and is $O(x^\nu)$, with $\text{Re}\,\nu > -1$, for $x \geq 0$ prove that

$$\lim_{a\to+0}(a+i\alpha)^{-\mu}\hat{G}(\alpha) = (i\alpha)^{-\mu}G(\alpha)$$

when $|\alpha|$ is large. (Hint: if $\Gamma(\alpha)$ vanishes for $|\alpha| \leq 1$, $\Gamma(\alpha)(a+i\alpha)^{-\mu}$ is good: use Theorem 3.9.)

12. If $g(x) \in K_+$ and is $O(x^\nu)$, with $\text{Re}\,\nu > -1$, for $x \geq 0$ prove that $I^\mu g$ (as defined in § 6.6) is absolutely integrable over any finite interval if $\text{Re}(\mu+\nu) > -2$ unless μ is a non-positive integer. If $\text{Re}\,\nu \geq -1+\eta$ and $\text{Re}(\mu) = -1-\varepsilon$ ($\eta > \varepsilon > 0$, μ not a non-positive integer) and if g is well behaved at infinity, show that $I^\mu g$ is well behaved at infinity. Deduce that $G(\alpha) = o(|\alpha|^{-\nu-1+\varepsilon})$ as $|\alpha| \to \infty$.

13. If g satisfies the conditions of Theorem 9.4 and if $xg(x)$ is well behaved at infinity prove that

$$G'(\alpha) = \sum_{m=1}^{M} G'_m(\alpha) + o(|\alpha|^{-N}).$$

It will be observed that Theorem 9.4 is not suitable for dealing with periodic generalised functions since such generalised functions must have either no singularity or an infinite number of them. The most obvious extension to cope with these is to permit a periodic generalised function to have a finite number of singularities in each period.

Definition 9.4. *A periodic generalised function, with period $2l$, is said to have a finite number of singularities in the period $-l < x \leq l$ if, in each of the $M+1$ intervals (M finite) $-l < x < -l(1-\varepsilon) < x_1$, $x_1 < x < x_2, \ldots, x_{M-1} < x < x_M, x_M < x < l(1+\varepsilon)$ for some $\varepsilon > 0$, the generalised function is equal to an infinitely differentiable function.*

Theorem 9.5. *If the periodic generalised function $g(x)$ has a finite number of singularities at $x_1, \ldots, x_M$ in the period $-l < x \leq l$ and if $g^{(N)}(x) - g_m^{(N)}(x)$ is absolutely integrable in some interval including x_m $(m = 1, \ldots, M)$ then c_q, the qth Fourier coefficient of g, satisfies*

$$c_q = \frac{1}{2l}\sum_{m=1}^{M} G_m\left(\frac{q\pi}{l}\right) + o(|q|^{-N})$$

as $|q| \to \infty$.

Proof. Recall that, on account of the remark at the end of §5.5,

$$c_q = \frac{1}{2l}\int_{-\infty}^{\infty} g(x)\tau_1\left(\frac{x}{2l}\right)\mathrm{e}^{-iq\pi x/l}\,\mathrm{d}x$$

where $\tau_1(x)$ is any fine function which is zero for $|x| \geq 1$ and such that $\tau_1(x) + \tau_1(x-1) = 1$ when $0 \leq x \leq 1$. Choose

$$\tau_1(x) = \begin{cases} 0 & (x \leq -\frac{1}{2} + \frac{1}{4}\varepsilon) \\ 1 - \tau\{2\varepsilon^{-1}(2x+1) - 1\} & (-\frac{1}{2} + \frac{1}{4}\varepsilon < x < -\frac{1}{2} + \frac{1}{2}\varepsilon) \\ 1 & (-\frac{1}{2} + \frac{1}{2}\varepsilon \leq x \leq \frac{1}{2} + \frac{1}{4}\varepsilon) \\ \tau\{2\,\varepsilon^{-1}(2x-1) - 1\} & (\frac{1}{2} + \frac{1}{4}\varepsilon < x < \frac{1}{2} + \frac{1}{2}\varepsilon) \\ 0 & (x \geq \frac{1}{2} + \frac{1}{2}\varepsilon) \end{cases}$$

where τ is defined in §5.5. Then it is easily verified that τ_1 has the required properties. Now, since $\tau_1(x/2l)$ vanishes outside $-l(1-\frac{1}{2}\varepsilon) \leq x \leq l(1+\varepsilon)$, $g(x)\tau_1(x/2l)$ and its derivatives are well behaved at infinity and their only singularities are at $x_1, \ldots, x_M$

where $\tau_1 = 1$. Therefore, Theorem 9.4 gives

$$\frac{1}{2l}\int_{-\infty}^{\infty} g(x)\tau_1(x/2l)\,\mathrm{e}^{-\mathrm{i}\alpha x}\,\mathrm{d}x = \frac{1}{2l}\sum_{m=1}^{M} G_m(\alpha) + o(|\alpha|^{-N})$$

as $|\alpha| \to \infty$. Since $g(x)\tau_1(x/2l) = \sum_{q=-\infty}^{\infty} c_q \mathrm{e}^{\mathrm{i}q\pi x/l}\tau_1(x/2l)$ the left-hand side is also $\sum_{q=-\infty}^{\infty} c_q T_1(l\alpha/\pi - q)$. This is an absolutely and uniformly convergent series for any finite α because $c_q = O(|q|^{N_1})$ for some N_1 (Corollary 5.13) and $T_1(l\alpha/\pi - q) = O(|q|^{-N_1-2})$ as $|q| \to \infty$. Since T_1 is continuous in α the series is also continuous so that, by putting $\alpha = q\pi/l$, the proof is completed. □

Example 9. The periodic generalised function $g(x)$ is defined as equal to e^{ax} in $-\pi < x < \pi$. In $-\pi < x \le \pi$, g has one singularity at $x = \pi$ where

$$\begin{aligned} g(x) = {} & \{\cosh a\pi - \sinh a\pi\,\operatorname{sgn}(x-\pi)\} \\ & \times \{1 + a(x-\pi) + \tfrac{1}{2}a^2(x-\pi)^2\} + O(|x-\pi|^3). \end{aligned}$$

Hence, with $g_1(x) = \sinh a\pi\,\operatorname{sgn}(x-\pi)\{1 + a(x-\pi) + \frac{1}{2}a^2(x-\pi)^2\}$, Theorem 9.5 and Corollary 4.3*b* give

$$\begin{aligned} c_q &= -(1/2\pi)\sinh a\pi\,\mathrm{e}^{-\mathrm{i}q\pi}(-2\mathrm{i}q^{-1} - 2aq^{-2} + 2\mathrm{i}a^2q^{-3}) + o(|q|^{-3}) \\ &= \frac{\mathrm{i}(-1)^q}{q\pi}\sinh a\pi(1 - (\mathrm{i}a/q) - a^2/q^2) + O(q|q|^{-4}) \end{aligned}$$

as $|q| \to \infty$, the order of the error being estimated from the order term in g.

Example 10. The odd periodic generalised function $g(x)$ is defined as equal to e^{ax} in $0 < x < \pi$. In this case a Fourier sine series is obtained for g, the coefficients b_m being $2\mathrm{i}$ times the Fourier coefficients for the periodic function which equals $\mathrm{e}^{a|x|}\operatorname{sgn} x$ in $-\pi < x \le \pi$. Here there are two singularities, $x = 0$ where

$$\mathrm{e}^{a|x|}\operatorname{sgn} x = (1 + \tfrac{1}{2}a^2x^2)\operatorname{sgn} x + ax + O(|x|^3)$$

and $x = \pi$ where

$$\begin{aligned} & \mathrm{e}^{a|x|}\operatorname{sgn} x \\ & = -\mathrm{e}^{a\pi}[\{1 + \tfrac{1}{2}a^2(x-\pi)^2\}\operatorname{sgn}(x-\pi) + a(x-\pi)] + O(|x-\pi|^3). \end{aligned}$$

Therefore, take

$$g_1(x) = (1 + \tfrac{1}{2}a^2 x^2)\,\mathrm{sgn}\,x,$$
$$g_2(x) = -\,e^{a\pi}\{1 + \tfrac{1}{2}a^2(x-\pi)^2\}\,\mathrm{sgn}(x-\pi)$$

and so

$$b_q = 2ic_q = (2/\pi q)\{1 - (-1)^q e^{a\pi}\}(1 - a^2/q^2) + O(|q|^{-4}).$$

Exercises

14. A periodic generalised function is equal to $(x+\pi)^{-\mu}$ in $-\pi < x \le \pi$. Show that its Fourier coefficient is asymptotic to

$$\{(-\mu)!/2\pi\}(-1)^q |q|^{\mu-1} e^{(\pi/2)i(\mu-1)\,\mathrm{sgn}\,q}$$

as $|q| \to \infty$ and estimate the error.

15. Find an asymptotic expression with error $o(|q|^{-3})$ for c_q in the Fourier expansion of the periodic function $e^{|\cos x|}$.

16. Find an asymptotic expression, with error $o(|q|^{-3})$, for c_q in the Fourier expansion of the periodic function equal to $|x| \ln |x|$ in $(-1, 1)$.

17. Show that the coefficient b_q in the Fourier sine series for the odd periodic function equal to $x^{1/2}$ in $0 < x < l$ equals $(-1)^{q-1} 2l^{1/2}/\pi q + O(|q|^{-3/2})$.

9.3. The method of stationary phase

A more general type of integral than the Fourier integral is $\int g(x) e^{-i\alpha f(x)} dx$. When f and g are conventional functions an estimate of the value of the integral for large $|\alpha|$ is obtained by arguing that in a small change of x, g will not alter much whereas the sinusoidal terms will go through many oscillations. Where this is true the contribution to the integral can be expected to be zero. The argument fails near a point where $f'(x) = 0$ since there the sinusoidal terms do not fluctuate so rapidly; so the main contribution comes from near points where f is stationary. If x_0 is such a point the approximate contribution is

$$\int_{-\infty}^{\infty} g(x_0) \exp[-i\alpha\{f(x_0) + \tfrac{1}{2}(x-x_0)^2 f''(x_0)\}]\,dx$$
$$= \{2\pi/\alpha|f''(x_0)|\}^{1/2} g(x_0) \exp\{-i\alpha f(x_0) - \tfrac{1}{4}\pi i\,\mathrm{sgn} f''(x_0)\}$$

and the asymptotic behaviour of the integral is obtained by adding the contributions of all stationary points. For this reason the method is often called *Kelvin's method of stationary phase*.

To generalise the method and also to place the theory on a more rigorous footing we shall suppose that the interval of integration can be split up into a finite number of sub-intervals in each of which $f(x)$ is either strictly increasing or strictly decreasing. Points at which $f'(x)$ is zero, infinite or discontinuous are permitted only at the ends of such sub-intervals. There is no loss of generality in restricting attention to f strictly increasing since the other possibility can be covered by changing the sign of α (or i). Accordingly we are concerned with integrals of the type

$$\int_a^b g(x)\mathrm{e}^{-\mathrm{i}\alpha f(x)}\,\mathrm{d}x$$

where $b > a$ and $f(x) > f(x_1)$ for $b \geq x > x_1 \geq a$.

Now suppose that

$$f'(x) = (x-a)^\rho (b-x)^\sigma f_1(x)$$

where f_1 is infinitely differentiable, positive and non-zero for $a \leq x \leq b$. At $x = a$, f' vanishes if $\rho > 0$ but is infinite if $\rho < 0$ though $f(a) = -\infty$ only if $\rho \leq -1$. Similar conclusions apply at $x = b$. If neither ρ nor σ is -1, f satisfies the conditions for change of variable given in §8.2 (see e.g. Definitions 8.5, 8.6). Therefore the substitution $y = f(x)$ gives (cf. (21) of Chapter 8)

$$\int_{f(a)}^{f(b)} \frac{g\{f^{-1}(y)\}}{f'(x)}\mathrm{e}^{-\mathrm{i}\alpha y}\,\mathrm{d}y$$

where $1/f'(x)$ can be replaced by $\mathrm{d}f^{-1}/\mathrm{d}y$. Now we can use the results of the preceding section to determine the asymptotic behaviour provided that $g(f^{-1})\mathrm{d}f^{-1}/\mathrm{d}y$ is a generalised function of y which comes within the scope of that theory.

To examine this possibility assume firstly that $g(x)$ has a singularity like $|x-x_m|^\beta$ where $a < x_m < b$. Near $x = x_m$, $y - y_m = (x-x_m)f'(x_m)$ where $y_m = f(x_m)$ and so, near $y = y_m$,

$$\frac{g\{f^{-1}(y)\}}{f'(x)} \approx \frac{|y-y_m|^\beta}{f'(x_m)|f'(x_m)|^\beta}.$$

Thus the transformed integrand has a singularity of the same type, though additional powers may be present. It may be shown similarly that the same is true for the other permitted singularities and, since integrability of derivatives is clearly preserved, it follows that interior singularities of g present no new problems.

The position with regard to the end-points must be considered separately, partly because of the special behaviour of f' there and partly because $g(f^{-1})$ must contain the factor $\mathrm{H}\{y-f(a)\}\mathrm{H}\{f(b)-y\}$. Suppose that $g(x)$ behaves as $(x-a)^{\beta}\mathrm{H}(x-a)$ near $x=a$ and that $\rho > -1$. Near $y=f(a)$,

$$y-f(a)=(x-a)^{\rho+1}(b-a)^{\sigma}f_1(a)/(\rho+1)=c(x-a)^{\rho+1},$$

say, because $\rho \neq -1$. Then

$$\frac{g\{f^{-1}(y)\}}{f'(x)} \approx c^{-(1+\beta)/(\rho+1)}\{y-f(a)\}^{(\beta-\rho)/(\rho+1)}\frac{\mathrm{H}\{y-f(a)\}}{(\rho+1)}$$

which is the first term of a series of the form

$$\sum_{m=0} a_m\{y-f(a)\}^{(\beta-\rho+m)/(\rho+1)}.$$

A singularity of this type comes within the scope of Theorem 9.4 provided that the other conditions in the theorem are satisfied. However, a better estimate of the error caused by truncating the series expansion about $y=f(a)$ is often obtained by employing the results of Exercise 12. Since $f(a)$ is finite it is clear that no problem at infinity arises when $\rho > -1$.

If $\rho < -1, f(a) = -\infty$ and now $g(f^{-1})/f'$ consists of a series of the form $\sum_{m=0} a_m(-y)^{(\beta-\rho+m)/(\rho+1)}$ which is well behaved at infinity and so presents no particular problem.

The general terms in the expansion are complicated to write down and it is doubtful whether, in a practical example, it is easier to substitute in the general formula or to work out the expansion directly. Accordingly the results obtained will be summarised in a way which indicates how the important parts of the asymptotic expansion arise.

Theorem 9.6. *Let* $f'(x)=(x-a)^{\rho}(b-x)^{\sigma}f_1(x)$ *where* $\rho \neq -1$, $\sigma \neq -1$ *and* f_1 *is positive and infinitely differentiable in* $a \leq x \leq b$. *Then* $\int_a^b g(x)e^{-i\alpha f(x)}\mathrm{d}x$ *is evaluated by the substitution* $y=f(x)$. *If* $g(x)$ *has a singularity* g_m *in* $a<x<b$ *and* $g^{(N)}-g_m^{(N)}$ *is absolutely integrable then an expansion as in Theorem 9.4 with error* $o(|\alpha|^{-N})$ *is obtained. The end-points give expansions in powers of*

$\alpha^{-1/(\rho+1)}$ and $\alpha^{-1/(\sigma+1)}$ with errors of $o(\alpha^{\varepsilon-\mu/(\rho+1)})$ and $o(\alpha^{\varepsilon-\nu/(\sigma+1)})$ if the last terms are $\alpha^{-(\mu-1)/(\rho+1)}$ and $\alpha^{-(\nu-1)/(\sigma+1)}$ respectively.

(*Note*: As for Theorem 9.4 we need only require that $g^{(N)}$ and $f^{(N+1)}$ be absolutely integrable away from singularities. For instance, if g behaves as $(x-a)^\beta \mathrm{H}(x-a)$ and $\rho > -1$, the asymptotic expansion from the end-point will involve $|\alpha|^{-(\beta+1)/(\rho+1)}$, $|\alpha|^{-(\beta+2)/(\rho+1)}, \dots,$ $|\alpha|^{-(\beta+N)/(\rho+1)}$ with an error which is $o(|\alpha|^{\varepsilon-(\beta+N+1)/(\rho+1)})$.)

Example 11. Consider $\int_0^1 x^\mu (1-x)^\nu \mathrm{e}^{-\mathrm{i}\alpha x^2} \mathrm{d}x$. Putting $y = x^2$ we have

$$\tfrac{1}{2}\int_0^1 y^{(\mu/2)-1/2}(1-y^{1/2})^\nu \mathrm{e}^{-\mathrm{i}\alpha y}\,\mathrm{d}y.$$

There are singularities only at the end-points where the integrand behaves as

$$\{y^{(\mu/2)-1/2}(1-\nu y^{1/2}) + O(y^{(\mu/2)+1/2})\}\,\mathrm{H}(y) \qquad (y \approx 0),$$

$$[\{\tfrac{1}{2}(1-y)\}^\nu\{1+\tfrac{1}{2}(\tfrac{1}{2}\nu-\mu+1)(1-y)\}+O((1-y)^{\nu+2})]\,\mathrm{H}(1-y) \qquad (y \approx 1).$$

Hence

$$\int_0^1 x^\mu(1-x)^\nu \mathrm{e}^{-\mathrm{i}\alpha x^2}\mathrm{d}x$$
$$= (\tfrac{1}{2}\mu-\tfrac{1}{2})!\,\tfrac{1}{2}|\alpha|^{-(\mu/2)-1/2}\mathrm{e}^{-(\pi/4)\mathrm{i}(\mu+1)\,\mathrm{sgn}\,\alpha}$$
$$-(\tfrac{1}{2}\mu)!\,\tfrac{1}{2}\nu|\alpha|^{-(\mu/2)-1}\mathrm{e}^{-(\pi/4)\mathrm{i}(\mu+2)\,\mathrm{sgn}\,\alpha}$$
$$+o(|\alpha|^{\varepsilon-(\mu/2)-3/2}) + \frac{\nu!}{2^{\nu+1}}|\alpha|^{-\nu-1}\mathrm{e}^{(\pi/2)\mathrm{i}(\nu+1)\,\mathrm{sgn}\,\alpha-\mathrm{i}\alpha}$$
$$+\frac{(\nu+1)!(\tfrac{1}{2}\nu-\mu+1)}{2^{\nu+2}}|\alpha|^{-\nu-2}\mathrm{e}^{(\pi/2)\mathrm{i}(\nu+2)\,\mathrm{sgn}\,\alpha-\mathrm{i}\alpha}+o(|\alpha|^{\varepsilon-\nu-3})$$

where ε is arbitrarily small and positive. Both error terms are included since it is not known which is the dominant one until the relative magnitudes of μ and ν are given.

Example 12. Consider $\int_0^1 x^\mu(1-x)^\nu(x-\tfrac{1}{2})^{-2}\mathrm{e}^{-\mathrm{i}\alpha x^2}\mathrm{d}x$. Put $y = x^2$ to obtain

$$\tfrac{1}{2}\int_0^1 y^{(\mu/2)-1/2}(1-y^{1/2})^\nu(y^{1/2}-\tfrac{1}{2})^{-2}\mathrm{e}^{-\mathrm{i}\alpha y}\,\mathrm{d}y.$$

There are singularities at the end-points, which can be handled as in the preceding example, and also at $y = \frac{1}{4}$ where the integrand is approximately $2^{1-\mu-\nu}(y - \frac{1}{4})^{-2}$. Hence

$$\int_0^1 x^\mu (1-x)^\nu (x - \tfrac{1}{2})^{-2} e^{-i\alpha x^2} dx$$
$$= (\tfrac{1}{2}\mu - \tfrac{1}{2})!\, 2|\alpha|^{-(\mu/2)-1/2} e^{-(\pi/4)i(\mu+1)\operatorname{sgn}\alpha} + o(|\alpha|^{\varepsilon-(\mu/2)-1})$$
$$+ \nu!\, 2^{1-\nu}|\alpha|^{-\nu-1} e^{-i\alpha+(\pi/2)i(\nu+1)\operatorname{sgn}\alpha} + o(|\alpha|^{\varepsilon-\nu-2})$$
$$+ 2^{-\mu-\nu}\pi\alpha \operatorname{sgn}\alpha + O(1).$$

Example 13. In $\int_0^1 x^\mu (1-x)^\nu e^{-i\alpha x^{-1}} dx$ put $y = x^{-1}$ to derive

$$\int_1^\infty y^{-\mu-\nu-2}(y-1)^\nu e^{-i\alpha y} dy.$$

The integrand is well behaved at infinity, so that it is only necessary to consider the behaviour at the singularity $y = 1$ where the integrand is $(y-1)^\nu H(y-1) + O(y-1)^{\nu+1}$. Hence

$$\int_0^1 x^\mu (1-x)^\nu e^{-i\alpha x^{-1}} dx = \nu!\,|\alpha|^{-\nu-1} e^{-i\alpha+(\pi/2)i(\nu+1)\operatorname{sgn}\alpha} + O(|\alpha|^{-\nu-2}).$$

Exercises

18. Prove that

$$\int_0^1 e^{-i\alpha x^3} dx = \tfrac{1}{3}!\, e^{-(\pi/6)i\operatorname{sgn}\alpha}|\alpha|^{-1/3} - \frac{(-\frac{1}{3})!}{(-\frac{4}{3})!} e^{-i\alpha+(\pi/2)i\operatorname{sgn}\alpha}|\alpha|^{-1} + o(|\alpha|^{\varepsilon-2}).$$

19. Prove that

$$\int_0^1 (1-x)^\mu e^{-i\alpha x^{-3}} dx = \frac{\mu!}{3^{\mu+1}}|\alpha|^{-\mu-1} e^{-i\alpha-(\pi/2)i(\mu+1)\operatorname{sgn}\alpha}$$
$$\times \{1 - \tfrac{2}{3}(\mu+1)(\mu+2)|\alpha|^{-1} e^{-(\pi/2)i\operatorname{sgn}\alpha}\} + O(|\alpha|^{-\mu-3}).$$

20. (i) If $g(x) = (x-a)^\beta g_1(x), f'(x) = (x-a)^\rho f_1(x) (\rho > -1)$ where g_1 and f_1 are N times continuously differentiable in $a \le x \le b, f_1$ being positive, show that the contribution from $x = a$ in $\int_a^b g(x) e^{-i\alpha f(x)} dx$ is

$$\sum_{m=0}^{N-1} \frac{\{(m+\beta-\rho)/(\rho+1)\}!\,a_m}{m!(\rho+1)}|\alpha|^{-(m+\beta+1)/(\rho+1)}$$
$$\times \exp\left\{-i\alpha f(a) - \tfrac{1}{2}\pi i \frac{m+\beta-\rho}{\rho+1}\operatorname{sgn}\alpha\right\} + o(|\alpha|^{\varepsilon-(\beta+N+1)/(\rho+1)})$$

where

$$a_m = \left[(\rho+1)\{y-f(a)\}^{\rho/(\rho+1)}\frac{\mathrm{d}}{\mathrm{d}y}\right]^m \frac{g(f^{-1})}{f'(x)}\{y-f(a)\}^{(\rho-\beta)/(\rho+1)}.$$

Verify that the same result is obtained from the integral with respect to x by expanding $g_1(x)\exp[-\mathrm{i}\alpha\{f(x)-f(a)-(x-a)^{\rho+1}f_1(a)\}]$ about $x=a$.

(ii) If in (i), $g(x)=(x-a)^\beta \ln(x-a)g_1(x)$ show that the same expression holds as $\alpha\to\infty$ provided that

$$a_m = h_1^{(m)}(0) + \left\{\psi\left(\frac{m+\beta-\rho}{\rho+1}\right) - \ln\alpha - \tfrac{1}{2}\pi\mathrm{i}\right\}h_0^{(m)}(0)/(\rho+1)$$

where $h_0(u)=g_1(x)(x-a)^\beta u^{-\beta}\,\mathrm{d}x/\mathrm{d}u$, $h_1(u)=h_0(u)\ln\{(x-a)/u\}$ and $u^{\rho+1}=f(x)-f(a)$.

21. Prove that

$$\frac{1}{2\pi}\int_0^{2\pi} \mathrm{e}^{-\mathrm{i}\alpha(\theta-\sin\theta)}\,\mathrm{d}\theta = \frac{|\alpha|^{-1/3}}{2^{2/3}3^{1/6}\pi}\left\{(-\tfrac{2}{3})! - \frac{\tfrac{2}{3}!2^{1/3}.3}{140}|\alpha|^{-4/3}\right\} + O(|\alpha|^{-7/3}).$$

22. Show that

$$\frac{1}{\pi}\int_0^{\pi}\cos\{\alpha(\theta-\sin\theta)\}\,\mathrm{d}\theta - \frac{\sin\alpha\pi}{\pi}\int_0^\infty \mathrm{e}^{-\alpha(t+\sinh t)}\,\mathrm{d}t$$

has the same asymptotic expression as given in Exercise 21.

23. Find an asymptotic development of $\int_{-\infty}^{\infty}\mathrm{e}^{-\mathrm{i}\alpha x^2}x^{-1}\ln|x-1|\,\mathrm{d}x$ with an error $o(|\alpha|^{-1})$.

24. Derive the first two terms of the asymptotic development of

$$\int_0^{\pi/2}\mathrm{e}^{-\mathrm{i}\alpha\,\mathrm{sgn}\,x}\ln x\,\mathrm{d}x.$$

9.4. Generalised functions with isolated singularities

The asymptotic behaviour of Fourier transforms in n dimensions is considerably more complicated than that in one dimension. Primarily, this is because singularities occur not only at isolated points but also on curves and, in general, on hypersurfaces which may be of any dimension up to $n-1$. Since these hypersurfaces may have cusps, conical points or other unpleasant deviations from smoothness it will be realised that any comprehensive treatment will be extremely complex. Such a discussion is beyond the scope of this book and attention will be limited to some cases of practical importance.

Firstly, let us deal with isolated singularities.

Definition 9.5. *A generalised function is said to have a finite number of isolated singularities* $\boldsymbol{x}_1, \dots, \boldsymbol{x}_M$ *if it is equal to an infinitely differentiable function except at these points.*

It is obvious that $\delta(\boldsymbol{x})$ and r^β have a finite number of isolated singularities; in both cases the origin is the only singularity. On the other hand neither $|x_1|$ nor $(x_1^2 - x_2^2)^{-1}$ has a finite number of isolated singularities when $n \geq 2$. Let

$$g_m(\boldsymbol{x}) = \sum_{p=1}^{P_m} \{c_{1p}|\boldsymbol{x} - \boldsymbol{x}_m|^{\beta_{1p}} + c_{2p}|\boldsymbol{x} - \boldsymbol{x}_m|^{\beta_{2p}} \ln|\boldsymbol{x} - \boldsymbol{x}_m|$$
$$+ c_{3p}\partial_1^{p_1} \dots \partial_n^{p_n} \delta(\boldsymbol{x} - \boldsymbol{x}_m)\}$$

where c_{1p}, β_{1p} are complex and need not be the same for every m. Then

Theorem 9.7. *If the generalised function* $g(\boldsymbol{x})$ *has a finite number of isolated singularities at* $\boldsymbol{x}_1, \dots, \boldsymbol{x}_M$ *and if*

$$(\partial_1^2 + \partial_2^2 + \dots + \partial_n^2)^N \{g(\boldsymbol{x}) - g_m(\boldsymbol{x})\}$$

is absolutely integrable in some sphere including $\boldsymbol{x}_m$ $(m = 1, \dots, M)$ *and if* $(\partial_1^2 + \dots + \partial_n^2)^N g(\boldsymbol{x})$ *is well behaved at infinity then*

$$G(\boldsymbol{\alpha}) = \sum_{m=1}^{M} G_m(\boldsymbol{\alpha}) + o(|\boldsymbol{\alpha}|^{-2N})$$

as $|\boldsymbol{\alpha}| \to \infty$ *along a radius.*

Proof. This is completely analogous to the proof of Theorem 9.4. □

The condition on the derivatives can be somewhat lightened by noting that if $|\boldsymbol{\alpha}| \to \infty$ implies that $|\alpha_s| \to \infty$ the proof of Theorem 9.4 still holds with a derivative with respect to x_s. This leads to

Corollary 9.7. *If* $|\boldsymbol{\alpha}| \to \infty$ *implies that* $|\alpha_s| \to \infty$ *the derivative* $(\partial_1^2 + \dots + \partial_n^2)^N$ *can be replaced by* ∂_s^N *without altering the validity of Theorem 9.7, the error term being* $o(|\alpha_s|^{-N})$.

Different values of N may correspond to different values of s, indicating that the magnitude of the error may be appreciably smaller in some parts of space than in others.

Example 14. $r^{\nu}J_{\nu}(r)$ has an isolated singularity at the origin and is well behaved at infinity. Since the behaviour near the origin is $r^{2\nu}/\nu!2^{\nu} + O(r^{2\nu+2})$ Corollary 9.7 shows that

$$\begin{aligned} G(\boldsymbol{\alpha}) &= \frac{(\nu+\frac{1}{2}n-1)!2^{\nu+n}\pi^{n/2}}{\nu!(-\nu-1)!|\boldsymbol{\alpha}|^{2\nu+n}} + o(|\boldsymbol{\alpha}|^{-N}) \\ &= -(\nu+\tfrac{1}{2}n-1)!2^{\nu+n}\pi^{(n/2)-1}|\boldsymbol{\alpha}|^{-2\nu-n}\sin\nu\pi + o(|\boldsymbol{\alpha}|^{-N}) \end{aligned}$$

where N is the integer satisfying $2\nu+2 \leq N-n < 2\nu+3$. As before we expect $o(|\boldsymbol{\alpha}|^{-N})$ to be $O(|\boldsymbol{\alpha}|^{-2\nu-2-n})$.

9.5. Integrals over a finite domain

The theory of the preceding section is not sufficiently embracing to include generalised functions which have singularities on curves or hypersurfaces instead of at a point. In particular, generalised functions which vanish outside a finite domain are excluded unless they are sufficiently smooth at the boundary. Such generalised functions will be considered in this section. The case of two dimensions will be discussed firstly in order to simplify the analysis and bring out clearly the kind of phenomenon which can arise. Our main purpose will be to show how the dominant asymptotic terms can be determined and to indicate how other terms could be derived if necessary. General formulae are usually so complicated and cumbersome that it is not very useful to have them.

Consider therefore $\int_D g(x_1, x_2)\mathrm{e}^{-\mathrm{i}(\alpha_1 x_1+\alpha_2 x_2)}\,\mathrm{d}x_1\,\mathrm{d}x_2$ where D is a connected domain in R_2 to be specified more precisely later on. Let $\phi(x)$ be a fine function which equals unity in $(-1, 1)$, vanishes outside $(-2, 2)$ and satisfies $0 \leq \phi(x) \leq 1$ everywhere. Then, if g possesses an isolated singularity g_0 at an interior point $\boldsymbol{x}_0$ of D consider $g(\boldsymbol{x}) - g_0(\boldsymbol{x})\phi(|\boldsymbol{x}-\boldsymbol{x}_0|/\varepsilon)$ and $g_0(\boldsymbol{x})\phi(|\boldsymbol{x}-\boldsymbol{x}_0|/\varepsilon)$ where $\varepsilon > 0$ is chosen so small that $|\boldsymbol{x}-\boldsymbol{x}_0| \leq 2\varepsilon$ lies entirely within D. The integral over D of $g_0\phi$ can be extended to the whole of R_2, because of the choice of ε, and so comes within the scope of the preceding section. On the other hand, $g - g_0\phi$ has no singularity at $\boldsymbol{x}_0$ or, at any rate, is significantly less singular there. This process enables the isolated singularities in the interior of D to be separated out and, consequently, there is no real loss of generality if it is assumed from now on that g, together with as many derivatives as are desired, is continuous inside D.

With regard to the behaviour at boundary points it will be supposed, for the moment, that g is continuous there; the effect of singularities at boundary points will be examined later.

Now fix the direction of (α_1, α_2) and rotate the axes of the variables of integration by means of $\boldsymbol{x} = \boldsymbol{L}\boldsymbol{y}$ until the y_1-axis coincides with this direction. Then the integral is $\int_D g_1(y_1, y_2)\mathrm{e}^{-\mathrm{i}|\alpha|y_1}\,\mathrm{d}y_1\,\mathrm{d}y_2$ where $g_1(y) = g(Ly)$. It is now proposed to evaluate this integral by integrating first with respect to y_2 and then with respect to y_1. This is certainly legitimate if D is finite and g continuous or, if D is not necessarily finite, when g is absolutely integrable over D. We shall assume that conditions of this type are satisfied.

The straight line $y_1 =$ constant may intersect the boundary of D in several points or in none. Let there be intersections in the interval $y_m \leq y_1 \leq y_M$ but none outside. Then $\int_D$ will be a sum of integrals of the type

$$\int_{-\infty}^{\infty} \mathrm{e}^{-\mathrm{i}|\alpha|y_1}\left\{\int_{h_1}^{h_2} g_1(y_1, y_2)\,\mathrm{d}y_2\right\}\mathrm{H}(y_1 - y_m)\,\mathrm{H}(y_M - y_1)\,\mathrm{d}y_1$$

where $y_2 = h_1(y_1)$ and $y_2 = h_2(y_1)$ are intersections of $y_1 =$ constant with the boundary of D such that $h_1 < y_2 < h_2$ consists of interior points of D. It follows from §9.2 that the asymptotic behaviour is dictated by the singularities in y_1 of $\int_{h_1}^{h_2} g_1(y_1, y_2)\,\mathrm{d}y_2 = h(y_1)$ (say) and those due to the presence of $\mathrm{H}(y_1 - y_m)\,\mathrm{H}(y_M - y_1)$.

Ignore, for the moment, the singularities of $h(y_1)$. When $y_1 = y_m$, $h_2 = h_1$ unless the boundary of D at y_m is a straight line parallel to the y_2-axis. It will now be assumed that near y_m

$$\left.\begin{aligned} h_2(y_1) - h_2(y_m) &= a_2(y_1 - y_m)^{\mu_2}(1 + \ldots), \\ h_1(y_1) - h_1(y_m) &= a_1(y_1 - y_m)^{\mu_1}(1 + \ldots) \end{aligned}\right\} \tag{3}$$

where $\mu_1 > 0$, $\mu_2 > 0$ and the dots represent positive integral powers of $(y_1 - y_m)$. The representation (3) covers many possible types of

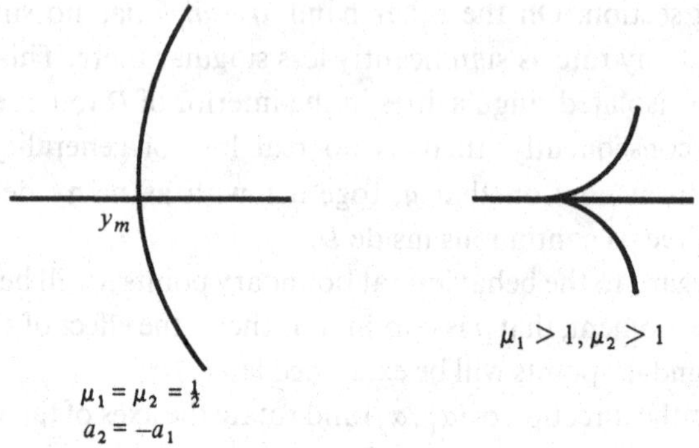

FIGURE 9.1

boundary; for example, $\mu_1 = \mu_2 = \frac{1}{2}$ and $a_2 = -a_1$ gives a smooth curve whereas $\mu_2 > 1$ and $\mu_1 > 1$ gives a cusp (see Fig. 9.1). With g_1 continuous there is a singularity at y_m of

$$g_1\{y_m, h_1(y_m)\}\{a_2(y_1 - y_m)^{\mu_2} - a_1(y_1 - y_m)^{\mu_1}\}\,\mathrm{H}(y_1 - y_m)$$

which contributes

$$-\mathrm{i}g_1\{y_m, h_1(y_m)\}\,\mathrm{e}^{-\mathrm{i}|\alpha|y_m} \times \{\mu_2!|\alpha|^{-\mu_2-1}a_2\,\mathrm{e}^{-\pi\mathrm{i}\mu_2/2} - \mu_1!a_1|\alpha|^{-\mu_1-1}\,\mathrm{e}^{-\pi\mathrm{i}\mu_1/2}\} \qquad (4)$$

to the Fourier transform. The error will depend upon the behaviour of g_1 near y_m but, if g_1 is continuously differentiable there, the error will be $o(|\alpha|^{-N})$ where N is the largest integer such that

$$N < 2 + \inf(\mu_1, 2\mu_1 - 1, 2\mu_2 - 1, \mu_2).$$

If the boundary at D is a straight line parallel to the y_2-axis the form (3) can still be employed provided that $\mu_1 = \mu_2 = 0$ and the terms $h_2(y_m)$ and $h_1(y_m)$ are removed. There is then a singularity

$$\mathrm{H}(y_1 - y_m)\int_{a_1}^{a_2} g_1(y_m, y_2)\,\mathrm{d}y_2$$

leading to a Fourier transform

$$-\mathrm{i}|\alpha|^{-1}\,\mathrm{e}^{-\mathrm{i}|\alpha|y_m}\int_{a_1}^{a_2} g_1(y_m, y_2)\,\mathrm{d}y_2.$$

The contribution from y_M may be dealt with similarly. Thus, if

$$h_2(y_1) - h_2(y_M) = b_2(y_M - y_1)^{\nu_2}(1 + \dots),$$
$$h_1(y_1) - h_1(y_M) = b_1(y_M - y_1)^{\nu_1}(1 + \dots)$$

when $y_1 \approx y_M$ there is a contribution to the Fourier transform of

$$\mathrm{i}g_1\{y_M, h_1(y_M)\}\{\nu_2! b_2 |\boldsymbol{\alpha}|^{-\nu_2 - 1} \mathrm{e}^{\pi \mathrm{i}\nu_2/2} - \nu_1! b_1 |\boldsymbol{\alpha}|^{-\nu_1 - 1} \mathrm{e}^{\pi \mathrm{i}\nu_1/2}\} \mathrm{e}^{-\mathrm{i}|\boldsymbol{\alpha}| y_M}.$$

Now convert these results back to the original variables. For simplicity of description it will be assumed that the boundary of D has no cusps or points where the slope is discontinuous – the reader will see easily what happens when these are present. Lines parallel to the y_1-axis become lines parallel to the direction of (α_1, α_2) so points such as y_m and y_M are boundary points where the tangent to the boundary is perpendicular to this direction, for example, the points $\boldsymbol{x}_0$ and $\boldsymbol{x}_1$ of Fig. 9.2.

Let (x, y) be a right-handed system of axes with origin at $\boldsymbol{x}_0$ and with the positive x-axis parallel to the direction of (α_1, α_2). Then what has been shown above can be formulated as

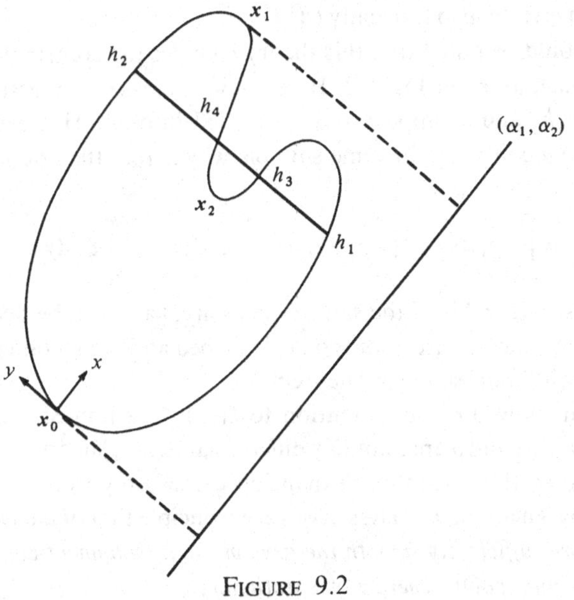

FIGURE 9.2

Theorem 9.8. *If $\boldsymbol{x}_0$ is a boundary point where $\boldsymbol{\alpha}.\boldsymbol{x}$ is a minimum, if in a neighbourhood of $\boldsymbol{x}_0$ the boundary has equation*

$$y = a_2 x^{\mu_2} \text{ (larger } y\text{)}, \quad y = a_1 x^{\mu_1} \text{ (smaller } y\text{)}, \quad \mu_1 > 0, \mu_2 > 0,$$

and if g has continuous first derivatives in this neighbourhood then the contribution to $\int_D g(\boldsymbol{x}) \mathrm{e}^{-\mathrm{i}\boldsymbol{\alpha}.\boldsymbol{x}} \, \mathrm{d}\boldsymbol{x}$ is

$$-\mathrm{i}g(\boldsymbol{x}_0)\{\mu_2! |\boldsymbol{\alpha}|^{-\mu_2 - 1} a_2 \mathrm{e}^{-\pi \mathrm{i}\mu_2/2} - \mu_1! |\boldsymbol{\alpha}|^{-\mu_1 - 1} a_1 \mathrm{e}^{-\pi \mathrm{i}\mu_1/2}\} \mathrm{e}^{-\mathrm{i}\boldsymbol{\alpha}.\boldsymbol{x}_0} + o(|\boldsymbol{\alpha}|^{-N})$$

where $N < 2 + \inf(\mu_1, 2\mu_1 - 1, 2\mu_2 - 1, \mu_2)$. If $\mu_1 = \mu_2 = 0$ the contribution is

$$-\mathrm{i}|\boldsymbol{\alpha}|^{-1} \mathrm{e}^{-\mathrm{i}\boldsymbol{\alpha}.\boldsymbol{x}_0} \int_{a_1}^{a_2} g \, \mathrm{d}y.$$

If $\boldsymbol{\alpha}.\boldsymbol{x}$ is a maximum at $\boldsymbol{x}_0$ change the sign of i *throughout the contributions (except for the factor $\mathrm{e}^{-\mathrm{i}\boldsymbol{\alpha}.\boldsymbol{x}_0}$).*

In the particular case when the boundary has a radius of curvature ρ at $\boldsymbol{x}_0$, $\mu_1 = \mu_2 = \frac{1}{2}$ and $a_2 = -a_1 = \sqrt{(2\rho)}$; then the contribution reduces to

$$(2\pi\rho)^{1/2} |\boldsymbol{\alpha}|^{-3/2} g(\boldsymbol{x}_0) \mathrm{e}^{-\mathrm{i}\boldsymbol{\alpha}.\boldsymbol{x}_0 - 3\pi \mathrm{i}/4}$$

with an error which is usually $O(|\boldsymbol{\alpha}|^{-2})$.

It should be noted that this theory includes the contribution of a point such as $\boldsymbol{x}_2$ in Fig. 9.2. Here, as y_1 increases, the extra intersections h_3 and h_4 appear as $\boldsymbol{x}_2$ is passed through. However, if we imagine g continued in some suitable way across the boundary we can write

$$\int_{h_1}^{h_3} g_1 \, \mathrm{d}y_2 + \int_{h_4}^{h_2} g_1 \, \mathrm{d}y_2 = \int_{h_1}^{h_2} g_1 \, \mathrm{d}y_2 - \int_{h_3}^{h_4} g_1 \, \mathrm{d}y_2.$$

The first integral is of the same type as $h(y_1)$ and can be dealt with by methods which are about to be described and the second integral comes within the scope of Theorem 9.8.

Let us now turn our attention to $h(y_1)$. It is immediately clear that, if h_1, h_2 and g are infinitely differentiable, $h(y_1)$ has no singularities and so the asymptotic expansion comes only from the points given by Theorem 9.8. Thus arises the principle that, *if the boundary and g are sufficiently smooth the asymptotic development comes from the boundary points where $\boldsymbol{\alpha}.\boldsymbol{x}$ is stationary.*

Contributions come from other boundary points only when there is a lack of smoothness at them. It is virtually self-evident that the more derivatives the curve possesses the less significant is the asymptotic behaviour. Therefore we shall concentrate primarily on the effect of the first derivative. Suppose then that near the boundary point $y_1 = y_b, y_2 = h_2(y_b)$ the boundary is given by

$$\begin{aligned} y_2 - h_2(y_b) &= c_1(y_b - y_1)^{\sigma_1} \quad (y_1 \le y_b) \\ &= c_2(y_1 - y_b)^{\sigma_2} \quad (y_1 \ge y_b) \end{aligned}$$

where $1 \ge \sigma_1 > 0$, $1 \ge \sigma_2 > 0$. Then, if $y_1 \le y_b$

$$h(y_1) - h(y_b) \approx c_1 g_1 \{y_b, h_2(y_b)\}(y_b - y_1)^{\sigma_1}$$

and, if $y_1 \ge y_b$,

$$h(y_1) - h(y_b) \approx c_2 g_1 \{y_b, h_2(y_b)\}(y_1 - y_b)^{\sigma_2}$$

so that there is a singularity involving the factors $(y_1 - y_b)^{\sigma_2} \mathrm{H}(y_1 - y_b)$ and $(y_b - y_1)^{\sigma_1} \mathrm{H}(y_b - y_1)$.

The consequent contribution to the Fourier transform is

$$-\mathrm{i}g_1 \{y_b, h_2(y_b)\} \{\sigma_2! c_2 |\alpha|^{-\sigma_2 - 1} \mathrm{e}^{-\pi \mathrm{i}\sigma_2/2} - \sigma_1! c_1 |\alpha|^{-\sigma_1 - 1} \mathrm{e}^{\pi \mathrm{i}\sigma_1/2}\} \mathrm{e}^{-\mathrm{i}|\alpha| y_b}.$$

The error in this expression will depend upon h_1, g_1 and the higher order terms in the equation for the curve; it can be estimated by methods already established.

Returning to our original variables and with (x, y) chosen as before we have

Theorem 9.9. *If $\boldsymbol{x}_0$ is a boundary point where the slope of the boundary has a singularity, the equation of the boundary being*

$$y = c_1(-x)^{\sigma_1} \quad (x \le 0), \quad y = c_2 x^{\sigma_2} \quad (x \ge 0)$$

where $0 < \sigma_1, \sigma_2 \le 1$ the dominant contribution to $\int_D g(\boldsymbol{x}) \mathrm{e}^{-\mathrm{i}\boldsymbol{\alpha}.\boldsymbol{x}} \, \mathrm{d}\boldsymbol{x}$ is

$$-\mathrm{i}g(\boldsymbol{x}_0)\{\sigma_2! c_2 |\boldsymbol{\alpha}|^{-\sigma_2 - 1} \mathrm{e}^{-\pi \mathrm{i}\sigma_2/2} - \sigma_1! c_1 |\boldsymbol{\alpha}|^{-\sigma_1 - 1} \mathrm{e}^{\pi \mathrm{i}\sigma_1/2}\} \mathrm{e}^{-\mathrm{i}\boldsymbol{\alpha}.\boldsymbol{x}_0}$$

for $y < 0$ in D.

It is, of course, assumed that $\boldsymbol{x}_0$ is not a point already covered by Theorem 9.8. The theory includes as a special case the intersection of two straight lines (neither being perpendicular to the direction of (α_1, α_2)) by putting $\sigma_1 = \sigma_2 = 1$ and gives

$$-(c_1 + c_2)g(\boldsymbol{x}_0)|\boldsymbol{\alpha}|^{-2}\mathrm{e}^{-i\boldsymbol{\alpha}.\boldsymbol{x}_0}.$$

This is generally less significant than the contribution of a stationary boundary point and disappears altogether when the two arcs are portions of the same straight line because then $c_1 = -c_2$.

Another special case is the point of inflexion where $\sigma_1 = \sigma_2 = \frac{1}{3}$, $c_1 = -c_2$; the contribution is

$$\tfrac{1}{3}!\, ic_1 g(\boldsymbol{x}_0)3^{1/2}|\boldsymbol{\alpha}|^{-4/3}\mathrm{e}^{-i\boldsymbol{\alpha}.\boldsymbol{x}_0}.$$

It must be emphasised that all the asymptotic formulae in this and the preceding section refer to what happens in a fixed direction. Thus, if $\alpha_1 = |\boldsymbol{\alpha}| \cos\theta$ and $\alpha_2 = |\boldsymbol{\alpha}| \sin\theta$, they can be expected to give good results if $|\boldsymbol{\alpha}| > A(\theta)$. But there is no presumption that A is independent of θ so that A may be different for differing directions. In other words *the formulae are not necessarily valid uniformly with respect to angle* (the coefficients c_1 and c_2 in Theorem 9.9 depend on θ). In some cases uniformly valid formulae can be derived though they are usually significantly less simple than the ones already obtained. The kind of difficulty which arises can be seen easily by considering the transition which must occur when a stationary boundary point has $\mu_1 = \mu_2 = 2$ for one angle and has $\mu_1 = \mu_2 = 3$ for a nearby angle (cf. also Exercise 25).

Exercises

25. Obtain the dominant asymptotic behaviour of $\int_D \mathrm{e}^{-i\boldsymbol{\alpha}.\boldsymbol{x}}\,\mathrm{d}\boldsymbol{x}$ by the methods of this section, D being the square $|x_1| \le 1$, $|x_2| \le 1$. Check with the formula derived by evaluating the integral exactly, paying special attention to what happens when $\boldsymbol{\alpha}$ is nearly parallel to a side of the square.

26. Obtain the dominant asymptotic behaviour of $\int_D \mathrm{e}^{-i\boldsymbol{\alpha}.\boldsymbol{x}}\,\mathrm{d}\boldsymbol{x}$ when D is (i) the circle $|\boldsymbol{x}| \le 1$, (ii) the ellipse $x_1^2/a^2 + x_2^2/b^2 \le 1$ and (iii) the parabola $x_2^2 \le 4ax_1$.

27. Find the first two terms of the asymptotic development of $\int_D \mathrm{e}^{-i\alpha_1 x_1}\,\mathrm{d}\boldsymbol{x}$ and the first term of $\int_D x_2 \mathrm{e}^{-i\alpha_1 x_1}\,\mathrm{d}\boldsymbol{x}$ where D consists of $|x_2| \le x_1^{1/2} + x_1^{3/2}$, $x_1 \ge 0$. Estimate the error in the first term of $\int_D g(\boldsymbol{x})\mathrm{e}^{-i\alpha_1 x_1}\,\mathrm{d}\boldsymbol{x}$ if g has a continuous first derivative.

28. If, in Theorem 9.8, the equation of the boundary is $y = a_2 x^{\mu_2} \ln x$ (larger y) and $y = a_1 x^{\mu_1} \ln x$ (smaller y) find the contribution to the asymptotic development.

29. Repeat Exercise 28 for Theorem 9.9.

30. If $g_1(y_1, y_2) = \{y_2^2 + (y_1 - y_m)^2\}^{\beta/2}$ show that (4) must be replaced by

$$-\mathrm{i}e^{-\mathrm{i}|\alpha|y_m - \pi\mathrm{i}\beta/2}\{(\beta+\mu_2)!a_2|\alpha|^{-\beta-\mu_2-1}\,e^{-\pi\mathrm{i}\mu_2/2} - (\beta+\mu_1)!a_1|\alpha|^{-\beta-\mu_1-1}\,e^{-\pi\mathrm{i}\mu_1/2}\}$$

when $\mu_1 > 1$ and $\mu_2 > 1$. If $0 < \mu_1 < 1$, $0 < \mu_2 < 1$ and $\mathrm{Re}\,\beta \neq -1$ the appropriate formula is

$$\frac{e^{-\mathrm{i}|\alpha|y_m}}{\beta+1}[\{(\beta+1)\mu_2\}!|a_2|^{\beta+1}|\alpha|^{-(\beta+1)\mu_2-1}\,e^{-(\pi/2)\mathrm{i}\{(\beta+1)\mu_2+\mathrm{sgn}\,a_2\}} - \{(\beta+1)\mu_1\}!|a_1|^{\beta+1}|\alpha|^{-(\beta+1)\mu_1-1}\,e^{-(\pi/2)\mathrm{i}\{(\beta+1)\mu_1+\mathrm{sgn}\,a_1\}}]$$

when $a_1 a_2 > 0$; when $a_1 a_2 < 0$ this is still true if $\mathrm{Re}\,\beta > -1$ but if $\mathrm{Re}\,\beta < -1$ the dominant behaviour is proportional to $|\alpha|^{-\beta-2}$. If $\mu_1 = \mu_2 = 1$ show that the main term is proportional to $|\alpha|^{-\beta-2}$. What happens if $\beta = -1$?

Explain how these results can be used to elucidate the contribution of an isolated singularity of g at a stationary boundary point.

31. Do an investigation parallel to Exercise 30 for a similar singularity at a boundary point of the type in Theorem 9.9.

32. Obtain the dominant terms of $\int_D g(\boldsymbol{x})e^{-\mathrm{i}\alpha_1 x_1}\,\mathrm{d}\boldsymbol{x}$, g being continuously differentiable, when D is the interior of the lemniscate $(x_1^2 + x_2^2)^2 = a^2(x_1^2 - x_2^2)$.

The same methods can be used in principle to elucidate what happens for Fourier transforms in n dimensions. For example, the singularities of g can still be eliminated in the same way and a rotation of axes will lead to a discussion of $\int e^{-\mathrm{i}|\alpha|y_1}\int_C g_1(\boldsymbol{y})\,\mathrm{d}y_2 \ldots \mathrm{d}y_n\,\mathrm{d}y_1$ where C is the intersection of D and the hyperplane $y_1 = \text{constant}$. But the detailed discussion of the integral over C is more difficult because of the highly complicated forms which can be taken by hypersurfaces in several dimensions. For instance, a hypersurface may touch a hyperplane in a point or in a hypersurface of several dimensions of which the parts cannot be dealt with separately. (Even in three dimensions a surface can be tangential to a plane along a continuous closed curve.) Therefore, discussion will be limited to some of the simpler cases which occur in practice.

Firstly we shall consider the case where a hyperplane $y_1 = $ a constant is tangential to D at an isolated boundary point. This point will always be called $(y_m, y_{2m}, \ldots, y_{nm})$ and we shall put $z_j = y_{j+1} - y_{j+1,m}$ $(j = 1, \ldots, n-1)$. Let the hypersurface be convex near the boundary point and lie in $y_1 \geq y_m$. Let its equation in this neighbourhood be $y_1 - y_m = \frac{1}{2}z^T Az$ where matrix notation has been employed and A is a positive definite matrix. Then, if $g_1(y_m)$ be an abbreviation for the value of g_1 at the boundary point, the integral over C is effectively $g_1(y_m)\int_C \mathrm{d}z$ where C is now the domain $\frac{1}{2}z^T Az \leq y_1 - y_m$. By means of a rotation of axes $w = Lz$ the quadratic form $z^T Az$ becomes $w^T LAL^T w$ and L can be chosen so that this is $\sum_{j=1}^{n-1} w_j^2/\rho_j$. Note that

$$1/\rho_1 \ldots \rho_{n-1} = \det LAL^T = \det A$$

since L is orthogonal. The positive numbers $\rho_1, \ldots, \rho_{n-1}$ are the principal radii of curvature of D at y_m and $1/\rho_1 \ldots \rho_{n-1}$ is known as the *Gaussian curvature*. If now we put $w_j = u_j\{2\rho_j(y_1 - y_m)\}^{1/2}$ the integral becomes $2^{(n-1)/2} g_1(y_m)(y_1 - y_m)^{(n-1)/2}(\rho_1 \ldots \rho_{n-1})^{1/2} \int \mathrm{d}\boldsymbol{u}$ taken over the unit sphere in $n - 1$ dimensions. Since the volume of this sphere is $\pi^{(n-1)/2}/(\frac{1}{2}n - \frac{1}{2})!$ the singularity in y_1 is

$$(2\pi)^{(n/2)-1/2} g_1(y_m)(\rho_1 \cdots \rho_{n-1})^{1/2}(y_1 - y_m)^{(n/2)-1/2} \mathrm{H}(y_1 - y_m)/(\tfrac{1}{2}n - \tfrac{1}{2})!$$

and the contribution to the asymptotic expansion is

$$(2\pi)^{(n/2)-1/2} g_1(y_m)(\rho_1 \cdots \rho_{n-1})^{1/2} |\boldsymbol{\alpha}|^{-(n+1)/2} \mathrm{e}^{-\mathrm{i}|\boldsymbol{\alpha}|y_m - \pi\mathrm{i}(n+1)/4}.$$

Theorem 9.10. *If $\boldsymbol{x}_0$ is a boundary point where $\boldsymbol{\alpha}.\boldsymbol{x}$ is a minimum and the boundary is synclastic with equation as above, the contribution to $\int_D g(\boldsymbol{x})\mathrm{e}^{-\mathrm{i}\boldsymbol{\alpha}.\boldsymbol{x}}\,\mathrm{d}\boldsymbol{x}$ is*

$$(2\pi)^{(n/2)-1/2} g(\boldsymbol{x}_0)(\rho_1 \cdots \rho_{n-1})^{1/2} |\boldsymbol{\alpha}|^{-(n+1)/2} \mathrm{e}^{-\mathrm{i}\boldsymbol{\alpha}.\boldsymbol{x}_0 - \pi\mathrm{i}(n+1)/4}.$$

If $\boldsymbol{\alpha}.\boldsymbol{x}$ is a maximum, change $\mathrm{e}^{-\pi\mathrm{i}(n+1)/4}$ to $\mathrm{e}^{\pi\mathrm{i}(n+1)/4}$.

Now consider a boundary point where the hypersurface is anticlastic, i.e. not all the principal radii of curvature are of the same sign, but all the radii of curvature are finite and non-zero. The domain of integration is limited to some fixed finite neighbourhood of y_m as well as being restricted to $\frac{1}{2}z^T Az \leq y_1 - y_m$. It is convenient to specify the precise domain later. Proceed as for the synclastic case by introducing the variable $\boldsymbol{w}$ but choose the coordinates so that $\rho_1, \ldots, \rho_p (1 \leq p < n - 1)$ are all positive and $\rho_{p+1}, \ldots, \rho_{n-1}$ are all negative. We put $w_j = u_j(2|\rho_j|)^{1/2}$ and then $\frac{1}{2}z^T Az \leq y_1 - y_m$

becomes $u_1^2+\dots+u_p^2-u_{p+1}^2-\dots-u_{n-1}^2\le y_1-y_m$ or $r_p^2-r_{n-p-1}^2 \le y_1-y_m$ in terms of spherical polar coordinates in spaces of p and $n-p-1$ dimensions respectively. The fixed domain is now selected as the union of the spheres of radii $2d$ and d in these two spaces respectively. Then our integral becomes

$$\frac{4(2\pi)^{(n/2)-1/2}|\rho_1\dots\rho_{n-1}|^{1/2}}{(\frac{1}{2}p-1)!(\frac{1}{2}n-\frac{1}{2}p-\frac{3}{2})!}g_1(y_m)$$

$$\times\int_{(y_m-y_1)\mathrm{H}(y_m-y_1)}^{d} r_{n-p-1}^{n-p-2}\int_0^{(y_1-y_m+r_{n-p-1}^2)^{1/2}} r_p^{p-1}\,\mathrm{d}r_p\,\mathrm{d}r_{n-p-1}$$

because the area of the unit sphere in m dimensions is $2\pi^{m/2}/(\frac{1}{2}m-1)!$. The repeated integral can be written as

$$\frac{1}{2p}\int_{-x\mathrm{H}(-x)}^{d^2} t^{(n-p-3)/2}(x+t)^{p/2}\,\mathrm{d}t$$

with $x=y_1-y_m$ and $r_{n-p-1}=t^{1/2}$. Integration by parts gives

$$\int_{-x\mathrm{H}(-x)}^{d^2} t^{(n-p-3)/2}(x+t)^{p/2}\,\mathrm{d}t$$

$$=\frac{2d^{n-p-1}(x+d^2)^{p/2}}{n-1}+\frac{px}{n-1}\int_{-x\mathrm{H}(-x)}^{d^2} t^{(n-p-3)/2}(x+t)^{(p/2)-1}\,\mathrm{d}t.$$

The first term has no singularity at $x=0$ and so can be ignored from now on. Thus we have a reduction formula which, when p is even, leads to

$$\frac{(\frac{1}{2}p)!(\frac{1}{2}n-\frac{1}{2}p-\frac{3}{2})!}{(\frac{1}{2}n-\frac{1}{2})!}(-x)^{(n/2)-1/2}\mathrm{H}(-x)\mathrm{e}^{-\pi i\{(p/2)+1\}}$$

for the singular part of the integral with respect to t. Hence the singularity, when p is even, is

$$\frac{(2\pi)^{(n/2)-1/2}|\rho_1\dots\rho_{n-1}|^{1/2}}{(\frac{1}{2}n-\frac{1}{2})!}g_1(y_m)(-x)^{(n/2)-1/2}\mathrm{H}(-x)\mathrm{e}^{-\pi i\{(p/2)+1\}}$$

with a contribution to the asymptotic development of

$$(2\pi)^{(n/2)-1/2}g_1(y_m)|\rho_1\dots\rho_{n-1}|^{1/2}|\boldsymbol{\alpha}|^{-(n/2)-1/2}\mathrm{e}^{-(\pi/4)i(2p-n+3)-i|\boldsymbol{\alpha}|y_m}. \tag{5}$$

If p is odd the reduction formula reaches finally

$$\frac{(\frac{1}{2}p)!(\frac{1}{2}n-\frac{1}{2}p-1)!}{(\frac{1}{2}n-\frac{1}{2})!\pi^{1/2}}x^{(p/2)+1/2}\int_{-x\mathrm{H}(-x)}^{d^2} t^{(n-p-3)/2}(x+t)^{-1/2}\,\mathrm{d}t$$

for the possible singular part. Now

$$\int_{-x\mathrm{H}(-x)}^{d^2}\frac{t^{\mu}}{(x+t)^{1/2}}\mathrm{d}t=\frac{2d^{2\mu}(x+d^2)^{1/2}}{1+2\mu}-\frac{2\mu x}{1+2\mu}\int\frac{t^{\mu-1}}{(x+t)^{1/2}}\mathrm{d}t$$

where once again the first term is not singular. Therefore this is another reduction formula which eventually arrives at $\int(x+t)^{-1/2}\mathrm{d}t$ or $\int t^{-1/2}(x+t)^{-1/2}\mathrm{d}t$ according as n is even or odd. Hence, when p is odd, the singularity is

$$\frac{(2\pi)^{(n/2)-1/2}|\rho_1\cdots\rho_{n-1}|^{1/2}}{(\frac{1}{2}n-\frac{1}{2})!}g_1(y_m)x^{(n/2)-1/2}\mathrm{H}(x)\mathrm{e}^{(\pi/2)\mathrm{i}(n-p-1)}\quad (n\text{ even})$$

$$\frac{(2\pi)^{(n/2)-1/2}|\rho_1\cdots\rho_{n-1}|^{1/2}}{(\frac{1}{2}n-\frac{1}{2})!}g_1(y_m)x^{(n/2)-1/2}\ln|x|\,\mathrm{e}^{(\pi/2)\mathrm{i}(n-p)}\quad (n\text{ odd}).$$

In either case the contribution to the asymptotic development turns out to be the same as (5). Combining these results with Theorem 9.10 we have

Theorem 9.11. *If $\boldsymbol{x}_0$ is a boundary point where $\boldsymbol{\alpha}.\boldsymbol{x}$ is stationary and the equation of the boundary nearby is $\boldsymbol{\alpha}.(\boldsymbol{x}-\boldsymbol{x}_0)/|\boldsymbol{\alpha}|=\frac{1}{2}\boldsymbol{z}^{\mathrm{T}}A\boldsymbol{z}$ the contribution to $\int_D g(\boldsymbol{x})\mathrm{e}^{-\mathrm{i}\boldsymbol{\alpha}.\boldsymbol{x}}\,\mathrm{d}\boldsymbol{x}$ is*

$$(2\pi)^{(n/2)-1/2}g(\boldsymbol{x}_0)|\rho_1\cdots\rho_{n-1}|^{1/2}|\boldsymbol{\alpha}|^{-(n/2)-1/2}\mathrm{e}^{-\mathrm{i}\boldsymbol{\alpha}.\boldsymbol{x}_0-\pi\mathrm{i}(\sigma+2)/4}$$

as $|\boldsymbol{\alpha}|\to\infty$, σ being the signature of A.

The signature of a quadratic form is the number of positive terms less the number of negative terms when it has been reduced to a sum of squares, i.e. $2p-n+1$.

Next consider what happens at a stationary point when the boundary is degenerate in the sense that one or more of the principal radii of curvature is infinite. The leading terms in the equation of the boundary will then contain other than quadratic elements. Suppose that with suitable coordinates w_j the equation of the boundary is

$$y_1-y_m=\sum_{j=1}^{n-1}c_j w_j^{m_j}$$

where $m_1, \dots, m_{n-1}$ are positive integers greater than 1. Consider firstly the behaviour of the integral when the w_j are restricted so that $w_j \geq 0 (j = 1, \dots, n-1)$. Let $c_1, \dots, c_p$ be positive and $c_{p+1}, \dots, c_{n-1}$ be negative and non-zero. Put $w_j = (u_j^2/|c_j|)^{1/m_j}$; then the integral to be evaluated is

$$\left(\prod_{j=1}^{n-1} \frac{2}{m_j|c_j|^{1/m_j}}\right) \int \left(\prod_{j=1}^{n-1} u_j^{-1+2/m_j}\right) d\boldsymbol{u}$$

subject to the restriction $u_1^2 + \dots + u_p^2 - u_{p+1}^2 - \dots - u_{n-1}^2 \leq y_1 - y_m$. As before we integrate over the spheres in p and $n - p - 1$ dimensions but with the constraint that the angles do not exceed $\frac{1}{2}\pi$ because w_j must not be negative. A typical angular integration takes the form

$$\prod_{j=1}^{p-1} \int_0^{\pi/2} (\cos\theta_j)^{-1+2/m_j} (\sin\theta_j)^{-1+2a_j} d\theta_j$$

where $a_j = \sum_{k=j+1}^{p} 1/m_k$. The integral is known (see §4.5) and so our integral reduces to

$$4 \prod_{j=1}^{p-1} \frac{(1/m_j)!(a_j-1)!}{(a_j - 1 + 1/m_j)!|c_j|^{1/m_j}}$$

$$\times \prod_{j=p+1}^{n-2} \frac{(1/m_j)!(b_j-1)!}{(b_j - 1 + 1/m_j)!|c_j|^{1/m_j}} \int r_{n-p-1}^{2b-1} r_p^{2a-1} dr_p dr_{n-p-1}$$

where $b_j = \sum_{k=j+1}^{n-1} 1/m_k$, $a = \sum_{j=1}^{p} 1/m_j$, $b = \sum_{j=p+1}^{n-1} 1/m_j$. Since

$$a_j + 1/m_j = a_{j-1} \quad (j = 2, \dots, p-1)$$

and

$$b_j + 1/m_j = b_{j-1} \quad (j = p+2, \dots, n-2)$$

this can be written as

$$\frac{1}{(a-1)!(b-1)!} \left\{\prod_{j=1}^{n-1} \frac{(1/m_j)!}{|c_j|^{1/m}}\right\} \frac{1}{a} \int_{-xH(-x)}^{d^2} t^{b-1}(x+t)^a dt.$$

The employment of the reduction formulae described earlier leads to

$$\int t^{b-1}(x+t)^a dt = (-1)^b \frac{(b-1)!a!}{(a+b)!}(-x)^{a+b} H(-x)$$

$$= (-1)^{a+1} \frac{(b-1)!a!}{(a+b)!} x^{a+b} H(x)$$

as far as singular behaviour is concerned, the first expression being valid if a is an integer and the second if a is not an integer but b is. If neither a nor b is an integer the formula

$$\int t^{b-1}(x+t)^a \mathrm{d}t$$

$$= \frac{(b-1)!\,a!\,(a+b-s-r)!}{(b-r-1)!\,(a-s)!\,(a+b)!}(-1)^r x^{r+s}\int_{-x\mathrm{H}(-x)}^{d^2} t^{b-r-1}(x+t)^{a-s}\mathrm{d}t$$

is reached, r and s being integers such that $0 > a-s > -1$, $0 > b-r-1 > -1$. Now, if $a+b \neq s+r$,

$$\begin{aligned}\int_0^{d^2} t^{b-r-1}(x+t)^{a-s}\mathrm{d}t &= \frac{(a-s)x}{a+b-s-r}\int_0^{d_2} t^{b-r-1}(x+t)^{a-s-1}\mathrm{d}t\\ &= \frac{(a-s)x}{a+b-s-r}\int_0^{\infty} t^{b-r-1}(x+t)^{a-s-1}\mathrm{d}t\\ &= \frac{(b-r-1)!\,(-a-b+r+s-1)!}{(s-a-1)!}x^{a+b-s-r}\end{aligned}$$

(from §4.5) as far so as singular behaviour is concerned. Similarly

$$\int_{-x}^{d^2} t^{b-r-1}(x+t)^{a-s}\mathrm{d}t = \frac{(a-s)!\,(-a-b+r+s-1)!}{(r-b)!}(-x)^{a+b-s-r}.$$

Hence, if $a+b \neq s+r$,

$$\int t^{b-1}(x+t)^a \mathrm{d}t$$

$$= \frac{(b-1)!\,a!}{(a+b)!\sin\pi(a+b)}\{x^{a+b}\mathrm{H}(x)\sin\pi a - (-x)^{a+b}\mathrm{H}(-x)\sin\pi b\}$$

as far as singular behaviour is concerned. Finally, if $a+b = s+r$,

$$\int_0^{d^2} t^{b-r-1}(x+t)^{a-s}\mathrm{d}t = \int_0^{\infty}\{t^{b-r-1}(x+t)^{a-s} - (x+t)^{-1}\}\mathrm{d}t - \ln x,$$

$$\int_{-x}^{d^2} t^{b-r-1}(x+t)^{a-s}\mathrm{d}t = \int_0^{\infty}\{(x+t)^{b-r-1}t^{a-s} - (x+t)^{-1}\}\mathrm{d}t - \ln(-x)$$

so that subtraction gives for the singularity

$$\begin{aligned}
&\mathrm{H}(x)\int_0^\infty \{t^{b-r-1}(1+t)^{a-s}-(1+t)^{b-r-1}t^{a-s}\}\,\mathrm{d}t-\ln|x| \\
&\quad=\mathrm{H}(x)\int_0^1 u^{-1}\{(1-u)^{b-r-1}-(1-u)^{r-b}\}\,\mathrm{d}u-\ln|x| \\
&\quad=\mathrm{H}(x)\lim_{\mu\to 0}\left\{\frac{(\mu-1)!(b-r-1)!}{(\mu+b-r-1)!}-\frac{(\mu-1)!(r-b)!}{(\mu+r-b)!}\right\}-\ln|x| \\
&\quad=\mathrm{H}(x)\{\psi(r-b)-\psi(b-r-1)\}-\ln|x| \\
&\quad=\mathrm{H}(x)\pi\cot\pi(b-r-1)-\ln|x|.
\end{aligned}$$

Therefore, in this case

$$\begin{aligned}
&\int t^{b-1}(x+t)^a\,\mathrm{d}t \\
&\qquad=\frac{a!(b-1)!(-1)^r x^{a+b+1}}{(a-s)!(a+b)!(b-r-1)!}\{\pi\cot\pi(b-r-1)\,\mathrm{H}(x)-\ln|x|\}
\end{aligned}$$

for the singular behaviour.

In all cases the asymptotic behaviour is found to be

$$-\mathrm{i}|\alpha|^{-1-\sum 1/m_j}\exp\{-\tfrac{1}{2}\pi\mathrm{i}\textstyle\sum(1/m_j)\operatorname{sgn}c_j\}\displaystyle\prod_{k=1}^{n-1}\frac{(1/m_k)!}{|c_k|^{1/m_k}}. \qquad (6)$$

So far the variables w_j have been forced to be non-negative. Suppose now that w_1 is negative whereas the others are still non-negative. A change in sign of w_1 returns to the previous case except that the sign of c_1 is altered if m_1 is odd. There is therefore a contribution which is the same as (6) apart from any necessary modification to c_1. Now allow w_1 to be unrestricted, w_2 to be negative and the others non-negative. Once again, by changing the sign of w_2, we see that there is the same contribution as when w_1 is unrestricted but with a change of sign for c_2 if m_2 is odd. Proceeding in this way we obtain

Theorem 9.12. *If the equation of the boundary is* $y_1-y_m=\sum_{j=1}^{n-1}c_jw_j^{m_j}$ *the first term in the asymptotic development is*

$$-\mathrm{i}g_1(y_m)2^{n-1}|\alpha|^{-1-\sum_{j=1}^{n-1}1/m_j}\left\{\prod_{k=1}^{n-1}\frac{(1/m_k)!d_k}{|c_k|^{1/m_k}}\right\}\mathrm{e}^{-\mathrm{i}|\alpha|y_m}$$

where $d_k = e^{-\{(\pi/2)i/m_k\}\,\mathrm{sgn}\, c_k}$ *(m_k even) and* $d_k = \cos(\pi/2m_k)$ *(m_k odd).*

One further boundary point will be discussed, namely a conical point in the neighbourhood of which the equation of the boundary is $y_1 - y_m = (z^{\mathrm{T}} Az)^{1/2}$ where A is positive definite. After a rotation of axes this becomes $y_1 - y_m = (\sum_{j=1}^{n-1} w_j^2/\rho_j)^{1/2}$ where $\rho_1, \dots, \rho_{n-1}$ are positive. Putting $w_j = u_j(y_1 - y_m)(2\rho_j)^{1/2}$ our integral is

$$2^{(n/2)-1/2} g_1(y_m)(y_1 - y_m)^{n-1}(\rho_1 \cdots \rho_{n-1})^{1/2} \int \mathrm{d}\boldsymbol{u}$$

taken over the unit sphere in $n-1$ dimensions. Thus the singularity is

$$(2\pi)^{(n/2)-1/2} g_1(y_m)(\rho_1 \cdots \rho_{n-1})^{1/2}(y_1 - y_m)^{n-1}\,\mathrm{H}(y_1 - y_m)/(\tfrac{1}{2}n - \tfrac{1}{2})!.$$

Theorem 9.13. *The contribution from a conical point of the type stated is*

$$(2\pi)^{(n/2)-1/2} g_1(y_m)(\rho_1 \cdots \rho_{n-1})^{1/2} |\boldsymbol{\alpha}|^{-n} e^{-i|\boldsymbol{\alpha}|y_m - \pi i n/2}.$$

Exercises

33. Estimate $\int_D e^{-i\boldsymbol{\alpha}.\boldsymbol{x}}\,\mathrm{d}\boldsymbol{x}$ when $|\boldsymbol{\alpha}|$ is large if D is (i) the sphere $|\boldsymbol{x}| \le a$, (ii) the 'quadric' $\sum_{j=1}^{n} c_j x_j^2 \le 1$. Compare with the exact value when D is $|x_1| \le 1, |x_2| \le 1, \dots, |x_n| \le 1$.

34. Find the dominant term of $\int_D r^2 e^{-i\alpha_1 x_1}\,\mathrm{d}\boldsymbol{x}$ if $n=3$ and D is

$$3x_2^2 - 2x_3 x_2 + 3x_3^2 \le x_1.$$

35. If $n=3$ and D is $c_1 x_2^2 + c_2 x_3^2 \le x_1$ $(c_1 > 0, c_2 > 0)$ estimate $\int_D e^{-i|\alpha|x_1}\,\mathrm{d}\boldsymbol{x}$.

36. If the hypersurface near y_m has equation $y_1 - y_m = \sum_{j=1}^{n-1} c_j z_1^{m_{1j}} \dots z_{n-1}^{m_{n-1,j}}$ where $\det(m_{jk}) \ne 0$ show that the contribution of the domain where $z_1, \dots, z_{n-1}$ are all non-negative to the asymptotic development is

$$\frac{-ig_1(y_m)|\boldsymbol{\alpha}|^{-1-\sum_j p_j}}{p_1 \cdots p_{n-1} \det(m_{jk})} \exp\{-i|\boldsymbol{\alpha}|y_m - \tfrac{1}{2}\pi i \sum_j p_j \,\mathrm{sgn}\, c_j\} \prod_{k=1}^{n-1} \frac{p_k!}{|c_k|^{p_k}}$$

where $p_j = \sum_{k=1}^{n-1} M_{jk}$ and $\sum_{k=1}^{n-1} M_{jk} m_{ki} = 0\,(j \ne i),\ 1\,(j = i)$.

What is the contribution if the restrictions on $z_1, \dots, z_{n-1}$ are dropped? (Hint: Use $z_1^{m_{1j}} \dots z_{n-1}^{m_{n-1,j}}$ as a new variable.)

37. Use Exercise 36 when $g=1$, $n=3$ and $y_1 - y_m = c_1 z_1^3 z_2 + c_2 z_1 z_2^5$.

38. If D is $(\sum_{j=2}^{n} c_j x_j^m)^{1/m} \le x_1$ where $c_2, \dots, c_n$ are positive show that $\int_D e^{-i|\alpha|x_1}\,\mathrm{d}\boldsymbol{x}$ is asymptotically $B|\boldsymbol{\alpha}|^{-n}$ and determine B.

9.6. Stationary phase in higher dimensions

In n dimensions the natural generalisation of the integral discussed in §9.3 is $\int_D g(\boldsymbol{x})\mathrm{e}^{-\mathrm{i}\alpha f(\boldsymbol{x})}\,\mathrm{d}\boldsymbol{x}$ where $|\alpha|$ is large. For simplicity it will be assumed that g has no singularities inside D. Surround each point where grad $f=0$ (or where at least one component of the grad is infinite or indeterminate) by a small sphere of radius ε. Denote by D_ε the domain D with these spheres excluded. Then if $f(\boldsymbol{x})$ is taken as a new variable y_1 in D_ε we obtain an integral of the type considered in the preceding section. An immediate deduction is that the contributions to the asymptotic development come from the points on the transformed boundary of D_ε where y_1 is stationary or where the boundary is not sufficiently smooth. However, the contributions from the portions of the spheres of radius ε are clearly cancelled by equal and opposite contributions from within (the same argument applies if D_ε is subdivided into a number of subdomains). Consequently it can be concluded that the only boundary points which need be considered are those which come from the boundary of D. These must be such that some hypersurface $f(\boldsymbol{x})=$ constant touches there or they are part of a portion of boundary with which $f=$ constant coincides or are points where the boundary lacks smoothness. As well as these points the contributions from the points so far excluded (some of which may be on the boundary) must also be determined. Recalling Exercise 20, in which it is indicated that the results in one dimension can be obtained by making an appropriate expansion of f about the point under consideration, we shall assume that we can adopt the same process in n dimensions. (For a more geometrical approach see D. S. Jones and M. Kline, *J. Math. Phys.* **37**, 1 (1958).)

It will be assumed that α is positive; results for α negative can be deduced by changing the sign of i and replacing α by $-\alpha$.

Consider firstly an interior point of D where grad $f=0$. Suppose that in a neighbourhood of this point $\boldsymbol{x}_0$,

$$f(\boldsymbol{x})=f(\boldsymbol{x}_0)+\tfrac{1}{2}(\boldsymbol{x}^{\mathrm{T}}-\boldsymbol{x}_0^{\mathrm{T}})A(\boldsymbol{x}-\boldsymbol{x}_0)+f_1(\boldsymbol{x})$$

where det $A\neq 0$ and f_1 vanishes more rapidly than $|\boldsymbol{x}-\boldsymbol{x}_0|^2$ as $\boldsymbol{x}\to\boldsymbol{x}_0$. Then the essence of the method consists in expanding $g(\boldsymbol{x})\mathrm{e}^{-\mathrm{i}\alpha f_1(\boldsymbol{x})}$ about $\boldsymbol{x}=\boldsymbol{x}_0$ and integrating term by term over the whole of space. The first term will be

$$\int_{-\infty}^{\infty} g(\boldsymbol{x}_0)\exp[-\mathrm{i}\alpha\{f(\boldsymbol{x}_0)+\tfrac{1}{2}(\boldsymbol{x}^{\mathrm{T}}-\boldsymbol{x}_0^{\mathrm{T}})A(\boldsymbol{x}-\boldsymbol{x}_0)\}]\,\mathrm{d}\boldsymbol{x}.$$

Change the origin to $\boldsymbol{x}_0$ and then rotate the axes until the quadratic form becomes $\sum_{j=1}^{n} y_j^2/\rho_j$ so that the integral is

$$g(\boldsymbol{x}_0)\,\mathrm{e}^{-\mathrm{i}\alpha f(\boldsymbol{x}_0)}\prod_{j=1}^{n}\int_{-\infty}^{\infty}\mathrm{e}^{-(\mathrm{i}/2)\alpha y_j^2/\rho_j}\,\mathrm{d}y_j$$
$$=(2\pi/\alpha)^{n/2}g(\boldsymbol{x}_0)|\rho_1\ldots\rho_n|^{1/2}\,\mathrm{e}^{-(\pi/4)\mathrm{i}\sigma-\mathrm{i}\alpha f(\boldsymbol{x}_0)} \qquad (7)$$

where σ is the signature of A. The numbers $\rho_1,\ldots,\rho_n$ can be regarded as principal radii of curvature; also $|\rho_1\ldots\rho_n|$ can be written as $1/|\det A|$.

If the next terms in the expansion of g are linear and quadratic respectively while f_1 starts off with cubic terms, then

$$g\,\mathrm{e}^{-\mathrm{i}\alpha f_1}=g(\boldsymbol{x}_0)+\sum a_j x_j+\sum a_{jk}x_j x_k+\ldots$$
$$+\alpha(\sum x_j x_k x_l+\sum x_j x_k x_l x_m)+\ldots$$

The linear terms can be ignored since they provide an odd function and therefore disappear from the integration. The quadratic terms produce an additional factor of $1/\alpha$. The cubic terms can be neglected but the quartic terms (which come from the cubic terms in f_1 and the linear in g or the quartic in f_1) produce a factor $1/\alpha$. It is easily checked that no power provides a larger factor so that, in these circumstances, the error in (7) is $O(\alpha^{-(n/2)-1})$. If g or f_1 contained fractional powers the error could be determined in the same way.

Theorem 9.14. *If, near the interior stationary point* $\boldsymbol{x}_0$,

$$f(\boldsymbol{x})\approx f(\boldsymbol{x}_0)+\tfrac{1}{2}(\boldsymbol{x}^{\mathrm{T}}-\boldsymbol{x}_0^{\mathrm{T}})A(\boldsymbol{x}-\boldsymbol{x}_0),$$

then

$$\int g(\boldsymbol{x})\,\mathrm{e}^{-\mathrm{i}\alpha f(\boldsymbol{x})}\,\mathrm{d}\boldsymbol{x}\sim\left(\frac{2\pi}{\alpha}\right)^{n/2}g(\boldsymbol{x}_0)|\rho_1\ldots\rho_n|^{1/2}\,\mathrm{e}^{-(\pi/4)\mathrm{i}\sigma-\mathrm{i}\alpha f(\boldsymbol{x}_0)}$$

as $\alpha\to\infty$.

More general forms for g and f can be dealt with in the same way. For example, if $g(\boldsymbol{x})\approx x_1^{m_1}\ldots x_n^{m_n}$ where $m_1,\ldots,m_n$ are non-negative integers and $f(\boldsymbol{x}) \approx f(\boldsymbol{x}_0) + \sum_{j=1}^{n} c_j x_j^{n_j}$ where $c_1, \ldots, c_n$ are non-zero and $n_1, \ldots, n_n$ are integers greater than 1, the dominant asymptotic term is

$$e^{-i\alpha f(\boldsymbol{x}_0)} \prod_{j=1}^{n} \frac{[\{(m_j+1)/n_j\} - 1]!}{n_j(\alpha|c_j|)^{(m_j+1)/n_j}} d_j e^{-(\pi/4)i(m_j+1)\,\mathrm{sgn}\, c_j}$$

where $d_j = 1 + (-1)^{m_j}$ (n_j even) and $d_j = 1 + (-1)^{m_j} e^{(\pi/2)i(m_j+1)\,\mathrm{sgn}\, c_j}$ (n_j odd).

An internal conical point can be handled similarly. If

$$f(\boldsymbol{x}) = f(\boldsymbol{x}_0) + \{(\boldsymbol{x}^{\mathrm{T}} - \boldsymbol{x}_0^{\mathrm{T}})A(\boldsymbol{x} - \boldsymbol{x}_0)\}^{1/2}$$

where A is positive-definite, take $\boldsymbol{x}_0$ as origin and rotate axes until the quadratic form is a sum of squares $\sum y_j^2/\rho_j$. Putting $y_j = \rho_j^{1/2} w_j$ we obtain

$$g(\boldsymbol{x}_0) e^{-i\alpha f(\boldsymbol{x}_0)} (\rho_1 \ldots \rho_n)^{1/2} \int_{-\infty}^{\infty} \exp\left\{-i\alpha\left(\sum_{j=1}^{n} w_j^2\right)^{1/2}\right\} d\boldsymbol{w}$$

$$= g(\boldsymbol{x}_0)(\rho_1 \ldots \rho_n)^{1/2} \frac{2\pi^{n/2}}{(\frac{1}{2}n - 1)!} e^{-i\alpha f(\boldsymbol{x}_0)} \int_0^{\infty} r^{n-1} e^{-i\alpha r}\, dr$$

on introducing spherical polar coordinates. There follows

Theorem 9.15. *If at an interior point*

$$f(\boldsymbol{x}) \approx f(\boldsymbol{x}_0) + \{(\boldsymbol{x}^{\mathrm{T}} - \boldsymbol{x}_0^{\mathrm{T}})A(\boldsymbol{x} - \boldsymbol{x}_0)\}^{1/2}$$

then

$$\int g(\boldsymbol{x}) e^{-i\alpha f(\boldsymbol{x})}\, d\boldsymbol{x}$$

$$\sim (\tfrac{1}{2}n - \tfrac{1}{2})! 2^n \pi^{(n/2)-1/2} g(\boldsymbol{x}_0)(\rho_1 \ldots \rho_n)^{1/2} \alpha^{-n} e^{-(\pi/2)in - i\alpha f(\boldsymbol{x}_0)}$$

as $\alpha \to \infty$.

The duplication formula $z!(z-\frac{1}{2})! = (2z)!\pi^{1/2}/2^{2z}$ has been used to put the expression in the given form.

Next, we examine the effect of the boundary point at which $f(\boldsymbol{x}) =$ constant is tangential to the boundary $h(\boldsymbol{x}) = 0$. If $\boldsymbol{x}_0$ is this point, expand f and h about the point to obtain

$$h(\boldsymbol{x}) = \sum_j (\partial h/\partial x_j)_0 (x_j - x_{0j}) + \tfrac{1}{2}(\boldsymbol{x}^{\mathrm{T}} - \boldsymbol{x}_0^{\mathrm{T}})B(\boldsymbol{x} - \boldsymbol{x}_0),$$

$$f(\boldsymbol{x}) = f(\boldsymbol{x}_0) + \sum_j (\partial f/\partial x_j)_0 (x_j - x_{0j}) + \tfrac{1}{2}(\boldsymbol{x}^{\mathrm{T}} - \boldsymbol{x}_0^{\mathrm{T}})A(\boldsymbol{x} - \boldsymbol{x}_0)$$

when only second orders are retained. The elements of A and B are the second partial derivatives of f and h respectively at $\boldsymbol{x}_0$. The condition of tangency implies that $(\partial h/\partial x_j)_0/(\partial f/\partial x_j)_0 = 1/k$ where k is independent of j. It can always be arranged that $k > 0$, by changing the sign of h if necessary. Therefore $k = |\operatorname{grad} f|_0/|\operatorname{grad} h|_0$. Let $(\partial h/\partial x_j)_0 = k_1 b_j$ where $k_1 = |\operatorname{grad} h|_0$; then $(\partial f/\partial x_j)_0 = kk_1 b_j$.

The object of the following manoeuvres is to produce a suitable system of coordinates in which $h = 0$ becomes $X_1 = 0$ and the positive X_1 axis is inside D. Make a rotation of axes $\boldsymbol{y} = L(\boldsymbol{x} - \boldsymbol{x}_0)$ of which the first line is to be $y_1 = \sum_{j=1}^n b_j(x_j - x_{0j})$ if $\sum_{j=1}^n b_j(x_j - x_{0j}) > 0$ corresponds to the inside of D; in the contrary case take $y_1 = -\sum_{j=1}^n b_j(x_j - x_{0j})$. It will be assumed that the first alternative is valid. Then

$$h(\boldsymbol{x}) = k_1 y_1 + \tfrac{1}{2}\boldsymbol{y}^{\mathrm{T}} LBL^{\mathrm{T}} \boldsymbol{y}, \tag{8}$$
$$f(\boldsymbol{x}) = f(\boldsymbol{x}_0) + kk_1 y_1 + \tfrac{1}{2}\boldsymbol{y}^{\mathrm{T}} LAL^{\mathrm{T}} \boldsymbol{y}.$$

Define $\boldsymbol{X}$ by $k_1 X_1 = h(\boldsymbol{x})$ and $X_j = y_j (j \neq 1)$. To express y_1 in terms of $\boldsymbol{X}$ it is necessary to invert equation (8). To the degree of approximation worked to here $y_1 = X_1 - \frac{1}{2}\boldsymbol{X}^{\mathrm{T}} LBL^{\mathrm{T}} \boldsymbol{X}/k_1$ and then

$$f(\boldsymbol{x}) = f(\boldsymbol{x}_0) + kk_1 X_1 + \tfrac{1}{2}\boldsymbol{X}^{\mathrm{T}} LCL^{\mathrm{T}} \boldsymbol{X}$$

where $C = A - kB$. Thus

$$\int g(\boldsymbol{x})\,\mathrm{e}^{-\mathrm{i}\alpha f(\boldsymbol{x})}\,\mathrm{d}\boldsymbol{x} \sim g(\boldsymbol{x}_0)\,\mathrm{e}^{-\mathrm{i}\alpha f(\boldsymbol{x}_0)} \int \exp[-\mathrm{i}\alpha\{kk_1 X_1 + \tfrac{1}{2}\boldsymbol{X}^{\mathrm{T}} LCL^{\mathrm{T}} \boldsymbol{X}\}]\,\mathrm{d}\boldsymbol{X} \tag{9}$$

the integration being over $X_1 \geq 0$. The reader will easily check that the terms involving X_1 in the quadratic form give a smaller contribution than the linear term in X_1. Consequently

$$\int g(\boldsymbol{x})\,\mathrm{e}^{-\mathrm{i}\alpha f(\boldsymbol{x})}\,\mathrm{d}\boldsymbol{x} \sim \frac{-\mathrm{i}(2\pi)^{(n/2)-1/2}}{kk_1\alpha^{(n/2)+1/2}} g(\boldsymbol{x}_0)|\rho_1 \cdots \rho_{n-1}|^{1/2}\,\mathrm{e}^{-(\pi/4)\mathrm{i}\sigma - \mathrm{i}\alpha f(\boldsymbol{x}_0)}$$

where $\rho_1, \ldots, \rho_{n-1}$ are the principal radii of curvature derived from the quadratic form $\boldsymbol{X}^{\mathrm{T}} LCL^{\mathrm{T}} \boldsymbol{X}$ with $X_1 = 0$ and σ is the corresponding signature.

To relate $\rho_1, \ldots, \rho_{n-1}$ more conveniently to f and h, note that the quadratic form is $y^T LCL^T y$ correct to the second order, subject to $y_1 = 0$ or, alternatively, $(x^T - x_0^T)C(x - x_0)$ subject to $b^T(x - x_0) = 0$, b having components $b_1, \ldots, b_n$. Hence $\rho_1, \ldots, \rho_{n-1}$ are the principal radii of curvature of $b^T(x - x_0) = \frac{1}{2}(x^T - x_0^T)C(x - x_0)$ at $x = x_0$. In other words $1/\rho_1 \ldots \rho_{n-1}$ is the Gaussian curvature of this hypersurface.

An explicit formula may be obtained as follows. Transfer the origin to x_0. Then $x^T Cx$ is stationary subject to $x^T x = 1$ and $b^T x = 0$ if

$$Cx - \lambda x - \mu b = 0.$$

Then $x^T Cx = \lambda$ in view of the conditions on x. Hence the values of λ correspond to $1/\rho_1, \ldots$ Now

$$x = (C - \lambda I)^{-1} \mu b$$

and therefore

$$b^T(C - \lambda I)^{-1} b = b^T x/\mu = 0$$

which is the equation to determine λ. The coefficient of λ^{n-1} is $(-1)^{n-1} b^T b = (-1)^{n-1}$ and the coefficient of λ^0 is $\sum_i \sum_j C_{ij} b_i b_j$ where $C_{ji} = C_{ij}$ is the co-factor of $c_{ij} = a_{ij} - kb_{ij}$ in det C. Hence the Gaussian curvature $1/\rho_1 \ldots \rho_{n-1}$ is $\sum_{i=1}^n \sum_{j=1}^n C_{ij} b_i b_j$.

Hence *the contribution of a point of tangential contact is given by*

$$\int g(x) e^{-i\alpha f(x)} dx \sim \frac{-i(2\pi)^{(n/2)-1/2} g(x_0) e^{-(\pi/4)i\sigma - i\alpha f(x_0)}}{|\sum_{i=1}^n \sum_{j=1}^n C_{ij} b_i b_j|^{1/2} |\mathrm{grad} f|_0 \alpha^{(n/2)+1/2}}. \tag{10}$$

In the particular case when $n = 2$, $C_{11} = a_{22} - kb_{22}$, $C_{22} = a_{11} - kb_{11}$ and $C_{12} = C_{21} = kb_{12} - a_{12}$. Then

$$\left| \sum_{i=1}^{2} \sum_{j=1}^{2} C_{ij} b_i b_j \right|$$
$$= |(a_{22} - kb_{22})b_1^2 - 2(a_{12} - kb_{12})b_1 b_2 + (a_{11} - kb_{11})b_2^2|$$
$$= |(1/\rho') - (1/\rho_0)| \, |\mathrm{grad} f|_0$$

where ρ_0 and ρ' are the radii of curvature of $f(x) = f(x_0)$ and $h = 0$ at x_0. If therefore the two curves have the same curvature (including sign) at x_0 (10) will break down and the integral will be asymptotically larger.

In general, if one of $\rho_1, \ldots, \rho_{n-1}$ is infinite, the same thing happens and cubic (or higher) terms must be retained in the expansion of f. But in that case similar terms must also be kept in h, (8) and in the expression for y_1 in terms $\boldsymbol{X}$. The corresponding terms will appear in (9). They can be handled as in preceding integrals but the results are more cumbersome.

It might happen that the boundary and $f(\boldsymbol{x}) = f(\boldsymbol{x}_0)$ coincide over a neighbourhood of $\boldsymbol{x}_0$. Then (10) will no longer be valid. The same method can still be applied but in general the integrals with respect to $X_2, \ldots, X_n$ will become integrals over the region of coincidence of $f(\boldsymbol{x}) = f(\boldsymbol{x}_0)$ and the boundary. (Cf. Theorem 9.8 when $\mu_1 = \mu_2 = 0$.)

With regard to the effect caused by lack of smoothness of the boundary we shall only consider what happens at a conical point. Take the origin at the conical point and suppose that, after a rotation of axes if necessary, the equation of the boundary takes the form $y_1 = \rho_1^{1/2}(\sum_{j=2}^n y_j^2/\rho_j)^{1/2}$ where $\rho_1, \ldots, \rho_n$ are all positive and the interior of D corresponds to $y_1 < \rho_1^{1/2}(\sum_{j=2}^n y_j^2/\rho_j)^{1/2}$. Assume that in the same coordinates $f(\boldsymbol{x}) = f(\boldsymbol{x}_0) + \boldsymbol{a}^{\mathrm{T}}\boldsymbol{y} + \ldots$, the dots indicating higher order terms. Put $y_j = \rho_j^{1/2} w_j$; then

$$\int g(\boldsymbol{x})\mathrm{e}^{-\mathrm{i}\alpha f(\boldsymbol{x})}\,\mathrm{d}\boldsymbol{x} \sim g(\boldsymbol{x}_0)\mathrm{e}^{-\mathrm{i}\alpha f(\boldsymbol{x}_0)}(\rho_2 \ldots \rho_n)^{1/2}\int \exp\{-\mathrm{i}\alpha \textstyle\sum a_j\rho_j^{1/2}w_j\}\,\mathrm{d}\boldsymbol{w}$$

taken over $w_1 \le (\sum_{j=2}^n w_j^2)^{1/2}$. The integral can be evaluated by first integrating over the sphere of radius w_1 in $w_2, \ldots, w_n$-space and then integrating with respect to w_1 from 0 to ∞. Thus the integral becomes (cf. Theorem 7.30)

$$\int_0^\infty \exp\{-\mathrm{i}\alpha a_1\rho_1^{1/2}w_1\}$$

$$\times \int_0^{w_1} (2\pi)^{(n/2)-1/2} w^{(n/2)-1/2}\frac{J_{(n-3)/2}(\alpha\rho w)\,\mathrm{d}w\,\mathrm{d}w_1}{(\alpha\rho)^{(n/2)-3/2}}$$

where $\rho = |\sum_{j=2}^n a_j^2\rho_j|^{1/2}$. Since $\int^z t^{\nu+1}J_\nu(t)\,\mathrm{d}t = z^{\nu+1}J_{\nu+1}(x)$, the integral is

$$(2\pi/\alpha\rho)^{(n/2)-1/2}\int_0^\infty \exp\{-\mathrm{i}\alpha a_1\rho_1^{1/2}w_1\}w_1^{(n/2)-1/2}J_{(n-1)/2}(\alpha\rho w_1)\,\mathrm{d}w_1$$
$$= \frac{(\frac{1}{2}n-1)!\,2^{n-1}\pi^{(n/2)-1}}{(\alpha\rho+\alpha a_1\rho_1^{1/2}-\mathrm{i}0)^{n/2}(\alpha\rho-\alpha a_1\rho_1^{1/2}+\mathrm{i}0)^{n/2}}$$

from Exercise 36 of Chapter 4. Hence, *if the boundary takes the form*

$$y_1 = \rho_1^{1/2}\left(\sum_{j=2}^{n} y_j^2/\rho_j\right)^{1/2}, \qquad f(\boldsymbol{x}) = f(\boldsymbol{x}_0) + \boldsymbol{a}^{\mathrm{T}}\boldsymbol{y} + \dots$$

and $\left|\sum_{j=2}^{n} a_j^2 \rho_j\right| \neq a_1^2 \rho_1$ *the contribution of the boundary point is given by*

$$\int g(\boldsymbol{x})\mathrm{e}^{-\mathrm{i}\alpha f(\boldsymbol{x})}\mathrm{d}\boldsymbol{x} \sim \frac{(\frac{1}{2}n-1)!\,2^{n-1}\pi^{(n/2)-1}}{\alpha^n\left\{\sum_{j=2}^{n} a_j^2\rho_j - a_1^2\rho_1\right\}^{n/2}} \tag{11}$$

where the radical is defined as $\left|\sum a_j^2\rho_j - a_1^2\rho_1\right|^{n/2}$ *when* $\sum a_j^2\rho_j > a_1^2\rho_1$ *and as* $\left|a_1^2\rho_1 - \sum a_j^2\rho_j\right|^{n/2}\mathrm{e}^{(\pi/2)\mathrm{i}n\,\mathrm{sgn}\,a_1}$ *when* $\sum a_j^2\rho_j < a_1^2\rho_1$.

If $\sum a_j^2\rho_j = a_1^2\rho_1$ higher order terms in f must be taken into account.

Exercises

39. Estimate $\int_D \mathrm{e}^{-\mathrm{i}\alpha f(\boldsymbol{x})}\mathrm{d}\boldsymbol{x}$ when $n = 2$, D is the unit circle and $f = \cos x_1 \cos x_2$.

40. When $n = 3$, S consists of that part of $x_1^2/a^2 + x_2^2/b^2 + x_3^2/c^2 = 1, a > b > 0$, for which $x_3 \geq 0$. R_0 and R_1 are the distances of a point of S from $(0, 0, x_{30})$ and $(0, 0, x_{31})$ respectively. If $x_{30} > c > 0$ and $x_{31} < 0$ show that

$$\int_S \mathrm{e}^{-\mathrm{i}\alpha(R_0+R_1)}\mathrm{d}S \sim \frac{2\pi(x_{30}-c)(c-x_{31})}{\mathrm{i}\alpha(x_{30}-x_{31})}\mathrm{e}^{-\mathrm{i}\alpha(x_{30}-x_{31})}.$$

If $x_{31} > c$ the right-hand side must be changed to $-2\pi\mathrm{i}\mathrm{e}^{-\mathrm{i}\alpha(x_{30}+x_{31}-2c)}/\alpha(a'b')^{1/2}$ where $a' = (x_{30}-c)^{-1} + (x_{31}-c)^{-1} + 2c/a^2$ and $b' = (x_{30}-c)^{-1} + (x_{31}-c)^{-1} + 2c/b^2$.

41. If, in Exercise 40, S is $x_1^2/a^2 + x_2^2/b^2 + x_3^2/c^2 - (dx_1^3 + ex_3^3)/3c = 1$ with $x_3 \leq 0$ examine the asymptotic behaviour of the integral in $x_{31} > -c$. In particular, show that when $(x_{31}+c)^{-1} + (x_{30}+c)^{-1} = 2c/b^2$ the integral is asymptotic to

$$(-\tfrac{2}{3})!\,\pi^{1/2}3^{-1/6}\alpha^{-5/6}c^{-1/2}ab(a^2-b^2)^{-1/2}|e|^{-1/3}\mathrm{e}^{-\mathrm{i}\alpha(x_{30}+x_{31}+2c)-\pi\mathrm{i}/4}.$$

If $a = b$ show that the corresponding result is $O(\alpha^{-2/3})$.

42. Calculate $\int_D \mathrm{e}^{-\mathrm{i}\alpha f(\boldsymbol{x})}\mathrm{d}\boldsymbol{x}$ correct to $O(\alpha^{-2})$ when $n = 2$, D is the square $0 < x_1 < \pi, 0 < x_2 < \pi$ and $f(\boldsymbol{x}) = \sin x_1 + 2\sin x_2$.

There is also the possibility that f is stationary at a boundary point. Suppose that near such a point

$$h(\boldsymbol{x}) = \boldsymbol{b}^{\mathrm{T}}(\boldsymbol{x}-\boldsymbol{x}_0) + \tfrac{1}{2}(\boldsymbol{x}^{\mathrm{T}}-\boldsymbol{x}_0^{\mathrm{T}})B(\boldsymbol{x}-\boldsymbol{x}_0),$$
$$f(\boldsymbol{x}) = f(\boldsymbol{x}_0) + \tfrac{1}{2}(\boldsymbol{x}^{\mathrm{T}}-\boldsymbol{x}_0^{\mathrm{T}})A(\boldsymbol{x}-\boldsymbol{x}_0).$$

Rotate the axes until the quadratic form in f is a sum of squares and $\boldsymbol{y} = L(\boldsymbol{x} - \boldsymbol{x}_0)$. Then

$$h(\boldsymbol{x}) = \boldsymbol{b}^{\mathrm{T}} L^{\mathrm{T}} \boldsymbol{y} + \tfrac{1}{2}\boldsymbol{y}^{\mathrm{T}} L B L^{\mathrm{T}} \boldsymbol{y},$$

$$f(\boldsymbol{x}) = f(\boldsymbol{x}_0) + \sum_{j=1}^{n} y_j^2/\rho_j.$$

Put $y_j = |\rho_j|^{1/2} w_j$ or $\boldsymbol{y} = C\boldsymbol{w}$ and then rotate the axes in $\boldsymbol{w}$-space via $\boldsymbol{w} = L_1 \boldsymbol{u}$ until $\boldsymbol{b}^{\mathrm{T}} L^{\mathrm{T}} C\boldsymbol{w} = k_0 u_1$; then

$$h(\boldsymbol{x}) = k_0 u_1 + \tfrac{1}{2}\boldsymbol{u}^{\mathrm{T}} L_1^{\mathrm{T}} C^{\mathrm{T}} L B L^{\mathrm{T}} C L_1 \boldsymbol{u},$$

$$f(\boldsymbol{x}) = f(\boldsymbol{x}_0) + (u_1^2 + \ldots + u_p^2 - u_{p+1}^2 - \ldots - u_n^2).$$

It is assumed that the interior of D corresponds to $u_1 > 0$. Define $\boldsymbol{X}$ by $k_0 X_1 = h(\boldsymbol{x})$ and $X_j = u_j (j \neq 1)$ so that $u_1 = X_1$ correct to the first order and $f(\boldsymbol{x}) = f(\boldsymbol{x}_0) + X_1^2 + \ldots + X_p^2 - X_{p+1}^2 - \ldots - X_n^2$ correct to second order. Hence

$$\int g(\boldsymbol{x}) \mathrm{e}^{-\mathrm{i}\alpha f(\boldsymbol{x})} \mathrm{d}\boldsymbol{x}$$

$$= |\rho_1 \ldots \rho_n|^{1/2} g(\boldsymbol{x}_0) \mathrm{e}^{-\mathrm{i}\alpha f(\boldsymbol{x}_0)} \int \mathrm{e}^{-\mathrm{i}\alpha(X_1^2 + \ldots + X_p^2 - \ldots - X_n^2)} \mathrm{d}\boldsymbol{X}$$

the integration being over $X_1 \geq 0$. Consequently *the contribution of a boundary stationary point is given by*

$$\int g(\boldsymbol{x}) \mathrm{e}^{-\mathrm{i}\alpha f(\boldsymbol{x})} \mathrm{d}\boldsymbol{x} = 2^{(n/2)-1} \left(\frac{\pi}{\alpha}\right)^{n/2} g(\boldsymbol{x}_0) |\rho_1 \ldots \rho_n|^{1/2} \mathrm{e}^{-(\pi/4)\mathrm{i}\sigma - \mathrm{i}\alpha f(\boldsymbol{x}_0)}. \tag{12}$$

9.7. Inverse polynomials

This section is concerned with the asymptotic evaluation of Fourier transforms of inverse polynomials. Although these have not been defined in full generality the meaning to be attached to them will become clearer in the following.

Consider $\int_{-\infty}^{\infty} \{P(\boldsymbol{x})\}^{-1} \mathrm{e}^{-\mathrm{i}\boldsymbol{\alpha}.\boldsymbol{x}} \mathrm{d}\boldsymbol{x}$ where P is a polynomial. Rotate the axes of integration until the y_1-axis coincides with the direction of $\boldsymbol{\alpha}$. Then the integral becomes $\int_{-\infty}^{\infty} \{P_1(\boldsymbol{y})\}^{-1} \mathrm{e}^{-\mathrm{i}|\boldsymbol{\alpha}|y_1} \mathrm{d}\boldsymbol{y}$ where P_1 is the polynomial into which P is transformed by the rotation. Regard

P_1 as a polynomial in y_1 of degree m (say) with roots $y_1 = Y_1(\hat{y})$, $Y_2(\hat{y}), \ldots, Y_m(\hat{y})$ where $\hat{y}$ is the $(n-1)$-dimensional vector with components $y_2, \ldots, y_n$. For simplicity it will be assumed that $Y_1, \ldots, Y_m$ are distinct, providing simple poles of the integrand. Then the integral with respect to y_1 can be evaluated by the method of residues; only those Y which are real or have a positive imaginary part will survive. Hence the integral is

$$\sum_{j=1}^{m} -\varepsilon\pi i \int_{-\infty}^{\infty} \frac{e^{-i|\alpha|Y_j} d\hat{y}}{[\partial P_1/\partial y_1]_{y_1=Y_j}}$$

where $\varepsilon = 0, 1, 2$ according as Y_j has a positive imaginary part, is real, or has a negative imaginary part.

For large $|\alpha|$ the terms in which Y_j has a negative imaginary part are clearly exponentially damped. They will therefore be ignored from now on.

The real Y_j come within the scope of the preceding section on stationary phase and there are contributions from the points $\hat{y}_0$ where $\partial Y_j/\partial y_2 = \ldots = \partial Y_j/\partial y_n = 0$. Assuming that such a point comes within the scope of Theorem 9.14 we see that its contribution is

$$-\pi i \left(\frac{2\pi}{|\alpha|}\right)^{(n/2)-1/2} \frac{|\rho_1 \cdots \rho_{n-1}|^{1/2} \exp\{-\tfrac{1}{4}\pi i\sigma - i|\alpha| Y_j(\hat{y}_0)\}}{[\partial P_1/\partial y_1]_{y_1=Y_j, \hat{y}=\hat{y}_0}}.$$

Now $P_1(Y_j, \hat{y}) = 0$ so that $(Y_j, \hat{y}_0)$ is a point on the hypersurface $P_1 = 0$. Also

$$\frac{\partial P_1}{\partial y_1}\frac{\partial Y_j}{\partial y_k} + \frac{\partial P_1}{\partial y_k} = 0$$

for $k = 2, \ldots, n$. Since $\partial Y_j/\partial y_k$ vanishes at $\hat{y}_0$ so does $\partial P_1/\partial y_k$. It follows that $(Y_j, \hat{y}_0)$ is a point of $P_1 = 0$ at which the normal is parallel to the y_1-axis. Furthermore at this point $[\partial P_1/\partial y_1] = \boldsymbol{i}.\text{grad } P_1$, $\boldsymbol{i}$ being a unit vector in the direction of the y_1-axis.

The interpretation of this in the original coordinates is as follows. Draw the hypersurface $P = 0$. There are contributions to the asymptotic development from all points of the hypersurface at which the normal is parallel to the direction of $\boldsymbol{\alpha}$. Let $\boldsymbol{x}_s$ be a typical one.

Then

$$\int_{-\infty}^{\infty} \{P(\boldsymbol{x})\}^{-1} \mathrm{e}^{-\mathrm{i}\boldsymbol{\alpha}.\boldsymbol{x}}\, \mathrm{d}\boldsymbol{x}$$

$$\sim \sum_s \frac{-\mathrm{i}2^{(n/2)-1/2}\pi^{(n/2)+1/2}|\rho_1 \cdots \rho_{n-1}|^{1/2}\mathrm{e}^{-(\pi/4)\mathrm{i}\sigma - \mathrm{i}\boldsymbol{\alpha}.\boldsymbol{x}_s}}{|\boldsymbol{\alpha}|^{(n/2)-3/2}[\boldsymbol{\alpha}.\mathrm{grad}\, P]} \tag{13}$$

where $\rho_1, \dots, \rho_{n-1}$ *are the principal radii of curvature of* $P = 0$ *at* $\boldsymbol{x}_s$ *and* σ *is the signature.*

Of course, if one of the radii of curvature is infinite this formula is invalid. Higher order terms in $P = 0$ must then be taken into account; the appropriate results may be determined from the preceding sections.

There is an alternative point of view which is of some interest in certain circumstances. Assume that P is a polynomial in $\boldsymbol{x}$ and ξ and the integral $\int_{-\infty}^{\infty}\int_{-\infty}^{\infty}\{P(\boldsymbol{x}, \xi)\}^{-1}\mathrm{e}^{-\mathrm{i}\boldsymbol{\alpha}.\boldsymbol{x}-\mathrm{i}\alpha_0\xi}\,\mathrm{d}\boldsymbol{x}\,\mathrm{d}\xi$ with $\alpha_0 > 0$ is to be estimated. Let P be such that

$$P(\boldsymbol{x}, \xi) = \{\xi - \Xi_1(\boldsymbol{x})\}\{\xi - \Xi_2(\boldsymbol{x})\}\dots\{\xi - \Xi_m(\boldsymbol{x})\}$$

where $\Xi_1, \dots, \Xi_m$ are all real. Then

$$\int_{-\infty}^{\infty}\int_{-\infty}^{\infty}\{P(\boldsymbol{x}, \xi)\}^{-1}\mathrm{e}^{-\mathrm{i}\boldsymbol{\alpha}.\boldsymbol{x}-\mathrm{i}\alpha_0\xi}\mathrm{d}\boldsymbol{x}\,\mathrm{d}\xi = -\pi\mathrm{i}\sum_{j=1}^{m}\int_{-\infty}^{\infty}\frac{\mathrm{e}^{-\mathrm{i}(\boldsymbol{\alpha}.\boldsymbol{x}+\alpha_0\Xi_j)}}{[\partial P/\partial\xi]_{\xi=\Xi_j}}\mathrm{d}\boldsymbol{x}. \tag{14}$$

On the other hand it is easily checked that

$$\int_0^1\int_0^{t_1}\dots\int_0^{t_{m-2}} f^{(m-1)}\{x_1(1-t_1) + x_2(t_1 - t_2) + \dots$$
$$+ x_m t_{m-1}\}\mathrm{d}t_1\dots\mathrm{d}t_{m-1} = \sum_{j=1}^{m} f(x_j)\Big/\prod_{k\neq j}(x_j - x_k)$$

so that (14) can also be written as

$$\int_{-\infty}^{\infty}\int_{-\infty}^{\infty}\{P(\boldsymbol{x}, \xi)\}^{-1}\mathrm{e}^{-\mathrm{i}\boldsymbol{\alpha}.\boldsymbol{x}-\mathrm{i}\alpha_0\xi}\mathrm{d}\boldsymbol{x}\,\mathrm{d}\xi$$
$$= (-\mathrm{i})^m\pi\alpha_0^{m-1}\int_{-\infty}^{\infty}\int_0^1\dots\int_0^{t_{m-2}}\exp[-\mathrm{i}\{\boldsymbol{\alpha}.\boldsymbol{x} + \alpha_0\Xi_1(1-t_1) + \dots$$
$$+ \alpha_0\Xi_m t_{m-1}\}]\mathrm{d}\boldsymbol{x}\,\mathrm{d}t_1\dots\mathrm{d}t_{m-1}. \tag{15}$$

As a matter of fact (15) is valid even if P has multiple zeros whereas (14) holds only if the zeros are simple.

It is now evident that, provided $(|\boldsymbol{\alpha}|^2 + \alpha_0^2)^{1/2} \gg 1$, the method of stationary phase may be applied.

10
Some applications

10.1. Integral equations

An equation of the form

$$\int f_1(t)k(x,t)\,\mathrm{d}t + f_1(x) = f_2(x) \tag{1}$$

in which k (the *kernel*) and f_2 are known is called an *integral equation* for f_1. In the form stated it is known as an *integral equation of the second kind*; if the second term on the left-hand side is absent it is an *integral equation of the first kind*. If $f_2 \equiv 0$ the integral equation is said to be *homogeneous*. Integral equations in which one limit of integration is fixed and the other is x are said to be of *Volterra type*. More general equations, in which derivatives of either the first or second term of the left-hand side of (1) occur, can arise and are sometimes known as *integro-differential equations*.

In principle there is no reason why t and x in (1) should not be replaced by $\boldsymbol{t}$ and $\boldsymbol{x}$ so that what has been said is applicable to integral equations in several variables.

We shall use the same terminology for integral equations involving generalised functions in so far as it is appropriate. Generalised functions are especially suitable for integral equations in which the integral is a convolution. Consider, for example,

$$\int_{-\infty}^{\infty} g_1(t)k(x-t)\,\mathrm{d}t + g_1(x) = g_2(x) \tag{2}$$

where g_1, g_2 and k are generalised functions whose properties will be specified more precisely later on.

Note firstly that the distinction between integral equations of the first and second kinds disappears in many circumstances. For (§6.6)

$$g_1(x) = \int_{-\infty}^{\infty} g_1(t)\delta(x-t)\,\mathrm{d}t$$

so that (2) can be written

$$\int_{-\infty}^{\infty} g_1(t)k_0(x-t)\mathrm{d}t = g_2(x) \tag{3}$$

where $k_0(t) = k(t) + \delta(t)$. Conversely, if (3) were given, an integral equation of the form (2) could be obtained by putting $k_0(t) = k(t) + \delta(t)$. Therefore, so long as the operation described does not affect the class of the kernel, it is sufficient to develop the theory for the integral equation of the first kind.

Suppose now that the kernel is the Fourier transform of a fairly good function. Then (3) can be written

$$\int_{-\infty}^{\infty} G_1(\beta)\Psi(\alpha-\beta)\mathrm{d}\beta = G_2(\alpha) \tag{4}$$

where g has been altered to G to simplify the notation. On account of Theorem 6.7, this implies that

$$2\pi g_1(-x)\psi(-x) = g_2(-x) \tag{5}$$

where G_1, G_2 and Ψ are the Fourier transforms of g_1, g_2 and ψ respectively. Hence, if ψ possesses an inverse, a solution of (5) is $g_1(x) = \{\psi(x)\}^{-1} g_2(x)/2\pi$ and a solution of (4) is

$$G_1(\alpha) = \frac{1}{2\pi}\int_{-\infty}^{\infty} \{\psi(x)\}^{-1} g_2(x)\mathrm{e}^{-\mathrm{i}\alpha x}\mathrm{d}x. \tag{6}$$

If $\psi(x)$ does not vanish then it is evident that (6) is the one and only solution of (4). On the other hand, if ψ has zeros for real values of x, (6) is not the general solution of (4). For arbitrary multiples of δ and its derivatives can be added, according to the rules of §6.2, to g_1 without affecting (5).

It is also clear that (4) may have no solutions which are generalised functions. For example, if $g_2(x) = 1$ and $\psi(x) = \mathrm{e}^{-x^2/2}$, we cannot take $g_1(x) = (2\pi)^{-1}\mathrm{e}^{x^2/2}$ since this is not a generalised function. Hence

$$\int_{-\infty}^{\infty} G_1(\beta)\exp\{-\tfrac{1}{2}(\alpha-\beta)^2\}\mathrm{d}\beta = (2\pi)^{1/2}\delta(\alpha)$$

does not have a solution which is a generalised function. Similarly, if ψ is a fine function (5) cannot be consistent unless g_2 vanishes wherever ψ is identically zero; even if this is true and the inverse of ψ can be found g_1 will still be arbitrary where ψ is identically zero.

There is no really simple way of characterising the ψ so that one can be sure of a solution but in most practical cases there is usually no difficulty in determining whether there is a solution.

The homogeneous equation

$$\int_{-\infty}^{\infty} G_1(\beta)\Psi(\alpha-\beta)\mathrm{d}\beta = 0$$

obviously has no solution, other than zero, which is a generalised function when ψ does not vanish at any x. If ψ is zero for some x then G_1 will be non-zero in general.

Example 1. Consider the integral equation

$$G_1(\alpha) - a\int_{-\infty}^{\infty} \mathrm{e}^{-|\alpha-\beta|}\, G_1(\beta)\mathrm{d}\beta = G_2(\alpha),$$

a being real. Then (Exercise 31 of Chapter 6)

$$g_1(x) - 2ag_1(x)/(1+x^2) = g_2(x)$$

or

$$g_1(x) = (1+x^2)(1+x^2-2a)^{-1}g_2(x) \tag{7}$$

together with terms which may arise from the zeros of $1+x^2-2a$. If $a<\frac{1}{2}$ there are no real zeros; if $a=\frac{1}{2}$ there is a double zero at $x=0$; if $a>\frac{1}{2}$ there are two simple zeros at $x=\pm(2a-1)^{1/2}$. Hence (7) is the general solution if $a<\frac{1}{2}$ but if $a\geq\frac{1}{2}$ there must be added $C_0\delta(x)+C_1\delta'(x)$ when $a=\frac{1}{2}$ and $C_0\delta\{x+(2a-1)^{1/2}\}+C_1\delta\{x-(2a-1)^{1/2}\}$ when $a>\frac{1}{2}$, C_0 and C_1 being arbitrary.

Now $(1+x^2)(1+x^2-2a)^{-1} = 1+2a(1+x^2-2a)^{-1}$ and therefore

$$G_1(\alpha) = G_2(\alpha) + \frac{a}{(1-2a)^{1/2}}\int_{-\infty}^{\infty} G_2(\beta)\exp\{-|\alpha-\beta|(1-2a)^{1/2}\}\,\mathrm{d}\beta \qquad (a<\tfrac{1}{2}). \tag{8}$$

If $a\geq\frac{1}{2}$, $(1+x^2-2a)^{-1}$ is not fairly good and so

$$G_1(\alpha) = G_2(\alpha) + 2a\int_{-\infty}^{\infty}(1+x^2-2a)^{-1}g_2(x)\mathrm{e}^{-\mathrm{i}\alpha x}\mathrm{d}x + G_3(\alpha)$$

where

$$G_3(\alpha) = C_0 + C_1\alpha \quad (a = \tfrac{1}{2})$$

and

$$G_3(\alpha) = C_0 \exp\{i\alpha(2a-1)^{1/2}\} + C_1 \exp\{-i\alpha(2a-1)^{1/2}\} \quad (a > \tfrac{1}{2}).$$

It should be remarked that (8) would not be correct if the growth of generalised functions were not restricted at infinity. Otherwise

$$A \exp\{\alpha(1-2a)^{1/2}\} + B \exp\{-\alpha(1-2a)^{1/2}\}$$

with A and B arbitrary could be added to the right-hand side of (8).

By taking $G_2 = 0$ we see that there are no generalised functions (except 0) which satisfy $G_1(\alpha) = a\int_{-\infty}^{\infty} e^{-|\alpha-\beta|} G_1(\beta)\,d\beta$ when $a < \frac{1}{2}$.

Exercises

1. If $G_1(\alpha)$ is a generalised function such that

$$G_1(\alpha) - a\int_{-\infty}^{\infty} e^{\alpha-\beta} H(\beta-\alpha) G_1(\beta)\,d\beta = e^{-|\alpha|},$$

a being real and $a \neq 2$, prove that

$$G_1(\alpha) = \begin{cases} \{2/(2-a)\}\{e^{-\alpha} - e^{(1-a)\alpha}\}H(\alpha) & (a > 1) \\ \{2/(2-a)\}\{e^{-\alpha}H(\alpha) + e^{(1-a)\alpha}H(-\alpha)\} & (a < 1) \\ 2e^{-\alpha}H(\alpha) - \operatorname{sgn}\alpha + C & (a = 1). \end{cases}$$

What is the solution when $a = 2$?

Verify that the conventional function $e^{(1-a)\alpha}$ is a solution of the homogeneous equation.

2. Show that the generalised function satisfying

$$2aG_1(\alpha) - \int_{-\infty}^{\infty} G_1(\beta)\operatorname{sech}\tfrac{1}{2}(\alpha-\beta)\,d\beta = 0$$

is $G_1(\alpha) = C_1 + C_2\alpha$ when $a = 2$. Are there solutions for other real values of a?

If $a = 0$ and the right-hand side were replaced by $G_2(\alpha)$ what could be said about G_1?

3. If $\int_{-\infty}^{\infty} e^{-|\alpha-\beta|} G_1(\beta)\,d\beta = G_2(\alpha)$ prove that $G_1(x) = \frac{1}{2}G_2(x) - \frac{1}{2}G_2''(x)$.

4. If g_1, g_2 are even and $\lambda^2 \neq 1/2\pi$, prove that the solution of

$$g_1(x) = g_2(x) + \lambda\int_{-\infty}^{\infty} e^{-ixy} g_1(y)\,dy$$

is

$$g_1(x) = \left\{ g_2(x) + \lambda \int_{-\infty}^{\infty} e^{-ixy} g_2(y) dy \right\} / (1 - 2\pi\lambda^2).$$

5. Prove that $\mathscr{H}_n(x)$ satisifies

$$(-i)^n (2\pi)^{1/2} g_1(x) = \int_{-\infty}^{\infty} e^{-ixy} g_1(y) dy.$$

Consider now

$$g_1(x) - \int_{-\infty}^{\infty} k(x-t) g_1(t) dt = g_2(x) \tag{9}$$

where k is not necessarily the Fourier transform of a fairly good function. It will then be necessary to impose restrictions on g_1 in order that the convolution exists and this constraint, in turn, will limit the possible g_2. Suppose, for example, that k, g_1 and g_2 are all members of L_2'. Then (9) gives

$$G_1(\alpha) - K(\alpha) G_1(\alpha) = G_2(\alpha) \tag{10}$$

where G_1, KG_1 and G_2 are all members of K_2. If $1 - K(\alpha)$ has an inverse a solution is

$$G_1(\alpha) = \{1 - K(\alpha)\}^{-1} G_2(\alpha) \tag{11}$$

provided that the right-hand is in K_2. Obviously a sufficient condition for (11) to give the solution of (9) is that $K(\alpha) < 1$. For then $G_1 \in K_2$ and $g_1 \in L_2'$. The condition $K(\alpha) < 1$ must hold for all α; a counter-example is supplied by taking $K(\alpha) = 1 - e^{-\alpha^2}$ when (10) has no solution which is a generalised function, for general G_2.

Theorem 10.1. *If g_1, g_2 and $k \in L_2'$ and the upper bound of $K(\alpha)$ is less than* 1, *the solution of* (9) *is*

$$g_1(x) = \frac{1}{2\pi} \int_{-\infty}^{\infty} \frac{G_2(\alpha) e^{i\alpha x} d\alpha}{\{1 - K(\alpha)\}}.$$

The effect of zeros of $1 - K(\alpha)$ can be determined by considering the homogeneous equation so that $G_2 \equiv 0$. Then the real zeros of $1 - K(\alpha)$ give rise to a G_1 of the form $\sum_{j=0}^{r} C_j \delta^{(j)} (\alpha - \alpha_s)$. However, no terms of this type are in K_2 and therefore the corresponding g_1 is not in L_2'. Hence, in this case, the only possibility is $g_1 \equiv 0$.

Example 2. Consider the integral equation

$$\int_{-\infty}^{x} (x-t)^{\beta} g_1(t)\,\mathrm{d}t = x^{\mu}\,\mathrm{H}(x)$$

where $\operatorname{Re}\beta < -\frac{1}{2}$, $\operatorname{Re}\mu < -\frac{1}{2}$, $\operatorname{Re}(\mu-\beta) < \frac{1}{2}$ and none of β, μ, $\mu-\beta-1$ is a negative integer. Since $x^{\beta}\,\mathrm{H}(x) \in L_2'$ for $\operatorname{Re}\beta < -\frac{1}{2}$,

$$\beta!\,\mathrm{e}^{-\pi\mathrm{i}(\beta+1)/2}(\alpha-\mathrm{i}0)^{-\beta-1}G_1(\alpha) = \mu!\,\mathrm{e}^{-\pi\mathrm{i}(\mu+1)/2}(\alpha-\mathrm{i}0)^{-\mu-1} \tag{12}$$

or

$$G_1(\alpha) = (\mu!/\beta!)\mathrm{e}^{\pi\mathrm{i}(\beta-\mu)/2}(\alpha-\mathrm{i}0)^{\beta-\mu}.$$

Hence

$$g_1(x) = \frac{\mu!\,x^{\mu-\beta-1}\,\mathrm{H}(x)}{\beta!(\mu-\beta-1)!}. \tag{13}$$

This is the only solution in L_2'; if this restriction were removed $\sum_{j=0}^{r} C_j x^j$, where $r < -1-\operatorname{Re}\beta$, could be added to g_1.

Different solutions of (9) can, of course, be obtained by working with the class L_p' instead of L_2'. Thus

Example 3. If, in Example 2, the restrictions on β and μ are altered to $\operatorname{Re}\beta < -1/p$, $\operatorname{Re}(\mu-\beta) < 1/p$ $(1 < p \le 2)$ and none of β, μ, $\mu-\beta-1$ is a negative integer (12) is still valid under the assumption $g_1 \in L_q'\{q = p/(p-1)\}$, $G_1 \in K_p$ (Theorem 6.14). It then follows that (13) is also correct.

Exercises

6. If $\operatorname{Re}\beta < -\frac{1}{2}$, $\operatorname{Re}\mu < -\frac{1}{2}$, $\operatorname{Re}(\mu-\beta) < \frac{1}{2}$ and none of $\beta, \mu, \mu-\beta-1$ is a negative integer solve

$$\int_{-\infty}^{\infty} |x-t|^{\beta} g_1(t)\,\mathrm{d}t = |x|^{\mu}$$

given that $g_1 \in L_2'$.

7. Show that a solution of

$$g_1(x) = \frac{1}{\pi}\int_{-\infty}^{\infty} g_1(t)\frac{\sin a(x-t)}{x-t}\mathrm{d}t$$

such that $g \in L'_2$ is $g_1(x) = (1/2\pi)\int_{-a}^{a} G(\alpha)\mathrm{e}^{\mathrm{i}\alpha x}\,\mathrm{d}\alpha$ where $G \in L_2$. Another solution, which is not in L_2', is $g_1(x) = \sin bx$ with $-a < b < a$.

8. If k, g_1 and g_2 all belong to L_2' and the upper bound of $K(\alpha)K(-\alpha)$ is less than 1, show that the solution of

$$g_1(x) = g_2(x) + \int_{-\infty}^{\infty} k(x-t)g_1(-t)\mathrm{d}t$$

is

$$g_1(x) = \frac{1}{2\pi}\int_{-\infty}^{\infty} \frac{G_2(\alpha) + K(\alpha)G_2(-\alpha)}{1 - K(\alpha)K(-\alpha)} e^{i\alpha x}\,\mathrm{d}\alpha.$$

(The integral in the integral equation may also be written $\int_{-\infty}^{\infty} k(x+t)g_1(t)\,\mathrm{d}t$.)

9. What can be said about g_1 in Example 2 if $\mu - \beta - 1$ is a negative integer?
10. Can the restriction $\mathrm{Re}\,\mu < -\frac{1}{2}$ be dropped in Example 2?
11. Show that (13) holds if $\mathrm{Re}\,\beta < -1 + 1/p$ $(1 < p \leq 2)$, $\mathrm{Re}(\mu - \beta) < -1 + 1/p$ and none of $\beta, \mu, \mu - \beta - 1$ is a negative integer under the assumption that $g_1 \in L_p'$.
12. If $-2 < \mathrm{Re}\,\beta < -1 + 1/p$ $(1 < p \leq 2)$ and $\beta \neq -1$ prove that the solution of

$$\int_{-\infty}^{x} (x-t)^{\beta} g_1(t)\mathrm{d}t = g_2(x)$$

such that $g_1 \in L_p'$, $g_2 \in L_p'$ and both $G_1, G_2 \in K_q$ is

$$\pi g_1(x) = (\beta + 1)\sin \pi\beta \int_{-\infty}^{x} (x-t)^{-\beta-2} g_2(t)\mathrm{d}t.$$

13. If $(1 + x^2)^{-1/4} g_1(x)$ is bounded and tends to zero as $|x| \to \infty$ prove that the solution of

$$g_1(x) - \frac{1}{\pi}\int_{-\infty}^{\infty} \frac{g_1(t)\,\mathrm{d}t}{1 + (x-t)^2} = \frac{x}{1 + x^2}$$

is

$$g_1(x) = (\pi \coth \pi x - x^{-1})\sqrt{(\tfrac{1}{2}\pi)} + C$$

where C is arbitrary. (Hint: Theorem 6.16.)

10.2. Hilbert transforms

There is one type of integral equation which is of such frequent occurrence in applied mathematics that it is worth considering in some detail. It takes the form

$$\frac{1}{\pi}\int_{-\infty}^{\infty} g_1(t)(x-t)^{-1}\mathrm{d}t = g_2(x). \tag{14}$$

This has been encountered already in §6.5 Example 9 where it is shown that

$$x^{-1} = h(x) - h''(x) \tag{15}$$

where

$$h(x) = \int_0^\infty \frac{\sin \alpha x}{(1+\alpha^2)} \mathrm{d}\alpha,$$

so that $x^{-1} \in L'_p$ for $p > 1$. The left-hand side of (14) therefore exists whenever $g_1 \in L'_{p_1}$ with $1 \le p_1 < \infty$. If $g_1 \in L'_{p_1}$ and $G_1 \in K_{q_1}$ where $q_1 = p_1/(p_1 - 1)$ (the condition on G_1 is automatically satisfied if $1 \le p_1 \le 2$) the Fourier transform of (14) is

$$-\mathrm{i}\, G_1(\alpha) \operatorname{sgn} \alpha = G_2(\alpha) \tag{16}$$

by Theorem 6.14. The left-hand side is in K_{q_1} and therefore $G_2 \in K_{q_1}$ so that G_2 is the product of a polynomial in α and a function in L_{q_1}. If $2 \le p_1 < \infty$ then $1 < q_1 \le 2$ and g_2 must belong to L'_{p_1}. Since (16) involves ordinary functions it can be multiplied by sgn α and

$$-\pi G_1(\alpha) = -\pi \mathrm{i} G_2(\alpha) \operatorname{sgn} \alpha.$$

The conclusion is that, if $g_1 \in L'_p$ $(2 \le p < \infty)$ and $G_1 \in K_q$ then $g_2 \in L'_p$ and

$$g_1(t) = -\frac{1}{\pi} \int_{-\infty}^\infty (t-x)^{-1} g_2(x) \mathrm{d}x. \tag{17}$$

Alternatively, if $G_2 \in K_q (1 < q \le 2)$ the unique solution of (14) with $G_1 \in K_q$ is given by (17). The generalised functions g_1 and g_2 connected by (14) and (17) are said to be *Hilbert transforms.*

Exercises

14. Solve $(1/\pi)\int_{-\infty}^\infty (x-t)^{-1} g_1(t) \mathrm{d}t = \delta'(x)$ with $g_1 \in L'_2$.

15. If $g_2(x)$ is (i) $\mathrm{H}(a^2 - x^2)(a > 0)$, (ii) $1/(1+x^2)$, (iii) $|x|^\nu J_\nu(|x|) \operatorname{sgn} x$ $(\operatorname{Re} \nu < -\frac{1}{2})$ show that $g_1(t)$ is (i)$(1/\pi) \ln|(a+t)/(a-t)|$, (ii) $-t/(1+t^2)$, (iii) $-|t|^\nu Y_\nu(|t|)$.

16. Show that the solution of

$$\frac{1}{\pi}\int_{-\infty}^\infty (x-t)^{-1} g_1(t) \mathrm{d}t = |x|^\beta \operatorname{sgn} x,$$

where $\operatorname{Re} \beta < 0$ and β is not a negative integer, is $|x|^\beta \tan \frac{1}{2}\pi(\beta + 1)$

provided that $g_1 \in L'_2$ if $\operatorname{Re}\beta < -\frac{1}{2}$ and $g_1 \in L'_p$ where $p > -\operatorname{Re}(1/\beta)$ if if $0 > \operatorname{Re}\beta > -1$.

17. If $g_1 \in L'_p (2 \le p < \infty)$, $G_1 \in K_q$ and

$$\int_{-\infty}^{\infty} (x-t)^{-m} g_1(t)\, dt = g_2(t)$$

prove that

$$\pi^2 g_1^{(m-1)}(t) = (m-1)!(-1)^m \int_{-\infty}^{\infty} (t-x)^{-1} g_2(x)\, dx.$$

18. If $g_1 \in L'_p (1 \le p < \infty)$ and $G_1 \in K_q$ prove that the solution of

$$cg_1(x) + \int_{-\infty}^{\infty} (x-t)^{-1} g_1(t)\, dt = g_2(x)$$

is

$$g_1(t) = \frac{c}{c^2+\pi^2} g_2(t) - \frac{1}{c^2+\pi^2} \int_{\infty} (t-x)^{-1} g_2(x)\, dx$$

where c is a complex constant such that $c^2 \neq -\pi^2$.

19. If $g \in L'_p (1 \le p \le \infty)$ and $G \in K_q$ prove that

$$cg(x) + \int_{-\infty}^{\infty} (x-t)^{-1} g(t)\, dt = 0$$

is satisfied, when $c = \pi i$, by any g such that $G = 0$ for $\alpha < 0$ and, when when $c = -\pi i$, by any g such that $G = 0$ for $\alpha > 0$.

The omission of the possibility $g_1 \in L'_\infty$ from (17) is due to a genuine difference between the connection between Hilbert transforms when $g_1 \in L'_p$ $(p < \infty)$ and when $g_1 \in L'_\infty$. Indeed it is not obvious that $\int_{-\infty}^{\infty} (x-t)^{-1} g_1(t)\, dt$ has a meaning when $g_1 \in L'_\infty$ (e.g. consider $g_1(t) = \operatorname{sgn} t$). Suppose that it does have; then it is the derivative, of some order, of $\int_{-\infty}^{\infty} (x-t)^{-1} f(t)\, dt$ where $f \in L_\infty$. Now

$$\int_{-\infty}^{\infty} (x-t)^{-1} f(t)\, dt = (x-i) \int_{-\infty}^{\infty} \frac{f(t)}{t-i} (x-t)^{-1}\, dt - \int_{-\infty}^{\infty} \frac{f(t)}{t-i}\, dt.$$

Since f is bounded, $f(t)/(t-i) \in L_p (p > 1)$ and the first term on the right-hand side has a meaning by what has gone before. Hence we must have $\left|\int_{-\infty}^{\infty} \{f(t)/(t-i)\}\, dt\right| < \infty$. On the other hand we can reverse the argument. Hence, *a necessary and sufficient condition that* $\int_{-\infty}^{\infty} (x-t)^{-1} g_1(t)\,\mathrm{d}t$ *exists is that*

$$\left| \int_{-\infty}^{\infty} \frac{f(t)}{t-\mathrm{i}} \mathrm{d}t \right| < \infty. \tag{18}$$

The integral here is understood in the sense of §6.1, Exercise 5, i.e.

$$\int_{-\infty}^{\infty} \frac{f(t)}{t-\mathrm{i}} \mathrm{d}t = \lim_{M\to\infty} \int_{-M}^{M} \frac{f(t)}{t-\mathrm{i}} \mathrm{d}t.$$

It is now convenient to introduce L''_∞, a subspace of L'_∞.

Definition 10.1. *The generalised function* $g \in L''_\infty$ *if, and only if,* $g = \sum_{m=1}^{M} f_m^{(r_m)}$ *where* $f_m \in L_\infty$ *and*

$$\left| \int_0^a \{f_m(t+x) - f_m(x-t)\}\,\mathrm{d}t \right| < K|a|^\mu \qquad (m = 1, \ldots, M) \tag{19}$$

for all real a, where M is finite, $0 < \mu < 1$ *and K is independent of a and x.*

Since $f \in L_\infty$ it is bounded and so, if $|a| < a_0$,

$$\left| \int_0^a \{f(t+x) - f(x-t)\}\,\mathrm{d}t \right| < B|a| < Ba_0^{1-\mu}|a|^\mu.$$

Therefore the condition in Definition 10.1 is satisfied for bounded a when $f \in L_\infty$. Thus the effect of this condition only applies for large a and imposes certain limits on the behaviour of f at infinity. These limits are not very restrictive since $f(x) = \mathrm{e}^{\mathrm{i}x}$ satisfies the condition. Also, if $f(t)$ and $f(-t)$ both have the same constant value in a neighbourhood of $t = \infty$ the condition is complied with; it fails, however, if $f(t)$ and $f(-t)$ have different constant values.

It can now be shown that, if $g \in L''_\infty$, $\int_{-\infty}^{\infty} (x-t)^{-1} g(t)\,\mathrm{d}t$ has a meaning. It is clearly sufficient to show that (18) is valid for an f satisfying Definition 10.1. Now

$$\int_{-M}^{M} \frac{f(t)}{t-\mathrm{i}} \mathrm{d}t = \int_0^M \frac{t}{t^2+1} \{f(t) - f(-t)\}\,\mathrm{d}t$$

$$= \left[\frac{t f_1(t)}{t^2+1} \right]_0^M - \int_0^M \frac{1-t^2}{(1+t^2)^2} f_1(t)\,\mathrm{d}t$$

where $f_1(t) = \int_0^t \{f(u) - f(-u)\}\,\mathrm{d}u$. From (19) with $x = 0$, $|f_1(t)| <$

$K|t|^\mu$ so that

$$\left|\int_{-M}^{M}\frac{f(t)}{t-\mathrm{i}}\mathrm{d}t\right| < KM^{\mu-1}+K\int_0^\infty \frac{|1-t^2|t^\mu}{(1+t^2)^2}\mathrm{d}t.$$

Allowing $M\to\infty$ and recalling that $0<\mu<1$ we see that (18) holds.

The next stage in dealing with (14) consists in showing that $g_2\in L''_\infty$ when $g_1\in L''_\infty$. It is obviously sufficient to show that

$$\int_{-\infty}^{\infty}(x-t)^{-1}f(t)\mathrm{d}t\in L''_\infty$$

when f satisfies (19). On account of (15) this will be true if

$$\int_{-\infty}^{\infty}h(x-t)f(t)\mathrm{d}t\in L''_\infty.$$

Now

$$\int_{-M}^{M}f(x-t)h(t)\,\mathrm{d}t=\int_0^M\{f(x-t)-f(x+t)\}h(t)\mathrm{d}t$$

since $h(t)$ is an odd function of t. Integration by parts gives

$$\int_{-M}^{M}f(x-t)h(t)\mathrm{d}t=[h(t)f_2(x,t)]_{-M}^{M}-\int_{-M}^{M}h'(t)f_2(x,t)\mathrm{d}t \quad (20)$$

where $f_2(x,t)=\int_0^t\{f(x-u)-f(x+u)\}\,\mathrm{d}u$. On account of (19) $|f_2|<K|t|^\mu$. Also, from the explicit formula for h,

$$h(x)\approx -x\ln|x|+Ax+O(x^2),\quad h'(x)\approx -\ln|x|+O(1)\qquad (x\approx 0)$$
$$h(x)=O(1/|x|),\quad h'(x)=O(1/|x|^2)\qquad (|x|\to\infty).$$

From this it is evident that the first term on the right-hand side of (20) tends to zero as $M\to\infty$ and that the magnitude of the second term does not exceed $K\int_{-\infty}^{\infty}|t|^\mu|h'(t)|\,\mathrm{d}t<\infty$. Thus $\int_{-\infty}^{\infty}f(x-t)h(t)\,\mathrm{d}t\in L_\infty$ and therefore $g_2\in L'_\infty$. Further

$$\int_0^a\int_{-\infty}^{\infty}f(u)\{h(t+x-u)-h(x-t-u)\}\,\mathrm{d}u\,\mathrm{d}t$$
$$=\int_0^a\int_0^\infty h(u)f_3(x,t,u)\,\mathrm{d}u\,\mathrm{d}t$$

where

$$f_3(x,t,u)=f(t+x-u)-f(t+x+u)-f(x-t-u)+f(x-t+u).$$

Now if $f_4(x,u) = \int_0^\alpha f_3(x,t,u)\mathrm{d}t$ the last integral can be written as

$$\int_0^\infty h(u)f_4(x,u)\mathrm{d}u = [h(u)f_5(x,u)]_0^\infty - \int_0^\infty h'(u)f_5(x,u)\mathrm{d}u \quad (21)$$

where $f_5(x,u) = \int_0^u f_4(x,v)\mathrm{d}v$. From (19), $|f_4(x,u)| < K|a|^\mu$ and so

$$|f_5(x,u)| < K|u|\,|a|^\mu. \quad (22)$$

But

$$|f_5(x,u)| = \left|\int_0^a \int_0^u f_3(x,t,v)\,\mathrm{d}v\,\mathrm{d}t\right| < \int_0^{|a|} K|u|^\mu \mathrm{d}t < K|u|^\mu|a|. \quad (23)$$

It follows that the first term in (21) vanishes; in the second term we use (22) in the interval $(0, |a|)$ and (23) in $(|a|, \infty)$. Then

$$\left|\int_0^\infty h(u)f_4(x,u)\mathrm{d}u\right| < K|a|^\mu \int_0^{|a|} u|h'(u)|\mathrm{d}u + K_1|a| \int_{|a|}^\infty u^{\mu-2}\mathrm{d}u$$
$$< K|a|^\mu \ln|a| + K_1|a|^\mu$$
$$< K|a|^{\mu'}$$

where $0 < \mu' < 1$. The proof that $g_2 \in L''_\infty$ is concluded.

Return now to the integral equation (14). If $g_1 \in L'_\infty$ then clearly $(t-\mathrm{i})^{-1}g_1(t) \in L'_p$ for $p > 1$ and, in particular, this is true if $g_1 \in L''_\infty$. Therefore $(1/\pi)\int_{-\infty}^\infty (t-\mathrm{i})^{-1}g_1(t)(x-t)^{-1}\mathrm{d}t$ is a generalised function. Hence $(1/\pi)\int_{-\infty}^\infty g_1(t)(x-t)^{-1}\mathrm{d}t - (1/\pi)(x-\mathrm{i})\int_{-\infty}^\infty (t-\mathrm{i})^{-1}g_1(t)$ $(x-t)^{-1}\mathrm{d}t$ or $-(1/\pi)\int_{-\infty}^\infty (t-\mathrm{i})^{-1}g_1(t)\mathrm{d}t$ must exist when $g_1 \in L''_\infty$. Consequently (14) can be expressed as

$$\frac{1}{\pi}(x-\mathrm{i})\int_{-\infty}^\infty (t-\mathrm{i})^{-1}g_1(t)(x-t)^{-1}\mathrm{d}t - A = g_2(x) \quad (24)$$

where $A = (1/\pi)\int_{-\infty}^\infty (t-\mathrm{i})^{-1}g_1(t)\mathrm{d}t$. Since $(t-\mathrm{i})^{-1}g_1 \in L'_p (p > 1)$ it follows that the solution of (24) is given by (17), i.e.

$$(t-\mathrm{i})^{-1}g_1(t) = -\frac{1}{\pi}\int_{-\infty}^\infty (t-x)^{-1}\frac{g_2(x)+A}{x-\mathrm{i}}\mathrm{d}x$$
$$= -\frac{1}{\pi}\int_{-\infty}^\infty \frac{(t-x)^{-1}}{x-\mathrm{i}}g_2(x)\mathrm{d}x - \frac{\mathrm{i}A}{t-\mathrm{i}}.$$

Integration with respect to t reveals that it is necessary that

$$A = -\frac{1}{\pi^2}\int_{-\infty}^{\infty}\int_{-\infty}^{\infty}\frac{(t-x)^{-1}}{x-\mathrm{i}}g_2(x)\mathrm{d}x\,\mathrm{d}t - \mathrm{i}A\int_{-\infty}^{\infty}\frac{\mathrm{d}t}{t-\mathrm{i}}$$

or

$$\int_{-\infty}^{\infty}\int_{-\infty}^{\infty}\frac{(t-x)^{-1}}{x-\mathrm{i}}g_2(x)\mathrm{d}x\,\mathrm{d}t = 0 \tag{25}$$

since $\int_{-\infty}^{\infty}(t-\mathrm{i})^{-1}\,\mathrm{d}t = \pi\mathrm{i}$.

Since $g_2 \in L''_\infty$ we may use the same argument for the splitting of $\int_{-\infty}^{\infty}(t-x)^{-1}g_2(x)\mathrm{d}x$ as was used for g_1. Thus, if (25) is satisfied,

$$g_1(t) = -\frac{1}{\pi}\int_{-\infty}^{\infty}(t-x)^{-1}g_2(x)\,\mathrm{d}x + B$$

when $B = -(1/\pi)\int_{-\infty}^{\infty}(x-\mathrm{i})^{-1}g_2(x)\mathrm{d}x + A$. In fact since A is indeterminate, B is arbitrary. This is also obvious because adding a constant to g_1 leaves it in L''_∞ and does not alter (14) since $\int_{-\infty}^{\infty}(x-t)^{-1}\,\mathrm{d}t = 0$.

Summarising we have

Theorem 10.2. *If*

$$\int_{-\infty}^{\infty}\int_{-\infty}^{\infty}\frac{(t-x)^{-1}}{x-\mathrm{i}}g_2(x)\mathrm{d}x\,\mathrm{d}t \neq 0$$

the integral equation

$$\frac{1}{\pi}\int_{-\infty}^{\infty}g_1(t)(x-t)^{-1}\,\mathrm{d}t = g_2(x)$$

has no solution with $g_1 \in L''_\infty$, $g_2 \in L''_\infty$. If

$$\int_{-\infty}^{\infty}\int_{-\infty}^{\infty}\frac{(t-x)^{-1}}{x-\mathrm{i}}g_2(x)\mathrm{d}x\,\mathrm{d}t = 0$$

the solution is

$$g_1(t) = -\frac{1}{\pi}\int_{-\infty}^{\infty}(t-x)^{-1}g_2(x)\mathrm{d}x + B$$

where B is an arbitrary constant.

Exercises

20. Show that Theorem 10.2 is still valid if $x-\mathrm{i}$ is replaced by $x-a\mathrm{i}$, a being real and non-zero.

21. If $g \in L'_p$ where $1 \le p < \infty$ show that $g \in L''_\infty$.
22. If $g_1, g_2 \in L''_\infty$, $\int_{-\infty}^{\infty} \int_{-\infty}^{\infty} (t-x)^{-1}(x-\mathrm{i})^{-1} g_2(x)\mathrm{d}x\,\mathrm{d}t = 0$ and

$$\frac{1}{\pi}\int_{-\infty}^{\infty} g_1(t)(x-t)^{-m}\mathrm{d}t = g_2(x)$$

show that

$$g_1^{(m-1)}(t) = (m-1)!\frac{(-1)^m}{\pi}\int_{-\infty}^{\infty}(t-x)^{-1} g_2(x)\mathrm{d}x + B$$

where B is an arbitrary constant.
23. If $g_2(x)$ is (i) $\cos x$, (ii) $|x|^\nu J_\nu(|x|)\,\mathrm{sgn}\,x$ $(-\frac{1}{2} \le \mathrm{Re}\,\nu \le \frac{1}{2})$ and $g_1 \in L''_\infty$ show that $g_1(t)$ is (i) $B - \sin t$, (ii) $B - |t|^\nu Y_\nu(|t|)$.
24. If $g_2(x) = \text{constant} \ne 0$ show that (14) has no solution with $g_1 \in L''_\infty$.
25. Examine the validity of the following scheme: If

$$\int_0^\infty (x-t)^{-1} g_1(t)\mathrm{d}t = g_2(x) \quad (x > 0)$$

put $t = u^2, x = y^2$ and then use $2y(y^2-u^2)^{-1} = (y-u)^{-1} + (y+u)^{-1}$ to provide a Hilbert kernel for $2ug_1(u^2)\,\mathrm{sgn}\,u$. Thus

$$\pi^2 u g_1(u^2)\,\mathrm{sgn}\,u = -\int_0^\infty 2y^2(u^2-y^2)^{-1} g_2(y^2)\mathrm{d}y$$

or

$$g_1(t) = -\frac{1}{\pi^2 t^{1/2}}\int_0^\infty (t-x)^{-1} x^{1/2} g_2(x)\mathrm{d}x.$$

What is the effect of adding $C\delta'(x)$ to g_2?
26. Do an investigation of

$$\int_0^1 (x-t)^{-1} g_1(t)\mathrm{d}t = g_2(x) \quad (0 < x < 1)$$

similar to Exercise 25, making the substitution $x = (y+1)^{-1}, t = (u+1)^{-1}$ leading to

$$g_1(t) = \frac{-t^{1/2}}{\pi^2(1-t)^{1/2}}\int_0^1 (t-x)^{-1}\left(\frac{1-x}{x}\right)^{1/2} g_2(x)\mathrm{d}x.$$

What is the effect of making the transformation $x = 1 - z, t = 1 - v$ first?

It is sometimes convenient, and especially for applications in electrical engineering, to work in the complex plane. Consider

$$f(z) = \int_{-\infty}^{\infty} (z-t)^{-1} g(t)\mathrm{d}t \qquad (\mathrm{Im}\,z > 0)$$

where $g \in L'_p$ $(1 \le p < \infty)$. If $f_0 \in L_p$, $\int_{-\infty}^{\infty} (z-t)^{-m} f_0(t)\,dt$ exists for $m = 1, 2 \ldots$ because $(z-t)^{-m} \in L_q$. Thus $\int_{-\infty}^{\infty} (z-t)^{-1} f_0(t)dt$ is analytic at all points $\operatorname{Im} z > 0$ and it follows, by taking derivatives, that $f(z)$ is a function which is analytic at all points $\operatorname{Im} z > 0$.

If $g_1 \in L'_p$ $(1 \le p < \infty)$ and $G_1 \in K_q$ the Fourier transform of $f_1(z)$ is, from Theorem 4.5, $-2\pi i e^{-\alpha y}\,H(\alpha)G_1(\alpha)$ where $y = \operatorname{Im} z$. This may also be written as $-2\pi i e^{-\alpha y} H(\alpha) G_1(\alpha) \operatorname{sgn} \alpha$ or $-2\pi i e^{-\alpha y}$ $H(\alpha) i G_2(\alpha)$ which is the Fourier transform of $i\int_{-\infty}^{\infty} (z-t)^{-1} g_2(t)dt$. Hence

$$\int_{-\infty}^{\infty} (z-t)^{-1} g_1(t)dt = i\int_{-\infty}^{\infty} (z-t)^{-1} g_2(t)dt \qquad (\operatorname{Im} > 0) \tag{26}$$

where $\pi g_2(x) = \int_{-\infty}^{\infty} (x-t)^{-1} g_1(t)dt$.

In particular (26) implies when $z = iy$ that

$$\pi i \int_{-\infty}^{\infty} (iy-t)^{-1} g_2(t)dt + \int_{-\infty}^{\infty} (iy-t)^{-1} \int_{-\infty}^{\infty} (t-x)^{-1} g_2(x)dx = 0$$

which is another way of writing

$$\int_{-\infty}^{\infty}\int_{-\infty}^{\infty} \frac{(t-x)^{-1}}{x-iy} g_2(x)dt = 0.$$

This can be identified with (25) on putting $y = 1$ and shows, together with Exercise 21, that the solution (17) of (14) is consistent with Theorem 10.2 when $g_1 \in L'_p$ $(1 \le p < \infty)$, the condition necessary for a solution being automatically complied with.

Since $\operatorname{Lim}_{y \to +0} (z-t)^{-1} = (x + i0 - t)^{-1} = (x-t)^{-1} - \pi i \delta(x-t)$ from (16) of Chapter 4, we deduce that, when $g_1 \in L'_p$ $(1 \le p < \infty)$

$$\lim_{y \to +0} \int_{-\infty}^{\infty} (z-t)^{-1} g_1(t)dt = \int_{-\infty}^{\infty} (x-t)^{-1} g_1(t)dt - \pi i g_1(x)$$
$$= -\pi i\{g_1(x) + i g_2(x)\}$$

on account of §6.6. This demonstrates the relation between Hilbert transforms and a function analytic in $\operatorname{Im} z > 0$.

Exercises

27. If $g_2 \in L''_\infty$ prove that

$$\int_{-\infty}^{\infty} \frac{g_2(t)}{t-z}\mathrm{d}t = \frac{\mathrm{i}}{\pi}\int_{-\infty}^{\infty}\frac{\mathrm{d}t}{z-t}\int_{-\infty}^{\infty}(t-u)^{-1}g_2(u)\mathrm{d}u$$

$$-\frac{1}{\pi \mathrm{i}}\int_{-\infty}^{\infty}\int_{-\infty}^{\infty}\frac{(t-u)^{-1}g_2(u)}{u-\mathrm{i}}\mathrm{d}u\,\mathrm{d}t$$

when $\operatorname{Im} z > 0$. Deduce the result corresponding to (26) when (25) is satisfied.

28. If $f(z) = (\mathrm{i}/\pi)\int_{-\infty}^{\infty} g_1(t)/(z-t)\,\mathrm{d}t$ and $g_1(t) = g_1(-t)$ prove that

$$f(z) = \frac{\mathrm{i}}{\pi}\int_{-\infty}^{\infty}\frac{z g_1(t)}{z^2-t^2}\mathrm{d}t.$$

Deduce that, if g_1 is real, $\overline{f(z)} = f(-\bar{z})$, the bar indicating a complex conjugate.

More generally, show that $\overline{f(z)} = f(-\bar{z})$ ($\operatorname{Im} z \neq 0$) if, and only if,

$$\overline{g_1(t)} = g_1(-t)$$

(cf. §10.6).

29. Determine $f(z)$ when $g_1(t)$ is (i) $\delta(t)$, (ii) $|t|^{-1/2}$, (iii) $1/(t^2+1)$.

30. In the *theories of coherence and communication* a *real signal* g_1 is such that $g_1 \in L'_p$ $(1 \le p < \infty)$, $G_1 \in K_q$ and g_1 is real. The corresponding *complex analytic signal* g_2 is defined by $g_2(x) = (1/\pi)\int_{-\infty}^{\infty} G_1(\alpha)\mathrm{H}(\alpha)\mathrm{e}^{\mathrm{i}\alpha x}\,\mathrm{d}\alpha$. Prove that $\operatorname{Re} g_2 = g_1$ and $\operatorname{Im} g_2 = (1/\pi)\int_{-\infty}^{\infty} g_1(t)(x-t)^{-1}\,\mathrm{d}t$; deduce that

$$g_1(t) = -\frac{1}{\pi}\int_{-\infty}^{\infty}(t-x)^{-1}\{\operatorname{Im} g_2(x)\}\,\mathrm{d}x.$$

10.3. Ordinary linear differential equations

In conventional analysis one frequently encounters the problem of determining a solution of

$$a_0(x)\frac{\mathrm{d}^m f_1}{\mathrm{d}x^m} + a_1(x)\frac{\mathrm{d}^{m-1}f_1}{\mathrm{d}x^{m-1}} + \ldots + a_m(x)f_1 = f_2(x) \quad (27)$$

in $x \ge x_0$ such that $f_1(x) = f_{10}$, $\mathrm{d}f_1/\mathrm{d}x = f_{11}, \ldots, \mathrm{d}^{m-1}f_1/\mathrm{d}x^{m-1} = f_{1,m-1}$ at $x = x_0$. The question arises as to what the corresponding problem is in terms of generalised functions. Since a simple change of origin will convert $x = x_0$ to $x = 0$ it will be sufficient if, in the following, attention is confined to the case where f_1 and its derivatives are given at $x = 0$. If we suppose that f_1 can be identified with a generalised function in $x \geq 0$ then we put $g_1(x) = f_1(x)\mathrm{H}(x)$. Consequently

$$g_1'(x) = \mathrm{H}(x)\mathrm{d}f_1/\mathrm{d}x + f_1(0)\delta(x),$$
$$g_1''(x) = \mathrm{H}(x)\mathrm{d}^2 f_1/\mathrm{d}x^2 + f_{11}\,\delta(x) + f_{10}\,\delta'(x),$$
$$\vdots$$
$$g_1^{(m)}(x) = \mathrm{H}(x)\mathrm{d}^m f_1/\mathrm{d}x^m + \textstyle\sum_{r=0}^{m-1} f_{1r}\,\delta^{(m-1-r)}(x).$$

Hence, account being taken of (27)

$$\sum_{r=0}^{m} a_{m-r}(x) g_1^{(r)}(x) = f_2(x)\mathrm{H}(x) + \sum_{r=0}^{m} a_{m-r}(x) \sum_{s=0}^{r-1} f_{1s}\delta^{(r-1-s)}(x)$$

it being assumed that the multiplication of a_{m-r} and $g_1^{(r)}$ is permissible. This is a particular case of

$$\sum_{r=0}^{m} a_{m-r}(x) g_1^{(r)}(x) = g_2(x). \tag{28}$$

Note that here the values of g_1 are not specified at the origin in accordance with the principle that the values of generalised functions at points are, in general, indeterminate. Of course, if the values of f_1 and its derivatives were not given at the origin we could still arrive at an equation of the type (28) by assigning them arbitrary values there.

However, the reader must be warned that if, given (27), he immediately solves (28) he may obtain a generalised function which is not the desired solution of (27). For example, let it be required to find f_1 so that

$$\mathrm{d}f_1/\mathrm{d}x - af_1 = 0 \quad (a > 0) \tag{29}$$

in $x \geq 0$ and $f_1(0) = 1$. The form corresponding to (28) is

$$g_1'(x) - ag_1(x) = \delta(x). \tag{30}$$

By taking a Fourier transform we have

$$(\mathrm{i}\alpha - a)G_1 = 1$$

whence

$$g_1(x) = -\,\mathrm{e}^{ax}\,\mathrm{H}(-x).$$

This is not the correct f_1, because it is zero for $x > 0$ and does not take the correct value at $x = 0$. The reason why (30) fails to supply the required solution of (29) is that $f_1 = e^{ax}$ is not a generalised function. On the other hand if we had asked for the solution of (29) in $x \leq 0$ such that $f_1(0) = -1$, (30) would be unaltered and would provide the correct f_1. Some care is therefore necessary in going from differential equations for functions to those for generalised functions.

When one turns to consider solutions of (28) there are features which do not arise in differential equations for functions. For example, the equation

$$xg'(x) + g(x) = 0 \tag{31}$$

can be written $(xg)' = 0$, whence $xg = C_1$ so that

$$g = C_1 x^{-1} + C_2 \delta(x).$$

This solution *involves two arbitrary constants* although (31) is a *differential equation of the first order.*

It is also possible to construct differential equations which have no non-trivial solution. Thus

$$x^3 g'(x) + 2g(x) = 0 \tag{32}$$

implies that

$$g = \begin{cases} C_1 e^{1/x^2} & (x > 0) \\ \\ C_2 e^{1/x^2} & (x < 0). \end{cases}$$

But since there is no generalised function which can take this form (§11.1) the only permissible solution of (32) is $g = 0$.

A general theory, indicating when (28) can be expected to supply satisfactory solutions, is postponed until Chapter 11 (§11.7). In the meantime we observe that if $e_1(x)$ is a generalised function such that

$$\sum_{r=0}^{m} a_{m-r}(x) e_1^{(r)}(x - t) = \delta(x - t)$$

then e_1 is known as an *elementary solution.* If g_2 is the Fourier transform of a fairly good function $e_1 * g_2$ is defined and

$$\sum_{r=0}^{m} a_{m-r}(e_1 * g_2)^{(r)} = \delta * g_2 = g_2$$

from equation (8) of Chapter 6. The same result is true if e_1 is the Fourier transform of a fairly good function.

Thus, if either e_1 or g_2 is the Fourier transform of a fairly good function a solution of (28) is provided by finding an elementary solution and then forming $e_1 * g_2$.

Naturally, there can be situations in which (28) can be solved although one may not be able to pass through the intermediate stage of using an elementary solution.

If the coefficients a_r are independent of x the differential equation becomes one with constant coefficients. In that case (28), after division by a_0, may be written as

$$g_1^{(m)}(x) + b_1 g_1^{(m-1)}(x) + \dots + b_m g_1(x) = g_2(x). \qquad (33)$$

The Fourier transform of this is

$$\{(\mathrm{i}\alpha)^m + b_1(\mathrm{i}\alpha)^{m-1} + \dots + b_m\}\, G_1(\alpha) = G_2(\alpha).$$

Denote the zeros of the polynomial factor of the left-hand side by $\alpha_1, \alpha_2, \dots, \alpha_m$. Then

$$\mathrm{i}^m \left\{ \prod_{r=1}^{m} (\alpha - \alpha_r) \right\} G_1(\alpha) = G_2(\alpha).$$

Since division by $\alpha - \alpha_r$ can always be carried out (§6.2), G_1 can always be determined. In fact we can write symbolically

$$G_1(\alpha) = \mathrm{i}^{-m} \left\{ \prod_{r=1}^{m} (\alpha - \alpha_r) \right\}^{-1} G_2(\alpha) + G_3(\alpha)$$

where G_3 consists of the additional terms containing arbitrary constants which arise from those α_r which are real. For each simple real zero G_3 will have a term such as $C\delta(\alpha - \alpha_r)$ and for each multiple real zero $(\alpha - \alpha_r)^s$ on the left-hand side G_3 will contain $\sum_{p=0}^{s-1} C_p \times \delta^{(p)}(\alpha - \alpha_r)$. The solution of (33) can now be obtained by taking a Fourier inverse.

It has thus been shown that a *linear differential equation with constant coefficients, of which the right-hand side is a generalised function, always possesses a solution.* In particular, a linear differential equation with constant coefficients always has an elementary solution. The practical utility of a such a solution requires further consideration.

More generally, the system in n unknowns

$$\sum_{r=0}^{m} A_r g_1^{(r)} = g_2,$$

where g_1 and g_2 are column vectors with n elements and A_r is an $n \times n$ matrix ($A_m = I$), has Fourier transform

$$\left\{ \sum_{r=0}^{m} A_r (\mathrm{i}\alpha)^r \right\} G_1 = G_2.$$

Since det $\{\sum_{r=0}^{m} (\mathrm{i}\alpha)^r A_r\}$ is a polynomial in α the inverse of the factor of G_1 exists and so G_1 can be found. Thus *a system of linear differential equations with constant coefficients and $A_m = I$ always possesses a solution which is a generalised function.*

Exercises

31. Show that the general solution of

$$xg'(x) - \beta g(x) = 0$$

where β is not a negative integer is

$$g(x) = C_1 x^{\beta} \mathrm{H}(x) + C_2(-x)^{\beta} \mathrm{H}(-x)$$

C_1 and C_2 being arbitrary.

If $\beta = -m$ ($m = 1, 2, \ldots$) show that

$$g(x) = C_1 x^{-m} + C_2 \delta^{(m-1)}(x).$$

32. Find the generalised functions which satisfy

(i) $g''(x) - 4g(x) = 0$, (iii) $g''(x) + 2g'(x) + 5g(x) = 0$,

(ii) $g''(x) + 4g(x) = 0$, (iv) $g'''(x) - 3g'(x) + 2g(x) = 0$.

33. Find the generalised functions which satisfy

(i) $g'(x) + 3g(x) = \mathrm{e}^{\mathrm{i}ax}$ (a real), (iii) $g''(x) + g(x) = \cos x$,

(ii) $g''(x) - 2g'(x) + g(x) = x^2$, (iv) $g'''(x) + 3g'(x) - 2\mathrm{i}g(x) = \mathrm{e}^{\mathrm{i}x}$.

34. Show that the solution of $g''(x) - g(x) = -\delta(x - \xi)$ is $g(x) = \frac{1}{2}\mathrm{e}^{-|x-\xi|}$.

35. Show that the general solution of $g''(x) = -\delta(x-\xi)$ is

$$g(x) = (\xi - x)\mathrm{H}(x-\xi) + Ax + B$$

where A and B are arbitrary constants. Deduce that, if $-1 < \xi < 1$ and $g(-1) = g(1) = 0$, $g(\xi) = -\tfrac{1}{2}\{|x-\xi| + x\xi - 1\}$.

10.4. Linear partial differential equations

The type of partial differential equation which is of common occurrence in applied mathematics is

$$P(\partial)f_1 = f_2 \tag{34}$$

where

$$P(\partial) \equiv \sum_{m_1} \dots \sum_{m_n} a_{m_1,\dots,m_n} \frac{\partial^{m_1}}{\partial x_1^{m_1}} \cdots \frac{\partial^{m_n}}{\partial x_n^{m_n}} \qquad (m_1 + \dots + m_n \le M)$$

which may be abbreviated conveniently as

$$P(\partial) = \sum_{m=1}^{M} a_m \partial^m.$$

In addition to satisfying (34) f_1 may also be required to comply with certain boundary conditions. These will usually take the form that combinations such as $\sum_{m=1}^{M-1} b_m(x)\partial^m f_1$ are prescribed on some hypersurface S. Usually, also, f_1 will be needed on only one side of S.

If we take a generalised function g_1, equal to f_1 on this side of S and equal to 0 on the other side, a derivative $\partial_1 g_1$ will consist of $\partial f_1/\partial x_1$ together with a generalised function which corresponds to a hypersurface layer on S. Then, if g_2 is defined in the same way for f_2 as g_1 for f_1,

$$P(\partial)g_1 = g_2 + g_3 \tag{35}$$

where $P(\partial)$ now involves generalised partial derivatives and g_3 is composed of terms which involve the values of the derivatives (up to order $M-1$) of f_1 on S.

One is then faced with the problem of finding a generalised function which solves (35). If one wishes to identify this with f_1 it will be necessary to verify that the solution found vanishes on the correct side of S and, possibly, that it is continuously differentiable a sufficient number of times.

Example 4. Given that

$$\frac{\partial^M f_1}{\partial x_n^M} + \sum_{m=1}^{M-1} a_m \partial^m f_1 = f_2$$

in $x_n > 0$ and that $\partial^m f_1/\partial x_n^m = f_{1m}(x_1, \ldots, x_{n-1})$ $(m = 0, 1, \ldots, M-1)$ then $g_1(\boldsymbol{x}) = f_1(\boldsymbol{x})\mathrm{H}(x_n)$ and $g_2(\boldsymbol{x}) = f_2(\boldsymbol{x})\mathrm{H}(x_n)$. Hence (35) is

$$\partial_n^M g_1 + \sum_{m=1}^{M-1} a_m \partial^m g_1 = g_2 + g_3$$

where

$$g_3 = \sum_{m=0}^{M-1} f_{1m} \delta^{(M-m-1)}(x_n)$$
$$+ \sum_{m_1} \cdots \sum_{m_n} \sum_{m=0}^{m_n - 1} a_{m_1, \ldots, m_n} \frac{\partial^{m_1}}{\partial x_1^{m_1}} \cdots \frac{\partial^{m_{n-1}}}{\partial x_{n-1}^{m_{n-1}}} f_{1m} \delta^{(m_n - m - 1)}(x_n).$$

Example 5. If $\partial^2 f_1/\partial x_1 \partial x_2 = f_2$ in $x_2 > 0$ and $f_1 = f_{10}(x_1)$, $\partial f_1/\partial x_2 = f_{11}(x_1)$ on $x_2 = 0$ $(n = 2)$ then

$$\partial_1 \partial_2 g_1 = g_2 + f'_{10} \delta(x_2).$$

The different properties of g_3 in Examples 4 and 5 should be noted. If $g_3 = 0$ in Example 4, it follows that $f_{10} = 0$ and then, successively, that $f_{11} = 0, \ldots, f_{1,M-1} = 0$. This shows that two different sets of values of f_1 and its derivatives on $x_n = 0$ will lead to different g_3 and consequently different g_1 and f_1. On the other hand, $g_3 = 0$ in Example 5 implies only that $f_{10} =$ constant and f_{11} is arbitrary; in this case different f_{11} do not lead to different g_1. Example 5 is known as a *characteristic boundary value problem.*

An *elementary solution* of (35) is a generalised function $e_1(\boldsymbol{x})$ such that

$$P(\partial) e_1(\boldsymbol{x}) = \delta(\boldsymbol{x}). \tag{36}$$

If g_2 or e_1 is the Fourier transform of a fairly good function then $g_1 = e_1 * g_2$ solves

$$P(\partial) g_1 = g_2. \tag{37}$$

Some examples of elementary solutions have been encountered

already. For instance, equations (16) and (18) of Chapter 7 show that

$$(\partial_1^2 + \partial_2^2 + \dots + \partial_n^2)e_1 = \delta(\boldsymbol{x})$$

has solution

$$e_1(\boldsymbol{x}) = \begin{cases} \{(\tfrac{1}{2}n - 2)!/4\pi^{n/2}\}r^{2-n} & (n \neq 2) \qquad (38) \\ (1/2\pi)\ln r & (n = 2). \qquad (39) \end{cases}$$

More generally

$$(\partial_1^2 + \dots + \partial_n^2)^k e_1 = \delta$$

is solved by

$$e_1 = \begin{cases} \dfrac{(\tfrac{1}{2}n - k - 1)!(-1)^k r^{2k-n}}{(k-1)!2^{2k}\pi^{n/2}} & (n \text{ odd or } 2k < n) \qquad (40) \\[2ex] \dfrac{(-1)^{1-n/2} r^{2k-n} \ln r}{(k-1)!(k-\tfrac{1}{2}n)!2^{2k-1}\pi^{n/2}} & (n \text{ even and } 2k - n \geq 0). \qquad (41) \end{cases}$$

An elementary solution is not, in general, unique. This is because, given any elementary solution, we can add to it any solution of the *homogeneous equation*

$$P(\partial)g_1 = 0 \qquad (42)$$

and still have an elementary solution; of course, if the only solution of the homogeneous equation is $g_1 \equiv 0$ any elementary solution is unique.

When the coefficients in the partial differential equation are constants a powerful method of finding both elementary solutions and solutions of the homogeneous equation (42) is by Fourier transforms.

Example 6. If

$$(\partial_1^2 + \dots + \partial_n^2 - 1)g_1 = 0 \qquad (43)$$

then

$$(1 + \alpha^2)G_1(\boldsymbol{\alpha}) = 0$$

which implies that $G_1 = 0$ and therefore $g_1 = 0$. Thus (43) is not satisfied by any generalised function other than zero. (There are functions satisfying (43) which are not generalised functions, e.g. e^{x_1}.)

If

$$(\partial_1^2 + \ldots + \partial_n^2 - 1)e_1 = \delta(\boldsymbol{x}) \tag{44}$$

then

$$(1 + \alpha^2)E_1(\boldsymbol{\alpha}) = -1$$

and

$$e_1(\boldsymbol{x}) = -\frac{1}{(2\pi)^{n/2}} r^{1-n/2} K_{(n/2)-1}(r)$$

from (7.37). Thus e_1 is the elementary solution of (44).

Since E_1 is a fairly good function the solution of

$$(\partial_1^2 + \ldots + \partial_n^2 - 1)g_1 = g_2$$

is

$$g_1(\boldsymbol{x}) = -\frac{1}{(2\pi)^{n/2}} \int_{-\infty}^{\infty} |\boldsymbol{x} - \boldsymbol{t}|^{1-n/2} K_{(n/2)-1}(|\boldsymbol{x} - \boldsymbol{t}|) g_2(\boldsymbol{t}) \, d\boldsymbol{t}. \tag{45}$$

Example 7. If

$$(\partial_1^2 + \ldots + \partial_n^2)g_1 = 0 \tag{46}$$

then

$$\alpha^2 G_1(\boldsymbol{\alpha}) = 0 \tag{47}$$

which, on account of Exercise 21 of Chapter 7, shows that G_1 is a linear combination of $\delta(\boldsymbol{\alpha})$ and its derivatives. Thus g_1 must be a polynomial. In fact it is just as easy to determine the polynomial solutions of (46) as the solutions of (47).

An elementary solution of

$$(\partial_1^2 + \ldots + \partial_n^2)e_1 = \delta(\boldsymbol{x})$$

is obtained from

$$E_1 = -\alpha^{-2}.$$

It follows from Theorems 7.31 and 7.33 that

$$e_1(x) = \begin{cases} -\{(\tfrac{1}{2}n-2)!/4\pi^{n/2}\}r^{2-n} & (n \neq 2) \\ (1/2\pi)\ln r & (n=2) \end{cases}$$

in agreement with (38) and (39). To this may be added any of the polynomial solutions of (46).

Example 8. The partial differential equation

$$(\partial_1^2 - \dots - \partial_n^2)e_1 = \delta(x) \tag{48}$$

leads to $\tilde{q}E_1 = -1$. A possible solution is $E_1 = -\tilde{q}^{-1}$ (others are $-(\tilde{q} \pm \mathrm{i}0)^{-1}$). With this choice of E_1 Exercise 97 of Chapter 8 reveals that

$$e_1(x) = q_+^{1-n/2}/(1-\tfrac{1}{2}n)!4\pi^{(n/2)-1} \tag{49}$$

when n is odd whereas Exercise 98 of Chapter 8 shows that, when n is even,

$$e_1(x) = \delta^{((n/2)-2)}(q)/4\pi^{(n/2)-1} \quad (n \neq 2) \tag{50}$$

and, if $n = 2$,

$$e_1(x) = \tfrac{1}{8}\operatorname{sgn} q.$$

Other elementary solutions can be constructed by, for example, adding multiples of $\delta(\tilde{q})$ and $\delta(\tilde{q})\mathrm{H}(\alpha_1)$ to E_1. In this way, Exercises 99 and 100 of Chapter 8 together with Theorem 8.13 supply

$$e_1(x) = \begin{cases} \dfrac{q_+^{1-n/2}\mathrm{H}(x_1)}{(1-\frac{1}{2}n)!2\pi^{(n/2)-1}} & (n \text{ odd}) \\ \delta^{((n/2)-2)}(q)\mathrm{H}(x_1)/2\pi^{(n/2)-1} & (n \text{ even but } n \neq 2) \\ \tfrac{1}{2}\mathrm{H}(q)\mathrm{H}(x_1) & (n=2). \end{cases} \tag{51}$$

Yet more may be obtained by this process through Theorem 8.13; for example,

$$e_1(x) = \frac{L^r q_+^{1-(n/2)+r}\mathrm{H}(x_1)}{(1-\frac{1}{2}n+r)!r!\pi^{(n/2)-1}2^{2r+1}} \tag{52}$$

when n is odd.

Of the various possibilities (51) and (52) are probably more useful than (49) and (50) so that succeeding formulae will be given in terms of them. For instance, a solution of

$$(\partial_1^2 - \dots - \partial_n^2)g_1 = g_2 \tag{53}$$

is, when n is odd.

$$g_1(x) = \frac{(\partial_1^2 - \dots - \partial_n^2)^{(n/2)-3/2}}{(\frac{1}{2}n - \frac{3}{2})!2^{n-2}\pi^{(n/2)-1/2}} \int_{-\infty}^{\infty} u_+^{-1/2}(x-t)g_2(t)\,dt \tag{54}$$

in the notation of Example 15 of Chapter 8. In (54) it is assumed that g_2 is such that the right-hand side exists. It might happen that the convolution of g_2 with $u_+^{-1/2}$ did not exist whereas that with $u_+^{1-n/2}$ did; in that case one of the other forms of the elementary solution should be employed. If n is even, a solution of (53) is

$$g_1(x) = \frac{(\partial_1^2 - \dots - \partial_n^2)^{(n/2)-1}}{(\frac{1}{2}n - 1)!2^{n-1}\pi^{(n/2)-1}} \times \int_{-\infty}^{\infty} g_2(x-t)\,\mathrm{H}(t_1^2 - t_2^2 - \dots - t_n^2)\mathrm{H}(t_1)\,dt \tag{55}$$

when account is taken of Exercises 90 and 91 of Chapter 8.

Example 9. Take a Fourier transform of (48) with respect to x_1; this is permissible by Theorem 7.23. Then (48) becomes

$$(\alpha_1^2 + \partial_2^2 + \dots + \partial_n^2)e_1 = -\delta(x_2)\dots\delta(x_n)$$

where now e_1 depends on $\alpha_1, x_2, \dots, x_n$. From (52) or (54),

$$e_1 = \frac{(-1)^{(n/2)-3/2}(\alpha_1^2 + \partial_2^2 + \dots + \partial_n^2)^{(n/2)-3/2}}{(\frac{1}{2}n - \frac{3}{2})!2^{n-2}\pi^{(n/2)-1/2}} \times \int_{-\infty}^{\infty} \frac{e^{-i\alpha_1 x_1}\,\mathrm{H}(x_1)\,\mathrm{H}(q)}{(x_1^2 - x_2^2 - \dots - x_n^2)^{1/2}}\,dx_1$$

when n is odd. If $x_2^2 + \dots + x_n^2$ is denoted by r_0^2 the integral is

$$\int_{r_0}^{\infty} \frac{e^{-i\alpha_1 x_1}}{(x_1^2 - r_0^2)^{1/2}}\,dx_1 = -\tfrac{1}{2}\pi i \mathrm{H}_0^{(2)}(\alpha_1 r_0)$$

when $\alpha_1 > 0$, $H_0^{(2)}$ being the Hankel function of order zero. Making a slight change of notation we see that a solution of

$$(\partial_1^2 + \dots + \partial_n^2 + k^2)e_1 = \delta(x), \tag{56}$$

k being real and positive, is

$$e_1 = \frac{(-1)^{(n/2)-1}\mathrm{i}(\partial_1^2 + \ldots + \partial_n^2 + k^2)^{(n/2)-1}}{(\frac{1}{2}n-1)!2^n\pi^{(n/2)-1}} H_0^{(2)}(kr) \qquad (57)$$

when n is even. Since

$$(\partial_1^2 + \ldots + \partial_n^2 + k^2)r^{-\nu}H_\nu^{(2)}(kr) = -k(n-2\nu-2)r^{-\nu-1}H_{\nu+1}^{(2)}(kr) \quad (58)$$

this may be written conveniently as

$$e_1 = \frac{\mathrm{i}k^{(n/2)-1}r^{1-n/2}}{2^{(n/2)+1}\pi^{(n/2)-1}} H_{(n/2)-1}^{(2)}(kr) \qquad (59)$$

with, however, the understanding that this is to be interpreted in the form (57).

Similarly, from Exercise 90 of Chapter 8 and (51), a solution of (56) when n is odd is

$$e_1 = \frac{(-1)^{(n/2)+1/2}(\partial_1^2 + \ldots + \partial_n^2 + k^2)^{(n/2)-1/2}}{(\frac{1}{2}n-\frac{1}{2})!2^n\pi^{(n/2)-1/2}} \int_r^\infty \mathrm{e}^{-\mathrm{i}kx_1}\,\mathrm{d}x_1$$

$$= \frac{(-1)^{(n/2)-1/2}(\partial_1^2 + \ldots + \partial_n^2 + k^2)^{(n/2)-1/2}}{(\frac{1}{2}n-\frac{1}{2})!2^n\pi^{(n/2)-1/2}k} \mathrm{e}^{-\mathrm{i}kr} \qquad (60)$$

since $\int_r^\infty \mathrm{e}^{-\mathrm{i}kx_1}\,\mathrm{d}x_1 = \mathrm{e}^{-\mathrm{i}kr}\{\pi\delta(k) - \mathrm{i}k^{-1}\}$ and $k > 0$. If we use the fact that $\mathrm{e}^{-\mathrm{i}z} = (\frac{1}{2}\pi z)^{1/2}H_{-1/2}^{(2)}(z)$ and (58) we find that (60) may also be abbreviated to (59). Thus, (59) *provides an elementary solution of* (56) *whether n be even or odd.*

To illustrate the effect of boundary conditions let us consider the *initial value problem* in which a solution of

$$\sum_{m=2}^{n} \frac{\partial^2 f}{\partial x_m^2} - \frac{\partial^2 f}{\partial x_1^2} = 0$$

is required in $x_1 > 0$ given that $f = f_1$, $\partial f/\partial x_1 = f_2$ on $x_1 = 0$. Define $g(\mathbf{x}) = f(\mathbf{x})\mathrm{H}(x_1)$ so that a solution of

$$(\partial_1^2 - \ldots - \partial_n^2)g = f_2\delta(x_1) + f_1\delta'(x_1)$$

must be found. This corresponds to (53). When n is even (55) shows that

$$g(\mathbf{x}) = \frac{(\partial_1^2 - \ldots - \partial_n^2)^{(n/2)-1}}{(\frac{1}{2}n-1)!2^{n-1}\pi^{(n/2)-1}} \int \{f_2\mathrm{H}(x_1^2 - t_2^2 - \ldots - t_n^2)\mathrm{H}(x_1)$$

$$+ f_1 2x_1\delta(x_1^2 - t_2^2 - \ldots - t_n^2)\mathrm{H}(x_1)\}\,\mathrm{d}t_2 \ldots \mathrm{d}t_n.$$

Denote the intersection of the semi-cone

$$(x_1 - t_1)^2 \geq (x_2 - t_2)^2 + \dots + (x_n - t_n)^2 \quad (t_1 \leq x_1)$$

and the hyperplane $t_1 = 0$ by S, and denote the boundary of S by Ω. Then the formula for g may be written

$$g(\boldsymbol{x}) = \frac{(\partial_1^2 - \dots - \partial_n^2)^{(n/2)-1}}{(\frac{1}{2}n - 1)!2^{n-1}\pi^{(n/2)-1}} \mathrm{H}(x_1)\left\{\int_S f_2 \, \mathrm{d}S + \int_\Omega f_1 \, \mathrm{d}\Omega\right\}. \quad (61)$$

This demonstrates the validity of *Huygens' principle* when n is even ($\neq 2$). For the integrals depend upon $\boldsymbol{x}$ only to the extent that S and Ω do. The effect of the differential operator is to ensure that only variations of the limits of these domains will contribute. Thus values of f_2 in the interior of S will not affect $g(\boldsymbol{x})$ which will, therefore, depend only upon values on the semi-cone.

If $n = 2$, (61) must be modified to

$$g(\boldsymbol{x}) = \tfrac{1}{2}\int_S f_2 \mathrm{d}S + f_1(P_1) + f_1(P_2)$$

where P_1 and P_2 are the values of t_2 where $t_2 = x_2 \pm x_1$.

When n is odd, (52) gives

$$g(\boldsymbol{x}) = \frac{(\partial_1^2 - \dots - \partial_n^2)^{(n/2)-1/2}}{(\frac{1}{2}n - \frac{1}{2})!\,2^{n-1}\,\pi^{(n/2)-1/2}}$$

$$\times \int_S \left\{ f_2 u_+^{1/2}(\boldsymbol{x} - \boldsymbol{t}) - f_1 \frac{\partial}{\partial t_1} u_+^{1/2}(\boldsymbol{x} - \boldsymbol{t}) \right\}_{t_1 = 0} \mathrm{d}S. \quad (62)$$

Huygens' principle is no longer valid.

Exercises

It is assumed in the following exercises that generalised functions satisfy conditions such that surface integrals exist.

36. Show that an elementary solution of

$$\partial_1 \partial_2 \dots \partial_n g = \delta(\boldsymbol{x})$$

is $g = \mathrm{H}(x_1)\mathrm{H}(x_2)\dots\mathrm{H}(x_n)$.

37. Show that the elementary solution of

$$(\partial_1^2 + \dots + \partial_n^2 - 1)^k e_1 = \delta(\boldsymbol{x})$$

can be written

$$e_1 = \frac{(-1)^k r^{k-n/2}}{2^{(n/2)+k-1}\pi^{n/2}} K_{(n/2)-k}(r).$$

38. Prove that $P(\partial)g_1 = g_2$ has a solution g_1 which is the Fourier transform of a fairly good function if and only if g_2 is the Fourier transform of a fairly good function and $G_2(\alpha)/P(i\alpha)$ is fairly good.

39. Show that an elementary solution of

$$(\partial_1^2 + \dots + \partial_p^2 - \partial_{p+1}^2 - \dots - \partial_n^2)^k e_1 = \delta(x)$$

is given by

$$e_1 = \begin{cases} \dfrac{(\frac{1}{2}n-k-1)!(-1)^k e^{\pm\pi i(n-p)/2}}{(k-1)!2^{2k}\pi^{n/2}}(q \pm i0)^{k-n/2} & (n \text{ odd or } 2k < n) \\[2ex] \dfrac{(-1)^{(n/2)-1} e^{\pm\pi i(n-p)/2}}{(k-\frac{1}{2}n)!(k-1)!2^{2k}\pi^{n/2}}(q \pm i0)^{k-n/2}\ln(q \pm i0) & (n \text{ even and } 2k \geq n). \end{cases}$$

Generalise this to the equation

$$\left(\sum_{i=1}^{n}\sum_{j=1}^{n} q_{ij}\partial_i\partial_j\right)^k e_1 = \delta(x)$$

where q_{ij} is real and $q_{ij} = q_{ji}$.

40. If n is odd prove that

$$\partial_1^{n-3} u_+^{(n/2)-2} = (\tfrac{1}{2}n-2)!(2\xi)^{n-3} u_+^{1-n/2}/(1-\tfrac{1}{2}n)!$$

where $\xi^2 = x_2^2 + \dots + x_n^2$. Deduce that a solution of

$$(\partial_1^2 - \dots - \partial_n^2)g_1 = g_2$$

is

$$g_1(x) = \frac{1}{(\frac{1}{2}n-2)!2^{n-2}\pi^{(n/2)-1}}\partial_1^{n-3}\int_{-\infty}^{\infty} u_+^{(n/2)-2}(x-t)g_2(t)\,dt.$$

Show that the same formula also holds if n is even ($n \geq 4$) by proving that

$$\partial_1^{n-3} u_+^{(n/2)-2} = (\tfrac{1}{2}n-2)!(2\xi)^{n-3} H(x_1)\delta^{((n/2)-2)}(q).$$

Thus, if $g_2 = f_2(x_2, \dots, x_n)\delta(x_1)$,

$$g_1(x) = \frac{H(x_1)}{(\frac{1}{2}n-2)!2^{n-2}\pi^{(n/2)-1}}\partial_1^{n-3}\int_0^{x_1}(x_1^2-\eta^2)^{(n/2)-2}\eta f_3\,d\eta$$

where $f_3 = \int_\Omega f_2(x_2 + \beta_2\eta, \dots, x_n + \beta_n\eta)\,d\Omega$, Ω being a hypersphere with centre $(x_2, \dots, x_n)$ and radius η with direction cosines $\beta_2, \dots, \beta_n$, and $\eta^{n-2}\,d\Omega$ an element of its area.

41. If $n = 3$ show that (62) can be written in *Volterra's form*

$$g(x) = -\frac{H(x_1)}{2\pi}\int_0^{x_1}\int_0^{2\pi}\frac{f_2(x_2+\eta\cos\phi, x_3+\eta\sin\phi)}{(x_1^2-\eta^2)^{1/2}}\eta\,d\eta\,d\phi$$
$$-\frac{H(x_1)}{2\pi}\partial_1\int_0^{x_1}\int_0^{2\pi}\frac{f_1(x_2+\eta\cos\phi, x_3+\eta\sin\phi)}{(x_1^2-\eta^2)^{1/2}}\eta\,d\eta\,d\phi.$$

42. If $n = 4$ show that (61) can be written in *Poisson's form*

$$g(x) = -\frac{x_1 H(x_1)}{4\pi}\int_0^{\pi}\int_0^{2\pi} f_2(x_2+x_1\sin\theta\cos\phi, x_3+x_1\sin\theta\sin\phi,$$
$$x_4+x_1\cos\theta)\sin\theta\,d\theta\,d\phi$$
$$-H(x_1)\partial_1\frac{x_1}{2\pi}\int_0^{\pi}\int_0^{2\pi} f_1(x_2+x_1\sin\theta\cos\phi, x_3+x_1\sin\theta\sin\phi,$$
$$x_4+x_1\cos\theta)\sin\theta\,d\theta\,d\phi.$$

43. By taking a Fourier transform of (48) with respect to x_n and then replacing n by $n+1$ show that an elementary solution of

$$(\partial_1^2 - \ldots - \partial_n^2 + k^2)e_1 = \delta(x) \qquad (k > 0)$$

is given by

$$e_1 = \begin{cases} \dfrac{(\partial_1^2 - \ldots - \partial_n^2 + k^2)^{(n/2)-1}}{(\frac{1}{2}n-1)!\,2^n\pi^{(n/2)-1}} H(x_1)H(q)J_0(k|q|^{1/2}) & (n \text{ even}) \\[2ex] \dfrac{(\partial_1^2 - \ldots - \partial_n^2 + k^2)^{(n/2)-1/2}}{(\frac{1}{2}n-1)!\,2^{n-1}\pi^{(n/2)-1/2}k} H(x_1)H(q)\sin k|q|^{1/2} & (n \text{ odd}). \end{cases}$$

Deduce that, if $n = 2$, a solution of

$$(\partial_1^2 - \partial_2^2 + k^2)g = 0$$

such that $g = 0$, $\partial_1 g = f(x_2)$ at $x_1 = 0$ is

$$g = \tfrac{1}{2}H(x_1)\int_0^{x_1}\{f(x_2+t)+f(x_2-t)\}\,J_0\{k(x_1^2-t^2)^{1/2}\}\,dt.$$

44. Find an elementary solution of

$$\left\{\prod_{s=1}^{m}(\partial_1^2 - \ldots - \partial_n^2 + k_s^2)\right\}e_1 = \delta(x) \qquad (k_m > k_{m-1} > \ldots > k_1 > 0).$$

45. If (x_1, x_2, x_3) is a point of the volume enclosed by the closed surface S and

$$(\partial_1^2 + \partial_2^2 - \partial_3^2)g = -g_2$$

show that

$$g(x)=\frac{1}{2\pi}\int_D\{(t_3-x_3)^2-r^2\}^{-1/2}g_2(t)dt$$

$$-\frac{1}{2\pi}\partial_3\int_{S_1}\frac{g_2}{\{(t_3-x_3)^2-r^2\}^{1/2}}\left\{\frac{t_3-x_3}{r}\frac{\partial r}{\partial v}+\frac{\partial t_3}{\partial v}\right\}dS$$

$$+\frac{1}{2\pi}\int_{S_1}\frac{1}{\{(t_3-x_3)^2-r^2\}^{1/2}}\left\{\frac{\partial g_2}{\partial t_1}\frac{\partial t_1}{\partial v}+\frac{\partial g_2}{\partial t_2}\frac{\partial t_2}{\partial v}-\frac{\partial g_3}{\partial t_3}\frac{\partial t_3}{\partial v}\right\}dS$$

where $r^2=(t_1-x_1)^2+(t_2-x_2)^2$, S_1 is the area cut out of S by the cone $r\le t_3-x_3$, D is the volume bounded by the cone and $t_3=0$, and $\partial/\partial v$ denotes a derivative along the normal to S.

46. If $n=4$ and $(\partial_1^2-\partial_2^2-\partial_3^2-\partial_4^2)g_1=g_2$ show that

$$g_1=\frac{1}{4\pi}\int_{-\infty}^{\infty}\int_{-\infty}^{\infty}\int_{-\infty}^{\infty}[g_2]t_0^{-1}\,dt_2\,dt_3\,dt_4$$

where $t_0^2=(x_2-t_2)^2+(x_3-t_3)^2+(x_4-t_4)^2$ and $[g_2]=g_2(x_1-t_0,t_2,t_3,t_4)$.

If S is a closed surface bounding the volume T in (x_2,x_3,x_4)-space and g_1 and g_2 vanish outside S prove *Kirchhoff's formula*

$$g_1=\frac{1}{4\pi}\int_T[g_2]t_0^{-1}d\tau+\frac{1}{4\pi}\int_S\left\{\left[\frac{\partial g_1}{\partial v}\right]\frac{1}{t_0}-[g_1]\frac{\partial}{\partial v}\frac{1}{t_0}+\frac{1}{t_0}\left[\frac{\partial g_1}{\partial x_1}\right]\frac{\partial t_0}{\partial v}\right\}dS$$

where $\partial/\partial v$ is along the outward normal to S and $[\partial g_1/\partial v]$ means calculate $\partial g_1(x_1,t_2,t_3,t_4)/\partial v$ and then replace x_1 by x_1-t_0.

47. Show that an elementary solution of

$$(\partial_1^2+\ldots+\partial_n^2)^k e_1=\delta(x)$$

is

$$e_1=\begin{cases}\dfrac{(\frac{1}{2}n-k-1)!(-1)^k r^{2k-n}}{(k-1)!2^{2k}\pi^{n/2}} & (n \text{ odd, or } n \text{ even and } 2k<n)\\[2ex] \dfrac{(-1)^{1-n/2}r^{2k-n}\ln r}{(k-1)!(k-\frac{1}{2}n)!2^{2k-1}\pi^{n/2}} & (n \text{ even and } 2k\ge n).\end{cases}$$

48. If $n=2$ and g_1,g_2 vanish outside the closed curve C show that

$$(\partial_1^2+\partial_2^2)^2g_1=g_2$$

is satisfied inside C by

$$g_1(x)=\frac{1}{2\pi}\int_C\left\{\frac{1}{4}\frac{\partial}{\partial v}(r_0^2\ln r_0)(\partial_1^2+\partial_2^2)g_1+g_1\frac{\partial}{\partial v}\ln r_0-\frac{\partial g_1}{\partial v}\ln r_0\right.$$

$$\left.-\frac{1}{4}r_0^2\ln r_0\frac{\partial}{\partial v}(\partial_1^2+\partial_2^2)g_1\right\}dS+\frac{1}{8\pi}\int r_0^2\ln r_0 g_2\,dx_1\,dx_2,$$

$\partial/\partial v$ being along the outward normal to C and r_0 the distance to the point of observation.

49. Show, by using (55) (with the derivative inside the integral) and doing the integration with respect to t_2, t_3 and t_4 first, that

$$(\partial_1^2 - \partial_2^2 - \partial_3^2 - \partial_4^2)g_1 = \delta\{x_2 - f_2(x_1)\}\delta\{x_3 - f_3(x_1)\}\delta\{x_4 - f_4(x_1)\}$$

has a solution

$$g_1 = (1/4\pi)\{t_1 - (x_2 - f_2)f'_2 - (x_3 - f_3)f'_3 - (x_4 - f_4)f'_4\}^{-1}$$

the argument of f_2, f_3, f_4 and their derivatives being $x_1 - t_1$. Here t_1 is the positive solution of

$$t_1 = |\boldsymbol{x} - \boldsymbol{f}(x_1 - t_1)| \tag{63}$$

where $\boldsymbol{x}$ and $\boldsymbol{f}$ are three-dimensional vectors with components (x_2, x_3, x_4) and (f_2, f_3, f_4) respectively. (It is assumed that there is only one t_1 satisfying (63) otherwise additional terms would be present.) If $\boldsymbol{R} = \boldsymbol{x} - \boldsymbol{f}(x_1 - t_1)$ and $v = f'(x_1 - t_1)$ the formula for g_1 can be written

$$g_1 = 1/4\pi(|\boldsymbol{R}| - \boldsymbol{R}.v)$$

in which form it is known as a *Liénard-Wiechert potential.*

50. If, in Exercise 49, $f_2(t) = vt, f_3 = f_4 = 0$ where v is constant and $v > 1$, show that

$$g_1 = (1/2\pi)\{(vx_1 - x_2)^2 - (x_3^2 + x_4^2)(v^2 - 1)\}^{-1/2}$$
$$\times \mathrm{H}\{vx_1 - x_2 - (x_3^2 + x_4^2)^{1/2}(v^2 - 1)^{1/2}\}.$$

51. If $(k_1^2 - \alpha^2)(k_2^2 - \alpha^2)G = 0$ and $k_1^2 \neq k_2^2$ show that $G = G_2 + G_1$ where $(k_1^2 - \alpha^2)\, G_1 = 0$ and $(k_2^2 - \alpha^2)G_2 = 0$. Deduce that

$$(\partial_1^2 + \dots + \partial_n^2 + k_1^2)(\partial_1^2 + \dots + \partial_n^2 + k_2^2)g = 0$$

is solved by $g = g_1 + g_2$ where $(\partial_1^2 + \dots + \partial_n^2 + k_1^2)g_1 = 0$, $(\partial_1^2 + \dots + \partial_n^2 + k_2^2)g_2 = 0$.

52. If $(k^2 - \alpha^2)\, G_1 = 0$ verify that $(k^2 - \alpha^2)G = G_1$ is satisfied by

$$G = \partial_1(\alpha_1 G_1) + \partial_2(\alpha_2 G_1) + \dots + \partial_n(\alpha_n G_1).$$

Deduce that $(\partial_1^2 + \dots + \partial_n^2 + k^2)^2 g = 0$ is solved by

$$g = g_0 + (x_1\partial_1 + x_2\partial_2 + \dots + x_n\partial_n)g_1$$

where $(\partial_1^2 + \dots + \partial_n^2 + k^2)g_0 = 0$ and $(\partial_1^2 + \dots + \partial_n^2 + k^2)g_1 = 0$.

Systems of partial differential equations are also frequently encountered in the form

$$\sum_{k=1}^{N} P_{jk}(\partial)g_k = g_{0j} \quad (j = 1, \dots, N)$$

where $P_{jk}(\partial)$ is of the same type as $P(\partial)$. In this case, an elementary solution can be defined by a matrix with elements e_{jk} where

$$\sum_{k=1}^{N} P_{jk}(\partial)e_{kl} = \delta_{jl}\delta(\boldsymbol{x})$$

provided that these equations are consistent. The reason for inserting the proviso can be understood by considering two of Maxwell's equations for electromagnetic waves, namely

$$\operatorname{curl} \boldsymbol{e} + \partial \boldsymbol{b}/\partial t = 0, \tag{64}$$

$$\operatorname{div} \boldsymbol{b} = 0. \tag{65}$$

If $\delta(x)\delta(y)\delta(z)\delta(t)\boldsymbol{i}$ were put on the right-hand side of (64) the divergence of (64) would not be consistent with the time derivative of (65).

There is therefore a good deal to be said for restricting one's attention only to the type of right-hand side which can be regarded as arising naturally from physical sources. Another way of avoiding the difficulty is to eliminate all unknowns but one and deal with the single equation so obtained. However, this may introduce the necessity of checking that any solution of the single equation that is obtained also provides a solution of the original equations.

Yet another method consists, not of finding an elementary solution, but of determining a solution with a convolution on the right-hand side.

Example 10. As an example consider Maxwell's equations in the form

$$\begin{aligned} &\operatorname{curl} \boldsymbol{e} + \partial \boldsymbol{b}/\partial t = 0, \quad \operatorname{div} \boldsymbol{b} = 0, \\ &\operatorname{curl} \boldsymbol{h} - \partial \boldsymbol{d}/\partial t = \boldsymbol{j}, \quad \operatorname{div} \boldsymbol{d} = \rho, \\ &\operatorname{div} \boldsymbol{j} + \partial \rho/\partial t = 0, \quad \boldsymbol{b} = \mu \boldsymbol{h}, \quad \boldsymbol{d} = \varepsilon \boldsymbol{e} \end{aligned}$$

where μ and ε are constants. In this case $n = 4$ and x_1, x_2, x_3 and x_4 are identified with t, x, y and z respectively; $\boldsymbol{e}$, $\boldsymbol{b}$, $\boldsymbol{h}$, $\boldsymbol{d}$ and $\boldsymbol{j}$ are vectors in (x, y, z)-space. The equations which will be discussed are

$$\begin{aligned} &\operatorname{curl} \boldsymbol{e} + \mu \partial \boldsymbol{h}/\partial t = \boldsymbol{a}, \quad \mu \operatorname{div} \boldsymbol{h} = p, \\ &\operatorname{curl} \boldsymbol{h} - \varepsilon \partial \boldsymbol{e}/\partial t = \boldsymbol{j} + \boldsymbol{c}, \quad \varepsilon \operatorname{div} \boldsymbol{e} = \rho + q, \\ &\operatorname{div} \boldsymbol{j} + \partial \rho/\partial t = 0 \end{aligned}$$

where div $\boldsymbol{a} = \partial p/\partial t$ and div $\boldsymbol{c} + \partial q/\partial t = 0$. On taking a Fourier transform we obtain

$$\mathrm{i}\xi \wedge \boldsymbol{E} + \mathrm{i}\alpha_1 \mu \boldsymbol{H} = \boldsymbol{A}, \qquad \mathrm{i}\mu\xi . \boldsymbol{H} = P,$$
$$\mathrm{i}\xi \wedge \boldsymbol{H} - \mathrm{i}\alpha_1 \varepsilon \boldsymbol{E} = \boldsymbol{J} + \boldsymbol{C}, \quad \mathrm{i}\varepsilon\xi . \boldsymbol{E} = R + Q$$

where ξ is a 3-vector with components $\alpha_2, \alpha_3, \alpha_4$. The elimination of $\boldsymbol{H}$ leads to

$$(|\xi|^2 - \alpha_1^2 \mu\varepsilon)\boldsymbol{E} = -\mathrm{i}\alpha_1 \mu(\boldsymbol{J} + \boldsymbol{C}) + \mathrm{i}\xi \wedge \boldsymbol{A} - \mathrm{i}(R + Q)\xi/\varepsilon.$$

It follows from Exercise 46 that

$$\boldsymbol{e} = -\frac{\partial}{\partial t}\frac{\mu}{4\pi}\int [\boldsymbol{j} + \boldsymbol{c}]t_0^{-1} + \mathrm{curl}\frac{1}{4\pi}\int [\boldsymbol{a}]t_0^{-1} - \frac{1}{\varepsilon}\mathrm{grad}\frac{1}{4\pi}\int [\rho + q]t_0^{-1}$$

where

$$t_0^2 = (x_2 - t_2)^2 + (x_3 - t_3)^2 + (x_4 - t_4)^2$$
$$= (x - t_2)^2 + (y - t_3)^2 + (z - t_4)^2,$$

$[g] = g(t - t_0/c, t_2, t_3, t_4)$, $\mu\varepsilon = 1/c^2$ and the integrals are over the whole of (t_2, t_3, t_4)-space.

Similarly,

$$\boldsymbol{h} = \mathrm{curl}\frac{1}{4\pi}\int [\boldsymbol{j} + \boldsymbol{c}]t_0^{-1} + \frac{\partial}{\partial t}\frac{\varepsilon}{4\pi}\int [\boldsymbol{a}]t_0^{-1} - \frac{1}{\mu}\mathrm{grad}\frac{1}{4\pi}\int [p]t_0^{-1}.$$

Hence, a solution of Maxwell's equations in their original form is,

$$\boldsymbol{e} = -\frac{\partial}{\partial t}\frac{\mu}{4\pi}\int [\boldsymbol{j}]t_0^{-1} - \mathrm{grad}\frac{1}{4\pi\varepsilon}\int [\rho]t_0^{-1},$$

$$\boldsymbol{h} = \mathrm{curl}\frac{1}{4\pi}\int [\boldsymbol{j}]t_0^{-1}.$$

If $\boldsymbol{e}$, $\boldsymbol{h}$, $\boldsymbol{j}$ and ρ all vanish outside the surface S in (x, y, z)-space then

$$\boldsymbol{a} = (\mathrm{grad}\, f) \wedge \boldsymbol{e}\delta(f),$$
$$\boldsymbol{e} = (\mathrm{grad}\, f) \wedge \boldsymbol{h}\delta(f),$$
$$p = \mu\delta(f)\boldsymbol{h}.\mathrm{grad}\, f$$

and

$$q = \varepsilon\delta(f)\boldsymbol{e}.\mathrm{grad}\, f$$

where $f>0$ inside $S, f<0$ outside S and $f=0$ on S. Consequently

$$\begin{aligned} \boldsymbol{e} = &-\frac{\partial}{\partial t}\frac{\mu}{4\pi}\int_T [\boldsymbol{j}]t_0^{-1}\,\mathrm{d}\tau - \operatorname{grad}\frac{1}{4\pi\varepsilon}\int_T [\rho]t_0^{-1}\,\mathrm{d}\tau \\ &+\frac{1}{4\pi}\frac{\partial}{\partial t}\int_S \boldsymbol{v}\wedge[\boldsymbol{b}]t_0^{-1}\,\mathrm{d}S \\ &-\operatorname{curl}\frac{1}{4\pi}\int_S \boldsymbol{v}\wedge[\boldsymbol{e}]t_0^{-1}\,\mathrm{d}S + \operatorname{grad}\frac{1}{4\pi}\int_S \boldsymbol{v}.[\boldsymbol{e}]t_0^{-1}\mathrm{d}S, \end{aligned} \tag{66}$$

$$\begin{aligned} \boldsymbol{h} = &\operatorname{curl}\frac{1}{4\pi}\int_T [\boldsymbol{j}]t_0^{-1}\,\mathrm{d}\tau - \frac{1}{4\pi}\int_S \boldsymbol{v}\wedge[\boldsymbol{d}]t_0^{-1}\,\mathrm{d}S \\ &-\operatorname{curl}\frac{1}{4\pi}\int_S \boldsymbol{v}\wedge[\boldsymbol{h}]t_0^{-1}\,\mathrm{d}S + \operatorname{grad}\frac{1}{4\pi}\int_S \boldsymbol{v}.[\boldsymbol{h}]t_0^{-1}\,\mathrm{d}S \end{aligned} \tag{67}$$

where T is the interior of S and $\boldsymbol{v}$ the outward normal to S.

Exercises

53. If, in Maxwell's equations, the dependence of $\boldsymbol{e}, \boldsymbol{h}, \boldsymbol{j}$ and ρ on t is $\mathrm{e}^{\mathrm{i}\omega t}$ show that the formulae corresponding to (66) and (67) are

$$\begin{aligned} \boldsymbol{e} = &\frac{\mathrm{i}\omega\mu}{4\pi}\int_T \boldsymbol{j}\psi\,\mathrm{d}\tau - \operatorname{grad}\frac{1}{4\pi\varepsilon}\int_T \rho\psi\,\mathrm{d}\tau - \frac{1}{4\pi}\operatorname{curl}\int_S \boldsymbol{v}\wedge\boldsymbol{e}\psi\,\mathrm{d}S \\ &-\frac{1}{4\pi\mathrm{i}\omega\varepsilon}(\operatorname{grad}\operatorname{div}+k^2)\int_S \boldsymbol{v}\wedge\boldsymbol{h}\psi\,\mathrm{d}S, \\ \boldsymbol{h} = &\operatorname{curl}\frac{1}{4\pi}\int_T \boldsymbol{j}\psi\,\mathrm{d}\tau + \frac{1}{4\pi\mathrm{i}\omega\mu}(\operatorname{grad}\operatorname{div}+k^2)\int_S \boldsymbol{v}\wedge\boldsymbol{e}\psi\,\mathrm{d}S \\ &-\frac{1}{4\pi}\operatorname{curl}\int_S \boldsymbol{v}\wedge\boldsymbol{h}\psi\,\mathrm{d}S \end{aligned}$$

where $k^2=\omega^2\mu\varepsilon, \psi=\mathrm{e}^{-\mathrm{i}kt_0}/t_0$.

54. If, in Example 10, $\rho=\delta\{x-f_2(t)\}\delta\{y-f_3(t)\}\delta\{z-f_4(t)\}$ and $\boldsymbol{j}=\rho\boldsymbol{f}'$ prove that

$$\boldsymbol{e}=\frac{\lambda^3}{4\pi\varepsilon}\left[\left(1-\frac{|\boldsymbol{v}|^2}{c^2}\right)\left(\boldsymbol{R}-\frac{\boldsymbol{v}|\boldsymbol{R}|}{c}\right)+\frac{1}{c^2}\boldsymbol{R}\wedge\left\{\left(\boldsymbol{R}-\frac{\boldsymbol{v}|\boldsymbol{R}|}{c}\right)\wedge\boldsymbol{v}'\right\}\right],$$

$$\boldsymbol{h}=c\varepsilon\boldsymbol{R}\wedge\boldsymbol{e}/|\boldsymbol{R}|$$

where $|v| < c$, the notation being that of Exercise 49, and $1/\lambda = |\boldsymbol{R}| - \boldsymbol{R}.v/c$. (This is the field radiated by a unit charge whose position at time t is $f(t)$.)

55. Examine the possibility of finding an elementary solution of

$$\sum_{j=1}^{4}\sum_{k=1}^{4} a_{lj}^{(k)}\left(\frac{\partial}{\partial x_k} - a_k\right)u_j - b\sum_{j=1}^{4} b_{lj}u_j = 0 \qquad (l = 1, 2, 3, 4)$$

where a_k and b are constants and

$$a_{lj}^{(1)} \equiv \begin{pmatrix} 0 & 0 & 0 & 1 \\ 0 & 0 & 1 & 0 \\ 0 & 1 & 0 & 0 \\ 1 & 0 & 0 & 0 \end{pmatrix}, \qquad a_{lj}^{(2)} \equiv \begin{pmatrix} 0 & 0 & 0 & -\mathrm{i} \\ 0 & 0 & \mathrm{i} & 0 \\ 0 & -\mathrm{i} & 0 & 0 \\ \mathrm{i} & 0 & 0 & 0 \end{pmatrix},$$

$$a_{lj}^{(3)} \equiv \begin{pmatrix} 0 & 0 & 1 & 0 \\ 0 & 0 & 0 & -1 \\ 1 & 0 & 0 & 0 \\ 0 & -1 & 0 & 0 \end{pmatrix}, \qquad a_{lj}^{(4)} \equiv \begin{pmatrix} -1 & 0 & 0 & 0 \\ 0 & -1 & 0 & 0 \\ 0 & 0 & -1 & 0 \\ 0 & 0 & 0 & -1 \end{pmatrix},$$

$$b_{lj} \equiv \begin{pmatrix} 1 & 0 & 0 & 0 \\ 0 & 0 & 0 & 0 \\ 0 & 0 & -1 & 0 \\ 0 & 0 & 0 & -1 \end{pmatrix},$$

56. The equations of motion in *Stokes flow* are

$$\mu\nabla^2 v - \operatorname{grad} p = 0, \qquad \operatorname{div} v = 0$$

in (x_1, x_2, x_3)-space. Show that a solution of

$$\mu(\partial_1^2 + \partial_2^2 + \partial_3^2)e_{jk} - \partial_j p_k = \delta_{jk}\,\delta(\boldsymbol{x}), \qquad \partial_j e_{jk} = 0$$

(the summation convention being employed) is given by

$$e_{jk} = -\frac{r^{-1}\delta_{jk}}{4\pi\mu} + \frac{1}{8\pi\mu}\partial_j\partial_k r, \qquad p_k = \frac{1}{4\pi}\partial_k r^{-1}.$$

Deduce that a Stokes flow which vanishes outside the closed surface S is given at interior points by

$$v_j(\boldsymbol{x}) = \int_S \left\{\left(p v_k - \mu\frac{\partial v_k}{\partial \nu}\right)e_{jk} + v_k\left(\mu\frac{\partial e_{jk}}{\partial \nu} + \nu_k p_j\right)\right\}\mathrm{d}S$$

where the argument of e_{jk} and p_j is $\boldsymbol{x} - \boldsymbol{t}$, $\boldsymbol{t}$ is a point of S and ν is the outward normal to S.

57. If a magnetic field is present in Stokes flow and the Hartmann number is much larger than the Reynolds number the equations of Exercise 56 are modified to

$$\nabla^2 v - \operatorname{grad} p + M^2(v \wedge i) \wedge i = 0, \qquad \operatorname{div} v = 0.$$

Show that a solution of

$$(\partial_1^2 + \partial_2^2 + \partial_3^2)e_{jk} - \partial_j p_k + M^2(e_{1k}\delta_{j1} - e_{jk}) = \delta_{jk}\delta(x), \qquad \partial_j e_{jk} = 0$$

gives

$$e_{1k} = \partial_k(g_1 - g_2) - M\delta_{ik}(g_1 + g_2), \quad p_k = M\partial_k(g_1 + g_2) - M^2\delta_{ik}(g_1 - g_2),$$
$$e_{j1} = \partial_j(g_1 - g_2) \quad (j \neq 1)$$

where $g_1 = r^{-1}\mathrm{e}^{(M/2)(x_1 - r)}/8\pi M$, $g_2 = r^{-1}\mathrm{e}^{-(M/2)(x_1 + r)}/8\pi M$.

If the flow vanishes outside the closed surface S,

$$v_j(x) = \int_S \left\{\left(p v_k - \frac{\partial v_k}{\partial \nu}\right)e_{jk} + v_k\left(\frac{\partial e_{jk}}{\partial \nu} + \nu_k p_j\right)\right\} \mathrm{d}S.$$

(Note that $(\partial_1^2 + \partial_2^2 + \partial_3^2 - M\partial_1)g_1 = -\delta(x)$ and $(\partial_1^2 + \partial_2^2 + \partial_3^2 + M\partial_1)g_2 = -\delta(x)$.)

There is an approach for finding an elementary solution which is of interest in certain circumstances. Suppose that the partial differential equation

$$P(\partial_1, \dots, \partial_n)e_1 = \delta(x) \tag{68}$$

is to be solved. From §7.9 it is known that

$$\delta(x) = \begin{cases} \dfrac{(-1)^{(n/2)-1/2}}{2^n \pi^{n-1}} \displaystyle\int_\Omega \delta^{(n-1)}(\omega . x)\,\mathrm{d}\Omega & (n \text{ odd}) \\[2ex] \dfrac{(n-1)!(-1)^{n/2}}{(2\pi)^n} \displaystyle\int_\Omega (\omega . x)^{-n}\,\mathrm{d}\Omega & (n \text{ even}) \end{cases}$$

where Ω is the unit sphere and ω a point of Ω. Therefore we examine possible solutions of

$$P(\partial_1, \dots, \partial_n)g_1 = \delta(\omega . x).$$

Since g_1 need depend only on $\omega . x$, this equation is

$$P(\omega_1 \mathrm{d}, \dots, \omega_n \mathrm{d})g_1(\xi, \omega) = \delta(\xi) \tag{69}$$

where $\xi = \omega . x$ and d is a derivative with respect to ξ. Equation (69) is an ordinary differential equation to determine g_1. Once it has been solved a solution of (68) is provided by

$$e_1(x)=\begin{cases}\dfrac{(-1)^{(n/2)-1/2}}{2^n\pi^{n-1}}\displaystyle\int_\Omega \frac{\partial^{n-1}}{\partial\xi^{n-1}}g_1(\xi,\omega)\mathrm{d}\Omega & (n \text{ odd})\\[2ex] \dfrac{(n-1)!(-1)^{n/2}}{(2\pi)^n}\displaystyle\int_\Omega\int_{-\infty}^{\infty} g_1(\xi-\eta,\omega)\eta^{-n}\mathrm{d}\eta\,\mathrm{d}\Omega & (n \text{ even}).\end{cases}$$

If P is a homogeneous polynomial of degree m the introduction of g_1 can be avoided. For it is only necessary to consider

$$P(\omega_1,\dots,\omega_n)\mathrm{d}^m g_2=\delta^{(n-1)}(\xi)$$

or ξ^{-n} according as n is odd or even. The solution of this equation is immediate. For example, if m is even,

$$P(\omega_1,\dots,\omega_n)g_2=\begin{cases}|\xi|^{m-n}/(m-n)!2 & (m\geq n)\\ \delta^{(n-m-1)}(\xi) & (m<n)\end{cases}$$

if n is odd, whereas if n is even

$$P(\omega_1,\dots,\omega_n)g_2=\begin{cases}\dfrac{(-1)^{n-1}\xi^{m-n}\ln|\xi|}{(m-n)!(n-1)!} & (m\geq n)\\[2ex] (n-m-1)!(-1)^m\xi^{m-n}/(n-1)! & (m<n).\end{cases}$$

The expression for e_1 can be deduced at once; for example, if m is even, n is odd and $m<n$

$$e_1(x)=\frac{(-1)^{(n/2)-1/2}}{2^n\pi^{n-1}}\int_\Omega\frac{\delta^{(n-m-1)}(\omega . x)}{P(\omega_1,\dots,\omega_n)}\mathrm{d}\Omega.$$

Such an expression has a definite meaning only if $P(\omega_1,\dots,\omega_n)$ does not vanish on Ω. If P does vanish, an interpretation such as a Cauchy principal value is necessary.

10.5. Approximate behaviour of elementary solutions

Although the Fourier transform provides a systematic way of finding elementary solutions of partial differential equations with constant coefficients, the solution so obtained is in the form of a Fourier integral. Often this integral cannot be evaluated easily.

Nevertheless it is frequently possible to deduce useful information by employing the results of Chapter 9.

Example 11. Consider in two dimensions,

$$(\partial_1^4 + \partial_2^4 - 1)e_1 = \delta(\boldsymbol{x}).$$

An elementary solution is

$$e_1 = \frac{1}{4\pi^2}\int_{-\infty}^{\infty} (\alpha_1^4 + \alpha_2^4 - 1)^{-1}\, \mathrm{e}^{\mathrm{i}(\alpha_1 x_1 + \alpha_2 x_2)}\, \mathrm{d}\boldsymbol{\alpha}.$$

According to §9.7 the asymptotic behaviour for large $|\boldsymbol{x}|$ may be determined from consideration of the curve $x^4 + y^4 = 1$. First find the points where the normal to this curve is parallel to the radius vector to (x_1, x_2); these satisfy $x^3/y^3 = x_1/x_2$. There are two such points and, if $x_2/x_1 \gg 1$, they are $(\eta, 1 - \frac{1}{4}\eta^4)$ and $(-\eta, -1 + \frac{1}{4}\eta^4)$ approximately where $\eta = (x_1/x_2)^{1/3}$. At $(\eta, 1 - \frac{1}{4}\eta^4)$ $\boldsymbol{\nu}$.grad is 4 and at the other point it is -4. The curvature has magnitude $3\eta^2$ at both points. Hence

$$e_1 \sim \frac{\mathrm{i}\mathrm{e}^{(\pi/4)\mathrm{i} + \mathrm{i}(\eta x_1 + x_2)}}{8(2\pi)^{1/2}3^{1/2}\eta(x_1^2 + x_2^2)^{1/4}} - \frac{\mathrm{i}\mathrm{e}^{(\pi/4)\mathrm{i} - \mathrm{i}(\eta x_1 + x_2)}}{8(2\pi)^{1/2}3^{1/2}\eta(x_1^2 + x_2^2)^{1/4}}$$

$$\sim -\frac{\mathrm{e}^{(\pi/4)\mathrm{i}}\sin\{(x_1^{4/3} + x_2^{4/3})x_2^{1/3}\}}{4(6\pi)^{1/2}x_1^{1/3}x_2^{1/6}}$$

as $x_2 \to \infty$ with $x_2 \gg x_1 > 0$.

Exercises

58. Repeat Example 11 but without making the approximation $x_2 \gg x_1$.

59. Find the asymptotic behaviour of a solution of

$$(a_1^2\partial_1^2 + \partial_2^2 + \partial_3^2 + 1)e_1 = \delta(\boldsymbol{x})$$

in three dimensions.

60. The electric field of a harmonic wave in a crystal satisfies

$$\nabla^2 \boldsymbol{e} - \text{grad div}\, \boldsymbol{e} + \omega^2 \underline{\varepsilon}.\boldsymbol{e} = 0$$

where the tensor $\underline{\varepsilon}$ has only diagonal elements $\varepsilon_1, \varepsilon_2$ and ε_3. If D_{jk} is the co-factor of d_{jk} in the determinant $|d_{jk}|$ where $d_{jk} = \delta_{jk}(\partial_1^2 + \partial_2^2 + \partial_3^2 + \varepsilon_j\omega^2) - \partial_j\partial_k$ and $e_j = D_{jk}u_k$ show that

$$\{(\partial_1^2 + \partial_2^2 + \partial_3^2)\psi(\partial) + \omega^2\phi(\partial) + \omega^4\}u_k = 0$$

where

$$\phi(\partial) = (\partial_2^2 + \partial_3^2)/\varepsilon_1 + (\partial_3^2 + \partial_1^2)/\varepsilon_2 + (\partial_1^2 + \partial_2^2)/\varepsilon_3,$$
$$\psi(\partial) = \partial_1^2/\varepsilon_2\varepsilon_3 + \partial_2^2/\varepsilon_3\varepsilon_1 + \partial_3^2/\varepsilon_1\varepsilon_2.$$

Show also that the points $\boldsymbol{\alpha}$ which contribute to the distant behaviour, in the direction of the unit vector $\boldsymbol{n}$, of an elementary solution satisfy

$$\sum_{j=1}^{3} \frac{n_j^2}{1 - \varepsilon_j \omega^2/(\boldsymbol{\alpha}.\boldsymbol{n})^2} = 0.$$

61. *Small amplitude harmonic magnetohydrodynamic waves* lead to

$$\omega^2\{\omega^2 - (a_0^2 + a_1^2)(\partial_1^2 + \partial_2^2 + \partial_3^2)\}e_1 + a_0^2 a_1^2 \partial_3^2(\partial_1^2 + \partial_2^2 + \partial_3^2)e_1 = \delta(\boldsymbol{x}).$$

If $a_1 < a_0$ show that, at a large distance, there may be two sets of waves or one set and indicate the region of space in which two sets may be expected.

62. In a *transversely isotropic elastic medium*, the following equations arise for harmonic waves:

$$a_3 \partial_3^2 g_2 + \{a_5(\partial_1^2 + \partial_2^2) + a_2 \partial_3^2\} g_3 + \omega^2 g_3 = 0,$$
$$a_5 \partial_3^2 g_1 + a_4(\partial_1^2 + \partial_2^2) g_1 - \omega^2 g_1 = \delta(x_1)\delta'(x_2)\delta(x_3),$$
$$a_3(\partial_1^2 + \partial_2^2) g_3 + \{a_5 \partial_3^2 + a_1(\partial_1^2 + \partial_2^2)\} g_2 + \omega^2 g_2 = -\delta'(x_1)\delta(x_2)\delta(x_3).$$

Examine the behaviour at large distances.

10.6. Covariance and generalised functions of positive type

In many branches of applied mathematics where random processes are involved one is concerned with the correlation between two variables. Usually the correlation is defined in terms of the *covariance* which is the expectation of the product of the two random functions.

Closely associated with the idea of covariance is the notion of *functions of positive type*. A function $f(t, t_1)$ is said to be of positive type on $T \times T$ if, for every *finite* subset T_m and every function $h(t)$ on T_m,

$$\sum_{t, t_1 \in T_m} f(t, t_1) h(t) \bar{h}(t_1) \geq 0.$$

It can be shown that $f(t, t_1)$ is a covariance if, and only if, it is of positive type.

In *stationary* random processes the covariance depends only on the difference between its arguments. From now on we shall restrict our attention to such processes and redefine a function of positive type so as to be particularly appropriate to such processes.

Definition 10.2. *A function $f(\boldsymbol{x})$ on R_n is said to be of positive type if for every finite set $\boldsymbol{x}_1, \ldots, \boldsymbol{x}_m$ and every function h,*

$$\sum_{j=1}^{m} \sum_{k=1}^{m} f(\boldsymbol{x}_j - \boldsymbol{x}_k) h(\boldsymbol{x}_j) \bar{h}(\boldsymbol{x}_k) \geq 0.$$

The notation $f \rhd 0$ will be used to signify a function of positive type.

Theorem 10.3. *If $f \rhd 0$ then*

$$f(\mathbf{0}) \geq 0, \qquad f(-\boldsymbol{x}) = \bar{f}(\boldsymbol{x}), \qquad |f(\boldsymbol{x})| \leq f(\mathbf{0}).$$

If, further, f is continuous at the origin, then f is continuous and bounded on R_n.

Proof. From Definition 10.2 with $m = 1$, $f(\mathbf{0}) \geq 0$. With $m = 2$, $\boldsymbol{x}_1 = \mathbf{0}$ and $\boldsymbol{x}_2 = \boldsymbol{x}$ we have

$$f(\mathbf{0})\{h(\mathbf{0})\bar{h}(\mathbf{0}) + h(\boldsymbol{x})\bar{h}(\boldsymbol{x})\} + f(\boldsymbol{x})h(\boldsymbol{x})\bar{h}(\mathbf{0}) + f(-\boldsymbol{x})h(\mathbf{0})\bar{h}(\boldsymbol{x}) \geq 0 \quad (70)$$

and therefore

$$\operatorname{Im}\{f(\boldsymbol{x})h(\boldsymbol{x})\bar{h}(\mathbf{0}) + f(-\boldsymbol{x})h(\mathbf{0})\bar{h}(\boldsymbol{x})\} = 0.$$

Choosing $h(\mathbf{0}) = 1$, $h(\boldsymbol{x}) = 1$ and then $h(\boldsymbol{x}) = \mathrm{i}$ we find

$$\operatorname{Im}\{f(\boldsymbol{x}) + f(-\boldsymbol{x})\} = 0, \qquad \operatorname{Re}\{f(\boldsymbol{x}) - f(-\boldsymbol{x})\} = 0,$$

i.e. $f(-\boldsymbol{x}) = \bar{f}(\boldsymbol{x})$. Also (70) can be written as

$$\left| h(\mathbf{0})\sqrt{f(\mathbf{0})} + \frac{h(\boldsymbol{x})f(\boldsymbol{x})}{\sqrt{f(\mathbf{0})}} \right|^2 + \left\{ f(\mathbf{0}) - \frac{f(\boldsymbol{x})\bar{f}(\boldsymbol{x})}{f(\mathbf{0})} \right\} |h(\boldsymbol{x})|^2 \geq 0$$

from which it is evident that $|f(\boldsymbol{x})| \leq f(\mathbf{0})$.

To prove the final result note that $f(\mathbf{0}) = 0$ (which implies $f \equiv 0$) can be excluded and, indeed, there is no loss of generality in taking $f(\mathbf{0}) = 1$ (by replacing f by $f/f(\mathbf{0})$). Then from Definition 10.2 with $m = 3$, $\boldsymbol{x}_1 = \mathbf{0}$, $\boldsymbol{x}_2 = \boldsymbol{x}$ and $\boldsymbol{x}_3 = \boldsymbol{y}$ we obtain a quadratic form in $h(\mathbf{0})$, $h(\boldsymbol{x})$ and $h(\boldsymbol{y})$. This can be non-negative only if its discriminant is non-negative, i.e. if

$$1 + f(\boldsymbol{x})f(-\boldsymbol{y})f(\boldsymbol{y} - \boldsymbol{x}) + f(\boldsymbol{y})f(-\boldsymbol{x})f(\boldsymbol{x} - \boldsymbol{y}) \\ - |f(\boldsymbol{x})|^2 - |f(\boldsymbol{y})|^2 - |f(\boldsymbol{x} - \boldsymbol{y})|^2 \geq 0$$

or if

$$|f(\boldsymbol{x}) - f(\boldsymbol{y})|^2 \leq 1 - |f(\boldsymbol{x} - \boldsymbol{y})|^2 - 2\operatorname{Re}[\bar{f}(\boldsymbol{x})f(\boldsymbol{y})\{1 - f(\boldsymbol{x} - \boldsymbol{y})\}].$$

Therefore, if f is continuous at the origin, i.e. if $f(x-y) \to f(0) = 1$ as $x \to y$, then $f(x) \to f(y)$. The proof is complete. □

The most obvious generalisation of Definition 10.2 so as to be applicable to generalised functions is to require that

$$\int_{-\infty}^{\infty}\int_{-\infty}^{\infty} g(x-t)\gamma(x)\bar{\gamma}(t)\,dx\,dt \geq 0$$

for every good γ. This may also be written as

$$\int_{-\infty}^{\infty} g(t)\int_{-\infty}^{\infty} \gamma(x)\bar{\gamma}(x-t)\,dx\,dt \geq 0.$$

Accordingly

Definition 10.3. *A generalised function g is said to be of positive type if*

$$\int_{-\infty}^{\infty} g(x)\int_{-\infty}^{\infty} \gamma(t)\bar{\gamma}(t-x)\,dt\,dx \geq 0 \tag{71}$$

for every good γ.

As with functions the notation $g \rhd 0$ will be employed to denote a generalised function of positive type.

Obviously a function $f \rhd 0$ can be identified, on account of Theorem 10.3, with a generalised function of positive type.

Theorem 10.4. *$g \rhd 0$ if, and only if, $G \geq 0$.*

Proof. By Theorem 7.13 and Theorem 7.38

$$\int_{-\infty}^{\infty} g(x)\int_{-\infty}^{\infty} \gamma(t)\bar{\gamma}(t-x)\,dt\,dx = \frac{1}{(2\pi)^n}\int_{-\infty}^{\infty} G(\alpha)\Gamma(-\alpha)\bar{\Gamma}(-\alpha)\,d\alpha. \tag{72}$$

Since $\Gamma\bar{\Gamma} \geq 0$ and is good it follows from the definition of §3.6 that, if $G \geq 0$, the right-hand side of (72) is non-negative and therefore (71) is satisfied, i.e. $g \rhd 0$.

Conversely, if $g \rhd 0$, (71) and (72) imply that

$$\int_{-\infty}^{\infty} G(\alpha)\Gamma(-\alpha)\bar{\Gamma}(-\alpha)\,d\alpha \geq 0. \tag{73}$$

It is not an immediate consequence that $G \geq 0$ because not every good $\Gamma_1 \geq 0$ can be expressed as $\Gamma\bar{\Gamma}$. Let $\{\phi_m\}$ be a sequence of real positive fine functions such that $\operatorname{Lim}_{m\to\infty} \phi_m G = G$ (such fine functions exist by Theorem 5.15). Let ϕ be a positive fine function which is unity wherever ϕ_m is non-zero. Then, if $\varepsilon > 0$,

$$(\varepsilon + \Gamma_1)\phi_m = (\varepsilon + \Gamma_1)\phi_m\phi^2 = \Gamma\bar{\Gamma}$$

where $\Gamma = \phi\{\phi_m(\varepsilon + \Gamma_1)\}^{1/2}$. Consequently (73) implies that

$$\int_{-\infty}^{\infty} G(\alpha)\phi_m(\alpha)\{\varepsilon + \Gamma_1(\boldsymbol{\alpha})\}\,\mathrm{d}\boldsymbol{\alpha} \geq 0$$

which, since ε can be chosen arbitrarily small, means that

$$\int_{-\infty}^{\infty} G(\boldsymbol{\alpha})\phi_m(\boldsymbol{\alpha})\Gamma_1(\boldsymbol{\alpha})\,\mathrm{d}\boldsymbol{\alpha} \geq 0.$$

On allowing $m \to \infty$ we deduce that $\int_{-\infty}^{\infty} G(\boldsymbol{\alpha})\Gamma_1(\boldsymbol{\alpha})\,\mathrm{d}\boldsymbol{\alpha} \geq 0$ for any good $\Gamma_1 \geq 0$. Hence $G \geq 0$ and the theorem is proved. □

Corollary 10.4. *If* $g \rhd 0$ *then* $\operatorname{Im}\{G(\boldsymbol{\alpha})\} = 0$ *and* $g(-\boldsymbol{x}) = \bar{g}(\boldsymbol{x})$.

Proof. If $g \rhd 0$ then $G \geq 0$ and hence, from Exercise 34 of Chapter 3, $\operatorname{Im} G = 0$. Further

$$g(-\boldsymbol{x}) = \frac{1}{(2\pi)^n}\int_{-\infty}^{\infty} G(\boldsymbol{\alpha})\mathrm{e}^{-\mathrm{i}\boldsymbol{\alpha}.\boldsymbol{x}}\,\mathrm{d}\boldsymbol{\alpha} = \bar{g}(\boldsymbol{x})$$

since $\operatorname{Im} G = 0$ and the corollary is demonstrated. □

There is another theorem of some interest:

Theorem 10.5. *If* $g \in \boldsymbol{L}_2'$ *then* $g * \tilde{g} \rhd 0$. *If* $g_1 \rhd 0$ *then* $g_1 = g * \tilde{g}$ $(g \in \boldsymbol{L}_2')$ *if, and only if,* $G_1 \in \boldsymbol{K}_1$.

Here $\tilde{g}(\boldsymbol{x}) = \bar{g}(-\boldsymbol{x})$.

Proof. If $g \in \boldsymbol{L}_2'$, Exercise 52 of Chapter 7 shows that the Fourier transform of $g * \tilde{g}$ is $G\bar{G}$ where $G \in \boldsymbol{K}_2$ and the first part of the theorem follows from Theorem 10.4. This also shows that $G_1 \in \boldsymbol{K}_1$ if $g_1 = g * \tilde{g}$.

Conversely, if $g_1 \rhd 0$ and $G_1 \in \boldsymbol{K}_1$ then $G_1 \geq 0$ and $\sqrt{G_1} \in \boldsymbol{K}_2$; the proof is complete. □

Example 12. $\delta(\boldsymbol{x}) \rhd 0$ because the Fourier transform of $\delta(\boldsymbol{x})$ is 1 and Theorem 10.4 applies. Similarly

$$(-\partial_1^2 - \partial_2^2 - \ldots - \partial_n^2)^k \delta(\boldsymbol{x}) \rhd 0. \tag{74}$$

On the other hand $r^\beta \geq 0$ if $\beta > -n$. Hence, from Theorem 10.4 and Theorem 7.31,

$$(\tfrac{1}{2}\beta + \tfrac{1}{2}n - 1)!\, r^{-\beta-n}/(-\tfrac{1}{2}\beta - 1)! \rhd 0$$

if $\beta \neq 2m$. The case $\beta = 2m$ is covered by (74). Since $x\,!$ is negative only in the intervals $(-1, -2)$, $(-3, -4)$, ... it follows that, if $2k > \beta > 2k - 2$,

$$(-1)^k r^{-\beta-n} \rhd 0.$$

Exercises

63. If $g \rhd 0$ prove that $\bar{g} \rhd 0$, $g(-\boldsymbol{x}) \rhd 0$ and $\bar{g}(-\boldsymbol{x}) \rhd 0$.
64. Show that $g \rhd 0$ if $\int_{-\infty}^{\infty} g(\boldsymbol{x})\gamma(\boldsymbol{x})\mathrm{d}\boldsymbol{x} \geq 0$ for every good $\gamma \rhd 0$.
65. If ψ is fairly good and $\rhd 0$, if $g \rhd 0$ prove that $\psi g \rhd 0$.
66. If $g_1 \in L'_p$ $(1 \leq p \leq \infty)$, $g_2 \in L'_q$ $\{q = p/(p-1)\}$, $G_1 \in \boldsymbol{K}_q$ and $G_2 \in \boldsymbol{K}_p$ show that $g_1 * g_2 \rhd 0$ when $g_1 \rhd 0$ and $g_2 \rhd 0$.
67. Show that $g \rhd 0$ if, and only if, $g = (1 - \partial_1^2 - \ldots - \partial_n^2)^k f$ where $f \rhd 0$ and is continuous on R_n. (Hint: for sufficiently large k, $G/(1+\alpha^2)^k$ is the Fourier transform of a function, continuous and bounded on R_n.)
68. If $g \rhd 0$ prove that $g \in L'_\infty$.

11
Weak functions

11.1. Weak functions in one dimension

The generalised function possesses the advantages of (i) always having a generalised derivative, (ii) always having a Fourier transform to which the Fourier inversion theorem is applicable. Its main disadvantage is that its growth at infinity is limited, essentially being bounded by some power of x. If we wish to remove this restriction without losing property (i) then we must expect that it may be necessary to pay the price of dispensing with (ii). The reason why the generalised function does not have unlimited increase at infinity is that $\int_{-\infty}^{\infty} g(x)\gamma(x)\mathrm{d}x$ must be finite for every good γ and γ need not tend to zero at infinity faster than exponentially. Thus to extend the range it is necessary to dispense with definitions via good functions and use functions which decay more rapidly at infinity. The obvious choice is to employ fine functions since these are zero outside a finite interval and so will allow a multiplying entity unrestricted growth at infinity. Accordingly

Definition 11.1. *A sequence* $\{\phi_m\}$ *of fine functions is said to be regular if, for every fine* ϕ, $\lim_{m\to\infty} \int_{-\infty}^{\infty} \phi_m(x)\phi(x)\mathrm{d}x$ *exists and is finite. Two regular sequences which have the same limit are said to be equivalent. An equivalence class of regular sequences is a weak function.*

A convenient notation is to write, if w is the weak function associated with the equivalence class of which $\{\phi_m\}$ is a typical member,

$$\lim_{m\to\infty} \int_{-\infty}^{\infty} \phi_m(x)\phi(x)\mathrm{d}x = \int_{-\infty}^{\infty} w(x)\phi(x)\mathrm{d}x.$$

As in §8.2, the behaviour of a weak function in $a < x < b$ can be determined by considering only ϕ and ϕ_m which vanish identically outside (a, b).

On account of Theorem 3.1, Definition 11.1 is capable of distinguishing between entities that are normally regarded as distinct.

Theorem 11.1. *Any generalised function is necessarily a weak function.*

Proof. It has been shown in §6.4 that any generalised function can be defined by a sequence of fine functions $\{\phi_m\}$ such that $\lim_{m\to\infty} \int_{-\infty}^{\infty} \phi_m(x)\gamma(x)\,dx$ exists for every good γ. Since the good functions include the fine functions the conditions of Definition 11.1 are satisfied and the theorem is proved. □

Consequently, any notation which has been used for a generalised function will be used for the corresponding weak function.

It is not of course true that every weak function is a generalised function. A graphic illustration of this is provided by the following theorem in which ϕ_{0m} is a fine function such that

$$\phi_{0m}(x) = \begin{cases} 1 & (|x| \le m) \\ 0 & (|x| \ge m+1) \end{cases}$$

and $0 \le \phi_{0m}(x) \le 1$.

Theorem 11.2. *If f is continuous on R_1 the sequence $\{\phi_m\}$, where $\phi_m(x) = \int_{-\infty}^{\infty} f(u)\rho\{m(u-x)\}m\phi_{0m}(u)\,du$, is regular and defines a weak function w such that*

$$\int_{-\infty}^{\infty} w(x)\phi(x)\,dx = \int_{-\infty}^{\infty} f(x)\phi(x)\,dx.$$

Proof. Since $\phi_{0m}(u) = 0$ for $|u| \ge m+1$ the integral in the definition of ϕ_m is actually over a finite interval and is uniformly convergent, together with its derivatives. Also, because $\rho\{m(u-x)\} = 0$ for $|x-u| \ge 1/m$, ϕ_m and its derivatives vanish for $|x| \ge m+1+1/m$. Therefore ϕ_m is a fine function.

To show that the sequence is regular we have

$$\int_{-\infty}^{\infty} \phi_m(x)\phi(x)\,dx = \int_{-\infty}^{\infty} f(u)\phi_{0m}(u)\left[\int_{-\infty}^{\infty} \phi(x)\rho\{m(u-x)\}m\,dx\right]du$$

since both integrals are actually over finite intervals and the integrand is continuous. On putting $x = u - y/m$, we obtain

$$\int_{-\infty}^{\infty} \phi_m(x)\phi(x)\,\mathrm{d}x = \int_{-1}^{1} \rho(y)\,\mathrm{d}y \left[\int_{-\infty}^{\infty} f(u)\{\phi(u - y/m) - \phi(u)\}\phi_{0m}(u)\,\mathrm{d}u + \int_{-\infty}^{\infty} f(u)\phi(u)\phi_{0m}(u)\,\mathrm{d}u \right].$$

Now, if $\phi(u)$ vanishes for $|u| \geq A$,

$$\int_{-\infty}^{\infty} f(u)\phi(u)\{\phi_{0m}(u) - 1\}\,\mathrm{d}u = \int_{-A}^{A} f(u)\phi(u)\{\phi_{0m}(u) - 1\}\,\mathrm{d}u \to 0$$

as $m \to \infty$ because $\phi_{0m}(u) = 1$ for $|u| \leq A$ when $m \geq A$. Also

$$\left| \int_{-\infty}^{\infty} f(u)\{\phi(u - y/m) - \phi(u)\}\phi_{0m}(u)\,\mathrm{d}u \right|$$

$$= \left| \int_{-A-|y|/m}^{A+|y|/m} f(u)\{\phi(u - y/m) - \phi(u)\}\phi_{0m}(u)\,\mathrm{d}u \right|$$

$$\leq \frac{|y|}{m} \int_{-A-|y|/m}^{A+|y|/m} |f|\,|\phi'|\,\mathrm{d}u = O(|y|/m)$$

since f and ϕ' are bounded. From this it is evident that

$$\lim_{m\to\infty} \int_{-\infty}^{\infty} \phi_m(x)\phi(x)\,\mathrm{d}x = \int_{-\infty}^{\infty} f(x)\phi(x)\,\mathrm{d}x$$

and the theorem is proved. □

One concludes from the theorem that $\mathrm{e}^x, \mathrm{e}^{x^2}, \mathrm{e}^{\mathrm{e}^x}, \ldots$ are weak functions. Since e^x is not a generalised function (§3.6) here is a definite example of a weak function which is not a generalised function. Indeed, it is evident that the class of weak functions is very much larger than the class of generalised functions. This is because weak functions are not restricted at infinity in the same way as generalised functions. On the other hand, weak functions are subject to the same limitations in the finite part of the plane as generalised functions. Thus the function equal to $\mathrm{e}^{2/x}$ in $x > 0$ and 0 in $x \leq 0$ is neither a weak function nor a generalised function. (Take $\phi(x) = \exp\{1/(x-1) - 1/x\}$ in $[0, 1]$ and zero elsewhere.)

Exercise

1. Show that Theorem 11.2 remains valid if the condition on f is replaced by $f \in L_1(a,b)$ for every finite (a,b).

11.2. The weak derivative and the weak limit

Many of the definitions and theorems for generalised functions are applicable to weak functions with little modification. Therefore the discussion will be concentrated on the main differences. Of course, proofs which depend on the use of the Fourier transform will have to be discarded because a weak function need not have a Fourier transform. (The Fourier transform of a fine function is not necessarily fine.)

Definition 11.2. *The weak functions* $w_1 + w_2$, $w(ax+b)$ *with* a *and* b *real,* $w'(x)$ *are defined by the regular sequences* $\{\phi_{1m} + \phi_{2m}\}$, $\{\phi_m(ax+b)\}$, $\{\phi'_m(x)\}$ *respectively. If* η *is infinitely differentiable on* R_1, *the weak function* $\eta(x)w(x)$ *is defined by the sequence* $\{\eta\phi_m\}$.

There is no difficulty in verifying, as in §3.3, that

$$\int_{-\infty}^{\infty} \{w_1(x) + w_2(x)\}\phi(x)\,\mathrm{d}x = \int_{-\infty}^{\infty} w_1(x)\phi(x)\,\mathrm{d}x + \int_{-\infty}^{\infty} w_2(x)\phi(x)\mathrm{d}x,$$

$$\int_{-\infty}^{\infty} w(ax+b)\phi(x)\,\mathrm{d}x = \int_{-\infty}^{\infty} w(x)\phi\left(\frac{x-b}{a}\right)\frac{\mathrm{d}x}{|a|},$$

$$\int_{-\infty}^{\infty} \{\eta(x)w(x)\}\phi(x)\,\mathrm{d}x = \int_{-\infty}^{\infty} w(x)\{\eta(x)\phi(x)\}\,\mathrm{d}x,$$

$$\int_{-\infty}^{\infty} w'(x)\phi(x)\,\mathrm{d}x = -\int_{-\infty}^{\infty} w(x)\phi'(x)\,\mathrm{d}x.$$

It is obvious that the derivative of a generalised function is the same whether it be regarded as a generalised or as a weak derivative. Higher order weak derivatives are defined in an obvious way.

Theorem 11.3. *If* f *is a function with conventional derivative* $\mathrm{d}f/\mathrm{d}x$ *and* $\mathrm{d}f/\mathrm{d}x \in L_1(a,b)$ *(*a *may be* $-\infty$ *and* b *may be* $+\infty$*), and if* $w(x) = f(x)$ *in* $a < x < b$ *then*

$$w'(x) = \mathrm{d}f(x)/\mathrm{d}x \quad (a < x < b).$$

Proof. This is similar to that of Theorem 3.8, noting that $\int_a^b f(x)\phi(x)\mathrm{d}x$ and $\int_a^b(\mathrm{d}f/\mathrm{d}x)\phi(x)\mathrm{d}x$ must both exist on account of the assumed conditions. □

There is thus equality between the weak and conventional derivatives when the latter exists.

Exercises

2. If η is infinitely differentiable on R_1 and w_1, w_2 weak functions such that $\eta(x)w_1(x) = w_2(x)$ prove that

$$\eta(ax+b)w_1(ax+b) = w_2(ax+b).$$

3. If η is infinitely differentiable on R_1 prove that

$$\eta(x)\delta(x) = \eta(0)\delta(x), \qquad \eta(x)\delta'(x) = -\eta'(0)\delta(x) + \eta(0)\delta'(x).$$

4. If w_1 and w_2 are weak functions show that

$$(w_1+w_2)' = w_1' + w_2', \qquad \{w_1(ax+b)\}' = aw_1'(ax+b),$$
$$\{\eta(x)w_1(x)\}' = \eta'(x)w_1(x) + \eta(x)w_1'(x),$$
$$\eta_1(x)\{\eta_2(x)w_1(x)\} = \{\eta_1(x)\eta_2(x)\}w_1(x) = \eta_2(x)\{\eta_1(x)w_1(x)\}$$

η, η_1, and η_2 being infinitely differentiable on R_1.

The weak limit is defined in a way analogous to that for generalised functions.

Definition 11.3. $\mathrm{Lim}_{\mu\to\mu_0}\, w_\mu = w$ *if, and only if,*

$$\lim_{\mu\to\mu_0}\int_{-\infty}^{\infty} w_\mu(x)\phi(x)\,\mathrm{d}x = \int_{-\infty}^{\infty} w(x)\phi(x)\,\mathrm{d}x$$

for every fine ϕ.

Theorem 11.4. *If* $\mathrm{Lim}_{\mu\to\mu_0}\, w_\mu = w$ *then*

(i) $\mathrm{Lim}_{\mu\to\mu_0}\, w'_\mu = w'$,

(ii) $\mathrm{Lim}_{\mu\to\mu_0}\, w_\mu(ax+b) = w(ax+b)$,

(iii) $\mathrm{Lim}_{\mu\to\mu_0}\, \eta(x)w_\mu(x) = \eta(x)w(x)$

if η *is infinitely differentiable on* R_1.

Proof. As for Theorem 3.14. □

Infinite series are handled in the same way as for generalised functions. They will be considered in more detail later.

Example 1. $\sum_{m=1}^{\infty} a_m \delta^{(m)}(x-m)$ is a weak function. For

$$\lim_{M\to\infty} \int_{-\infty}^{\infty} \sum_{m=1}^{M} a_m \delta^{(m)}(x-m)\phi(x)\,dx = \sum_{m=1}^{N} (-1)^m a_m \phi^{(m)}(m)$$

where N is the smallest integer such that $\phi(x)=0$ for $|x|>N$, and the series on the right-hand side contains only a finite number of terms.

Exercises

5. Prove that $w'(x) = \mathrm{Lim}_{h\to 0}\,\{w(x+h)-w(x)\}/h$.
6. If f_μ, f are continuous on R_1 and f_μ tends to f uniformly in x as $\mu \to \mu_0$ show that $\mathrm{Lim}_{\mu\to\mu_0} f_\mu = f$ in the weak sense. Deduce the corresponding result if $f_\mu, f \in L_1(a,b)$ and $\int_a^b |f_\mu(x) - f(x)|\,dx \to 0$.
7. If $\partial w_\mu/\partial\mu$ exists prove that $(\partial w_\mu/\partial\mu)' = \partial w'_\mu/\partial\mu$.
8. Is the weak function in Example 1 a generalised function?
9. Show that the following are weak functions:

   (i) $\sum_{m=1}^{\infty} a_m \delta^{(m)}(x-x_m)$, $x_m > x_{m-1}$ and $x_m \to \infty$ as $m\to\infty$,

   (ii) $\mathrm{Lim}_{M\to\infty}\,\{\sum_{m=1}^{M} \delta(x-1/m) - M\delta(x) + \delta'(x)\ln M\}$.

11.3. The classification of weak functions

The aim of this section is to show that, in a finite interval, any weak function is the weak derivative of some order of a continuous function.

Theorem 11.4a. *If w is a weak function and a and b are finite there are finite numbers K and r such that*

$$\left|\int_{-\infty}^{\infty} w(x)\phi(x)\,dx\right| = \left|\int_a^b w(x)\phi(x)\,dx\right| \le K \max_{a\le x\le b} |\phi^{(r)}(x)|$$

for all fine ϕ which vanish outside (a, b).

Proof. Let $\{\phi_m\}$ be a sequence of fine functions, vanishing outside (a, b), which define w in (a, b). Then there is a constant $K'(\phi)$ which depends on ϕ such that $|\int_a^b \phi_m(x)\phi(x)\,dx| \le K'(\phi)$ for all m, and

$$\left|\int_a^b w(x)\phi(x)\,dx\right| \le K'(\phi).$$

Also

$$\left|\int_a^b \phi_m(x)\phi(x)\,dx\right| \le \left\{\int_a^b |\phi_m(x)|\,dx\right\} \max_{a\le x\le b} |\phi| \le K_{m0} \max|\phi|$$

where K_{m0} is independent of ϕ. Now

$$\phi^{(m)}(x) = \int_a^x \frac{(x-t)^{r-m-1}}{(r-m-1)!}\phi^{(r)}(t)\,dt$$

so that $\max|\phi^{(m)}| \le (b-a)^{r-m-1} \max|\phi^{(r)}|/(r-m-1)!$ for $m < r$. Thus there is a finite K_{ms} such that

$$\left|\int_a^b \phi_m(x)\phi(x)\,dx\right| \le K_{ms} \max|\phi^{(s)}|$$

for every finite m.

We want now to show that, for some finite s, K_{ms} is bounded as $m \to \infty$. Suppose that the contrary is true, namely that for any given s an m can be found such that $|\int_a^b \phi_m(x)\phi(x)\,dx| > L_m \max|\phi^{(s)}|$ no matter how large L_m. Then, as in Theorem 3.16, a subsequence $\{\phi_{m_n}\}$ of $\{\phi_m\}$ and a sequence $\{\phi_{m0}\}$ can be found such that

(i) $\max|\phi_{m0}^{(p)}(x)| < 2^{-m}$ for $p = 0, 1, \ldots, m$,

(ii) $|\int_a^b \phi_{m_p}\phi_{m0}(x)\,dx| < 2^{-m}$ for $p = 1, 2, \ldots, m-1$,

(iii) $\int_{-\infty}^{\infty} \phi_{m_n}\phi_{n0}(x)\,dx > 1 + n + \sum_{r=1}^{n-1} K'(\phi_{r0})$.

Hence $\sum_{p=1}^{\infty} \phi_{p0}(x)$ is fine and

$$\int_{-\infty}^{\infty} \phi_{m_n}(x) \sum_{p=1}^{\infty} \phi_{p0}(x)\,dx > n.$$

It is evident that $\{\phi_{m_n}\}$ is not regular as $n \to \infty$, which is contrary to $\{\phi_m\}$ being the defining sequence of w. Hence K_{ms} must be bounded as $m \to \infty$ and the theorem is proved. □

Theorem 11.5. *If w is a weak function and a and b are finite*

$$w(x) = f^{(r)}(x) \quad (a < x < b)$$

where f is continuous and vanishes outside $a - \varepsilon \le x \le b + \varepsilon$, ε being an arbitrary positive number.

Proof. Define the linear functional F over the space of fine functions which vanish outside (a, b) by

$$F(\phi^{(r)}) = \int_a^b w(x)\phi(x)\,\mathrm{d}x.$$

With $\|\phi^{(r)}\| = \max|\phi^{(r)}|$, F is a bounded linear functional with $\|F\| = K$ on account of Theorem 11.4*a*. By the Hahn–Banach theorem (Theorem 1.11) F can be extended to a bounded linear functional over the space of continuous functions without increasing the norm of F. The Riesz representation theorem (§1.11) then gives

$$F(\phi^{(r)}) = \int_a^b \phi^{(r)}(x)\,\mathrm{d}h(x)$$

where the total variation of h is K and, indeed, we can take $\max|h| \le K$. Thus

$$\int_{-\infty}^{\infty} w(x)\phi(x)\,\mathrm{d}x = (-1)^r \int_a^b h^{(r+1)}(x)\phi(x)\,\mathrm{d}x$$

which shows that $w = (-1)^r h^{(r+1)} (a < x < b)$. If now h be replaced by any one of its indefinite integrals, say $h_0, w = (-1)^r h_0^{(r+2)}$ $(a < x < b)$ where h_0 is continuous. However, if $\phi_0(x)$ is a fine function which equals 1 in $a \le x \le b$ and vanishes outside $(a - \varepsilon, b + \varepsilon)$ we can also write

$$w = (-1)^r(\phi_0 h_0)^{(r+2)} \quad (a < x < b)$$

and the theorem is proved. □

Corollary 11.5. *If $w(x) = 0$ in $x < a$ and in $x > b > a$, a and b finite, then*

$$w(x) = \sum_{m=1}^{M} f_m^{(r_m)}(x)$$

where f_m is continuous and vanishes outside $a - \varepsilon \le x \le b + \varepsilon$, and M is finite.

Proof. Let ϕ_1 be a fine function which is unity for $a \le x \le b$ and vanishes outside $[a - \frac{1}{2}\varepsilon, b + \frac{1}{2}\varepsilon]$. Then $\phi_1 w = w$. Let ϕ be any fine function. By Theorem 11.5 there is an f which vanishes outside $[a - \varepsilon, b + \varepsilon]$ such that $w = f^{(r)}(x)$ in $a - \frac{1}{2}\varepsilon < x < b + \frac{1}{2}\varepsilon$ and

$$\int_{-\infty}^{\infty} w(x)\phi_2(x)\,dx = (-1)^r \int_{-\infty}^{\infty} f(x)\phi_2^{(r)}(x)\,dx$$

for any fine ϕ_2 which vanishes outside $[a - \frac{1}{2}\varepsilon, b + \frac{1}{2}\varepsilon]$. Hence

$$\int_{-\infty}^{\infty} w(x)\phi(x)\,dx = \int_{-\infty}^{\infty} w(x)\phi_1(x)\phi(x)\,dx = (-1)^r \int_{-\infty}^{\infty} f(x)(\phi_1\phi)^{(r)}\,dx$$

on identifying $\phi_1\phi$ with ϕ_2. Since $(\phi_1\phi)^{(r)} = \sum_{m=0}^{r} \dfrac{r!}{m!(r-m)!} \times \phi_1^{(r-m)}\phi^{(m)}$,

$$\int_{-\infty}^{\infty} w(x)\phi(x)\,dx = \sum_{m=0}^{r} \frac{r!(-1)^{r+m}}{m!(r-m)!} \int_{-\infty}^{\infty} \{\phi_1^{(r-m)} f\}^{(m)} \phi(x)\,dx$$

and the corollary follows at once because $\phi_1^{(r-m)} f$ is continuous and vanishes outside $a - \varepsilon \le x \le b + \varepsilon$. □

Theorem 11.5 gives a local picture of the structure of a weak function whereas Corollary 11.5 indicates the form, on R_1, of a weak function which vanishes outside a finite inverval. A global view of an arbitrary weak function can be obtained from the two following theorems:

Theorem 11.6. $w(x) = \sum_{m=-\infty}^{\infty} w(x)\tau(x - m)$.

Proof.

$$\int_{-\infty}^{\infty} \left\{ w(x) - \sum_{m=-\infty}^{\infty} w(x)\tau(x-m) \right\} \phi(x)\,dx$$

$$= \lim_{M \to \infty} \int_{-\infty}^{\infty} w(x) \left\{ 1 - \sum_{m=-M}^{M} \tau(x-m) \right\} \phi(x)\,dx. \quad (1)$$

Now $1 - \sum_{m=-M}^{M} \tau(x-m)$ vanishes for $|x| \le M$ and $\phi(x)$ is zero outside some finite interval. Hence, if M is sufficiently large, the fine function multiplying $w(x)$ on the right-hand side of (1) must be identically zero and therefore the right-hand side of (1) must be zero. The proof is complete. □

Theorem 11.7. *Any weak function $w(x)$ can be expressed in the form*

$$w(x) = \sum_{m=-\infty}^{\infty} f_m^{(r_m)}(x)$$

where f_m is continuous and vanishes outside $m-1-\varepsilon \le x \le m+1+\varepsilon$, ε being an arbitrary positive number.

Proof. By Corollary 11.5 $w(x)\tau(x-m)$ can be expressed as the sum of a finite number of continuous functions which vanish outside $m-1-\varepsilon \le x \le m+1+\varepsilon$ since $\tau(x-m)=0$ for $|x-m| \ge 1$. The theorem is now an immediate consequence of Theorem 11.6. □

Theorem 11.8. *If $w(x)$ is zero for $x>0$ and for $x<0$ then*

$$w(x) = \sum_{m=0}^{M} a_m \delta^{(m)}(x)$$

where M is finite.

Proof. By Theorem 11.5 w is the derivative of a continuous function which vanishes outside $(-\varepsilon, \varepsilon)$ and such a function must be a polynomial since w is zero in $x>0$ and $x<0$. The rest of the proof runs along lines similar to Theorem 3.21. □

Positive weak functions are defined in a manner analogous to that for generalised functions, i.e. $w \ge 0$ if $\int_{-\infty}^{\infty} w(x)\phi(x)\,dx \ge 0$ for all fine $\phi \ge 0$. It can be shown, as in Theorem 3.22, that if $w \ge 0$ $\int_{-\infty}^{\infty} w(x)\phi_m(x)\,dx \to 0$ when $\max|\phi_m(x)| \to 0$, ϕ_m vanishing outside (a, b).

Exercises

10. Verify that $\sum_{m=1}^{\infty} \delta^{(m)}(x-m)$ is a weak function, which demonstrates that Theorem 11.5 is not true if $b=\infty$. It also indicates that we cannot improve on Theorem 11.7.
11. Show that $\{w_m\}$ is a convergent sequence of weak functions, i.e. $\operatorname{Lim}_{m\to\infty} w_m = w$, if, and only if, $w_m = f_m^{(r)}$ where $\{f_m\}$ is a sequence of continuous functions which converge uniformly on (a,b), a and b finite. Show also that $w = (\operatorname{Lim}_{m\to\infty} f_m)^{(r)}$.

11.4. Sundry results

Odd and even weak functions are defined in exactly the same way as odd and even generalised functions, namely by $w(x) = -w(-x)$ and $w(x) = w(-x)$ respectively. There is no difficulty in showing that w' is odd (even) when w is even (odd) (cf. Theorem 4.1).

Singular integrals involving weak functions may be handled in the same way as for generalised functions and it is therefore not necessary to go into further detail.

However there is one result, Corollary 4.2, which must be proved again since the proof given so far depends upon Fourier transforms.

Theorem 11.9. *If $w'(x) = 0$ then $w(x) = C$, where C is an arbitrary constant.*

Proof. Let ϕ be any fine function. Then, if ρ is the fine function defined just before Theorem 3.3, $\int_{-\infty}^{x} \{\phi(y) - \rho(y)\int_{-\infty}^{\infty}\phi(t)\,dt\}\,dy$ is infinitely differentiable, vanishes for sufficiently negative x and, since $\int_{-1}^{1}\rho(y)\,dy = 1$, is zero for sufficiently positive x. Hence it is fine and, since $w' = 0$,

$$0 = \int_{-\infty}^{\infty} w'(x) \int_{-\infty}^{x} \left\{\phi(y) - \rho(y)\int_{-\infty}^{\infty}\phi(t)\,dt\right\} dy\,dx$$

$$= -\int_{-\infty}^{\infty} w(x)\left\{\phi(x) - \rho(x)\int_{-\infty}^{\infty}\phi(t)\,dt\right\} dx$$

or

$$\int_{-\infty}^{\infty} w(x)\phi(x)\,dx = \int_{-\infty}^{\infty} w(x)\,\rho(x)\,dx \int_{-\infty}^{\infty} \phi(t)\,dt.$$

Since the right-hand side is a constant multiple of $\int_{-\infty}^{\infty}\phi(t)\,dt$ we deduce that w is a constant and the proof is finished. □

Weak functions on a half space may be defined in a similar way to Definition 4.11. Once again the properties are not sufficiently different from generalised functions to justify detailed description.

The definition of a series has already been mentioned in §11.3. It is an immediate consequence of Theorem 11.4 (i) that

Theorem 11.10. *If $\sum_{m=1}^{\infty} w_m(x)$ is a weak function*

$$\left\{\sum_{m=1}^{\infty} w_m(x)\right\}' = \sum_{m=1}^{\infty} w_m'(x).$$

Thus the derivative of a series is obtained by taking derivatives term by term.

The representation of a weak function in a series has been obtained in Theorem 11.7. Using the notation of that theorem, let f_{1m} be defined by $f_{1m} = \int_{-\infty}^{x} f_m(y)\,dy$ $(m \geq 0)$ and $f_{1m} = \int_{\infty}^{x} f_m(y)\,dy$ $(m < 0)$. Then $f_{1m}' = f_m$ and f_{1m} is continuous; also f_{1m} vanishes for $x \leq m - 1 - \varepsilon$ if $m \geq 0$, though not necessarily for $x \geq m + 1 + \varepsilon$, and, if $m < 0, f_{1m}$ vanishes for $x \geq m + 1 + \varepsilon$. Now a meaning can certainly be attached to

$$\int_{-\infty}^{\infty} \left\{\sum_{m=-\infty}^{\infty} f_{1m}^{(r_m)}(x)\right\} \rho(x)\,dx$$

since at most a finite number of f_{1m} are non-zero where ρ is non-zero. Also, if ϕ is any fine function,

$$\int_{-\infty}^{\infty} \left\{\sum_{m=-M}^{M} f_m^{(r_m)}(x)\right\} \int_{-\infty}^{x} \left\{\phi(y) - \rho(y)\int_{-\infty}^{\infty} \phi(t)\,dt\right\} dy\,dx$$

$$= -\int_{-\infty}^{\infty} \left\{\sum_{m=-M}^{M} f_{1m}^{(r_m)}(x)\right\}\left\{\phi(x) - \rho(x)\int_{-\infty}^{\infty} \phi(t)\,dt\right\} dx.$$

Now the left-hand side certainly exists, as $M \to \infty$, because the integrand is the product of a weak function and a fine function. The term involving $\rho(x)$ exists by what has been stated above. Hence the remaining term must exist, i.e. $\sum_{m=-\infty}^{\infty} f_{1m}^{(r_m)}(x)$ is a weak function w_0. By Theorem 11.10 $w_0' = w$. Hence *given any weak function w there is a weak function w_0 such that $w_0' = w$.* That w_0 is not unique can be seen from Theorem 11.9.

Suppose now that $\sum_{m=1}^{\infty} w_m(x)$ exists. Let w_{m0} be any weak function such that $w_{m0}'(x) = w_m(x)$. Choose $w_{m1} = w_{m0} + a_m$ where a_m is a constant such that

$$\int_{-\infty}^{\infty} \{w_{m0}(x) + a_m\}\rho(x)\,dx = 0.$$

Then

$$\int_{-\infty}^{\infty}\sum_{m=1}^{M} w_{m1}(x)\phi(x)\,dx$$

$$= \sum_{m=1}^{M}\int_{-\infty}^{\infty} w_{m1}(x)\rho(x)\,dx \int_{-\infty}^{\infty}\phi(t)\,dt$$

$$+ \sum_{m=1}^{M}\int_{-\infty}^{\infty} w_{m1}(x)\left[\int_{-\infty}^{x}\left\{\phi(y)-\rho(y)\int_{-\infty}^{\infty}\phi(t)\,dt\right\}dy\right]dx$$

$$= -\sum_{m=1}^{M}\int_{-\infty}^{\infty} w_{m}(x)\int_{-\infty}^{x}\left\{\phi(y)-\rho(y)\int_{-\infty}^{\infty}\phi(t)\,dt\right\}dy\,dx$$

since $w'_{m1} = w_m$. The right-hand side tends to a limit as $M \to \infty$ because $\sum_{m=1}^{\infty} w_m$ is a weak function. Hence $\sum_{m=1}^{\infty} w_{m1}$ exists and, by Theorem 11.10, $(\sum_{m=1}^{\infty} w_{m1})' = \sum_{m=1}^{\infty} w_m$. Therefore

Theorem 11.11. *If $\sum_{m=1}^{\infty} w_m$ is a weak function there is a weak function $\sum_{m=1}^{\infty} w_{m1}$ such that $(\sum_{m=1}^{\infty} w_{m1})' = \sum_{m=1}^{\infty} w_m$ and $w'_{m1} = w_m$.*

Theorem 11.12. *If $\sum_{m=1}^{\infty} f_m$ converges uniformly on R_1, the f_m being continuous functions, then $\sum_{m=1}^{\infty} f_m^{(r)}$ exists as a weak function.*

Proof. By Theorem 1.24, $\sum_{m=1}^{\infty} f_m$ is continuous and therefore a weak function on account of Theorem 11.2. Application of Theorem 11.10 completes the proof. □

A converse to this theorem is provided by

Theorem 11.13. *If $\sum_{m=1}^{\infty} w_m$ is a weak function then, on any finite interval (a, b), there is an integer r (independent of m) such that $w_m = f_m^{(r)}$ where f_m is continuous and $\sum_{m=1}^{\infty} f_m$ converges uniformly on R_1.*

Proof. Consider the sequence $\{\sum_{m=1}^{M} w_m\}$. On account of Exercise 11 there is f_{1m} such that $\sum_{m=1}^{M} w_m = f_{1M}^{(r-1)}$ in $a < x < b$ where f_{1M} is continuous and $\{f_{1M}\}$ converges uniformly on (a, b) as $M \to \infty$. Let now $f_{2M} = 0$ $(x < a)$ and $f_{2M}(x) = \int_a^{\min(x,b)} f_{1M}(t)\,dt$ when $x \geq a$. Then f_{2M} is continuous on R_1 and $f'_{2M} = f_{1M}$ in $a < x < b$. Also

$$|f_{2N} - f_{2M}| \leq \int_a^b |f_{1N} - f_{1M}|\,dt \to 0$$

as $N, M \to \infty$ since $\{f_{1M}\}$ converges uniformly on (a, b). Hence $\{f_{2M}\}$ is a sequence of functions, continuous and converging uniformly on R_1, such that $\sum_{m=1}^{M} w_m = f_{2M}^{(r)}$ in $a < x < b$. The theorem now follows on taking $f_m = -f_{2m} - f_{2,m-1}$. □

Exercise

12. If $\lim_{M\to\infty} \int_{-\infty}^{\infty} |\sum_{m=1}^{p} f_{M+m} x)| \mathrm{d}x = 0$ for every integer $p \geq 1$ show that $\sum_{m=1}^{\infty} f_m^{(r)}$ exists as a weak function. Show also that there is a weak function w such that $\mathrm{Lim}_{m\to\infty} w_m = w$ if $\{\int_{-\infty}^{\infty} w_m(x)\phi(x)\mathrm{d}x\}$ is a convergent sequence.

11.5. Multiplication

As for generalised functions the analytical simplicity of weak functions engenders a corresponding algebraic complexity. In general, the product of two weak functions cannot be defined and when the product can be defined it is not necessarily associative.

We commence with a relatively simple case and gradually build up to more complicated ones.

Definition 11.4. *If $f_1 \in L_1(a, b), f_2 \in L_1(a, b)$ and $f_1 f_2 \in L_1(a, b)$ for every finite (a, b) the weak function determined by $f_1 f_2$ will also be denoted by $f_1 f_2$ (or $f_2 f_1$) and called the product of the weak functions f_1 and f_2.*

The definition makes sense because f_1, f_2 and $f_1 f_2$ are all weak functions on account of Exercise 1. This exercise also shows that the product $(f_1 f_2)$ of weak functions is related to the product $f_1 f_2$ of conventional functions by

$$\int_{-\infty}^{\infty} (f_1 f_2)\phi \,\mathrm{d}x = \int_{-\infty}^{\infty} f_1 f_2 \phi \,\mathrm{d}x. \tag{2}$$

Therefore the standard rules for the multiplication of conventional functions continue to apply, for example $\mathrm{e}^x \mathrm{e}^x = \mathrm{e}^{2x}$.

In particular, it should be observed that $f_1 f_2$ exists if either factor is continuous provided that the other factor belongs to $L_1(a, b)$ for every finite (a, b).

We may note that if f_2 is continuous and the integral on the right-hand side of (2) is understood as a Stieltjes integral (§1.12), $f_1\, \mathrm{d}x$

being replaced by $d\mu$ where μ is a measure (§1.11), a meaning could be attached to $(f_1 f_2)$ when $f_1 = \mu'$. However, for most practical applications, it will be sufficient to employ Definition 11.7 below.

The range of entities which may be included in a product is extended by requiring that Leibnitz's rule be complied with. This is certainly satisfied by any conventional functions within the scope of Definition 11.4 which possess conventional derivatives.

Definition 11.5. *If $w_1 w_2$ and $w_1' w_2$ exist the product $w_1 w_2'$ $(= w_2' w_1)$ is defined by $w_1 w_2' = (w_1 w_2)' - w_1' w_2$.*

If f and its first $m - 1$ derivatives are continuous and $f^{(m)} \in L_1(a, b)$ for every finite (a, b) we shall write $f \in L^m$. It is also convenient to employ $f \in L^0$ to mean $f \in L_1(a, b)$ for every finite (a, b). Clearly $L^m \subset L^{m-1}$.

Theorem 11.14. *If $w_1 = f_1'$ where $f_1 \in L^0$, if $f_2 \in L^1$ and if $f_1 f_2'$ exists then $f_2 w_1$ exists as a weak function.*

Proof. Since f_2 is continuous $f_1 f_2$ satisfies Definition 11.4. By hypothesis $f_1 f_2'$ exists and therefore $f_1' f_2$ or $w_1 f_2$ exists by Definition 11.5. □

Corollary 11.14a. *If $w_1 = f_1^{(m)}$ where $f_1 \in L^0$, if $f_2 \in L^m$ and if $f_1 f_2^{(m)}$ exists the product $f_2 w_1$ exists.*

Proof. It is assumed that $m > 1$ otherwise the corollary is the same as the preceding theorem. Since $f_2 \in L^m$ the products $f_1 f_2, f_1 f_2', f_1 f_2'', \ldots, f_1 f_2^{(m-1)}$ all exist. By hypothesis $f_1 f_2^{(m)}$ exists and therefore $f_1' f_2, f_1' f_2', \ldots, f_1' f_2^{(m-1)}$ all exist. The existence of $f_1' f_2$ and $f_1' f_2'$ implies that of $f_1'' f_2$. Similarly the products $f_1'' f_2', \ldots, f_1'' f_2^{(m-2)}$ exist. Continuing in this way we prove that $f_1^{(m)} f_2$ exists and the corollary is complete. □

It should be remarked that the product $f_1^{(m)} f_2$ in this corollary has the property

$$\int_{-\infty}^{\infty} f_1^{(m)} f_2 \phi \, dx = (-1)^m \int_{-\infty}^{\infty} f_1 (f_2 \phi)^{(m)} \, dx \tag{3}$$

the derivative on the right-hand side being calculated in the usual way on account of Exercise 4. Equation (3) can be proved by induction, for

$$\int_{-\infty}^{\infty} f_1^{(m)} f_2 \phi \, \mathrm{d}x = \int_{-\infty}^{\infty} \{(f_1^{(m-1)} f_2)' - f_1^{(m-1)} f_2'\} \phi \, \mathrm{d}x$$

$$= -\int_{-\infty}^{\infty} \{f_1^{(m-1)} f_2 \phi' + f_1^{(m-1)} f_2' \phi\} \, \mathrm{d}x$$

which shows that the result is true for m if it is true for $m - 1$. But with $m = 1$ it is clearly true and so (3) is valid.

Corollary 11.14b. *If* w_1 *vanishes outside a finite interval and* $f_2 \in L^m$, $f_2 w_1$ *exists provided that* m *be sufficiently large.*

Proof. On account of Theorem 11.5, $w_1 = f_1^{(r)}$ where f_1 is continuous and vanishes outside a finite interval. Therefore $f_1 f_2^{(m)}$ exists and, by Corollary 11.14*a*, $f_2 w_1$ certainly exists if $m \geq r$. □

The theory given so far shows that, if η is infinitely differentiable, i.e. $\eta \in L^\infty$ and w is the finite derivative of a continuous function, ηw exists. In fact, it is evident from (3) that

$$\int_{-\infty}^{\infty} (\eta w)\phi \, \mathrm{d}x = \int_{-\infty}^{\infty} w(\eta\phi) \, \mathrm{d}x. \tag{4}$$

Therefore, for this class of weak functions the multiplication of this section is consistent with the multiplication by infinitely differentiable functions defined in §11.2. However, not all weak functions are of this type (Theorem 11.7); clearly the ones which are excluded are those in which $r_m \to \infty$ as $m \to \infty$. To allow for this possibility we introduce

Definition 11.6. *Let* $\eta_1, \eta_2, \ldots,$ *be infinitely differentiable functions such that* $\sum_{m=1}^{\infty} \eta_m w_1 = w_1$. *Then, if* $(w_1 \eta_m) w_2$ *exists for every* m, *and* $\sum_{m=1}^{\infty} (w_1 \eta_m) w_2$ *exists,* $w_1 w_2$ *is defined by* $w_1 w_2 = \sum_{m=1}^{\infty} \{(w_1 \eta_m) w_2\}$.

Now, if in Definition 11.6 $w_2 = \eta$ and $\eta_1, \eta_2, \ldots$ are chosen as fine functions, $(w_1\eta_m)\eta$ exists by Corollary 11.14*b*. Also, by (4),

$$\sum_{m=1}^{M}\int_{-\infty}^{\infty}\{(w_1\eta_m)\eta\}\phi\,dx = \sum_{m=1}^{M}\int_{-\infty}^{\infty}(w_1\eta_m)(\eta\phi)\,dx$$

$$\rightarrow \int_{-\infty}^{\infty} w_1(\eta\phi)\,dx$$

as $M \to \infty$, since $\eta\phi$ is fine. In this way, consistency for multiplication by an infinitely differentiable function is obtained.

Definition 11.6 also permits the multiplication of weak functions whose singularities prevent their being subsumed under the previous theorems, at any rate when the singularities are finite in number and can be separated from one another by suitable $\eta_1, \eta_2, \ldots$

One further definition is necessary:

Definition 11.7. *If, in some interval with the origin as an interior point, w is equal to a continuous function, $w(x)\delta(x)$ is defined by*

$$w(x)\delta(x) = w(0)\delta(x).$$

Exercises

13. If $f_1 \in L_p(a,b)$ $(p \geq 1)$, $f_2 \in L_q(a,b)$ $\{q = p/(p-1)\}$ for every finite (a,b) show that $f_1 f_2$ exists as a weak function.

14. If $w_{1r} = f_{1r}^{(m)}$ where $f_{1r} \in L^0$, if $f_{2r} \in C^m$ and if $\mathrm{Lim}_{r\to\infty} w_{1r} = w_1 = f_1^{(m)}$, $\mathrm{Lim}_{r\to\infty} f_{2r} = f_2$ where $f_1 \in L^0, f_2 \in C^m$ show that $\mathrm{Lim}_{r\to\infty} f_{2r} w_{1r} = f_2 w_1$. Would this still be true if $f_{2r} \in L^m, f_2 \in L^m$ and $f_{1r} f_{2r}^{(m)}, f_1 f_2^{(m)}$ existed?

15. If $w_1 w_2$ exists and η is infinitely differentiable show that

$$\eta(w_1 w_2) = (\eta w_1)w_2 = w_1(\eta w_2).$$

16. If $f \in L^1$ show that $\delta(x)f(x) = f(0)\delta(x)$ without using Definition 11.7.

17. If $f_1 \in L_p(a,b)(p \geq 1)$, $f_2^{(m)} \in L_q(a,b)$ $\{q = p/(p-1)\}$, $f_3^{(m)} \in L_\infty(a,b)$ for every finite (a,b) show that $f_1^{(m)}(f_2 f_3) = (f_1^{(m)} f_2)f_3$.

The problem of division is no easier for weak functions than for generalised functions. The results are indeed comparable. We have

Theorem 11.15. *For any weak function w_2 the equation $xw_1(x) = w_2(x)$ has a solution; when $w_2 \in L^1$ it is given by $w_1(x) = x^{-1}w_2(x) + C\delta(x)$.*

Proof. The proof is similar to that of Theorem 6.5, with $\gamma(x) - \gamma(0)\mathrm{e}^{-x^2}$ being replaced by $\phi(x) - \phi(0)\sigma(x)$ where the fine function σ is such that $\sigma(0) = 1$. □

The methods of §6.2 may also be employed to solve $\eta(x)w_1(x) = w_2(x)$ where $\eta(x)$ is infinitely differentiable and possesses only isolated zeros. (The restriction at infinity is no longer necessary.) It can also be shown that $A(x)w_1(x) = w_2(x)$ always has a solution when A is analytic.

Exercise

18. If $x^k w_1(x) = f$ and $f \in L^k$ show that

$$w_1(x) = x^{-k} f(x) + \sum_{r=0}^{k-1} C_r \delta^{(r-1)}(x).$$

11.6. The convolution

The type of product known as the convolution has already been considered for generalised functions in §§6.3–6.6. The spaces L'_p (§6.5) were defined by means of derivatives of functions in L_p. Since such functions are generalised functions, these spaces are the same whether one considers them in terms of generalised functions or weak functions. Therefore they will not be discussed further here.

The convolution dealt with in §6.3 needs modification since it involves Fourier transforms and not all weak functions have Fourier transforms. In fact, the natural analogue of the Fourier transform of a fairly good function is a weak function which vanishes outside a finite interval (cf. Exercise 68 of Chapter 6). Indeed, we have

Definition 11.8. *If $\{\phi_m\}$ is a regular sequence of fine functions defining the weak function w_1, and w_2 is a weak function which vanishes outside a finite interval then $\{\int_{-\infty}^{\infty} w_2(t)\phi_m(x-t)\,dt\}$ is a regular sequence which defines a weak function to be denoted by $\int_{-\infty}^{\infty} w_2(t)w_1(x-t)\,dt$ or $\int_{-\infty}^{\infty} w_2(x-t)w_1(t)\,dt$ or $w_2 * w_1$ or $w_1 * w_2$.*

To justify the definition suppose that w_2 vanishes outside (a, b). Then, essentially $w_2 = f^{(r)}$ where f is continuous (Corollary 11.5) and

$$\int_{-\infty}^{\infty} w_2(t)\phi_m(x-t)\,dt = \int_a^b f(t)\phi_m^{(r)}(x-t)\,dt. \qquad (5)$$

The function on the right-hand side is clearly continuous and vanishes for $x > b_1 + b$ and for $x < a_1 + a$ if ϕ_m is zero outside (a_1, b_1). The properties of the kth derivative $\int_a^b f(t)\phi_m^{(r+k)}(x-t)\,dt$ are similar. Hence $w_2 * \phi_m$ is a fine function. We must now show that $\{w_2 * \phi_m\}$ is regular.

The right-hand side of (5) could be written, equally well, as $\int_{a_1}^{b_1} f(x-t)\phi_m^{(r)}(t)dt$. Let ϕ be any fine function. Then

$$\int_{-\infty}^{\infty} \phi(x)w_2 * \phi_m \,dx = \int_{-\infty}^{\infty} \phi(x) \int_{a_1}^{b_1} f(x-t)\phi_m^{(r)}(t)dt\,dx$$

$$= \int_{a_1}^{b_1} \phi_m^{(r)}(t) \int_{-\infty}^{\infty} f(u)\phi(u+t)\,du\,dt$$

$$= (-1)^r \int_{a_1}^{b_1} \phi_m(t) \int_{-\infty}^{\infty} f(u)\phi^{(r)}(u+t)\,du\,dt$$

since the integrand is composed entirely of continuous functions which vanish outside a finite interval. The right-hand side can be expressed as

$$\int_{-\infty}^{\infty} \phi_m(t) \int_{-\infty}^{\infty} w_2(u)\phi(u+t)\,du\,dt$$

$$\to \int_{-\infty}^{\infty} w_1(t) \int_{-\infty}^{\infty} w_2(u)\phi(u+t)\,du\,dt$$

as $m \to \infty$ because $\{\phi_m\}$ is regular and $\int_{-\infty}^{\infty} w_2(u)\phi(u+t)\,du$ is fine. It has therefore been shown that $\{w_2 * \phi_m\}$ is regular; in the course of the proof there has been demonstrated:

Theorem 11.16. *If w_2 vanishes outside a finite interval*

$$\int_{-\infty}^{\infty} \phi(x)w_1 * w_2 \,dx = \int_{-\infty}^{\infty} w_1(x) \int_{-\infty}^{\infty} w_2(t)\phi(t+x)\,dt\,dx.$$

Example 2. Since $\delta(x)$ vanishes outside a finite interval

$$\int_{-\infty}^{\infty} \phi(x)w_1 * \delta \,dx = \int_{-\infty}^{\infty} w_1(x) \int_{-\infty}^{\infty} \delta(t)\phi(t+x)\,dt\,dx$$

$$= \int_{-\infty}^{\infty} w_1(x)\phi(x)\,dx$$

which can be interpreted as

$$w_1 * \delta = w_1.$$

It may be shown similarly that

$$w_1 * \delta^{(k)} = w_1^{(k)} * \delta = w_1^{(k)}.$$

Theorem 11.17. *If w_2 and w_3 both vanish outside a finite interval $w_3 * (w_2 * w_1)$ exists and*

$$w_3 * (w_2 * w_1) = w_2 * (w_3 * w_1) = (w_3 * w_2) * w_1.$$

Proof. By Definition 11.8, $w_2 * w_1$ is a weak function and hence $w_3 * (w_2 * w_1)$ exists. Also the defining sequence is $w_3 * (w_2 * \phi_m)$ and

$$\int_{-\infty}^{\infty} \phi(x) w_3 * (w_2 * \phi_m) dx$$

$$= (-1)^{r+s} \int_{a_1}^{b_1} \phi_m(t) \int_{-\infty}^{\infty} f_2(u) \int_{-\infty}^{\infty} f_3(v) \phi^{(r+s)}(u+v+t) \, dv \, du \, dt$$

where $w_2 = f_2^{(r)}, w_3 = f_3^{(s)}$. Since the factors can be written in any of the orders stated in the theorem the proof is complete. □

Theorem 11.18. *If w_2 vanishes outside a finite interval*

$$(w_1 * w_2)' = w_1 * w_2' = w_1' * w_2.$$

Proof. The regular sequence defining $(w_1 * w_2)'$ is $\{(w_2 * \phi_m)'\}$ which may be written in the alternative forms $\{w_2 * \phi_m'\}$ or $\{w_2' * \phi_m\}$ and the proof is finished. □

Exercises

19. If $\{w_{2m}\}$ is a sequence of weak functions such that $\mathrm{Lim}_{m\to\infty} w_{2m} = w_2$ and all of $w_2, w_{21}, w_{22}, \ldots$ vanish outside the same finite interval show that

$$\mathrm{Lim}_{m\to\infty} w_1 * w_{2m} = w_1 * w_2.$$

(Hint: Exercise 11.)

20. If $\mathrm{Lim}_{m\to\infty} w_{1m} = w_1$ and w_2 vanishes outside a finite interval show that $\mathrm{Lim}_{m\to\infty} w_{1m} * w_2 = w_1 * w_2$.

21. If w_2 vanishes outside a finite interval show that

$$(w_1 * w_2)'' = w_1'' * w_2 = w_1 * w_2'' = w_1' * w_2'.$$

To lift the restriction that one of the factors in the convolution vanishes outside a finite interval it is necessary to apply some restriction to both factors. One of the most useful constraints is obtained by limiting attention to weak functions which are zero on the negative real axis.

Definition 11.9. $w \in W_+$ *if, and only if,* $w(x) = 0$ *for* $x < 0$.

Obviously the space W_+ contains K_+ and is more extensive than it.

Suppose now that $\{\phi_m\}$ defines $w \in W_+$. Then, if ϕ vanishes for $x \geq 0$,

$$\lim_{m\to\infty} \int_{-\infty}^{\infty} \phi_m \phi \, dx = \int_{-\infty}^{\infty} w\phi \, dx = 0$$

from Definition 11.9. Let $\eta(x)$ be an infinitely differentiable function which equals 1 for $x \geq 0$ and is zero for $x \leq -\varepsilon < 0$. For any fine function ϕ the function $(1-\eta)\phi$ is fine and vanishes for $x \geq 0$. Hence

$$\lim_{m\to\infty} \int_{-\infty}^{\infty} (1-\eta)\phi_m \phi \, dx = 0.$$

Therefore, when $w \in W_+$, we can replace $\{\phi_m\}$ by $\{\eta\phi_m\}$ without loss or, to put it another way, $\eta w = w$.

Let $w_1 \in W_+$, $w_2 \in W_+$ and let $\{\phi_{2m}\}$ be a sequence of fine functions defining w_2. Then $w_1 * \phi_{2m}$ is a weak function and, since ϕ_{2m} vanishes outside a finite interval, Theorem 11.16 gives

$$\begin{aligned}\int_{-\infty}^{\infty} \phi(x) w_1 * \phi_{2m} \, dx &= \int_{-\infty}^{\infty} w_1(x) \int_{-\infty}^{\infty} \phi_{2m}(t)\phi(t+x) \, dt \, dx \\ &= \int_{-\infty}^{\infty} w_1(x)\eta(x) \int_{-\infty}^{\infty} \eta(t)\phi_{2m}(t)\phi(t+x) \, dt \, dx \end{aligned} \tag{6}$$

on taking advantage of the remarks of the preceding paragraph.

Now $\eta(x)\int_{-\infty}^{\infty} w_2(t)\phi(x+t)\,dt$ vanishes outside a finite interval because $\phi(t) = 0$ for $t > b$ (say) and $w_2(t) = 0$ for $t < 0$. For x in this interval $\phi(x+t)$ is non-zero for only a finite interval of t and in that interval we can take $w_2 = f_2^{(r)}$ and obtain $\eta(x)(-1)^r\int_{-\infty}^{\infty} f_2(t)\phi^{(r)}(x+t)\mathrm{d}t$. In this form it is obviously infinitely differentiable and so $\eta(x)\int_{-\infty}^{\infty} w_2(t)\phi(x+t)\mathrm{d}t$ is a fine function for any fine ϕ. Thus $\{\eta(x)\int_{-\infty}^{\infty}\eta(t)\phi_{2m}(t)\phi(t+x)\mathrm{d}t\}$ is a sequence of fine functions defining a fine function. By Theorem 3.1 the convergence must be pointwise and therefore must be uniform since the limit is bounded. The same is true for its derivatives.

With this information we can replace w_1 by $f_1^{(s)}$ on the right-hand side of (6), transfer the s derivatives to the fine function and proceed to the limit as $m \to \infty$. Moving the derivatives back from the fine function to f_1 we conclude that

$$\lim_{m\to\infty}\int_{-\infty}^{\infty}\phi(x)w_1 * \phi_{2m}\,\mathrm{d}x = \int_{-\infty}^{\infty} w_1(x)\eta(x)\int_{-\infty}^{\infty} w_2(t)\eta(t)\phi(t+x)\,\mathrm{d}t\,\mathrm{d}x. \tag{7}$$

If $\{\phi_{1n}\}$ is a sequence defining w_1 the right-hand side of (7) can be expressed as

$$\begin{aligned}\lim_{n\to\infty}&\int_{-\infty}^{\infty}\phi_{1n}(x)\eta(x)\int_{-\infty}^{\infty} w_2(t)\eta(t)\phi(t+x)\,\mathrm{d}t\,\mathrm{d}x\\ &= \lim_{n\to\infty}\int_{-\infty}^{\infty} w_2(t)\eta(t)\int_{-\infty}^{\infty}\phi_{1n}(x)\eta(x)\phi(t+x)\,\mathrm{d}x\,\mathrm{d}t \end{aligned}\tag{8}$$

by means of Theorem 11.16. The argument for the right-hand side travels along similar lines to that for (6) and the limit obtained is the same as (7) with w_1 and w_2 interchanged. Accordingly

Definition 11.10. *If* $w_1 \in W_+$ *and* $w_2 \in W_+$ *the convolution* $w_1 * w_2$ *is defined by* $w_1 * w_2 = \operatorname{Lim}_{m\to\infty} w_1 * \phi_{2m} = \operatorname{Lim}_{n\to\infty}\phi_{1n} * w_1$, $w_1 * w_2 = w_2 * w_1$ *and*

$$\begin{aligned}\int_{-\infty}^{\infty}\phi(x)w_1 * w_2\,\mathrm{d}x &= \int_{-\infty}^{\infty} w_1(x)\eta(x)\int_{-\infty}^{\infty} w_2(t)\eta(t)\phi(t+x)\mathrm{d}t\,\mathrm{d}x\\ &= \int_{-\infty}^{\infty} w_2(t)\eta(t)\int_{-\infty}^{\infty} w_1(x)\eta(x)\phi(t+x)\mathrm{d}x\,\mathrm{d}t.\end{aligned}$$

It is possible to drop the factor η by adopting the convention of §4.4. Clearly $w_1 * w_2 \in W_+$.

Theorem 11.19. *If* w_1, w_2 *and* w_3 *all belong to* W_+ *then*

$$w_1 * (w_2 * w_3) = (w_1 * w_2) * w_3.$$

Proof. By Definition 11.10

$$\begin{aligned} w_1 * (w_2 * w_3) &= \operatorname*{Lim}_{m\to\infty} \phi_{1m} * (w_2 * w_3) = \operatorname*{Lim}_{m\to\infty} \operatorname*{Lim}_{n\to\infty} \phi_{1m} * (\phi_{2n} * w_3) \\ &= \operatorname*{Lim}_{m\to\infty} \operatorname*{Lim}_{n\to\infty} (\phi_{1m} * \phi_{2n}) * w_3 \\ &= (w_1 * w_2) * w_3 \end{aligned}$$

where the associativity proved in Theorem 11.17 has been employed. □

Another way of stating this is that

$$\int_{-\infty}^{\infty} \phi(x) w_1 * (w_2 * w_3)\, dx$$

$$= \int_{-\infty}^{\infty} w_1(u) \int_{-\infty}^{\infty} w_2(x) \int_{-\infty}^{\infty} w_3(t)\phi(t + x + u)\, dt\, dx\, du$$

and that the order of the weak functions on the right-hand side is immaterial.

Theorem 11.20. *If* $w_1 \in W_+, w_2 \in W_+$ *and* $w_1 * w_2 = 0$ *then either* $w_1 = 0$ *or* $w_2 = 0$.

Proof. Let ϕ be any fine function which is zero for $x < 0$. Then, from Theorem 11.19,

$$(w_1 * \phi) * (w_2 * \phi) = (\phi * \phi) * (w_1 * w_2) = 0.$$

Since $w_1 * \phi$ and $w_2 * \phi$ are continuous functions on $x \geq 0$ it follows from a theorem of Titchmarsh (for a proof see the appendix to this chapter) that either $w_1 * \phi = 0$ or $w_2 * \phi = 0$ on $x \geq 0$.

Let $\phi_0(x)$ be any fine function; then $\phi_0(x_0 - x)$ is a fine function which is zero for $x < 0$ for sufficiently large $x_0 \geq 0$. Put

$\phi(t) = \phi_0(x_0 - t)$. Then, if $w_1 * \phi = 0$ for $x \geq 0$,

$$\int_{-\infty}^{\infty} w_1(t)\phi_0(x_0 - x + t)\,dt = 0 \quad (x \geq 0).$$

Choose $x = x_0$ (permissible since $w_1 * \phi$ is a continuous function and $x_0 \geq 0$) and then $\int_{-\infty}^{\infty} w_1(t)\phi_0(t)\,dt = 0$. Since ϕ_0 is an arbitrary fine function $w_1 = 0$. Similarly it may be shown that $w_2 * \phi = 0$ implies $w_2 = 0$ and the theorem is proved. □

The fractional integral I^β, defined in §6.6, is also applicable to weak functions. It has the properties

$$I^\beta(I^\gamma w) = (I^\beta I^\gamma)w = I^{\beta+\gamma}w, \quad I^0 w = w$$

if $w \in W_+$. If β is not zero or a negative integer

$$I^\beta w = \int_0^x \frac{(x-t)^{\beta-1} w(t)}{(\beta-1)!}\,dt.$$

Exercises

22. If w_1 and w_2 belong to W_+ and

$$\int_0^x w_1(t)(x-t)^\beta\,dt = w_2(x)$$

show that

$$w_1(t) = \frac{1}{\pi}(\beta+1)\sin\pi\beta \int_0^t w_2(x)(t-x)^{-\beta-2}\,dx$$

when β is not an integer.

23. If w_μ, w and w_2 all belong to W_+ and $\operatorname{Lim}_{\mu\to\mu_0} w_\mu = w$ show that

$$\operatorname*{Lim}_{\mu\to\mu_0} w_\mu * w_2 = w * w_2.$$

Deduce that, if $\partial w_\mu/\partial\mu$ exists,

$$\frac{\partial}{\partial\mu}(w_\mu * w_2) = \left(\frac{\partial w_\mu}{\partial\mu}\right) * w_2.$$

24. If f is continuous for $x \geq 0$ and zero for $x < 0$ show that

$$\operatorname*{Lim}_{m\to\infty} f * f * \ldots (m \text{ terms}) = 0.$$

25. If $w \in W_+$ show that $w + w*w + w*w*w + \ldots$ is a weak function if and only if $\operatorname{Lim}_{m\to\infty} w*w*\ldots(m \text{ terms}) = 0$. (Hint: take the convolution of the the first m terms of the series with $w - \delta$ and use Theorem 11.20.)

26. If $w_1 = 1$, $w_2 = \delta'$ show that $w_1 * w_2 = 0$ which indicates that Theorem 11.20 need not be true when one of the factors is not in W_+.

11.7. Volterra integral equations

If w_1, w_2 and k all belong to W_+

$$w_1 - k * w_1 = w_2 \tag{9}$$

constitutes a Volterra *integral equation of the second kind.* Consider the weak function $w_2 + k * w_2 + k * k * w_2 + \ldots + (k*)^m * w_2$. We have

$$\begin{aligned}(\delta - k) * \{w_2 + \ldots + (k*)^m * w_2\} &= w_2 - (k*)^{m+1} * w_2 \\ &\to w_2\end{aligned}$$

as $m \to \infty$, by Exercise 23, if $\operatorname{Lim}_{m\to\infty} (k*)^m = 0$. Hence, *if* $\operatorname{Lim}_{m\to\infty} (k*)^m = 0$ *a solution of* (9) *is*

$$w_1 = w_2 + k * w_2 + k * k * w_2 + \ldots \tag{10}$$

Conversely, if (10) constitutes the solution of (9) it is necessary for the convergence of (10) that $\operatorname{Lim}_{m\to\infty} (k*)^m * w_2 = 0$. Assuming $\operatorname{Lim}_{m\to\infty} (k*)^m$ exists, Theorem 11.20 demands that either $w_2 = 0$ or $\operatorname{Lim}_{m\to\infty} (k*)^m = 0$. Thus, if $w_2 \neq 0$, (10) is the solution of (9) if, and only if, $\operatorname{Lim}_{m\to\infty} (k*)^m = 0$. In fact, it is the only solution when $\operatorname{Lim}_{m\to\infty} (k*)^m = 0$. For suppose $w_3 \in W_+$ were another solution of (9). Then $w_0 = w_1 - w_3$ would satisfy $w_0 = k * w_0$ and applying this to itself

$$w_0 = k * k * w_0 = \ldots = (k*)^m * w_0 \to 0$$

as $m \to \infty$. Consequently, if $\operatorname{Lim}_{m\to\infty} (k*)^m = 0$ (10) *is the unique solution of* (9).

Example 3. If k is continuous, Exercise 24 shows that $\operatorname{Lim}_{m\to\infty} (k*)^m = 0$ and the unique solution of

$$w_1(x) - \int_0^x k(x-t) w_1(t)\,dt = w_2(x)$$

is given by

$$\begin{aligned}w_1(x) = w_2(x) &+ \int_0^x k(x-t) w_2(t)\,dt \\ &+ \int_0^x k(x-t) \int_0^t k(t-u) w_2(u)\,du\,dt + \ldots\end{aligned} \tag{11}$$

Indeed this is valid for a much wider class of k. Suppose that $k(x)$ is integrable over every finite interval and $|k(x)| = O(x^{c-1})$ as $x \to 0$ for $c > 0$. Then, for any finite interval $0 \le x \le b$, $\int_0^k |k(t)|\,dt \le Ax^c$ and

$$\left|\int_0^x k * k\,dt\right| = \left|\int_0^x k(t) \int_0^{x-t} k(u)\,du\,dt\right|$$

$$\le \int_0^x |k(t)|\, A(x-t)^c\,dt$$

$$\le Ac \int_0^x (x-t)^{c-1} \int_0^t |k(u)|\,du\,dt$$

by an integration by parts. Hence

$$\left|\int_0^x k * k\,dt\right| \le A^2 c \int_0^x (x-t)^{c-1}\, t^c\,dt \le c\,!c\,!A^2 x^{2c}/(2c)!.$$

Similarly

$$\left|\int_0^x k * k * k\,dt\right| = \left|\int_0^x k(t) \int_0^{x-t} k * k\,du\,dt\right|$$

$$\le (c\,!Ax^c)^3/(3c)!$$

and, in general, $|\int_0^x (k*)^m\,dt| \le (c\,!Ax^c)^m/(mc)!$ as may be checked by induction. Hence

$$\left|\int_{-\infty}^{\infty} \phi(x) \int_0^x (k*)^m\,dt\,dx\right| \le \frac{(c\,!Ab^c)^m}{(mc)!} \max|\phi| \to 0$$

as $m \to \infty$. Therefore $\operatorname{Lim}_{m\to\infty} \int_0^x (k*)^m\,dt = 0$ and it follows from Theorem 11.4(i) that $\operatorname{Lim}_{m\to\infty} (k*)^m = 0$. Consequently (11) still holds. (For the validity of (11) under somewhat different conditions on k see § 12.3.)

Exercises

27. Show that (11) is valid if k is bounded on every finite interval, but not if $k(x) = \delta(x)$.

28. If k is integrable over every finite interval and $\int_0^x |k(t)|\,dt = O(x^c)$, $c > 0$, as $x \to 0$ show that (11) is valid.

29. Show that the solution of

$$w_1(x) - \lambda \int_0^x e^{x-t} w_1(t) dt = w_2(x)$$

in W_+ is

$$w_1(x) = w_2(x) + \lambda \int_0^x e^{(1+\lambda)(x-t)} w_2(t) dt.$$

30. If k is continuous on $x > 0$, with conventional derivative dk/dx, but $k(0) \neq 0$ obtain the solution of

$$\int_0^x k(x-t) w_1(t) dt = w_2(x),$$

when $\text{Lim}_{m\to\infty} (dk/dx *)^m = 0$, by taking a derivative with respect to x. Deduce the solution of

$$\int_0^x k^{(r)}(x-t) w_1(t) dt = w_2(x).$$

11.8. Ordinary linear differential equations

Consider the ordinary differential equation

$$\frac{d^m f_1}{dx^m} + a_1(x) \frac{d^{m-1} f_1}{dx^{m-1}} + \ldots + a_m(x) f_1 = f_2 \tag{12}$$

where $a_1, \ldots, a_m$ are bounded on the finite interval $[a, b]$ and f_2 is absolutely integrable over the same interval.

A point-function f_1 is called a *solution* of this differential equation if $f_1, df_1/dx, \ldots, d^{m-1}f_1/dx^{m-1}$ are absolutely continuous and $f_1, df_1/dx, \ldots, d^m f_1/dx^m$ satisfy the equation for almost all x. Such a solution can be constructed by writing the differential equation in the form

$$\frac{d^m f_1}{dx^m} = f_2 - a_1(x) \frac{d^{m-1} f_1}{dx^{m-1}} - \ldots - a_m(x) f_1$$

and using an iteration procedure in which the preceding approximation is inserted on the right-hand side. Denote this solution by f_0. Then it can be shown, if $a_r, da_r/dx, \ldots, d^p a_r/dx^p$ are all absolutely continuous and if $d^{p+1}a_r/dx^{p+1}$ is bounded, that $f_0, df_0/dx, \ldots, d^{m+p}f_0/dx^{m+p}$ are all absolutely continuous. In particular, if a_r and f_2 are infinitely differentiable, f_0 is infinitely differentiable.

To f_0 may be added any solutions of the homogeneous equation, i.e. (12) with $f_2 \equiv 0$. (These solutions can also be generated by iteration starting from the approximations $x^p/p!$ $(p = 0, 1, \ldots, m-1)$.) Solutions of the homogeneous equation have differentiability properties which depend only upon the coefficients a_r. Denote the linearly independent solutions of the homogeneous equation by $f_{11}, f_{12}, \ldots, f_{1m}$; then the general solution of (12) is

$$f_1 = f_0 + c_1 f_{11} + \ldots + c_m f_{1m} \tag{13}$$

where $c_1, \ldots, c_m$ are arbitrary constants. There are no other solutions which are point-functions.

Turn now to the corresponding equation in weak functions, namely

$$w_1^{(m)} + a_1(x) w_1^{(m-1)} + \ldots + a_m(x) w_1 = f_2. \tag{14}$$

A solution is given by $w_1 = f_1$ as expressed by (13) since f_1 can be regarded as a weak function. In fact, there are no other weak functions which satisfy the equation. For there are no other point-function solutions and so, in any other weak solution, $w_1^{(m)}$ could not be integrable; therefore $w_1^{(m)}$ would be expressible in the form $f^{(r)}$ where f is integrable and r is the least integer for which the representation is valid (cf. Theorem 11.5). But (14) could be written

$$w_1^{(m)} = f_2 - a_1(x) w_1^{(m-1)} - \ldots - a_m(x) w_1$$

and all the terms on the right-hand side involve a smaller number of derivatives than r. Consequently, there is a contradiction unless $r = 0$, i.e. w_1 is a point-function. Hence *the weak-function and point-function solutions of* (14) *are the same.*

For the more general equation

$$w_1^{(m)} + a_1(x) w_1^{(m-1)} + \ldots + a_m(x) w_1 = w_2 \tag{15}$$

note that, since a finite interval is involved, $w_2 = f_2^{(r)}$ where f_2 is integrable and r is the smallest integer for which the representation is feasible. Let w_3 be a weak function such that $w_3^{(m)} = w_2$. Either w_3 is integrable $(r \leq m)$ or, if $r > m$, it is at most the $(r-m)$th derivative of an integrable function. Equation (15) may be expressed in the form

$$(w_1 - w_3)^{(m)} + a_1(x)(w_1 - w_3)^{(m-1)} + \ldots + a_m(x)(w_1 - w_3)$$
$$= -a_1 w_3^{(m-1)} - \ldots - a_m w_3.$$

The highest derivative of an integrable function which occurs on the right-hand side is $r-1$. Thus repeated reduction of this type eventually produces an integrable function on the right-hand side. The process needs the multiplication of weak functions and is therefore valid only if the a_r are such that the products $a_r w_3^{(m-r)}$ $(r=1,\dots,m)$ exist (which is certainly true if the a_r are infinitely differentiable). Use of the information concerning point-function solutions now yields: *under suitable conditions on the a_r (in particular, if they are infinitely differentiable) the differential equation* (15) *has a solution*

$$w_1 = w_0 + c_1 f_{11} + \dots + c_m f_{1m}$$

where $c_1,\dots,c_m$ are arbitrary, and it is the only solution which is a weak function.

The theory can be extended to the differential equation

$$a_0(x)w_1^{(m)} + a_1(x)w_1^{(m-1)} + \dots + a_m(x)w_1 = w_2$$

provided that division by a_0 is permissible so that the equation can be reduced to the previous type. The division is not permitted if a_0 has a zero in the interval, for then the peculiarities which were mentioned in §10.3 (equation (31) of Chapter 10) recur.

11.9. Linear differential equations with constant coefficients

Consider the linear differential equation

$$w_1' - cw_1 = w_2 \tag{16}$$

where c is a constant (which may be complex) and $w_1 \in W_+$, $w_2 \in W_+$. To solve this equation introduce the weak function e_β defined by

$$e_\beta = e^{\beta x}\,\mathrm{H}(x).$$

Observe that $e_\beta' = \beta e_\beta + \delta(x)$. Thus if $w \in W_+$,

$$(e_\beta * w)' = e_\beta' * w = \beta e_\beta * w + w.$$

We deduce that a solution of (16) is $w_1 = e_c * w_2$. It follows, from the result proved at the end of the preceding section, that the solution of (16) is

$$w_1 = e_c * w_2 + C_1 e^{cx}$$

where C_1 is arbitrary. However, since $w_1 \in W_+$ and $e_c * w_2 \in W_+$ we must have $C_1 = 0$. Hence

Theorem 11.21. *If* $w_1 \in W_+$ *and* $w_2 \in W_+$ *the solution of* (16) *is*

$$w_1 = e_c * w_2.$$

This theorem is a tool which may be employed to solve the more general differential equation

$$w_1^{(m)} + a_1 w_1^{(m-1)} + \dots + a_m w_1 = w_2 \tag{17}$$

where $a_1, \dots, a_m$ are constants. Let $p_1, \dots, p_m$ be the roots of the equation

$$p^m + a_1 p^{m-1} + \dots + a_m = 0.$$

Then (17) may be written as

$$(D - p_1)(D - p_2) \dots (D - p_m) w_1 = w_2$$

where D indicates a generalised derivative. By Theorem 11.21,

$$(D - p_2) \dots (D - p_m) w_1 = e_{p_1} * w_2.$$

Repeating the argument we have

$$(D - p_3) \dots (D - p_m) w_1 = e_{p_2} * e_{p_1} * w_2$$

and thus

Corollary 11.21. *If* $w_1 \in W_+$ *and* $w_2 \in W_+$ *the solution of* (17) *is*

$$w_1 = e_{p_m} * \dots * e_{p_1} * w_2.$$

The order of the factors is of no significance on account of Theorem 11.19.

In calculating the convolution of Corollary 11.21 it is sometimes helpful to remark that

$$\begin{aligned} e_\beta * e_\gamma &= \mathrm{H}(x) \int_0^x \mathrm{e}^{\beta t} \mathrm{e}^{\gamma(x-t)} \, \mathrm{d}t \\ &= (\mathrm{e}^{\beta x} - \mathrm{e}^{\gamma x}) \, \mathrm{H}(x)/(\beta - \gamma) \quad (\beta \neq \gamma). \end{aligned}$$

Thus, if $\beta \neq \gamma$,

$$e_\beta * e_\gamma = (e_\beta - e_\gamma)/(\beta - \gamma). \tag{18}$$

On the other hand $e_\beta * e_\beta = x\mathrm{e}^{\beta x}\,\mathrm{H}(x)$,

$$e_\beta * x\,\mathrm{e}^{\beta x}\,\mathrm{H}(x) = \int_0^x t\,\mathrm{e}^{\beta t}.\mathrm{e}^{\beta(x-t)}\,\mathrm{d}t\,\mathrm{H}(x) = \tfrac{1}{2}x^2\,\mathrm{e}^{\beta x}\,\mathrm{H}(x)$$

and, in general, $e_\beta * x^m\,\mathrm{e}^{\beta x}\,\mathrm{H}(x)/m! = x^{m+1}\,\mathrm{e}^{\beta x}\,\mathrm{H}(x)/(m+1)!$. Hence

$$(e_\beta *)^m = x^{m-1}\,\mathrm{e}^{\beta x}\,\mathrm{H}(x)/(m-1)!. \tag{19}$$

For many purposes (18) and (19) are sufficient but on some occasions one requires, when $\beta \neq \gamma$,

$$(e_\beta *)^m * (e_\gamma *)^r = \mathrm{H}(x)\,\mathrm{e}^{\gamma x} \sum_{s=0}^{r} \frac{r!(m+s)!(-1)^{m-1}x^{r-s}}{s!(r-s)!(\beta-\gamma)^{m+s+1}}$$

$$+ \mathrm{H}(x)\,\mathrm{e}^{\beta x} \sum_{s=0}^{m} \frac{m!(r+s)!(-1)^s x^{m-s}}{s!(m-s)!(\beta-\gamma)^{r+s+1}} \tag{20}$$

a formula which may be checked by induction via the reduction

$$t^m(x-t)^r = xt^m(x-t)^{r-1} - t^{m+1}(x-t)^{r-1}.$$

Example 4. The solution of

$$w_1''' - 6w_1'' + 11w_1' - 6w_1 = w_2$$

when $w_1 \in W_+$, $w_2 \in W_+$ is, from Corollary 11.21,

$$w_1 = e_1 * e_2 * e_3 * w_2.$$

From (18)

$$e_1 * e_2 * e_3 = e_1 * (e_3 - e_2)$$
$$= \tfrac{1}{2}(e_3 - e_1) - (e_2 - e_1) = \tfrac{1}{2}e_3 - e_2 + \tfrac{1}{2}e_1.$$

Hence the required solution is

$$w_1 = \int_0^x (\tfrac{1}{2}\mathrm{e}^{3t} - \mathrm{e}^{2t} + \tfrac{1}{2}\mathrm{e}^t)w_2(x-t)\,\mathrm{d}t.$$

Example 5. The solution of

$$w_1'' + w_1 = w_2$$

is $w_1 = e_\mathrm{i} * e_{-\mathrm{i}} * w_2$. Since

$$e_\mathrm{i} * e_{-\mathrm{i}} = (e_\mathrm{i} - e_{-\mathrm{i}})/2\mathrm{i} = \sin x,$$

$$w_1 = \int_0^x w_2(t)\sin(x-t)\,\mathrm{d}t.$$

Example 6. The solution of

$$w_1''' - 4w_1'' + 5w_1' - 2w_1 = w_2$$

is $w_1 = e_1 * e_1 * e_2 * w_2$. From (18) and (19)

$$e_1 * e_1 * e_2 = e_1 * (e_2 - e_1) = e_2 - e_1 - x\,\mathrm{e}^x\,\mathrm{H}(x)$$

and so

$$w_1 = \int_0^x \{\mathrm{e}^{2t} - (1-t)\mathrm{e}^t\} w_2(x-t)\,\mathrm{d}t.$$

Exercises

31. If $w \in W_+$ and $x_0 \geq 0$ prove that the solutions of

(i) $w^{(m)}(x) = \mathrm{H}(x - x_0)$,

(ii) $(D - c)^m w(x) = \mathrm{H}(x - x_0)$,

(iii) $aw''(x) + 2bw'(x) + cw(x) = \delta'(x - x_0) \qquad (b^2 < ac)$

are

(i) $w = (x - x_0)^m \mathrm{H}(x - x_0)/m!$,

(ii) $w = \dfrac{(-1)^m}{c^m}\left\{1 - \mathrm{e}^{c(x-x_0)} \sum_{s=0}^{m-1} \dfrac{(-1)^s c^s}{s!}(x - x_0)^s\right\} \mathrm{H}(x - x_0)$

(iii) $w = (1/a)\mathrm{e}^{-(b/a)(x-x_0)}\{\cos \omega(x - x_0) - (b/a\omega)\sin \omega(x - x_0)\}$

where $\omega = (ac - b^2)^{1/2}/a$.

32. Show that, if $ac > b^2$,

$$aw''(x) + 2bw'(x) + cw(x) = \sum_{m=0}^{\infty} \delta(x - mx_0)$$

is satisfied by

$$w(x) = \sum_{m=0}^{\infty} \frac{\mathrm{e}^{-(b/a)(x-mx_0)}}{a\omega} \sin \omega(x - mx_0)\mathrm{H}(x - mx_0)$$

where $\omega = (ac - b^2)^{1/2}/a$.

Examine the behaviour of w as $x \to \infty$ when $b = 0$ and $\omega x_0 = 2\pi$.

33. Solve

(i) $w'' + 3w' + 2w = w_2$,

(ii) $w'''' - w = w_2$,

(iii) $w''' + 4w'' + 5w' + 2w = w_2$,

(iv) $w'''' - 8w'' + 16w = w_2$,

(v) $w^{(m)}(x) = x^\beta \mathrm{H}(x) \ln x$.

34. Verify that

$$(e_\beta *)^m = \frac{1}{(m-1)!}\frac{\partial^{m-1}}{\partial\beta^{m-1}}e_\beta,$$

$$(e_\beta *)^m * (e_\gamma *)^r = \frac{1}{(m-1)!(r-1)!}\frac{\partial^{m+r-2}}{\partial\beta^{m-1}\partial\gamma^{r-1}}e_\beta * e_\gamma.$$

35. If the solution of

$$w_1^{(m)} + a_1 w_1^{(m-1)} + \dots + a_m w_1 = \delta(x - x_0),$$

$x_0 \geq 0$, be denoted by $G(x - x_0)$ (the *Green's function* or elementary solution) when $w_1 \in W_+$ show that the solution of (17) is $w_1(x) = \int_0^x G(x-t)w_2(t)\,dt$.

The initial value problem is the problem of solving

$$\frac{d^m f_1}{dx^m} + \dots + a_m f_1 = f_2$$

in $x > 0$, subject to the conditions $f_1 = f_{10}$, $df_1/dx = f_{11}, \dots$, $d^{m-1}f_1/dx^{m-1} = f_{1,m-1}$ at $x = 0$. (If the conditions are given at $x = x_0$ we change the origin to x_0.) Putting $w_1(x) = f_1(x)\,H(x)$ we are led to

$$w_1^{(m)} + \dots + a_m w_1 = f_2 H + \sum_{r=0}^{m-1} f_{1r} \sum_{s=0}^{m-r-1} a_s \delta^{(m-r-1-s)}. \tag{21}$$

The solution of this in W_+ is

$$w_1 = e_{p_m} * \dots * e_{p_1} * \left\{ f_2 H + \sum_{r=0}^{m-1} f_{1r} \sum_{s=0}^{m-r-1} a_s \delta^{(m-1-r-s)} \right\}$$

where $a_0 = 1$.

There is no difficulty in verifying that w_1 and its first $m-1$ derivatives are continuous on $x > 0$, and w_1 will provide the solution of the initial value problem provided that the initial conditions are satisfied.

We shall show that they are satisfied in the sense that

$$\lim_{x \to +0} w_1^{(r)}(x) = f_{1r} \quad (r = 0, 1, \dots, m-1).$$

The convolution involves m integrations starting at 0. Therefore the first term on the right-hand side and its first $m-1$ derivatives must vanish at the origin.

Hence the first term can be ignored in so far as the initial conditions are concerned. With regard to the second term we can, on account of the general theory, write it as $f(x)\,\mathrm{H}(x)$ where f is a point function solution of the homogeneous equation. As a consequence

$$(f\mathrm{H})^{(m)} + \ldots + a_m f\mathrm{H} = \sum_{r=0}^{m-1} f^{(r)}(+0) \sum_{s=0}^{m-r-1} a_s \delta^{(m-r-1-s)}.$$

But this must be the same as the right-hand side of (21) with $f_2\,\mathrm{H}$ absent. Therefore, by Theorems 8.4 and 8.3,

$$\sum_{s=0}^{r} a_s f^{(r-s)}(+0) = \sum_{s=0}^{r} a_s f_{1,r-s} \qquad (r = 0, 1, \ldots, m-1).$$

Taking $r = 0, 1, \ldots$ we deduce successively that $f(+0) = f_{10}$, $f'(+0) = f_{11}, \ldots$ and the initial conditions are verified.

Exercise

36. Obtain the solutions in $x > 0$ of

(i) $f_1'' - 3f_1' + 2f_1 = 0, f_{10} = f_{11} = 1$,

(ii) $f_1'''' - 5f_1''' + 6f_1'' + 4f_1' - 4f_1 = 0, f_{10} = 0, f_{11} = f_{12} = 1, f_{13} = 3$,

(iii) $f_1'' + 6f_1' + 9f_1 = x^2 e^{-3x}, f_{10} = f_{11} = 0$,

(iv) $f_1'' + 6f_1' + 25f_1 = 0, f_{10} = f_{11} = 1$,

(v) $f_1''' - 3f_1'' + 3f_1' - f_1 = 54e^{4x}, f_{10} = 1, f_{11} = 1, f_{11} = 2, f_{12} = 3$

(Answers: in $x > 0$ (i) e^x, (ii) $\frac{1}{3}(e^{2x} - e^{-x})$, (iii) $\frac{1}{12}x^4 e^{-4x}$, (iv) $e^{-3x}(\cos 4x + \sin 4x)$ (v) $2e^{4x} - (9x^2 + 5x + 1)e^x$.)

11.10. The operational method

The aim of this section is to devise a technique which reduces the procedure of the preceding section to algebraic manipulations, i.e. to produce an *operational method*. To do this it is necessary to introduce an operational representation which has appropriate algebraic properties. The representation we choose is to make $p/(p - \beta)$ correspond to e_β and regard p as algebraic. Symbolically, we write

$$\frac{p}{p-\beta} \risingdotseq e_\beta \qquad \text{or} \qquad e_\beta \risingdotseq \frac{p}{p-\beta}. \tag{22}$$

This definition gives $1 \risingdotseq \mathrm{H}(x)$ and so has the advantage that 1 corresponds to 1 as far as $x > 0$ is concerned.

We shall now use E (with or without suffixes) to signify a series of the form $\sum_{m=1}^{M} \sum_{r=1}^{R} c_{mr}(e_{\beta_r}*)^m$ where M and R are finite; the operational representation of E will be denoted by $\mathscr{E}$. Then the rules for the combination of operational representations are: if c, c_1, c_2 are any complex constants

$$c\mathscr{E} \risingdotseq cE, \tag{23}$$

$$c_1\mathscr{E}_1 + c_2\mathscr{E}_2 \risingdotseq c_1 E_1 + c_2 E_2, \tag{24}$$

$$(1/p)\mathscr{E}_1\mathscr{E}_2 \risingdotseq E_1 * E_2. \tag{25}$$

The third rule states that the convolution of E_1 and E_2 is to be represented by the product of the operational representations multiplied by $1/p$, for example,

$$e_\beta * e_\gamma \risingdotseq \frac{1}{p}\frac{p}{p-\beta}\frac{p}{p-\gamma} = \frac{p}{(p-\beta)(p-\gamma)} \tag{26}$$

since we are regarding p as an algebraic quantity.

On the other hand, the right-hand side of (26), considered algebraically, can be expanded in partial fractions if $\beta \neq \gamma$; in fact

$$\frac{p}{(p-\beta)(p-\gamma)} = \frac{1}{\beta-\gamma}\left\{\frac{p}{p-\beta} - \frac{p}{p-\gamma}\right\} \risingdotseq \frac{1}{\beta-\gamma}(e_\beta - e_\gamma)$$

from (24), (23) and (22). Hence (26) implies

$$e_\beta * e_\gamma = \frac{1}{\beta-\gamma}(e_\beta - e_\gamma) \qquad (\beta \neq \gamma)$$

which is consistent with (18).

Further

$$\frac{p}{(p-\beta)^m} = \frac{1}{p^{m-1}}\left(\frac{p}{p-\beta}\right)^m \risingdotseq (e_\beta *)^m = \frac{x^{m-1}\mathrm{e}^{\beta x}\mathrm{H}(x)}{(m-1)!}$$

from (19). Thus

$$\frac{p}{(p-\beta)^m} \risingdotseq \frac{x^{m-1}\mathrm{e}^{\beta x}\mathrm{H}(x)}{(m-1)!}.$$

On account of Exercise 34 we deduce that

$$\frac{p}{(p-\beta)^m} \doteqdot \frac{1}{(m-1)!}\frac{\partial^{m-1}}{\partial\beta^{m-1}}\frac{p}{p-\beta}$$

in agreement with the ordinary rule for taking a derivative. One concludes that (20) is consistent with

$$\frac{p}{(p-\beta)^m(p-\gamma)^r} \doteqdot (e_\beta *)^m * (e_\gamma *)^r.$$

Exercise

37. Show that

(i) $\dfrac{p}{p^2+\beta^2} \doteqdot \dfrac{1}{\beta}\sin\beta x\,\mathrm{H}(x)$,

(ii) $\dfrac{p^2}{p^2+\beta^2} \doteqdot \mathrm{H}(x)\cos\beta x$,

(iii) $\dfrac{p}{(p-\beta)^2+\gamma^2} \doteqdot \dfrac{1}{\gamma}e^{\beta x}\mathrm{H}(x)\sin\gamma x$,

(iv) $\dfrac{p(p-\beta)}{(p-\beta)^2+\gamma^2} \doteqdot e^{\beta x}\mathrm{H}(x)\cos\gamma x$,

(v) $\dfrac{p^2}{(p^2-\gamma^2)^2} \doteqdot \dfrac{x\,\mathrm{H}(x)}{2\gamma}\sinh\gamma x$,

(vi) $\dfrac{p(p-\beta)}{\{(p-\beta)^2+\gamma^2\}^2} \doteqdot \dfrac{xe^{\beta x}}{2\gamma}\mathrm{H}(x)\sin\gamma x$.

By means of these rules it can be seen that

$$e_{p_m} * \ldots * e_{p_1} \doteqdot \frac{p}{p^m+a_1p^{m-1}+\ldots+a_m}.$$

A convenient reduction of the left-hand side can be achieved by expanding the right-hand side in partial fractions and interpreting this expansion.

Example 7. Find f in $x>0$ such that

$$\frac{\mathrm{d}^2f}{\mathrm{d}x^2}+3\frac{\mathrm{d}f}{\mathrm{d}x}+2f=4$$

with $f_0 = f_1 = 0$. In this case

$$e_{p_2} * e_{p_1} \doteqdot \frac{p}{p^2+3p+2} = \frac{p}{(p+2)(p+1)} = \frac{p}{p+1} - \frac{p}{p+2} = \mathrm{e}^{-t} - \mathrm{e}^{-2t}.$$

Hence the required solution is given by

$$f = \int_0^x (\mathrm{e}^{-t} - \mathrm{e}^{-2t}) 4\,\mathrm{d}t = 2 - 4\mathrm{e}^{-x} + 2\mathrm{e}^{-2x}.$$

Initial conditions may also be handled relatively simply by introducing the additional identification

$$p^m \doteqdot \delta^{(m-1)}(x) \qquad (m = 1, 2, \ldots) \tag{27}$$

and continuing with the previous rules. We then have

Theorem 11.22. *If $\mathscr{E} \doteqdot E$ then $p^r \mathscr{E} \doteqdot E^{(r)}$.*

Proof. If $E = e_\beta$, $\mathscr{E} = p/(p-\beta)$ and

$$p\mathscr{E} = \frac{p^2}{p-\beta} = p + \frac{\beta p}{p-\beta} \doteqdot \delta(x) + \beta e_\beta = e'_\beta$$

which proves the result for this particular E and $r = 1$. If $E = (e_\beta *)^m$ $(m \geq 2)$, then $\mathscr{E} = p/(p-\beta)^m$ and

$$p\mathscr{E} = \frac{p^2}{(p-\beta)^m} = \frac{p(p-\beta)+\beta p}{(p-\beta)^m} = \frac{p}{(p-\beta)^{m-1}} + \frac{\beta p}{(p-\beta)^m}$$
$$= (e_\beta *)^{m-1} + \beta(e_\beta *)^m = E',$$

Thus the result is true for general E and $r = 1$. The result for general r follows by induction.□

There is an interesting corollary to this theorem which is a consequence of expanding $\mathscr{E}$ in powers of $1/p$. Since $\mathscr{E}$ is the ratio of two polynomials the expansion consists of a polynomial in p plus a series of powers of $1/p$. Denote by $\langle \mathscr{E} \rangle$ the result of subtracting from $\mathscr{E}$ the positive powers in the polynomial, for example,

$$\langle p/(p-\beta) \rangle = p/(p-\beta),$$
$$\langle p^2/(p-\beta) \rangle = \langle p + \beta p/(p-\beta) \rangle = \beta p/(p-\beta).$$

Then

Corollary 11.22. *If* $\mathscr{E} \doteqdot E$ *then* $\lim_{x\to+0} E^{(r)} = \lim_{p\to\infty}\langle p^r\mathscr{E}\rangle$.

Proof. If $E = \sum_{m=1}^{M}\sum_{r=1}^{R} c_{mr}(e_{\beta r}*)^m$, $\mathscr{E} = \sum_{m=1}^{M}\sum_{r=1}^{R} c_{mr}p/(p-\beta_r)^m$. Therefore $\langle\mathscr{E}\rangle = \mathscr{E}$ and $\lim_{p\to\infty}\langle\mathscr{E}\rangle = \sum_{r=1}^{R} c_{1r} = \lim_{x\to+0} E$. Thus the corollary is true for $r = 0$ and follows for general r from Theorem 11.22. □

The definition (27) means that the terms in (21) arising from the initial conditions can be represented by $\sum_{r=0}^{M-1} f_{1r}\sum_{s=0}^{m-r-1} a_s p^{m-r-s}$ and their contribution to the solution will be given by the operational interpretation of

$$\frac{\sum_{r=0}^{m-1} f_{1r}\sum_{s=0}^{m-r-1} a_s p^{m-r-s}}{p^m + a_1 p^{m-1} + \dots + a_m}. \tag{28}$$

The numerator can be constructed very easily by observing that it contains only the terms with positive powers of p in the product

$$(p^m + a_1 p^{m-1} + \dots + a_m)\left(f_{10} + \frac{f_{11}}{p} + \dots + \frac{f_{1,m-1}}{p^{m-1}}\right).$$

Example 8. Solve $\mathrm{d}^2 f/\mathrm{d}x^2 + \omega^2 f = 0$ in $x > 0$ with arbitrary initial conditions.

The solution is the operational interpretation of

$$\frac{p^2 f_{10} + p f_{11}}{p^2 + \omega^2} \doteqdot \left\{f_0 \cos\omega x + \frac{f_1}{\omega}\sin\omega x\right\}\mathrm{H}(x)$$

from Exercise 37.

If f_2 can be written as an E, then it can be replaced by its operational representation and the whole solution carried out operationally. In this case we can formulate the following rule: *if* $f_2 \doteqdot \mathscr{F}_2$ *the operational representation of* f_1 *is*

$$(\mathscr{F}_2 + \mathscr{C})/\mathscr{L}$$

where $\mathscr{L} = p^m + a_1 p^{m-1} + \dots + a_m$,

$$\mathscr{C} = \text{terms with positive powers of } p \text{ in } \mathscr{L}\left(f_{10} + \frac{f_{11}}{p} + \dots + \frac{f_{1,m-1}}{p^{m-1}}\right).$$

When f_2 is not an E and therefore does not have an operational representation this rule must be modified. The contribution of the terms providing the initial conditions can be still calculated by $\mathscr{C}/\mathscr{L}$. But for the term involving f_2 the convolution $e_{p_m} * \dots * e_{p_1}$ is evaluated either directly or by operational means and then the convolution $e_{p_m} * \dots * e_{p_1} * f_2$ is formed

Example 9. If $d^2f/dx^2 + 4\,df/dx + 13f = 10e^{-x}$ and $f_0 = 0, f_1 = 2$ the right-hand side has an operational representation and so the operational representation of f in $x > 0$ is

$$\frac{1}{p^2+4p+13}\left\{\frac{10p}{p+1} + 2p\right\} = \frac{p}{p+1} - \frac{p^2+p}{p^2+4p+13}$$

$$\mathrel{\dot{=}} e^{-x} - e^{-2x}(\cos 3x - \tfrac{1}{3}\sin 3x).$$

On the other hand, if the $10e^{-x}$ were replaced by an f_2 which was not an E, for example, $f_2 = e^{x^2}$, we should first interpret

$$\frac{p}{p^2+4p+13} \mathrel{\dot{=}} \tfrac{1}{3}e^{-2x}\,H(x)\sin 3x$$

and then give the solution as

$$f = \int_0^x \tfrac{1}{3}e^{-2t}\sin 3t\, f_2(x-t)\,dt + \tfrac{2}{3}e^{-2x}\sin 3x.$$

Exercises

38. Show, from (27) and (25), that $e_\beta * \delta' \mathrel{\dot{=}} p^2/(p-\beta)$.
39. Obtain the solution of Exercise 36 by operational means.
40. Obtain solutions in $x > 0$ of

(i) $f_1'' + 4f_1' + 8f_1 = e^{-x}(8\cos x + 6\sin x),\ f_{10} = 2,\ f_{11} = 0,$
(ii) $f_1'' + 2af_1' + (a^2+b^2)f_1 = e^{x^2},\ f_{10} = f_{11} = 0,$
(iii) $f_1''' - 3f_1' + 2f_1 = e^{-x^2},\ f_{10} = f_{11} = 3,$
(iv) $f_1'''' - f_1 = 15\sinh 2x,\ f_{10} = f_{11} = 0,\ f_{12} = 2,\ f_{13} = 10.$

41. By means of Corollary 11.22 show that the interpretation of (28) satisfies the initial conditions.

The operational technique can be extended to simultaneous differential equations. The aim is once again to employ algebraic manipulations. However, simultaneous linear algebraic equations may be incompatible (e.g. $x + y = 3$ and $x + y = 5$) or not independent (e.g. $x + y = 3$ and $2x + 2y = 6$). There is a similar state of affairs for differential equations with the added complication that the initial conditions may turn out to be connected. We shall consider firstly the cases in which there is a natural generalisation of the process for a single equation, leaving the awkward cases until later.

We give details only for two equations; the procedure for a greater number will then be clear. Suppose that the differential equations are

$$L_1\left(\frac{d}{dx}\right)f_1 + M_1\left(\frac{d}{dx}\right)f_2 = f_3, \tag{29}$$

$$L_2\left(\frac{d}{dx}\right)f_1 + M_2\left(\frac{d}{dx}\right)f_2 = f_4 \tag{30}$$

where

$$\begin{aligned} L_1(t) &= a_{10}t^m + a_{11}t^{m-1} + \dots + a_{1m}, \\ M_1(t) &= b_{10}t^r + b_{11}t^{r-1} + \dots + b_{1r}, \\ L_2(t) &= a_{20}t^m + a_{21}t^{m-1} + \dots + a_{2m}, \\ M_2(t) &= b_{20}t^r + b_{21}t^{r-1} + \dots + b_{2r} \end{aligned}$$

where one at least of a_{10}, a_{20} and one at least of b_{10}, b_{20} are non-zero. Let the initial conditions be $\lim_{x\to+0} d^k f_1/dx^k = f_{1k}$ $(k = 0, 1, \dots, m-1)$ and $\lim_{x\to+0} d^k f_2/dx^k = f_{2k} (k=0, 1, \dots, r-1)$. Let

$$\mathscr{A} = f_{10} + \frac{f_{11}}{p} + \dots + \frac{f_{1,m-1}}{p^{m-1}}, \quad \mathscr{B} = f_{20} + \frac{f_{21}}{p} + \dots + \frac{f_{2,r-1}}{p^{r-1}}$$

and let $\mathscr{C}_1, \mathscr{C}_2$ be the positive powers of p in

$$L_1(p)\mathscr{A} + M_1(p)\mathscr{B}, \quad L_2(p)\mathscr{A} + M_2(p)\mathscr{B}$$

respectively. Then the operational rule is: *if* $f_3 \risingdotseq \mathscr{F}_3, f_4 \risingdotseq \mathscr{F}_4$ *and*

$$\begin{vmatrix} a_{10} & b_{10} \\ a_{20} & b_{20} \end{vmatrix} \neq 0$$

the algebraic solutions $\mathscr{F}_1$ *and* $\mathscr{F}_2$ *of*

$$L_1(p)\mathscr{F}_1 + M_1(p)\mathscr{F}_2 = \mathscr{F}_3 + \mathscr{C}_1, \tag{31}$$

$$L_2(p)\mathscr{F}_1 + M_2(p)\mathscr{F}_2 = \mathscr{F}_4 + \mathscr{C}_2, \tag{32}$$

have interpretations f_1 *and* f_2 *which are the unique solutions of* (29) *and* (30) *in* $x > 0$ *and satisfy the initial conditions.*

It will be observed that the rule for the construction of the algebraic equations can be stated as: replace f by $\mathscr{F}$, d/dx by p and add to the right-hand side the terms from the initial values of f_1 (as if f_2 and its derivatives were absent) and the terms from the initial values of f_2 (as if f_1 and its derivatives were absent).

$\mathscr{F}_1$ and $\mathscr{F}_2$ exist and are unique since

$$\Delta\left(=\begin{vmatrix} L_1(p) & M_1(p) \\ L_2(p) & M_2(p)\end{vmatrix}\right)$$

does not vanish identically.

Example 10. The algebraic equations corresponding to

$$7\frac{d^2 f_1}{dx^2} + 23f_1 - 8f_2 = 0,$$

$$3\frac{d^2 f_1}{dx^2} + 2\frac{d^2 f_2}{dx^2} - 13f_1 + 10f_2 = 0,$$

with $f_{10} = f_{20} = 0, f_{11} = f_{21} = 1$ are

$$(7p^2 + 23)\mathscr{F}_1 - 8\mathscr{F}_2 = 7p,$$

$$(3p^2 - 13)\mathscr{F}_1 + (2p^2 + 10)\mathscr{F}_2 = 5p.$$

It is easy to deduce that

$$7\mathscr{F}_1 = \frac{6p}{p^2+1} + \frac{p}{p^2+9}, \qquad 7\mathscr{F}_2 = \frac{21p}{p^2+1} - \frac{5p}{p^2+9}.$$

Hence

$$f_1 = \tfrac{1}{7}(6 \sin x + \tfrac{1}{3}\sin 3x), \qquad f_2 = \tfrac{1}{7}(12 \sin x - \tfrac{5}{3}\sin 3x).$$

The proof of the rule enunciated above requires three stages, (i) that the interpretations of $\mathscr{F}_1$ and $\mathscr{F}_2$ satisfy (29) and (30), (ii) that the interpretations satisfy the initial conditions and (iii) that the solution is unique. Now, by Theorem 11.22, $p\mathscr{F}_1 - p \mathrel{\overline{\overline{\cdot}}} (f_1 \mathrm{H})' - \delta = (\mathrm{d}f_1/\mathrm{d}x)\mathrm{H}$ so that

$$L_1\left(\frac{d}{dx}\right)f_1 + M_1\left(\frac{d}{dx}\right)f_2 \mathrel{\overline{\overline{\cdot}}} L_1(p)\mathscr{F}_1 + M_1(p)\mathscr{F}_2 - \mathscr{C}_1 = \mathscr{F}_3$$

from (31). Thus (29) is satisfied and (30) is dealt with similarly.

Also, by Corollary 11.22,

$$\lim_{x\to+0}\frac{\mathrm{d}^s f_1}{\mathrm{d}x^s}=\lim_{p\to\infty}\langle p^s\mathscr{F}_1\rangle$$

$$=\lim_{p\to\infty}\left\langle\frac{p^s}{\Delta}\{(\mathscr{F}_3+\mathscr{C}_1)M_2-(\mathscr{F}_4+\mathscr{C}_2)M_1\}\right\rangle$$

$$=\lim_{p\to\infty}\left\langle p^s\mathscr{A}+\frac{p^s}{\Delta}\{\mathscr{F}_3M_2-M_2\langle L_1\mathscr{A}+M_1\mathscr{B}\rangle\right.$$

$$\left.-\mathscr{F}_4M_1+M_2\langle L_2\mathscr{A}+M_2\mathscr{B}\rangle\}\right\rangle$$

since $\mathscr{C}_1=L_1\mathscr{A}+M_1\mathscr{B}-\langle L_1\mathscr{A}+M_1\mathscr{B}\rangle$ and similarly for $\mathscr{C}_2$. By definition $\langle L_1\mathscr{A}+M_1\mathscr{B}\rangle$ contains no positive powers of p as $p\to\infty$. Nor does $\mathscr{F}_3$, while M_2 contains no power higher than p^r. Since Δ contains p^{m+r} the highest power in $\mathscr{F}_3M_2-M_2\langle L_1\mathscr{A}+M_1\mathscr{B}\rangle$ is p^{-m}. Hence

$$\lim_{x\to+0}\frac{\mathrm{d}^s f_1}{\mathrm{d}x^s}=\lim_{p\to\infty}\langle p^s\mathscr{A}\rangle\qquad(s<m)$$

$$=f_{1s}.$$

We may show similarly that $\lim_{x\to+0}\mathrm{d}^s f_2/\mathrm{d}x^s=f_{2s}(s<r)$. Consequently the initial conditions are satisfied.

Suppose now that there were two solutions of (29) and (30) satisfying the initial conditions, say f_1,f_2 and f_{01},f_{02}. Then $\xi=f_1-f_{01}$, $\eta=f_2-f_{02}$ satisfy

$$L_1\left(\frac{\mathrm{d}}{\mathrm{d}x}\right)\xi+M_1\left(\frac{\mathrm{d}}{\mathrm{d}x}\right)\eta=0,\qquad L_2\left(\frac{\mathrm{d}}{\mathrm{d}x}\right)\xi+M_2\left(\frac{\mathrm{d}}{\mathrm{d}x}\right)\eta=0\quad(33)$$

with the initial conditions $\lim_{x\to+0}\mathrm{d}^s\xi/\mathrm{d}x^s=0\ (s<m)$, $\lim_{x\to+0}\mathrm{d}^s\eta/\mathrm{d}x^s=0\ (s<r)$. It follows from (33) that

$$a_{10}\lim_{x\to+0}\mathrm{d}^m\xi/\mathrm{d}x^m+b_{10}\lim_{x\to+0}\mathrm{d}^r\eta/\mathrm{d}x^r=0,$$

$$a_{20}\lim_{x\to+0}\mathrm{d}^m\xi/\mathrm{d}x^m+b_{20}\lim_{x\to+0}\mathrm{d}^r\eta/\mathrm{d}x^r=0.$$

Hence $\lim_{x\to+0}\mathrm{d}^m\xi/\mathrm{d}x^m=0$, $\lim_{x\to+0}\mathrm{d}^r\eta/\mathrm{d}x^r=0$ because

$$\begin{vmatrix}a_{10} & b_{10}\\ a_{20} & b_{20}\end{vmatrix}\neq 0.$$

By taking repeated derivatives of (33) we deduce that all derivatives of ξ and η vanish at $x = 0$.

However, it is clear from (33) that $\Delta(d/dx)\xi = 0$. The theory of a single differential equation shows that the only solution such that all derivatives vanish at the origin is identically zero, i.e. $\xi \equiv 0$. Similarly $\eta \equiv 0$ and it has been shown that (29) and (30) possess a unique solution.

If f_3 and f_4 do not have operational representations the above procedure is still valid in a formal way but now the term $(M_2 \mathscr{F}_3 - M_1 \mathscr{F}_4)/\Delta$ in $\mathscr{F}_1$ must be interpreted as $m_2 * f_3 - m_1 * f_4$ where $m_2 \doteqdot pM_2/\Delta, m_1 \doteqdot pM_1/\Delta$.

Exercise

42. Solve

(i) $\dfrac{df_1}{dx} + 5f_1 + 2f_2 = e^{-x}, \dfrac{df_2}{dx} + 2f_1 + 2f_2 = 0, f_{10} = 1, f_{20} = 0$;

(ii) $\dfrac{d^2f_1}{dx^2} + 2a\dfrac{df_2}{dx} + b^2f_1 = 0, \dfrac{d^2f_2}{dx^2} - 2a\dfrac{df_1}{dx} + b^2f_2 = 0,$

$$f_{10} = f_{11} = f_{21} = 0, f_{20} = 1;$$

(iii) $\dfrac{d^2f_1}{dx^2} + 6\dfrac{d^2f_2}{dx^2} + \dfrac{df_2}{dx} - 4f_1 - f_2 = 0,$

$$6\dfrac{d^2f_1}{dx^2} + \dfrac{d^2f_2}{dx^2} - \dfrac{df_1}{dx} - f_1 - 9f_2 = 0, f_{10} = f_{20} = 1, f_{11} = f_{21} = 0;$$

(iv) $9\dfrac{d^2f_1}{dx^2} + 3f_1 + 11\dfrac{df_2}{dx} + 9f_2 = e^{x^2}, 3\dfrac{d^2f_1}{dx^2} + 9f_1 + 5\dfrac{df_2}{dx} + 3f_2 = 0,$

$$f_{10} = 1, f_{11} = 2, f_{20} = 0;$$

(v) $\dfrac{d^2f_1}{dx^2} + a\dfrac{df_2}{dx} = \cos \omega x, \dfrac{d^2f_2}{dx^2} - a\dfrac{df_1}{dx} = 0, f_{10} = 3, f_{20} = 5, f_{11} = f_{21} = 0$;

(vi) $\dfrac{d^2f_1}{dx^2} + 3f_1 + f_2 + 3f_3 = 0, \dfrac{d^2f_2}{dx^2} + \tfrac{1}{2}f_1 + 2f_2 = 0, \dfrac{d^2f_3}{dx^2} + \tfrac{1}{2}f_1 + 2f_3 = 0,$

$$f_{10} = f_{20} = f_{30} = f_{11} = f_{31} = 0, f_{21} = 1.$$

(In forming algebraic equations for three differential equations add to the right-hand side the sum of terms for initial conditions which would be added for each f separately in the absence of the others.)

When

$$\begin{vmatrix} a_{10} & b_{10} \\ a_{20} & b_{20} \end{vmatrix} = 0$$

the situation is very different. One possibility is that (30) is just a constant multiple of (29). In that case f_2 can be chosen as any function (with appropriate continuity and differentiability) which satisfies the initial conditions and then f_1 is determined by the resulting differential equation. In this way a whole family of solutions is generated; there is no longer a unique solution to the simultaneous differential equations.

If $a_{20} = b_{20} = 0$ the terms $\mathrm{d}^m f_1/\mathrm{d}x^m$ and $\mathrm{d}^r f_2/\mathrm{d}x^r$ are absent from the left-hand side of (30). By allowing $x \to +0$ we see that a relation must exist between the initial values (unless the differential equation has the degenerate form $0 = f_4$). If this relation is not met the differential equations possess no solution.

If $b_{20}/a_{20} = b_{10}/a_{10}$ subtract a_{10}/a_{20} times (30) from (29) and suppose that the highest derivative of f_1 in the resulting differential equation is the sth ($s < m$). (If neither f_1 nor f_2 occurs on the left-hand side the equations are incompatible unless f_3 and f_4 are connected by a special relation when we return to the first possibility considered.) Then we have to solve (29) and

$$a\frac{\mathrm{d}^s f_1}{\mathrm{d}x^s} + \ldots = f_5. \tag{34}$$

Equation (30) may now be replaced by $(30) - (a_{10}/a)\mathrm{d}^{m-s}(34)/\mathrm{d}x^{m-s}$ which contains no derivative of f_1 as high as m. The process can be repeated so as to remove the highest derivative of f_1 at each step until we arrive at

$$N_1\left(\frac{\mathrm{d}}{\mathrm{d}x}\right)f_1 + N_2\left(\frac{\mathrm{d}}{\mathrm{d}x}\right)f_2 = f_6,$$

$$N_3\left(\frac{\mathrm{d}}{\mathrm{d}x}\right)f_2 = f_7.$$

If $N_3 \not\equiv 0$ we can solve for f_2 and then, if $N_1 \not\equiv 0$, find f_1. At each stage the number of arbitrary initial values is known, the total number being the sum of the degrees of N_1 and N_3. If $N_3 \equiv 0$ the equations are incompatible unless $f_7 \equiv 0$ when f_2 is arbitrary. If $N_1 \equiv 0$ the equations are either incompatible or f_1 is arbitrary.

Example 11.

$$\frac{d^3 f_1}{dx^3} + \frac{df_2}{dx} - f_2 = 0, \tag{35}$$

$$\frac{d^2 f_1}{dx^2} + 4f_1 + f_2 = 1. \tag{36}$$

Subtracting (d/dx) (36) from (35) gives

$$4\frac{df_1}{dx} + f_2 = 0. \tag{37}$$

From (36) and (37) follows

$$\frac{d^2 f_1}{dx^2} - 4\frac{df_1}{dx} + 4f_1 = 1$$

whence

$$f_1 = \tfrac{1}{4} + e^{2x}(c_1 + c_2 x)$$

which shows at once that only two of f_{10}, f_{11}, f_{12} can be selected arbitrarily. Moreover, (37) gives

$$f_2 = -4e^{2x}(2c_1 + c_2 + 2c_2 x)$$

so that f_{20} and f_{21} are known and cannot be chosen arbitrarily once f_1 has been determined. Conversely if f_{20} and f_{21} are given then f_{10}, f_{11}, f_{12} are fixed. In other words only two of $f_{20}, f_{21}, f_{10}, f_{11}$ and f_{12} are arbitrary.

11.11. Weak functions of several variables

This section is concerned with the weak function of more than one variable. The extension from the generalised functions of Chapter 7 is completely analogous to the extension in the case of a single variable. Therefore, formal definitions will be omitted for the most part and only the main properties will be described.

The weak function $w(\mathbf{x})$ on R_n is defined by means of a regular sequence $\{\phi_m\}$ of functions ϕ_m which are fine on R_n, and

$$\int_{-\infty}^{\infty} w(\boldsymbol{x})\phi(\boldsymbol{x})\,\mathrm{d}\boldsymbol{x} = \lim_{m\to\infty}\int_{-\infty}^{\infty} \phi_m(\boldsymbol{x})\phi(\boldsymbol{x})\,\mathrm{d}\boldsymbol{x}.$$

Any generalised function on R_n is also a weak function. The growth of weak functions is not limited at infinity.

Any f, continuous on R_n, defines a weak function w such that

$$\int_{-\infty}^{\infty} w(\boldsymbol{x})\phi(\boldsymbol{x})\,\mathrm{d}\boldsymbol{x} = \int_{-\infty}^{\infty} f(\boldsymbol{x})\phi(\boldsymbol{x})\,\mathrm{d}\boldsymbol{x};$$

therefore the symbol f is used to denote the corresponding weak function also.

The weak functions w_1+w_2, $w(a_1x_1+b_1,\dots,a_nx_n+b_n)$ $(a_1, b_1,\dots, a_n, b_n$ real), $\eta(x)w(x)$ (η infinitely differentiable on R_n), $\partial_p w$ are defined by the sequences $\{\phi_{1m}+\phi_{2m}\}$, $\{\phi_m(a_1x_1+b_1,\dots,a_nx_n+b_n)\}$ $\{\eta\phi_m\}$ and $\{\partial\phi_m/\partial x_p\}$ respectively. They have similar properties to the corresponding generalised functions and, in particular,

$$\partial_p\partial_q w = \partial_q\partial_p w.$$

If f is a continuous function with ordinary partial derivative $\partial f/\partial x_p$ and $\partial f/\partial x_p \in \boldsymbol{L}_1(\boldsymbol{a}, \boldsymbol{b})$, i.e. $\int_{a_1}^{b_1}\dots\int_{a_n}^{b_n}|\partial f/\partial x_p|\mathrm{d}x_1\dots\mathrm{d}x_n < \infty$, and if $w=f$ in $(\boldsymbol{a}, \boldsymbol{b})$ then $\partial_p w = \partial f/\partial x_p$ in $(\boldsymbol{a}, \boldsymbol{b})$. Thus the weak and conventional partial derivative can be identified when the latter exists.

It may be shown, as for generalised functions, that if $w=f$ inside the volume T and vanishes outside

$$\int_{-\infty}^{\infty}(\partial_1 w)\phi(\boldsymbol{x})\,\mathrm{d}\boldsymbol{x} = \int_T \frac{\partial f}{\partial x_1}\phi(\boldsymbol{x})\,\mathrm{d}\boldsymbol{x} - \int_S f(\boldsymbol{x})\phi(\boldsymbol{x})\cos\theta_1\,\mathrm{d}S$$

when f has a continuous partial derivative on $T+S$, S is the boundary of T and θ_1 is the angle between the outward normal to S and the x_1-axis. When f has second partial derivatives

$$\int_{-\infty}^{\infty}\{(\partial_1^2+\partial_2^2+\dots+\partial_n^2)w\}\phi(\boldsymbol{x})\,\mathrm{d}\boldsymbol{x}$$
$$= \int_T (\nabla^2 f)\phi(\boldsymbol{x})\,\mathrm{d}\boldsymbol{x} + \int_S\left(f\frac{\partial\phi}{\partial\nu} - \frac{\partial f}{\partial\nu}\phi\right)\mathrm{d}S.$$

The definition of limit is unaltered and $\mathrm{Lim}_{\mu\to\mu_0}\eta w_\mu = \eta w$ if $\mathrm{Lim}_{\mu\to\mu_0} w_\mu = w$.

If $(\boldsymbol{a}, \boldsymbol{b})$ is finite there are finite K and p such that

$$\left|\int_{-\infty}^{\infty} w(\boldsymbol{x})\phi(\boldsymbol{x})\mathrm{d}\boldsymbol{x}\right| \leq K \max_{\boldsymbol{x}\in[\boldsymbol{a},\, \boldsymbol{b}]} |\partial^{p}\phi(\boldsymbol{x})|$$

for all fine ϕ which vanish outside $(\boldsymbol{a}, \boldsymbol{b})$. Also there is a continuous f, which vanishes outside $[\boldsymbol{a}-\boldsymbol{\varepsilon}, \boldsymbol{b}+\boldsymbol{\varepsilon}]$ $(\varepsilon_1, \dots, \varepsilon_n$ all arbitrary and positive), such that

$$w(\boldsymbol{x}) = \partial_1^{p_1}\partial_2^{p_2}\dots\partial_n^{p_n} f(\boldsymbol{x}) \qquad (\boldsymbol{x}\in(\boldsymbol{a}, \boldsymbol{b})).$$

Furthermore $\mathrm{Lim}_{m\to\infty} w_m = w$ on $(\boldsymbol{a}, \boldsymbol{b})$ if, and only if, $w_m = \partial_1^{p_1}\dots\partial_n^{p_n} f_m$ where $\{f_m\}$ is a sequence of continuous functions which converge uniformly on $(\boldsymbol{a}, \boldsymbol{b})$.

One consequence of this is that if $w(x_1)$ is a weak function on R_1 then it is one on R_n. Another is that if $w = 0$ for $r > 0$ then

$$w(\boldsymbol{x}) = \sum_{m=0}^{M} a_m \partial_1^{m_1}\dots\partial_n^{m_n}\delta(\boldsymbol{x})$$

where M is finite and $m_1 + \dots + m_n = m$.

The proof in Corollary 7.21*a* is not applicable to weak functions and therefore a new proof must be provided to obtain a similar result.

Theorem 11.23. *$\int_{-\infty}^{\infty} w(\boldsymbol{x})\phi_1(x_1)\dots\phi_n(x_n)\mathrm{d}\boldsymbol{x} = 0$ for all possible $\phi_1, \phi_2, \dots, \phi_n$ each fine on R_1 if, and only if, $w = 0$.*

Proof. The statement is obviously true if $w = 0$ since $\phi_1\phi_2\dots\phi_n$ is fine on R_n.

Conversely, if the integral is zero, we can use the method of Theorem 5.7 to show that $w = 0$ provided that any fine ϕ can be approximated sufficiently closely by functions of the form $\sum a_m \phi_{m_1}(x_1)\dots\phi_{m_n}(x_n)$. Suppose that ϕ vanishes outside $(\boldsymbol{a}, \boldsymbol{b})$ and let $\phi_{01}(x_1)\dots\phi_{0n}(x_n)$ be a fine function which equals 1 in $(\boldsymbol{a}, \boldsymbol{b})$. Since ϕ is infinitely differentiable a well-known theorem of Weierstrass states that there is a sequence of polynomials $\{P_m\}$ which converges uniformly to ϕ on $[\boldsymbol{a}, \boldsymbol{b}]$ while each derivative converges uniformly to the corresponding derivative of ϕ. Then $\{P_m\phi_{01}\dots\phi_{0n}\}$ is a sequence of fine functions with the same property and of the required form. Using this sequence in the same way as the sequence of Hermite polynomials in Theorem 5.7 we deduce $\int_{-\infty}^{\infty} w(\boldsymbol{x})\phi(\boldsymbol{x})\mathrm{d}\boldsymbol{x} = 0$, i.e. $w = 0$. The proof is complete. □

By means of this theorem we see that it is sufficient to work with fine functions which are the product of n fine functions of a single variable.

If $\phi_{(n)}(x_n)$ is fine, and if $\{\phi_m\}$ defines w then $\mathrm{Lim}_{m\to\infty}\int_{-\infty}^{\infty}\phi_m(\boldsymbol{x})\phi_{(n)}(x_n)\mathrm{d}x_n$ is a weak function on R_{n-1} which is denoted by $\int_{-\infty}^{\infty} w(\boldsymbol{x})\phi_{(n)}(x_n)\,\mathrm{d}x_n$. As in Theorem 7.22 it has the property

$$\partial_1\int_{-\infty}^{\infty} w(\boldsymbol{x})\phi_{(n)}(x_n)\,\mathrm{d}x_n = \int_{-\infty}^{\infty}\{\partial_1 w(\boldsymbol{x})\}\phi_{(n)}(x_n)\,\mathrm{d}x_n$$

and

$$\int_{-\infty}^{\infty} w(\boldsymbol{x})\phi_{(1)}(x_1)\ldots\phi_{(n)}(x_n)\mathrm{d}\boldsymbol{x}$$

$$= \int_{-\infty}^{\infty}\phi_{(1)}(x_1)\int_{-\infty}^{\infty}\phi_{(2)}(x_2)\ldots\int_{-\infty}^{\infty} w(\boldsymbol{x})\phi_{(n)}(x_n)\,\mathrm{d}x_n\ldots\mathrm{d}x_2\,\mathrm{d}x_1.$$

The *direct product* of $w_1(\boldsymbol{x})$ on R_n and $w_2(\boldsymbol{y})$ on R_m is defined on R_{n+m} by

$$\int_{-\infty}^{\infty} w_1(\boldsymbol{x})w_2(\boldsymbol{y})\phi_1(\boldsymbol{x})\phi_2(\boldsymbol{y})\,\mathrm{d}\tau$$

$$= \int_{-\infty}^{\infty} w_1(\boldsymbol{x})\phi_1(\boldsymbol{x})\,\mathrm{d}\boldsymbol{x}\int_{-\infty}^{\infty} w_2(\boldsymbol{y})\phi_2(\boldsymbol{y})\,\mathrm{d}\boldsymbol{y}.$$

Clearly

$$w_1(\boldsymbol{x})w_2(\boldsymbol{y}) = w_2(\boldsymbol{y})w_1(\boldsymbol{x}),$$
$$w_1(\boldsymbol{x})\{w_2(\boldsymbol{y})w_3(\boldsymbol{z})\} = \{w_1(\boldsymbol{x})w_2(\boldsymbol{y})\}w_3(\boldsymbol{z})$$

and

$$\partial_1\{w_1(\boldsymbol{x})w_2(\boldsymbol{y})\} = \{\partial_1 w_1(\boldsymbol{x})\}w_2(\boldsymbol{y}).$$

Definition 7.14, Theorems 7.26, 7.27, 7.28 and Corollary 7.27 may continue to be used with weak functions substituted for generalised functions. The same remark applies to Exercises 28, 30 and the first halves of Exercises 25 and 26 of Chapter 7.

With regard to multiplication we generalise Definition 11.4 as follows: *if* $f_1 \in \boldsymbol{L}_1(\boldsymbol{a}, \boldsymbol{b})$, $f_2 \in \boldsymbol{L}_1(\boldsymbol{a}, \boldsymbol{b})$ *and* $f_1 f_2 \in \boldsymbol{L}_1(\boldsymbol{a}, \boldsymbol{b})$ *for every finite* $(\boldsymbol{a}, \boldsymbol{b})$ *the weak function determined by* $f_1 f_2$ *is also denoted by* $f_1 f_2$ *(or* $f_2 f_1$*). If* $w_1 w_2$ *and* $(\partial_1 w_1) w_2$ *exist the product* $w_1 \partial_1 w_2$ *is defined by*

$$w_1 \partial_1 w_2 = \partial_1(w_1 w_2) - (\partial_1 w_1) w_2.$$

If f and $\partial^{r_1 + r_2 + \dots + r_n} f / \partial x_1^{r_1} \dots \partial x_n^{r_n}$ $(0 \leq r_1 \leq p_1 - 1, \dots, 0 \leq r_n \leq p_{n-1})$ are continuous and $\partial^{p_1 + \dots + p_n} f / \partial x_1^{p_1} \dots \partial x_n^{p_n} \in \boldsymbol{L}_1(\boldsymbol{a}, \boldsymbol{b})$ for every finite $(\boldsymbol{a}, \boldsymbol{b})$ we write $f \in \boldsymbol{L}^{\boldsymbol{p}}$. We use $f \in \boldsymbol{L}^0$ to indicate that $f \in \boldsymbol{L}_1(\boldsymbol{a}, \boldsymbol{b})$ for every finite $(\boldsymbol{a}, \boldsymbol{b})$. Then it can be shown, as for Corollary 11.14*a*, if $w_1 = \partial_1^{p_1} \dots \partial_n^{p_n} f_1$ where $f_1 \in \boldsymbol{L}^0$, if $f_2 \in \boldsymbol{L}^{\boldsymbol{p}}$ and if $f_1 \partial_1^{p_1} \dots \partial_n^{p_n} f_2$ exists, that $f_2 w_1$ exists. In particular, if w_1 vanishes outside a finite interval and $f_2 \in \boldsymbol{L}^{\boldsymbol{p}}$, $f_2 w_1$ exists if p is large enough. Consequently ηw exists under these definitions and has the same properties as previously when η is infinitely differentiable.

If w is continuous in some neighbourhood surrounding the origin we define $w(\boldsymbol{x})\delta(\boldsymbol{x})$ by

$$w(\boldsymbol{x})\delta(\boldsymbol{x}) = w(\boldsymbol{0})\delta(\boldsymbol{x}).$$

As far as convolution is concerned the spaces $\boldsymbol{L}'_p$ involve no new features. Convolution in which one factor vanishes outside a finite $(\boldsymbol{a}, \boldsymbol{b})$ may continue to be handled by Definition 11.8 (with the small modification of replacing x and t by $\boldsymbol{x}$ and $\boldsymbol{t}$ respectively). Theorems 11.16, 11.17 and 11.18 remain valid.

There are also generalisations along the lines of §8.10. Let K be the cone introduced in that section.

Definition 11.11. $w \in \boldsymbol{W}_+$ *if, and only if,* $w(\boldsymbol{x}) = 0$ *outside* K.

If $w_1, w_2 \in \boldsymbol{W}_+$ the convolution $w_1 * w_2$ can be defined as in Definition 11.10 and Theorem 11.19 still holds. It can also be shown that Theorem 11.20 is still valid.

Let I^β be the operator defined at the end of §8.10. Then, if $w \in \boldsymbol{W}_+$,

$$I^\beta(I^\gamma w) = (I^\beta I^\gamma) w = I^{\beta+\gamma} w, \quad I^0 w = w.$$

Since

$$(\partial_1^2 - \dots - \partial_n^2)^k w_1 = w_2 \tag{38}$$

can be written, if $w_1, w_2 \in W_+$, as

$$I^{-2k}w_1 = w_2$$

it follows that the unique solution of (38) in W_+ is

$$\begin{aligned} w_1 &= I^{2k}w_2 \\ &= u_{2k} * w_2. \end{aligned}$$

This result generalises in some respects the formulae of Example 8 of Chapter 10.

Consider now the series $\sum_{m=0}^{\infty}\lambda^m I^{2m+2}w$ where λ is a complex number. For sufficiently large k, $I^k w$ differs from $u_+^{(k/2)-n/2} * w$ only by a constant factor and is a continuous function. Hence $I^{mk}w \to 0$ as $m \to \infty$ (cf. Exercise 24). Similarly $I^{mk+1}w, I^{mk+2}w, \ldots, I^{mk+k-1}w$ all tend to zero as $m \to \infty$. Consequently the given series converges weakly and is a weak function. Therefore

$$\begin{aligned} (I^{-2} - \lambda I^0) \sum_{m=0}^{\infty} \lambda^m I^{2m+2}w &= \sum_{m=0}^{\infty} \lambda^m I^{2m}w - \sum_{m=0}^{\infty} \lambda^{m+1} I^{2m+2}w \\ &= I^0 w = w. \end{aligned}$$

Consequently the unique solution of

$$(\partial_1^2 - \partial_2^2 - \ldots - \partial_n^2 - \lambda)w_1 = w_2$$

with $w_1, w_2 \in W_+$ is

$$w_1 = \sum_{m=0}^{\infty} \lambda^m I^{2m+2} w_2. \tag{39}$$

If n is odd (39) can be put in the form

$$w_1 = \sum_{m=0}^{\infty} \frac{\lambda^m u_+^{m+1-n/2} * w_2}{(m+1-\frac{1}{2}n)!\,m!\,\pi^{(n/2)-1}2^{2m+1}}.$$

Formally the series is the same as occurs in a Bessel function, which could be signified by writing

$$w_1 = \frac{\lambda^{(n/4)-1/2}}{\pi^{(n/2)-1}2^{n/2}}\{u_+^{(1/2)-n/4} I_{1-n/2}(\lambda^{1/2}u_+^{1/2})\} * w_2.$$

When $\lambda < 0$ it may be more convenient to convert this to

$$w_1 = \frac{|\lambda|^{(n/4)-1/2}}{\pi^{(n/2)-1}2^{n/2}}\{u_+^{(1/2)-n/4} J_{1-n/2}(|\lambda|^{1/2}u_+^{1/2})\} * w_2.$$

If n is even the formula looks more complicated because of the presence of additional terms since then I^2 does not involve $u^{1-n/2}$ but $\mathrm{H}(x_1)\delta^{((n/2)-2)}(q)$.

More generally, it can be shown by the same method that the solution of

$$(\partial_1^2 - \partial_2^2 - \dots - \partial_n^2 - \lambda)^k w_1 = w_2$$

with $w_1, w_2 \in W_+$ is

$$w_1 = \sum_{m=0}^{\infty} \frac{(m+k-1)!}{m!(k-1)} \lambda^m I^{2m+2k} w_2. \tag{40}$$

Exercises

43. If n is odd show that an abbreviation for (40) is

$$w_1 = \frac{\lambda^{(n/4)-k/2}}{\pi^{(n/2)-1}2^{(n/2)+k-1}} \{u_+^{(k/2)-n/4} I_{k-n/2}(\lambda^{1/2}u_+^{1/2})\} * w_2.$$

44. If $w_1, w_2 \in W_+$ prove that the solution of

$$(\partial_1^2 - \partial_2^2 - \dots - \partial_n^2 - \lambda_1)(\partial_1^2 - \partial_2^2 - \dots - \partial_n^2 - \lambda_2)w_1 = w_2 \qquad (\lambda_1 \neq \lambda_2)$$

is

$$w_1 = \frac{1}{\lambda_1 - \lambda_2} \sum_{m=0}^{\infty} (\lambda_1^m - \lambda_2^m) I^{2m+2} w_2.$$

11.12. Integration with respect to a parameter

If $w(\mu, \boldsymbol{x})$ is a weak function of $\boldsymbol{x}$ for all μ in some domain M and $\int_M \{\int_{-\infty}^{\infty} w(\mu, \boldsymbol{x})\phi(\boldsymbol{x})\,\mathrm{d}\boldsymbol{x}\}\,\mathrm{d}\mu$ exists for every fine ϕ, then $\int_M w(\mu, \boldsymbol{x})\,\mathrm{d}\mu$ is said to exist as a weak function and is defined by

$$\int_{-\infty}^{\infty} \phi(\boldsymbol{x}) \left\{ \int_M w(\mu, \boldsymbol{x})\,\mathrm{d}\mu \right\} \mathrm{d}\boldsymbol{x} = \int_M \left\{ \int_{-\infty}^{\infty} w(\mu, \boldsymbol{x})\phi(\boldsymbol{x})\,\mathrm{d}\boldsymbol{x} \right\} \mathrm{d}\mu.$$

It may be shown, as in §7.9, that if $\int_M w(\mu, \boldsymbol{x})\,\mathrm{d}\mu$ exists then so does $\int_M \partial_1 w(\mu, \boldsymbol{x})\,\mathrm{d}\mu$ and

$$\partial_1 \int_M w(\mu, \boldsymbol{x})\,\mathrm{d}\mu = \int_M \partial_1 w(\mu, \boldsymbol{x})\,\mathrm{d}\mu.$$

It can always be asserted that $\int_M w(\mu, \boldsymbol{x})\,\mathrm{d}\mu$ exists if M is a bounded domain and if, for every $\mu_0 \in M$, $\operatorname{Lim}_{\mu\to\mu_0} w(\mu, x) = w(\mu_0, x)$ (Theorem 7.41).

Exercise

45. Are the following statements true for every weak function w and any finite real number h,

$$\int_{-\infty}^{\infty} w(x+h)\phi(x)\,dx = \sum_{m=0}^{M-1}\int_{-\infty}^{\infty}\frac{h^m}{m!}w^{(m)}(x)\phi(x)\,dx + \frac{h^M}{(M-1)!}$$
$$\times \int_{-\infty}^{\infty} w^{(M)}(x)\int_0^1 (1-t)^{M-1}\phi(x-th)\,dt\,dx,$$

$$w(x+h) = \sum_{m=0}^{M-1}\frac{h^m}{m!}w^{(m)}(x) + \frac{h^M}{(M-1)!}\int_0^1 (1-t)^{M-1}w^{(M)}(x+th)\,dt,$$

$$w(x+h) = \sum_{m=0}^{M-1}\frac{h^m}{m!}w^{(m)}(x) + \frac{h^M}{M!}w^{(M)}(x+\xi h)$$

where $0 \le \xi \le 1$? (cf. Corollary 1.30c).

11.13. Change of variable: single variable case

The problem of changing variables in a weak function is similar to that for a generalised function but, because of the different behaviour at infinity, can be achieved with less restriction on the transformation.

Definition 11.12. *Let $\eta(x)$ be a real-valued infinitely differentiable function, with $\eta'(x) \neq 0$, which is a one-to-one mapping from R_1 to R_1. Then $\{\phi_m(\eta(x))\}$ is a regular sequence which defines a weak function to be denoted by $w\{\eta(x)\}$.*

Evidently $\phi_m(\eta)$ is infinitely differentiable and vanishes outside a finite interval of x, since the mapping is one-to-one and η is bounded for finite x. Hence $\phi_m(\eta)$ is a fine function of x. Conversely, if $\phi(x)$ is a fine function of x, $\phi(\eta^{-1}(y))$ is a fine function of y for similar reasons. Therefore

$$\int_{-\infty}^{\infty}\phi_m\{\eta(x)\}\phi(x)\,dx = \int_{\infty}^{\infty}\phi_m(y)\frac{\phi\{\eta^{-1}(y)\}}{|\eta'(x)|}dy$$
$$\to \int_{-\infty}^{\infty} w(y)\,\frac{\phi\{\eta^{-1}(y)\}}{|\eta'(x)|}dy.$$

This shows that the sequence $\phi_m(\eta)$ is regular and that

$$\int_{-\infty}^{\infty} w\{\eta(x)\}\phi(x)\mathrm{d}x = \int_{-\infty}^{\infty} w(y)\frac{\phi\{\eta^{-1}(y)\}}{|\eta'(x)|}\mathrm{d}y. \tag{41}$$

As in Theorem 8.6, it can be proved that

$$[w\{\eta(x)\}]' = \eta'(x)w'\{\eta(x)\}. \tag{42}$$

As in Example 3 of Chapter 8, one can demonstrate that

$$\delta\{\eta(x)\} = \delta(x - x_0)/|\eta'(x_0)|$$

where $\eta(x_0) = 0$.

For finite intervals we have

Definition 11.13. *Let $\eta(x)$ be a real-valued infinitely differentiable function which goes steadily from c to d as x goes from a to b (a, b, c, d all finite) with $\eta'(x) \neq 0$ in $a \leq x \leq b$. Then, if $\{\phi_m(y)\}$ is a regular sequence defining w(y) on (c, d), $\{\phi_m(\eta(x))\}$ is a regular sequence which defines $w\{\eta(x)\}$ on (a, b).*

It should be emphasised that $w(\eta)$ is defined only on (a, b) and that there may be no weak function on R_1 which coincides with it on (a, b). (Cf. remarks at the end of §11.1.)

Since we are now concerned only with fine functions which vanish outside (a, b) for the variable x or (c, d) for the variable y there is no difficulty in verifying that the definition has the requisite properties and that

$$\int_a^b w\{\eta(x)\}\phi(x)\,\mathrm{d}x = \int_c^d w(y)\frac{\phi\{\eta^{-1}(y)\}}{\eta'(x)}\mathrm{d}y \tag{43}$$

for ϕ vanishing outside (a, b). Also

$$[w\{\eta(x)\}]' = \eta'(x)w'\{\eta(x)\} \quad (a < x < b). \tag{44}$$

If one of a, b, c, d is not finite Definition 11.13 may no longer be valid. Suppose $a = 0$, $b = 1$, $\eta(x) = \mathrm{e}^{1/x}$. Then any fine function on (a, b) which behaves like $\mathrm{e}^{-1/x}$ near $x = 0$ will become $1/y$ near $y = \infty$ and will consequently not be fine. Thus fine functions may not remain fine under the mapping. Of course, if we are dealing with generalised functions on the infinite interval we can transform from the finite to infinite intervals provided that the conditions of Definitions 8.5 or 8.6 are satisfied. If this is not so and $w(y)$ is given on an infinite interval we can take (a, b) finite, in general, only when $w(y)$ vanishes outside a finite interval of y. On the other hand, if (c, d) is finite $\phi(\eta^{-1})$ will be fine on it whether (a, b) is finite or infinite and only a limited range of ϕ_m will be involved in the integration over (a, b) in (43); in this case $w(\eta)$ will be defined via (43).

Exercises

46. Suppose that $w\{\eta(x)\}$ is defined by requiring (43) to hold whenever the right-hand side has a meaning. If $w = f^{(m)}$, where $f(y)$ is continuous on $[c,d]$ and $f', \ldots, f^{(M-1)}$ are finite at $y = c$ and $y = d$, show that $w\{\eta(x)\}$ can be defined if $\eta \in L^m$ and $\mathrm{d}\eta/\mathrm{d}x$ is finite and non-zero on $[a,b]$. Will (44) still be valid?

47. If $y = f(x)$ investigate conditions under which

$$w(x)\frac{\mathrm{d}x}{\mathrm{d}y} = \sum_{m=0}^{M-1} \frac{1}{m!}\frac{\mathrm{d}^m}{\mathrm{d}y^m}[\{y - f(y)\}^m w(y)]$$

$$+ \frac{1}{(M-1)!}\frac{\mathrm{d}^M}{\mathrm{d}y^M}\int_0^1 \{u - f(u)\}^M w(u)(1-t)^{M-1}\,\mathrm{d}t$$

where $u + t\{f(u) - u\} = y$.

11.14. Change of variables: several variables

When there are several variables consider firstly the linear mapping specified by the matrix A with $\det A \neq 0$. Then we have

Definition 11.14. *If $\{\phi_m\}$ is a regular sequence defining w, the sequence $\{\phi_m(A\boldsymbol{x})\}$ is regular and defines a weak function which is denoted by $w(A\boldsymbol{x})$.*

The justification is similar to that for Definition 8.1 but using fine functions instead of good functions, and

$$\int_{-\infty}^{\infty} w(A\boldsymbol{x})\phi(\boldsymbol{x})\,\mathrm{d}\boldsymbol{x} = \int_{-\infty}^{\infty} w(\boldsymbol{y})\phi(A^{-1}\boldsymbol{y})\frac{\mathrm{d}\boldsymbol{y}}{|\det A|}. \tag{45}$$

Definition 8.4 may still be used for the definition of a homogeneous weak function and Theorems 8.3, 8.4 and 8.5 remain valid.

The modifications to Definitions 8.7 and 8.8 for weak functions are similar to those incorporated in Definitions 11.12 and 11.13.

We note that if $\eta(x)$ is a real-valued infinitely differentiable function which gives a one-to-one mapping from T to U, with the Jacobian J not vanishing on the closure of T, then

$$\int_T w\{\eta(x)\}\phi(x)\,\mathrm{d}x = \int_U w(x)\frac{\phi\{\eta^{-1}(y)\}}{|J|}\,\mathrm{d}y$$

provided that T and U are both finite or both infinite. Special consideration is necessary if one of T and U is finite and the other infinite.

Further

$$\partial_1 w\{\eta(x)\} = \left[\sum_{j=1}^{n} \partial_{yj} w(y)\partial_1 y_1\right]_{y=\eta} \qquad (x\in T)$$

and

$$\partial_2\partial_1 w(\eta) = \partial_1\partial_2 w(\eta).$$

Most of the formulae involving integrals in Chapter 8, for example, those in Example 11 and §§8.4 and 8.6 are unaltered if f, γ and g are replaced by η, ϕ and w.

11.15. Ultradistributions and Fourier transforms

It has already been mentioned that in order to permit weak functions unlimited growth at infinity the Fourier transform has to be abandoned. This is a pity because the Fourier transform is a valuable weapon in tackling problems. To recover it one has to be prepared to work with entities which are neither weak functions nor generalised functions.

Suppose that $\phi(t)$ is a fine function which vanishes outside $(-a, a)$ with $a > 0$. Define $\Phi(z)$ when z is the complex variable $x + \mathrm{i}y$ by

$$\Phi(z) = \int_{-a}^{a} \phi(t)\mathrm{e}^{-\mathrm{i}zt}\,\mathrm{d}t.$$

The integral converges uniformly over every bounded domain of z. The integrand is a continuous function of (z, t) for every complex z and t. Also the integrand is a regular function of z for every real t. Therefore $\Phi(z)$ is a regular function of z for all finite z. Moreover, integration by parts gives

$$(\mathrm{i}z)^k \Phi(z) = \int_{-a}^{a} \phi^{(k)}(t)\mathrm{e}^{-\mathrm{i}zt}\,\mathrm{d}t.$$

Thus $\Phi(z)$ is an entire function of z satisfying

$$|z^k \Phi(z)| \leq C_k \mathrm{e}^{a|y|}$$

for all z.

The converse can also be proved and is incorporated in

Theorem 11.24. *In order that $\phi(t)$ be fine and vanish outside $(-a, a)$ it is necessary and sufficient that*

$$\phi(t) = \frac{1}{2\pi}\int_{-\infty}^{\infty} \Phi(x)\mathrm{e}^{\mathrm{i}xt}\,\mathrm{d}x$$

where $\Phi(z)$ is entire and

$$|z^k \Phi(z)| \leq C_k \mathrm{e}^{a|y|} \qquad (k = 0, 1, 2, \ldots)$$

for all z.

Note that this theorem prevents the Fourier transform of a fine function being fine. For, if $\Phi(x)$ were fine, $\Phi(z)$ would disappear because an entire function which vanishes on a non-zero length of the x-axis is identically zero.

The space of entire functions Φ satisfying the conditions of Theorem 11.24 is called Z. Obviously, if $\Phi \in Z$,

$$\frac{1}{2\pi}\int_{-\infty}^{\infty} (\mathrm{i}z)^r \Phi(z)\mathrm{e}^{\mathrm{i}zt}\,\mathrm{d}x = \phi^{(r)}(t), \tag{46}$$

$$\frac{1}{2\pi}\int_{-\infty}^{\infty} \Phi^{(r)}(z)\mathrm{e}^{\mathrm{i}zt}\,\mathrm{d}x = (-\mathrm{i}t)^r \phi(t). \tag{47}$$

From (47) it is evident that $\Phi^{(r)} \in Z$ if $\Phi \in Z$. An immediate deduction from the inequalities of Theorem 11.24 is that $\Phi(z)$ is a good function of x for fixed y.

Definition 11.15. *A sequence $\{\Phi_m\}$ with $\Phi_m \in Z$ is said to be regular if, for every $\Phi \in Z$, $\lim_{m\to\infty} \int_{-\infty}^{\infty} \Phi_m(z)\Phi(x)\,\mathrm{d}x$ exists for each fixed y and is finite. Two regular sequences with the same limit are said to be* *equivalent. An equivalence class of regular sequences is an ultradistribution.*

Perhaps it might be more precise to say an ultradistribution of x for each fixed y. Denoting an ultradistribution by $u(z)$ or $u(x+\mathrm{i}y)$ we write

$$\lim_{m\to\infty}\int_{-\infty}^{\infty}\Phi_m(z)\Phi(x)\,\mathrm{d}x=\int_{-\infty}^{\infty}u(z)\Phi(x)\,\mathrm{d}x.$$

The equality of two ultradistributions and addition may be defined in an analogous way to that for weak functions; so may $u(az+c)$ where a is a non-zero real constant and c is complex. Here we may note that

$$\begin{aligned}\int_{-\infty}^{\infty}u(az+c)\Phi(x)\,\mathrm{d}x &= \lim_{m\to\infty}\int_{-\infty}^{\infty}\Phi_m(az+c)\Phi(x)\,\mathrm{d}x\\ &= \lim_{m\to\infty}\int_{-\infty}^{\infty}\Phi_m(x)\Phi\left(\frac{x-c}{a}-\mathrm{i}y\right)\frac{\mathrm{d}x}{|a|}\\ &= \int_{-\infty}^{\infty}u(x)\Phi\left(\frac{x-c}{a}-\mathrm{i}y\right)\frac{\mathrm{d}x}{|a|} \qquad (48)\end{aligned}$$

since $\Phi\{(x-c)/a-\mathrm{i}y\}\in Z$. Equation (48) with $a=1$ and $c=0$ shows that a knowledge of $u(x)$ is sufficient to determine $u(z)$. In other words (48) provides a kind of continuation of an ultradistribution from the real axis to the remainder of the complex plane.

For multiplication we need

Definition 11.16. *The entire function $\psi(z)$ is said to belong to $\mathscr{Y}$ if there are finite b and l such that*

$$|\psi(z)|\le C\mathrm{e}^{b|y|}(1+|z|^l)$$

for all z.

Then the product $\psi(z)u(z)$ may be defined by $\{\psi\Phi_m\}$.

The derivative $u'(z)$ comes from $\{\mathrm{d}\Phi_m/\mathrm{d}z\}$ and has the property

$$\int_{-\infty}^{\infty}u'(z)\Phi(x)\,\mathrm{d}x=-\int_{-\infty}^{\infty}u(z)\Phi'(x)\,\mathrm{d}x.$$

It has been pointed out that (48) enables $u(z)$ to be found from $u(x)$. Now $u(x)$ has a derivative $u'(x)$ in terms of x and so $u'(x)$ will furnish an ultradistribution $v(z)$. The notation is reasonable so long as $v(z)=u'(z)$. But, from (48),

$$\int_{-\infty}^{\infty} v(z)\Phi(x)\,dx = \int_{-\infty}^{\infty} u'(x)\Phi(x-iy)\,dx = -\int_{-\infty}^{\infty} u(x)\Phi'(x-iy)\,dx$$

$$= -\int_{-\infty}^{\infty} u(z)\Phi'(x)\,dx$$

$$= \int_{-\infty}^{\infty} u'(z)\Phi(x)\,dx$$

to that $v(z)=u'(z)$ and the notation is acceptable.

Definition 11.17. *If $\{\phi_m(t)\}$ is a sequence of fine functions defining the weak function $w(t)$, the sequence $\{\int_{-\infty}^{\infty}\phi_m(t)e^{-izt}dt\}$ defines an ultradistribution called the complex Fourier transform of w and denoted by $w_c(z)$. We write*

$$w_c(z) = \int_{-\infty}^{\infty} w(t)e^{-izt}\,dt.$$

As justification we observe that, if $\Phi_m(z)=\int_{-\infty}^{\infty}\phi_m(t)e^{-izt}dt$, $\Phi_m\in Z$ by Theorem 11.24. By the same theorem, if $\Phi\in Z$, $\phi(t)=(1/2\pi)\int_{-\infty}^{\infty}\Phi(x)e^{ixt}dx$ is fine. Hence

$$\int_{-\infty}^{\infty}\Phi_m(z)\Phi(x)\,dx = \int_{-\infty}^{\infty}\Phi(x)\int_{-\infty}^{\infty}\phi_m(t)e^{-izt}\,dt\,dx$$

$$= 2\pi\int_{-\infty}^{\infty}\phi_m(t)e^{yt}\phi(-t)\,dt$$

$$\to 2\pi\int_{-\infty}^{\infty} w(t)e^{yt}\phi(-t)\,dt$$

as $m\to\infty$ since $e^{yt}\phi(-t)$ is fine. Thus the sequence $\{\Phi_m\}$ is regular and defines an ultradistribution. In fact, we have demonstrated

Theorem 11.25. $\int_{-\infty}^{\infty} w_c(z)\Phi(x)\,dx = 2\pi\int_{-\infty}^{\infty} w(t)e^{yt}\phi(-t)\,dt$

as well as confirming that every weak function possesses a complex Fourier transform.

Definition 11.18. *If* $\{\Phi_m\}$ *defines the ultradistribution* $u(z)$ *the sequence* $\{\int_{-\infty}^{\infty}\Phi_m(z)\mathrm{e}^{-izt}\mathrm{d}x\}$ *with t real defines a weak function* $U(t)$ *called the Fourier transform of u. We write*

$$U(t)=\int_{-\infty}^{\infty}u(z)\mathrm{e}^{-izt}\mathrm{d}x=\int_{-\infty}^{\infty}u(x)\mathrm{e}^{-ixt}\mathrm{d}x.$$

According to Theorem 11.24 the sequence is one of fine functions and

$$\int_{-\infty}^{\infty}\phi(t)\int_{-\infty}^{\infty}\Phi_m(z)\mathrm{e}^{-izt}\mathrm{d}x\,\mathrm{d}t=\int_{-\infty}^{\infty}\Phi_m(z)\Phi(z)\mathrm{d}x$$

$$\to\int_{-\infty}^{\infty}u(z)\Phi(z)\mathrm{d}x$$

as $m\to\infty$. Consequently, every ultradistribution has a Fourier transform and we have

Theorem 11.26. $\int_{-\infty}^{\infty}U(t)\phi(t)\mathrm{d}t=\int_{-\infty}^{\infty}u(z)\Phi(z)\mathrm{d}x.$

$U(t)$ is a weak function and therefore has a complex Fourier transform $U_c(z)$ according to Definition 11.17. From Theorem 11.25

$$\int_{-\infty}^{\infty}U(t)\phi(t)\mathrm{d}t=\frac{1}{2\pi}\int_{-\infty}^{\infty}U_c(x)\Phi(-x)\mathrm{d}x.$$

The invocation of Theorem 11.26 and (48) supplies

$$\frac{1}{2\pi}\int_{-\infty}^{\infty}U_c(-x)\Phi(x)\mathrm{d}x=\int_{-\infty}^{\infty}u(x)\Phi(x)\mathrm{d}x$$

whence $u(x)=U_c(-x)/2\pi$. On account of (48) $u(z)=U_c(-z)/2\pi$ and so we have

Theorem 11.27. *If* $U(t)$ *is the Fourier transform of* $u(z)$ *then*

$$u(z)=\frac{1}{2\pi}\int_{-\infty}^{\infty}U(t)\mathrm{e}^{izt}\mathrm{d}t.$$

If $w_c(z)$ *is the complex Fourier transform of* $w(t)$ *then*

$$w(t)=\frac{1}{2\pi}\int_{-\infty}^{\infty}w_c(z)\mathrm{e}^{izt}\mathrm{d}x.$$

In other words the Fourier inversion theorem is valid both for the

Fourier transform of an ultradistribution and for the complex Fourier transform of a weak function.

Imagine now that $w(t)$ is a generalised function $g(t)$. Then $g(t)$ possesses a Fourier transform G as a generalised function and a complex Fourier transform g_c which is an ultradistribution. Since $\Phi(x)$ is good in x the sequence defining $g_c(x)$ also defines $G(x)$ so that $G(x)$ and $g_c(x)$ may be identified. This suggests that the notation $G(z)$ might be employed for the complex Fourier transform. It is intended to imply that $G(z)$ is the generalised function $G(x)$ when $y = 0$ but, nevertheless, $G(z)$ is an ultradistribution whose meaning when $y \neq 0$ is to be inferred from (48) i.e.

$$\int_{-\infty}^{\infty} G(z)\Phi(x)\,\mathrm{d}x = \int_{-\infty}^{\infty} G(x)\Phi(x - \mathrm{i}y)\,\mathrm{d}x \qquad (49)$$

where the right-hand side can be calculated by the theory of generalised functions. Since every generalised function is a generalised Fourier transform of a generalised function each generalised function $g(x)$ originates an ultradistribution $g(z)$.

For example,

$$\int_{-\infty}^{\infty} \delta(z)\Phi(x)\,\mathrm{d}x = \int_{-\infty}^{\infty} \delta(x)\Phi(x - \mathrm{i}y)\,\mathrm{d}x = \Phi(-\mathrm{i}y). \qquad (50)$$

Again, through (49), the ultradistribution z^{-1} has the property

$$\int_{-\infty}^{\infty} z^{-1}\Phi(x)\,\mathrm{d}x = \int_{-\infty}^{\infty} x^{-1}\Phi(x - \mathrm{i}y)\,\mathrm{d}x = -\int_{-\infty}^{\infty} \ln|x|\Phi'(x - \mathrm{i}y)\mathrm{d}x.$$

Let $\ln(x + \mathrm{i}y)$ have its principal value, with a cut along the negative real axis. Then, if $y > 0$,

$$\int_{-\infty}^{\infty} z^{-1}\Phi(x)\,\mathrm{d}x = -\int_{-\infty}^{\infty} \ln(x + \mathrm{i}y)\Phi'(x)\mathrm{d}x + \pi\mathrm{i}\Phi(-\mathrm{i}y)$$

$$= \int_{-\infty}^{\infty} \frac{\Phi(x)}{x + \mathrm{i}y}\mathrm{d}x + \pi\mathrm{i}\,\Phi(-\mathrm{i}y)$$

since $\mathrm{d}\{\ln(x + \mathrm{i}y)\}/\mathrm{d}z = 1/(x + \mathrm{i}y)$ for $y > 0$. Remembering (50) we deduce that

$$z^{-1} = 1/(x + \mathrm{i}y) + \pi\mathrm{i}\,\delta(z) \quad (y > 0). \qquad (51)$$

Similarly

$$z^{-1} = 1/(x + iy) - \pi i\delta(z) \quad (y < 0). \tag{52}$$

Equations (51) and (52) reveal that there is a clear-cut distinction between z^{-1} and $1/z$ when $y \neq 0$.

The complex Fourier transforms of $w'(t)$, $tw(t)$ and $w(at)$ (a real and non-zero) may be shown, as for generalised functions, to be $izw_c(z)$, $iw'_c(z)$ and $w_c(z/a)/|a|$ respectively.

There is one noteworthy theorem for ultradistributions which has no counterpart for weak or generalised functions.

Theorem 11.28. *If c is any complex constant and u(z) is an ultra-distribution*

$$u(z + c) = \sum_{m=0}^{\infty} \frac{c^m u^{(m)}(z)}{m!}.$$

Proof. By Theorems 11.26 and 11.27

$$\lim_{M\to\infty} \sum_{m=0}^{M} \frac{c^m}{m!} \int_{-\infty}^{\infty} u^{(m)}(z)\Phi(z)\,dx = \lim_{M\to\infty} \sum_{m=0}^{M} \frac{c^m}{m!} \int_{-\infty}^{\infty} (it)^m U(t)\phi(t)\,dt.$$

On account of Theorem 11.4*a*

$$\left| \int_{-\infty}^{\infty} \left\{ \sum_{m=0}^{M} \frac{(itc)^m}{m!} - e^{itc} \right\} U(t)\phi(t)\,dt \right|$$

$$\leq K \max_{a\leq t\leq b} \left| \left[\left\{ \sum_{m=0}^{M} \frac{(itc)^m}{m!} - e^{itc} \right\} \phi(t) \right]^{(r)} \right|$$

for some finite r when ϕ vanishes outside (a, b). The right-hand side disappears as $M \to \infty$ and so we obtain for our limit $\int_{-\infty}^{\infty} e^{itc} U(t)\phi(t)\,dt$. But this may be written as $\int_{-\infty}^{\infty} u(z + c)\Phi(z)\,dx$ and the proof is complete. □

While e^{t^2} is a weak function, e^{z^2} is not an ultradistribution, basically because its growth at infinity is too large. On the other hand $\delta(x)$ is both a weak function and an ultradistribution. Notwithstanding, $\delta(z)$ is an ultradistribution which is not a weak function if $y \neq 0$ because, if $\phi(x)$ is fine, so is $\phi(x)/(x + iy)$ but $\delta(z)$ would not assign a meaning on it. Thus some weak functions are not ultradistributions and some ultradistributions are not weak functions although generalised functions of x are both weak functions and ultradistributions. However, to each weak function corresponds an ultradistribution and conversely, the connecting mapping being the Fourier transform.

Exercises

48. Show that the complex Fourier transform of $w(t)\mathrm{e}^{-\mathrm{i}dt}$, where d is a complex constant, is $w_c(z+d)$ and that the Fourier transform of $u(z+d)$ is $\mathrm{e}^{\mathrm{i}dt}U(t)$.

49. If $\psi \in \mathscr{Y}$, prove that $\psi(z)\delta(z+c) = \psi(-c)\delta(z+c)$, c being a complex constant.

50. If c is a complex constant and $y > \operatorname{Im} c$ prove that

$$(z-c)^{-1} = 1/(z-c) + \pi\mathrm{i}\delta(z-c).$$

What is the corresponding result when $y < \operatorname{Im} c$?

51. Show that the complex Fourier transform of sgn t is $-2\mathrm{i}z^{-1}$. Show also that the Fourier transform of $1/z$ when $y \neq 0$ is $-2\pi\mathrm{i}\mathrm{H}(yt)\operatorname{sgn} y$.

52. By expanding e^{ct}, c a complex constant, in a series show that its complex Fourier transform is $\sum_{m=0}^{\infty} 2\pi(\mathrm{i}c)^m \delta^m(z)/m!$. Invoke Theorem 11.28 to show that this is $2\pi\delta(z+\mathrm{i}c)$. More generally, the complex Fourier transform of $t^m \mathrm{e}^{ct}$ is $2\pi\mathrm{i}^m\delta^{(m)}(z+\mathrm{i}c)$ and of cosh ct is $\pi\{\delta(z+\mathrm{i}c)+\delta(z-\mathrm{i}c)\}$.

53. If c is a complex number and $w(t)$ a weak function which vanishes outside a finite interval show that the complex Fourier transform of $\mathrm{e}^{ct}\sum_{m=-\infty}^{\infty} w(t-m)$ is $2\pi\sum_{m=-\infty}^{\infty} w_c(2\pi m)\delta(z+\mathrm{i}c-2\pi m)$.

54. If $-\pi < \operatorname{ph} z < 0$ show that the complex Fourier transform of t_+^β is $\beta!\,\mathrm{e}^{-\pi\mathrm{i}(\beta+1)/2} z^{-\beta-1}$. What is the corresponding result when $0 < \operatorname{ph} z < \pi$?

55. If $z^m u(z) = 0$ prove that $u(z) = \sum_{n=0}^{m-1} C_n \delta^{(n)}(z)$ where $C_0, \dots, C_{m-1}$ are arbitrary complex constants.

56. The weak function w_1 vanishes outside a finite interval. If w is any weak function prove that the complex Fourier transform of $w * w_1$ is $w_c(z)w_{1c}(z)$.

57. The weak function w satisfies

$$a_m w^{(m)} + a_{m-1} w^{(m-1)} + \dots + a_0 w = 0$$

where $a_0, \dots, a_m$ are complex constants. By taking a complex Fourier transform show that

$$w(t) = \sum_{p=1}^{r} \sum_{q=1}^{k_r} C_{pq} t^{q-1} \mathrm{e}^{\mathrm{i}c_p t}.$$

58. Extend the theory of this section to several variables.

11.16. The relation between weak functions and distributions

Distributions were defined by L. Schwartz in *Théorie des distributions* (Hermann, Paris (1957)) by means of linear functionals of fine functions in the following way. (Only the single variable case will be discussed but the argument carries over to several variables with hardly any modification.) A sequence $\{\phi_n\}$ of fine functions is said to be null if

(i) for every n, ϕ_n vanishes outside a fixed finite interval which is independent of n,

(ii) $\mathrm{d}^p\phi_n/\mathrm{d}x^p$ tends uniformly to zero in the interval of (i) as $n \to \infty$, for each value of $p\,(p = 0, 1, 2 \ldots)$.

A linear functional $F(\phi)$, defined for every fine ϕ, is said to be continuous when $F(\phi_n) \to 0$ as $n \to \infty$ for every null sequence $\{\phi_n\}$ of fine functions. Such a continuous linear functional is termed by Schwartz a *distribution.*

A weak function, as defined in this chapter, specifies a linear functional through

$$F(\phi) = \lim_{m \to \infty} \int_{-\infty}^{\infty} \phi_m(x)\phi(x)\,\mathrm{d}x$$

where $\{\phi_m\}$ is the defining sequence of the weak function. This linear functional is continuous in the sense of Schwartz because of Theorem 11.4*a* and conditions (i) and (ii) above. Therefore every weak function is a distribution.

Conversely, suppose that $F(\phi)$ is a distribution. Let $\rho(x)$ be the fine function defined in §3.2. Then $\phi_m(y) = \int_{-\infty}^{\infty} \phi(x) m\rho(mx - my)\,\mathrm{d}x$ is a fine function such that $\{\phi - \phi_m\}$ is a null sequence (cf. Theorem 3.3 and §6.4). Hence $F(\phi_m) \to F(\phi)$ as $m \to \infty$. However, if $w_m(y) = F\{m\rho(mx - my)\}$,

$$\int_{-\infty}^{\infty} w_m(y)\phi(y)\,\mathrm{d}y = F(\phi_m).$$

Thus $\lim_{m \to \infty} \int_{-\infty}^{\infty} w_m(y)\phi(y)\,\mathrm{d}y$ exists and defines a weak function w such that

$$\int_{-\infty}^{\infty} w(y)\phi(y)\,\mathrm{d}y = F(\phi)$$

(cf. Theorem 5.3 footnote). Thus every distribution is a weak function.

Consequently, there is complete equivalence between the theory of weak functions in this book and the theory of distributions.

The remarks at the end of §3.1 also indicate that there is equivalence between the generalised functions of this book and Schwartz's distributions of slow growth.

Appendix. Titchmarsh's theorem

In this appendix we prove the result: *if* f_1, f_2 *are continuous on* $x \geq 0$ *and* $f_1 * f_2 = 0$ *then* $f_1 = 0$ *or* $f_2 = 0$ *on* $x \geq 0$.

There does not seem to be a short proof, although there are many different ways of demonstration. We shall give one that uses relatively elementary processes.

Denote $x^n f_1(x)$ and $x^n f_2(x)$ by f_{1n} and f_{2n} respectively. Then

$$f_{11} * f_2 + f_1 * f_{21} = x(f_1 * f_2) = 0.$$

Therefore

$$(f_1 * f_{21}) * (f_{11} * f_2 + f_1 * f_{21}) = 0$$

which implies, on account of Theorem 11.19 and $f_1 * f_2 = 0$, that $(f_1 * f_{21}) * (f_1 * f_{21}) = 0$. Hence, if $f * f = 0$ implies $f = 0$, we have $f_1 * f_{21} = 0$. Repeating the process we find $f_1 * f_{2n} = 0$. We shall see that this necessitates $f_1(x-t) f_2(t) = 0$ for $0 \leq t \leq x$ and all $x \geq 0$. If there is a t_0 for which $f_2(t_0) \neq 0$ then $f_1(x - t_0) = 0$ for all $x \geq t_0$, i.e. $f_1(x) = 0$ for all $x \geq 0$. Hence either $f_1 = 0$ or $f_2 = 0$ for all $x \geq 0$.

It remains to prove the properties used concerning $f * f$ and $f_1 * f_{2n}$. If $f * f = 0$ then $f * f = 0$ for $0 \leq x \leq 2X$. Therefore, if T is the triangle $x + y \geq 0$, $x \leq X$, $y \leq X$,

$$\int_T e^{n(x+y)} f(X-x) f(X-y) \mathrm{d}x \, \mathrm{d}y = \int_0^{2X} e^{n(2X-x)} \int_0^x f(t) f(x-t) \mathrm{d}t \, \mathrm{d}x$$
$$= 0.$$

Hence, if $T + T_1$ is the square $|x| \leq X$, $|y| \leq X$,

$$\left| \left\{ \int_{-X}^{X} e^{nx} f(X-x) \mathrm{d}x \right\}^2 \right| = \left| \int_{-X}^{X} \int_{-X}^{X} e^{n(x+y)} f(X-x) f(X-y) \mathrm{d}x \, \mathrm{d}y \right|$$
$$= \left| \int_{T_1} e^{n(x+y)} f(X-x) f(X-y) \mathrm{d}x \, \mathrm{d}y \right|$$
$$\leq \int_{T_1} |f(X-x) f(X-y)| \mathrm{d}x \, \mathrm{d}y \leq 2X^2 B^2$$

where B is an upper bound for f in $0 \le x \le 2X$. Consequently

$$\left|\int_{-X}^{X} e^{nx} f(X-x)\,dx\right| \le \sqrt{2}\,XB$$

whence

$$\left|\int_{0}^{X} e^{nx} f(X-x)\,dx\right| \le 3XB \tag{A.1}$$

since

$$\left|\int_{-X}^{0} e^{nx} f(X-x)\,dx\right| \le XB.$$

Next, note that

$$\left|\sum_{m=1}^{\infty} \frac{(-1)^{m-1}}{m!} \int_0^x e^{mn(x-t)} f(t)\,dt\right| \le \sum_{m=1}^{\infty} \frac{e^{mn(x-X)}}{m!} \left|\int_0^X e^{mn(X-t)} f(t)\,dt\right|$$
$$\le 3XB[\exp\{e^{n(x-X)}\} - 1]$$

on account of (A.1). Hence, allowing $n \to \infty$, we have

$$\lim_{n\to\infty} \int_0^X [1 - \exp\{-e^{n(x-t)}\}] f(t)\,dt = 0 \quad (0 \le x < X).$$

It is immediately evident that there is no contribution from $t \ge x+\varepsilon$, $\varepsilon > 0$, and that the exponential term disappears for $t \le x - \varepsilon$. The contribution of the interval $(x-\varepsilon, x+\varepsilon)$ is $O(\varepsilon)$ so that if we choose ε small enough first and then n large enough we are obliged to conclude that $\int_0^x f(t)\,dt = 0$ $(0 \le x < X)$. From the continuity of f it follows that $f(x) = 0$ $(0 \le x \le X)$. Since this holds for every $X > 0$ it follows that $f = 0$.

Turning now to $f_1 * f_{2n} = 0$ we observe that this can be written $\int_0^x t^n f(t)\,dt = 0$ where $f(t)$ is continuous for $0 \le t \le x$. Assuming that $0 < x_0 < x$ we have

$$\left|\int_{x_0}^{x} \left(\frac{t}{x_0}\right)^n f(t)\,dt\right| = \left|-\int_0^{x_0} \left(\frac{t}{x_0}\right)^n f(t)\,dt\right| \le Mx_0.$$

The substitution $t = x_0 e^u$, $x = x_0 e^X$ now gives

$$\left|\int_0^X e^{nu} f_0(u)\,du\right| \le M$$

where $f_0(u) = f(x_0 e^u)x_0 e^u$. We can now use the same argument as that following (A.1) to demonstrate that $f_0(u) = 0$ for $0 \leq u \leq X$, i.e. $f(t) = 0$ for $x_0 \leq t \leq x$. If we take x_0 small enough, $f(t) = 0$ for $0 < t \leq x$ and, by continuity, $f(t) = 0$ for $0 \leq t \leq x$.

12
The Laplace transform

12.1. The Laplace transform

The value of Laplace transforms in applied mathematics is well known. Their advantage stems from the conversion of differentiation to multiplication. It is therefore desirable to generalise the transform so that it is applicable to weak functions without losing the basic properties which make the transform so useful for conventional functions. In fact we shall be able to do this only for a restricted class of weak functions.

The starting point is

Definition 12.1. *$f \in \mathscr{L}_+(c)$ if, and only if, $f = 0$ for $x < 0$, f is integrable over every finite interval and $f(x) = O(e^{cx})$ as $x \to \infty$, c being a real constant.*

Obviously if $f \in \mathscr{L}_+(c)$ then $f \in \mathscr{L}_+(c_0)$ for $c_0 \geq c$.

Let $s = \sigma + i\omega$ where σ and ω are real. Then if $f \in \mathscr{L}_+(c)$ the Laplace transform $\mathscr{F}$ of f is defined by

$$\mathscr{F}(s) = \int_0^\infty f(x)e^{-sx}\,dx. \tag{1}$$

Its existence is assured only when $\sigma > c$.

To generalise these ideas to a weak function w introduce

Definition 12.2. *$w \in \mathscr{L}'_+(c)$ if, and only if, $w = f^{(r)}$ for some finite r and $f \in \mathscr{L}_+(c)$.*

In other words $\mathscr{L}'_+(c)$ is a space of weak derivatives of functions which possess Laplace transforms. Since $H(x) \in \mathscr{L}_+(0)$, $\delta(x) \in \mathscr{L}'_+(0)$.

Definition 12.3. *If $w \in \mathscr{L}'_+(c)$ the Laplace transform $\mathscr{W}(s)$ is defined to be the Fourier transform with respect to ω of $e^{-\sigma x}w(x)$ $(\sigma \geq c)$, i.e.*

$$\mathscr{W}(s) = \int_{-\infty}^{\infty} \{w(x)e^{-\sigma x}\}e^{-i\omega x}\,dx.$$

It is necessary to check that the definition has a meaning. Observe firstly that, if w can be identified with $f \in \mathscr{L}_+(c)$, Definition 12.3 gives the same as (1) when $\sigma > c$ on account of Theorem 3.13. Accordingly the Laplace transform of a weak function can be identified with the conventional transform when the latter exists. Therefore all existing tables of Laplace transforms may continue to be used. For this reason it is convenient to write

$$\mathscr{W}(s) = \int_0^\infty w(x) e^{-sx} dx.$$

Now suppose that $f \in \mathscr{L}_+(c)$. Then

$$\{e^{-\sigma x} f(x)\}' = e^{-\sigma x} f'(x) - \sigma e^{-\sigma x} f(x).$$

Since $e^{-\sigma x} f(x)$ is a generalised function for $\sigma \geq c$ (being, in fact, bounded at infinity) it follows that $e^{-\sigma x} f'(x)$ is also. Further, if $e^{-\sigma x} f^{(r)}(x)$ is a generalised function so is $e^{-\sigma x} f^{(r+1)}(x)$ because

$$\{e^{-\sigma x} f^{(r)}(x)\}' = e^{-\sigma x} f^{(r+1)}(x) - \sigma e^{-\sigma x} f^{(r)}(x).$$

Hence, by induction, $e^{-\sigma x} f^{(r)}(x)$ is a generalised function for any finite r. Therefore Definition 12.2 shows, if $w \in \mathscr{L}'_+(c)$, that $e^{-\sigma x}$ *$w(x)$ is a generalised function for $\sigma \geq c$.* Consequently its Fourier transform is defined and $\mathscr{W}$ has a meaning.

Example 1. Since $e^{-\sigma x} \delta(x) = \delta(x)$ the Laplace transform of $\delta(x)$ is 1.

Example 2. The Laplace transform of $H(x)$ is $1/s$ $(\sigma > 0)$ and $\pi\delta(\omega) - i\omega^{-1}$ $(\sigma = 0)$ from Exercise 34 of Chapter 4. This is consistent with

$$\lim_{\sigma \to +0} 1/s = -i(\omega - i0)^{-1}.$$

More generally the Laplace transform of $H(x)e^{(c+id)x}$ is $1/(s - c - id)$ $(\sigma > c)$ and $\pi\delta(\omega - d) - i(\omega - d)^{-1}$ $(\sigma = c)$.

Theorem 12.1. *If $w \in \mathscr{L}'_+(c)$, the Laplace transform of $w(ax)$ $(a > 0)$ is $(1/a)\,\mathscr{W}(s/a)$ and of w' is $s\mathscr{W}(s)$ $(\sigma \geq c)$.*

Proof. By Theorem 3.10 the Fourier transform of $w(ax)e^{-\sigma x}$ is $(1/a)\mathscr{W}(\sigma/a + i\omega/a)$ which is the result stated. Also

$$e^{-\sigma x} w'(x) = \{e^{-\sigma x} w(x)\}' + \sigma e^{-\sigma x} w(x)$$

so that the Fourier transform of $e^{-\sigma x} w'(x)$ is $i\omega \mathscr{W}(s) + \sigma \mathscr{W}(s)$ by Theorem 3.12 and the proof is complete. □

Theorem 12.2. *If* $w_\mu \in \mathscr{L}'_+(c)$, $w \in \mathscr{L}'_+(c)$ *and* $\text{Lim}_{\mu\to\mu_0} w_\mu = w$ *then* $\text{Lim}_{\mu\to\mu_0} \mathscr{W}_\mu = \mathscr{W}$.

Proof.

$$\underset{\mu\to\mu_0}{\text{Lim}} \mathscr{W}_\mu = \underset{\mu\to\mu_0}{\text{Lim}} \int_{-\infty}^{\infty} \{w_\mu(x)e^{-\sigma x}\} e^{-i\omega x} dx$$

$$= \int_{-\infty}^{\infty} \underset{\mu\to\mu_0}{\text{Lim}} \{w_\mu(x)e^{-\sigma x}\} e^{-i\omega x} dx$$

by Theorem 3.14 (iv). Since $\text{Lim}_{\mu\to\mu_0} \{w_\mu(x)e^{-\sigma x}\} = w(x)e^{-\sigma x}$ by Theorem 11.4 (iii) the result follows. □

Example 3. By Theorem 12.1 and Example 2 the Laplace transform of $\{H(x)e^{\beta x}\}'$ is $s/(s-\beta)(\sigma > \text{Re}\,\beta)$ and $\pi\beta\delta(\omega - \text{Im}\,\beta) + s(s-\beta)^{-1}$ $(\sigma = \text{Re}\,\beta)$. Taking the limit as $\beta \to 0$ we see from Theorem 12.2 that the Laplace transform of $\delta(x)$ is $1\,(\sigma \geq 0)$ in consistence with Example 1.

Similarly Theorem 12.1 shows that the Laplace transform of $\delta^{(m)}(x)$ is s^m.

Theorem 12.3. *If* $w \in \mathscr{L}'_+(c)$, $\mathscr{W}(s)$ *is a regular function of s in* $\sigma > c$, $\mathscr{W}(s) = O(|s|^k)$ *for some finite k in* $\sigma > c$ *and* $\text{Lim}_{\sigma\to c+0} \mathscr{W}(s) = \mathscr{W}(c + i\omega)$.

Proof. Since $e^{-cx} w(x)$ is a generalised function in K_+, Theorem 6.19 shows that $\mathscr{W}(s)$ is a regular function of s in $\sigma > c$ and that $\text{Lim}_{\sigma\to c+0} \mathscr{W}(s) = \mathscr{W}(c + i\omega)$.

For the second part of the theorem note that $w = f^{(r)}$ where $f \in \mathscr{L}_+(c)$ and r is finite so that Theorem 12.1 gives $\mathscr{W}(s) = s^r \mathscr{F}(s)$. If $\sigma \geq c + \varepsilon > c$,

$$|\mathscr{F}(s)| \leq \int_0^{\infty} |f(x)| e^{-(c+\varepsilon)x} dx < \infty$$

since $f = O(e^{cx})$ as $x \to \infty$. The proof is complete. □

One can say that $\mathscr{W}(s)$ is a fairly good function of ω for each $\sigma > c$.

Examples 1, 2 and 3 are all in agreement with Theorem 12.3. Observe that, although Theorem 12.3 ensures that $\mathscr{W}$ is regular in $\sigma > c$, there is no reason why particular $\mathscr{W}$ should not be regular in a larger domain; $\delta(x)$, which gives $\mathscr{W} = 1$, is an example. On the other hand, $\mathrm{H}(x)$ is a counter-example which demonstrates that there are some weak functions whose Laplace transforms have a strip of regularity which does not go beyond $\sigma > c$. One can also see from these two examples that the domain of regularity of the Laplace transform of a weak function may be smaller than that of the transform of its derivative.

In some cases analytic continuation will supply an extension of $\mathscr{W}$ though not necessarily in the form of a strip. Thus $(s - c)^{1/2}$ could be continued analytically into the whole complex plane cut by a branch-line from c to $-\infty$.

Theorem 12.4. *If* $w \in \mathscr{L}_+(c)$,

$$w(x) = \begin{cases} \dfrac{e^{\sigma x}}{2\pi} \displaystyle\int_{-\infty}^{\infty} \mathscr{W}(s) e^{i\omega x} \, d\omega & (\sigma \geq c) \\[2ex] \dfrac{1}{2\pi i} \mathscr{D}^m \displaystyle\int_{\sigma - i\infty}^{\sigma + i\infty} \dfrac{\mathscr{W}(s) e^{sx}}{(s - c)^m} \, ds & (\sigma > c) \end{cases}$$

for sufficiently large m, and $\mathscr{D}w = w' - cw$.

Proof. Since $\mathscr{W}(s)$ is the Fourier transform of $e^{-\sigma x} w(x)$ with respect to ω,

$$e^{-\sigma x} w(x) = \frac{1}{2\pi} \int_{-\infty}^{\infty} \mathscr{W}(s) e^{i\omega x} \, d\omega \quad (\sigma \geq c)$$

from Theorem 3.11. The first result follows from multiplication by the infinitely differentiable $e^{\sigma x}$.

By Theorem 12.3 $\mathscr{W}$ is regular and $O(|s|^k)$ for some finite k in $\sigma > c$. Therefore $\mathscr{W}(s)/(s - c)^m$ is absolutely integrable with respect to ω for sufficiently large m. Hence (Theorem 3.13) $\int_{\sigma - i\infty}^{\sigma + i\infty} \{\mathscr{W}(s)/(s - c)^m\} e^{i\omega x} \, ds$ is a bounded function of x which is zero for $x < 0$, as can be seen by deforming the contour to the right and using the regularity of $\mathscr{W}$. Consequently $(\mathscr{D}^m/2\pi i)\int_{\sigma-i\infty}^{\sigma+i\infty}\{\mathscr{W}(s)/(s-c)^m\}\,e^{sx}\,ds \in \mathscr{L}'_+(c)$ and, by Theorem 12.1, has a Laplace transform which is $\mathscr{W}(s)$ in $\sigma > c$. The second half of the theorem now follows from the first half. □

In view of this theorem it is very convenient to adopt the notation

$$w(x) = \frac{1}{2\pi i}\int_{\sigma-i\infty}^{\sigma+i\infty} \mathscr{W}(s)\,e^{sx}\,ds \quad (\sigma \geq c)$$

with the understanding that the right-hand side is to be interpreted in the sense of Theorem 12.4.

In the course of proving Theorem 12.4 we have shown the converse of Theorem 12.3, namely

Theorem 12.5. *If $\mathscr{W}(s)$ is a regular function of s in $\sigma > c$ and $\mathscr{W}(s) = O(|s|^k)$ for some finite k in $\sigma > c$ then $\mathscr{W}$ is the Laplace transform of a $w \in \mathscr{L}'_+(c)$ and there is a $\mathscr{W}(c+i\omega)$ such that* $\operatorname{Lim}_{\sigma\to c+0} \mathscr{W}(s) = \mathscr{W}(c+i\omega)$.

The last statement is an immediate consequence of Theorem 12.3 and the fact that $w \in \mathscr{L}'_+(c)$.

Example 4. From Example 3 and Theorem 12.4 we see that, if $\sigma \geq 0$,

$$\frac{1}{2\pi i}\int_{\sigma-i\infty}^{\sigma+i\infty} s^m\,e^{sx}\,ds = \delta^{(m)}(x) \quad (m = 0, 1, \ldots).$$

Exercises

1. If $w_1, w_2 \in \mathscr{L}'_+(c)$ and c_1, c_2 are any complex constants show that $c_1 w_1 + c_2 w_2$ has Laplace transform $c_1 \mathscr{W}_1 + c_2 \mathscr{W}_2$.
2. Prove that the Laplace transforms of $H(x)\cos\beta x$, $H(x)\cosh\beta x$, $H(x)\sin\beta x$, $H(x)\sinh\beta x$, $H(x)|\sin ax|$ $(a > 0)$ are

$$\frac{s}{s^2+\beta^2}\quad(\sigma > |\operatorname{Im}\beta|), \qquad \frac{s}{s^2-\beta^2}\quad(\sigma > |\operatorname{Re}\beta|),$$

$$\frac{\beta}{s^2+\beta^2}\quad(\sigma > |\operatorname{Im}\beta|), \qquad \frac{\beta}{s^2-\beta^2}\quad(\sigma > |\operatorname{Re}\beta|),$$

$$\frac{a\coth(\pi s/2a)}{s^2+a^2}\quad(\sigma > 0)$$

respectively. What are the corresponding results when $\sigma = |\mathrm{Im}\,\beta|$ and $\sigma = |\mathrm{Re}\,\beta|$ in the first four cases?

3. Show that the Laplace transform of $x^\beta \mathrm{H}(x)$ is $\beta!/s^{\beta+1}$ $(\sigma > 0)$ for all complex β except negative integer values. Prove that the Laplace transform of $\mathrm{H}(x)\ln x$ is $-(\gamma + \ln s)/s$ $(\sigma > 0)$, the logarithm having its principal value, and that the Laplace transform of $x^{-m}\mathrm{H}(x)$ $(m = 1, 2, \dots)$ is $\{(-s)^{m-1}/(m-1)!\}\{\psi(m-1) - \ln s\}$.
4. Prove that the Laplace transform of $x^\beta \mathrm{H}(x)\ln x$ is $(\beta!/s^{\beta+1})\{\psi(\beta) - \ln s\}$ $(\sigma > 0)$ for all complex β except negative integer values.
5. If $w \in \mathscr{L}'_+(c)$ show that the Laplace transform of $w(x-a)$, $a \geq 0$, is $\mathrm{e}^{-as}\mathscr{W}(s)$.
6. If β is any complex constant and $w \in \mathscr{L}'_+(c)$ show that the Laplace transform of $\mathrm{e}^{\beta x}w(x)$ is $\mathscr{W}(s-\beta)$, $(\sigma \geq c + \mathrm{Re}\,\beta)$.
7. If $w \in \mathscr{L}'_+(c)$ show that the Laplace transform of $x^m w(x)$ $(m = 1, 2, \dots)$ is $(-1)^m \mathrm{d}^m \mathscr{W}(s)/\mathrm{d}s^m$ $(\sigma > c)$.
8. Find the Laplace transform of $\sum_{m=1}^{M} \delta^{(m)}(x-m)$, M finite, and use Theorems 12.2 and 12.3 to show that there is no c for which $\sum_{m=1}^{\infty} \delta^{(m)}(x-m) \in \mathscr{L}'_+(c)$.
9. Prove that $(s-1)^{1/3}$ is not the Laplace transform of a weak function which belongs to L'_2 and K_+.
10. Assuming that

$$\int_0^\infty \mathrm{e}^{-sx} J_\nu(\beta x) x^\nu \,\mathrm{d}x = \frac{(\nu - \frac{1}{2})!(2\beta)^\nu}{(s^2 + \beta^2)^{\nu + 1/2}\pi^{1/2}}$$

for $\sigma > |\mathrm{Im}\,\beta|$, $\mathrm{Re}\,\nu > -\frac{1}{2}$ examine the possibility that the formula is true for a wider range of ν.

Theorem 12.6. *If $w_1, w_2 \in \mathscr{L}'_+(c)$ the Laplace transform of $w_1 * w_2$ is $\mathscr{W}_1(s)\mathscr{W}_2(s)$ $(\sigma > c)$ and $\mathrm{Lim}_{\sigma \to c+0}\, \mathscr{W}_1(s)\mathscr{W}_2(s)$ $(\sigma = c)$.*

Proof. Since $w_1, w_2 \in \mathscr{L}'_+(c)$, $w_1, w_2 \in W_+$ and $w_1 * w_2$ exists by Definition 11.10. In fact $w_1 * w_2$ can be expressed in terms of derivatives of $f_1 * f_2$ where $w_1 = f_1^{(r)}$, $w_2 = f_2^{(k)}$ with $f_1, f_2 \in \mathscr{L}_+(c)$. Now

$$\mathrm{e}^{-\sigma x}(f_1 * f_2) = \mathrm{H}(x)\int_0^x \mathrm{e}^{-\sigma t} f_1(t)\,\mathrm{e}^{-\sigma(x-t)} f_2(t)\,\mathrm{d}t$$

and the right-hand side is the convolution of $\mathrm{e}^{-\sigma x}f_1(x)$ and $\mathrm{e}^{-\sigma x}f_2(x)$. But $\mathrm{e}^{-\sigma x}f_1(x)$ and $\mathrm{e}^{-\sigma x}f_2(x)$ are both in L_1 for $\sigma > c$ and so, by Theorem 6.13, the transform of their convolution is $\mathscr{F}_1(s)\mathscr{F}_2(s)$. The application of Theorem 12.1 now shows that the Laplace transform of $w_1 * w_2$ is $s^{r+k}\mathscr{F}_1(s)\mathscr{F}_2(s)$ or $\mathscr{W}_1(s)\mathscr{W}_2(s)$ $(\sigma > c)$.

Theorem 12.3 now gives the transform when $\sigma = c$. The theorem is proved. □

Alternatively we can say

$$\frac{1}{2\pi i}\int_{\sigma - i\infty}^{\sigma + i\infty} \mathscr{W}_1(s)\mathscr{W}_2(s)\,e^{sx}\,ds = w_1 * w_2 \quad (\sigma > c).$$

Theorem 12.7. *If the infinitely differentiable* $\eta \in \mathscr{L}'_+(c_1)$ *and* $w \in \mathscr{L}'_+(c_2)$, *the Laplace transform of* ηw *in* $\sigma > c_1 + c_2$ *is*

$$\frac{1}{2\pi i}\int_{\sigma_1 - i\infty}^{\sigma_1 + i\infty} \mathscr{H}(s_1)\mathscr{W}(s - s_1)\,ds_1 \quad (\sigma - c_2 > \sigma_1 > c_1)$$

where $\mathscr{H}$ *is the Laplace transform of* η.

Proof. If $\sigma > c_1 + c_2$ the Laplace transform of ηw is the Fourier transform with respect to ω of $e^{-\sigma x}\eta(x)w(x)$, i.e. of $e^{-\sigma_1 x}\eta(x).e^{-(\sigma-\sigma_1)x}w(x)$ with σ_1 chosen so that $c_1 < \sigma_1 < \sigma - c_2$. Since $e^{-\sigma_1 x}\eta(x)$ and $e^{-(\sigma-\sigma_1)x}w(x)$ are both in L'_1 Theorem 6.13 indicates that the Fourier transform of their product is the convolution of their transforms apart from a multiplicative constant, i.e.

$$\int_{-\infty}^{\infty} e^{-\sigma x}\eta(x)w(x)\,e^{-i\omega x}\,dx$$

$$= \frac{1}{2\pi}\int_{-\infty}^{\infty} \mathscr{H}(\sigma_1 + it)\,\mathscr{W}\{\sigma - \sigma_1 + i(\omega - t)\}\,dt$$

$$= \frac{1}{2\pi i}\int_{\sigma_1 - i\infty}^{\sigma_1 + i\infty} \mathscr{H}(s_1)\mathscr{W}(s - s_1)\,ds_1$$

and the proof is complete. □

Exercise

11. Can you replace η in Theorem 12.7 by a weak function by imposing some condition on w?

12.2. Ordinary differential equations

The ordinary differential equation

$$w_1^{(m)} + a_1 w_1^{(m-1)} + \ldots + a_m w_1 = w_2 \qquad (2)$$

with $w_1, w_2 \in \mathscr{L}'_+(c)$ can be handled by means of Laplace transforms. For Theorem 12.1 implies that

$$(s^m + a_1 s^{m-1} + \ldots + a_m)\mathscr{W}_1(s) = \mathscr{W}_2(s) \quad (\sigma \geq c).$$

Now c can always be increased if necessary (since $\mathscr{L}'_+(c) \supset \mathscr{L}'_+(c_0)$ for $c \leq c_0$) so that the zeros of the polynomial $s^m + a_1 s^{m-1} + \ldots + a_m$ have their real parts less than c. Then division by the polynomial is permissible and

$$\mathscr{W}_1(s) = \frac{\mathscr{W}_2(s)}{s^m + a_1 s^{m-1} + \ldots + a_m} \quad (\sigma \geq c).$$

Then $\mathscr{W}_1$ satisfies the conditions of Theorem 12.5 and w_1 can be found by the inversion formula (Theorem 12.4).

An alternative way of describing this conclusion is: if $w_2 \in \mathscr{L}'_+(c)$ there is a $c_0 \geq c$ such that $w_1 \in \mathscr{L}'_+(c_0)$, satisfies (2) and

$$w_1(x) = \frac{1}{2\pi \mathrm{i}} \int_{\sigma - \mathrm{i}\infty}^{\sigma + \mathrm{i}\infty} \frac{\mathscr{W}_2(s)\, \mathrm{e}^{sx}}{s^m + a_1 s^{m-1} + \ldots + a_m} \mathrm{d}s \quad (\sigma \geq c_0). \tag{3}$$

Comparison with the operational method of §11.10 is interesting, because Laplace transforms can be manipulated algebraically in their domains of regularity. Essentially the Laplace transform makes e_β or $\mathrm{e}^{\beta x}\mathrm{H}(x)$ correspond to $1/(s - \beta)$. Also Theorem 12.6 makes the convolution correspond to the multiplication of the transforms, which should be compared with equation (23) of Chapter 11. There are thus many similarities between the p-operators and Laplace transforms. However, the p-operator gives a *complete* operational method only when w_2 is of the type E (see §11.10) whereas the Laplace transform gives one under the much lighter restriction that $w_2 \in \mathscr{L}_+(c)$. Nevertheless it was pointed out in the remarks after Example 8 of Chapter 11 that the p-operator method could be used for a wider class of w_2 provided that the interpretation was carried out before the final convolution. A similar conclusion can be derived for (3) by writing it in the form (Theorem 12.6)

$$w_1(x) = f * w_2 \tag{4}$$

where

$$f(x) = \frac{1}{2\pi \mathrm{i}} \int_{\sigma - \mathrm{i}\infty}^{\sigma + \mathrm{i}\infty} \frac{\mathrm{e}^{sx}\, \mathrm{d}s}{s^m + \ldots + a_m} \quad (\sigma \geq c_0).$$

The form (4) solves (2) subject only to the restriction $w_2 \in W_+$.

Laplace transforms can also be used for differential equations with polynomial coefficients. Consider, for example,

$$xw'' + (1 - 2\nu)w' + xw = 0$$

with $w \in \mathscr{L}'_+(c)$. Then, from Exercise 7,

$$(s^2 + 1)\frac{\mathrm{d}\mathscr{W}}{\mathrm{d}s} + (2\nu + 1)s\mathscr{W} = 0 \quad (\sigma > c)$$

whence $\mathscr{W}(s) = C(s^2 + 1)^{-\nu - 1/2}$ and is a Laplace transform provided that $c = 0$. It follows from Exercise 10 that

$$w(x) = x^\nu J_\nu(x)\mathrm{H}(x)$$

at any rate when $\mathrm{Re}\,\nu > -\frac{1}{2}$. If $\nu = -\frac{1}{2} - m$ $(m = 0, 1, \ldots)$ $\mathscr{W}$ is a polynomial in s^2 and w is a linear combination of $\delta(x)$ and its derivatives.

Differential-difference equations also yield to Laplace transforms in suitable circumstances. Consider, for example,

$$w'(x) - w(x - a) = w_2(x) \quad (a \geq 0)$$

with $w_2 \in \mathscr{L}'_+(c)$. If $w \in \mathscr{L}'_+(c_0)$ $(c_0 \geq c)$ Exercise 5 gives

$$(s - \mathrm{e}^{-as})\mathscr{W}(s) = \mathscr{W}_2(s) \quad (\sigma > c_0)$$

which makes $\mathscr{W}$ a permissible transform certainly if $c_0 > 1$. Then

$$w(x) = \frac{1}{2\pi\mathrm{i}}\int_{\sigma - \mathrm{i}\infty}^{\sigma + \mathrm{i}\infty} \frac{\mathscr{W}_2(s)\,\mathrm{e}^{sx}}{s - \mathrm{e}^{-as}}\,\mathrm{d}s \quad (\sigma > c_0 > 1).$$

Exercises

12. Find a solution of

$$xw'' + (1 - \nu)w' + w = 0$$

by Laplace transforms.

13. Obtain a solution of

$$w'(x) - (1/a)\{w(x) - w(x - a)\} = w_2 \qquad (a > 0)$$

with $w, w_2 \in \mathscr{L}'_+(c)$.

12.3. Integral equations

The Volterra integral equation of the second kind

$$w_1 - k * w_1 = w_2$$

can be tackled by means of Laplace transforms. Let $k \in \mathscr{L}_+(c)$, $w_2 \in \mathscr{L}'_+(c)$ and suppose that $w_1 \in \mathscr{L}'_+(c_0)$ for some $c_0 \geq c$. Then

$$\{1 - \mathscr{K}(s)\}\mathscr{W}_1(s) = \mathscr{W}_2(s) \quad (\sigma > c_0).$$

Now

$$|\mathscr{K}(s)| = \left|\int_0^\infty k(x)\,\mathrm{e}^{-sx}\,\mathrm{d}x\right|$$

$$\leq \int_0^\infty |k(x)|\,\mathrm{e}^{-\sigma x}\,\mathrm{d}x < C\int_0^\infty \mathrm{e}^{(c-\sigma)x}\,\mathrm{d}x < 1$$

if σ is chosen sufficiently large. Hence, if c_0 is sufficiently large,

$$\mathscr{W}_1(s) = \mathscr{W}_2(s)/\{1 - \mathscr{K}(s)\} \quad (\sigma > c_0) \tag{5}$$

and w_1 follows by inversion.

One possibility is to expand $\{1 - \mathscr{K}(s)\}^{-1}$ in powers of $\mathscr{K}$ (permissible since $|\mathscr{K}| < 1$) and thereby recover equation (11) of Chapter 11. The reader should compare the results of this section and those of §11.7.

If the condition on k is lightened to $k \in \mathscr{L}'_+(c)$ it cannot be asserted in general that (5) supplies $\mathscr{W}_1$ because it is not certain that a c_0 can be found so that $\mathscr{K}(s) \neq 1$ in $\sigma > c_0$. Of course, for any particular $k \in \mathscr{L}'_+(c)$ for which a c_0 can be found for which $\mathscr{K}(s)$ does not approach 1 for any $\sigma > c_0$ (5) will provide a solution of the integral equation.

Similarly a solution of the integral equation of the first kind

$$k * w_1 = w_2$$

can be derived from

$$\mathscr{W}_1(s) = \mathscr{W}_2(s)/\mathscr{K}(s)$$

under any conditions on $\mathscr{K}$ which render division by it possible. Roughly speaking, these are that $\mathscr{K}$ does not vanish in $\sigma > c_0$ and does not tend to zero too rapidly as $|\omega| \to \infty$; otherwise, the conditions of Theorem 12.5 cannot be fulfilled.

Exercises

14. If

$$\int_0^x w_1(t)\frac{\cos a(x-t)^{1/2}}{(x-t)^{1/2}}\mathrm{d}t = w_2(x) \qquad (a>0)$$

with $w_2 \in \mathscr{L}'_+(c)$ show that

$$w_1(x) = \frac{1}{\pi}\int_0^x w_2'(t)\frac{\cosh a(x-t)^{1/2}}{(x-t)^{1/2}}\mathrm{d}t.$$

Can the condition on w_2 at infinity be lightened by using Exercise 28 of Chapter 11?

By making a change of origin examine the possibility that

$$\int_x^1 w_1(t)\frac{\cos a(t-x)^{1/2}}{(t-x)^{1/2}}\mathrm{d}t = w_2(x)$$

has a solution $w_1(x) = -\dfrac{1}{\pi}\displaystyle\int_x^1 w_2'(t)\frac{\cosh a(t-x)^{1/2}}{(t-x)^{1/2}}\mathrm{d}t.$

15. Solve $w * w = (1/2\beta)\mathrm{H}(x)\sinh\beta x + \frac{1}{2}x\mathrm{H}(x)\cosh\beta x$ if $w \in \mathscr{L}'_+(c)$ for some c.

12.4. Laplace transform of a weak function of several variables

The use of the Laplace transform in connection with partial differential equations requires the extension of the theory to weak functions of several variables.

Definition 12.4. *$f \in \mathscr{L}_1(c)$ if, and only if, $f = 0$ for $x_1 < 0$, $f \in \mathbf{L}^0$ and $f(\mathbf{x}) = O(\mathrm{e}^{cx_1})$ as $x_1 \to \infty$ for fixed $x_2, \ldots, x_n$, c being a real constant.*

The notation is chosen so as to be capable of natural generalisation for several Laplace transforms. Thus we would use $\mathscr{L}_2(c_1; c_2)$ to indicate an f with similar properties to those in Definition 12.4 in both x_1 and x_2.

If $f \in \mathscr{L}_1(c)$ the existence of the Laplace transform

$$\mathscr{F}(s, \mathbf{y}) = \int_0^\infty f(\mathbf{x})\mathrm{e}^{-sx_1}\mathrm{d}x_1,$$

$\mathbf{y}$ denoting $x_2, \ldots, x_n$, is guaranteed for $\sigma > c$.

Definition 12.5. *$w \in \mathscr{L}'_1(c)$ if, and only if, $w = \partial_1^{p_1} \ldots \partial_n^{p_n} f$ for some finite $p_1, \ldots, p_n$ and $f \in \mathscr{L}_1(c)$.*

Definition 12.6. *If* $w \in \mathscr{L}'_1(c)$ *the Laplace transform* $\mathscr{W}(s, \boldsymbol{y})$ *is defined by*

$$\mathscr{W}(s, \boldsymbol{y}) = \int_{-\infty}^{\infty} \{w(\boldsymbol{x})\,\mathrm{e}^{-\sigma x_1}\}\,\mathrm{e}^{-\mathrm{i}\omega x_1}\,\mathrm{d}x_1 \quad (\sigma \geq c).$$

Some clarification of this definition is necessary because, although a Fourier transform with respect to one variable of a generalised function of $\boldsymbol{x}$ has been given in Theorem 7.23, $w(\boldsymbol{x})\,\mathrm{e}^{-\sigma x_1}$ cannot be regarded as a generalised function of $\boldsymbol{x}$. In a sense it is a generalised function of x_1 and a weak function of the other variables. To cover this possibility the following meaning will be attached to the Fourier transform: *if for any fine (on* R_1*) functions* $\phi_{(2)}, \ldots, \phi_{(n)}$,

$$\int_{-\infty}^{\infty} \cdots \int_{-\infty}^{\infty} \{w(\boldsymbol{x})\,\mathrm{e}^{-\sigma x_1}\}\,\phi_{(2)}(x_2) \ldots \phi_{(n)}(x_n)\,\mathrm{d}x_2 \ldots \mathrm{d}x_n$$

is a generalised function of x_1 *the Fourier transform is defined by*

$$\int_{-\infty}^{\infty} \cdots \int_{-\infty}^{\infty} \phi_{(2)}(x_2) \ldots \phi_{(n)}(x_n) \int_{-\infty}^{\infty} \{w(\boldsymbol{x})\,\mathrm{e}^{-\sigma x_1}\}\,\mathrm{e}^{-\mathrm{i}\omega x_1}\,\mathrm{d}x_1\,\mathrm{d}x_2 \ldots \mathrm{d}x_n$$

$$= \int_{-\infty}^{\infty} \mathrm{e}^{-\mathrm{i}\omega x_1} \int_{-\infty}^{\infty} \cdots \int_{-\infty}^{\infty} \{w(\boldsymbol{x})\,\mathrm{e}^{-\sigma x_1}\}\,\phi_{(2)}(x_2) \ldots \times \phi_{(n)}(x_n)\,\mathrm{d}x_2 \ldots \mathrm{d}x_n\,\mathrm{d}x_1;$$

the right-hand side has a definite meaning on account of the assumption made (cf. §11.12) and identifies a weak function of ω and $\boldsymbol{y}$ because of Theorem 11.23.

It is now necessary to verify that $w \in \mathscr{L}'_1(c)$ implies the given condition. This will certainly be true if

$$\int_{-\infty}^{\infty} \{w(\boldsymbol{x})\,\mathrm{e}^{-\sigma x_1}\}\,\gamma(x_1)\phi_{(2)}(x_2) \ldots \phi_{(n)}(x_n)\,\mathrm{d}\boldsymbol{x}$$

exists for any good γ. Now, if $f \in \mathscr{L}_1(c)$ we can say that $|f| < C\,\mathrm{e}^{cx_1}$ for sufficiently large x_1 and $\boldsymbol{y}$ in a finite domain. Since

$$\int_{-\infty}^{\infty} \{f(\boldsymbol{x})\,\mathrm{e}^{-\sigma x_1}\}\,\gamma(x_1)\phi_{(2)}(x_2) \ldots \phi_{(n)}(x_n)\,\mathrm{d}\boldsymbol{x}$$

involves integration with respect to $\boldsymbol{y}$ only over a finite domain, the integral exists for $\sigma > c$ and $f(\boldsymbol{x})\mathrm{e}^{-\sigma x_1}$ satisfies the given condition. By replacing γ by γ' we see that $\partial_1\{f(\boldsymbol{x})\mathrm{e}^{-\sigma x_1}\}$ satisfies the given condition and consequently so does $\{\partial_1 f(\boldsymbol{x})\}\mathrm{e}^{-\sigma x_1}$. Similarly, by replacing $\phi_{(2)}$ by $\partial_2\phi_{(2)}$ it is found that $\{\partial_2 f(\boldsymbol{x})\}\mathrm{e}^{-\sigma x_1}$ satisfies the given condition. By taking as many derivatives as necessary in this way it is deduced that $w(\boldsymbol{x})\mathrm{e}^{-\sigma x_1}$ satisfies the given condition when $w \in \mathscr{L}'_1(c)$. Hence Definition 12.6 is justified.

Theorem 12.8. *If* $w \in \mathscr{L}'_1(c)$ *the Laplace transform* (i) *of* $\partial_1 w$ *is* $s\mathscr{W}(s, \boldsymbol{y})$, (ii) *of* $\partial_2 w$ *is* $\partial_2 \mathscr{W}(s, \boldsymbol{y})$ $(\sigma > c)$.

Proof. Since $\partial_1 w \in \mathscr{L}'_1(c)$ it has a Laplace transform and, as in §11.12, it is determined by the Fourier transform of

$$(\partial_1 + \sigma)\int_{-\infty}^{\infty} \cdots \int_{-\infty}^{\infty} \{w(\boldsymbol{x})\mathrm{e}^{-\sigma x_1}\}\phi_{(2)}(x_2)\ldots\phi_{(n)}(x_n)\,\mathrm{d}x_2\ldots\mathrm{d}x_n.$$

Consequently it is equal to $(\mathrm{i}\omega + \sigma)\mathscr{W}$ and (i) is proved.

With regard to (ii)

$$\int_{-\infty}^{\infty} \cdots \int_{-\infty}^{\infty} \phi_{(2)}(x_2)\ldots\phi_{(n)}(x_n)\partial_2\mathscr{W}(s,\boldsymbol{y})\,\mathrm{d}x_2\ldots\mathrm{d}x_n$$

$$= -\int_{-\infty}^{\infty} \cdots \int_{-\infty}^{\infty} \partial_2\phi_{(2)}(x_2)\ldots\phi_{(n)}(x_n)\mathscr{W}(s,\boldsymbol{y})\,\mathrm{d}x_2\ldots\mathrm{d}x_n$$

$$= -\int_{-\infty}^{\infty} \mathrm{e}^{-\mathrm{i}\omega x_1}\int_{-\infty}^{\infty} \cdots \int_{-\infty}^{\infty} \{w(\boldsymbol{x})\mathrm{e}^{-\sigma x_1}\}\partial_2\phi_{(2)}\ldots \times \phi_{(n)}\,\mathrm{d}x_2\ldots\mathrm{d}x_n\,\mathrm{d}x_1$$

$$= \int_{-\infty}^{\infty} \mathrm{e}^{-\mathrm{i}\omega x_1}\int_{-\infty}^{\infty} \cdots \int_{-\infty}^{\infty} \{\partial_2 w(\boldsymbol{x})\mathrm{e}^{-\sigma x_1}\}\phi_{(2)}\ldots \times \phi_{(n)}\,\mathrm{d}x_n\,\mathrm{d}x_2\ldots\mathrm{d}x_n\,\mathrm{d}x_1$$

and the theorem is complete. □

An immediate consequence of the limit properties of generalised and weak functions is

Theorem 12.9. *If* $w, w_\mu \in \mathscr{L}'_1(c)$ *and* $\operatorname{Lim}_{\mu\to\mu_0} w_\mu = w$ *then*

$$\operatorname*{Lim}_{\mu\to\mu_0} \mathscr{W}_\mu(s,\boldsymbol{y}) = \mathscr{W}(s,\boldsymbol{y}).$$

Further

Theorem 12.10. *If* $w \in \mathscr{L}'_1(c)$,

$$w(\boldsymbol{x}) = \frac{1}{2\pi i}\int_{\sigma - i\infty}^{\sigma + i\infty} \mathscr{W}(s, \boldsymbol{y}) e^{sx_1}\, ds \qquad (\sigma > c).$$

Obviously $\mathscr{W}$ must not grow too fast with respect to the variable s.

12.5. Partial differential equations

As an illustration of the use of Laplace transforms in partial differential equations consider the typical problem of finding f_1 in $x_1 > 0$, $x_2 > 0$ such that

$$\frac{\partial^2 f_1}{\partial x_1^2} - \frac{\partial^2 f_1}{\partial x_2^2} = 0$$

with $f_1 = 0$ on $x_2 = 0$, $f_1 = e^{cx_2}$ on $x_1 = 0$, $\partial f_1/\partial x_1 = 0$ on $x_1 = 0$. Introduce the weak function $w = f_1\, H(x_1) H(x_2)$; then

$$\begin{aligned}\partial_1^2 w - \partial_2^2 w &= [f_1]_{x_1=0}\, \delta'(x_1) H(x_2) + [\partial f_1/\partial x_1]_{x_1=0}\, \delta(x_1) H(x_2) \\ &\quad - [f_1]_{x_2=0}\, H(x_1)\delta'(x_2) - [\partial f_1/\partial x_2]_{x_2=0}\, H(x_1)\delta(x_2) \\ &= e^{cx_2} H(x_2)\delta'(x_1) - f_0(x_1) H(x_1)\delta(x_2)\end{aligned}$$

where $f_0(x_1) = [\partial f_1/\partial x_2]_{x_2=0}$.

Assume that $w \in \mathscr{L}'_1(c_0)$ for some c_0 and that $f_0 \in \mathscr{L}'_+(c_0)$; then

$$(s^2 - \partial_2^2)\, \mathscr{W}(s, x_2) = s\, e^{cx_2} H(x_2) - \mathscr{F}_0(s)\delta(x_2).$$

Now $\mathscr{W} = 0$ for $x_2 < 0$ and so, treating s as a parameter, the theory of §§11.9 and 11.10 gives

$$\mathscr{W}(s, x_2) = \left\{\frac{s e^{cx_2}}{s^2 - c^2} + \frac{\frac{1}{2} e^{sx_2}}{c - s} - \frac{\frac{1}{2} e^{-sx_2}}{s + c} + \frac{1}{s}\mathscr{F}_0(s) \sinh sx_2\right\} H(x_2).$$

Evidently $w \notin \mathscr{L}'_1(c_0)$ unless the terms involving e^{sx_2} disappear. Since $\mathscr{F}_0$ is at our disposal choose it to be $s/(s - c)$ (which complies with Theorem 12.5) and then

$$\mathscr{W}(s, x_2) = \frac{s}{s^2 - c^2}(e^{cx_2} - e^{-sx_2}) H(x_2)$$

which makes $w \in \mathscr{L}'_1(c_0)$ for $c_0 > c$. From Exercises 2 and 5,

$$w(\boldsymbol{x}) = \{e^{cx_2}\cosh cx_1\, H(x_1) - H(x_1 - x_2)\cosh c(x_1 - x_2)\}\, H(x_2)$$

$$= \begin{cases} \tfrac{1}{2}\{e^{c(x_2+x_1)} - e^{c(x_1-x_2)}\}\, H(x_2) & (x_2 < x_1) \\ \tfrac{1}{2}\{e^{c(x_2+x_1)} + e^{c(x_2-x_1)}\}\, H(x_1) & (x_2 > x_1). \end{cases}$$

The reader can easily check that this satisfies the conditions on f_1, except that its conventional derivatives may fail to exist on $x_1 = x_2$. Therefore the formula may be considered as a satisfactory solution to the problem. The question as to whether there might be weak solutions which are not in $\mathscr{L}'_1(c_0)$ for any finite c_0 will be left on one side.

As another example consider, on $x_1 > 0, x_2 > 0$

$$\frac{\partial f_1}{\partial x_1} - \frac{\partial^2 f_1}{\partial x_2^2} = 0$$

subject to $f_1 = 0$ on $x_1 = 0, f_1 = f(x_1)$ on $x_2 = 0$ and $\lim_{x_2 \to \infty} f_1 = 0$. As before introduce $w = f_1 H(x_1)H(x_2)$; then

$$\begin{aligned} \partial_1 w - \partial_2^2 w &= [f_1]_{x_1=0}\,\delta(x_1)H(x_2) - [f_1]_{x_2=0}\,H(x_1)\delta'(x_2) \\ &\quad - [\partial f_1/\partial x_2]_{x_2=0}\, H(x_1)\delta(x_2) \\ &= -f(x_1)H(x_1)\delta'(x_2) - f_0(x_1)H(x_1)\delta(x_2) \end{aligned}$$

where $f_0(x_1) = [\partial f_1/\partial x_2]_{x_2=0}$.

Assume that $w \in \mathscr{L}'_1(c_0), f, f_0 \in \mathscr{L}_+(c_0)$. Then

$$(s - \partial_2^2)\,\mathscr{W} = -\mathscr{F}(s)\delta'(x_2) - \mathscr{F}_0(s)\delta(x_2)$$

and hence

$$\mathscr{W}(s, x_2) = \{s^{-1/2}\mathscr{F}_0(s)\sinh s^{1/2}x_2 + \mathscr{F}(s)\cosh s^{1/2}x_2\}\, H(x_2).$$

Since $\lim_{x_2 \to \infty} f_1 = 0$, $\lim_{x_2 \to \infty} \mathscr{W}(s, x_2) = 0$ which requires that $\mathscr{F}_0(s)$ be selected so that $\mathscr{F}_0(s) = -s^{1/2}\mathscr{F}(s)$. Then

$$\mathscr{W}(s, x_2) = \mathscr{F}(s)e^{-s^{1/2}x_2}H(x_2).$$

Now

$$\int_0^\infty \frac{a}{2\pi^{1/2}x^{3/2}}e^{-sx - a^2/4x}\,dx = e^{-as^{1/2}} \qquad (a > 0);$$

consequently

$$w(x) = \frac{x_2 H(x_2) H(x_1)}{2\pi^{1/2}} \int_0^{x_1} \frac{f(t) e^{-x_2^2/4(x_1 - t)}}{(x_1 - t)^{3/2}} dt$$

from Theorem 12.6

Exercises

16. Try the first example above with $f_1 = f(x_2)$, $f \in \mathscr{L}_+(c)$, instead of e^{cx_2} on $x_1 = 0$.

17. Can you do the second example above by taking a Laplace transform with respect to x_2 instead of x_1?

18. In $x_1 > 0, x_2 > 0$

$$\frac{\partial^2 f_1}{\partial x_1^2} = \frac{\partial^2 f_1}{\partial x_2^2} + a \frac{\partial^3 f_1}{\partial x_2^2 \partial x_1} \qquad (a > 0)$$

and $f_1 = \partial f_1/\partial x_1 = 0$ on $x_1 = 0$, $f_1 = 1$ on $x_2 = 0$, and $\lim_{x_2 \to \infty} f_1 = 0$. Find a solution in the form of a Laplace transform and show that

$$\partial f_1/\partial x_2 = e^{-x_1/a} H(x_1)/(\pi a x_1)^{1/2} \text{ on } x_2 = 0.$$

19. In $x_1 > 0, x_2 > 0$

$$\frac{\partial^2 f_1}{\partial x_2^2} - 5a^2 f_1 = \frac{\partial f_1}{\partial x_1} \qquad (a > 0)$$

and $f_1 = 0$ for $x_1 = 0$, $f_1 = e^{-a^2 x_1}$ for $x_2 = 0$ and $\lim_{x_2 \to \infty} f_1 = 0$. Show that

$$f_1 = e^{-a^2 x_1 - 2a x_2} + O(e^{-5a^2 x_1})$$

as $x_1 \to \infty$.

20. Find f_1 in $x_1 > 0, x_2 > 0$ such that

$$\frac{\partial f_1}{\partial x_2} + \frac{\partial f_1}{\partial x_1} = x_2 e^{-x_1}$$

and $f_1 = 0$ on $x_2 = 0$, $f_1 = 1 - e^{-x_2}$ on $x_1 = 0$.

21. Find f_1 in $x_1 > 0, x_2 > 0$ such that

$$\frac{\partial^2 f_1}{\partial x_2^2} - \frac{\partial^2 f_1}{\partial x_1^2} = e^{-x_1 - x_2}$$

with $f_1 = 0$ on $x_2 = 0$, $f_1 = 0$ and $\partial f_1/\partial x_1 = x_2$ on $x_1 = 0$.

22. If

$$\frac{\partial^2 f_1}{\partial x_1^2} - \frac{\partial^2 f_1}{\partial x_2^2} - f_1 = 0 \qquad (x_1 > 0, x_2 > 0)$$

and $f_1 = \partial f_1/\partial x_1 = 0$ on $x_1 = 0$, $f_1 = x_1$ on $x_2 = 0$ prove that f_1 approaches $-x_2 e^{x_1}(2/\pi x_1)^{1/2}$ as $x_1 \to \infty$.

12.6. The bilateral Laplace transform

So far the Laplace transform has been defined only for weak functions which are zero in a half-space. In some problems it is convenient to be able to take a Laplace transform of a weak function which is not necessarily zero in a half-space. In order to do this it is necessary to put restrictions on the behaviour at both negative and positive infinity.

It will always be assumed in the following that c_1 and c_2 are real numbers such that $c_2 \geq c_1$. Then

Definition 12.7. *$f \in \mathscr{L}(c_1, c_2)$ if, and only if, f is integrable over every finite interval, $f = O(e^{c_1 x})$ as $x \to \infty$ and $f = O(e^{c_2 x})$ as $x \to -\infty$.*

It is immediately evident that $\mathscr{L}_+(c)$ could be regarded as a $\mathscr{L}(c, \cdot)$ in which c_2 was irrelevant. For that reason the same notation for Laplace transforms can be employed and

$$\mathscr{F}(s) = \int_{-\infty}^{\infty} f(x) e^{-sx} dx$$

exists if $f \in \mathscr{L}(c_1, c_2)$ and $c_1 < \sigma < c_2$.

Definition 12.8. *$w \in \mathscr{L}'(c_1, c_2)$ if, and only if, $w = f^{(r)}$ for some finite r and $f \in \mathscr{L}(c_1, c_2)$. The bilateral Laplace transform $\mathscr{W}(s)$ is the Fourier transform with respect to ω of $e^{-\sigma x} w(x)$ $(c_1 \leq \sigma \leq c_2)$.*

The argument justifying this definition is completely analogous to that used for Definition 12.3.

Since sgn $x \in \mathscr{L}(0, 0)$ it has a bilateral Laplace transform as a weak function. By Definition 12.8 it is the same as its Fourier transform. Consequently the bilateral transform has little further interest when $c_1 = c_2 = 0$. The case $c_1 = c_2 \neq 0$ is a little different because the weak function may then have a bilateral Laplace transform but not a Fourier transform.

Theorem 12.11. *If $w \in \mathscr{L}'(c_1, c_2)$ the bilateral Laplace transform of (i) $w(ax)(a > 0)$ is $(1/a)\mathscr{W}(s/a)$, (ii) w' is $s\mathscr{W}(s)(c_1 \leq \sigma \leq c_2)$. Furthermore, if $w_\mu \in \mathscr{L}'(c_1, c_2)$ and $\operatorname{Lim}_{\mu \to \mu_0} w_\mu = w$, then $\operatorname{Lim}_{\mu \to \mu_0} \mathscr{W}_\mu = \mathscr{W}$.*

Proof. This is the same as for Theorems 12.1 and 12.2. □

Theorem 12.12. *If* $w \in \mathscr{L}'(c_1, c_2)(c_2 > c_1)$, (i) $\mathscr{W}(s)$ *is a regular function of* s *in* $c_1 < \sigma < c_2$, (ii) $\mathscr{W}(s) = O(|s|^k)$ *for some finite* k *in* $c_1 < \sigma < c_2$, (iii) $\mathrm{Lim}_{\sigma \to c_1 + 0} \mathscr{W}(s) = \mathscr{W}(c_1 + i\omega)$, (iv) $\mathrm{Lim}_{\sigma \to c_2 - 0} \mathscr{W}(s) = \mathscr{W}(c_2 + i\omega)$.

Proof. The proof is analogous to that of Theorem 12.3. □

Theorem 12.13. *If* $w \in \mathscr{L}'(c_1, c_2)$,

$$w(x) = \frac{e^{\sigma x}}{2\pi i} \int_{-\infty}^{\infty} \mathscr{W}(s) e^{i\omega x} d\omega \qquad (c_1 \le \sigma \le c_2)$$

$$= \frac{1}{2\pi i} \mathscr{D}_1^m \int_{\sigma - i\infty}^{\sigma + i\infty} \frac{\mathscr{W}(s) e^{sx}}{(s - c_1)^m} ds \qquad (c_1 < \sigma < c_2)$$

for sufficiently large m, *and* $\mathscr{D}_1 w = w' - c_1 w$.

Proof. This is along the lines of Theorem 12.4. □

In the second formula for w, $s - c_1$ could be replaced by $s - c_2$ provided that $\mathscr{D}_1$ were replaced by $\mathscr{D}_2$.

As for the single-sided Laplace transform we write

$$w(x) = \frac{1}{2\pi i} \int_{\sigma - i\infty}^{\sigma + i\infty} \mathscr{W}(s) e^{sx} ds \qquad (c_1 \le \sigma \le c_2)$$

with the interpretation of Theorem 12.13 understood.

Analogous to Theorem 12.5 and with a similar proof is

Theorem 12.14. *If* $c_2 > c_1$, *if* $\mathscr{W}(s)$ *is a regular function of* s *in* $c_1 < \sigma < c_2$ *and* $\mathscr{W}(s) = O(|s|^k)$ *for some finite* k *in* $c_1 < \sigma < c_2$ *then* $\mathscr{W}(s)$ *is the bilateral transform of a* $w \in \mathscr{L}'(c_1, c_2)$.

Let $c_2 > c_1$ and assume that f_1 and f_2 are continuous as well as being in $\mathscr{L}(c_1, c_2)$. Then $f_1(t) f_2(x - t)$ is integrable over any finite interval of t. Also, as $t \to \infty$, $f_1(t) f_2(x - t) \sim e^{(c_1 - c_2)t}$ and, as $t \to -\infty$, $f_1(t) f_2(x - t) \sim e^{(c_2 - c_1)t}$. Hence $\int_{-\infty}^{\infty} f_1(t) f_2(x - t) dt$ or $f_1 * f_2$ exists for such f_1 and f_2; in fact $f_1 * f_2 \in \mathscr{L}(c_1 + \varepsilon, c_2 - \varepsilon)$ where $0 < \varepsilon < \frac{1}{2}(c_2 - c_1)$ but ε is otherwise arbitrary. For, in $t > 0$ both f_1 and f_2

are $O(e^{c_1x})$ and in $t<0$ both are $O(e^{c_2x})$. Therefore, when $x>0$,

$$\int_x^\infty f_1(t)f_2(x-t)\mathrm{d}t = O\left(\int_x^\infty e^{c_1t}.e^{c_2(x-t)}\mathrm{d}t\right) = O(e^{c_1x}),$$

$$\int_0^x f_1(t)f_2(x-t)\mathrm{d}t = O\left(\int_0^x e^{c_1t}.e^{c_1(x-t)}\mathrm{d}t\right) = O(xe^{c_1x}),$$

$$\int_{-\infty}^0 f_1(t)f_2(x-t)\mathrm{d}t = O\left(\int_{-\infty}^0 e^{c_2t}.e^{c_1(x-t)}\mathrm{d}t\right) = O(e^{c_1x}).$$

Addition now shows that $f_1 * f_2 = O(e^{(c_1+\varepsilon)x})$ as $x\to\infty$. Similarly, $f_1 * f_2 = O(e^{(c_2-\varepsilon)x})$ as $x\to-\infty$.

Now extend the definition of convolution to derivatives of f_1 and f_2 by the rule

$$f_1^{(m)} * f_2^{(k)} = (f_1 * f_2)^{(m+k)}. \tag{6}$$

Suppose now that $f\in\mathscr{L}(c_1, c_2)$ and consider

$$(\mathscr{D}-\tfrac{1}{2}\varepsilon)\,e^{(c_1+\varepsilon/2)x}\int_{-\infty}^x f(t)e^{-(c_1+\varepsilon/2)t}\mathrm{d}t = f(x).$$

As $x\to-\infty$

$$e^{(c_1+\varepsilon/2)x}\int_{-\infty}^x f(t)e^{-(c_1+\varepsilon/2)t}\mathrm{d}t = O\left(e^{(c_1+\varepsilon/2)x}\int_{-\infty}^x e^{(c_2-c_1-\varepsilon/2)t}\mathrm{d}t\right)$$
$$= O(e^{c_2x})$$

and, as $x\to\infty$,

$$e^{(c_1+\varepsilon/2)x}\int_{-\infty}^x f(t)e^{-(c_1+\varepsilon/2)t}\mathrm{d}t = O\left(e^{(c_1+\varepsilon/2)x}\int_{-\infty}^x f(t)e^{-(c_1+\varepsilon/2)t}\mathrm{d}t\right)$$
$$= O(e^{(c_1+\varepsilon/2)x}).$$

Hence f has been represented as the derivative of a continuous function which is in $\mathscr{L}(c_1+\tfrac{1}{2}\varepsilon, c_2)$. Consequently, by the rule (6), convolution can be defined if $f_1, f_2\in\mathscr{L}(c_1, c_2)$ and, by a further application of rule (6), $w_1 * w_2$ is defined for $w_1, w_2\in\mathscr{L}'(c_1, c_2)$. The above process shows that $w_1 * w_2\in\mathscr{L}'(c_1+\varepsilon, c_2-\varepsilon)$.

Theorem 12.15. *If $c_2 > c_1$, if $w_1, w_2 \in \mathscr{L}'(c_1, c_2)$ the bilateral Laplace transform of $w_1 * w_2$ is $\mathscr{W}_1(s)\mathscr{W}_2(s)$ in $c_1 + \varepsilon < \sigma < c_2 - \varepsilon$.*

Proof. This is the same as Theorem 12.6. □

Exercise

23. Is there an analogue to Theorem 12.7 for the bilateral Laplace transform?

12.7. Integral equations

The integral equation

$$w_1(x) - \int_{-\infty}^{\infty} k(x-t)w_1(t)\,dt = w_2(x)$$

can be solved by means of Laplace transforms under suitable circumstances. Suppose w_1, w_2, k are all in $\mathscr{L}'(c_1, c_2)$ with $c_1 < c_2$. Then, from Theorem 12.15,

$$\{1 - \mathscr{K}(s)\}\mathscr{W}_1(s) = \mathscr{W}_2(s)$$

and $\mathscr{W}_1$ can be found by division provided that $1 - \mathscr{K}(s)$ has no zeros in a strip within (c_1, c_2). The inversion Theorem 12.13 then gives w_1.

If $1 - \mathscr{K}(s)$ has zeros inside the strip then solutions of the homogeneous integral equation are involved. The discussion of the homogeneous integral equation will be given under the assumptions that

$$w_1, w_2 \in \mathscr{L}'(c_1, c_2) \quad \text{and} \quad k \in \mathscr{L}'(c_{01}, c_{02})$$

where $c_{01} < c_1$, $c_{02} > c_2$. The Laplace transform corresponding to

$$w_1(x) = \int_{-\infty}^{\infty} k(x-t)w_1(t)\,dt \tag{7}$$

is

$$\mathscr{W}_1(s)\{1 - \mathscr{K}(s)\} = 0$$

and holds within (c_1, c_2). Now $\mathscr{W}_1$ can be written as $\mathscr{W}_+ + \mathscr{W}_-$ where $\mathscr{W}_+$ is regular in $\sigma > c_1$ and $\mathscr{W}_-$ is regular in $\sigma < c_2$. Such a decomposition can always be constructed by $w_1 = w_{11} + w_{12}$ where $w_{11} = 0$ for $x < 0$ and $w_{12} = 0$ for $x > 0$. In general the decomposition is not unique. Alternatively we can take

$$\mathscr{W}_+(s) = -\frac{(s-c_1)^m}{2\pi i}\int_{c_1+\varepsilon-i\infty}^{c_1+\varepsilon+i\infty}\frac{\mathscr{W}_1(z)\,dz}{(z-c_1)^m(z-s)},$$

$$\mathscr{W}_-(s) = \frac{(s-c_1)^m}{2\pi i}\int_{c_2-\varepsilon-i\infty}^{c_2-\varepsilon+i\infty}\frac{\mathscr{W}_1(z)\,dz}{(z-c_1)^m(z-s)}$$

with m chosen sufficiently large for the integrals to converge.

The decomposition implies that

$$\mathscr{W}_+(s)\{1-\mathscr{K}(s)\} = -\mathscr{W}_-(s)\{1-\mathscr{K}(s)\}.$$

Now $\mathscr{K}(s)$ is regular within (c_{01}, c_{02}) and so the left-hand side is regular in (c_1, c_{02}) whereas the right-hand side is regular in (c_{01}, c_2). The two sides therefore are analytic continuations of one another and the only possible singularities of $\mathscr{W}_+$ and $\mathscr{W}_-$ in (c_{01}, c_{02}) are poles at the zeros of $1-\mathscr{K}(s)$. Moreover $\mathscr{W}_+(s) = -\mathscr{W}_-(s)$. Consequently w_1 vanishes, i.e. the homogeneous equation has only a trivial solution under the given conditions.

Exercises

24. Examine the integral equations of §10.1 with a view to solving them by Laplace transforms.

25. Show that the homogeneous integral equation (7) has a non-trivial solution when $k(x) = \delta(x)$.

26. If $k \in \mathscr{L}(c, -c)$ and $f = O(e^{c_0|x|})$ where $0 < c_0 < c$, show that $f = k * f$ may have a non-trivial solution. Is this still true for $k \in \mathscr{L}'(c, -c)$? (Try $f = x^m e^{-sx}$.)

12.8. The Wiener–Hopf integral equation

The Wiener–Hopf integral equation of the first kind is

$$\int_0^\infty k(x-t) f_1(t)\,dt = f_2(x) \quad (x > 0).$$

This may be put in terms of weak functions in the following way. Let $k \in \mathscr{L}'(c_1, c_2)$ and let $w_1, w_2 \in \mathscr{L}'_+(c_1)$. Then the integral equation is written as

$$\int_{-\infty}^\infty k(x-t) w_1(t)\,dt = w_2(x) \quad (x > 0).$$

By this is meant that $k * w_1$, which is a weak function for all x, can be identified with w_2 in $x > 0$.

Suppose now that we knew w_1 and could calculate $k * w_1$ in $x < 0$, say $k * w_1 = w_3(x < 0)$. Then, $w_3 = O(e^{(c_2 - \varepsilon)x})$ as $x \to -\infty$ and we may take $w_3 = 0$ $(x > 0)$. Then Theorem 11.8 gives

$$\int_{-\infty}^{\infty} k(x-t) w_1(t) \mathrm{d}t = w_2(x) + w_3(x) + \sum_{m=0}^{M} a_m \delta^{(m)}(x).$$

Applying a bilateral Laplace transform we obtain

$$\mathscr{K}(s)\mathscr{W}_1(s) = \mathscr{W}_2(s) + \mathscr{W}_3(s) + \sum_{m=0}^{M} a_m s^m \tag{8}$$

where $\mathscr{K}(s)$ is regular in $c_1 < \sigma < c_2$, $\mathscr{W}_2$ in $\sigma > c_1$ and $\mathscr{W}_3$ in $\sigma < c_2 - \varepsilon$.

In order to solve (8) additional restrictions are imposed on $\mathscr{K}$. Suppose that in $c_1 < \sigma < c_2$, $\mathscr{K}(s) \to e^{ia} s^{\mu}$ (a real) as $\omega \to \infty$, $\mathscr{K}(s) \to e^{ib} s^{\mu}$ (b real) as $\omega \to -\infty$ and that $|1 - |\mathscr{K}(s)/s^{\mu}|| < C|\omega|^{-\nu}$, $\nu > 0$ as $|\omega| \to \infty$, the inequality holding uniformly in any interior strip. Then $\mathscr{K}(s)(s - c_1)^{-\mu}$ approaches e^{ia}, e^{ib} as $\omega \to \infty, -\infty$ and its modulus tends to 1 uniformly in $c_1 + \varepsilon \le \sigma \le c_2 - \varepsilon$ as $|\omega| \to \infty$.

Hence it has at most a finite number of zeros in the interior strip. Let them be $s_1, s_2, \ldots, s_m$ and let

$$\mathscr{M}(s) = e^{-ia} \mathscr{K}(s) \frac{(s - c_1)^{m - (1/2\pi)(a-b) - \mu} (s - c_2)^{(1/2\pi)(a-b)}}{(s - s_1)(s - s_2) \ldots (s - s_m)} \tag{9}$$

the branch lines of $(s - c_1)^{\alpha}$ and $(s - c_2)^{\beta}$ being drawn so that $(s - c_1)^{\alpha}$ is regular in $\sigma > c_1$ and its phase tends to $\frac{1}{2}\pi\alpha, -\frac{1}{2}\pi\alpha$ as $\omega \to \infty, -\infty$ while $(s - c_2)^{\beta}$ is regular in $\sigma < c_2$ and its phase tends to $\frac{1}{2}\pi\beta, \frac{3}{2}\pi\beta$ as $\omega \to \infty, -\infty$. $\mathscr{M}(s)$ has no zeros in $c_1 < \sigma < c_2$ and tends to 1 as $|\omega| \to \infty$. Therefore $\ln \mathscr{M}$ is regular in $c_1 + \varepsilon \le \sigma \le c_2 - \varepsilon$ and behaves like $|\omega|^{-\nu}$ as $|\omega| \to \infty$. Consequently Cauchy's theorem gives

$$\ln \mathscr{M}(s) = \frac{1}{2\pi i} \int_{c_2 - \varepsilon - i\infty}^{c_2 - \varepsilon + i\infty} \frac{\ln \mathscr{M}(z)}{z - s} \mathrm{d}z - \frac{1}{2\pi i} \int_{c_1 + \varepsilon - i\infty}^{c_1 + \varepsilon + i\infty} \frac{\ln \mathscr{M}(z)}{z - s} \mathrm{d}z \tag{10}$$

for $c_1 + \varepsilon < \sigma < c_2 - \varepsilon$. The first integral is actually regular and bounded for $\sigma < c_2 - \varepsilon$; the second has similar properties in $\sigma > c_1 + \varepsilon$. By choosing ε small enough we can say

$$\ln \mathscr{M}(s) = \mathscr{N}_-(s) + \mathscr{N}_+(s)$$

where the suffix + indicates regularity in $\sigma > c_1$ and the suffix – regularity in $\sigma < c_2$. Hence it has been shown that $\mathscr{M}(s) = \mathscr{M}_+(s)\mathscr{M}_-(s)$ in $c_1 < \sigma < c_2$, where $\mathscr{M}_+$ is regular, bounded and *non-zero* in $\sigma > c_1$ whereas $\mathscr{M}_-$ has the same properties in $\sigma < c_2$. It follows from (9) that

$$\mathscr{K}(s) = \mathscr{K}_+(s)\mathscr{K}_-(s)(s - s_1)\ldots(s - s_m)$$

where $\mathscr{K}_+$ is regular and non-zero in $\sigma > c_1$ and behaves like $|\omega|^{-m-(1/2\pi)(a-b)+\mu}$ as $|\omega| \to \infty$ in $\sigma > c_1$; $\mathscr{K}_-$ is regular and non-zero in $\sigma < c_2$ and behaves like $|\omega|^{(1/2\pi)(a-b)}$ as $|\omega| \to \infty$ in $\sigma < c_2$.

Now rewrite (8) in the form

$$\mathscr{K}_+(s)\mathscr{W}_1(s)(s - s_1)\ldots(s - s_m) = \frac{1}{\mathscr{K}_-(s)}\left\{\mathscr{W}_2(s) + \mathscr{W}_3(s) + \sum_{m=0}^{M} a_m s^m\right\}.$$

$\mathscr{W}_2(s)/\mathscr{K}_-(s)$ is regular in $c_1 < \sigma < c_2$, since $\mathscr{K}_-$ has no zeros in the strip, and is $O\{|\omega|^{k-(1/2\pi)(a-b)}\}$ as $|\omega| \to \infty$. Hence, for sufficiently large p,

$$\frac{\mathscr{W}_2(s)}{(s - c_1)^p \mathscr{K}_-(s)} = \frac{1}{2\pi i}\int_{c_2-\varepsilon-i\infty}^{c_2-\varepsilon+i\infty} \frac{\mathscr{W}_2(z)\,dz}{(z - c_1)^p \mathscr{K}_-(z)(z - s)} - \frac{1}{2\pi i}\int_{c_1+\varepsilon-i\infty}^{c_1+\varepsilon+i\infty} \frac{\mathscr{W}_2(z)\,dz}{(z - c_1)^p \mathscr{K}_-(z)(z - s)}$$

from which it follows that

$$\mathscr{W}_2(s)/\mathscr{K}_-(s) = \mathscr{W}_+(s) + \mathscr{W}_-(s)$$

where $\mathscr{W}_+$ is regular in $\sigma > c_1$ and $O\{|\omega|^{k-(1/2\pi)(a-b)}\}$ while $\mathscr{W}_-$ behaves similarly in $\sigma < c_2$. Therefore

$$\mathscr{K}_+(s)\mathscr{W}_1(s)(s - s_1)\ldots(s - s_m) - \mathscr{W}_+(s) = \mathscr{W}_-(s) + \left\{\mathscr{W}_3(s) + \sum_{m=0}^{M} a_m s^m\right\}\Big/ \mathscr{K}_-(s). \tag{11}$$

The left-hand side is regular in $\sigma > c_1$ and the right-hand side is regular in $\sigma < c_2$. They are therefore analytic continuations of the same integral function. However the growth at infinity of the terms on the left-hand side is limited to $|s|^{k_0-(1/2\pi)(a-b)}$ where k_0 is the greater of k and $k_1+\mu$, k_1 being the limitation on $\mathscr{W}_1$. The growth of terms on the right-hand side is similarly limited though the k_0 may be different. Consequently, by the extension of Liouville's theorem, the integral function must be a polynomial whose highest degree does not exceed the greater of $k_0-(1/2\pi)(a-b)$ on the two sides of (11). Since, however, M is arbitrary k_0 cannot be specified more closely than by saying it is finite. Denoting the polynomial by $\mathscr{P}(s)$ we have shown that

$$\mathscr{K}_+(s)\mathscr{W}_1(s)(s-s_1)\dots(s-s_m)-\mathscr{W}_+(s)=\mathscr{P}(s).$$

There are now two possibilities. Either $\mathscr{W}_+ + \mathscr{P}$ does not have the zeros $s_1,\dots,s_m$ in which case $\mathscr{W}_1$ cannot be regular in the appropriate domain and no solution of the required type exists. Or $\mathscr{W}_+ + \mathscr{P}$ does have the same zeros (by special choice of the coefficients of $\mathscr{P}$ if necessary) and then

$$\mathscr{W}_1(s)=\frac{\mathscr{W}_+(s)+\mathscr{P}(s)}{(s-s_1)\dots(s-s_m)\mathscr{K}_+(s)}. \qquad (12)$$

The formula for w_1 now follows by inversion and must have the requisite properties since the conditions of Theorem 12.5 are satisfied.

In particular, (12) will hold if $\mathscr{K}$ has no zeros in the strip $c_1 < \sigma < c_2$, i.e. $m=0$.

It will be observed that $\mathscr{W}_+$ can be calculated from the known k and w_2 so that the only unknown features of (12) are the coefficients of $\mathscr{P}$. If these do not have to be determined so that $\mathscr{W}_1$ has no poles in $\sigma > c_1$, they will either have to remain completely arbitrary or be fixed by other considerations such as the behaviour of w_1 near the origin.

Example 5. In the equation

$$\int_0^\infty e^{-|x-t|}w_1(t)\,dt=\delta(x-c) \quad (x>0, c>0)$$

$k(x)=e^{-|x|}$ so $\mathscr{K}(s)=-2/(s^2-1)$. In this case $\mathscr{K}$ has no zeros in

$-1 < \sigma < 1$ and simple inspection reveals that $\mathscr{K}_+(s) = 1/(s+1)$, $\mathscr{K}_-(s) = -2/(s-1)$. Hence

$$\frac{\mathscr{W}_1(s)}{s+1} = -\tfrac{1}{2}(s-1)\left\{e^{-sc} + \mathscr{W}_3(s) + \sum_{m=0}^{M} a_m s^m\right\}$$

or

$$\frac{\mathscr{W}_1(s)}{s+1} + \tfrac{1}{2}(s-1)e^{-sc} = -\tfrac{1}{2}(s-1)\left\{\mathscr{W}_3(s) + \sum_{m=0}^{M} a_m s^m\right\}.$$

Proceeding as above

$$\frac{\mathscr{W}_1(s)}{s+1} + \tfrac{1}{2}(s-1)e^{-sc} = \mathscr{P}(s)$$

or

$$\mathscr{W}_1(s) = -\tfrac{1}{2}(s^2-1)e^{-sc} + (s+1)\mathscr{P}(s)$$

whence

$$w_1(x) = \tfrac{1}{2}\{\delta(x-c) - \delta''(x-c)\} + \sum_{p=0}^{P} b_p\{\delta^{(p+1)}(x) + \delta^{(p)}(x)\}.$$

The coefficients b_p are arbitrary and cannot be determined without knowledge of the behaviour of w_1 at the origin.

12.9. A related partial differential equation

The technique used to resolve (8) may also be applied to partial differential equations sometimes, employing a generalisation of the bilateral transform for several variables similar to that of § 12.4 for the one-sided transform. As an illustration, suppose that f satisfies

$$\frac{\partial^2 f}{\partial x_1^2} + \frac{\partial^2 f}{\partial x_2^2} + \beta^2 f = 0 \qquad (\beta = \beta_r - i\beta_i, \beta_r > 0, \beta_i > 0)$$

except possibly on $x_1 \geq 0$, $x_2 = 0$ where it is given that $f = e^{-i\beta x_1 \cos\theta}$ $(0 < \theta < \frac{1}{2}\pi)$. The function f is continuous everywhere and so are its first partial derivatives except possibly across $x_1 \geq 0$, $x_2 = 0$. It is also assumed that $\lim_{|x_2| \to \infty} f = 0$.

Let $w_1 = fH(x_2)$, $w_2 = fH(-x_2)$. Then

$$\partial_1^2 w_1 + \partial_2^2 w_1 + \beta^2 w_1 = f(+0)\delta'(x_2) + f'(+0)\delta(x_2),$$
$$\partial_1^2 w_2 + \partial_2^2 w_2 + \beta^2 w_2 = -f(-0)\delta'(x_2) - f'(-0)\delta(x_2)$$

where $f(\pm 0) = \lim_{x_2 \to \pm 0} f$, $f'(\pm 0) = \lim_{x_2 \to \pm 0} \partial f / \partial x_2$. By assumption

$$f(+0) = f(-0) = -\,e^{-i\beta x_1 \cos\theta}$$

on $x_1 > 0$. Evidently, $w = w_1 + w_2$ satisfies

$$\partial_1^2 w + \partial_2^2 w + \beta^2 w = \{f'(+0) - f'(-0)\}\,\delta(x_2).$$

Take a bilateral Laplace transform with respect to x_1, assuming that w_1, w_2 and w all belong to $\mathscr{L}'(-\beta_i \cos\theta, \beta_i)$. Then

$$(\partial_2^2 + \kappa^2)\,\mathscr{W} = \{\mathscr{F}'(+0) - \mathscr{F}'(-0)\}\,\delta(x_2)$$

where $\kappa^2 = s^2 + \beta^2$.

Define κ by choosing that branch of $(s^2 + \beta^2)^{1/2}$ which reduces to β when $s = 0$. Then κ has a negative imaginary part in $-\beta_i < \sigma < \beta_i$. Since $\lim_{|x_2| \to \infty} f = 0$, $\lim_{|x_2| \to \infty} \mathscr{W} = 0$ and so

$$\mathscr{W} = -(1/2i\kappa)\,\{\mathscr{F}'(+0) - \mathscr{F}'(-0)\}\,e^{-i\kappa|x_2|}. \tag{13}$$

If the analysis were repeated for w_1 it would be found that

$$\mathscr{W}_1 = \mathscr{F}(+0)\cos\kappa x_2 + (1/\kappa)\mathscr{F}'(+0)\sin\kappa x_2$$

which corresponds to a w_1 which is zero in $x_2 < 0$ only if

$$\mathscr{F}(+0) + \mathscr{F}'(+0)/i\kappa = 0 \tag{14}$$

and then $\mathscr{W}_1 = \mathscr{F}(+0)e^{-i\kappa x_2}$. Similarly $\mathscr{W}_2 = \mathscr{F}(-0)e^{i\kappa x_2}$ and

$$\mathscr{F}(-0) - \mathscr{F}'(-0)/i\kappa = 0. \tag{15}$$

The subtraction of (15) from (14) gives

$$\mathscr{F}'(-0) + \mathscr{F}'(+0) = 0. \tag{16}$$

Now

$$\begin{aligned}\mathscr{F}'(+0) &= \int_{-\infty}^{\infty} \left(\frac{\partial f}{\partial x_2}\right)_{x_2 = +0} e^{-sx_1}\,dx_1 \\ &= \left\{\int_0^{\infty} + \int_{-\infty}^{0}\right\}\left(\frac{\partial f}{\partial x_2}\right)_{x_2 = +0} e^{-sx_1}\,dx_1 \\ &= \mathscr{F}'_+(+0) + \mathscr{F}'_-(+0)\end{aligned}$$

say, where $\mathscr{F}'_+$ and $\mathscr{F}'_-$ are regular in $\sigma > -\beta_i \cos\theta$ and $\sigma < \beta_i$

respectively. Further $\mathscr{F}'_-(-0) = \mathscr{F}'_-(+0)$ and so (16) becomes

$$2\mathscr{F}'_-(0) = -\mathscr{F}'_+(+0) - \mathscr{F}'_+(-0).$$

The left-hand side is regular in $\sigma < \beta_i$, the right-hand side in $\sigma > -\beta_i \cos\theta$. Therefore they are analytic continuations of the same integral function. If we assume that $f'(\pm 0)$ is $O(|x|^{\mu-1})$ $(\mu > 0)$ as $|x| \to 0$ both $\mathscr{F}'_-$ and $\mathscr{F}'_+$ must tend to zero as $|s| \to \infty$ in the respective domains of regularity. Therefore the integral function must be zero, i.e.

$$\mathscr{F}'_-(0) = 0, \qquad \mathscr{F}'_+(+0) = -\mathscr{F}'_+(-0). \tag{17}$$

Thus (13) can be written

$$\begin{aligned}\mathscr{F}'_+(+0)/\kappa &= -\mathrm{i}\mathscr{F}(+0) = -\mathrm{i}\mathscr{F}_+(+0) - \mathrm{i}\mathscr{F}_-(+0)\\ &= \mathrm{i}/(s + \mathrm{i}\beta\cos\theta) - \mathrm{i}\mathscr{F}_-(+0)\end{aligned}$$

on account of the known form of $f(+0)$ on $x_1 > 0$. Hence

$$\mathscr{F}'_+(+0)/(s+\mathrm{i}\beta)^{1/2} = \mathrm{i}(s-\mathrm{i}\beta)^{1/2}/(s+\mathrm{i}\beta\cos\theta) - \mathrm{i}(s-\mathrm{i}\beta)^{1/2}\mathscr{F}_-(+0). \tag{18}$$

The first term on the right-hand side has now to be split into the sum of + and − terms. The technique used in the preceding section to separate $\mathscr{W}_2/\mathscr{K}_-$ can be employed but is unnecessary here because the only singularity on one side is a simple pole so that the split can be achieved by inspection. In fact, the required separation is

$$\frac{(s-\mathrm{i}\beta)^{1/2}}{s+\mathrm{i}\beta\cos\theta} = \frac{(s-\mathrm{i}\beta)^{1/2} - (-\mathrm{i}\beta\cos\theta - \mathrm{i}\beta)^{1/2}}{s+\mathrm{i}\beta\cos\theta} + \frac{(-\mathrm{i}\beta\cos\theta - \mathrm{i}\beta)^{1/2}}{s+\mathrm{i}\beta\cos\theta}$$

the first term being regular in $\sigma < \beta_i$ and the second in $\sigma > -\beta_i\cos\theta$. Therefore (18) can be expressed as

$$\frac{\mathscr{F}'_+(+0)}{(s+\mathrm{i}\beta)^{1/2}} - \frac{\mathrm{i}(-\mathrm{i}\beta\cos\theta - \mathrm{i}\beta)^{1/2}}{s+\mathrm{i}\beta\cos\theta} = \frac{\mathrm{i}(s-\mathrm{i}\beta)^{1/2} - \mathrm{i}(-\mathrm{i}\beta\cos\theta - \mathrm{i}\beta)^{1/2}}{s+\mathrm{i}\beta\cos\theta} - \mathrm{i}(s-\mathrm{i}\beta)^{1/2}\mathscr{F}_-(+0). \tag{19}$$

Once again the argument about the respective domains of regularity of the two sides leads one to conclude that both sides are equal to the same integral function. The assumption already made that $f'(+0)$ is $O(|x|^{\mu-1})$ as $|x| \to 0$ indicates that $\mathscr{F}'_+(+0)$ is $O(|s|^{-\mu})$ as

$|s| \to \infty$ in $\sigma > -\beta_i \cos\theta$ (Theorem 9.4). Therefore the left-hand side of (19) shows that the integral function tends to zero as $|s| \to \infty$ in $\sigma > -\beta_i \cos\theta$. Likewise the right-hand side shows that the integral function is $o(|s|^{1/2})$ as $|s| \to \infty$ in $\sigma < \beta_i$, provided that $f(+0)$ is integrable at the origin. Liouville's theorem consequently dictates that the integral function must be a constant, which must be zero on account of the behaviour as $|s| \to \infty$ in $\sigma > -\beta_i \cos\theta$. Hence

$$\mathscr{F}'_+(+0) = \frac{\mathrm{i}(-\mathrm{i}\beta\cos\theta - \mathrm{i}\beta)^{1/2}(s+\mathrm{i}\beta)^{1/2}}{s+\mathrm{i}\beta\cos\theta}. \tag{20}$$

It follows from (17) and (13) that

$$\mathscr{W} = -\frac{(-\mathrm{i}\beta\cos\theta - \mathrm{i}\beta)^{1/2}}{(s+\mathrm{i}\beta\cos\theta)(s-\mathrm{i}\beta)^{1/2}}\mathrm{e}^{-\mathrm{i}\kappa|x_2|}.$$

Hence

$$f = -\frac{1}{2\pi\mathrm{i}}\int_{\sigma-\mathrm{i}\infty}^{\sigma+\mathrm{i}\infty}\frac{(-\mathrm{i}\beta\cos\theta - \mathrm{i}\beta)^{1/2}}{(s+\mathrm{i}\beta\cos\theta)(s-\mathrm{i}\beta)^{1/2}}\mathrm{e}^{sx_1-\mathrm{i}\kappa|x_2|}\,\mathrm{d}s \tag{21}$$

with $-\beta_i\cos\theta < \sigma < \beta_i$.

The evaluation of this formula needs some manipulation. We quote only

$$\lim_{x_2\to+0}\frac{\partial f}{\partial x_2} = \frac{\mathrm{i}(-\mathrm{i}\beta\cos\theta - \mathrm{i}\beta)^{1/2}}{\pi^{1/2}x_1^{1/2}} + \frac{\mathrm{i}\beta^2\sin^2\theta\,\mathrm{e}^{-\mathrm{i}\beta x_1\cos\theta}}{\pi^{1/2}(-\mathrm{i}\beta\cos\theta - \mathrm{i}\beta)^{1/2}}\int_0^{x_1}\frac{\mathrm{e}^{\mathrm{i}\beta u(\cos\theta-1)}\,\mathrm{d}u}{u^{1/2}}$$

in $x_1 > 0$. The method of stationary phase (§9.3) can be used to develop approximate formulae from (21) when $x_1^2 + x_2^2$ is large, there being a point of stationary phase at $s = -\mathrm{i}\beta x_1/(x_1^2+x_2^2)^{1/2}$.

Exercises

27. Repeat the problem of this section with the boundary conditions on f changed to

$$\lim_{x_2\to+0}\frac{\partial f}{\partial x_2} = \lim_{x_2\to-0}\frac{\partial f}{\partial x_2} = \mathrm{i}\beta\sin\theta\,\mathrm{e}^{-\mathrm{i}\beta x_1\cos\theta} \text{ on } x_1 > 0$$

and $\lim_{x_2\to+0} f$ not necessarily equal to $\lim_{x_2\to-0} f$ on $x_1 > 0$.

28. If the limitation on the growth of $\mathscr{F}'_+(+0)$ and $\mathscr{F}_-(+0)$ be dropped show that the term $C(s+i\beta)^{1/2}$ can be added to (20) without violating (19), or (17). The consequent f vanishes on $x_2=0$, $x_1>0$ but $\lim_{x_2\to+0}\partial f/\partial x_2 = -Ce^{-i\beta x_1}/2\pi^{1/2}x^{3/2}$ so that the problem does not have a unique solution unless conditions are imposed on f and its derivatives at the origin. Are there any other terms which could be added to (20) without violating (19) when there is no restriction on the growth at infinity?

29. Investigate the possibility that the operation $\text{Lim}_{\beta i\to+0}$ applied to (21) would provide the solution to the problem in which $\beta_i=0$.

Table of Fourier transforms

In the following table the notation of the main text is employed, β is a complex number, $m = 1, 2, \ldots$ and $k = 0, 1, 2, \ldots$

$g(x)$	$G(\alpha) = \int_{-\infty}^{\infty} g(x) e^{-i\alpha x} \mathrm{d}x$
$\delta^{(k)}(x-b)$	$(i\alpha)^k e^{-ib\alpha}$
$x^k e^{ibx}$	$2\pi i^k \delta^{(k)}(\alpha - b)$
$\mathrm{H}(x)$	$\pi\delta(\alpha) - i\alpha^{-1}$
$\operatorname{sgn} x$	$-2i\alpha^{-1}$
x^{-m}	$\dfrac{(-i)^m \pi}{(m-1)!}\alpha^{m-1} \operatorname{sgn} \alpha$
$x^{\beta}\mathrm{H}(x)$ (β not an integer)	$\beta! e^{-\pi i(\beta+1)/2}(\alpha - i0)^{-\beta-1}$
$\lvert x\rvert^{\beta}$ (β not an integer)	$\beta! \lvert\alpha\rvert^{-\beta-1} 2\cos\frac{1}{2}\pi(\beta+1)$
$x^{-m} \operatorname{sgn} x$	$2e^{-\pi i(m-1)/2}\alpha^{m-1}\{\psi(m-1) - \ln\lvert\alpha\rvert\}/(m-1)!$
$\ln\lvert x\rvert$	$\pi\{2\psi(0)\delta(\alpha) - \alpha^{-1}\operatorname{sgn}\alpha\}$
$\lvert x\rvert^{\beta}\operatorname{sgn} x$ (β not an integer)	$\beta!\lvert\alpha\rvert^{-\beta-1}\operatorname{sgn}\alpha\,(-2i)\sin\frac{1}{2}\pi(\beta+1)$
$(x + ic)^{-m}$ (c real and non-zero)	$-2\pi i e^{-c\alpha}\mathrm{H}(\alpha c)(-i\alpha)^{m-1}\operatorname{sgn} c/(m-1)!$
$(x \pm i0)^{\beta}$ (β not 0 or a positive integer)	$2\pi e^{\pm \pi i\beta/2}(\pm\alpha)^{-\beta-1}\mathrm{H}(\pm\alpha)/(-\beta-1)!$
$x^{\beta}\mathrm{H}(x)\ln x$ (β not an integer)	$\beta! e^{-\pi i(\beta+1)/2}[\{\psi(\beta) - \frac{1}{2}\pi i\}(\alpha - i0)^{-\beta-1} - (\alpha - i0)^{-\beta-1}\ln(\alpha - i0)]$
$\lvert x\rvert^{\beta}\ln\lvert x\rvert$ (β not an integer)	$\beta! 2\cos\frac{1}{2}\pi(\beta+1)[\{\psi(\beta) - \frac{1}{2}\pi\tan\frac{1}{2}\pi(\beta+1)\}\lvert\alpha\rvert^{-\beta-1} - \lvert\alpha\rvert^{-\beta-1}\ln\lvert\alpha\rvert]$
$\lvert x\rvert^{\beta}\ln\lvert x\rvert\operatorname{sgn} x$ (β not an integer)	$\beta!(-2i)\sin\frac{1}{2}\pi(\beta+1)[\{\psi(\beta) + \frac{1}{2}\pi\cot\frac{1}{2}\pi(\beta+1)\}\lvert\alpha\rvert^{-\beta-1} \times \operatorname{sgn}\alpha - \lvert\alpha\rvert^{-\beta-1}\ln\lvert\alpha\rvert\operatorname{sgn}\alpha]$
$x^k\mathrm{H}(x)\ln x$	$k! e^{-\pi i(k+1)/2}[\pi i(-1)^{k+1}\alpha^{-k-1} + \{\psi(k) - \frac{1}{2}\pi i\}\alpha^{-k-1} - \alpha^{-k-1}\ln\lvert\alpha\rvert + (-1)^k\pi i\psi(k)\delta^{(k)}(\alpha)/k!]$
$x^k\ln\lvert x\rvert$	$\pi e^{-\pi ik/2}\{(-1)^k 2\psi(k)\delta^{(k)}(\alpha) - k!\alpha^{-k-1} \times \operatorname{sgn}\alpha\}$

$g(x)$	$G(\alpha) = \int_{-\infty}^{\infty} g(x) e^{-i\alpha x} dx$
$x^{-m} H(x)$	$(-i)^{m-1} \alpha^{m-1} \{\psi(m-1) - \ln\|\alpha\| - \frac{1}{2}\pi i \operatorname{sgn} \alpha\} / (m-1)!$
$x^{-m} H(x) \ln x$	$(-i\alpha)^{m-1} [\frac{1}{3}\pi^2 - \psi'(m-1) + \{\psi(m-1) - \ln\|\alpha\| - \frac{1}{2}\pi i \operatorname{sgn} \alpha\}^2] / (m-1)! 2$
$x^{-m} \operatorname{sgn} x \ln \|x\|$	$(-i\alpha)^{m-1} [\{\ln\|\alpha\| - \psi(m-1)\}^2 - \frac{1}{6}\pi^2 - \psi'(m-1)] / (m-1)!$
$x^k \{\ln\|x\| - \psi(k)\}^2$	$k! 2\pi (-i)^k \alpha^{-k-1} \operatorname{sgn} \alpha \ln\|\alpha\| + 2\pi i^k \{\frac{1}{6}\pi^2 + \psi'(k)\} \delta^{(k)}(\alpha)$

$g(\boldsymbol{x})$	$G(\boldsymbol{\alpha}) = \int_{-\infty}^{\infty} g(\boldsymbol{x}) e^{-i\boldsymbol{\alpha}.\boldsymbol{x}} d\boldsymbol{x}$
$e^{-r^2/2}$	$(2\pi)^{n/2} e^{-\alpha^2/2}$
$\exp(-\frac{1}{2}\boldsymbol{x}^T A \boldsymbol{x})$	$\frac{(2\pi)^{n/2}}{(\det A)^{1/2}} \exp(-\frac{1}{2}\boldsymbol{\alpha}^T A^{-1} \boldsymbol{\alpha})$
$r^\beta \quad (\beta \neq 2k \text{ or } -n-2k)$	$\frac{(\frac{1}{2}\beta + \frac{1}{2}n - 1)!}{(-\frac{1}{2}\beta - 1)!} 2^{\beta+n} \pi^{n/2} \alpha^{-\beta-n}$
r^{2k}	$(2\pi)^n (-1)^k (\partial_1^2 + \ldots + \partial_n^2)^k \delta(\boldsymbol{\alpha})$
r^{-n-2k}	$\frac{(-1)^k \pi^{n/2} \alpha^{2k}}{(\frac{1}{2}n + k - 1)! k! 2^{2k-1}} \{\frac{1}{2}\psi(\frac{1}{2}n + k - 1) + \frac{1}{2}\psi(k) + \ln 2 - \ln \alpha\}$
$\ln(1/r)$	$(\frac{1}{2}n - 1)! 2^{n-1} \pi^{n/2} \alpha^{-n} - (2\pi)^n \{\frac{1}{2}\psi(n-1) + \frac{1}{2}\psi(0) + \ln 2\} \delta(\boldsymbol{\alpha})$
$r^\beta \ln r \quad (\beta \neq 2k \text{ or } -n-2k)$	$\frac{(\frac{1}{2}\beta + \frac{1}{2}n - 1)!}{(-\frac{1}{2}\beta - 1)!} 2^{\beta+n} \pi^{n/2} [\alpha^{-\beta-n} \{\frac{1}{2}\psi(\frac{1}{2}\beta + \frac{1}{2}n - 1) + \frac{1}{2}\psi(-\frac{1}{2}\beta - 1)\} - \alpha^{-\beta-n} \ln \frac{1}{2}\alpha]$
$r^{2k} \ln r$	$(\frac{1}{2}n + k - 1)! k! 2^{n+2k-1} \pi^{n/2} (-1)^{k+1} \alpha^{-n-2k} + (-1)^k (2\pi)^n \{\frac{1}{2}\psi(\frac{1}{2}n + k - 1) + \frac{1}{2}\psi(k) + \ln 2\} \times (\partial_1^2 + \ldots + \partial_n^2)^k \delta(\boldsymbol{\alpha})$

$g(x)$	$G(\alpha)=\int_{-\infty}^{\infty} g(x)e^{-i\alpha:x}dx$
$r^{-n-2k}\ln r$	$\frac{(-1)^k\pi^{n/2}\alpha^{2k}}{(\frac{1}{2}n+k-1)!k!2^{2k}}[\{\frac{1}{2}\psi(\frac{1}{2}n+k-1)+\frac{1}{2}\psi(k)-\ln\frac{1}{2}\alpha\}^2-\frac{1}{4}\{\psi'(\frac{1}{2}n+k-1)+\psi'(k)\}+\frac{1}{12}\pi^2]$

In the following $q=x_1^2+\ldots+x_p^2-\ldots-x_n^2$, $\tilde{q}$ the same in terms of α and $\tilde{L}$ the operator defined in the text.

$(q\pm i0)^\beta$ $(\beta\neq k$ or $-\frac{1}{2}n-k)$	$(\beta+\frac{1}{2}n-1)!\pi^{n/2}2^{2\beta+n}e^{\pm\pi i(p-n)/2}\times(\tilde{q}\mp i0)^{-\beta-n/2}/(-\beta-1)!$
$(q\pm i0)^k=q^k$	$(2\pi)^n(-\tilde{L})^k\delta(\alpha)$
$(q\pm i0)^{-(n/2)-k}$	$\frac{(-1)^k\pi^{n/2}e^{\pm\pi i(p-n)/2}}{(\frac{1}{2}n+k-1)!k!2^{2k}}[\{2\ln 2+\psi(\frac{1}{2}n+k-1)+\psi(k)\}\tilde{q}^k-\tilde{q}^k\ln(\tilde{q}\mp i0)]$
q_+^β $(\beta\neq\pm k$ or $-\frac{1}{2}n-k)$	$\beta!(\beta+\frac{1}{2}n-1)!\pi^{(n/2)-1}2^{2\beta+n}\{\tilde{q}_-^{-\beta-n/2}\sin\frac{1}{2}p\pi-\tilde{q}_+^{-\beta-n/2}\sin(\beta-\frac{1}{2}p+\frac{1}{2}n)\pi\}$
q_+^k	$(\frac{1}{2}n+k-1)!k!(-1)^k\pi^{(n/2)-1}i2^{n+2k-1}\times\{e^{-\pi i(p-n)/2}(\tilde{q}+i0)^{-(n/2)-k}-e^{\pi i(p-n)/2}(\tilde{q}-i0)^{-(n/2)-k}\}-(-1)^k2^n i\pi^{n-1}\{i\pi+\psi(k)+\psi(\frac{1}{2}n+k-1)-\psi(\frac{1}{2}n-1)-\psi(0)\}\tilde{L}^k\delta(\alpha)$
q_+^{-m}	See Corollary 8.10
$q_+^{-(n/2)-k}$	See Theorem 8.12
q^{-m}	$\frac{(\frac{1}{2}n-m-1)!}{(m-1)!}\pi^{n/2}2^{n-2m}\{(-1)^{(n/2)-1/2}\tilde{q}_+^{m-n/2}\times\sin\frac{1}{2}p\pi+(-1)^m\tilde{q}_-^{m-n/2}\cos\frac{1}{2}p\pi\}$ (n odd) $\frac{(\frac{1}{2}n-m-1)!}{(m-1)!}\pi^{n/2}2^{n-2m}(-1)^{n/2}\tilde{q}^{m-n/2}\cos\frac{1}{2}p\pi+\frac{2^{n-2m}\pi^{(n/2)+1}}{(m-1)!}(-1)^m\delta^{((n/2)-m-1)}(\tilde{q})\times\sin\frac{1}{2}p\pi$ (n even and $m<\frac{1}{2}n-1$)
$q_+^\beta\ln q$	See Theorem 8.11 and Corollary 8.11

$g(\boldsymbol{x})$	$G(\boldsymbol{\alpha}) = \int_{-\infty}^{\infty} g(\boldsymbol{x}) \mathrm{e}^{-\mathrm{i}\boldsymbol{\alpha}.\boldsymbol{x}} \mathrm{d}\boldsymbol{x}$
$\delta^{(k)}(q)$ (n odd or n even and $k < \frac{1}{2}n - 1$)	$(\frac{1}{2}n - k - 2)!\pi^{(n/2)-1}\mathrm{i}2^{n-2k-3}(-1)^{k+1}$ $\times\{\mathrm{e}^{-\pi\mathrm{i}(p-n)/2}(\tilde{q}+\mathrm{i}0)^{k+1-n/2}$ $-\mathrm{e}^{\pi\mathrm{i}(p-n)/2}(\tilde{q}-\mathrm{i}0)^{k+1-n/2}\}$
$q_+^\beta \mathrm{H}(x_1)$ ($\beta \neq \pm k$ or $-\frac{1}{2}n - k$)	$\beta!(\beta + \frac{1}{2}n - 1)!\pi^{(n/2)-1}2^{2\beta+n-1}$ $\{\tilde{q}_-^{-\beta-n/2} + \mathrm{e}^{\mathrm{i}\pi(\beta+n/2)}\tilde{q}_+^{-\beta-n/2} - \tilde{q}_+^{-\beta-n/2}$ $\mathrm{H}(\alpha_1) \times 2\mathrm{i}\sin(\beta + \frac{1}{2}n)\pi\}$
$\delta^{(m-1)}(q)\mathrm{H}(x_1)$	$(\frac{1}{2}n - m - 1)!\pi^{(n/2)-1}2^{n-2m-1}(-1)^m$ $\times\{(-1)^m\tilde{q}_-^{m-n/2} + \mathrm{e}^{\mathrm{i}n\pi/2}\tilde{q}_+^{m-n/2}$ $-\tilde{q}_+^{m-n/2}\mathrm{H}(\alpha_1)2\mathrm{i}\sin\frac{1}{2}n\pi\}$ (n odd) $(\frac{1}{2}n - m - 1)!\pi^{(n/2)-1}2^{n-2m-1}(-1)^{m-n/2}$ $(\tilde{q}+\mathrm{i}0)^{m-n/2} - 2^{n-2m}\pi^{n/2}\mathrm{i}\delta^{((n/2)-m-1)}(\tilde{q})\mathrm{H}(\alpha_1)$ (n even and $m \leq \frac{1}{2}n - 1$)
$(Q \pm \mathrm{i}0)^\beta$ ($\beta \neq k$ or $-\frac{1}{2}n - k$)	$\dfrac{(\beta + \frac{1}{2}n - 1)!\pi^{n/2}}{(-\beta-1)!\lvert\det Q\rvert^{1/2}}2^{2\beta+n}\mathrm{e}^{\pm\pi\mathrm{i}(p-n)/2}$ $\times(\tilde{Q} \mp \mathrm{i}0)^{-\beta-n/2}$
$(Q + a^2)^\beta$ ($\beta \neq k$ or $-\frac{1}{2}n - k$)	$\dfrac{\beta!\mathrm{i}\pi^{(n/2)-1}}{\lvert\det Q\rvert^{1/2}}(2a)^{\beta+n/2}\mathrm{e}^{\pi\mathrm{i}(p-n-2\beta)/2}$ $\times(\tilde{Q} - \mathrm{i}0)^{-(\beta/2)-n/4}K_{\beta+n/2}\{\alpha(\tilde{Q} - \mathrm{i}0)^{1/2}\}$

Table of Laplace transforms

$w(x)$	$\mathscr{W}(s) = \int_0^\infty w(x)\,\mathrm{e}^{-sx}\,\mathrm{d}x$
$\delta^{(m)}(x)$	$s^m \quad (\sigma \geq 0)$
$x^\beta \mathrm{H}(x) \quad (\beta \neq -1, -2, \dots)$	$\beta!/s^{\beta+1} \quad (\sigma > 0)$
$x^{-m}\mathrm{H}(x)$	$\dfrac{(-s)^{m-1}}{(m-1)!}\{\psi(m-1) - \ln s\}$
$\mathrm{H}(x)\cos \beta x$	$s/(s^2+\beta^2) \quad (\sigma > \lvert\mathrm{Im}\,\beta\rvert)$
$\mathrm{H}(x)\sin \beta x$	$\beta/(s^2+\beta^2) \quad (\sigma > \lvert\mathrm{Im}\,\beta\rvert)$
$\mathrm{H}(x)\cosh \beta x$	$s/(s^2-\beta^2) \quad (\sigma > \lvert\mathrm{Re}\,\beta\rvert)$
$\mathrm{H}(x)\sinh \beta x$	$\beta/(s^2-\beta^2) \quad (\sigma > \lvert\mathrm{Re}\,\beta\rvert)$
$\mathrm{H}(x)\lvert\sin ax\rvert \quad (a > 0)$	$\dfrac{a\coth(\pi s/2a)}{s^2+a^2} \quad (\sigma > 0)$
$x^\beta \mathrm{H}(x)\ln x \quad (\beta \neq -1, -2, \dots)$	$\dfrac{\beta!}{s^{\beta+1}}\{\psi(\beta) - \ln s\} \quad (\sigma > 0)$
$x^\nu J_\nu(\beta x)$	$\dfrac{(\nu-\frac{1}{2})!(2\beta)^\nu}{(s^2+\beta^2)^{\nu+1/2}\pi^{1/2}} \quad (\sigma > \lvert\mathrm{Im}\,\beta\rvert, \mathrm{Re}\,\nu > -\frac{1}{2})$
$x^{\nu/2} I_\nu(2x^{1/2})$	$s^{-\nu-1}\mathrm{e}^{1/s} \quad (\sigma > 0, \mathrm{Re}\,\nu > -1)$

Index